WileyPLUS

WileyPLUS is a research-based online environment for effective teaching and learning.

WileyPLUS builds students' confidence because it takes the guesswork out of studying by providing students with a clear roadmap:

- what to do
- how to do it
- if they did it right

It offers interactive resources along with a complete digital textbook that help students learn more. With *WileyPLUS*, students take more initiative so you'll have greater impact on their achievement in the classroom and beyond.

Now available for

Bb
Blackboard

For more information, visit www.wileyplus.com

WileyPLUS

ALL THE HELP, RESOURCES, AND PERSONAL SUPPORT YOU AND YOUR STUDENTS NEED!

www.wileyplus.com/resources

1st DAY OF CLASS ... AND BEYOND!

2-Minute Tutorials and all of the resources you and your students need to get started

WileyPLUS

Student Partner Program

Student support from an experienced student user

Wiley Faculty Network

Collaborate with your colleagues, find a mentor, attend virtual and live events, and view resources
www.WhereFacultyConnect.com

WileyPLUS

Quick Start

Pre-loaded, ready-to-use assignments and presentations created by subject matter experts

Technical Support 24/7
FAQs, online chat, and phone support
www.wileyplus.com/support

Your *WileyPLUS* Account Manager, providing personal training and support

Sixth Edition

Introducing Physical Geography

Alan Strahler

BOSTON UNIVERSITY

WILEY

Vice President and Executive Publisher	Jay O'Callaghan
Executive Editor	Ryan Flahive
Product Designer	Beth Tripmacher
Editorial Operations Manager	Lynn Cohen
Assistant Content Editor	Darnell Sessoms
Editorial Assistant	Julia Nollen
Senior Content Manager	Micheline Frederick
Senior Production Editor	Sandra Rigby
Media Specialist	Anita Castro
Creative Director	Harry Nolan
Senior Designer	Wendy Lai
Senior Marketing Manager	Margaret Barrett
Photo Manager	Hilary Newman
Senior Photo Editor	Jennifer Atkins
Cover Photo and Chapter Openers	Yann Arthus-Bertrand/Altitude

This book was typeset in 10/12 New Baskerville at MPS Limited and printed and bound by Quadgraphics/Versailles. The cover was printed by Quadgraphics/Versailles.

The paper in this book was manufactured by a mill whose forest management programs include sustained-yield harvesting of its timberlands. Sustained-yield harvesting principles ensure that the number of trees cut each year does not exceed the amount of new growth.

This book is printed on acid-free paper. ∞

ISBN 13   978-111-839620-9
BRV ISBN 13   978-111-829193-1

Printed in the United States of America.

10   9   8   7   6   5   4   3   2   1

Preface

Welcome to the Sixth Edition of *Introducing Physical Geography*! Our latest edition takes a fresh look at the book's second half, covering soils, Earth materials, surface processes, and landforms, while retaining the features adopted in the Fifth Edition. Chapters 10 through 17 have been largely revised and restructured to provide both a better flow of concepts and updated treatments in many areas. Here's what's new:

- Chapter 10, *Global Soils*, now begins with sections focused on soil chemistry and soil moisture, and presents soil orders grouped by maturity, climate, parent materials, and organic matter.
- Chapter 11, *Earth Materials and Plate Tectonics*, has been reorganized to proceed from geologic time to Earth's inner structure, rock types, and the cycle of rock change. We follow that discussion with our distinctive coverage of plate tectonics.
- Chapter 12, *Tectonic and Volcanic Landforms*, places tectonic landforms first in this edition, rooting the discussion more firmly in the context of plate tectonics. Our treatment of earthquakes is also revised and restructured.
- Chapter 13, *Weathering and Mass Wasting*, sorts weathering processes more specifically in this edition, according to their physical and chemical nature, and presents types of mass wasting organized by substrate moisture (dry to wet) and speed of action (fast to slow).
- Chapter 14, *Freshwater of the Continents*, focuses on a more direct flow of concepts, from groundwater to surface water, streamflow, and flooding.
- Chapter 15, *Landforms Made by Running Water*, organizes the discussion of fluvial processes by erosion, transportation, and deposition, and breaks out stream gradation and evolution at a top level to lead into fluvial landforms.
- Chapter 16, *Landforms Made by Wind and Waves*, separates processes and landforms more directly, leading to a better understanding of coastline development and specific dune types. Tsunamis are covered here as a wave phenomenon.
- Chapter 17, *Glacial and Periglacial Landforms*, has been restructured to ensure a better flow of information, from types of glaciers to glacial processes and on to landforms, integrating both alpine and continental glaciers. Periglacial processes now are discussed in this chapter, including permafrost and mass wasting.

For this edition we have also revised and refined these features:

- We updated the *Eye on Global Change* feature, which leads off many chapters, with new data and information. Likewise, the *Focus on Remote Sensing* feature, appearing in several chapters, has been updated, as needed.
- We restyled the art, and added many new photos.
- To maintain relevance and student interest, we updated our examples of natural phenomena, to include more recent hurricanes, tornados, volcanic eruptions, landslides, earthquakes, floods, and wildfires.
- Finally, we renewed our chapter opening photos, from the collection of the renowned photographer Yann Arthus-Bertrand, featuring images from his new book *The New Earth From Above 365 Days*. Each provides a fresh viewpoint on the contents of the chapter at hand.

Even as we have moved forward in the areas just noted, we have maintained our emphasis in the Fifth Edition on visual learning. By juxtaposing graphics and photos, we effectively and synergistically develop and illustrate facts and concepts. Many of these art pieces also incorporate text blocks, highlighted with callout lines to particular parts of the photos or graphics; these aid learning more effectively than in-narrative explanations alone. We have also retained the use of "windows," text boxes within certain columns to reinforce important key concepts and add visual interest . The *In Review* section at the end of each chapter reads like an abstract, encapsulating the key concepts and relevant terms in short bullet points. The review questions, visualizing exercises, and essay questions all offer opportunities for students to reinforce and demonstrate the knowledge they have acquired.

Geo Media Library

This easy-to-use website features animations, videos, and interactive exercises, making it possible to quickly reinforce and illustrate key concepts from the text. Students can use these resources for tutorials, as well as a self-quizzing device to complement the textbook and enhance their understanding of geography. Easy integration of this content into course management systems and homework assignments enables instructors to conveniently integrate multimedia with their syllabi and more traditional reading and writing assignments.

Resources in the Geo Media Library include:

Animations: Key diagrams and drawings from our rich signature art program have been animated to produce a virtual experience of difficult concepts. Such animations are crucial to the understanding of this content for visual learners.

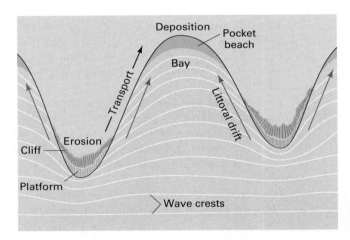

Videos: Brief video clips provide real-world examples of geographic features, and put these examples into context with the concepts covered in the text.

Simulations: Computer-based models of geographic processes allow students to manipulate data and variables so they may explore and interact with virtual environments.

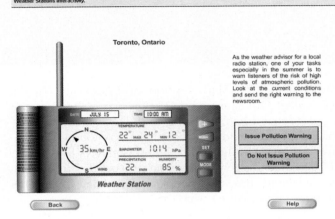

Interactive Exercises: Learning activities and games build off our presentation material, giving students an opportunity to test their understanding of key concepts and explore additional visual resources.

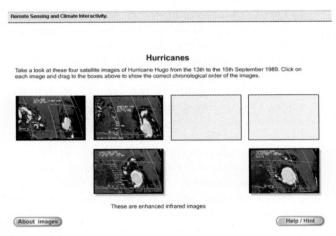

WileyPLUS

WileyPLUS is an online teaching and learning environment that integrates the entire digital textbook with the most effective instructor and student resources to accommodate every learning style. It can be used with or in place of the textbook.

With *WileyPLUS*:

- Students can achieve concept mastery with the help of additional study tools such as online self-assessments, videos, animations, and flashcards, all presented in a media-rich, structured environment that is available 24/7.

■ Instructors can personalize and manage their courses more effectively by tracking content and assessments to learning objectives, assignments, grade tracking, ability to upload assessments, and more.

In addition to the instructor and student resources provided on our book's companion website, *WileyPLUS* offers these additional features:

1. Google Earth Lecture Tours: New to this edition, interactive Google Earth tours, developed by Alan Halfen, University of Kansas, can be used by instructors as supplemental lecture material or as stand-alone lectures. Written for Chapters 11 through 17, each tour follows the basic chapter layout and contains links at each stop to key terms and definitions. Instructors can edit the tours as well as manipulate them using the navigation control to better present the specific topic under instruction. The tours also include an instructor guide and a note-taking guide for students.

2. Google Earth Topical Tours: Topic-specific tours developed by Professor Randy Rutberg, CUNY Hunter, on glaciers, igneous rocks, plate boundaries, and sedimentary rocks give students an in-depth review and an engaging look at the world around them.

3. ## contextualized animations and videos, tagged to learning objectives chapter content, including 12 new animations.

Google Earth™

For Introducing Physical Geography Sixth Edition

The Sixth Edition of *Introducing Physical Geography* incorporates the diverse resource of Google Earth™. Through satellite imagery shown in the text and online tours available for each chapter using Google Earth™, instructors and students can view and interact with landforms and landscapes anywhere on the globe, to better demonstrate and learn how the processes of physical geography work. Completing the online resources available when taking a chapter tour in Google Earth™ are study tools such as new Google Earth™ lecture tours, topical tours, and bonus practice with interactivities.

Earth from Above
An Aerial Portrait of Our Planet

Since 1990, photographer Yann Arthus-Bertrand has flown over hundreds of countries to compile an aerial portrait of our planet. His photographs invite all of us to reflect upon the future of Earth and its inhabitants. Over the past 50 years, we humans have changed ecosystems more rapidly and extensively than in any comparable period of time in human history. Everywhere, under the assaults of humankind, our planet's ecosystem appears to be deteriorating—freshwater, oceans, forests, air, arable land, open spaces, cities. . . . In whatever form of media—books, exhibitions, websites, films, posters—"Earth from Above" reminds us that each and every one of us is responsible for the future of the Earth. And because each one of us plays a part, we all have the duty to act. For more on the work of Yann Arthus-Bertrand go to www.yannarthusbertrand.org; and for more on how you can participate in the recovery of Earth's ecosystem, go to www.goodplanet.org.

www.yannarthusbertrand.org
www.goodplanet.org

Toward a Sustainable Development

Since 1950, worldwide economic growth has been considerable, and global production of goods and services has multiplied by a factor of 8. During this same period, while the world's population has a little more than doubled, the volume of fish caught has multiplied by 5, and the volume of meat produced by 6. The demand for energy has multiplied by 5; oil consumption has multiplied by 7; and carbon dioxide emissions, the main cause of the greenhouse effect and global warming, by 5. Since 1900, freshwater consumption has multiplied by 6, chiefly to provide for agriculture.

And yet, 20 percent of the world's population has no access to sources of drinking water, 25 percent are without electricity, and 40 percent have no sanitary installation; 820 million people are underfed, and half of humanity lives on less than $2 a day. In other words, a fifth of the world's population lives in industrialized countries, consuming and producing in excess and generating massive pollution. The remaining four-fifths live in developing countries and, for the most part, in poverty.

Overexploitation of resources leads to the constant degradation of our planet's ecosystem, and severely limits supplies of freshwater, ocean water, forests, air, arable land, and much more.

That's not all. By 2050, the Earth will have close to 3 billion additional inhabitants. These people will live, for the most part, in developing countries. As these countries develop, their economic growth will jockey for position with that of industrialized nations—within the limits of ecosystem Earth.

If every individual living on the planet were to consume as much as a person living in the Western world, we would need three planets the size of Earth to satisfy all their needs. Fortunately, there is a way we can meet everyone's needs while preserving natural resources for future generations: We must actively promote and support technologies that are less polluting and less water and energy-consuming. Referred to as *sustainable development*, this approach to living on Earth represents progress for humanity: to consume not *less*, but *better*.

The current situation on Earth is not irreversible, but we need to begin making changes immediately. We have the opportunity to turn toward more sustainable development, one that allows us to improve the living conditions of the world's citizens and to satisfy the needs of generations to come. This development would be based on an economic growth that is respectful both of humans and the natural resources of our unique planet.

Such development requires that we improve production methods and change our consumption habits. With the active participation of all the world's citizens, each and every one of us can make a valuable contribution to the future of the Earth and humankind, starting right now.

The Earth from Above

Introduction
Physical Geography and the Tools Geographers Use

Chapter 1
The Earth as a Rotating Planet

Chapter 2
The Earth's Global Energy Balance

Chapter 3
Air Temperature

Chapter 4
Atmospheric Moisture and Precipitation

Chapter 5
Winds and Global Circulation

Chapter 6
Weather Systems

Chapter 7
Global Climates and Climate Change

Chapter 8
Biogeographic Processes

Chapter 9
Global Biogeography

Chapter 10
Global Soils

Chapter 11
Earth Materials and Plate Tectonics

Chapter 12
Tectonic and Volcanic Landforms

Chapter 13
Weathering and Mass Wasting

Chapter 14
Freshwater of the Continents

Chapter 15
Landforms Made by Running Water

Chapter 16
Landforms Made by Waves and Wind

Chapter 17
Glacial and Periglacial Landforms

Book Companion Website

On the companion website to this text— www.wiley.com/college/strahler—you will find a wealth of study and practice materials, including:

Student Online Resources

- *Self-quizzes*: Chapter-based multiple-choice and fill-in-the-blank questions.
- *Annotated Weblinks*: Useful weblinks selected to enhance chapter topics and content.
- *Lecture Note Handouts*: Key images and slides from the instructor PowerPoint presentations are made available so that, when in class, students can focus on the lecture, annotate figures, and add their own notes.
- *Media Library*: Link to the media library for students to explore key concepts in greater depth using videos, animations, and interactive exercises.

Concept Caching

- This online database of photographs explores what a physical feature looks like. Photographs and GPS coordinates are "cached" and categorized along with core concepts of geography. Professors can access the images or submit their own by visiting www.ConceptCaching.com.

Instructor Resources

- This section includes all student resources, plus:
  - *PowerPoint Lecture Slides*: Chapter-oriented slides, along with lecture notes and text art.
  - *Computerized Test Bank*: Multiple-choice, fill-in-the blank, and essay questions, available in both Respondus and Diploma.
  - *Instructor's Manual*: Lecture notes, learning objectives, guides to additional resources, and teaching tips for enhancing the classroom experience.
  - *Clicker Questions*: A set of questions for each chapter that can be used during lectures to check understanding using PRS, HITT, or CPS clicker systems.
  - *Image Gallery*: Both line art and photos from the text.

Acknowledgments

The preparation of *Introducing Physical Geography*, Sixth Edition, was greatly aided by the following reviewers who read and evaluated various parts of the manuscript:

William M. Buhay, University of Winnipeg
Christopher Justice, University of Maryland, College Park
Robert M. Hordon, Rutgers University
Francis A. Galgano, United States Military Academy, West Point
Jeffrey P. Schaffer, Napa Valley College
Thomas A. Terich, Western Washington University

We would also like to thank the reviewers who read and commented on prior editions:

Tanya Allison, Montgomery College
Joseph M. Ashley, Montana State University
Roger Balm, Rutgers State University
David Bixler, Chaffey College
Peter Blanken, University of Colorado, Boulder
Kenneth L. Bowden, Northern Illinois University
David Butler, University of North Carolina, Chapel Hill
Les Dean, Riverside College
Christopher H. Exline, University of Nevada
Jerry Green, Miami University
Clarence Head, University of Central Florida
Peter W. Knightes, Central Texas College
Christopher S. Larsen, State University of New York, Buffalo
Robert E. Lee, Seattle Central Community College
Denyse Lemaire, Rowen University
Eugene J. Palka, U.S. Military Academy, West Point
James F. Petersen, Southwest Texas State University
Clinton Rowe, University of Nebraska-Lincoln
Carol Shears, Ball State University
Carol I. Zinser, State University of New York, Plattsburgh

It is with particular pleasure that I thank the staff at Wiley for their long hours of careful work in the preparation and production of our Sixth Edition. They include our Executive Editor Ryan Flahive, Editorial Assistant Julia Nollen, Photo Editor Jennifer Atkins, Senior Content Manager Micheline Frederick, Designer Wendy Lai, Marketing Manager Margaret Barrett, Product Designer Beth Tripmacher, and especially our Senior Production Editor Sandra Rigby, who made it all come together. Particular thanks go to Sandra and Micheline for making every page look so great!

I am also indebted to Dr. Timothy Foresman, coauthor of the Second Edition of *Visualizing Physical Geography*, and our Development Editor for that edition, Rebecca Heider, whose efforts in the restructuring of the latter chapters greatly strengthened the book. Their vision was extremely helpful as I put this new edition together.

In closing, I would like to acknowledge the debt I owe to my former coauthor, Arthur Strahler, who contributed so much to the field of education in physical geography in the 49 years of his collaboration with John Wiley & Sons, and his 29 years of collaboration with me. I deeply wish he could still be at my side today as the coauthor of *Introducing Physical Geography*, Sixth Edition.

Alan Strahler
Boston, Massachusetts
December, 2012

About the Author

Alan Strahler, presently Professor of Geography at Boston University, earned his PhD degree in Geography from Johns Hopkins, in 1969. He has published more than 250 articles in the refereed scientific literature, largely on the theory of remote sensing of vegetation; he has also contributed to the fields of plant geography, forest ecology, and quantitative methods. In 2011, Professor Strahler received the William T. Pecora Award from the Department of the Interior and the National Aeronautics and Space Administration for his outstanding contributions to understanding the Earth by means of remote sensing. In 2009, his work in geographic education was recognized with an award from the SAIC Estes Memorial Teaching Award from the American Society of Photogrammetry and Remote Sensing. He also received the Association of American Geographers/Remote Sensing Specialty Group Medal for Outstanding Contributions to Remote Sensing, in 1993.

Professor Strahler holds the honorary degree DSHC from the Université Catholique de Louvain, Belgium, awarded in 2000, and is a Fellow of the American Association for the Advancement of Science. With the late Arthur Strahler, he is the coauthor of 7 textbook titles, with 13 revised editions, on physical geography and environmental science. He is also the author or coauthor of *Visualizing Physical Geography*, first and second editions, and *Visualizing Weather and Climate*, both titles in the Wiley Visualizing Series.

Brief Contents

Contents

Sixth Edition

Introducing Physical Geography

Introduction
Physical Geography and the Tools Geographers Use

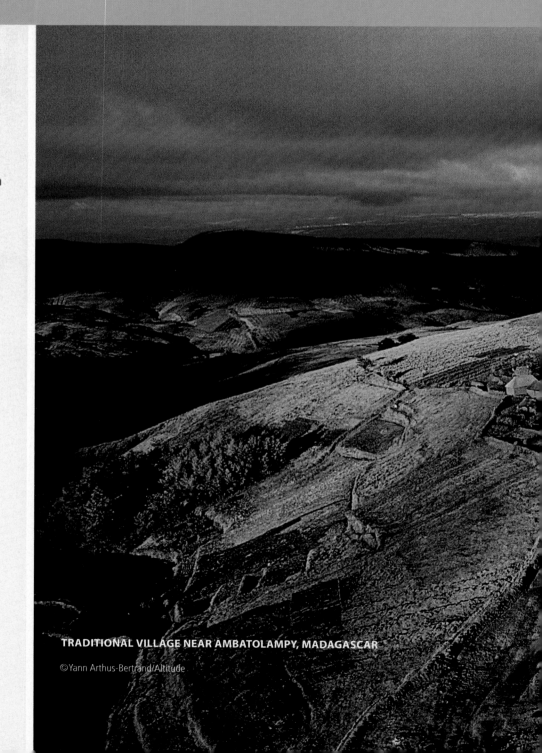

This traditional village in Madagascar, in a fertile highland area south of the capital, has a unique geography and relation to its surroundings. A cluster of houses on the brow of a hill occupies the village center, with walled kitchen gardens and crops that need frequent tending nearby. Field crops occupy the hilltop and gentler slopes, while forest grows on the steeper slopes. Human settlements have a physical setting that places bounds on the kinds of human activities that take place there. In this book, we will focus on the natural processes that shape the physical landscape and provide the habitat of the human species.

TRADITIONAL VILLAGE NEAR AMBATOLAMPY, MADAGASCAR
©Yann Arthus-Bertrand/Altitude

Physical Geography and the Tools Geographers Use

Geography is a modern discipline with ancient roots. But what is geography? What are the big ideas of physical geography? How is physical geography related to global climate change? Biodiversity? Extreme events? Geographers use special tools to study the Earth. How do maps depict the Earth's curved surface on a flat piece of paper? How does a geographic information system (GIS) work? How do geographers use remote sensing? These are some of the questions we will answer in our Introduction.

Introducing Geography

What is geography? Put simply, **geography** is the study of the evolving character and organization of the Earth's surface. It is about how, why, and where human and natural activities occur and how these activities are interconnected.

To get a better understanding of geography, think of it as having two sides. One side, which we can term **regional geography**, is concerned with how the Earth's surface is differentiated into unique *places*. Take Vancouver, British Columbia, for example (Figure I.1) What makes Vancouver unique? Is it its spectacular setting where the Pacific Ranges meet the Pacific Ocean? The marine west-coast climate that provides its mild and rainy winters and blue summer skies? Its position as a seaport gateway to Asia? Its English roots combined with dashes of French, Asian, and even Russian culture? In fact, all of these attributes contribute to making Vancouver the unique place that it is.

Although places are unique, the physical, economic, and social processes that form them are not. Thus, geographers are concerned with discovering, understanding, and modeling the processes that differentiate the Earth's surface into places. This is the other side of geography, which we can term **systematic geography**. Why are pineapples cheap in Hawaii and expensive in Toronto? Oranges cheap in Florida and expensive in North Dakota? Steak cheap in Kansas City and expensive in Boston? These are examples of a simple principle of economic geography—that prices include transportation costs and that when goods travel a longer distance, they are usually more expensive. Discovering such principles and extending them to model and predict spatial phenomena is the domain of systematic geography. To summarize, geographers study both the "vertical" integration of characteristics that define a place and the "horizontal" connections between places.

What makes geography different from other disciplines? Geography adopts a unique set of perspectives to analyze the world and its human and natural phenomena. These perspectives include the spatial viewpoint of geographers, the interest of geographers in the synthesis of ideas across the boundaries of conventional studies, and geographers' usage of tools to represent and manipulate spatial information and spatial phenomena. Figure I.2 shows these perspectives in the form of a cube, with each perspective displayed on a different face.

The first unique perspective of geography is its spatial **viewpoint**. Geographers are interested not only in how something happens, but also where it happens and how it is related to other happenings nearby and far away. The spatial viewpoint can focus at three levels. At the *place* level, geographers study how processes are integrated at a single location or within a single region. For example, a physical geographer may study the ecology, climate, and soils of a national park. At the *space* level, geographers look at how places are interdependent. An economic geographer may examine how flows of goods, information, or money connect cities and towns that are of different sizes and at different distances apart. Geographers also look at human and natural activities at different *scales*, sometimes zooming in for a close look at

Al Harvey/All Canada Photos/©Corbis

I.1 Vancouver, British Columbia

This cosmopolitan city enjoys a spectacular setting on the Strait of Georgia, flanked by the Pacific and Vancouver Island Ranges.

EYE ON THE LANDSCAPE **What else would the geographer see?**

A For the physical geographer, Vancouver's environment combines snow-capped peaks eroded by glaciers, conifer forests adapted to the cool, maritime climate, and an arm of the ocean that erodes the coast by wave and tidal action. **B** For the human geographer, the image shows a center of economic activity marked by Vancouver's office and residential towers. Areas of low buildings document the differentiation of the city into districts with diverse characters and history. The road and freeway network demonstrates the city's reliance on cars and trucks to move people and goods within the city. **C** The physical environment interacts with human activity through Vancouver's role as a port city, where land- and waterborne transportation modes meet. Large commercial vessels in the bay mingle with sailboats and powerboats, highlighting the importance of the city's marine setting to both shipping and recreation.

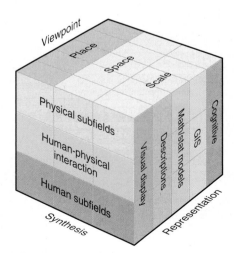

I.2 Perspectives of geography

The three unique perspectives of geography—its spatial viewpoint, its synthesis of related fields, and its representation of spatial processes and information—are diagrammed as three dimensions occupying the sides of a cube.

something small, or pulling back for an overview of something large. Often, what looks important at one scale is less important at another.

The second perspective of geography is **synthesis**. Geographers are very interested in putting together ideas from different fields and assembling them in new ways—a process called synthesis. Of particular interest to geographers are studies that link conventional areas of study. In physical geography, for example, a biogeographer may investigate how streamside vegetation affects the flood flow of rivers, thus merging the physical geography subfields of ecology and hydrology. The many connections between environmental processes and human activities are also subjects of geographic synthesis. For example, a classic study area in geography is perception of hazards: Why do people build houses next to rivers or beaches when it is only a matter of time before floods or storms will wash their homes away? Here, geographers study the interaction of hydrology with perception and cognitive learning.

The third perspective of geography is geographic **representation**. Here, geographers develop and perfect tools for representing and manipulating information spatially. *Cartography*—the art and science of making and drawing *maps*—is a subfield of geography that focuses on *visual display* of spatial relationships. Visual display also includes remote sensing—acquiring images of the Earth from aircraft or spacecraft and enhancing them to better display spatial information. *Verbal descriptions* use the power of words to explain or evoke geographic phenomena. *Mathematical and statistical models* predict how a phenomenon of interest varies over space and through time. *Geographic information systems* store, manipulate, and display spatial information in very flexible ways. *Cognitive representation* refers to spatial relationships as they are stored in the human brain—mental mapping of real space into the subjective space that people experience.

Taken together, the perspectives of viewpoint, synthesis, and representation define geography as a unique discipline that focuses on how the natural

> Geography as a discipline has a unique set of perspectives. Geographers look at the world from the viewpoint of geographic space, focus on synthesizing ideas from different disciplines, and develop and use special techniques to represent and manipulate spatial information.

and human patterns of the Earth's physical and cultural landscape change and interact in space and time.

HUMAN AND PHYSICAL GEOGRAPHY

Like many other areas of study, geography has a number of subfields, each with a different focus but often overlapping and interlocking with other subfields. We can organize these subfields into two broad realms: **human geography**, which deals with social, economic and behavioral processes that differentiate places; and **physical geography**, which examines the natural processes occurring at the Earth's surface that provide the physical setting for human activities. Figure I.3 is a diagram showing the principal fields of physical and human geography. Reading downward from the left, we see five fields of physical geography, from climatology to biogeography, which are illustrated in Figure I.4. These topics are the main focus of this text.

Climatology is the science that describes and explains the variability in space and time of the heat and moisture states of the Earth's surface, especially its land surfaces. Since heat and moisture states are part of what we call weather, we can think of climate as a description of average weather and its variation at places around the world. Chapters 1–7 will familiarize you with the essentials of climatology, including the processes that control the weather we experience daily. Climatology is also concerned with climate

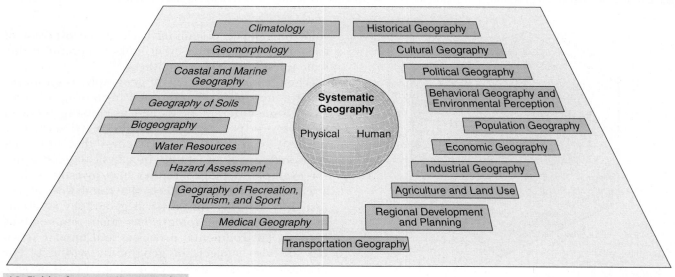

I.3 Fields of systematic geography

Physical and human geography have many interrelated subfields.

change, past, present, and future. One of the most rapidly expanding and challenging areas of climatology is global climate modeling, which we touch on in several chapters. This field attempts to predict how human activities, such as converting forestland to agricultural use, or releasing CO_2 by burning fossil fuels, will change global climate.

Geomorphology is the science of Earth surface processes and landforms. The Earth's surface is constantly being altered under the combined influence of human and natural factors. The work of gravity in the collapse and movement of Earth materials, as well as the work of flowing water, blowing wind, breaking waves, and moving ice, acts to remove and transport soil and rock and to sculpt a surface that is constantly being renewed through volcanic and tectonic activity. The closing chapters of our book (Chapters 12–17) describe these geomorphic processes, while the basic geologic processes that provide the raw material are covered in Chapters 11–12. Modern geomorphology also focuses on modeling landform-shaping processes to predict both short-term, rapid changes, such as landslides, floods, or coastal storm erosion, and long-term, slower changes, such as soil erosion in agricultural areas or as a result of strip mining.

The field of **coastal and marine geography** combines the study of geomorphic processes that shape shores and coastlines with their application to coastal development and marine resource utilization. Chapter 16 describes these processes and provides some perspectives on problems of human occupation of the coastal zone.

Geography of soils includes the study of the distribution of soil types and properties and the processes of soil formation. It is related to both geomorphic processes of rock breakup and weathering, and to biological processes of growth, activity, and decay of organisms living in the soil (Chapter 10). Since both geomorphic and biologic processes are influenced by the surface temperature and availability of moisture, broad-scale soil patterns are often related to climate.

Biogeography, covered in Chapters 8 and 9, is the study of the distributions of organisms at varying spatial and temporal scales, as well as the processes that produce these distribution patterns. Local distributions of plants and animals typically depend on the suitability of the habitat that supports them. In this application, biogeography is closely aligned with *ecology*, which is the study of

> Five major fields of physical geography are: climatology, geomorphology, coastal and marine geography, geography of soils, and biogeography.

the relationship between organisms and environment. Over broader scales and time periods, the migration, evolution, and extinction of plants and animals are key processes that determine their spatial distribution patterns. Thus, biogeographers often seek to reconstruct past patterns of plant and animal communities from fossil evidence of various kinds. *Biodiversity*—the assessment of biological diversity from the perspective of maintaining the diversity of life and life-forms on Earth—is a biogeographic topic of increasing importance, due to ongoing and increasing human impact on the environment. The present global-scale distribution of life-forms as the great biomes of the Earth provides a basic context for biodiversity.

In addition to these five main fields of physical geography, two others are strongly involved with applications of physical geography: water resources and hazards assessment. **Water resources** is a broad field that couples basic study of the location, distribution, and movement of water, for example, in river systems or as groundwater, with the utilization and quality of water for human use. This field involves many aspects of human geography, including regional development and planning, political geography, and agriculture and land use. We touch on water resources briefly in this book, in Chapters 14 and 15, where we discuss water wells, dams, and water quality.

Hazards assessment is another field that blends physical and human geography. What are the risks of living next to a river, and how do inhabitants perceive those risks? What is the role of government in protecting citizens from floods, or assisting them in recovery from flood damages? Answering questions such as these requires not only knowledge of how physical systems work, but also how humans, as both individuals and societies, perceive and interact with their physical environment. In this text, we develop an understanding of the physical processes of floods, earthquakes, landslides, and other disaster-causing natural events as a background for appreciating hazards to humans and their activities.

Many of the remaining fields of human geography have linkages with physical geography. For example, climatic and biogeographic factors may determine the spread of disease-carrying mosquitoes (medical geography). Mountain barriers may isolate populations and increase the cost of transporting goods from one place to another (cultural geography, transportation geography). Unique landforms and landscapes may be destinations for tourism (geography of recreation, tourism, and sport). Nearly all human activities take place in a physical environment that varies in space and time, so the physical processes that we examine in this text provide a background useful for further learning in any of geography's fields.

1.4 Fields of physical geography

The principal subfields of physical geography covered in this book are climatology, biogeography, geography of soils, geomorphology, and coastal and marine geography.

CP Photo Art/Getty Images, Inc.

▲ **CLIMATOLOGY**
Climatology studies the transfers of energy and matter between the surface and atmosphere that control weather and climate.

▼ **GEOMORPHOLOGY**
Geomorphology is the study of landform-making processes.

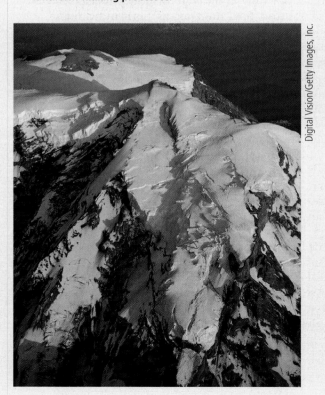

Digital Vision/Getty Images, Inc.

James Randklev/Getty Images, Inc.

Elena Kalistratova/Getty Images, Inc.

▲ BIOGEOGRAPHY
Biogeography examines the distribution patterns of plants and animals
and relates them to environment,
migration, evolution, and extinction.

◄ COASTAL AND MARINE GEOGRAPHY
Coastal and marine geography examines
coastal processes, marine resources, and
their human interface.

Fletcher & Baylis/Photo Researchers, Inc.

GEOGRAPHY OF SOILS ►
Soils are influenced by their parent
material, climate, biota, and time.

Spheres, Systems, and Cycles

As a part of your introduction to physical geography, it will be useful to take a look at the big picture and examine some ideas that arch over all of physical geography—that is, spheres, systems, and cycles. The first of these ideas is that of the four great physical realms, or *spheres* of Earth—atmosphere, lithosphere, hydrosphere, and biosphere. These realms are distinctive parts of our planet with unique components and properties. The second big idea is that of *systems*—viewing the processes that shape our landscape as a set of interrelated components that comprise a system. The systems viewpoint stresses linkages and interactions and helps us to understand complex problems, such as global climate change or loss of biodiversity. The last big idea is that of *cycles*—regular changes in systems that reoccur through time.

THE SPHERES—FOUR GREAT EARTH REALMS

The natural systems that we will encounter in the study of physical geography operate within the four great realms, or **spheres**, of the Earth. These are the atmosphere, the lithosphere, the hydrosphere, and the biosphere (Figure I.5).

The **atmosphere** is a gaseous layer that surrounds the Earth. It receives heat and moisture from the surface and redistributes them, returning some heat and all

I.5 The Earth realms

The natural systems of physical geography operate in four great Earth realms.

I.6 The life layer

As this sketch shows, the life layer is the layer of the Earth's surface that supports nearly all of the Earth's life. It includes the land and ocean surfaces and the atmosphere in contact with them.

the moisture to the surface. The atmosphere also supplies vital elements—carbon, hydrogen, oxygen, and nitrogen—that are needed to sustain life.

The outermost solid layer of the Earth, or **lithosphere**, establishes the platform for most Earthly life-forms. The solid rock of the lithosphere bears a shallow layer of soil in which nutrient elements become available to organisms. The surface of the lithosphere is sculpted into landforms. These features—such as mountains, hills, and plains—provide varied habitats for plants, animals, and humans.

The liquid realm of the Earth is the **hydrosphere**, which principally comprises the mass of water in the world's oceans. It also includes solid ice in mountain and continental glaciers, which, like liquid ocean and freshwater, is subject to flow under the influence of gravity. Within the atmosphere, water occurs as gaseous vapor, liquid droplets, and solid ice crystals. In the lithosphere, water is found in the uppermost layers in soils and in groundwater reservoirs.

The **biosphere** encompasses all living organisms of the Earth. Life-forms on Earth utilize the gases of the atmosphere, the water of the hydrosphere, and the nutrients of the lithosphere, and so the biosphere is dependent on all three of the other great realms. Figure I.5 diagrams this relationship.

Most of the biosphere is contained in the shallow surface zone called the **life layer**. It includes the surface of the lands and the upper 100 m or so (about 300 ft) of the ocean (Figure I.6). On land, the life layer is the zone of interactions among the biosphere, lithosphere, and atmosphere, with the hydrosphere represented by rain, snow, stillwater in ponds and lakes, and running water in rivers. In the ocean, the life layer is the zone of interactions among the hydrosphere, biosphere, and atmosphere, with the lithosphere represented by nutrients dissolved in the upper layer of seawater.

SCALE, PATTERN, AND PROCESS

As we saw earlier, geographers have unique perspectives on understanding the physical and human organization of the Earth's surface. Three interrelated themes that often arise in geographic study are scale, pattern, and process. **Scale** refers to the level of structure or organization at which a phenomenon is studied. **Pattern** refers to the variation in a phenomenon that is seen at a particular scale. **Process** describes how the factors that affect a phenomenon act to produce a pattern at a particular scale.

To make these ideas more real, imagine yourself as an astronaut, returning to Earth from a voyage to the Moon. As you approach the Earth and finally touch down on land, your view of our planet takes in scales ranging from

global to local. As the scale changes, so do the patterns and processes that you observe (Figure I.7). At the *global scale*, you see the Earth's major physical features—oceans of blue water; continents of brown Earth, green vegetation, and white snow and ice; and an atmosphere of white clouds and clear air. The pattern of land and ocean is created by the processes of plate tectonics, which shape landmasses and ocean basins across the eons of geologic time. The pattern of white clouds, which includes a band of persistent clouds near the Earth's Equator and spirals of clouds moving across the globe, is created

I.7 Scale, pattern, and process

As the Earth is viewed at increasingly finer scales, various patterns, created by different processes, emerge.

▲ GLOBAL SCALE

At the global scale, the major surface features of the Earth and atmospheric circulation are readily visible.

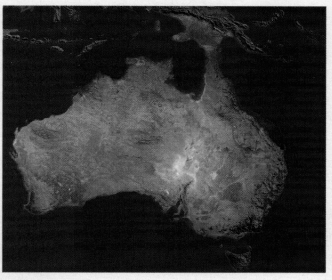

▲ CONTINENTAL SCALE

At the continental scale, climate determines the pattern of vegetation. In this image of Australia, green colors indicate healthy vegetation, with reds, browns, and whites showing sparse vegetation cover and desert.

Jacques Descloitres, MODIS Land Rapid Response Team, Courtesy NASA, Visible Earth.

▲ REGIONAL SCALE

At the regional scale, broad patterns of human activity are visible, such as this example of deforestation in Rondonia, in the Brazilian Amazon. The herringbone patterns show conversion of forest to cropland on a grid pattern.

Earth Satellite Corporation/ Photo Researchers, Inc.

▲ LOCAL SCALE

At the local scale, shown in this Landsat image of the San Francisco Bay region, the details of development emerge, as well as the shapes of individual landforms.

by atmospheric circulation processes that depend on solar heating coupled with the Earth's slow rotation on its axis. These processes act much more quickly and on a finer spatial scale than those of plate tectonics.

At the *continental scale*, you see the broad differentiation of landmasses into regions of dry desert and moister vegetated regions, a pattern caused by atmospheric processes that provide some areas with more precipitation than others. In some regions, air temperatures keep water frozen, producing sea ice and glaciers. Air temperature and precipitation are the basic elements of climate, and so we may regard climate as a major factor affecting the landscape on a continental level.

At the *regional scale*, mountain ranges, deserts, lakes, and rivers create a varied pattern generated by the interaction between the geologic processes that raise mountains and lower valleys, and the atmospheric processes that produce the water that runs off the continents while supporting the growth of vegetation. Also evident at the regional scale are broad patterns of human activity, such as the deforestation of the Amazon. Agricultural regions

Scale, pattern, and process are three interrelated geographic themes. Scale refers to the level of structure or organization; pattern refers to the variation seen at a particular scale; and process describes how the pattern at a particular scale is produced.

are clearly visible, distinguished by repeating geometric patterns of fields.

At the *local scale*, we zoom in on a landscape showing a distinctive pattern in fine detail. For example, our image of the San Francisco Bay region reveals both the natural processes that carve hillslopes and canyons from mountain masses and the human processes that superimpose city and suburb on the natural landscape. At the finest scale, we see individual-scale landscape features, such as sand dunes, bogs, or freeways, each of which is the result of a different process.

These examples illustrate the themes of scale, pattern, and process as they apply to the landscapes of our planet. Keep in mind, however, that these themes are quite general ones. Throughout this book, we will see many examples of scale, pattern, and process applied to such diverse phenomena as climate, vegetation, soils, and landforms. We will zoom in and out, examining processes at local scales and applying them to regions to create and explain broad patterns observed at continental and global scales. In this way, you will gain a better understanding of how the Earth's surface changes and evolves in response to natural and human activities.

SYSTEMS IN PHYSICAL GEOGRAPHY

The processes that interact within the four realms to shape the life layer and differentiate global environments

1.8 Dimensions of global change

The dimensions of global change touch on many human activities.

CARBON CYCLE
Clear-cutting of timber, shown here on the Olympic Peninsula, Washington, removes carbon from the landscape, while regrowth returns carbon through photosynthesis.

POLLUTION
Human activity can cause pollution of air and water, leading to changes in natural habitats, as well as impacts on human health. The discharge from this Canadian pulp and paper mill is largely water vapor, but pulp mill pollutants often include harmful sulfur oxides.

GLOBAL CLIMATE CHANGE
Is the Earth's climate changing? Nearly all global change scientists have concluded that human activities have resulted in climate warming, and that weather patterns, shown here in this satellite image of clouds and weather systems over the Pacific Ocean, are changing.

are varied and complex. A helpful way to understand the relationships among these processes is to study them as **systems**. System, a common English word used in everyday speech, typically refers to a set or collection of things that are somehow related or organized. An example is the solar system—a collection of planets that revolve around the Sun. In the text, we will use the word system in this way quite often. We will also use the word to define a scheme for naming things. For example, we will introduce a climate system, in Chapter 7, and a soil classification system, in Chapter 10. Finally, we will use system to mean a group of interrelated processes that operate simultaneously in the physical landscape.

When we study physical geography using a *systems approach,* we look for linkages and interactions among processes. For example, global warming should enhance the process of evaporation of water from oceans and moist land surfaces, generating more clouds. But an increase in clouds also affects the process of solar reflection, in which white, fleecy clouds reflect solar radiation back out to space. This leaves less radiation to be absorbed by the atmosphere and surface and so should tend to cool our planet, reducing global warming. This is actually an example of negative feedback, in which one process counteracts another process to reduce its impact. (We'll present more information about this topic in Chapter 6.) More examples of this systems viewpoint in physical geography will be given throughout the text.

TIME CYCLES

Many natural systems show **time cycles**, rhythms in which processes change in a regular and repeatable fashion. For example, the annual revolution of the Earth around the Sun generates a time cycle of incoming solar energy flow. We speak of this cycle as the rhythm of the seasons. The rotation of the Earth on its axis sets up the night-and-day cycle of darkness and light. The Moon, in its monthly orbit around the Earth, sets up its own time cycle, which we see in ocean tides.

The astronomical time cycles of Earth rotation and solar revolution are described in several places in our early chapters. Other time cycles, with durations of tens to hundreds of thousands of years, describe the alternate growth and shrinkage of the great continental ice sheets. Still others, with durations of millions of years, describe cycles of the solid Earth in which supercontinents form, break apart, and reform anew.

Physical Geography, Environment, and Global Change

Physical geography is concerned with the natural world around us—in short, with the human environment. Because natural processes are constantly active, the Earth's environments are constantly changing (Figure I.8).

Keith Myers/Kansas City Star/MCT via Getty Images

◀ **EXTREME EVENTS**
The frequency of hurricanes, severe storms, droughts, and floods may be increasing as the global climate warms. A tornado flattened this neighborhood in Joplin, Missouri, on May 24, 2011.

BIODIVERSITY ▶
Reduction in the area and degradation of the quality of natural habitats is decreasing biodiversity. The banks of this stream in the rainforest of Costa Rica are lined with many different plant species.

imagebroker/Alamy Limited

Sometimes the changes are slow and subtle, as when crustal plates move over geologic time to create continents and ocean basins. At other times, the changes are rapid, as when hurricane winds flatten vast areas of forests or even tracts of houses and homes.

Environmental change is now produced not only by the natural processes that have acted on our planet for millions of years, but also by human activity. The human race has populated our planet so thoroughly that few places remain free of some form of human impact. Global change, then, involves not only natural processes, but also human processes that interact with them. Physical geography is the key to understanding this interaction.

Environment and global change are so important that we open nearly every chapter with a special section titled "Eye on Global Change" to introduce change-related topics. What are some of the important topics of global change that lie within the scope of physical geography? Let's examine a few.

GLOBAL CLIMATE CHANGE

Are human activities changing global climate? It seems that almost every year we hear that it has been the hottest year, or one of the hottest years, on record. But climate is notoriously variable. Could such a string of hot years be part of the normal variation? This is the key question facing scientists studying global climate change. Over the past decade, nearly all scientists have come to the opinion that human activity has, indeed, begun to change our climate. How has this happened?

The answer lies in the greenhouse effect. As human activities continue to release gases that block heat radiation from leaving the Earth, the greenhouse effect intensifies. The most prominent of these gases is CO_2, which is released by fossil fuel burning. Others include methane (CH_4), nitrous oxide (NO), and the chlorofluorocarbons that, until recently, served as coolants in refrigeration and air conditioning systems and as aerosol spray propellants. Taken with other gases, they act to raise the Earth's surface temperature, with consequences including dislocation of agricultural areas, rise in sea level, and increased frequency of extreme weather events, such as severe storms or record droughts.

Climate change is a recurring theme throughout this book, ranging from the urban heat island effect that tends to raise city temperatures (Chapter 3), to the El Niño phenomenon that alters global atmospheric and ocean circulation (Chapter 5), to the effect of clouds on global warming (Chapter 6), and to a rising sea level due to the expansion of seawater with increasing temperatures (Chapter 16).

THE CARBON CYCLE

One way to reduce human impact on the greenhouse effect is to slow the release of CO_2 from fossil fuel burning. But modern civilization depends on the energy of fossil fuels to carry out almost every task, so reducing fossil fuel consumption to stabilize the increasing concentration of CO_2 in the atmosphere is not easy. Fortunately, some natural processes do reduce atmospheric CO_2. Plants, for example, withdraw CO_2 from the atmosphere by taking it up in photosynthesis to construct plant tissues, such as cell walls and wood. In addition, CO_2 is soluble in seawater. These two important pathways, by which carbon flows from the atmosphere to lands and oceans, are part of the carbon cycle. Biogeographers and ecologists are now focusing in detail on the global carbon cycle in order to better understand the pathways and magnitudes of carbon flow. They hope that this understanding will suggest alternative actions that can reduce the rate of CO_2 buildup without penalizing economic growth. The processes of the carbon cycle are described in Chapter 8.

> Environmental change is produced by both natural and human processes. Human activities are currently changing both the Earth's climate and the global flows of carbon from land to ocean to atmosphere.

BIODIVERSITY

Among scientists, environmentalists, and the public, there is growing awareness that the diversity in the plant and animal forms harbored by our planet—the Earth's biodiversity—is an immensely valuable resource that, if protected, will be cherished by future generations. One important reason for preserving as many natural species as possible is that, over time, species have evolved natural biochemical defense mechanisms against diseases and predators. These defense mechanisms involve bioactive compounds that can sometimes be very useful, ranging from natural pesticides that increase crop yields to medicines that fight human cancer.

Another important reason for maintaining biodiversity is that complex ecosystems with many species tend to be more stable and to respond better to environmental change. If human activities inadvertently reduce biodiversity, there is a greater risk of unexpected and unintended human effects on natural environments. Biogeographers focus on both the existing biodiversity of the Earth's many natural habitats and the processes that create and maintain biodiversity. These topics are treated in Chapters 8 and 9.

Human activity is reducing the biodiversity of many of the Earth's natural habitats. Environmental pollution degrades habitat quality for plants and animals, as well as humans. Extreme weather events, which will become more frequent as a result of human-induced climate change, as well as other rare natural events, are increasingly destructive to our expanding human population.

POLLUTION

As we all know, unchecked human activity can degrade environmental quality. In addition to releasing CO_2, fuel burning can yield hazardous gases that react to form toxic compounds such as ozone and nitric acid in photochemical smog. Water pollution from fertilizer runoff, toxic wastes of industrial production, and acid mine drainage can severely degrade water quality. This degradation impacts not only the ecosystems of streams and rivers, but also the human populations that depend on rivers and streams as sources of water supply. Groundwater reservoirs can also be polluted or turn salty in coastal zones when drawn down excessively.

Environmental pollution, its causes, its effects, and the technologies used to reduce pollution, comprise a subject in its own right. As a text in physical geography that emphasizes the natural processes of the Earth's land surface, we touch on air and water pollution in two chapters: Chapter 4 for air pollution and Chapter 14 for surface water pollution, irrigation effects, and groundwater contamination.

EXTREME EVENTS

Catastrophic events—floods, fires, hurricanes, earthquakes, and the like—can have great and long-lasting impacts on both human and natural systems. Are human activities increasing the frequency of these extreme events? As our planet warms in response to changes in the greenhouse effect, global climate modelers predict that weather extremes will become more severe and more frequent. Droughts and consequent wildfires and crop failures will happen more often, as will spells of rain and flood runoff. In the last decade, we have seen numerous examples of extreme weather events, from Hurricane Katrina in 2005—the most costly storm in U.S. history—to the devastating Texas drought of 2011–2012. Is human activity responsible for the increased occurrence of these extreme events? Significant evidence now points in that direction.

Other extreme events, such as earthquakes, volcanic eruptions, and tsunamis, are produced by forces deep within the Earth that are not affected by human activity. But as the human population continues to expand and rely increasingly on a technological infrastructure, ranging from skyscrapers to the Internet, we are becoming more sensitive to damage and disruption caused by extreme events.

This text describes many types of extreme events and their causes. In Chapters 4 and 6, we discuss thunderstorms, tornadoes, cyclonic storms, and hurricanes. Droughts in the African Sahel are presented in Chapter 7. Earthquakes, volcanic eruptions, and tsunamis are covered in Chapter 12. Floods are described in Chapter 14.

Tools in Physical Geography

Geographers use a number of specialized tools to examine, explore, and interact with spatial data (Figure I.9). One of the oldest tools is the *map*, a representation of space showing where things are. While maps will never go out of style, computers have enhanced our ability to store, retrieve, and analyze spatial data through the development of the *geographic information system* (*GIS*). Acquiring geographic information for input to GIS has recently been made much easier through use of the *global positioning system* (GPS), which allows hand-held electronic equipment, linked to signals from orbiting spacecraft, to easily determine the latitude, longitude, and elevation of any point on the Earth's surface to within a few meters.

Satellites carrying imaging instruments have provided a wealth of information about the Earth's surface layers, including land, oceans, and atmosphere, that is vital to geographic study. The field of processing, enhancing, and analyzing images and measurements made from aircraft and spacecraft is known as *remote sensing*. Recent developments linking remote sensing, GIS, and GPS with the Internet have produced new Earth visualization tools, such as Google Earth, that are also of great interest to geographers.

Tools in geography also include *mathematical modeling* and *statistics*. Using math

> Maps, geographic information systems (GISs), and remote sensing are important geographic tools to acquire, display, and manipulate spatial data. Mathematical modeling and statistics are also helpful tools for the geographer.

and computers to model geographic processes is a powerful approach to understanding both natural and human phenomena. Statistics provides methods that can be used to manipulate geographic data so that we can ask and answer questions about differences, trends, and patterns. But because these tools rely heavily on specialized knowledge, they are not included here. Our text does, however, present many examples of geographic information obtained using modeling and statistics.

I.9 Tools of Physical Geography

Geographers rely on specialized tools to analyze spatial data.

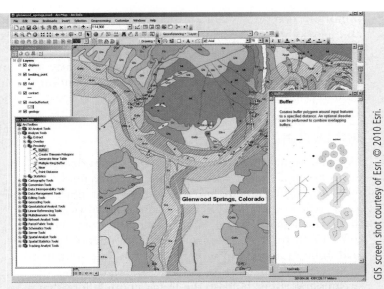

▲ GEOGRAPHIC INFORMATION SYSTEMS (GIS)

Computer programs that store and manipulate geographic data are essential to modern applications of geography. This screen from the ARCGIS program package shows a geologic map of Glenwood Springs, Colorado, with an added buffer corridor surrounding the Colorado River and a tributary.

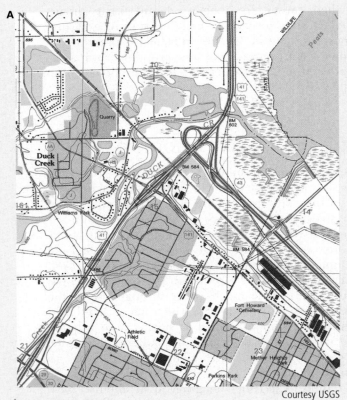

Courtesy USGS

▲ CARTOGRAPHY

A portion of the U.S. Geological Survey 1:24,000 topographic map of Green Bay, Wisconsin. Using symbols, the map shows creeks and rivers; a bay; swampy regions; urban developed land; streets, roads, and highways.

MATHEMATICAL MODELING ▶

By describing a phenomenon using a mathematical model, a geographer can predict outcomes and examine what-if scenarios. These equations demonstrate the calculation of an exponential growth factor.

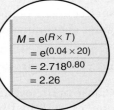

$$M = e^{(R \times T)}$$
$$= e^{(0.04 \times 20)}$$
$$= 2.718^{0.80}$$
$$= 2.26$$

REMOTE SENSING

Remote sensing includes observing the Earth from the perspective of an aircraft or spacecraft. Wildfires on the Greek island of Peloponnesus, seen in a Landsat image, provide an example.

▼

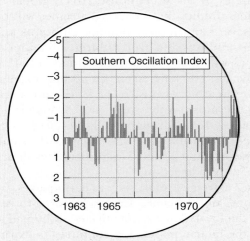

▲ STATISTICS

Statistical tools, such as this graph, allow the exploration of geographic data to determine trends and develop mathematical models. The plot shows the value of the Southern Oscillation Index, an indicator of El Niño conditions.

Maps and Cartography

Cartography is the field of geography concerned with making maps. A **map** is a paper representation of space showing point, line, or area data—that is, locations, connections, and regions. It typically displays a set of characteristics or features of the Earth's surface that are positioned on the map in much the same way that they occur on the surface. The map's scale links the true distance between places with the distance on the map.

Maps play an essential role in the study of physical geography because much of the information content of geography is stored and displayed on maps. Map literacy—the ability to read and understand what a map shows—is a basic requirement for day-to-day functioning in our society. Maps appear widely in newspapers and magazines and in nearly every television newscast. Most people routinely use highway maps, street maps, and maplike displays on navigation systems. Maps also pop up on web sites to show locations. The purpose of this part of the chapter is to provide additional information on the art and science of maps.

MAP PROJECTIONS

Cartographers record position on the Earth's surface using latitude and longitude. You'll read more about latitude and longitude in Chapter 1, but for now, all you need to know is that latitude measures position in a north-south direction and that longitude measures position in an east-west direction. Lines of equal latitude are *parallels*, and lines of equal longitude are *meridians*.

A **map projection** is an orderly system of lines of latitude and longitude used as a base for drawing a map on a flat surface. A projection is needed because the Earth's surface is not flat but, rather, curved in a shape that is very close to the surface of a sphere. All map projections misstate the shape of the Earth in some way. It's simply impossible to transform a spherical surface to a flat (planar) surface without violating the true surface.

Perhaps the simplest of all map projections is a grid of perfect squares. In this simple map, horizontal lines are parallels, and vertical lines are meridians. They are equally spaced in degrees, so this projection is sometimes called an *equal-angle grid*. A grid of this kind can show the true spacing (approximately) of the parallels, but it fails to show how the meridians converge toward the two poles. This convergence causes the grid to fail dismally in high latitudes, and the map usually has to be terminated at about $70° - 80°$ north and south.

Early attempts to find satisfactory map projections made use of a simple concept. Imagine the spherical Earth grid as a cage of wires located on meridians and parallels. A tiny light source is placed at the center of the cage, and the image of the wire grid is cast upon a surface outside the sphere. This situation is like a reading lamp with a lampshade. Basically, three kinds of "lampshades" can be used, as shown in Figure I.10.

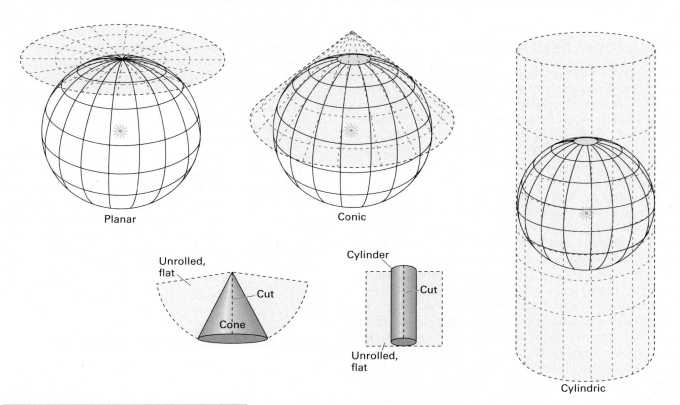

Planar

Unrolled, flat

Cut

Cone

Conic

Cylinder

Cut

Unrolled, flat

Cylindric

I.10 Simple ways to generate map projections

Rays from a central light source cast shadows of the spherical geographic grid on target screens. The conical and cylindrical screens can be unrolled to become flat maps.

First is a flat paper disk perched on the North Pole. The shadow of the wire grid on this plane surface will appear as a combination of concentric circles (parallels) and radial straight lines (meridians). Here we have a polar-centered, or *polar projection*. Second is a cone of paper resting point-up on the wire grid. The cone can be slit down the side, unrolled, and laid flat to produce a map that is some part of a full circle. This is called a *conic projection*. Parallels are arcs of circles, and meridians are radiating straight lines. Third, a cylinder of paper can be wrapped around the wire sphere so as to be touching all around the Equator. When slit down the side along a meridian, the cylinder can be unrolled to produce a *cylindrical projection*, which is a true rectangular grid.

> Map projections allow the curved surface of the Earth to be displayed on a flat map.

None of these three projection methods can show the entire Earth grid, no matter how large a sheet of paper is used to receive the image. Obviously, if the entire Earth grid, or large parts of it, are to be shown, some quite different system must be devised. In Chapter 1, we describe three types of projections used throughout the book: the polar projection; the Mercator projection, which is a cylindrical projection; and the Winkel Tripel projection, which uses special mathematics that provide minimum distortion in a global map.

WileyPLUS Map Projections
Watch an animation showing how map projections are constructed.

SCALES OF GLOBES AND MAPS

All globes and maps depict the Earth's features in much smaller size than the true features they represent. Globes are intended in principle to be perfect scale models of the Earth itself, differing from the Earth only in size. The scale of a globe is the ratio between the size of the globe and the size of the Earth, where "size" is some measure of length or distance (but not of area or volume).

Take, for example, a globe 20 cm (about 8 in.) in diameter, representing the Earth, which has a diameter of about 13,000 km. The scale of the globe is the ratio between 20 cm and 13,000 km. Dividing 13,000 by 20, we see that 1 centimeter on the globe represents 650 kilometers on the Earth. This relationship holds true for distances between any two points on the globe.

Scale is often stated as a simple fraction, termed the **scale fraction**. It can be obtained by reducing both Earth and globe distances to the same unit of measure, which in this case is centimeters. (There are 100,000 centimeters in 1 kilometer.) The advantage of the scale fraction is that it is entirely free of any specified units of measure, such as the foot, mile, meter, or kilometer. It is usually written as a fraction with a numerator of 1, either followed by a colon or with the numerator above the denominator, separated by a slash (/).For the example of the globe shown above, the scale fraction is obtained by reducing 20/1300000000 to 1/65000000, or 1:65,000,000.

In contrast to a globe, a flat map cannot have a constant scale. In flattening the curved surface of the sphere to conform to a plane surface, all map projections stretch the Earth's surface in a nonuniform manner, so that the map scale changes from place to place on the map. However, it is usually possible to select a meridian or parallel—the Equator, for example—for which a scale fraction can be given, relating the map to the globe it represents.

SMALL-SCALE AND LARGE-SCALE MAPS

When geographers refer to small-scale and large-scale maps, they mean the value of the scale fraction. For example, a global map at a scale of 1:65,000,000 has a scale fraction value of 0.00000001534, which is obtained by dividing 1 by 65,000,000. A hiker's topographic map might have a scale of 1:25,000, for a scale value of 0.000040. Since the global-scale value is smaller, it is a *small-scale map*, while the hiker's map is a *large-scale map*. Note that this contrasts with common use of the terms *large-scale* and *small-scale*. When we refer in conversation to a large-scale phenomenon or effect, we typically mean something that takes place over a large area and that is usually best presented on a small-scale map.

Maps of large scale show only small sections of the Earth's surface. Because they "zoom in," they are capable of carrying an enormous amount of geographic information in a convenient and effective manner. Most large-scale maps feature a graphic scale, which is a line marked off into units representing kilometers or miles. Figure I.11 shows a portion of a large-scale map on which sample graphic scales in miles, feet, and kilometers are superimposed. Graphic scales make it easy to measure ground distances.

For practical reasons, maps are printed on sheets of paper usually less than a meter (3 ft) wide, as in the case of the ordinary highway map or navigation chart. Bound books of maps—atlases—usually have pages no larger than 30 by 40 cm (about 12 by 16 in.), whereas maps found in textbooks and scientific journals are even smaller. Maps displayed on a computer screen are often more flexible, allowing the viewer to scroll over the map in any direction, and to zoom in and zoom out as well. In many cases, the details displayed change with the scale of viewing.

CONFORMAL AND EQUAL-AREA MAPS

As we can see from Figure I.10, the shape and area of a small feature, such as an island or peninsula, will change as the feature is projected from the surface of the globe to a map. With some projections, the area will change,

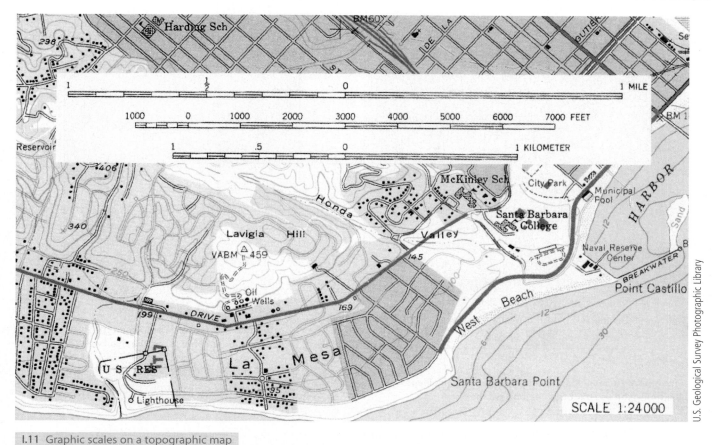

I.11 Graphic scales on a topographic map

A portion of a modern, large-scale topographic map, for which three graphic scales have been provided.

but the shape will be preserved. Such a projection is referred to as *conformal*. The Mercator projection shown in Chapter 1, Figure 1.11 is an example. Here, every small twist and turn of the shoreline of each continent is shown in its proper shape. However, the growth of the continents with increasing latitude shows that the Mercator projection does not depict land areas uniformly. A projection that does show area uniformly is referred to as *equal-area*. Here, continents show their relative areas correctly, but their shapes are distorted. No projection can be both conformal and equal-area; only a globe has that property.

> Conformal map projections show shapes correctly, whereas equal-area maps show areas correctly.

INFORMATION CONTENT OF MAPS

The information conveyed by a map projection grid system is limited to one category only: absolute location of points on the Earth's surface. To be more useful, maps also carry other types of information. Figure I.11 is a portion of a large-scale *multipurpose map*. Map sheets published by national governments, such as this one, are usually multipurpose maps. Using a great variety of symbols, patterns, and colors, these maps have high informa-tion content. Appendix 3 shows a larger example of a multipurpose map, a portion of a U.S. Geological Survey topographic quadrangle map for San Rafael, California.

In contrast to the multipurpose map is the *thematic map*, which shows only one type of information, or theme. We use many thematic maps in this text. Some examples include Figure 4.26, which maps the frequency of severe hailstorms in the United States; Figure 5.17, atmospheric surface pressures; Figure 7.4, mean annual precipitation of the world; and Figure 7.8, world climates.

MAP SYMBOLS

Symbols on maps associate information with points, lines, and areas. To show information at a point, we use a *dot*, defined as a small symbol to show point location. It might be a closed circle, an open circle, a letter, a numeral, or a graphic symbol of the object it represents (see "church with tower" in Figure I.12). A *line* can vary in width and be single or double, colored, dashed, or dotted. A *patch* denotes a particular area, typically using a distinctive pattern or color or a line marking its edge.

Figure I.12 shows symbols applied to a map. There are two kinds of dot symbols (both symbolic of churches), three kinds of line symbols, and three kinds of patch

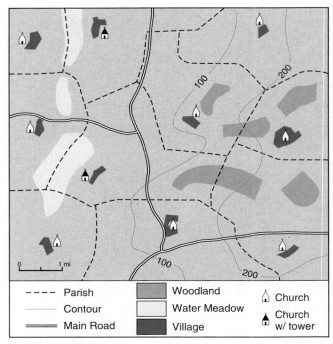

---- Parish	Woodland	⌂ Church
Contour	Water Meadow	Church w/ tower
Main Road	Village	

I.12 Map symbols

A multipurpose map of an imaginary area, with 10 villages illustrating the use of dots, lines, and patches.

symbols. Altogether, eight types of information are present. Line symbols freely cross patches, and dots can appear within patches. Two different kinds of patches can overlap. For more examples of symbols, consult the display of topographic map symbols facing the map of San Rafael in Appendix 3.

Map symbols can vary with map scale. Maps of very large scale—for example, a plot plan of a house—can show objects in their true outline form. As map scale is decreased, representation becomes more and more generalized. In physical geography, an excellent example is the depiction of a river, such as the lower Mississippi, shown in Figure I.13. The level of depiction of fine detail in a map is described by the term *resolution*. Maps of large scale have much greater resolving power than maps of small scale.

PRESENTING NUMERICAL DATA ON THEMATIC MAPS

In physical geography, we often need to display numerical information on maps. For example, a weather map typically shows air temperature, air pressure, and wind speed. Another category of information consists simply of the presence or absence of something. In this case, we can simply place a dot to mean "present," so that when entries are completed, the map shows a field of scattered dots (Figure I.14).

In some scientific programs, measurements are taken uniformly; for example, at the centers of grid squares, laid over a map. For many classes of data, however, the locations of the observation points are predetermined

Maps of the Mississippi River on three scales (slightly enlarged for reproduction).

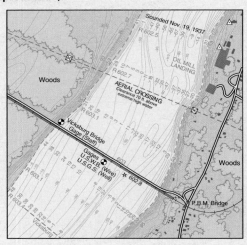

▲ **1:20,000 SCALE**
This map shows a detailed plan of the river, which even includes contours on the riverbed.

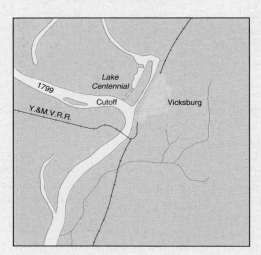

▲ **1:250,000 SCALE**
At this scale, the river is depicted by two lines showing its banks, and color to indicate the area of the river.

▲ **1:3,000,000 SCALE**
At a very small scale, the river is shown as a solid line.

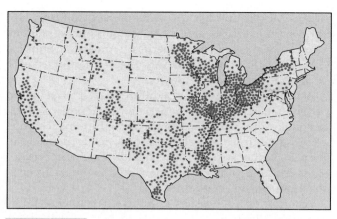

I.14 Dot map

A dot map showing the distribution of soils of the order Alfisols in the United States.

by a fixed and nonuniform set of observing stations. For example, weather and climate data are often collected at airports. Whatever the sampling method used, we end up with an array of numbers and dots indicating their location on the base map.

Although the numbers and locations may be accurate, it may be difficult to see the spatial pattern present in the data being displayed. For this reason, cartographers often simplify arrays of point values into isopleth maps. An **isopleth** is a line of equal value (from the Greek *isos*, "equal," and *plethos*, "fullness" or "quantity"). Figure 3.21 shows how an isopleth map is constructed for temperature data. In this case, the isopleth is an *isotherm*, or line of constant temperature. In drawing an isopleth, the line is routed among the points in a way that best indicates a uniform value, given the observations at hand.

Isopleth maps are important in various branches of physical geography. Table I.1 gives a partial list of isopleths of various kinds used in the earth sciences, together with their names and the kinds of information they display. A special kind of isopleth, the *topographic contour* (or isohypse), is shown on the maps in Figures I.11, I.13 (1:20,000 scale), and in the portion of the San Rafael topographic map in Appendix 3. Topographic contours show the configuration of land surface features, such as hills, valleys, and basins.

In contrast to the isopleth map is the *choropleth* map, which identifies information in categories. Our global maps of vegetation (Figure 9.7) and soils (Figure 10.18)

are examples of thematic choropleth maps.

Cartography is a rich and varied field of geography with a long history of conveying geographic information accurately and efficiently. If you are interested in maps and mapmaking, you might want to investigate cartography further.

> Isopleth maps show lines of equal value. Choropleth maps show categorical information associated with particular areas.

The Global Positioning System

The latitude and longitude coordinates of a point on the Earth's surface describe its position exactly. But how are those coordinates determined? For the last few hundred years, we have known how to use the position of the stars in the sky, coupled with an accurate clock, to determine the latitude and longitude of any point. Linked with advances in mapping and surveying, these techniques became highly accurate, but they remained impractical for precisely determining locations in a short period of time.

Thanks to new technology originally developed by the U.S. Naval Observatory for military applications, there is now in place a **global positioning system (GPS)** that can provide accurate location information anywhere in the world within a minute or two. The system uses about 30 satellites that orbit the Earth every 12 hours, continuously broadcasting their position and a highly accurate time signal (Figure I.15).

To determine location, a receiver listens simultaneously to radio signals from four or more satellites. The receiver compares the time readings transmitted by each satellite with the receiver's own clock to determine how long it took for each signal to reach the receiver. Radio signals travel at a known rate of speed, which enables the receiver to then convert the travel time into the distance between the receiver and the satellite. Coupling the distance to each satellite with the position of the satellite in its orbit at the time of the transmission, the receiver calculates its position on the ground to within about 20 m (66 ft) horizontally and 30 m (98 ft) vertically.

The accuracy of the location may, however, be disrupted by several types of errors. One of the major causes of error is the effect of the atmosphere on the speed of the

Table I.1 Examples of Isopleths			
Name of Isopleth	Greek Root	Property Described	Examples in Figures
Isobar	*barros*, weight	Barometric pressure	5.17
Isotherm	*therme*, heat	Temperature of air, water, or soil	3.21
Isotach	*tachos*, swift	Fluid velocity	5.24
Isohyet	*hyetos*, rain	Precipitation	4.19
Isohypse (topographic contour)	*hypso*, height	Elevation	Appendix 3

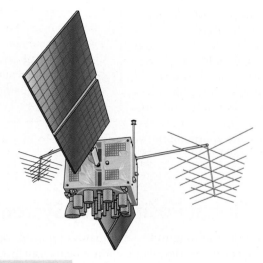

I.15 GPS satellite

A GPS satellite as it might look in orbit high above the Earth. The U.S. Navy NAVSTAR GPS satellite system consists of about 30 orbiting satellites.

radio waves as they pass from the satellite to the receiver. Charged particles at the outer edge of the atmosphere (ionosphere) and water vapor in the lowest atmospheric layer (troposphere) act to slow the radio waves. Since the conditions in these layers can change within a matter of minutes, the speed of the radio waves varies in an unpredictable way. Another transmission problem is that the radio waves may bounce off local obstructions before reaching the receiver, causing two slightly different signals to arrive at the receiver at the same time. This "multipath error" is a source of noise that confuses the receiver.

There is a way, nevertheless, to determine location within about 1 m (3.3 ft) horizontally and 2 m (6.6 ft) vertically. The method uses two GPS units, one at a base station and one that is mobile and used to determine the desired locations. The base station unit is placed at a position that is known to a very high degree of accuracy.

> The global geographic positioning system uses signals from a constellation of orbiting satellites to locate points on the Earth with high accuracy.

By comparing its exact position with that calculated from each satellite signal, it determines the small deviations from orbit of each satellite, any small variations in each satellite's clock, and the exact speed of that satellite's radio signal through the atmosphere at that moment. It then broadcasts that information to the GPS field unit, where it is used to calculate the position more accurately. Because this method compares two sets of signals, it is known as *differential GPS*.

In North America, differential GPS information is now available everywhere using the *Wide Area Augmentation System (WAAS),* provided by the U.S. Federal Aviation Administration and the Department of Transportation. The system includes about 25 ground receiving stations that monitor the signals of GPS satellites and yield a stream of differential correction information. This information is uploaded to a geostationary satellite, where it is rebroadcast to receivers on the ground. A GPS unit with a built-in WAAS receiver can determine position to within a few meters.

The enhanced accuracy of differential GPS is required for coastal navigation, where a few meters in position can make the difference between a shipping channel and a shoal. It is also required for a new generation of aircraft landing systems that will allow much safer instrument landings, using equipment that is much lower in cost than existing systems.

As GPS technology has developed, costs have fallen exponentially. It is now possible to buy a small, hand-held GPS receiver for less than $100. GPS chips now routinely provide location information in smartphones for various applications. Besides plotting your progress on a computer-generated map as you drive your car or sail your boat, GPS technology can even help parents keep track of their children at a theme park.

WileyPLUS Global Positioning Systems
Watch an animation on the Navistar Global Positioning System to learn more about how the system works.

Geographic Information Systems

Maps, like books, are very useful devices for storing information, but they have limitations. Recent advances in computing capability have enabled geographers to develop a powerful new tool to work with spatial data: the **geographic information system (GIS)**. A GIS is a computer-based system for acquiring, processing, storing, querying, creating, analyzing, and displaying spatial data. Geographic information systems have allowed geographers, geologists, geophysicists, ecologists, planners, landscape architects, and others to develop applications of spatial data processing ranging from planning land subdivisions on the fringes of suburbia to monitoring the deforestation of the Amazon Basin.

WileyPLUS Geographic Information Systems
Watch a narrated animation to explore the key ideas behind geographic information systems and see some examples.

SPATIAL OBJECTS IN GEOGRAPHIC INFORMATION SYSTEMS

Geographic information systems are designed to manipulate spatial objects. A **spatial object** is a geographic point, line, or area to which some information is attached. This information may be as simple as a place name or as complicated as a large data table with many types of information. Some spatial objects are illustrated in Figure I.16.

A *point* is a spatial object without an area, only a location. A *line* is also a spatial object with no area, but it has two points associated with it, one for each end of the line. These special points are often referred to as *nodes*.

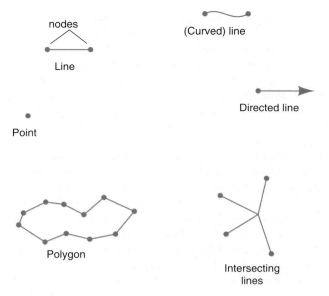

I.16 Spatial objects

Spatial objects in a GIS can include points, lines of various types, intersecting lines, and polygons.

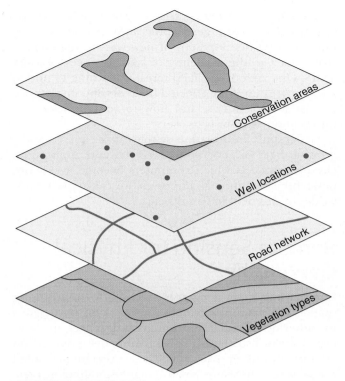

I.17 Data layers in a GIS

A GIS allows easy overlay of spatial data layers for such queries as "Identify all wells on conservation land."

Normally a line is straight, but it can also be defined as a smooth curve having a certain shape. If the two nodes marking the ends of the line are differentiated as starting and ending, then the line has a direction. If the line has a direction, then its two sides can be distinguished. This allows information to be attached to each side—for example, labels for land on one side and water on the other. Lines connect to other lines when they share a common node. A series of connected lines that form a closed chain is a *polygon*. A polygon identifies an *area*, the last type of spatial object.

By defining spatial objects in this way, computer-based geographic information systems allow easy manipulation of the objects and permit many different types of operations to compare objects and generate new objects. As an example, suppose we have a GIS data layer composed of conservation land in a region represented as polygons, and another layer containing the location of preexisting water wells as points within the region (Figure I.17). It is very simple to use the GIS to identify the wells that are on conservation land. Or the conservation polygons containing wells may also be identified, and even output, as a new data layer. By comparing the conservation layer with a road network layer portrayed as a series of lines, we could identify the conservation polygons containing roads.

> Geographic information systems use computers to store, process, analyze, and display spatial data.

We could also compare the conservation layer to a layer of polygons showing vegetation type, and tabulate the amount of conservation land in forest, grassland, brush, and so forth. We could even calculate distance zones around a spatial object, for example, to create a map of buffer zones that are located within, say, 100 m

(328 ft) of conservation land. Many other possible manipulations exist.

KEY ELEMENTS OF A GIS

A geographic information system consists of five elements: data acquisition, preprocessing, data management, data manipulation and analysis, and product generation. Each is a component or process needed to ensure the functioning of the system as a whole.

In the *data acquisition* process, data are gathered together for the particular application. These may include maps, air photos, tabular data, and other forms as well. In *preprocessing*, the assembled spatial data are converted to forms that can be ingested by the GIS to produce data layers of spatial objects and their associated information.

The *data management* component creates, stores, retrieves, and modifies data layers and spatial objects. It is essential to proper functioning of all parts of the GIS. The *manipulation and analysis* component is the real workhorse of the GIS. Utilizing this component, the user asks and answers questions about spatial data and creates new data layers of derived information.

The last component of the GIS, *product generation*, generates output products in the form of maps, graphics, tabulations, or statistical reports that are the end products desired by the users. Taken together, these components comprise a system that can serve many geographic applications at many scales.

Many new and exciting areas of geographic research are associated with geographic information systems, ranging from development of new ways to manipulate spatial data to the modeling of spatial processes using a GIS. One area of special interest is learning to understand how outputs are affected by errors and uncertainty in spatial data inputs, and how to communicate this information effectively to users.

Geographic information systems make up a rapidly growing field of geographic research and application. Given the rate at which computers become ever more powerful as technology improves, we can expect great strides in this field in future years.

Remote Sensing for Physical Geography

Another important geographic technique for acquiring spatial information is **remote sensing**. This term refers to gathering information from great distances and over broad areas, usually through instruments mounted on aircraft or orbiting spacecraft. These instruments, called *remote sensors*, measure electromagnetic radiation coming from the Earth's surface and atmosphere as received at the aircraft or spacecraft platform. The data acquired by remote sensors are typically displayed as images—photographs or similar depictions on a computer screen or color printer—but are often processed further to provide other types of outputs, such as maps of vegetation condition or extent, or of land-cover class. Information obtained can range from fine local detail (such as the arrangement of cars in a parking lot) to a global-scale picture (for example, the "greenness" of vegetation for an entire continent). As you read this textbook, you will see many examples of remote sensing, especially images from orbiting satellites.

All substances, whether naturally occurring or synthetic, are capable of reflecting, transmitting, absorbing, and emitting *electromagnetic radiation*. For remote sensing, however, we are only concerned with energy that is reflected or emitted by an object and reaches the remote sensor. For remote sensing of reflected energy, the Sun is the source of radiation in many applications. As we will see in Chapter 2, solar radiation reaching the Earth's surface is largely in the form of light energy that includes visible, near-infrared, and shortwave infrared light. Remote sensors are commonly constructed to measure radiation reflected from the Earth in all or part of this range of light energy. For remote sensing of emitted energy, the object or substance itself is the source of the radiation, which is related largely to its temperature.

COLORS AND SPECTRAL SIGNATURES

To the human eye, most objects or substances at the Earth's surface possess color. This means that they reflect radiation variously in different parts of the visible spectrum. Figure I.18 shows how the reflectance of water, vegetation, and soil varies, with light ranging in wavelength from visible to shortwave infrared. Water surfaces are always dark but are slightly more reflective in the blue and green regions of the visible spectrum. Thus, clear water appears blue or blue-green to our eyes. Beyond the visible region, water absorbs nearly all radiation it receives and so looks black in images acquired in the near-infrared and shortwave infrared regions.

Vegetation appears dark green to the human eye, which means that it reflects more energy in the green portion of the visible spectrum while reflecting somewhat less in the blue and red portions. But vegetation also reflects very strongly in near-infrared wavelengths, which the human eye cannot see. Because of this property, vegetation appears very bright in near-infrared images. This distinctive behavior of vegetation—appearing dark in visible bands and bright in the near-infrared—is the basis for much of vegetation remote sensing, as we will see in many examples of remotely sensed images throughout this book.

The soil spectrum shows a slow increase of reflectance across the visible and near-infrared spectral regions. Looking at the visible part of the spectrum, we see that soil is brighter overall than vegetation and is somewhat more reflective in the orange and red portions. Thus, it appears brown. (Note that this is just a "typical" spectrum; soil color can actually range from black to bright yellow or red.)

We refer to the pattern of relative brightness within the spectrum as the **spectral signature** of an object or type of surface. Spectral signatures can be used to recognize objects or surfaces in remotely sensed images in much the same way that we recognize objects by their colors. In computer processing of remotely sensed images, spectral signatures can be used to make classification maps, showing, for example, water, vegetation, and soil.

> Different types of remotely sensed objects or surfaces often reflect the electromagnetic spectrum differently, providing characteristic spectral signatures.

THERMAL INFRARED SENSING

While objects reflect some of the solar energy they receive, they also emit internal energy as heat that can be remotely sensed. Warm objects emit more *thermal radiation* than cold ones, so warmer objects appear brighter in thermal infrared images. You can look ahead to Chapter 2 (Figure 2.4) to see a night scene of a suburban street imaged by a thermal infrared camera. In Chapter 3 (Figure 3.8), you can see a thermal image showing how the downtown area of a city is hotter than the surrounding area.

Besides temperature, the intensity of infrared emission depends on the *emissivity* of an object or a

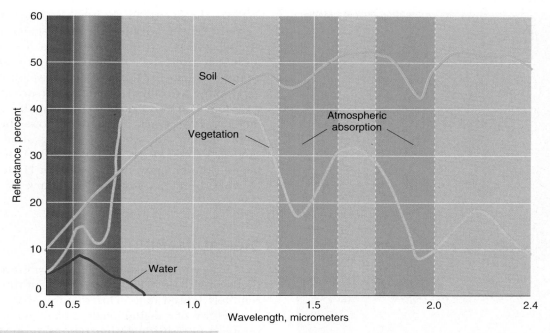

I.18 Reflectance spectra of vegetation, soil, and water

The amount of energy reflected by surfaces of vegetation, soil, and water depends on the wavelength of the light. At the left is the visible spectrum, from about 0.4 to 0.65 micrometers (μm). This is the portion of the spectrum to which our eyes are sensitive. The graph also shows two spectral regions where the atmosphere absorbs radiation. In these regions, a spaceborne sensor can "see" the Earth's surface only dimly, if at all.

substance. Objects with higher emissivity appear brighter at a given temperature than objects with lower emissivities. Differences in emissivity affect thermal images. For example, two different surfaces might be at the same temperature, but the one with the higher emissivity will look brighter because it emits more energy. Crystalline minerals often show varying emissivities at different wavelengths in the thermal infrared spectrum. In a way, this is like having a particular color, or spectral signature, in the thermal infrared spectral region. In Chapter 11 we will see examples of how some rock types can be distinguished and mapped using thermal infrared images.

ACTIVE SENSING—RADAR AND LIDAR

There are two classes of remote sensor systems: passive and active. *Passive systems* acquire images without providing a source of wave energy. The most familiar passive system is the camera, which uses electronic detectors or photographic film to sense solar energy reflected from the scene. *Active systems* use a beam of wave energy as a source, sending the beam toward an object or surface. Part of the energy is reflected back to the source, where it is recorded by a detector.

Radar is an example of an active sensing system that is often deployed on aircraft or spacecraft. (The acronym comes from RAdio Detection And Ranging.) Radar systems in remote sensing use the *microwave* portion of the electromagnetic spectrum, so named because the waves have a short wavelength compared to other types of radio waves. Radar systems emit short pulses of microwave radiation and then "listen" for a returning microwave echo. By analyzing the strength of each return pulse and the exact time it is received, an image is created showing the surface as it is illuminated by the radar beam.

> Radar is an active remote sensing system that emits a pulse of microwaves and then measures the time and strength of the response scattered back to the radar instrument from the ground or atmosphere.

Radar systems used for land imaging emit microwave energy that is not significantly absorbed by atmospheric water. This means that radar systems can penetrate clouds to provide images of the Earth's surface in any weather. In contrast, ground-based weather radars use microwaves that are scattered by water droplets or ice crystals and produce an image of precipitation over a region. They detect rain, snow, and hail, and are used in local weather forecasting.

Radar images are produced by radar instruments aboard an aircraft or spacecraft that send pulses of radio waves downward and sideward as the craft flies

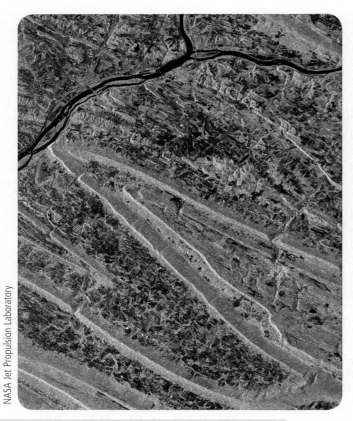

NASA Jet Propulsion Laboratory

I.19 Side-looking radar image from central Pennsylvania

Zigzag ridges in the Ridge and Valley regions of central Pennsylvania are the most prominent feature of this radar image centered just southeast of Sunbury. The ridges running from upper left to lower right curve and turn sharply, revealing the geologic structure of the region. Between the ridges are agricultural lands, with varying patterns of forests and fields. The top of the image shows the confluence of the Susquehanna River with its West Branch at the town of Sunbury. This image was acquired by a NASA radar instrument aboard the space shuttle Endeavor. The area shown is about 30 by 38 km (19 by 24 mi).

forward. Surfaces oriented most nearly at right angles to the slanting radar beam will return the strongest echo and therefore appear lightest in tone. In contrast, those surfaces facing away from the beam will appear darkest. The effect is to produce an image resembling a three-dimensional model of the landscape illuminated at a strong angle (Figure I.19).

Another type of active imaging uses **lidar** (Light Detection And Ranging). In a typical lidar system, a laser generates rapid pulses of light that are focused into a beam, and the beam is directed toward the ground from an aircraft flying overhead. When a pulse hits the ground, a portion of the light is reflected back toward the aircraft. A light detector, also focused in the direction of the beam, measures the strength of the returning pulse and the time it has taken for the pulse to travel out and back.

Since the speed of light is known, the travel time of the pulse return can determine the distance between the aircraft and the ground to within a few centimeters in many cases. By sweeping the ground with the beam, an image of the distance between the aircraft and ground is acquired. From this distance measure and knowledge of the exact position of the aircraft, an image can be constructed that shows the elevation of the ground and objects on the ground very accurately. Figure I.20 shows how a NASA aircraft lidar imaged the changes in a portion of Hatteras Island, North Carolina, following the attack of Hurricane Isabel in 2003.

DIGITAL IMAGING

Modern remote sensing relies heavily on computer processing to extract and enhance information from remotely sensed data. This requires that the data be in the form of a **digital image**. In a digital image, large numbers of individual observations, termed *pixels*, are arranged in a systematic way related to the Earth's position from which the observations were acquired. These images are quite similar to the digital images provided by digital cameras and cellular phones, although they are often acquired using scanning technology, described below. *Image processing* refers to the manipulation of digital images to extract, enhance, and display the information that they contain. In remote sensing, image processing is a very broad field that includes many methods and techniques for processing remotely sensed data.

Many remotely sensed digital images are acquired by scanning systems, which may be mounted in aircraft or on orbiting space vehicles. *Scanning* is the process of receiving information instantaneously from only a very small portion of the area being imaged (Figure I.21). The scanning instrument senses a very small field of view that runs rapidly across the ground scene. Light from the field of view is focused on a detector that responds very quickly to small changes in light intensity. Electronic circuits read out the detector at very short time intervals and record the intensities. Later, a computer reconstructs a digital image of the ground scene from the measurements acquired by the scanning system. Most scanning systems in common use are **multispectral scanners**. These devices have multiple detectors and measure brightness in several wavelength regions simultaneously.

An alternative to scanning is *direct digital imaging* using large numbers of detectors arranged in a two-dimensional array (Figure I.22). This technology is in common use in digital cameras as well as in scientific applications. The array has millions of tiny detectors arranged in rows and columns that

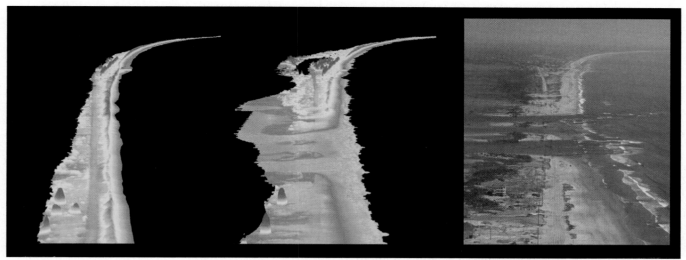

U.S. Geological Survey

I.20 Lidar images of Hatteras Island, before and after Hurricane Isabel

These two images, acquired by NASA's Experimental Advanced Airborne Research Lidar (EAARL), show a portion of Hatteras Island, part of North Carolina's outermost barrier beach, before and after Hurricane Isabel, which struck in September 2003. The color scale shows the elevation measured by the lidar pulses, with highest elevations in red and lowest in green. The area of low dunes in the middle of the left image gave way to the wave attack of the storm, and the island was breached, as shown in the right image. The main road down the island, visible on the left as stripe of constant elevation behind the dunes, was washed away. In the foreground, several buildings are identified by the height of their roofs. Two buildings were washed away by the storm.

individually measure the amount of light they receive during an exposure. Electronic circuitry reads out the measurement made by each detector, composing the entire image rapidly. Advanced digital cameras now record detail as finely as film cameras.

SATELLITE ORBITS

With the development of orbiting Earth satellites carrying remote sensing systems, this technology has expanded into a major branch of geographic research. Because orbiting satellites can image and monitor large geographic areas, or even the entire Earth, we can now carry out global and regional studies that cannot be done in any other way.

Courtesy Fairchild Imaging

I.22 An area array of detectors

The center part of this computer chip is covered by an array of about 4.2 million tiny light detectors arranged in a square of 2048 rows and columns. Although similar in concept to the area arrays used in common digital cameras and smartphones, this array is specially designed for advanced scientific, space, medical, industrial, and commercial applications.

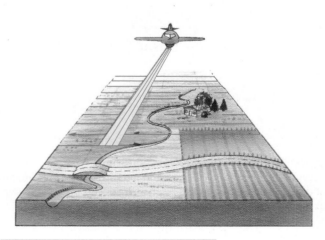

I.21 Multispectral scanning from aircraft

As the aircraft flies forward, the scanner sweeps from side to side. The result is a digital image covering the overflight area.

Most satellites designed for remote sensing use a **Sun-synchronous orbit** (Figure I.23). As the satellite circles the Earth, passing near each pole, the Earth rotates underneath it, allowing all of the Earth to be imaged after repeated passes. The orbit is designed so that the images of a location acquired on different days are taken at the same hour of the day. In this way, the solar lighting conditions remain about the same from one image to the next. Typical Sun-synchronous orbits take 90 to 100 minutes to circle the Earth and are located at heights of about 700 to 800 km (430 to 500 mi) above the Earth's surface.

Another orbit used in remote sensing is the **geostationary orbit**. Instead of orbiting above the poles, a satellite in geostationary orbit constantly revolves above the Equator. The orbit height, about 35,800 km (22,200 mi), is set so that the satellite makes one revolution in exactly 24 hours in the same direction that the Earth

> A Sun-synchronous orbit allows a remote imager to cover nearly all the Earth's surface, with only slowly varying illumination conditions. A geostationary orbit places the imager above a single point on the Equator, viewing an area of about half the Earth.

turns. Thus, the satellite always remains above the same point on the Equator. From its high vantage point, the geostationary orbiter provides a view of nearly half of the Earth at any moment.

Geostationary orbits are ideal for observing weather, and the weather satellite images readily available on television and the Internet are obtained from geostationary remote sensors. Geostationary orbits are also used by communications satellites. Since a geostationary orbiter remains at a fixed position in the sky for an Earthbound observer, a high-gain antenna can be pointed at the satellite and fixed in place permanently, enabling high-quality, continuous communications. Satellite television systems also use geostationary orbits.

LANDSAT, MODIS, AND THE EARTH OBSERVING SYSTEM

The launch of the first *Landsat* satellite by NASA in 1972 provided nearly complete coverage of the Earth's surface for the first time in history, acquiring simultaneous images drawn from four wavelength bands in the visible and near-infrared parts of the spectrum. Since that time, six additional Landsat imagers have contributed to a record

---

I.23 Satellite orbits

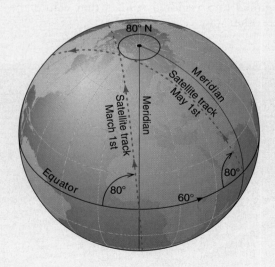

▲ EARTH TRACK OF A SUN-SYNCHRONOUS ORBIT
The satellite circles the globe about every 90 minutes, imaging the Earth as it turns eastward underneath the satellite track. With the Earth track inclined at about 80° to the Equator, the orbit slowly swings eastward at about 1° per day, maintaining its relative position with respect to the Sun from season to season. This keeps the solar lighting conditions similar from one image of a location to the next. Between March 1 and May 1 (shown) the orbit moves about 60°.

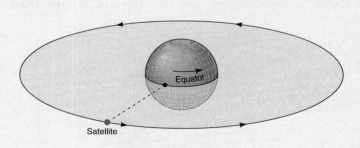

▲ MOTION OF A GEOSTATIONARY SATELLITE
Because the satellite revolves around the Earth above the Equator at the same rate as the Earth's rotation, it appears fixed in the sky above a single point on the Equator.

Scientists use comparisons of satellite data over time to show changes to the Earth's surface.

US Geological Survey (USGS) Courtesy NASA/Goddard Space Flight Center

US Geological Survey (USGS), Goddard Space Fligh Center/Courtesy NASA

▲ LAS VEGAS IN 1973
In 1973, approximately 500,000 people lived in the desert city of Las Vegas.

▲ LAS VEGAS IN 2006
By 2006, the population had grown to 1.5 million. Las Vegas sprawled in every direction, especially to the northwest and south. This unprecedented rate of growth has put a strain on water and energy supplies.

of four decades of images of our planet. As the Landsat sensor system was refined, later imagers added more spectral bands and provided a basic spatial resolution of 30 m (98.4 ft) per pixel, with global coverage every 16 days. Remotely sensed data from Landsat are used across a range of geographic studies, from crop analysis to transportation corridor planning to sea ice measurements to mapping of the expansion of deserts. Thanks to lengthy records from Landsat, coupled with other satellite data, scientists and policymakers now can study changes to the Earth's surface to evaluate how humans are altering our planet and determine what our environmental response to those changes should be (Figure I.24).

In 1999, NASA launched the first *MODIS* imager aboard the Terra spacecraft; it was followed, in 2002, by a second MODIS imager on the Aqua spacecraft.

Unlike Landsat imagers, MODIS provides daily images covering nearly the entire globe at a basic spatial resolution of 1 km (0.62 mi) per pixel (Figure I.25). Acquiring image data in 36 spectral bands for studies of ocean, land, and atmosphere, MODIS is an ideal instrument for viewing large regions, and even continents, many times per year. MODIS (for Moderate-Resolution Imaging Spectroradiometer) is being used to monitor global change in many ways, ranging from expansion of agricultural areas to tropical deforestation and shrinking arctic sea ice cover. We will see a number of other examples of MODIS images in later chapters.

MODIS and Landsat are part of NASA's Earth Observing System (EOS), an integrated array of satellite platforms and instruments that make integrated measurements of Earth system processes from long-term

Jacques Descloitres, MODIS Rapid Response Team, NASA/GSFC

I.25 MODIS views of eastern North America

The **MODIS (Moderate-Resolution Imaging Spectroradiometer)** instrument is ideal for imaging large areas, such as eastern North America, shown here in an image from April 14, 2003. At this time of year, plants in the southern and central regions are leafing out and appear green. To the north, darker tones indicate leafless trees and fields. Snow cover is largely gone, but still visible in northern New England and New York State.

EYE ON THE LANDSCAPE What else would the geographer see? **A** The rich agricultural land along the Mississippi River, on the left side of the image, is still largely fallow, judging by the brown colors. **B** The Smoky Mountains along the border of Tennessee and North Carolina are visible as a "tongue" of brown colors surrounded by greener hills and lowlands. **C** The green and brown tones of shallow Lake Erie indicate sediment and phytoplankton at or near the surface while the water of the other Great Lakes is largely clear.

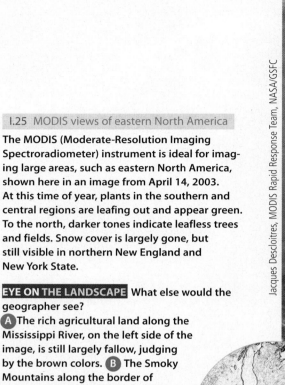

observations of the land surface, biosphere, atmosphere, and oceans (Figure I.26). The system focuses on observing and studying clouds, water and energy cycles, oceans, chemistry of the atmosphere, changes to the land surface, water and ecosystem processes, glaciers and polar ice sheets, and the solid Earth.

Remote sensing is an exciting, expanding field within physical geography and the geosciences in general. Throughout this text, you will see many more examples of remotely sensed images.

Earth Visualization Tools

Within the last few years, remote sensing, geographic information systems, and GPS technology have been integrated into new and exciting Internet tools for visualizing the Earth. Google Earth and World Wind are outstanding examples of these **Earth visualization tools**.

GOOGLE EARTH

Google Earth is a program for personal computers that allows users to roam the Earth's surface at will and zoom in on images showing the surface in detail. The program uses the Internet to access a large database of images maintained by Google (Figure I.27). The spatial resolution of the images varies from location to location, depending on the Earth imagery available. Basic coverage is provided largely by Landsat satellite data, with pixels 15 or 30 m (50 or 100 ft) on a side. But many areas of much higher spatial detail are present, using other satellite sources as well as air photos. Some states and a number of European countries are covered by images

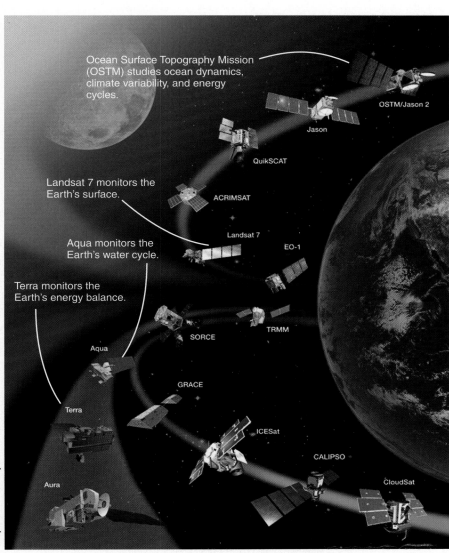

Jenny Mottar/Courtesy NASA

Ocean Surface Topography Mission (OSTM) studies ocean dynamics, climate variability, and energy cycles.

OSTM/Jason 2

Jason

QuikSCAT

Landsat 7 monitors the Earth's surface.

ACRIMSAT

Aqua monitors the Earth's water cycle.

Landsat 7

EO-1

Terra monitors the Earth's energy balance.

SORCE

TRMM

Aqua

GRACE

Terra

ICESat

CALIPSO

Aura

CloudSat

I.26 Earth Observing System satellites

The constellation of Earth Observing System (EOS) satellites depicted in this image highlight a portion of NASA's many remote-sensing research programs that target daily events and long-term changes for our planet, including biodiversity protection, natural resources depletion, human effects on the planet, and climate changes. Together these satellites provide continuous data about all of the Earth's systems.

at 1-m spatial resolution. A few locations are covered at resolutions as fine as 15 cm (6 in.). Most of the images are less than three years old.

The images in Google Earth are linked to an elevation database that was also made possible by an application of remote sensing using radar mapping technology. As a result, the elevation of each pixel is known to an accuracy of 10–30 m (33–98 ft), depending on the location. This allows computation of synthetic three-dimensional views of the landscape, including simulated fly-overs. By placing the viewer in motion over the landscape, the fly-over gives a strong visual impression of three-dimensional terrain, even though it is viewed on a flat computer screen.

Also linked to the image database are layers of GIS information. These include natural features, such as rivers and peaks, as well as political land boundaries and place names. A road network can also be superimposed. A search capability allows the user to type in a location (for example, the name of a city or town) and have

the program zoom in to a close view. You can even ask to see restaurants, lodgings, parks, and recreation areas through the GIS linkages.

Because Google Earth provides a view of nearly every point on land, it is a very useful tool for studying physical geography. The book's web site provides fly-over tours for each chapter. These are files of placemarks locating views of interest that can be downloaded and opened with the Google Earth application.

> Earth visualization tools merge remote sensing, GIS, and GPS technology to provide interactive viewing of the Earth's surface and its physical and cultural features.

OTHER EARTH VISUALIZATION TOOLS

Although Google Earth is at present the most technologically advanced of web-based Earth visualization

I.27 Google Earth images

These four images show screens for Google Earth.

OPENING SCREEN ▶

The opening screen shows a view of the globe centered on the United States. At the left are saved locations, called placemarks, and a list of GIS features that can be enabled.

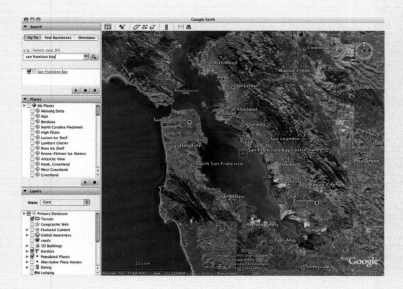

◀ **ZOOMING IN**

Zooming in to the San Francisco Bay region, we can see many physical and cultural features identified by name.

CHANGING THE VIEWPOINT ▶

By changing the viewpoint, the program provides an oblique image of the city of San Francisco looking northward.

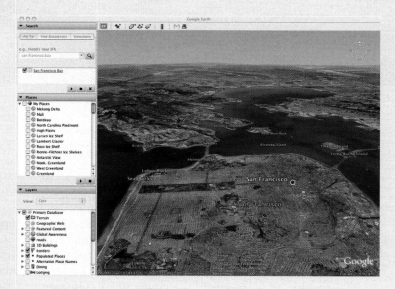

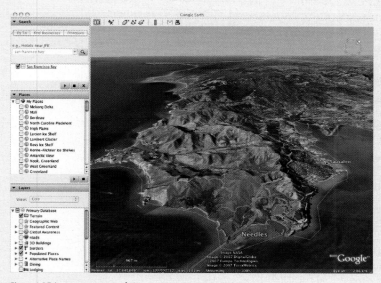

◀ **LOOKING AROUND**

Moving the field of view and choosing a new viewpoint, the town of Sausalito, with Mount Tamalpais in the background, comes into view.

Figure I.27 images courtesy of © Image NASA, Image © 2007 TeraMetrics
http://www.truearth.com, Image © 2007 Europa Technologies, Image © 2007 Digital Globe

tools, several other tools are readily available. NASA's World Wind opens a similar window on the Earth, starting with a global view and zooming in for fine detail. It also uses an elevation database, so it can provide three-dimensional renderings of the surface, as well as realistic fly-overs. World Wind's major strength is that users have easy access to computer program commands for development of custom applications, such as viewing the Grand Canyon or Mount Everest in true 3-D, or plotting a GPS track to visualize a trail hike or road trip.

TerraServer.com offers overhead high-resolution photos, with spatial resolutions as fine as 30 cm (11.8 in.) per pixel over many urbanized areas of the United States. Outside of the United States, most of the coverage is at 15 m (50 ft) per pixel, with some inset areas in higher resolution. The service requires a subscription and payment for downloading images.

TerraServer–USA, a service of the U.S. Geological Survey, provides online topographic maps for the United States. Many of these are orthophoto maps that use high-resolution air photos as a base. The Geological Survey also hosts the National Map on the web, an extensive database with a large number of map layers, including administrative boundaries, geographic names, geology, land use and cover, natural hazards, topography, and transportation.

A Look Ahead

In this introduction we have presented the fundamental concepts related to geography, physical geography, and some of the overarching ideas of physical geography. We have introduced some of the key environmental and global change topics that we will discuss throughout our text. We have also described some of the special tools that geographers use, including maps, geographic information systems, and remote sensing. Armed with these tools and ideas, we are ready to proceed to the subject itself. We will start with weather and climate, where we will see how solar energy drives a vast circulation of atmosphere and oceans that changes our physical environment from day to day, week to week, and year to year.

Humans are now the dominant species on the planet. Nearly every part of the Earth has felt human impact in some way. As the human population continues to grow and rely more heavily on natural resources, our impact on natural systems will continue to increase. Each of us is charged with the responsibility to treat the Earth well and respect its finite nature. Understanding the processes that shape our habitat as they are described by physical geography helps us all to become better citizens of the Earth, our home planet.

IN REVIEW INTRODUCING PHYSICAL GEOGRAPHY

- **Geography** is the study of the evolving character and organization of the Earth's surface. Geography has a unique set of perspectives. Geographers look at the world from the **viewpoint** of geographic space, focus on the **synthesis** of ideas from different disciplines, and develop and use special techniques for the **representation** and manipulation of spatial information.

- **Human geography** deals with social, economic, and behavioral processes that differentiate places; and **physical geography** examines the natural processes occurring at the Earth's surface that provide the physical setting for human activities.

- **Climatology** is the science that describes and explains the variability in space and time of the heat and moisture states of the Earth's surface, especially its land surfaces. **Geomorphology** is the science of Earth surface processes and landforms. **Coastal and marine geography** is a field that combines the study of the geomorphic processes that shape shores and coastlines with their application to coastal development and marine resource utilization. **Geography of soils** includes the study of the distribution of soil types and properties and the processes of soil formation. **Biogeography** is the study of the distribution of organisms at varying spatial and temporal scales, as well as the processes that

produce these distribution patterns. **Water resources** and **hazards assessment** are applied fields that blend both physical and human geography.

- **Spheres**, **systems**, and **time cycles** are three overarching themes that appear in physical geography. The four great Earth realms are **atmosphere**, **hydrosphere**, **lithosphere**, and **biosphere**. The **life layer** is the shallow surface layer where lands and oceans meet the atmosphere and where most forms of life are found.

- Scale, pattern, and process are three interrelated geographic themes. **Scale** refers to the level of structure or organization at which a phenomenon is studied; **pattern** refers to the variation in a phenomenon seen at a particular scale; and **process** describes how the factors that affect a phenomenon act to produce a pattern at a particular scale. The processes of physical geography operate at multiple scales, including *global, continental, regional,* and *local.*

- Physical processes often act together in an organized way that we can view as a **system**. A *systems* approach to physical geography looks for linkages and interactions between processes.

- Natural systems may undergo periodic, repeating changes that constitute **time cycles**. Important time cycles in physical geography range in length from hours to millions of years.

- Physical geography is concerned with the natural world around us—the human environment. Natural and human processes are constantly changing that environment.

- Global climate is changing in response to human impacts on the greenhouse effect. Global pathways of carbon flow can influence the greenhouse effect and, therefore, are the subject of intense research interest.

- Maintaining global biodiversity is important both for maintaining the stability of ecosystems and guarding a potential resource of bioactive compounds for human benefit. Unchecked human activity can degrade environmental quality and cause pollution.

- Extreme events take ever-higher tolls on life and property as populations expand. Extreme weather—storms and droughts, for example—will be more frequent with global warming caused by human activity.

- Important tools for studying the fields of physical geography include **maps**, **geographic information systems**, **remote sensing**, *mathematical modeling*, and *statistics*.

- **Cartography** is the art and science of making **maps**, which depict objects, properties, or activities as they are located on the Earth's surface.

- A **map projection** is an orderly system for displaying the curved surface of the Earth on a flat map. Common map projections include *polar, conic*, and *cylindrical*.

- The scale of a map relates distance on the Earth to distance on a globe or flat map. It is expressed by the **scale fraction**. *Large-scale maps* show small areas, while *small-scale maps* show large areas.

- *Conformal* maps preserve the shapes of geographic features, but not their areas. *Equal-area* maps show the areas of geographic features correctly, but distort their shape. Only a globe is both conformal and equal-area.

- *Multipurpose maps* use symbols, patterns, and colors to convey different types of information on the same map. *Thematic maps* display a single class of information, or theme.

- Map symbols include *dots, lines*, and *patches. Resolution* describes the level of detail shown on a map.

- **Isopleth** maps show lines of equal value for a continuously varying property. They are constructed from individual observations at points. A temperature map of isotherms is an example. A *choropleth* map shows categories of information, such as soil type or rock type, as areas on a map.

- The **global positioning system (GPS)** locates the position of an observer on the Earth using signals from Earth satellites. The fine accuracy of location is affected by the atmosphere, which is constantly changing.

- *Differential GPS* significantly improves location accuracy. It uses two GPS receivers, one at a base station and one at nearby locations to be plotted. In North America, the *Wide Area Augmentation System* provides differential GPS information by geostationary satellite broadcast.

- A **geographic information system (GIS)** is a computer-based tool for working with spatial data. It works with **spatial objects**, which include *points, lines*, and *polygons*, and manipulates information associated with spatial objects.

- A geographic information system has five elements: *data acquisition, preprocessing, data management, data manipulation and analysis*, and *product generation*.

- **Remote sensing** refers to acquiring information from a distance, usually of large areas, by instruments called *remote sensors* flown on aircraft or spacecraft. The information is obtained by measuring *electromagnetic radiation* reflected or emitted by an object or type of Earth surface.

- Most objects or surfaces reflect the colors of the spectrum differently, creating distinctive **spectral signatures**. Vegetation is dark green in the visible spectrum, but bright in the near-infrared.

- Objects or surfaces emit *thermal radiation* in proportion to their temperature. *Emissivity*, which affects the amount of radiation emitted at a given temperature, varies with the object or surface type.

- *Passive* sensing systems rely on environmental illumination or internal emission, while *active* systems provide their own source of energy. **Radar** is an active sensing system that uses microwave radiation to image landforms or track storms. **Lidar** is an active system that uses laser light pulses to map surface elevation in very fine detail.

- Remote sensing uses **digital images** that are processed by computer. *Image processing* is used to extract and enhance the information content of digital images. Digital images are typically acquired by a **multispectral scanner** or by a *direct digital imager* that uses an array of detectors.

- Satellite orbits used for remote sensing include Sun-synchronous and geostationary. The **Sun-synchronous orbit** covers most of the Earth while maintaining similar illumination conditions for repeat images. The **geostationary orbit** keeps the imager always above the same point on the Equator.

- NASA's Earth Observing System (EOS) is a constellation of orbiting satellites and instruments for making long-term observations of land, oceans, and atmosphere. *Landsat* and *MODIS* are examples of imagers that are very useful in studying physical geography and global change.

- **Earth visualization tools** integrate remote sensing, GIS, and GPS technology to create a visual simulation of the Earth's surface viewable on a networked computer. Google Earth is a prominent example.

KEY TERMS

geography, p. 4
regional geography, p. 4
systemic geography, p.4
viewpoint, p. 4
synthesis, p. 5
representation, p. 6
human geography, p. 6
physical geography, p. 6
climatology, p. 6
geomorphology, p. 7
coastal and marine
   geography, p. 7
geography of soils, p. 7

biogeography, p. 7
water resources, p. 7
hazards assessment, p. 7
spheres, p. 10
atmosphere, p. 10
lithosphere, p. 10
hydrosphere, p. 10
biosphere, p. 10
life layer, p. 10
scale, p. 11
pattern, p. 11
process, p. 11
systems, p. 13

time cycles, p. 13
cartography, p. 17
map, p. 17
map projection, p. 17
scale fraction, p. 18
isopleth, p. 21
global positioning system
   (GPS), p. 21
geographic information
   system (GIS), p. 21
spatial object, p. 22
remote sensing, p. 24
spectral signature, p. 24

radar, p. 25
lidar, p. 26
digital image, p. 26
multispectral
   scanners, p. 27
Sun-synchronous
   orbit, p. 28
geostationary orbit, p. 28
Earth visualization
   tools, p.30

REVIEW QUESTIONS

1. What is **geography?** Identify three perspectives used by geographers in studying the physical and human characteristics of the Earth's surface.
2. How does **human geography** differ from **physical geography?**
3. Identify and define five important subfields of science within physical geography.
4. Identify and define three interrelated themes that often arise in geographic study.
5. Name and describe each of the four great physical realms of Earth. What is the **life layer?**
6. Provide two examples of processes or systems that operate at each of the following scales: *global, continental, regional,* and *local.*
7. How is the word "system" used in physical geography? What is a *systems approach?*
8. What is a **time cycle** as applied to a system? Give an example of a time cycle evident in natural systems.
9. Identify and describe two interacting components of global change.
10. How is global climate change influenced by human activity?
11. Why are current research efforts focused on the carbon cycle?
12. Why is loss of biodiversity a concern of biogeographers and ecologists?
13. How does human activity degrade environmental quality? Provide a few examples.
14. How do extreme events affect human activity? Is human activity influencing the size or reoccurrence rate of extreme events?
15. Describe three types of **map projections** as they might occur, by projecting a wire globe onto a flat sheet of paper.
16. What is the **scale fraction** of a map or globe? Can the scale of a flat map be uniform everywhere on the map? Do large-scale maps show large areas or small areas?
17. How do *conformal* and *equal-area* maps differ?
18. What types of symbols are found on maps, and what types of information do they carry?
19. How are numerical data represented on maps? Identify three types of **isopleths**. What is a *choropleth* map?
20. What is the **global positioning system?** How does it work? What factors cause errors in determining ground locations?
21. What is *differential GPS*, and why is it important?
22. What is a **geographic information system?**
23. Identify and describe three types of **spatial objects**.
24. What are the key elements of a GIS?
25. What is **remote sensing?** What is a remote sensor?
26. Compare the reflectance spectra of water, vegetation, and a typical soil. How do they differ in the visible spectrum? In near-infrared and shortwave infrared wavelengths?
27. What is *emissivity*, and how does it affect the amount of energy emitted by an object?
28. Is **radar** an example of an active or a passive remote sensing system? Why?
29. How might **lidar** be used to detect change in a coastal environment?
30. What is a **digital image?** What advantage does a digital image have over a photographic image? Describe two ways of acquiring a digital image.
31. How does a **Sun-synchronous orbit** differ from a **geostationary** orbit? What are the advantages of each type?
32. Compare Landsat and MODIS. Which has the finer spatial resolution? Which images the globe more frequently?
33. What technologies are involved in **Earth visualization tools?** How do these tools make use of the Internet?

Chapter 1
The Earth as a Rotating Planet

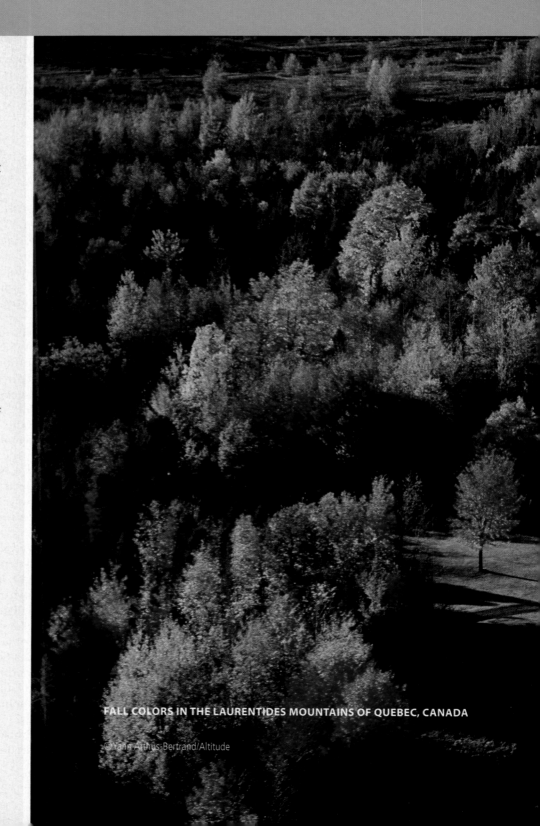

The march of the seasons is nowhere more prominent than in the deciduous forests of the northern hemisphere. Here in Canada, sugar maples, red maples, and birches provide vibrant fall colors as their leaves turn to red and gold. Canada is a country of forests, with half of its land area covered by trees. Although the trees are a renewable resource, Canada is taking steps to reduce the felling of natural forest for pulp and paper and to keep its timber harvesting sustainable. These measures help preserve the ability of the forest to take up atmospheric carbon dioxide, reducing the rate of increase.

FALL COLORS IN THE LAURENTIDES MOUNTAINS OF QUEBEC, CANADA

©Yann Arthus-Bertrand/Altitude

The Earth as a Rotating Planet

T his chapter is concerned with the motion of the Earth as a planet—both its rotation around its polar axis and its revolution around the Sun. What are the environmental effects of the Earth's rotation? How does the rotation naturally lead to the geographic grid of parallels and meridians? How is the curved geographic grid projected to construct flat maps? How does our global system of timekeeping work? What is the cause of the seasons, in which the length of the daylight period varies with latitude through the year? These are some of the questions we will answer in this chapter.

The Shape of the Earth

As we all learn early in school, the Earth's shape is very close to a sphere (Figure 1.1). Pictures taken from space by astronauts and by orbiting satellites also show us that the Earth is a ball rotating in space.

Today it seems almost nonsensical that many of our ancestors thought the world was flat. But to ancient sailors voyaging across the Mediterranean Sea, the shape and breadth of the Earth's oceans and lands were hidden. Imagine standing on one of their ships, looking out at the vast ocean, with no land in sight. The surface

1.1 Our spherical Earth

PHOTO OF EARTH'S CURVATURE
This astronaut photo shows the Earth's curved horizon from low-Earth orbit.
▼

Courtesy NASA

▲
DISTANT SHIP
Seen through a telescope, the decks of a distant ship seem to be under water. This phenomenon is easily explained by a curved Earth surface that appears to rise up between the observer and the ship.

of the sea would seem perfectly flat, stretching out and meeting the sky along a circular horizon. Given this view, perhaps it is not so surprising that early sailors, believing the Earth was a flat disk, feared their ships would fall off its edge if they ventured too far from land.

We also gain information about the shape of the Earth when we watch the Sun set when there are clouds in the sky. The clouds continue to receive the direct light of the Sun, although it has already set, as seen at ground level. The movement of solar illumination across the clouds is easily explained by a rotating spherical Earth.

Actually, the Earth is not perfectly spherical. The Earth's equatorial diameter, at about 12,756 km (7926 mi), is very slightly larger than the polar diameter, which is about 12,714 km (7900 mi). As the Earth spins, the outward force of rotation causes it to bulge slightly at the equator and flatten at the poles. The difference is very small—about three-tenths of 1 percent—but strictly speaking the Earth's squashed shape is closer to what is known as an *oblate ellipsoid*, not a sphere.

An even more accurate representation of the Earth's shape is the *geoid*, which is a reference surface based on the pull of gravity over the globe (Figure 1.2). It is defined by a set of mathematical equations and has many applications in mapmaking, as well as navigation.

courtesy Mike Sandiford, School of Earth Sciences, University of Melbourne, Victoria, 3010, Australia. University of Melbourne, mikes@unimelb.edu.au.

1.2 The geoid

Pictured here is a greatly exaggerated geoid, in which small departures from a sphere are shown as very large deviations.

The Earth's Rotation

The Earth spins slowly on its *axis*—an imaginary straight line through its center and poles—a motion we refer to as **rotation**. We define a solar day by one complete rotation, and for centuries have chosen to divide the solar day into exactly 24 hours. The North and South **Poles** are defined as the two points on the Earth's surface where the axis of rotation emerges. The direction of the Earth's rotation is shown in Figure 1.3.

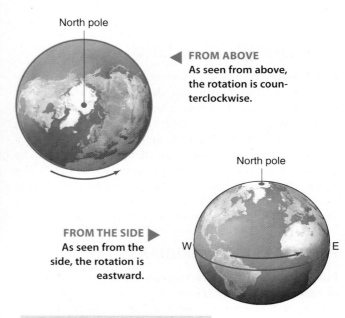

North pole

FROM ABOVE
As seen from above, the rotation is counterclockwise.

North pole

FROM THE SIDE
As seen from the side, the rotation is eastward.

W   E

1.3 Direction of the Earth's rotation

You can picture the direction of the Earth's rotation in two ways.

Richard L. Carlton/Photo Researchers, Inc.

CLOUD ILLUMINATION

As you watch a sunset from the ground, the Sun lies below the horizon, no longer illuminating the land around you. But at the height of the clouds, the Sun has not yet dipped below the horizon, so it continues to bathe them in red and pinkish rays. As the Sun descends, the red band of light slowly moves farther toward the horizon. In this dramatic sunset photo, the far distant clouds are still directly illuminated by the Sun's last rays. For the clouds directly overhead, however, the Sun has left the sky.

The Earth's rotation is important for three reasons. First, the axis of rotation serves as a reference in setting up the geographic grid of latitude and longitude, which we will discuss later in the chapter. Second, it provides the day as a convenient measure of the passage of time, with the day in turn divided into hours, minutes, and seconds. Third, it has important effects on the physical and life processes on Earth.

ENVIRONMENTAL EFFECTS OF THE EARTH'S ROTATION

All forms of life on the planet's surface are governed by the daily rhythms of the Sun. Green plants receive and store solar energy during the day and consume some of it at night. Among animals, some are active during the day, others at night. The day–night cycle also sets in motion the daily air temperature cycle that is observed in most places on the Earth.

The directions of large motions of the atmosphere and oceans are also affected, as the turning of the planet makes their paths curve. As we will see in Chapter 5, weather systems and ocean currents respond to this phenomenon, which is known as the *Coriolis effect*.

Finally, the Earth's rotation, combined with the Moon's gravitational pull on the planet, causes the rhythmic rise and fall of the ocean surface, which we know as the *tides*. The ebb and flow of tidal currents is a life-giving pulse for many plants and animals and provides a clock that regulates many daily human activities in the coastal zone. When we examine the tide and its currents in greater detail in Chapter 16, we will see that the Sun also has an influence on the tides.

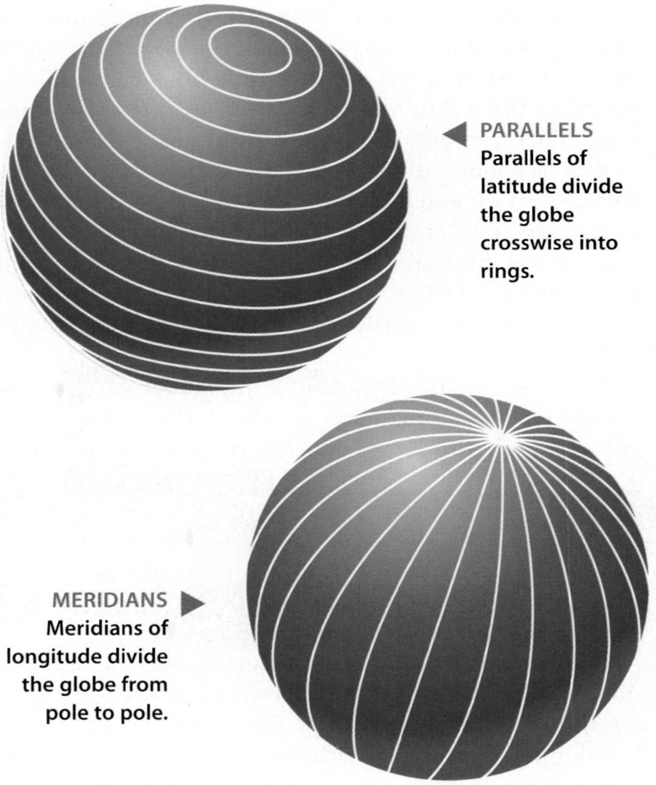

The Geographic Grid

It is impossible to lay a flat sheet of paper over a sphere without creasing, folding, or cutting it—as you know if you have tried to gift-wrap a ball. This simple fact has caused mapmakers problems for centuries. Because the Earth's surface is curved, we cannot divide it into a rectangular grid any more than we could smoothly wrap a globe in a sheet of graph paper. Instead, we divide the Earth into what is known as the *geographic grid*. This is made up of a system of imaginary circles, called parallels and meridians, which are shown in Figure 1.4.

PARALLELS AND MERIDIANS

Imagine cutting the globe just as you might slice an onion to make onion rings (Figure 1.4). Lay the globe on its side, so that the axis joining the North and South Poles runs perpendicular to your imaginary knife and begin to slice. Each cut creates a circular outline that passes

PARALLELS
Parallels of latitude divide the globe crosswise into rings.

MERIDIANS
Meridians of longitude divide the globe from pole to pole.

1.4 Parallels and meridians

around the surface of the globe. This circle is known as a parallel of latitude, or a **parallel**. The Earth's longest parallel of latitude is the **Equator**, which lies midway between the two poles. We use the Equator as a fundamental reference line for measuring position.

Now imagine slicing the Earth through the axis of rotation instead of across it, just as you would cut up a lemon to produce wedges. The outlines of the cuts form circles on the globe, each of which passes through both poles. Half of this circular outline, connecting one pole to the other, is known as a meridian of longitude, or, more simply, a **meridian**.

Meridians and parallels define geographic directions. When you walk directly north or south, you follow a meridian; when you walk east or west, you follow a parallel. There are an infinite number of parallels and meridians that can be drawn on the Earth's surface, just as there are an infinite number of positions on the globe. Every point on the Earth is associated with a unique combination of one parallel and one meridian. The position of the point is defined by their intersection.

> The geographic grid consists of an orderly system of circles—meridians and parallels—that are used to locate position on the globe.

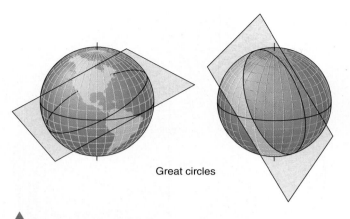

Great circles

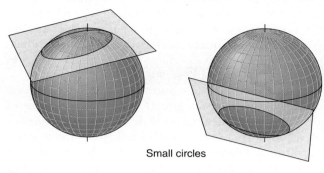

Small circles

GREAT CIRCLES
A great circle is created when a plane passes through the Earth, intersecting the Earth's center.

SMALL CIRCLES
Small circles are created when a plane passes through the Earth but does not intersect the center point.

1.5 Great and small circles

Meridians and parallels are made up of two types of circles: great and small (Figure 1.5). A **great circle** is created when a plane passing through the center of the Earth intersects the Earth's surface. It bisects the globe into two equal halves. A **small circle** is created when a plane passing through the Earth, but not through the Earth's center, intersects the Earth's surface. Meridians are actually halves of great circles, while all parallels except the Equator are small circles.

Because great circles can be aligned in any direction on the globe, we can always find a great circle that passes through two points on the globe. As we will see shortly, in our discussion of map projections, the portion of the great circle between two points is the shortest distance between them.

LATITUDE AND LONGITUDE

We label parallels and meridians by their **latitude** and **longitude**. (Figure 1.6). The Equator divides the globe into two equal portions: the *northern hemisphere* and the *southern hemisphere*. Parallels are identified by their angular distance from the Equator, which ranges from 0° to 90°. All parallels in the northern hemisphere are described by a north latitude (N), and all parallels south of the Equator are given as a south latitude (S).

Meridians are identified by longitude, which is an angular measure of how far eastward or westward the meridian is from a reference meridian, called the *prime meridian*. The prime meridian is sometimes known as

> Latitude and longitude uniquely determine the position of a point on the globe. Latitude records the parallel, and longitude the meridian, associated with the point.

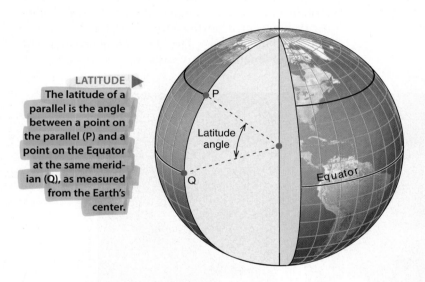

LATITUDE ▶
The latitude of a parallel is the angle between a point on the parallel (P) and a point on the Equator at the same meridian (Q), as measured from the Earth's center.

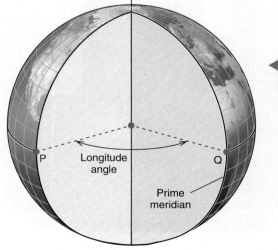

LONGITUDE
The longitude of a meridian is the angle between a point on that meridian at the Equator (P) and a point on the prime meridian at the Equator (Q), as measured at the Earth's center.

1.6 Latitude and longitude angles

the *Greenwich meridian* because it passes through the old Royal Observatory at Greenwich, near London, England (Figure 1.7). It has a longitude value of 0°. The longitude of a meridian on the globe is measured eastward or westward from the prime meridian, depending on which direction gives the smaller angle. Longitude then ranges from 0° to 180°, east or west (E or W).

Used together, latitude and longitude pinpoint locations on the geographic grid (Figure 1.8). Fractions of latitude or longitude angles are described using minutes and seconds. A *minute* is 1/60 of a degree, and a *second* is 1/60 of a minute, or 1/3600 of a degree. So, the latitude 41°, 27 minutes ('), and 41 seconds (") north (lat. 41°27′ 41″ N) means 41° north plus 27/60 of a degree plus 41/3600 of a degree. This cumbersome system has now largely been replaced by decimal notation. In this example, the latitude 41° 27′ 41″ N translates to 41.4614° N.

Degrees of latitude and longitude can also be used as distance measures. A degree of latitude, which measures distance in a north-south direction, is equal to about 111 km (69 mi). The distance associated with a degree of longitude, however, will be progressively

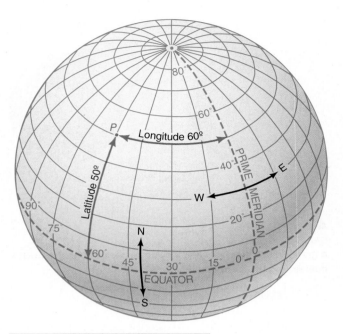

1.8 Latitude and longitude of a point

The point, *P*, lies on the parallel of latitude at 50° north (50° from the Equator) and on the meridian at 60° west (60° from the prime meridian). Its location is therefore lat. 50° N, long. 60° W.

reduced with latitude because meridians converge toward the poles. For example, at 60° latitude, a degree of longitude has a length exactly half of that at the Equator, or 55.5 km (34.5 mi).

A third axis exists for defining locations. Elevation—that is, height or altitude—complements latitude and longitude. Elevation is usually measured from sea level, but can also be expressed as a distance from the Earth's center. Three-dimensional mapping and analysis rely on the use of elevation to create terrain maps and images that can represent the Earth more realistically.

Map Projections

The problem of how to best display the Earth's surface has puzzled cartographers, or mapmakers, throughout history (Figure 1.9). The oldest maps were limited by a lack of knowledge of the world, rather than by difficulties caused by the Earth's curvature. They tended to represent political or religious views rather than geographic reality. Ancient Greek maps from the sixth century BC show the world as an island, with Greece at its center, while medieval maps from the fourteenth century placed Jerusalem at the locus.

But by the fifteenth century, ocean-faring explorers such as Columbus and Magellan were extending the reaches of the known world. These voyagers took mapmakers with them to record the new lands that they discovered, and navigation charts were highly valued.

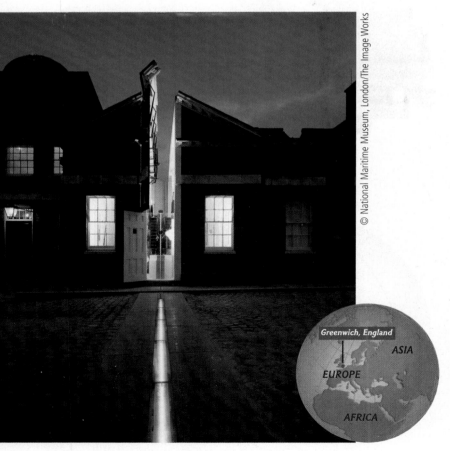

1.7 The prime meridian

This stripe in the forecourt of the old Royal Observatory at Greenwich, England, marks the prime meridian.

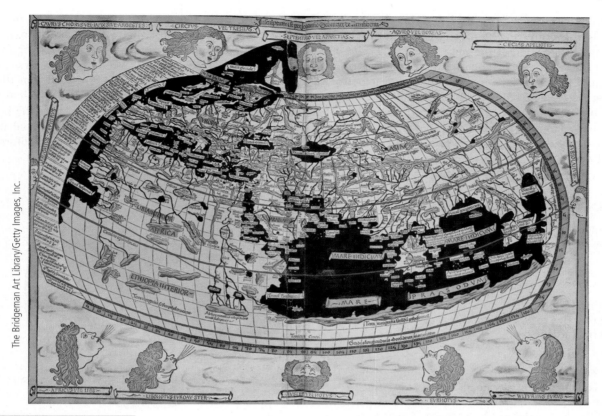

1.9 Ptolemy's map of the world

This atlas page shows a reproduction of a map of the world as it was known in ancient Greece.

Mapmakers, who now had a great deal of information about the world to set down, were challenged by the difficulty of representing the curved surface of the Earth on a flat page.

One of the earliest attempts to tackle the curvature problem for large-scale maps was made by the Belgian cartographer, Gerardus Mercator, in the sixteenth century, and it is still used today. A number of other systems, or **map projections**, have been developed to translate the curved geographic grid to a flat one. We will concentrate on the three most useful types, including Mercator's. Each has its advantages and drawbacks. (You can read more about maps and map projections in the introductory chapter.)

WileyPLUS Map Projections
Watch an animation showing how map projections are constructed.

POLAR PROJECTION

The *polar projection* (Figure 1.10) is normally centered on either the North or South Pole. Meridians are straight lines radiating outward from the pole, and parallels are

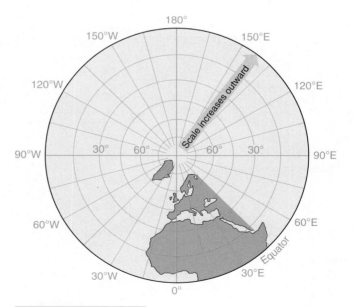

1.10 A polar projection

The map is centered on the North (or South) Pole. All meridians are straight lines radiating from the center point, and all parallels are concentric circles. The scale fraction increases in an outward direction, making shapes toward the edges of the map appear larger.

nested circles centered on the pole. The map is usually cut off to show only one hemisphere so that the Equator forms the outer edge of the map. Because the intersections of the parallels with the meridians always form true right angles, this projection shows the true shapes of all small areas. That is, the shape of a small island would always be shown correctly, no matter where it appeared on the map. However, because the scale fraction increases in an outward direction, the island would look larger toward the edge of the map than near the center.

MERCATOR PROJECTION

In the *Mercator projection*, the meridians form a rectangular grid of straight vertical lines, while the parallels form

straight horizontal lines (Figure 1.11). The meridians are evenly spaced, but the spacing between parallels increases at higher latitude so that the spacing at 60° is double that at the Equator. As the map reaches closer to the poles, the spacing increases so much that the map must be cut off at some arbitrary parallel, such as 80° N. This change of scale enlarges features near the pole.

The Mercator projection has several special properties. Mercator's goal was to create a map that sailors could

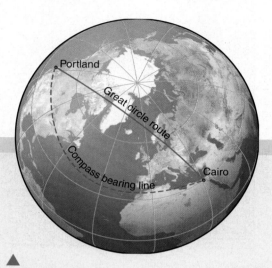

GREAT CIRCLE DISTANCE
The true shortest distance, drawn directly the globe as the crow flies, appears as a curved line on the Mercator projection.

1.11 The Mercator projection

The Mercator projection consists of a grid of parallels and meridians intersecting at right angles.

COMPASS BEARING
The compass line connecting two locations, such as Portland and Cairo, shows the compass bearing of a course directly connecting them. However, the shortest distance between them lies on a great circle, which is a longer, curving line on this map projection.
The diagram at the right side of the map shows how rapidly the map scale increases at higher latitudes. At lat. 60°, the scale is double the equatorial scale. At lat. 80°, the scale is six times greater than at the Equator.

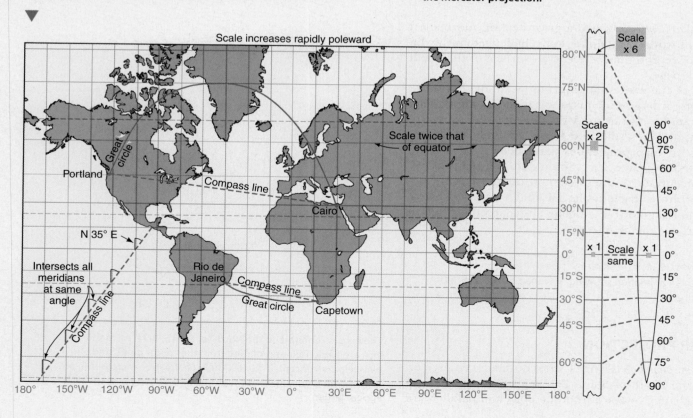

use to determine their course. A straight line drawn anywhere on his map gives you a line of constant compass direction. So a navigator can simply draw a line between any two points on the map and measure the *bearing*, or direction angle of the line, with respect to a nearby meridian on the map. Since the meridian is a true north-south line, the angle will give the compass bearing to be followed. Once aimed in that compass direction, a ship or an airplane can be held to the same compass bearing to reach the final point or destination (Figure 1.11).

But this line does not necessarily follow the shortest actual distance between two points, which we can easily plot out on a globe. We have to be careful—Mercator's map can falsely make the shortest distance between two points appear much longer than the compass line joining them.

> The Mercator projection shows a line of constant compass bearing as a straight line and so is used to display directional features, such as wind direction.

Because the Mercator projection shows the true compass direction of any straight line on the map, it is used to show many types of straight-line features. Among these features are flow lines of winds and ocean currents, directions of crustal features (such as chains of volcanoes), and lines of equal values, such as lines of equal air temperature or equal air pressure. That's why the Mercator projection is chosen for maps of temperatures, winds, and pressures.

CONFORMAL AND EQUAL-AREA MAPS

The shape and area of a small feature, such as an island or peninsula, will change as the feature is projected from the surface of the globe to a map. With some projections, the area will change, but the shape will be preserved. Such a projection is referred to as *conformal*. The Mercator projection (Figure 1.11) is an example. Here, every small twist and turn of the shoreline of each continent is shown in its proper shape. However, with increasing latitude, the growth of the continents shows that the Mercator projection does not depict land areas uniformly. A projection that does show area uniformly is referred to as *equal-area*. Here, continents show their relative areas correctly, but their shapes are distorted. No projection can be both conformal and equal-area; only a globe has that property.

> Conformal map projections show shapes correctly, whereas equal-area maps show areas correctly.

WINKEL TRIPEL PROJECTION

The *Winkel Tripel projection* (Figure 1.12) is named after its inventor, Oswald Winkel (1873–1953). The German

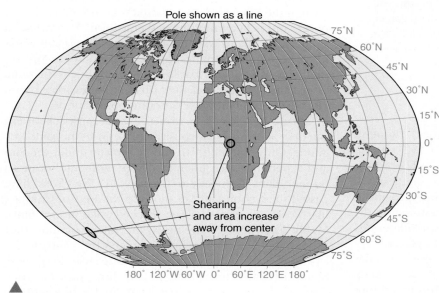

▲ SHAPE DISTORTION
This projection is very useful for displaying world maps because it shows the true shapes and areas of countries and continents, with only small distortions as compared to the Mercator or other global projections. Shearing and relative area increase toward the map's east and west edges and near the poles.

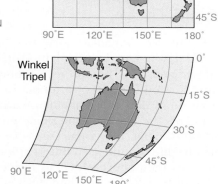

▲ SHEARING AND AREA
Shearing occurs when meridians and parallels are curved, distorting true shape. The Mercator map shows the shape of Australia correctly, but the area of Australia is shown more accurately in the Winkel Tripel map.

1.12 The Winkel Tripel projection

The Winkel Tripel projection minimizes distortion in both shape and area.

word *tripel*, translated as "triplet," refers to the property that the projection minimizes the sum of distortions to area, distance, and direction. The projection has parallels that are nearly straight, curving slightly toward the edges of the map. The meridians are increasingly curved with distance from the central meridian.

The Winkel Tripel projection is neither perfectly conformal nor perfectly equal-area. Compared to the Mercator map, the shapes of coastlines and continents are somewhat distorted by shearing, which increases away from the central meridian and toward the poles. However, the areas of continents and countries are shown much more accurately than in the Mercator map. Only in the polar regions near the east and west edges of the map do areas grow significantly with latitude.

Because it shows areas and shapes with only a small amount of distortion, the Winkel Tripel projection is well suited to displaying global data. It is an ideal choice for world maps showing the world's climate, soils, and vegetation, and we use it in many places in this book.

Maps are in wide use today for many applications as a simple and an efficient way of compiling and storing spatial information. However, in the past few decades, maps have been supplemented by more powerful computer-based methods for acquiring, storing, processing, analyzing, and outputting spatial data. These are contained within *geographic information systems (GISs)*. The "Tools in Physical Geography" section in our Introduction presents some basic concepts of geographic information systems and how they work.

> The Winkel Tripel projection shows the countries and continents of the globe with minimal distortion of shape, area, and scale.

Global Time

There's an old Canadian joke that goes, "Repent! The world will end at midnight!—or, 12:30 A.M. in Newfoundland." It's humorous because independent-minded Newfoundlanders use a time zone that is a half hour ahead of the other Canadian maritime provinces. It highlights the fact that one single instant across the world—no matter how cataclysmic—is simultaneously labeled by different times in different local places.

Humans long ago decided to divide the solar day into 24 units, called hours, and devised clocks to keep track of hours in groups of 12. Yet, different regions set their clocks differently—when it is 10:03 A.M. in New York, it is 9:03 A.M. in Chicago, 8:03 A.M. in Denver, and 7:03 A.M. in Los Angeles. These times differ by exactly one hour. How did this system come about? How does it work?

Even in today's advanced age, our global time system is oriented to the Sun. Think for a moment about the Sun moving across the sky. In the morning, the Sun is low on the eastern horizon, and as the day progresses, it rises higher until at *solar noon* it reaches its highest point

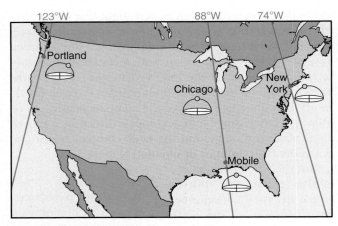

1.13 Time and the Sun

When it is noon in Chicago, it is 1:00 P.M. in New York and only 10:00 A.M. in Portland. Yet in Mobile, about 1600 km (1000 mi) away, it is also noon. This is because time is determined by longitude, not latitude.

in the sky. If you check your watch at that moment, it will read a time somewhere near twelve o'clock (noon). After solar noon, the Sun's elevation in the sky decreases. By late afternoon, the Sun hangs low in the sky, and at sunset it rests on the western horizon.

Imagine for a moment that you are in Chicago (Figure 1.13). The time is noon, and the Sun is at or near its highest point in the sky. You call a friend in New York and ask about the position of the Sun. Your friend will say that the Sun has already passed solar noon, its highest point, and is beginning its descent down. Meanwhile, a friend in Portland will report that the Sun is still working its way up to its highest point. But a friend in Mobile, Alabama, will tell you that the time in Mobile is the same as in Chicago, and that the Sun is at about solar noon. How do we explain these different observations?

The difference in time makes sense because solar noon can only occur simultaneously in places with the same longitude. Only one meridian can be directly under the Sun and experience solar noon at a given moment. Locations on meridians to the east of Chicago, like New York, have passed solar noon, and locations to the west of Chicago, like Portland, have not yet reached solar noon. Since Mobile and Chicago have nearly the same longitude, they experience solar noon at approximately the same time.

Figure 1.14 indicates how time varies with longitude. Since the Earth turns 360° in a 24-hour day, the rotation rate is 360°/24=15° per hour. So 15° of longitude equates to one hour of time.

STANDARD TIME

We've just seen that locations with different longitudes experience solar noon at different times. But what would happen if each town or city set its clocks to read

> In the standard time system, we keep time by standard meridians that normally differ by one hour from each other.

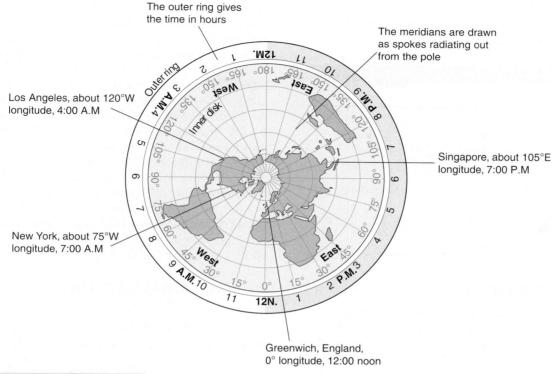

The outer ring gives
the time in hours

The meridians are drawn
as spokes radiating out
from the pole

Los Angeles, about 120°W
longitude, 4:00 A.M

Singapore, about 105°E
longitude, 7:00 P.M

New York, about 75°W
longitude, 7:00 A.M

Greenwich, England,
0° longitude, 12:00 noon

1.14 The relation of longitude to time

This polar projection illustrates how longitude is related to time for an example of noon on the prime meridian. The alignment of meridians with hours shows the time at other locations.

12:00 at its own local solar noon? All cities and towns on different meridians would have different local time systems. With today's instantaneous global communication, chaos would soon result.

Standard time simplifies the global timekeeping problem. In the **standard time system** the globe is divided into **time zones.** People within a zone keep time according to a *standard meridian* that passes through their zone. Since the standard meridians are usually 15° apart, the difference in time between adjacent zones is normally one hour. In some geographic regions, however, the difference is only one half hour.

Figure 1.15 shows the time zones observed in northern North America. The United States and its Caribbean

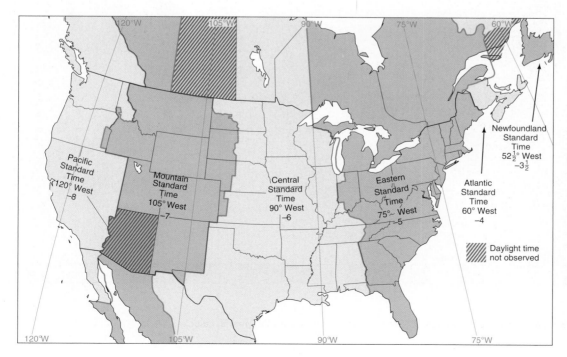

1.15 Time zones of the contiguous United States and southern Canada

The name, standard meridian, and number code are shown for each time zone. Time zone boundaries often follow preexisting natural or political boundaries. For example, the Eastern time-Central time boundary line follows Lake Michigan down its center, and the Mountain time-Pacific time boundary follows a ridge-crest line also used by the Idaho-Montana state boundary.

possessions fall within seven time zones. Six zones cover Canada. Their names and standard meridians of longitude are shown in Table 1.1:

U.S. Zones	Meridian	Canadian Zones
	52½°	Newfoundland
Atlantic	60°	Atlantic
Eastern	75°	Eastern
Central	90°	Central
Mountain	105°	Mountain
Pacific	120°	Pacific-Yukon
Alaska-Bering	135°	
Hawaii	150°	

Table 1.1 Examples of North American Time Zones

WORLD TIME ZONES

According to our map of the world's time zones (Figure 1.16), the country spanning the greatest number of time zones is Russia. From east to west, Russia

spans 11 zones, but groups them into 8 standard time zones. China covers 5 time zones but runs on a single national time using the standard meridian of Beijing.

A few countries, such as India and Iran, keep time using a meridian that is positioned midway between standard meridians, so that their clocks depart from those of their neighbors by 30 or 90 minutes. Some states or provinces within countries also keep time by 7½° meridians, such as the Canadian province of Newfoundland and the interior Australian states of South Australia and Northern Territory.

World time zones are often referred to by number to indicate the difference in hours between time in a zone and time in Greenwich. A number of −7, for example, indicates that local time is seven hours behind Greenwich time, while a +3 indicates that local time is three hours ahead of Greenwich time.

INTERNATIONAL DATE LINE

Take a world map or globe with 15° meridians. Start at the Greenwich 0° meridian and count along the 15° meridians in an eastward direction. You will find that the

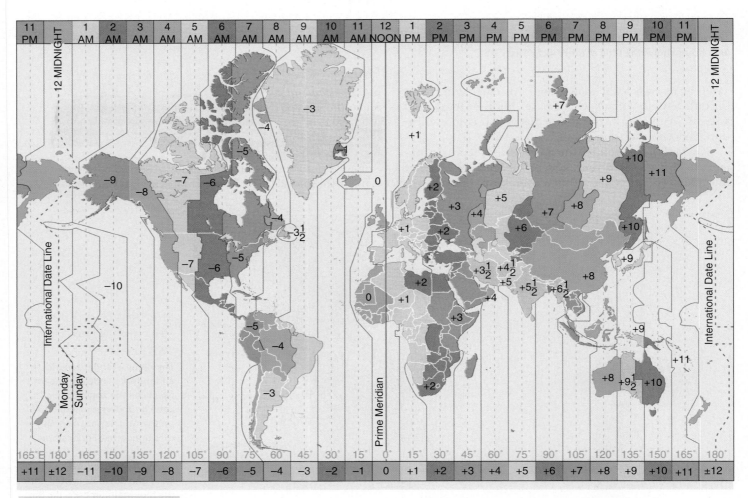

1.16 Time zones of the world

Dashed lines represent 15° meridians, and bold lines represent 7½° meridians. Alternate zones appear in color.

180th meridian is number 12 and that the time at this meridian is, therefore, 12 hours later than Greenwich time. Counting in a similar manner westward from the Greenwich meridian, we find that the 180th meridian is again number 12 but that the time is 12 hours earlier than Greenwich time. We seem to have a paradox: How can the same meridian be both 12 hours ahead of Greenwich time and 12 hours behind it? The answer is that each side of this meridian is experiencing a different day.

Imagine that you are on the 180° meridian on June 26. At the exact instant of midnight, the same 24-hour calendar day covers the entire globe. Stepping east will place you in the very early morning of June 26, while stepping west will place you very late in the evening of June 26. You are on the same calendar day on both sides of the meridian but 24 hours apart in time.

Doing the same experiment an hour later, at 1:00 A.M., stepping east you will find that you are in the early morning of June 26. But if you step west, you will find that midnight of June 26 has passed, and it is now the early morning of June 27. So on the west side of the 180th meridian, it is also 1:00 A.M. but it is one day later than on the east side. For this reason, the 180th meridian serves as the *International Date Line*. This means that if you travel westward across the date line, you must advance your calendar by one day. If traveling eastward, you set your calendar back by a day.

Air travelers on Pacific routes between North America and Asia cross the date line. For example, flying westward from Los Angeles to Sydney, Australia, you may depart on a Tuesday evening and arrive on a Thursday morning after a flight that lasts only 14 hours. On an eastward flight from Tokyo to San Francisco, you may actually arrive the day before you take off, taking the date change into account!

> When crossing the International Date Line in an eastward direction, travelers turn back their calendars back one day.

Actually, the International Date Line does not follow the 180th meridian exactly. Like many time zone boundaries, it deviates from the meridian for practical reasons. As shown in Figure 1.16, it has a zigzag offset between Asia and North America, as well as an eastward offset in the South Pacific to keep clear of New Zealand and several island groups.

DAYLIGHT SAVING TIME

The United States and many other countries observe some form of **daylight saving time**, in which clocks are set ahead by an hour (sometimes two) for part of the year. Although it was once thought that adding daylight hours to the end of the workday would save electricity and reduce traffic accidents and crime, the evidence now shows that the primary effects are economic—allowing more retail shopping and recreation, for example. Although something of a mixed blessing, daylight saving time is now a part of normal life in most places.

In the United States, daylight saving time comes into effect on the second Sunday in March and is discontinued on the first Sunday of November. Arizona (except the Navajo Nation), Puerto Rico, Hawaii, U.S. Virgin Islands, Guam, the Northern Mariana Islands, and American Samoa do not observe daylight saving time. Although many other nations observe daylight saving time, they do not always begin and end it on the same days of the year. In the European Union, daylight saving time is called *summer time*. It begins on the last Sunday in March and ends on the last Sunday in October.

PRECISE TIMEKEEPING

Since the 1950s, the most accurate time has been kept using atomic clocks, which are based on the frequency of microwave energy emission from atoms of the element cesium cooled to near absolute zero. These very accurate clocks keep time to better than one part in 1 trillion. Atomic time is a universal standard that is not related to the Earth's rotation. Civil time sources use Coordinated Universal Time (UTC), which is derived from atomic time and provides a day of 86,400 seconds (24 hours) in length to match the Earth's mean rotation rate with respect to the Sun. Coordinated Universal Time is administered by the Bureau International de l'Heure, located near Paris.

Our Earth is a much less precise timekeeper, exhibiting small changes in the angular velocity of its rotation on its axis and variations in the time it takes to complete one circuit around the Sun. As a result, constant adjustments to the timekeeping system are necessary.

The Earth's Revolution Around the Sun

So far, we have discussed the importance of the Earth's rotation on its axis. But what about the Earth's movement as it orbits the Sun? We refer to this motion as the Earth's **revolution** around the Sun. The Earth takes 365.242 days to travel around the Sun—almost a quarter of a day longer than the calendar year of 365 days. Every four years, this time adds up to nearly one extra day, which we account for by inserting a 29th day into February in leap years. Further minor corrections—such as omitting the extra day in century years—are necessary to keep the calendar on track.

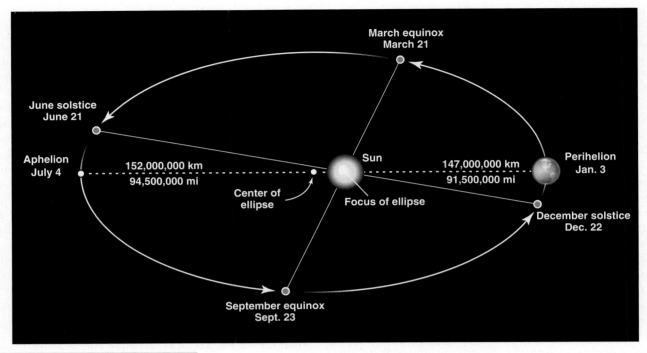

1.17 Orbit of the Earth around the Sun

The Earth's orbit around the Sun is not quite circular; rather, it is in the shape of an ellipse (drawn here in an exaggerated way). As a result, the distance between the Sun and the Earth varies with the time of year.

The Earth's orbit around the Sun is shaped like an ellipse, or oval (Figure 1.17). This means that the distance between the Earth and Sun varies somewhat through the year. The Earth is nearest to the Sun at *perihelion*, which occurs on or near January 3. It is farthest away from the Sun at *aphelion*, on or near July 4. However, the distance between Sun and Earth varies only by about 3 percent during one revolution because the elliptical orbit is shaped very much like a circle. For most purposes we can regard the orbit as circular.

Which way does the Earth revolve? Imagine yourself in space, looking down on the North Pole. From this viewpoint, the Earth travels counterclockwise around the Sun (Figure 1.18). This is the same direction as the Earth's rotation.

MOTIONS OF THE MOON

The Moon rotates on its axis and revolves about the Earth in the same direction as the Earth rotates and revolves around the Sun. But the Moon's rate of rotation is synchronized with the Earth's rotation so that one side of the Moon is permanently directed toward the Earth while the opposite side of the Moon remains hidden. It was only in 1959, when a Soviet spacecraft passing the Moon transmitted photos back to Earth, that we caught our first glimpse of the far side.

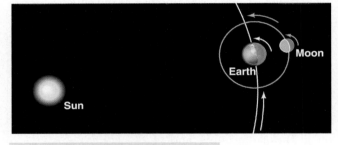

1.18 Revolution of the Earth and Moon

Viewed from a point over the Earth's North Pole, the Earth both rotates and revolves in a counterclockwise direction. From this viewpoint, the Moon also rotates counterclockwise.

The phases of the Moon are determined by the position of the Moon in its orbit around the Earth, which in turn determines how much of the sunlit Moon is seen from the Earth. It takes about 29.5 days for the Moon to go from one full Moon to the next. In the twilight photo of a moonlit scene in Figure 1.19, the Moon is about half full. From the way that the Sun illuminates the Moon as a sphere, it is easy to see that the Sun is to the left and just below the horizon.

WileyPLUS The Earth's Revolution Around the Sun
Watch a narrated animation to see how the Earth revolves around the Sun to cause the seasons.

© Aline Hopkins/Alamy Limited

1.19 Midnight in July in East Greenland

Although it is midnight, the Sun is only just below the horizon, bathing the scene in soft twilight. The way the spherical Moon is lit by the Sun also shows that the Sun is located below the horizon and to the left.

EYE ON THE LANDSCAPE What else would the geographer see? These mountain landforms (A) show the effects of glacial ice and frost action during the most recent Ice Age. The patches of snow mark the sites where small glaciers once formed, carving shallow basins in the bedrock of the peaks. The glacier in the foreground (B) flows into the sea, and is called a tidewater glacier. Its surface is dotted with rocks and debris that have melted out of the glacier. Crevasses fracture the ice as it descends.

TILT OF THE EARTH'S AXIS

Depending on where you live in the world, the effects of the changing seasons can be large. But why do we experience seasons on Earth? And why do the hours of daylight change throughout the year—most extremely at the poles, and less so near the Equator?

Seasons arise because the Earth's axis is not perpendicular to the plane containing the Earth's orbit around the Sun, which is known as the *plane of the ecliptic*. Figure 1.20 shows this plane as it intersects the Earth. If we extend the imaginary axis out of the North Pole into space, it always

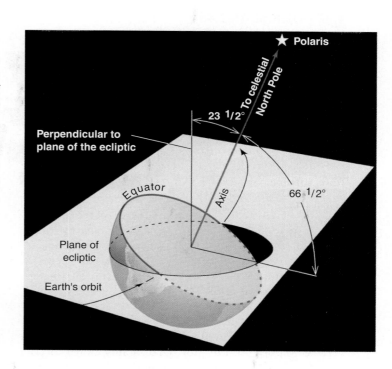

1.20 The tilt of the Earth's axis of rotation with respect to its orbital plane

As the Earth moves in its orbit on the plane of the ecliptic around the Sun, its rotational axis remains pointed toward Polaris, the North Star, and makes an angle of 66½° with the ecliptic plane. The axis of the Earth is thus tilted at an angle of 23½° away from a right angle to the plane of the ecliptic.

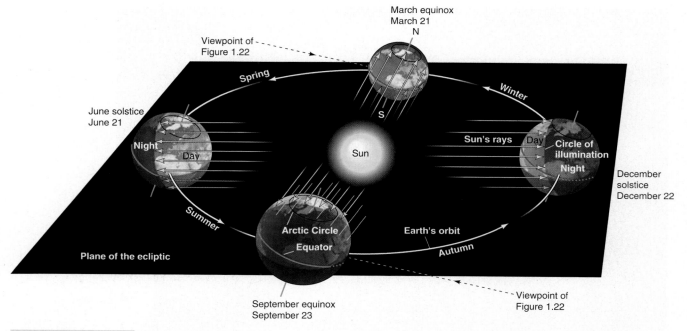

1.21 The four seasons

The Earth revolves once around the Sun in a year, passing through each of its four seasons. The four seasons occur because the Earth's tilted axis is always pointed toward the same point in space, very close to the North Star. That is, a line through the Earth's axis of rotation at each season is parallel to a line through the axis at any of the other seasons. Because of this fixed direction of rotation, the northern hemisphere is tipped toward the Sun for the June solstice and away from the Sun for the December solstice. Both hemispheres are illuminated equally at the equinoxes

aims toward Polaris, the North Star. The direction of the axis does not change as the Earth revolves around the Sun. Let's investigate this phenomenon in more detail.

THE FOUR SEASONS

Figure 1.21 shows the full Earth orbit traced on the plane of the ecliptic. On December 22, the north polar end of the Earth's axis leans at the maximum angle away from the Sun, 23½°. This event is called the **December solstice**, or *winter solstice* in the northern hemisphere. At this time, the southern hemisphere is tilted toward the Sun and enjoys strong solar heating.

Six months later, on June 21, the Earth has traveled to the opposite side of its orbit. This is known as the **June solstice**, or *summer solstice* in the northern hemisphere. The north polar end of the axis is tilted at 23½° toward the Sun, while the South Pole and southern hemisphere are tilted away.

The equinoxes occur midway between the solstice dates. At an equinox, the Earth's axis is not tilted toward the Sun or away from it. The **March equinox** (*vernal equinox* in the northern hemisphere) occurs near March 21, and the **September equinox** (*autumnal equinox*) occurs near September 23. The conditions at

> The axis of the Earth's rotation is tilted by 23½° away from a perpendicular to the plane of the ecliptic. This tilt causes the seasons.

the two equinoxes are identical as far as the Earth–Sun relationship is concerned. The date of any solstice or equinox in a particular year may vary by a day or so, since the revolution period is not exactly 365 days.

EQUINOX CONDITIONS

The Sun's rays always divide the Earth into two hemispheres—one that is bathed in light and one that is shrouded in darkness. The *circle of illumination* is the circle that separates the day hemisphere from the night hemisphere. The *subsolar point* is the single point on the Earth's surface where the Sun is directly overhead at a particular moment.

At equinox, the circle of illumination passes through the North and South Poles, as we see in Figure 1.22. The Sun's rays graze the surface at both poles, so the surfaces at the poles receive very little solar energy. The subsolar point falls on the Equator. Here, the angle between the Sun's rays and the Earth's surface is 90°, so that point receives the full force of solar illumination. At noon at latitudes in between, such as 40° N, the Sun strikes the surface at an angle that is less than 90°. The angle that marks the Sun's elevation above the horizon is known as the *noon angle*. Simple geometry shows that for equinox conditions, the noon angle is equal to 90° minus the latitude, so that at 40° N, the noon angle is 50°.

One important feature of the equinox is that day and night are of equal length everywhere on the globe.

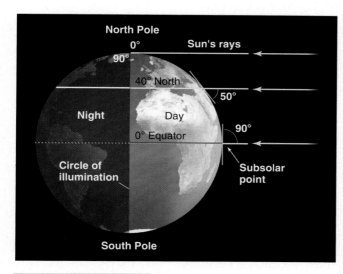

1.22 Equinox conditions

At equinox, the Earth's axis of rotation is exactly at right angles to the direction of solar illumination. The circle of illumination passes through the North and South Poles. The subsolar point lies on the Equator. At both poles, the Sun is seen at the horizon. The viewpoint for this diagram, shown by the dashed arrow lines in Figure 1.21, is away from the plane of the ecliptic.

You can see this by imagining yourself located at a point on the 40° N parallel. As the world turns, you will be in daylight for exactly half the day and in night for the other half.

SOLSTICE CONDITIONS

Now let's examine the solstice conditions in Figure 1.23. The June solstice is shown on the left. Imagine that you are back at a point on the lat. 40° N parallel. Unlike at equinox, the circle of illumination no longer divides your parallel into equal halves because of the tilt of the northern hemisphere toward the Sun. Instead, daylight

covers most of the parallel, with a smaller amount passing through twilight and darkness. For you, the day is now considerably longer (about 15 hours) than the night (about 9 hours). Now step onto the Equator. You can see that this is the only parallel that is divided exactly into two. On the Equator, daylight and nighttime hours will be equal throughout the year.

The farther north you go, the more the effect increases. Once you move north of lat. 66½°, the day continues unbroken for 24 hours. Looking at Figure 1.23, we can see that is because the lat. 66½° parallel is positioned entirely within the daylight side of the circle of illumination. This parallel is known as the **Arctic Circle**. Even though the Earth rotates through a full cycle during a 24-hour period, the area north of the Arctic Circle will remain in continuous daylight. We can also see that the subsolar point is at a latitude of 23½° N. This parallel is known as the **Tropic of Cancer**. Because the Sun is directly over the Tropic of Cancer at this solstice, solar energy is most intense here.

The conditions are reversed at the December solstice. Back at 40° N lat., the night is about 15 hours long, while daylight lasts about 9 hours. All the area south of 66½° S lat. lies under the Sun's rays, inundated with 24 hours of daylight. This parallel is known as the **Antarctic Circle**. The subsolar point has shifted to a point on the parallel at 23½° S lat., known as the **Tropic of Capricorn**.

We have carefully used the term *daylight* to describe the period of the day during which the Sun is above the horizon. When the Sun is not too far below the horizon, the sky is still lit by *twilight*. At high latitudes during the polar night, twilight can be several hours long and

> At the June solstice, the North Pole is tilted toward the Sun. At the December solstice, it is tilted away from the Sun.

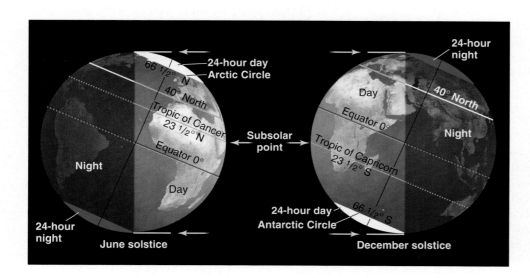

1.23 Solstice conditions

At the solstice, the north end of the Earth's axis of rotation is fully tilted either toward or away from the Sun. Because of the tilt, polar regions experience either a 24-hour day or a 24-hour night. The subsolar point lies on one of the tropics, at lat. 23½° N or S.

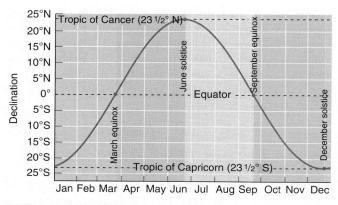

1.24 The Sun's declination throughout the year

The latitude of the subsolar point marks the Sun's declination, which changes slowly throughout the year, from 23½° S to 23½° N to 23½° S.

provide enough illumination for many outdoor activities.

The solstices and equinoxes are four special events that occur only once during the year. Between these times, the latitude of the subsolar point travels northward and southward in an annual cycle, looping between the Tropics of Cancer and Capricorn. We call the latitude of the subsolar point the Sun's *declination* (Figure 1.24).

As the seasonal cycle progresses, the polar regions that are bathed in 24-hour daylight, or shadowed in 24-hour

> The Sun's declination describes the latitude of the subsolar point as it ranges from 23½° S (December solstice) to 23½° N (June solstice) throughout the year.

night, shrink and then grow. At other latitudes, the length of daylight changes slightly from one day to the next, except at the Equator, where it remains the same. In this way, the Earth experiences the rhythm of the seasons as it continues its revolution around the Sun.

WileyPLUS Observing Earth–Sun Relationships
Watch a video to learn how ancient peoples used the position of the rising and setting Sun on the horizon to devise annual calendars to predict wet and dry seasons, planting, and harvesting.

WileyPLUS Web Quiz
Take a quick quiz on the key concepts of this chapter.

A Look Ahead

This chapter has focused on the daily rotation of the Earth on its axis and the annual revolution of the Earth around the Sun. As we will see in Chapter 2, the daily and annual rhythms of the Earth's motion create a global pattern of energy flow from the Sun to the Earth that changes from minute to minute, day to day, and season to season. This flow powers most of the natural processes that we experience every day, from changes in the weather to the work of streams in carving the landscape.

WileyPLUS Web Links
Visit web sites to find out more about longitude, maps, and globes. Find the time at any location in the world. Find the position of the Sun in the sky for any location, time, and date. Learn more about GPS.

IN REVIEW THE EARTH AS A ROTATING PLANET

- Although the Earth appears in space photos to be a sphere, it is slightly flattened at the poles into a shape resembling an *oblate ellipsoid*. The *geoid* is a closer approximation of the Earth's true shape.
- The Earth rotates on its *axis* once in 24 hours. The intersection of the axis of rotation with the Earth's surface marks the North and South **Poles**. The direction of rotation is counterclockwise when viewed from above the North Pole.
- The Earth's **rotation** provides the daily alternation of sunlight and darkness, the tides, and a sideward turning of the paths of ocean and air currents.
- The *geographic grid*, which consists of **parallels** and **meridians**, helps us mark locations on the globe. *Great circles* always bisect the globe, but *small circles* do not.
- Geographic location is labeled using **latitude** and **longitude**. The **Equator** and the *prime meridian* act as references to locate any point on Earth.
- **Map projections** display the Earth's curved surface on a flat page.

- The *polar projection* pictures the globe as we might view it from the North or South Pole. Meridians radiate outward from the poles, like spokes in a wheel.
- The *Mercator projection* converts the curved geographic grid into a flat, rectangular grid and best displays directional features. A straight line on a Mercator projection is a constant compass bearing.
- *Conformal* maps preserve the shape of individual features, such as islands or coastlines, while *equal-area* maps show the areas of regions and continents accurately.
- The *Winkel Tripel projection* shows the entire globe while minimizing distortion in shape, area, and scale, and is well suited to showing world maps of climate, vegetation, and soils.
- Our system of timekeeping is oriented to the Sun's apparent motion in the sky. *Solar noon* marks the position of the Sun at its highest point in the sky each day. All locations on the same meridian experience solar noon at the same instant.

- We keep *standard time* in **time zones** according to *standard meridians* that are normally 15° apart. Since the Earth rotates by 15° each hour, time zones normally differ by one hour.
- At the *International Date Line*, the calendar day changes—advancing a day for westward travel, dropping back a day for eastward travel.
- *Daylight saving time* advances the clock by one hour. Most nations observe daylight saving time, although starting and ending dates may differ.
- Atomic clocks provide a very accurate basis for global timekeeping. Civil time is kept in **Coordinated Universal Time (UTC),** which uses a day length of 86,400 seconds (24 hours) to match the Earth's rotation rate with respect to the Sun.
- The Moon rotates and revolves about the Earth in the same direction that the Earth revolves around the Sun. The Moon's rotation is synchronized with its revolution so that one side always faces the Earth while the other side remains unseen.
- The Earth's axis of rotation is tilted at an angle of 23½° from a perpendicular to the *plane of the ecliptic* and points to a fixed location in space. The seasons arise from the **revolution** of the Earth in its orbit around the Sun and this tilt of the Earth's axis. The solstices and equinoxes mark the cycle of this revolution.
- At the **June** (summer) **solstice**, the *northern hemisphere* is tilted toward the Sun. At the **December** (winter) **solstice**, the *southern hemisphere* is tilted toward the Sun. At the **March** and **September equinoxes**, the Earth is tilted neither toward nor away from the Sun, and day and night are of equal length.

KEY TERMS

rotation p. 39	small circle p. 41	daylight saving time p. 49	September equinox p. 52
pole p. 39	latitude p. 41	revolution p. 49	Arctic Circle p. 53
parallel p. 40	longitude p. 41	December solstice p. 52	Tropic of Cancer p. 53
Equator p. 40	map projection p. 43	June solstice p. 52	Antarctic Circle p. 53
meridian p. 40	standard time system p. 47	March equinox p. 52	Tropic of Capricorn p. 53
great circle p. 41	time zones p. 47		

REVIEW QUESTIONS

1. How do we know that the Earth is "round"? What is the approximate shape of the Earth? Define the term *geoid*.
2. What is meant by the Earth's *rotation*? Describe three environmental effects of the Earth's rotation.
3. Describe the geographic grid, including parallels and meridians. Distinguish between great and small circles and apply these terms to parallels and meridians.
4. How do latitude and longitude determine position on the globe? In what units are they measured? What function do the Equator and the prime meridian serve in determining latitude and longitude?
5. Identify three types of map projections, and describe each briefly. Give reasons why you might choose alternate map projections to display different types of geographical information.
6. Explain the global timekeeping system. Define and use the terms *standard time, standard meridian*, and *time zone* in your answer.
7. What is the *International Date Line?* Where is it found? Why is it necessary?
8. What is meant by the "tilt of the Earth's axis"? How is the tilt responsible for the seasons?

VISUALIZING EXERCISES

1. Sketch a diagram of the Earth at an equinox. Show the North and South Poles, the Equator, and the circle of illumination. Indicate the direction of the Sun's incoming rays and shade the night portion of the globe.
2. Sketch a diagram of the Earth at the June (summer) solstice, showing the same features. Include the Tropics of Cancer and Capricorn, and the Arctic and Antarctic Circles.

ESSAY QUESTION

1. Suppose that the Earth's axis were tilted at 40° to the plane of the ecliptic instead of 23½°. What would be the global effects of this change? How would the seasons change at your location?

Chapter 2
The Earth's Global Energy Balance

Qatar, located on the Persian Gulf, joins other Arab nations in a love of camel racing. Careful breeding has turned the lowly pack animal of the desert into a sleek, strong racer capable of galloping at 18 m/s (40 mi/hr) in sprints and 11 m/s (25 mi/hr) for an hour or more. The shadows of this line of racing camels in training demonstrates how the intensity of the Sun depends on the angle of the Sun in the sky. When the Sun is low, its energy is spread across a larger amount of surface, like the shadows of the camels. When the Sun is overhead, its energy is most intense.

TRAINING RACING CAMELS, AR RAYYAN, QATAR

©Yann Arthus-Bertrand/Altitude

The Earth's Global Energy Balance

O ur planet receives a nearly constant flow of solar energy that powers all life processes and most processes of the atmosphere and Earth's surface. What are the characteristics of this energy? How and where does the Earth and atmosphere absorb this solar energy? How is solar energy converted to heat that is ultimately radiated back to space? How does the Earth-atmosphere system trap heat to produce the greenhouse effect? These are some of the questions we will answer in this chapter.

EYE ON GLOBAL CHANGE

The Ozone Layer—Shield to Life

High above the Earth's surface lies an atmospheric layer rich in **ozone**, a form of oxygen in which three oxygen atoms are bonded together (O_3). Ozone is a highly reactive gas that can be toxic to life and damaging to materials; but high in the atmosphere it serves an essential purpose: to shelter life on the Earth's surface from powerful ultraviolet radiation emitted by the Sun. Without the ozone layer to absorb this radiation, bacteria exposed at the Earth's surface would be destroyed, and unprotected animal tissues would be severely damaged.

The ozone layer is presently under attack by air pollutant gases produced by human activity. The most important gases are **chlorofluorocarbons (CFCs)**, synthetic industrial chemical compounds containing chlorine, fluorine, and carbon atoms. Although CFCs were banned in aerosol sprays in the United States beginning in 1976, they are still used as cooling fluids in some refrigeration systems. When appliances containing CFCs leak or are discarded, their CFCs are released into the air.

Ozone is constantly being formed and destroyed by chemical reactions in the upper atmosphere, and the balance between formation and destruction determines the concentration of ozone. CFC molecules move up to the ozone layer where they decompose to chlorine oxide (ClO), which attacks ozone, converting it to ordinary oxygen (O_2) by a chain reaction. This lowers the concentration of ozone, and with less ozone, there is less absorption of ultraviolet radiation.

A hole in the ozone layer was discovered over the continent of Antarctica in the mid-1980s (Figure 2.1). In recent years, the ozone layer there has been found to thin during the early spring of the southern hemisphere, reaching a minimum during the month of September or October. Thereafter, the ozone hole typically shrinks slowly and ultimately disappears in early December.

In the northern hemisphere, the formation of an ozone hole is less likely to occur, due to unfavorable conditions. Still, arctic ozone holes have occurred several times in the past decade, with well-developed holes recorded in 2005 and 2011. Atmospheric computer models are projecting more such events in the period 2012–2019.

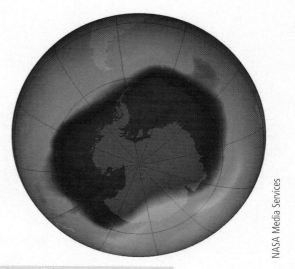

NASA Media Services

2.1 Ozone hole, September 24, 2006

The Antarctic ozone hole of 2006 was the largest on record, covering about 29.5 million km² (about 11.4 million mi²). Low values of ozone are shown in purple, ranging through blue, green, and yellow. Ozone concentration, which is measured in Dobson units, saw its lowest value—85 units—on October 8, 2006.

Aerosols introduced into the stratosphere by volcanic activity also can act to reduce ozone concentrations. The June 1991 eruption of Mount Pinatubo, in the Philippines, reduced global ozone in the stratosphere by 4 percent during the following year, with reductions over midlatitudes of up to 9 percent.

Since 1978, surface-level ultraviolet radiation has been increasing. Over most of North America, the increase has been about 4 percent per decade. This trend is expected to multiply the number of skin cancer cases. Crop yields and some forms of aquatic life may also suffer. Today, we are all aware of the dangers of harmful ultraviolet rays to our skin, and the importance of using sunscreen before going outdoors.

In response to the global threat of ozone depletion, in 1987, 23 nations signed a treaty to cut global CFC consumption by 50 percent by 1999. The treaty proved effective: by 1997, stratospheric chlorine concentrations had topped out and started to fall. In 2003, scientists, using three NASA satellite instruments and three international ground stations, confirmed a slowing in the rate of ozone depletion starting in 1997.

In operation for more than two decades, the international agreement has had an effect. Though not yet a reversal of ozone loss, the trend is encouraging. Current predictions show that the ozone layer will be restored by the middle of the century.

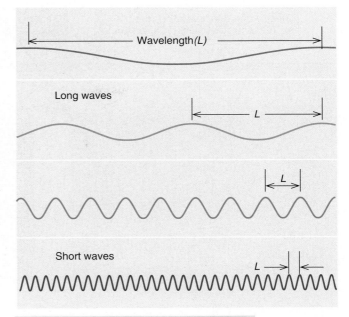

2.2 Wavelength of electromagnetic radiation

Electromagnetic radiation can be described as a collection of energy waves of different lengths. Wavelength is the distance from one wave crest to the next.

Electromagnetic Radiation

All surfaces—from the fiery Sun in the sky to the skin covering our bodies—constantly emit radiation. Very hot objects, such as the Sun or a lightbulb filament, give off radiation that is nearly all in the form of light. Most of this energy is visible light, which we perceive as having the colors of the rainbow; but the Sun also emits ultraviolet and infrared light that cannot be seen directly.

Cooler objects than the Sun, such as Earth surfaces and even our own bodies, emit heat radiation. So, our planet's surface and its atmosphere constantly emit heat. Over the long run, the Earth emits as much energy as it absorbs from the Sun, creating a global **energy balance**.

You can think of **electromagnetic radiation** as a collection of energy waves that travel away from the surface of an object. Electromagnetic energy occurs across a wide range of wavelengths. Light, radio waves, and infrared heat waves are all forms of electromagnetic radiation of different wavelengths.

Wavelength is the distance separating one wave crest from the next wave crest, as you can see in Figure 2.2. In this book, we will measure wavelength in micrometers. A *micrometer* is one millionth of a meter (10^{-6} m). The tip of your little finger is about 15,000 micrometers wide. We use the abbreviation μm for the micrometer. The first letter is the Greek letter μ, or mu.

Electromagnetic waves differ in wavelength throughout their entire range, or *spectrum* (Figure 2.3). *Gamma rays* and *X rays* lie at the short-wavelength end of the spectrum. Their wavelengths are normally expressed in *nanometers*. A nanometer is 1 one-thousandth of a micrometer, or 10^{-9} m, and is abbreviated nm. Gamma and X rays have high energies and can be hazardous to health. *Ultraviolet radiation* begins at about 10 nm and extends to 400 nm (or 0.4 μm). It can also damage living tissues.

Visible light begins at about 0.4 μm with the color violet. Colors then gradually change, through blue, green, yellow, orange, and red, until we reach the end of the visible spectrum at about 0.7 μm. Next is *near-infrared* radiation, with wavelengths from 0.7 to 1.2 μm. This radiation is very similar to visible light—most of it comes from the Sun. We can't see near-infrared light because our eyes are not sensitive to radiation beyond about 0.7 μm.

Shortwave infrared radiation also mostly comes from the Sun; it lies between 1.2 and 3.0 μm. *Middle-infrared* radiation, from 3.0 μm to 6 μm, can come from the Sun or from very hot sources on the Earth, such as forest fires and gas-well flames.

Next we have *thermal infrared* radiation, between 6 μm and 300 μm. This is given off by bodies at temperatures normally found at the Earth's surface. Figure 2.4 shows a thermal infrared

Visible light includes colors ranging from violet to red and spans the wavelength range of about 0.4 to 0.7 μm.

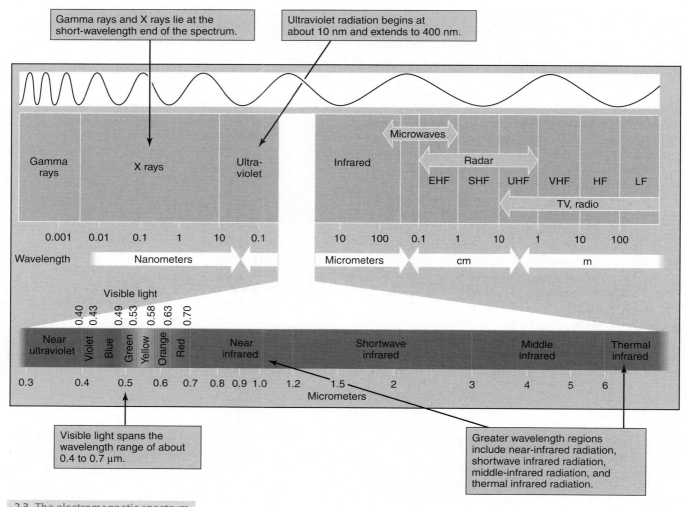

Gamma rays and X rays lie at the short-wavelength end of the spectrum.

Ultraviolet radiation begins at about 10 nm and extends to 400 nm.

Visible light spans the wavelength range of about 0.4 to 0.7 µm.

Greater wavelength regions include near-infrared radiation, shortwave infrared radiation, middle-infrared radiation, and thermal infrared radiation.

2.3 The electromagnetic spectrum

Electromagnetic radiation can exist at any wavelength. By convention, names are assigned to specific wavelength regions.

image of a suburban scene obtained at night using a special sensor. Here, yellow tones indicate the warmest temperatures, and black tones the coldest. Windows appear yellow because they are warm and radiate more intensely. House walls are intermediate in temperature and appear red. The lawn, driveway and roof appears cool, in tones of blue and violet. The sky is coldest and is shown in black.

WileyPLUS The Electromagnetic Spectrum
Expand your vision! Go to this animation and click on parts of the electromagnetic spectrum to reveal images that can't be sensed directly with your eyes.

RADIATION AND TEMPERATURE

There are two important physical principles to remember about the emission of electromagnetic radiation. The first is that hot objects radiate more energy than cooler objects. The flow of radiant energy emitted from the surface of an object is directly related to the absolute temperature of the surface, measured on the Kelvin absolute temperature scale, raised to the fourth power. So if you double the absolute temperature of an object, it will emit 16 times more energy from its surface. Even a small rise in temperature can mean a large increase in the rate at which radiation is given off by an object or surface.

The second principle is that the hotter the object, the shorter the wavelengths of radiation it emits. This inverse relationship between wavelength and temperature means that very hot objects like the Sun emit radiation at short wavelengths. Because the Earth is a much cooler object, it emits radiation with longer wavelengths. This principle explains why the Sun emits energy as light and the Earth emits energy as thermal infrared radiation.

> Hotter objects radiate substantially more energy than cooler objects. Hotter objects also radiate energy at shorter wavelengths.

2.4 A thermal infrared image

This suburban home was imaged at night in the thermal infrared spectral region. Black and blue tones indicate lower temperatures; red and yellow tones, higher temperatures. Ground and sky are coldest, while the windows of the heated home are warmest.

SOLAR RADIATION

Our Sun is a ball of constantly churning gases that are heated by continuous nuclear reactions. It is about average in size compared to other stars, and it has a surface temperature of about 6000°C (about 11,000°F). The Sun's energy travels outward in straight lines, or rays, at a speed of about 300,000 km (about 186,000 mi) per second—the speed of light. At that rate, it takes the energy about 8½ minutes to travel the 150 million km (93 million mi) from the Sun to the Earth.

The rays of solar radiation spread apart as they move away from the Sun. This means that a square meter on Mars will intercept less radiation than on Venus because Mars lies farther from the Sun. The Earth receives only about one half of one billionth of the Sun's total energy output.

Solar energy is generated by nuclear fusion reactions inside the Sun, as hydrogen is converted to helium at very high temperatures and pressures. A vast quantity of energy is generated this way, which finds its way to the Sun's surface. The rate of solar energy production is nearly constant, so the output of solar radiation also remains nearly constant, as does the amount of solar energy received by the Earth. The rate of incoming energy, known as the *solar constant*, is measured beyond the outer limits of the Earth's atmosphere, before any energy has been lost in the atmosphere.

You've probably seen the *watt* (W) used to describe the *power*, or rate of energy flow, of a lightbulb or other home appliance. When we talk about the intensity of received (or emitted) radiation, we must take into account both the power of the radiation and the surface area being hit by (or giving off) energy. So we use units of watts per square meter (W/m^2). The solar constant has a value of about 1361 W/m^2. Because there are no common equivalents for this energy flow rate in the English system, we will use only metric units.

CHARACTERISTICS OF SOLAR ENERGY

Let's look in more detail at the Sun's output as it is received by the Earth (illustrated in Figure 2.5). Energy intensity is shown on the graph on the vertical scale. Note that it is a logarithmic scale—that is, each whole unit marks an intensity 10 times greater than the one below. Wavelength is shown on the horizontal axis, also on a logarithmic scale.

The left side of Figure 2.5 shows how the Sun's incoming electromagnetic radiation varies with wavelength. The uppermost line indicates how a "perfect" Sun would supply solar energy at the top of the atmosphere. By "perfect," we mean a Sun at a temperature of 6000 K radiating as a *blackbody*, an ideal surface that follows physical theory exactly. The solid line shows the actual output of the Sun as measured at the top of the atmosphere. It is quite close to the "perfect" Sun, except for ultraviolet wavelengths, where the real Sun emits less energy. The Sun's output peaks in the visible part of the spectrum. We can see that human vision is adjusted to the wavelengths where solar light energy is highest.

The solar radiation actually reaching the Earth's surface is quite different from the solar radiation measured

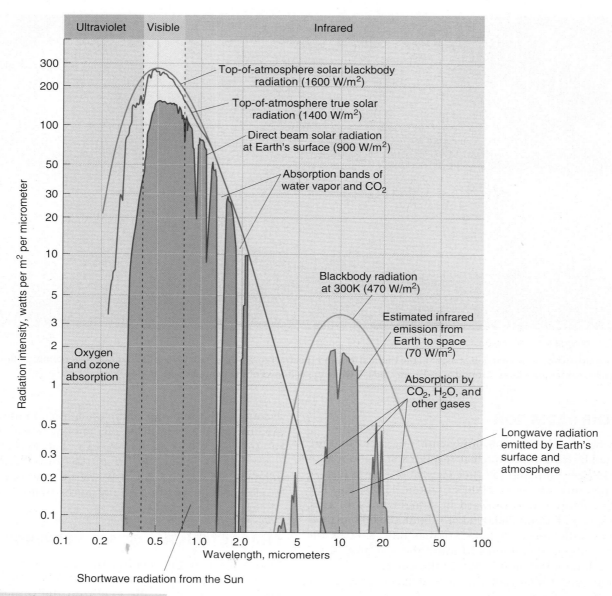

Ultraviolet Visible   Infrared

Top-of-atmosphere solar blackbody
radiation (1600 W/m^2)

Top-of-atmosphere true solar
radiation (1400 W/m^2)

Direct beam solar radiation
at Earth's surface (900 W/m^2)

Absorption bands of
water vapor and CO_2

Blackbody radiation
at 300K (470 W/m^2)

Estimated infrared
emission from
Earth to space
(70 W/m^2)

Oxygen
and ozone
absorption

Absorption by
CO_2, H_2O, and
other gases

Longwave radiation
emitted by Earth's
surface and
atmosphere

Shortwave radiation from the Sun

Radiation intensity, watts per m^2 per micrometer

Wavelength, micrometers

2.5 Spectra of solar and Earth radiation

The Earth radiates less energy than the Sun, and this energy is emitted at longer wavelengths. This figure plots both shortwave radiation, which comes from the Sun (left side), and longwave radiation, which is emitted by the Earth's surface and atmosphere (right side). Note that radiation intensity is shown on a logarithmic scale. We have not taken scattering into account in this illustration.

above the Earth's atmosphere. This is because solar radiation is both absorbed and scattered by varying amounts at different wavelengths as it passes through the atmosphere.

Molecules and particles in the atmosphere intercept and absorb radiation at particular wavelengths. This atmospheric **absorption** directly warms the atmosphere in a way that affects the global energy balance, as we will discuss toward the end of this chapter. Solar rays can also be scattered into different directions when they collide with molecules or particles in the atmosphere.

Scattering that turns solar rays back toward space is called **reflection**.

Solar energy received at the surface ranges from about 0.3 μm to 3 μm. This is known as **shortwave radiation**. We will now turn to the longer wavelengths of energy that are emitted by the Earth and atmosphere.

Shortwave radiation refers to wavelengths emitted by the Sun, which are in the range of about 0.3 to 3 μm. Longwave radiation refers to wavelengths emitted by cooler objects, such as Earth surfaces, which range from about 3 to 30 μm.

LONGWAVE RADIATION FROM THE EARTH

Remember that both the range of wavelengths and the intensity of radiation emitted by an object depend on the object's temperature. Because the Earth's surface and atmosphere are much colder than the Sun, our planet radiates less energy than the Sun, and this energy is emitted at longer wavelengths.

The right side of Figure 2.5 shows exactly that. The upper line shows the radiation of a blackbody at a temperature of about 300 K (23°C; 73°F), which is a good approximation for the Earth as a whole. At this temperature, radiation ranges from about 3 to 30 μm and peaks at about 10 μm in the thermal infrared region. This thermal infrared radiation emitted by the Earth is **longwave radiation**.

Beneath the blackbody curve is an irregular series of peaks that show upwelling energy emitted by the Earth and atmosphere as measured at the top of the atmosphere. Some wavelengths in this range seem to be missing, especially between 6 and 8 μm, 14 and 17 μm, and above 21 μm. The reason is that these wavelengths are almost completely absorbed by the atmosphere before they can escape. Water vapor and carbon dioxide are the main absorbers, and play a large part in the greenhouse effect, which we will discuss shortly.

There are still three regions where outgoing energy flow from the Earth to space is significant: 4–6 μm, 8–14 μm, and 17–21 μm. We call these *windows* through which longwave radiation leaves the Earth and flows to space.

THE GLOBAL RADIATION BALANCE

The Earth constantly absorbs solar shortwave radiation and emits longwave radiation. Figure 2.6 presents a simple diagram of this energy flow process, which we refer to as the Earth's **global radiation balance**.

The Sun provides a nearly constant flow of shortwave radiation that is intercepted by the Earth. Scattering by atmospheric particles and Earth surfaces reflects part of this radiation back into space without absorption. The remaining energy is absorbed by atmosphere, land, or ocean, and is ultimately emitted as longwave radiation to space. In the long run, absorbed incoming radiation is balanced by emitted outgoing radiation. Since the temperature of a surface is determined by the amount of energy it absorbs and emits, the Earth's overall temperature tends to remain constant.

Insolation over the Globe

Most natural phenomena on the Earth's surface—from the downhill flow of a river to the movement of a sand dune to the growth of a forest—are powered by the Sun, either directly or indirectly. It is the power source for wind, waves, weather, rivers, and ocean currents, as we will see here and in later chapters.

Although the flow of solar radiation to the Earth as a whole remains constant, different places on the planet receive energy at different rates and at different times. What causes this variation?

The flow rate of incoming solar radiation, observed at the top of the atmosphere, is known as **insolation**. It is measured in units of watts per square meter (W/m²). Insolation at a particular moment and particular location depends on the angle of the Sun above the horizon. It is greatest when

> Insolation refers to the flow rate of incoming solar radiation. It is high when the Sun is high in the sky.

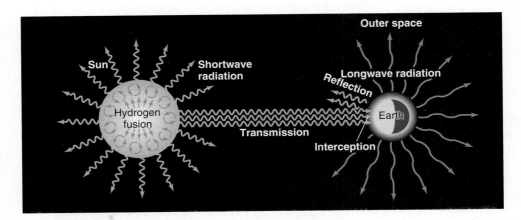

2.6 The global radiation balance

Shortwave radiation from the Sun is transmitted through space, where it is intercepted by the Earth. A portion of the intercepted radiation is reflected back to space, while the remainder is absorbed. The absorbed radiation is then ultimately emitted as longwave radiation to outer space.

the Sun is directly overhead, and it decreases when the Sun is low in the sky, since the same amount of solar energy is spread out over a greater area of ground surface (Figure 2.7).

WileyPLUS The Sun's Noon Angle and the Length of Day

Imagine yourself, through this animation, watching the Earth from a point far out in space, where it is easy to see how both the Sun's angle at noon and the length of day vary with the seasons and latitude for any point on Earth.

DAILY INSOLATION THROUGH THE YEAR

Daily insolation at a location is the average insolation rate taken over a 24-hour day. Daily insolation depends on two factors: (1) the angles at which the Sun's rays strike the surface during that day, and (2) how long the location is exposed to the rays. In Chapter 1 we saw that both of these factors are controlled by latitude and the time of year. At midlatitude locations in summer, for example, days are long and the Sun rises to a position high in the sky, heating the surface more intensely.

2.7 Solar intensity and Sun angle

The intensity of the solar beam depends on the angle between the beam and the surface.

1 unit of surface area

One unit of light is concentrated over one unit of surface area.

 VERTICAL RAYS
Sunlight, represented by the flashlight, is most intense when the beam is vertical.

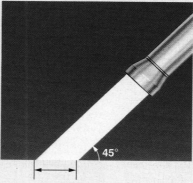

1.4 units of surface area

One unit of light is dispersed over 1.4 units of surface area.

 RAYS AT 45° ANGLE
When the beam strikes the surface at an angle of 45°, it covers a larger surface, and so is less intense.

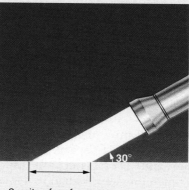

2 units of surface area

One unit of light is dispersed over 2 units of surface area.

 RAYS AT 30° ANGLE
At 30°, the beam covers an even greater surface and is even weaker.

BEAM ANGLE AND LATITUDE ▶
Because the angle of the solar beam striking the Earth varies with latitude, insolation is strongest near the Equator and weakest near the poles.

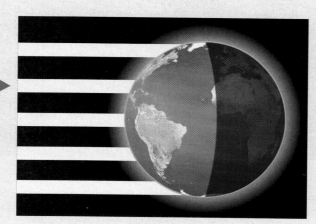

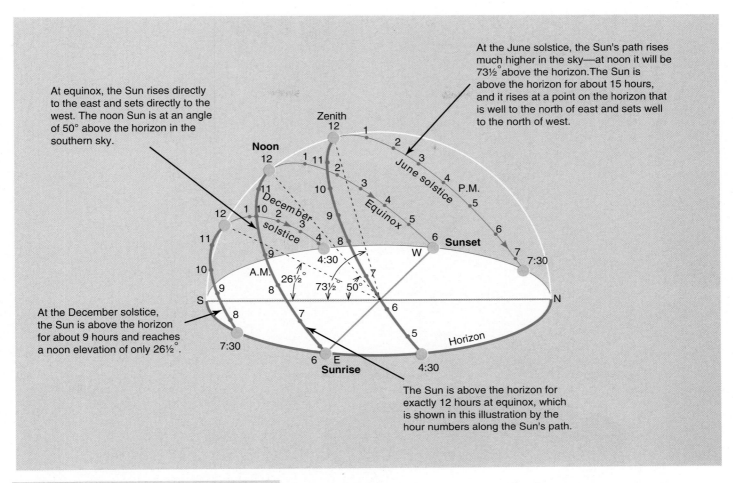

At the June solstice, the Sun's path rises much higher in the sky—at noon it will be 73½° above the horizon. The Sun is above the horizon for about 15 hours, and it rises at a point on the horizon that is well to the north of east and sets well to the north of west.

At equinox, the Sun rises directly to the east and sets directly to the west. The noon Sun is at an angle of 50° above the horizon in the southern sky.

At the December solstice, the Sun is above the horizon for about 9 hours and reaches a noon elevation of only 26½°.

The Sun is above the horizon for exactly 12 hours at equinox, which is shown in this illustration by the hour numbers along the Sun's path.

2.8 The path of the Sun in the sky at 40° N latitude

Through the seasons, the Sun's path changes greatly in position and height above the horizon.

How does the angle of the Sun vary during the day? It depends on the Sun's path. Near noon, the Sun is high above the horizon—the Sun's angle is greater, and so insolation is higher. Figure 2.8 shows the typical conditions found in midlatitudes in the northern hemisphere, for example, at New York or Denver. An observer standing on a wide plain will see a small area of the Earth's surface bounded by a circular horizon. The Earth's surface appears flat, and the Sun seems to travel inside a vast dome in the sky.

Comparing the three paths shown in the figure, we find that both the length of time the Sun is in the sky and the angle of the Sun during the main part of the day change with the time of year. At the June solstice, average daily insolation will be greatest, since the Sun is in the sky longer and reaches higher elevations. At the December solstice, daily insolation will be lowest, with a shorter daily path and lower elevations. At the equinox, the insolation will be intermediate.

Figure 2.9 shows the Sun's path for three other latitudes. At the North Pole, the Sun moves in a circle in the sky at an elevation that changes with the seasons. At the Equator, the Sun is always in the sky for 12 hours, but its noon angle varies through the year. At the Tropic of Capricorn, the Sun is in the sky longest and reaches its highest elevations at the December solstice.

Based on this analysis, daily insolation will vary widely with the seasons at most latitudes. As shown in Figure 2.10, daily insolation at 40° will range from about 160 W/m² at the December solstice to about 460 W/m² at the June solstice. Insolation drops to zero at the North Pole at the September equinox, when the Sun's circular path sinks below the horizon, and does not rise again until the March equinox. However, the peak insolation at the June solstice is greater at the North Pole—about 500 W/m²—than at any other latitude.

At the Equator, daily insolation varies from about 380 W/m² to about 430 W/m², and there are two maximums. Each is near the time of an equinox, when the Sun is directly overhead at noon. At the solstices,

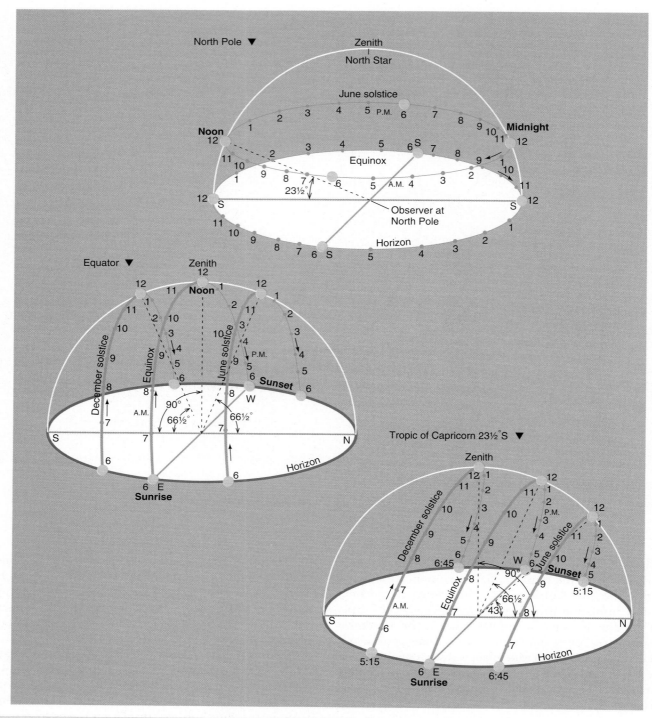

2.9 The Sun's path at the North Pole, Equator, and Tropic of Capricorn

insolation is lower because the Sun's path is lower in the sky, as shown in Figure 2.9.

WileyPLUS The Path of the Sun in the Sky
View this animation to follow the daily path of the Sun in the sky. It shows how both latitude and season affect the Sun's motion, as seen by an observer on the ground.

ANNUAL INSOLATION BY LATITUDE

How does latitude affect *annual insolation*—the rate of insolation averaged over an entire year? Figure 2.11 shows two curves of annual insolation by latitude: one for the actual case of the Earth's axis tilted at 23½° and the other for an Earth with an untilted axis.

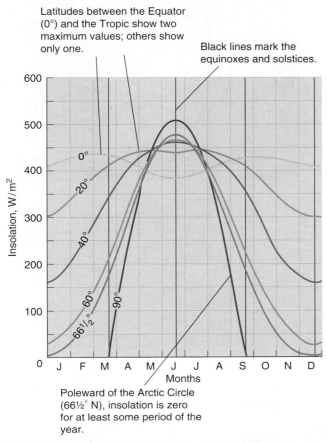

Latitudes between the Equator (0°) and the Tropic show two maximum values; others show only one.

Black lines mark the equinoxes and solstices.

Poleward of the Arctic Circle (66½° N), insolation is zero for at least some period of the year.

2.10 Daily insolation throughout the year at various latitudes (northern hemisphere)

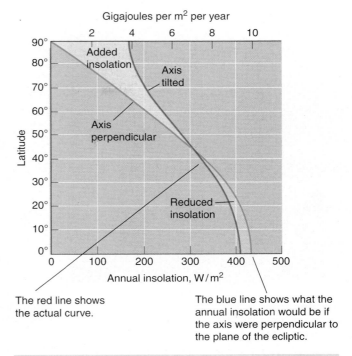

The red line shows the actual curve.

The blue line shows what the annual insolation would be if the axis were perpendicular to the plane of the ecliptic.

2.11 Annual insolation from the Equator to either pole for the Earth

Let's look first at the real case of a tilted axis. We can see that annual insolation varies smoothly from the Equator to the pole and is greater at lower latitudes. But high latitudes still receive a considerable flow of solar energy—the annual insolation value at the pole is about 40 percent of the value at the Equator.

Now let's look at what would happen if the Earth's axis were not tilted. With the axis perpendicular to the plane of the ecliptic, there are no seasons. Annual insolation is very high at the Equator because the Sun passes directly overhead at noon every day throughout the year. Annual insolation at the poles is zero because the Sun's rays always skirt the horizon.

We can see that without a tilted axis our planet would be a very different place. The tilt redistributes a very significant portion of the Earth's insolation from the equatorial regions toward the poles. So even though the poles do not receive direct sunlight for six months of the year, they still receive nearly half the amount of annual solar radiation as the Equator.

WORLD LATITUDE ZONES

The seasonal pattern of daily insolation provides a convenient way to divide the globe into broad latitude zones (Figure 2.12) that we will use in this book. The *equatorial zone* encompasses the Equator and covers the latitude belt roughly 10° N lat. to 10° S lat. Here the Sun provides intense insolation throughout most of the year, and the days and nights are of roughly equal length. Spanning the Tropics of Cancer and Capricorn are the *tropical zones*, ranging from 10°–25° N and S lat. A marked seasonal cycle exists in these zones, combined with high annual insolation.

Moving toward the poles, we come to the *subtropical zones*, which lie roughly between 25°–35° N and S lat. These zones have a strong seasonal cycle and a high annual insolation. The *midlatitude zones* are next, between 35°–55° N and S lat. The length of daylight varies significantly from winter to summer here, so seasonal contrasts in insolation are quite striking. As a result, these regions experience a wide range in annual surface temperature.

The *subarctic* and *subantarctic zones* border the midlatitude zones at 55°–60° N and S lat. The *arctic* and *antarctic zones* lie between 65°–70° N and S lat., astride the Arctic and Antarctic Circles. These zones have an extremely large yearly variation in day lengths, yielding enormous contrasts in insolation over the year. Finally, the *north* and *south polar zones* range from about 75° latitude to the poles. They experience the greatest seasonal insolation contrast of all, and have 24-hour days or nights for much of the year.

Richard Nowitz/NG Image Collection

Kike Calvo/V&W/The Image Works

▲ **SUBARCTIC ZONE**
Much of the subarctic zone is covered by evergreen forest, seen here with a ground cover of snow, near Churchill, Hudson Bay region, Canada.

▲ **MIDLATITUDE ZONE**
A summer midlatitude land-scape in the Tuscany region of Italy.

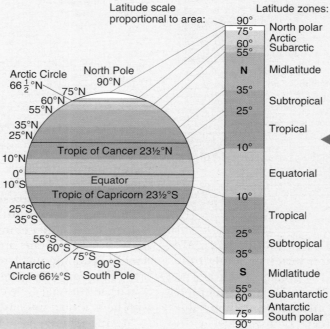

Latitude scale proportional to area:

Latitude zones:

90°	North polar
75°	Arctic
60°	
55°	Subarctic
N	Midlatitude
35°	
25°	Subtropical
10°	Tropical
	Equatorial
10°	
25°	Tropical
35°	Subtropical
S	Midlatitude
55°	Subantarctic
60°	Antarctic
75°	South polar
90°	

Arctic Circle 66½°N
North Pole 90°N
75°N
60°N
55°N
35°N
25°N
10°N
0°
10°S
Tropic of Cancer 23½°N
Equator
Tropic of Capricorn 23½°S
25°S
35°S
55°S
60°S
75°S
90°S
South Pole
Antarctic Circle 66½°S

◄ **LATITUDE ZONES**
A geographer's system of latitude zones, based on the seasonal patterns of daily inso-lation observed over the globe.

▼ **TROPICAL ZONE**
The tropical zone is the home of the world's driest deserts. Pictured here are sand dunes in the Namib Desert, Namibia.

Tierbild Okapia/Photo Researchers, Inc.

▼ **EQUATORIAL ZONE**
An equatorial rainforest, as seen along a stream in the Gunung Palung National Park, Borneo, Indonesia.

Tim Laman/NG Image Collection

Composition of the Atmosphere

Thus far, we've mentioned a few of the gases found in the Earth's atmosphere. A longer look here at the composition of the atmosphere will be useful as we continue exploring the Earth's energy balance.

The Earth is surrounded by air, a mixture of various gases that reaches up to a height of many kilometers. This envelope of air makes up our atmosphere (Figure 2.13). It is held in place by the Earth's gravity. Almost all the atmosphere (97 percent) lies within 30 km (19 mi) of the Earth's surface. The upper limit of the atmosphere is at a height of approximately 10,000 km (about 6000 mi) above the Earth's surface—a distance that is nearly as large as Earth's diameter.

The proportion of gases in dry air is highly uniform up to an altitude of about 80 km (50 mi). About 99 percent of pure, dry air is nitrogen (about 78 percent by volume) and oxygen (about 21 percent). These two main component gases of the lower atmosphere are perfectly mixed, so pure, dry air behaves as if it is a single gas with very definite physical properties.

Nitrogen gas is a molecule consisting of two nitrogen atoms (N_2). It does not easily react with other substances. Soil bacteria do take up very small amounts of nitrogen, which can be used by plants, but otherwise, nitrogen is largely a "filler," adding inert bulk to the atmosphere.

In contrast, *oxygen gas* (O_2) is chemically very active, combining readily with other elements in the process of *oxidation*. Fuel combustion is a rapid form of oxidation, while certain types of rock decay (weathering) are very slow forms of oxidation. Living tissues require oxygen to convert foods into energy.

The remaining 1 percent of dry air is mostly argon, an inactive gas of little importance in natural processes, with a very small amount of *carbon dioxide* (CO_2), amounting to about 0.0385 percent. Although the amount of CO_2 is small, it is a very important atmospheric gas because it absorbs much of the incoming shortwave radiation from the Sun and outgoing long-wave radiation from the Earth. This contributes to the *greenhouse effect*, which we will return to in a later section. Carbon dioxide is also used by green plants, which convert it to its chemical compounds to build up their tissues, organs, and supporting structures during photosynthesis.

Water vapor is another important atmospheric gas. Individual water vapor molecules mix freely with other atmospheric gases, but water vapor can vary highly in concentration. Water vapor usually makes up less than 1 percent of the atmosphere, but under very warm, moist conditions, as much as 2 percent of the air can be water vapor. Since it is a good absorber of heat radiation, like carbon dioxide, it plays a major role in warming the lower atmosphere and enhancing the greenhouse effect.

Another small, but important, constituent of the atmosphere is *ozone*, which we described in our opening feature, "Eye on Global Change." Ozone in the upper atmosphere is beneficial because it shields life at the Earth's surface from harmful solar ultraviolet radiation. But in the lowest layers of the atmosphere, ozone is an air pollutant that damages lung tissue and aggravates bronchitis, emphysema, and asthma.

Energy Transfer

We've used the familiar word *temperature* several times so far in this chapter. But what is temperature? Particles of mass can have many kinds of energy. One kind is the kinetic energy of molecules within a substance, in which the molecules are moving or vibrating. Temperature measures the amount of this energy; the more rapid the motion, the higher the temperature.

Physicists define the term *heat* as a flow of internal energy transferred from one substance to another. Objects can't contain heat; they only contain a certain level of internal energy. But as they lose or gain internal energy, heat flows. As we saw earlier, all objects lose internal energy to their surroundings by radiating electromagnetic energy in proportion to their temperature. This radiant energy flow is thus a form of heat. Heat flow also occurs by conduction, convection, and latent heat transfer.

When substances with different temperatures are placed in contact, some of kinetic energy of motion of the molecules in the warmer substance is transferred to the cooler substance. This transfer of internal energy is called *conduction*. Neighboring particles exchange energy until all particles have the same level of internal energy. In other words, the cooler object becomes warmer and the warmer object becomes cooler.

Convection is a flow of internal energy that occurs when matter moves from one place to another. This

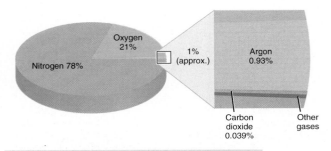

2.13 Component gases of the lower atmosphere

Values show percentage by volume for dry air. Nitrogen and oxygen form 99 percent of the air on Earth, with other gases, principally argon and carbon dioxide, accounting for the final 1 percent.

motion happens spontaneously in fluids—liquids or gases—when parts or regions of the fluid are at different temperatures and therefore have different densities. For example, air heated by the burner on a stove will expand and rise, creating an upward flow of heat. As we will see later in this chapter, convection helps drive the circulation of the Earth's atmosphere and oceans and redirects much of the solar heat flow received between the tropics to higher latitudes.

SENSIBLE AND LATENT HEAT

The flow of internal energy between two substances resulting in a temperature change is referred to as **sensible heat**. But a flow of heat to or from an object can also occur without a change in temperature, when a change of state (solid, liquid, or gas) occurs—for example, when liquid water evaporates to become water vapor. An energy flow that changes the state of a substance is known as **latent heat**. The latent heat changes the potential energy that is held in the positioning of atoms or molecules within a substance. Atoms or molecules moving freely in a gas have the highest potential energy, while atoms or molecules in a liquid state have lower potential energy. In a solid, the atoms or molecules have the lowest potential energy.

Latent heat taken up in a change of state is completely released when the change of state is reversed. For example, the amount of energy required to evaporate 1 kilogram (kg) of water liquid at 25°C to water vapor is 2441 kilojoules (kJ). If that water vapor is then condensed to liquid water at the same temperature, 2441 kJ of energy is released to the surroundings. For freezing and thawing, the latent heat flow required to melt 1 kg of ice is 334 kJ. When the water refreezes, it will provide a latent heat flow to the surroundings of 334 kJ.

In the Earth-atmosphere system, *sensible heat transfer* occurs when air is heated or cooled by ocean or land surfaces and when currents of warm water and air mix with cooler air and water. *Latent heat transfer* occurs when water evaporates from a moist land surface or from open water, transferring energy from the surface to the atmosphere. That latent heat is later released as sensible heat, often far away, when the water vapor condenses to form water droplets or snow crystals. On a global scale, latent heat transfer is a very important mechanism for transporting large amounts of heat from one region of the Earth to another.

> Sensible heat transfer refers to the flow of heat between the Earth's surface and the atmosphere by conduction or convection. Latent heat transfer refers to the flow of heat carried by changes of state of water.

WileyPLUS Latent Heat
Brush up on how water changes state between solid, liquid, and gaseous forms, and how potential energy of state is either absorbed or released to provide latent heat transfer. An animation.

The Global Energy System

Human activity around the globe has changed the planet's surface cover and added carbon dioxide to the atmosphere. Have we irrevocably shifted the balance of energy flows? Is our Earth absorbing more solar energy and becoming warmer? Or is it absorbing less and becoming cooler? If we want to understand human impact on the Earth-atmosphere system, then we need to examine the global energy balance in detail.

The flow of energy from the Sun to the Earth and then back out into space is a complex system. Solar energy is the ultimate power source for the Earth's surface processes, so when we trace the energy flows between the Sun, surface, and atmosphere, we are really studying how these processes are driven.

SOLAR ENERGY LOSSES IN THE ATMOSPHERE

Let's examine the flow of insolation through the atmosphere on its way to the surface. Figure 2.14 gives typical

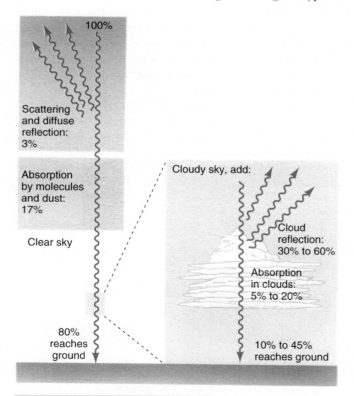

2.14 Fate of incoming solar radiation

Losses of incoming solar energy are much lower when skies are clear (left) than when there is cloud cover (right).

values for losses of incoming shortwave radiation in the solar beam as it penetrates the atmosphere. Gamma rays and X rays from the Sun are almost completely absorbed by the thin outer layers of the atmosphere, while much of the ultraviolet radiation is also absorbed, particularly by ozone.

As the radiation moves deeper through denser layers of the atmosphere, it can be scattered by gas molecules, dust, or other particles in the air, deflecting it in any direction. Apart from this change in direction, it is unchanged. Scattered radiation moving in all directions through the atmosphere is known as *diffuse radiation*. Some scattered radiation flows down to the Earth's surface, while some flows upward. This upward flow of diffuse radiation escaping back to space, also known as *diffuse reflection*, amounts to about 3 percent of incoming solar radiation.

What about absorption? As we saw earlier, molecules and particles can absorb radiation as it passes through the atmosphere. Carbon dioxide and water vapor are the most important absorbers, but because the water vapor content of air can vary widely, absorption also varies from one global environment to another. About 17 percent of incoming solar radiation is absorbed, raising the temperature of atmospheric layers. After taking into account absorption and scattering, about 80 percent of the incoming solar radiation reaches the ground.

Clouds can greatly increase the amount of incoming solar radiation reflected back to space. Reflection from thick, low clouds deflects about 30 to 60 percent of incoming radiation back into space. Clouds also absorb as much as 5 to 20 percent of radiation.

When accounting for both cloudy and clear skies on a global scale, only about half of the total insolation at the top of the atmosphere reaches the surface. When this energy strikes the surface, it can be either absorbed or scattered upward. Absorption heats the surface, raising the surface temperature. The scattered radiation reenters the atmosphere, and much of it passes through, directly to space.

ALBEDO

The proportion of shortwave radiant energy scattered upward by a surface is called its **albedo**. Snow and ice have high albedos (0.45 to 0.85), reflecting most of the solar radiation that hits them and absorbing only a small amount. In contrast, a black pavement, which has a low albedo (0.03), absorbs nearly all the incoming solar energy (Figure 2.15). Water also has a low albedo (0.02), unless the Sun illuminates it at a low angle, producing Sun glint. The energy absorbed by a surface warms the air immediately above it by conduction and convection, so surface temperatures are warmer over low-albedo than over high-albedo surfaces. Fields, forests, and bare ground have intermediate albedos, ranging from 0.03 to 0.25.

The Earth and atmosphere system, taken as a whole, has an albedo of between 0.29 and 0.34. This means that our planet sends back to space slightly less than one-third of the solar radiation it receives. It also means that our planet absorbs slightly more than two-thirds of the solar radiation it receives. This balance between reflected and absorbed solar radiation is what determines the overall temperature of Earth.

COUNTERRADIATION AND THE GREENHOUSE EFFECT

As well as being warmed by shortwave radiation from the Sun, the Earth's surface is significantly heated by the longwave radiation emitted by the atmosphere and absorbed by the ground. Let's look at this in more detail.

Figure 2.16 shows the energy flows between the surface, atmosphere, and space. On the left we can see the flow of shortwave radiation from the Sun to the surface. Some of this radiation is reflected back to space, but much is absorbed, warming the surface.

Meanwhile, the Earth's surface emits longwave radiation upward. Some of this radiation escapes directly to space, while the remainder is absorbed by the atmosphere.

What about longwave radiation emitted by the atmosphere? Although the atmosphere is colder than the surface, it also emits longwave radiation, which is emitted in all directions, and so some radiates upward to space while the remainder radiates downward toward the Earth's surface. We call this downward flow **counterradiation**. It replaces some of the heat emitted by the surface.

This mechanism, in which the atmosphere absorbs longwave radiation leaving the surface and returns it to the surface through counterradiation is termed the **greenhouse effect** (Figure 2.17). The term *greenhouse* is not, however, quite accurate. Like the atmosphere, the window glass in a greenhouse is transparent to solar shortwave radiation, while absorbing and reradiating longwave radiation. But unlike the atmosphere, a greenhouse is warmed mainly by keeping the warm air inside the greenhouse from mixing with the outside air, not by counterradiation from the glass.

Counterradiation from the atmosphere to the Earth's surface helps warm the surface and creates the greenhouse effect. It is enhanced by carbon dioxide and water in the atmosphere.

Counterradiation depends greatly on the presence of carbon dioxide and water vapor in the atmosphere. Remember that much of the longwave radiation emitted upward from the Earth's surface is absorbed by these two gases. This absorbed energy raises the temperature of the atmosphere, causing it to emit more counterradiation. So, the lower atmosphere, with its

2.15 Albedo contrasts

The albedo of a surface depends on the nature of the surface material.

John Dunn/Arctic Light/NGImage Collection

Jeremy Woodhouse/Masterfile

BRIGHT SNOW
A layer of new, fresh snow has a high albedo, reflecting most of the sunlight it receives. Only a small portion is absorbed.

BLACKTOP ROAD
Asphalt paving reflects little light, so it appears dark or black and has a low albedo. It absorbs nearly all of the solar radiation it receives.

Altrendo/Getty Images, Inc.

WATER
Water absorbs solar radiation and has a low albedo, unless the radiation strikes the water surface at a low angle. In that case, Sun glint raises the albedo.

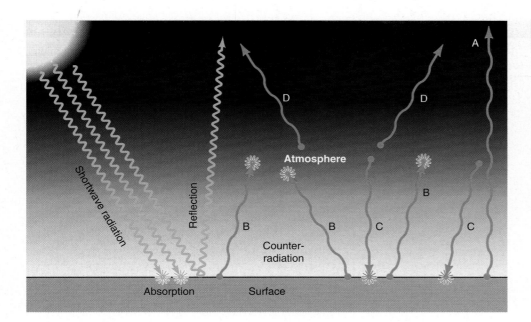

Shortwave radiation

Reflection

Atmosphere

D

D

A

B

B

B

B

C

C

Counter-
radiation

Absorption   Surface

2.16 Counterradiation and the greenhouse effect

Shortwave radiation passes through the
atmosphere and is absorbed or reflected
at the surface. Absorption warms the
surface, which emits longwave radiation.
Some of this flow passes directly to space
(A), but most is absorbed by the atmosphere (B). In turn, the atmosphere radiates longwave energy back to the surface
as counterradiation (C) and also to space
(D). The counterradiation produces the
greenhouse effect.

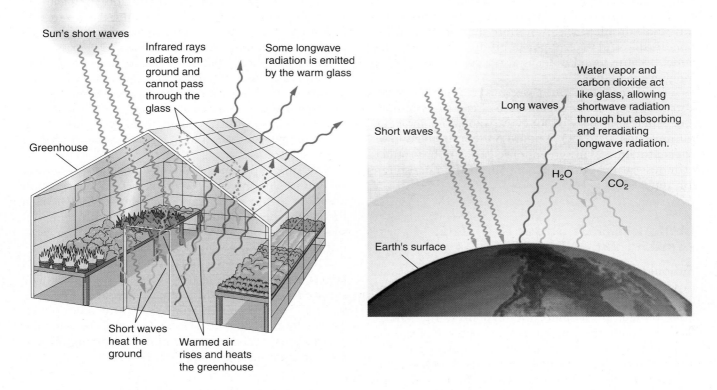

Sun's short waves

Infrared rays
radiate from
ground and
cannot pass
through the
glass

Some longwave
radiation is emitted
by the warm glass

Greenhouse

Short waves
heat the
ground

Warmed air
rises and heats
the greenhouse

Water vapor and
carbon dioxide act
like glass, allowing
shortwave radiation
through but absorbing
and reradiating
longwave radiation.

Long waves

Short waves

H_2O CO_2

Earth's surface

2.17 A greenhouse and the greenhouse effect

Water vapor and carbon dioxide act like glass, allowing shortwave radiation through, but absorbing and radiating longwave radiation.

Table 2.1 Greenhouse gases characterization

Greenhouse gases	Chemical formula	Pre-industrial concentration ppb*	2010 concentration	Anthropogenic sources	Global warming potential**
Carbon dioxide	CO_2	278,000	389,000	Fossil fuel, combustion, land use conversion, cement production	1
Methane	CH_4	700	1866	Fossil fuels, rice paddies, waste dumps, livestock	21
Nitrous oxide	N_2O	275	323	Fertilizer, industrial processes, combustion	310
CFC-12	CCl_2F_2	0	0.537	Liquid coolants, foams	7000
HCFC-22	$CHClF_2$	0	0.210	Liquid coolants	1350
Sulfur hexafluoride	SF_6	0	0.684	Dielectric fluid	23,900

*Parts per billion per volume of air in the atmosphere.

**Global warming potential for 100-year time frame, based on CO_2 value of 1.

longwave-absorbing gases, acts like a blanket that keeps the surface warm. Cloud layers, which are composed of tiny water droplets, are even more important than carbon dioxide and water vapor in producing a blanketing effect because liquid water is also a strong absorber of longwave radiation.

A number of other *greenhouse gases,* shown in Table 2.1, also contribute to the greenhouse effect. Many of these are better absorbers of longwave radiation, molecule for molecule, than CO_2, but are present in much lower concentrations. Many greenhouse gases are by-products of industrial processes and remain in the atmosphere for hundreds of years.

GLOBAL ENERGY BUDGETS OF THE ATMOSPHERE AND SURFACE

Although energy may change its form from shortwave to longwave radiation, or to sensible heat or latent heat, it cannot be created or destroyed. Like a household budget of income and expenses, the energy flows between the Sun and the Earth's atmosphere and surface must balance over the long term. The global energy budget shown in Figure 2.18 takes into account all the important energy flows, and helps us to understand how changes in these flows might affect the Earth's climate. It uses a scale in which the amount of incoming solar radiation is represented as 100 units.

Let's look first at the top of the atmosphere, where we see the balance for the Earth-atmosphere system as a whole. Incoming solar radiation (100 units) is balanced by exiting shortwave reflection from the Earth's surface and atmosphere, and outgoing longwave radiation coming from the atmosphere and surface.

The atmosphere's budget is also balanced: it receives 152 units and loses 152 units. Received energy includes absorbed incoming solar radiation, absorbed longwave radiation from the surface, and latent and sensible heat transfer from the surface. The atmosphere loses longwave energy by radiation to space and counterradiation to the surface.

The surface receives 144 units and loses 144 units. Incoming energy consists of direct solar radiation absorbed at the surface and longwave radiation from the atmosphere. Exiting energy includes latent and sensible heat transfer to the atmosphere and longwave radiation to the atmosphere and space.

The greenhouse effect is readily visible where the two largest arrows appear, at the center of the figure. The surface loses 102 units of longwave energy but receives 95 units of counterradiation from the atmosphere. These flows amount to a loop that absorbs but then returns most of the longwave radiation leaving the surface, keeping surface temperatures warm.

CLIMATE AND GLOBAL CHANGE

The global energy budget helps us understand how global change might affect the Earth's climate. For example, suppose that clearing forests for agriculture, and turning agricultural lands into urban and suburban

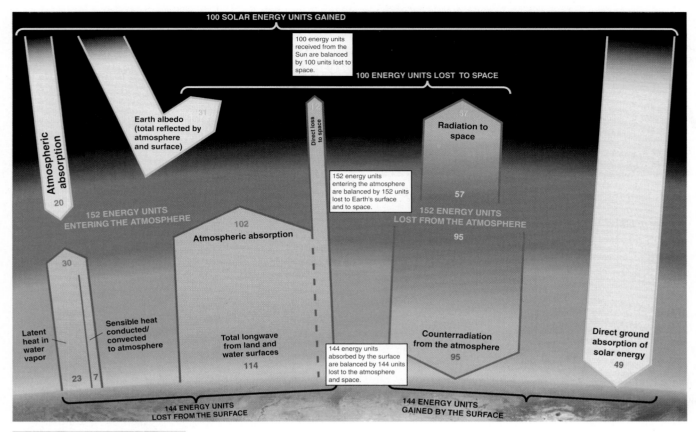

100 SOLAR ENERGY UNITS GAINED

100 energy units received from the Sun are balanced by 100 units lost to space.

100 ENERGY UNITS LOST TO SPACE

Earth albedo (total reflected by atmosphere and surface)   31

Atmospheric absorption   20

Direct loss to space

Radiation to space   57

152 ENERGY UNITS ENTERING THE ATMOSPHERE

152 energy units entering the atmosphere are balanced by 152 units lost to Earth's surface and to space.

152 ENERGY UNITS LOST FROM THE ATMOSPHERE   95

Atmospheric absorption   102

Latent heat in water vapor   23

Sensible heat conducted/ convected to atmosphere   7

30

Total longwave from land and water surfaces   114

144 energy units absorbed by the surface are balanced by 144 units lost to the atmosphere and space.

Counterradiation from the atmosphere   95

Direct ground absorption of solar energy   49

144 ENERGY UNITS LOST FROM THE SURFACE

144 ENERGY UNITS GAINED BY THE SURFACE

2.18 The global energy balance

Energy flows continuously among the Earth's surface, atmosphere, and space. The relative size of each flow is based on an arbitrary 100 units of solar energy reaching the top of the Earth's atmosphere. The difference between solar energy absorbed by the Earth system (100 units incoming − 31 reflected = 69 absorbed) and the energy absorbed at the surface (144 units) is the energy (75 units) that is recycled within the Earth system (144−69=75). The larger this number, the warmer the Earth system's climate.

areas, decreases surface albedo. In that case, more energy would be absorbed by the ground, raising its temperature. That, in turn, would increase the flow of surface longwave radiation to the atmosphere, which would be absorbed and would then boost counterradiation. The total effect would probably be to amplify warming through the greenhouse effect.

What if industrial aerosols caused more low, thick clouds to form? Low clouds would increase shortwave reflection back to space, causing the Earth's surface and atmosphere to cool. What about increasing condensation trails from jet aircraft? These could cause more high, thin clouds to form, which absorb more longwave energy and make the atmosphere warmer, thereby boosting counterradiation and intensifying the greenhouse effect. The energy flow linkages between the Sun, surface, atmosphere, and space are critical components of our climate system, and human activities can modify these flows significantly.

Net Radiation, Latitude, and the Energy Balance

Although the energy budgets of the Earth's surface and atmosphere are in balance overall, they do not have to balance at each particular place on the Earth, nor do they have to balance at all times. At night, for example, there is no incoming radiation from the Sun, yet the Earth's surface and atmosphere still emit outgoing radiation.

Net radiation is the difference between all incoming radiation and all outgoing radiation. In places where radiant energy flows in faster than it flows out, net radiation is positive, providing an energy surplus. In other places, net radiation can be negative. For the entire Earth and atmosphere, the net radiation is zero over a year.

We saw earlier that solar energy input varies strongly with latitude. What is the effect of this variation on net radiation? To answer this question, let's look at Figure 2.19, which shows the net radiation profile from

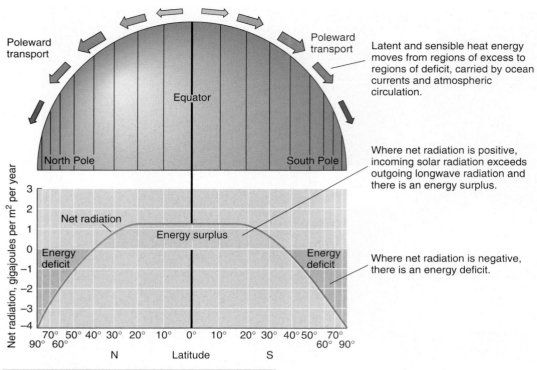

Poleward transport

Poleward transport

Latent and sensible heat energy moves from regions of excess to regions of deficit, carried by ocean currents and atmospheric circulation.

Equator

North Pole

South Pole

Net radiation

Energy surplus

Energy deficit

Energy deficit

Net radiation, gigajoules per m² per year

Where net radiation is positive, incoming solar radiation exceeds outgoing longwave radiation and there is an energy surplus.

Where net radiation is negative, there is an energy deficit.

90° 70° 50° 40° 30° 20° 10° 0° 10° 20° 30° 40° 50° 70° 90°
   60°   60°
N   Latitude   S

2.19 Annual surface net radiation from pole to pole

A surplus of net radiation occurs between the tropics; a deficiency is found between the tropics and the poles.

2.20 Solar power

Solar energy powers wind and water motion, and can also generate electricity.

Simon Fraser/Science Photo Library/Photo Researchers, Inc.

EMERGENCY
STATE OF HAWAII
CALL BOX

▲ **SOLAR-POWERED CALL BOX**
This emergency telephone is powered by the solar cell atop its pole.

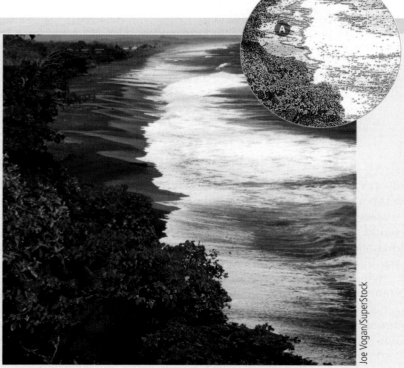

Joe Vogan/SuperStock

WAVE EROSION ▲
Ocean waves, powered by the Sun through the Earth's wind system, attack and erode the coastline, carving distinctive coastal landforms.

EYE ON THE LANDSCAPE **What else would the geographer see?**
Why is the sand along this beach Ⓐ such a dark color?. One reason might be that the beach is formed from eroded particles of lava, produced by past volcanic eruptions. As waves and salt spray attack volcanic layers in cliffs and headlands, grains of dark-colored lava are released. Wind-driven waves sweep the dark sand along the coast, producing beaches where the sand accumulates.

pole to pole. Between about 40° N and 40° S there is a net radiant energy gain, labeled "Energy surplus." In other words, incoming solar radiation exceeds outgoing longwave radiation. Poleward of 40° N and 40° S, the net radiation is negative and is labeled "Energy deficit," meaning that outgoing longwave radiation exceeds incoming shortwave radiation.

If you examine the graph carefully, you will find that the area labeled "Energy surplus" is equal in size to the combined areas labeled "Energy deficit." So the net radiation for the Earth's surface as a whole is zero, as expected, with global incoming shortwave radiation balancing exactly global outgoing longwave radiation.

Because there is an energy surplus at low latitudes, and an energy deficit at high latitudes, energy will flow from low latitudes to high. This energy is transferred poleward as latent and sensible heat; warm ocean water and warm, moist air move poleward, while cooler water and cooler, drier air move toward the Equator.

We'll return to these flows in later chapters. But keep in mind that this **poleward energy transfer**, driven by the imbalance in net radiation between low and high latitudes, is the power source for broad-scale atmospheric circulation patterns and ocean currents. Without this circulation, low latitudes would heat up and high latitudes would cool down until a radiative balance was achieved, leaving the Earth with much more extreme temperature contrasts—very different from the planet that we are familiar with now. The images in Figure 2.20 illustrate some of the ways that natural processes and human uses are driven by solar power.

> Poleward energy transfer moves latent and sensible heat from the low latitudes toward the poles. Warm ocean water and warm, moist air flow poleward, and are replaced by cooler water and cooler, drier air flowing toward the Equator.

Galaxy Contact/Explorer/Photo Researchers, Inc.

▲ TROPICAL CYCLONE
Solar power also indirectly powers severe storms like Typhoon Odessa, shown here in a space photo.

EYE ON THE LANDSCAPE **What else would the geographer see?**
This photo, taken by an astronaut in orbit, shows the structure of the hurricane, or typhoon, very nicely. You can easily see the central eye Ⓐ, where air descends rapidly and ground wind speeds are light and variable. The radiating arms of the storm Ⓑ are formed by bands of severe thunderstorms and rain spiraling inward toward the center of the storm.

Greg Vaughn/Alamy Limited

▲ WATER POWER
The hydrologic cycle, powered by solar evaporation of water over oceans, generates runoff from rainfall that erodes and deposits sediment.

FOCUS ON REMOTE SENSING

CERES—Clouds and The Earth's Radiant Energy System

The Earth's global radiation balance is the primary determinant of long-term surface temperature, which is of great importance to life on Earth. Because this balance can be affected by human activities, such as converting forests to pasturelands or releasing greenhouse gases into the atmosphere, it is important to monitor the Earth's radiation budget over time, as accurately as possible.

For more than 20 years, NASA has studied the Earth's radiation budget from space. An ongoing NASA experiment entitled *CERES—Clouds and the Earth's Radiant Energy System*—is sending a new generation of instruments into orbit to scan the Earth and measure the amount of shortwave and longwave radiation leaving the Earth at the top of the atmosphere.

Figure 2.21 shows global reflected solar energy and emitted longwave energy averaged over the month of March 2000 as obtained by CERES. The top image shows average shortwave flux ("flux" means "flow"), ranging from 0 to 210 W/m². The largest flows occur over regions of thick clouds near the Equator, where the bright, white clouds reflect much of the solar radiation back to space. In the midlatitudes, persistent cloudiness during this month also shows up as light tones. Tropical deserts, the Sahara for example, are also bright. Snow and ice surfaces in polar regions are quite reflective; but in March, the amount of radiation received in polar regions is low. As a result, they don't appear as bright in this image. Oceans, especially where skies are clear, absorb solar radiation and thus show low shortwave fluxes.

Longwave flux is shown in the bottom image on a scale from 100 to 320 W/m². Cloudy equatorial regions have low values, showing the blanketing effect of thick clouds that trap longwave radiation beneath them. Warm tropical oceans in

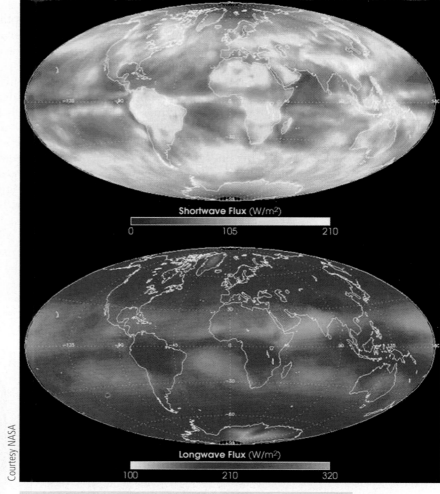

Courtesy NASA

2.21 Global shortwave and longwave energy fluxes from CERES

These images show average shortwave and longwave energy flows from Earth for March 2000, as measured by the CERES instrument on NASA's Terra satellite platform.

regions of clear sky emit the most longwave flux. Poleward, surface and atmospheric temperatures drop, so longwave energy emission also drops significantly.

As you can see from these images, clouds are very important determiners of the global radiation balance. A primary goal of the CERES experiment is to learn more about the Earth's cloud cover, which changes from minute to minute and hour to hour. This knowledge can be used to

improve global climate models that predict the impact of human and natural change on the Earth's climate.

The most important contribution of CERES, however, is continuous and careful monitoring of the Earth's radiant energy flows. In this way, small, long-term changes, induced by human or natural change processes, can be detected, in spite of large variations in energy flows from place to place and time to time caused by clouds.

A Look Ahead

The Earth's energy balance is a sensitive one involving many factors that determine how energy is transmitted and absorbed. Have human activities already altered the components of the planetary radiation balance? Scientists have shown convincingly that industrial releases of certain gases, such as carbon dioxide, have enhanced the greenhouse effect, causing global temperatures to warm. Human habitation, through cultivation and urbanization of land, has raised surface albedo and affected the transfer of latent and sensible heat to the atmosphere, modifying the global energy balance. But to understand these effects and others fully requires further study of the processes of heating and cooling of the Earth's atmosphere, lands, and oceans. Our next chapter concerns air temperature, addressing how and why it varies daily and annually depending on the surface energy balance.

IN REVIEW   THE EARTH'S GLOBAL ENERGY BALANCE

- The **ozone** (O_3) layer in the upper atmosphere absorbs solar ultraviolet radiation, shielding surface life from these harmful rays. Industrial chlorofluorocarbons (CFCs) speed the breakdown of ozone, reducing the amount of shielding. During certain conditions, ozone holes of reduced ozone concentration form over the Antarctic continent and, less frequently, over the Arctic.
- **Electromagnetic radiation** is a form of energy emitted by all objects. The wavelength of the radiation determines its characteristics. The hotter an object, the shorter the *wavelengths* of the radiation and the greater the amount of radiation that it emits.
- Radiation emitted by the Sun includes *ultraviolet, visible, near-infrared,* and *shortwave infrared* radiation. *Thermal infrared* radiation is emitted by Earth surfaces. The atmosphere absorbs and scatters radiation in certain wavelength regions.
- The flow of radiation emitted by an object increases very rapidly with its temperature. The wavelengths emitted decrease with increasing temperature.
- Continuous nuclear reactions within the Sun emit vast quantities of energy, largely in the form of light. The Earth receives solar radiation at a near-constant rate known as the *solar constant.* Solar radiation is strongest in the wavelength range of visible light. Radiation flows are measured in *watts* per square meter.
- Molecules and particles in the atmosphere both absorb and **scatter** incoming **shortwave radiation**. Upward scattering is also called **reflection**.
- The Earth's surface emits **longwave radiation**. While much of this emitted radiation is absorbed by the atmosphere, some radiation at particular wavelengths, called *windows*, passes through the atmosphere and out to space.
- The Earth continuously absorbs and scatters solar shortwave radiation and emits longwave radiation. In the long run, the gain and loss of radiant energy remains in a **global radiation balance**, and the Earth's average temperature remains constant.
- Solar energy powers most natural phenomena, directly or indirectly. Insolation, the rate of solar radiation flow available at a location at a given moment, is greater when the Sun is higher in the sky.
- *Daily insolation,* the average insolation value for a 24-hour day, depends on the range of Sun angles and length of the daylight period at the location. Near the Equator, daily insolation is greater at the equinoxes than at the solstices. Between the tropics and poles, the Sun rises higher in the sky and stays longer in the sky at the summer solstice than at the equinox, and longer at the equinox than at the winter solstice.
- *Annual insolation* is greatest at the Equator and lowest at the poles. However, the poles still receive 40 percent of the annual radiation received at the Equator.
- The pattern of annual insolation with latitude leads to a natural naming convention for latitude zones: *equatorial, tropical, subtropical, midlatitude, subarctic (subantarctic), arctic (antarctic),* and *polar.*
- The Earth's atmosphere is dominated by *nitrogen* and *oxygen* gases. *Carbon dioxide* and *water vapor* are only small constituents by volume, but are very important because they absorb longwave radiation and enhance the **greenhouse effect**.
- Heat is a flow of internal energy from one substance to another. Internal energy can flow by *conduction* and *convection.*

- **Sensible heat** is a flow of internal energy between substances that results in a temperature change. **Latent heat** is an energy flow that changes one state of a substance—solid, liquid, or gas—to another. *Sensible heat transfer* refers to the flow of heat between the Earth's surface and the atmosphere by conduction or convection. *Latent heat transfer* refers to the flow of energy carried by changes of state of water.
- Part of the solar radiation passing through the atmosphere is absorbed or scattered by molecules, dust, and larger particles. Some of the scattered radiation returns to space as *diffuse reflection*. The land surfaces, ocean surfaces, and clouds also reflect some solar radiation back to space.
- The proportion of radiation that a surface absorbs is termed its albedo. The albedo of the Earth and atmosphere as a whole planet is about 30 percent.
- Water vapor, CO_2, and other *greenhouse gases* absorb longwave energy emitted by the Earth's surface, causing the atmosphere to counterradiate some of that longwave radiation back to Earth, thereby creating the greenhouse effect. Because of this counterradiation, the Earth's surface temperature is considerably

warmer than we might expect for an Earth without an atmosphere.
- Flows of energy to and from the Earth-atmosphere system, as well as the atmosphere and surface taken individually, must balance over the long run. Energy flows within the Earth-atmosphere system include shortwave radiation, longwave radiation, sensible heat, and latent heat.
- Humans can affect the Earth's energy balance and global climate by such activities as changing surface albedos or causing more cloud formation.
- **Net radiation** describes the balance between incoming and outgoing radiation. At latitudes lower than 40°, annual net radiation is positive, while it is negative at higher latitudes. This imbalance creates **poleward energy transfer** of latent and sensible heat in the motions of warm water and warm, moist air, which provides the power that drives ocean currents and broad-scale atmospheric circulation patterns.
- NASA scientists monitor and map the upward flows of shortwave and longwave radiation over the globe to detect small, long-term changes that could affect global climate.

KEY TERMS

ozone, p. 58
chlorofluorocarbons (CFCs), p. 58
energy balance, p. 59
electromagnetic radiation, p. 59

absorption, p. 62
scattering, p. 62
reflection, p. 62
shortwave radiation, p. 62
longwave radiation, p. 63

global radiation balance, p. 63
insolation, p. 63
sensible heat, p. 70
latent heat, p. 70
albedo, p. 71

counterradiation, p. 71
greenhouse effect, p. 71
net radiation, p. 75
poleward energy transfer, p. 77

REVIEW QUESTIONS

1. What are **CFCs,** and how do they impact the ozone layer?
2. When and where have **ozone** reductions been reported? What action has occurred to help restore the ozone layer?
3. What is **electromagnetic radiation?** How is it characterized? Identify the major regions of the electromagnetic spectrum.
4. How does the temperature of an object influence the nature and amount of electromagnetic radiation that it emits?
5. What is the *solar constant?* What is its value? What are the units with which it is measured?
6. How does solar radiation received at the top of the atmosphere differ from solar radiation received at the Earth's surface? What are the roles of **absorption** and **scattering?**

7. Compare the terms **shortwave radiation** and **longwave radiation**. What are their sources?
8. How does the atmosphere affect the flow of longwave energy from the Earth's surface to space?
9. What is the Earth's global energy balance, and how are shortwave and longwave radiation involved?
10. How does the Sun's path in the sky influence *daily insolation* at a location? Compare the summer solstice and equinox paths of the Sun in the sky for 40° N lat. and the Equator.
11. What influence does latitude have on the annual cycle of daily insolation? On *annual insolation?*
12. Identify the two largest components of dry air. Why are *carbon dioxide* and *water vapor* important atmospheric constituents?
13. Explain how the terms **latent heat** *transfer* and **sensible heat** *transfer* apply to the Earth–atmosphere system.

14. What is the fate of incoming solar radiation? Discuss absorption, scattering, and reflection, including the role of clouds.
15. Define **albedo** and give two examples.
16. Describe the **counterradiation** process and how it relates to the **greenhouse effect**.
17. Discuss the energy balance of the Earth's surface. Identify the types and sources of energy flows that the surface receives. Do the same for the energy flows it loses.
18. Discuss the energy balance of the atmosphere. Identify the types and sources of energy flows that the atmosphere receives. Do the same for the energy flows it loses.
19. What is **net radiation?** How does it vary with latitude?
20. What is the role of **poleward energy transfer** in balancing the net radiation budget by latitude?
21. Using CERES as an example, explain the effect of clouds on shortwave and longwave radiation leaving the Earth−atmosphere system.

VISUALIZING EXERCISES

1. Place yourself in Figure 2.8. Imagine that you are standing in the center of the figure where the N–S and E–W lines intersect. Turn so that you face south. Using your arm to point at the Sun's position, trace the path of the Sun in the sky at the equinox. It will rise exactly to your left, swing upward to about a 50° angle, and then descend to the horizon exactly at your right. Repeat for the summer and winter solstices, using the figure as a guide. Then try it for the North Pole, the Equator, and the Tropic of Capricorn.
2. Sketch the world latitude zones on a circle representing the globe and give their approximate latitude ranges.
3. Sketch a simple diagram of the Sun above a layer of atmosphere above the Earth's surface. Using Figure 2.18 as a guide, draw arrows indicating flows of energy among the Sun, atmosphere, and surface. Label each arrow using terms from Figure 2.18.

ESSAY QUESTIONS

1. Suppose the Earth's axis of rotation were perpendicular to the orbital plane instead of tilted at 23½° away from perpendicular. How would global insolation be affected? How would insolation vary with latitude? How would the path of the Sun in the sky change with the seasons?
2. Imagine that you are following a beam of either (a) shortwave solar radiation entering the Earth's atmosphere heading toward the surface, or (b) a beam of longwave radiation emitted from the surface heading toward space. How will the atmosphere influence the beam?

Chapter 3
Air Temperature

McMurdo Research Station, located on Ross Island at the edge of the Ross Sea, is America's main research base in Antarctica. It is also a very cold place—in fact, the high, low, and mean temperature of every month is below freezing. One of the main determiners of air temperature is latitude, and at about 78° S, the Sun never rises higher than about 35° above the horizon. On this summer's day, the low Sun casts a thin, reddish glow on the base and Mount Erebus, in the far distance.

MCMURDO STATION, ROSS ISLAND, ANTARCTICA

©Yann Arthus-Bertrand/Altitude

Air Temperature

O ne of the first things we notice when stepping outdoors is the temperature of the air. But why does air temperature vary from day to day and from season to season? Why are cities warmer than suburbs? Why are mountains cooler than surrounding plains? Why is it always warm at the Equator? Why does Siberia get so cold in winter? Have human activities caused global warming? If so, what are the effects? This chapter will answer these questions and more.

Carbon Dioxide—On the Increase

One of the most important factors affecting air temperatures over the long run is the greenhouse effect, in which atmospheric gases absorb outgoing longwave radiation and reradiate a portion back to the surface. This makes surface temperatures warmer. Apart from water vapor, carbon dioxide gas plays the largest role in the greenhouse effect, and CO_2 concentration is increasing.

In the centuries before global industrialization, carbon dioxide concentration in the atmosphere was at a level below 300 parts per million (ppm) by volume (Figure 3.1). Since then, the amount has increased substantially, to 392 ppm in 2011. Why? When fossil fuels are burned, they yield water vapor and carbon dioxide. Water vapor does not present a problem because a large amount of water vapor is normally present in the atmosphere. But because the normal amount of CO_2 was so small, fossil fuel burning has raised the level substantially. According to studies of bubbles of atmospheric gases trapped in glacial ice, the present level of CO_2 is the highest attained in the last 800,000 years, and nearly double the amount present during glaciations of the most recent Ice Age.

Even with future concerted global action to reduce CO_2 emissions, scientists estimate that levels will stabilize at a value not lower than about 550 ppm by the late twenty-first century. This will nearly double preindustrial levels and is very likely to cause a significant increase in global temperatures.

Predicting the future buildup of CO_2 is difficult because not all the carbon dioxide emitted into the air by fossil fuel burning remains there. Plants take up CO_2 in photosynthesis to build their tissues, and under present conditions, global plant matter

is accumulating. Phytoplankton in ocean waters also take up CO_2, converting it into carbonate that sinks to the ocean floor. Together, these processes remove about half the CO_2 released to the atmosphere by fossil fuel burning.

Forecasting fossil fuel consumption is another uncertain enterprise. Future energy needs depend on global economic growth, which is difficult to predict, as is the future efficiency of energy use. The amount of CO_2 released also depends on the effectiveness of global action to reduce the rate of emissions through conservation and use of alternative energy sources.

Although there is a great deal of uncertainty about future atmospheric concentrations of carbon dioxide, one thing is certain: Fuel consumption will continue to release carbon dioxide, and its effect on climate will continue to increase.

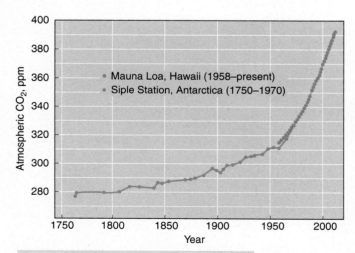

3.1 Increases in atmospheric carbon dioxide

Atmospheric concentrations of CO_2 have increased from 280 in 1750 to 393 ppm in 2012.

84

Surface and Air Temperature

This chapter focuses on **air temperature**—that is, the temperature of the air as observed at 1.2 m (4 ft) above the ground surface. Air temperature conditions many aspects of human life, from the clothing we wear to the fuel costs we pay. Air temperature and air temperature cycles also act to select the plants and animals that make up the biological landscape of a region. And air temperature, along with precipitation, is a key determiner of climate, which we will explore in more depth in Chapter 7.

> Five key factors influence a station's air temperature and its variation: latitude, surface type, coastal or interior location, elevation, and atmospheric and oceanic circulations.

Five important factors influence air temperature (Figure 3.2):

1. **Latitude.** Daily and annual cycles of insolation vary systematically with latitude, causing air temperatures and air temperature cycles to vary as well. Yearly insolation decreases toward the poles, so less energy is available to heat the air. But because the seasonal cycle of insolation becomes more intense with latitude, high latitudes experience a much greater range in air temperatures throughout the year.

2. **Surface type.** Urban air temperatures are generally higher than rural temperatures. City surface materials—asphalt, roofing shingles, stone, brick—hold little water, compared to the moist soil surfaces of rural areas and forests, so there is little cooling through evaporation. Urban materials are also darker and absorb a greater portion of the Sun's energy than vegetation-covered surfaces. The same is true for areas of barren or rocky soil surfaces, such as those of deserts.

3. **Coastal or interior location.** Locations near the ocean experience a narrower range of air temperatures than locations in continental interiors. Because water heats and cools more slowly than land, air temperatures over water are less extreme than temperatures over land. When air flows from water to land, a coastal location will feel the influence of the adjacent water.

4. **Elevation.** Temperature decreases with elevation. At high elevation there is less atmosphere above the surface, and greenhouse gases provide a less effective insulating blanket. More surface heat is lost to space. On high peaks, snow accumulates and remains longer. The reduced greenhouse effect also results in greater daily temperature variation.

5. **Atmospheric and oceanic circulations**. Local temperatures can rise or fall rapidly when air from one region is brought into another. Temperatures of coastal regions can be influenced by warm or cold coastal currents. (We will investigate this factor more fully in Chapter 5.)

We will return to these factors in pages to come. But first, we will look at surface and air temperature in more detail.

SURFACE TEMPERATURE

Temperature is a familiar concept. It is a measure of the level of kinetic energy of the atoms in a substance, whether it is a gas, liquid, or solid. When a substance receives a flow of radiant energy, such as sunlight, the kinetic energy level increases, and its temperature rises. Similarly, if a substance loses energy by radiation, its temperature falls. This energy flow moves in and out of a solid or liquid substance at its **surface**—for example, the very thin surface layer of soil that actually absorbs

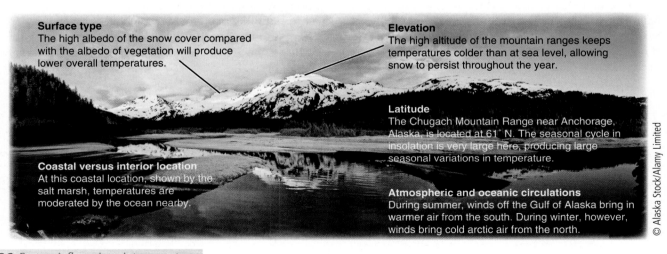

Surface type
The high albedo of the snow cover compared with the albedo of vegetation will produce lower overall temperatures.

Elevation
The high altitude of the mountain ranges keeps temperatures colder than at sea level, allowing snow to persist throughout the year.

Latitude
The Chugach Mountain Range near Anchorage, Alaska, is located at 61° N. The seasonal cycle in insolation is very large here, producing large seasonal variations in temperature.

Coastal versus interior location
At this coastal location, shown by the salt marsh, temperatures are moderated by the ocean nearby.

Atmospheric and oceanic circulations
During summer, winds off the Gulf of Alaska bring in warmer air from the south. During winter, however, winds bring cold arctic air from the north.

© Alaska Stock/Alamy Limited

3.2 Factors influencing air temperature

Five main factors affect temperature and its variability at a given location. All are visible in this view of the Chugash National Forest, Alaska.

3.3 Solar energy flow

Solar light energy strikes the Earth's surface and is largely absorbed, warming the surface.

solar shortwave radiation and radiates longwave radiation out to space.

The temperature of a surface is determined by the balance among the various energy flows that move across it. **Net radiation**—the balance between incoming shortwave radiation and outgoing longwave radiation—produces a radiant energy flow that can heat or cool

a surface. During the day, incoming solar radiation normally exceeds outgoing longwave radiation, so the net radiation balance is positive and the surface warms (Figure 3.3). Energy flows through the surface into the cooler soil below. At night, net radiation is negative, and the soil loses energy as the surface temperature falls and the surface radiates longwave energy to space.

Energy may also move to or from a surface in other ways. **Conduction** describes the flow of sensible heat from a warmer substance to a colder one through direct contact. When energy flows into the soil from its warm surface during the day, it flows by conduction. At night, the energy is conducted back to the colder soil surface.

Latent heat transfer is also important. When water evaporates at a surface, it removes the energy stored in the change of state from liquid to vapor, thus cooling the surface. When water condenses at a surface, latent heat energy is released, warming the surface.

Another form of energy transfer is **convection**, in which energy is distributed in a fluid by mixing. If the surface is in contact with a fluid, such as a soil surface with air above, upward- and downward-flowing currents can act to warm or cool the surface.

AIR TEMPERATURE

In contrast to surface temperature is *air temperature*, which is measured at a standard height of 1.2 m (4.0 ft) above the ground surface. Air temperature can be quite different from surface temperature. When you walk across a parking lot on a clear summer day, you will notice that the pavement is a lot hotter than the air against the upper part of your body. In general, air temperatures above a surface reflect the same trends as ground surface temperatures, but ground temperatures are likely to be more extreme.

In the United States, temperature is still widely measured and reported using the Fahrenheit scale. In this book, we use the Celsius temperature scale, which is the international standard. On the Celsius scale, the freezing point of water is 0°C and the boiling point is 100°C. Conversion formulas between these two scales are given in Figure 3.4.

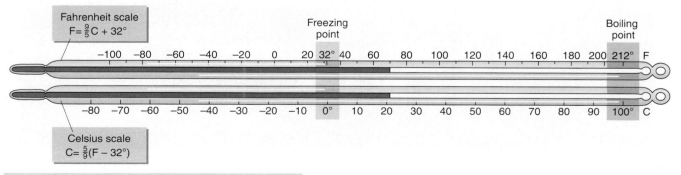

Fahrenheit scale
$F = \frac{9}{5}C + 32°$

Freezing point

Boiling point

Celsius scale
$C = \frac{5}{9}(F - 32°)$

3.4 Celsius and Fahrenheit temperature scales compared

At sea level, the freezing point of water on the Celsius (C) scale is 0°, while it is 32° on the Fahrenheit (F) scale. Boiling occurs at 100°C, or 212°F.

Air temperature measurements are made routinely at weather stations (Figure 3.5). Although some weather stations report temperatures hourly, most only report the maximum and minimum temperatures recorded during a 24-hour period. These are the most important values in observing long-term trends in temperature.

Temperature measurements are reported to governmental agencies charged with weather forecasting, such as the U.S. Weather Service or the Meteorological Service of Canada. These agencies typically make available daily, monthly, and yearly temperature statistics for each station using the daily maximum, minimum, and mean temperatures. The *mean daily air temperature* is defined as the average of the maximum and minimum daily values. The *mean monthly air temperature* is the average of mean daily temperatures in a month. These statistics, along with others such as daily precipitation, are used to describe the climate of the station and its surrounding area.

> Surface air temperature is measured under standard conditions at 1.2 m (4 ft) above the ground. Maximum and minimum temperatures are typically recorded. The mean daily temperature is taken as the average of the minimum and maximum temperatures.

WileyPLUS Weather Station Interactivity
Check out a modern automatic weather station. See how it measures air and ground temperatures.

TEMPERATURES CLOSE TO THE GROUND

Soil, surface, and air temperatures within a few meters of the ground change throughout the day (Figure 3.6). The daily temperature variation is greatest just above the surface. The air temperature at standard height is far less variable. In the soil, the daily cycle becomes gradually less pronounced with depth, until we reach a point where daily temperature variations on the surface cause no change at all.

ENVIRONMENTAL CONTRASTS: URBAN AND RURAL TEMPERATURES

On a hot day, rural environments will feel cooler than urban environments (Figure 3.7). In rural areas, water is taken up by plant roots; it then moves to the leaves and evaporates from leaf pores in a process called **transpiration**. This process cools leaf surfaces, which in turn cool nearby air. Evaporation also occurs from moist soil surfaces, cooling the air close to the surface. We refer to the combined effects of transpiration and evaporation as **evapotranspiration**.

There are other reasons why urban surfaces are hotter than rural ones. Many city surfaces are dark, and strongly absorb solar energy. In fact, asphalt paving absorbs more than twice as much solar energy as vegetation. Rain runs off the roofs, sidewalks, and streets into storm sewer

GIPhotoStock/Photo Researchers, Inc.

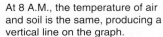
3.5 Weather station

This weather station at Furnace Creek, in Death Valley National Park, has an evaporation pan in the foreground that measures the rate of evaporation in this desert climate. Just to the left of the pan is a 3-cup anemometer that measures wind speed very near the evaporating water surface. The tall cylinder behind the pan and to the right is a rain gauge. The white louvered box, called a Stevenson shelter, will contain thermometers measuring temperatures and possibly relative humidity.

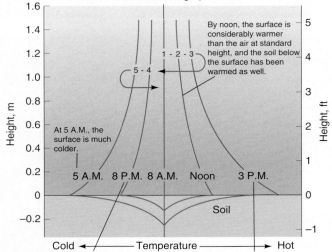
3.6 Daily temperature profiles close to the ground

The red curves in the figure show a set of temperature profiles for a bare, dry soil surface from about 30 cm (12 in.) below the surface to 1.5 m (4.9 ft) above it at five times of day (1–5).

3.7 Rural and urban surfaces

Rural and urban surfaces have different characteristics that affect surface temperatures.

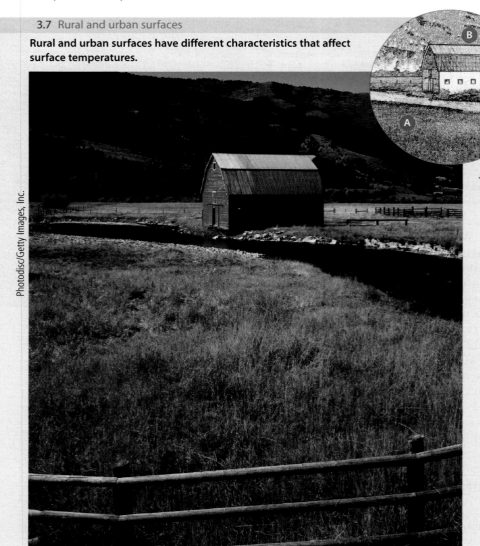

◀ **RURAL SURFACES**

As we see here in this view of Ruby River Creek, near Alder, Montana, rural surfaces are composed of moist soil, covered largely by vegetation. Evapotranspiration keeps these surfaces cooler.

EYE ON THE LANDSCAPE **What else would the geographer see?** The flat landscape in the foreground Ⓐ, stretching to the hillslopes in the background, is a river floodplain. During the Ice Age, the climate in this region was a lot wetter, and a much larger Ruby River would have swept back and forth across this wide plain, leaving a surface now ideally suited to farming with irrigation. The patchy vegetation on the hillslopes Ⓑ suggests a semiarid climate, with bands of drought-resistant conifers in the creek valleys and the drier slopes covered by grasses and shrubs.

▼ **URBAN SURFACES**

As demonstrated by this photo of a shopping center in Pennsylvania, urban surfaces are largely composed of asphalt paving and roofing, concrete, building stone, and similar materials. Sewers drain away rainwater, keeping urban surfaces dry. Because evapotranspiration is limited, surface and air temperatures are hotter.

Photodisc/Getty Images, Inc.

Jeff Greenberg/Getty Images, Inc.

systems. Because the city surfaces are dry, there is little evaporation to help lower temperatures.

Another important factor is waste heat. In summer, city air temperatures are raised by air conditioning, which pumps thermal energy out of buildings and releases it to the air. In winter, thermal energy from buildings and structures is conducted directly to the urban environment.

> Urban surfaces lack moisture and so are warmer than rural surfaces during the day. At night, urban materials conduct stored thermal energy to the surface, which also keeps temperatures warmer.

THE URBAN HEAT ISLAND

As a result of these effects, air temperatures in the central region of a city are typically several degrees warmer than those of the surrounding suburbs and countryside, as shown in Figure 3.8. The sketch of a temperature profile across an urban area in the late afternoon shows this effect. We call the central area an **urban heat island**, because it has a significantly elevated temperature. At night, the thermal energy conducted into the ground and urban structures flows outward, warming the night air so that the heat island remains warmer than its surroundings during the night, too. The thermal infrared

3.8 Urban heat island effect

The nature of urban surfaces and materials, coupled with waste heat generation, create an urban heat island.

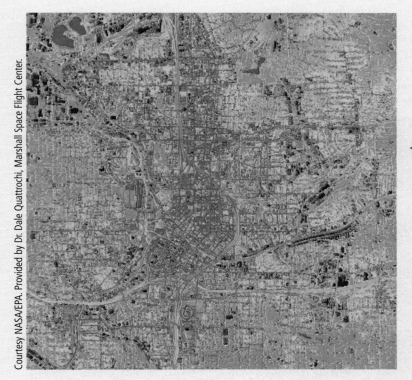

Courtesy NASA/EPA. Provided by Dr. Dale Quattrochi, Marshall Space Flight Center.

◀ **ATLANTA, GEORGIA**
This image, taken at night in May, over downtown Atlanta, Georgia, shows the urban heat island. The main city area, in tones of red and yellow, is clearly warmer than the suburban area, in blue and green. The street pattern of asphalt pavement is shown very clearly as a red grid, with many of the downtown squares filled with red.

TEMPERATURE DIAGRAM ▶
This diagram shows how air temperatures might vary across the urban and rural areas during the late afternoon. Downtown and commercial areas are warmest, while rural farmland is coolest.

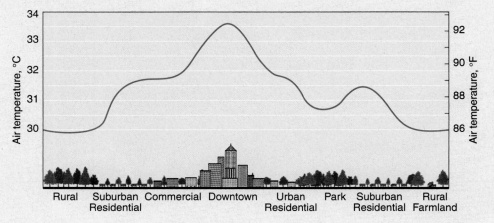

image of the Atlanta central business district at night demonstrates the heat island effect.

The urban heat island effect has important economic consequences. Higher temperatures in the summer raise the demand for air conditioning and electric power. The fossil fuel burned to generate this power contributes CO_2 and air pollutants to the air. The increased temperatures can escalate smog formation, which is unhealthy and damaging to materials. To reduce these effects, many cities are planting more vegetation and using more reflective surfaces, such as concrete or bright roofing materials, to reflect solar energy back to space.

The heat island effect does not necessarily apply to cities in desert climates. In the desert, the evapotranspiration of the irrigated vegetation of the city may actually keep the city cooler than the surrounding barren region.

HIGH-MOUNTAIN ENVIRONMENTS

We have seen that the ground surface affects the temperature of the air directly above it. But what happens as you travel to higher elevations (Figure 3.9)? For example, as you climb higher on a mountain, you may become short of breath, and you might notice that you sunburn more easily. You also feel the temperature drop, as you ascend. If you camp out, you'll see that the nighttime temperature drops lower than you might expect, even given that temperatures are generally cooler the farther up you go.

What causes these effects? At high elevations there is significantly less air above you, so air pressure is low. This makes it more difficult to catch your breath, simply because of the reduced oxygen pressure in your lungs. And with fewer molecules to scatter and absorb the Sun's light, the Sun's rays will feel stronger.

3.9 High elevation

These mountain peaks in the Wind River Range, Wyoming, climb to about 3500 m (about 12,000 ft). At higher elevations, there is less air to absorb solar radiation. The sky is deep blue and temperatures are cooler.

Raymond Gehman/NG Image Collection

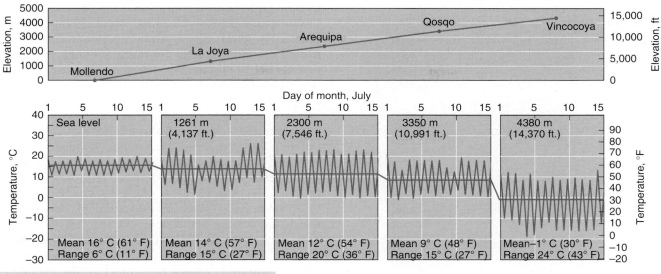

3.10 The effect of elevation on air temperature cycles

The graph shows the mean air temperature for mountain stations in Peru, lat. 15° S, during the same 15 days in July. As elevation rises, the mean daily temperature decreases and the temperature range increases.

There is less carbon dioxide and water vapor, and so the greenhouse effect is reduced. With less warming, temperatures will tend to drop even lower at night. Later in this chapter, we will see how this pattern of decreasing air temperature extends high up into the atmosphere.

Figure 3.10 shows temperature graphs for five stations at different heights in the Andes mountain range in Peru. Mean temperatures clearly decrease with elevation, from 16°C (61°F) at sea level to −1°C (30°F) at 4380 m (14,370 ft). The range between maximum and minimum temperatures also increases with elevation, except for Qosqo. Temperatures in this large city do not dip as low as you might expect because of its urban heat island.

> At high elevations, air temperatures are generally cooler and have a greater day-to-night range. Both effects occur because the thickness and density of the air column above decrease with elevation; therefore, the greenhouse effect is weaker at high elevations.

TEMPERATURE INVERSION

So far, air temperatures seem to decrease with height. But is this always true? Think about what happens on a clear, calm night. The ground surface radiates longwave energy to the sky, and net radiation becomes negative. The surface cools. This means that air near the surface also cools, as we saw in Figure 3.6. If the surface stays cold, a layer of cooler air above the ground will build up under a layer of warmer air, as shown in Figure 3.11. This is a **temperature inversion**.

In a temperature inversion, the temperature of the air near the ground can fall below the freezing point. This temperature condition is called a *killing frost*—even though actual frost may not form—because of its effect on sensitive plants during the growing season.

Growers of fruit trees or other crops use several methods to break up an inversion. Large fans can be used to mix the cool air at the surface with the warmer air above, and oil-burning heaters are sometimes used to warm the surface air layer.

> In a temperature inversion, air temperature does not decrease with height, but increases instead. Temperature inversions commonly appear on clear, calm nights.

TEMPERATURE INDEXES

Temperature can also be used with other weather and climate data to produce *temperature indexes*, indicators of the temperature's impact upon environmental and human conditions. Two of the more familiar indexes are the wind chill index and the heat index.

The *wind chill index* is used to determine how cold temperatures feel to us, based on not only the actual

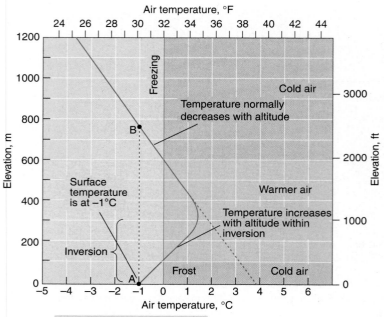

3.11 Temperature inversion

While air temperature normally decreases with altitude (dashed line), in an inversion, temperature increases with altitude. Air temperature is the same at the surface (A) as at 760 m (2500 ft) aloft (B).

temperature but also the wind speed. Air is actually a very good insulator, so when the air is still, our skin temperature can be very different from the temperature of the surrounding environment. However, as air moves across our skin, it removes sensible and latent

heat and transports it away from our bodies. During the summer, this process keeps us cool as sweat is evaporated away, lowering our skin temperature. During the winter, it conducts away the thermal energy necessary to keep our bodies warm, cooling our skin and making conditions feel much colder than the actual measured temperature.

The wind chill index, which is used in the United States and measured in degrees Fahrenheit (°F), can be quite different from the actual temperature (Figure 3.12). For example, an actual temperature of 30°F (−1°C) and a wind speed of 30 mi/hr (13.45 m/s) produce a wind chill of 15°F (−26°C).

The *heat index* gives an indication of how hot we feel based on the actual temperature and the *relative humidity*. Relative humidity is the humidity given in most weather reports and indicates how much water vapor is in the atmosphere as a percentage of the maximum amount possible. Low relative humidity indicates relatively dry atmospheric conditions, while high relative humidity indicates relatively humid atmospheric conditions. When the relative humidity is high, less water evaporates from the skin because the surrounding atmosphere is already relatively moist, leaving the skin surface hotter.

Like the wind chill index, the heat index is given in °F; it, too, can be quite different from the actual temperature (Figure 3.13). For example, if the actual temperature is 90°F (32°C) and the relative humidity is 90 percent, the heat index indicates that the temperature will feel like 122°F (50°C)—a difference of 32°F (18°C)!

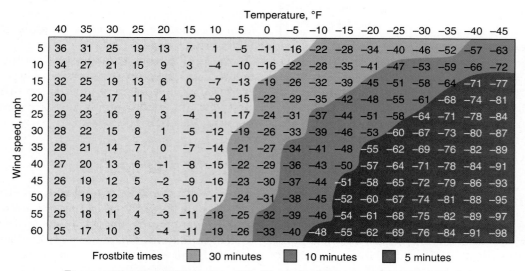

3.12 Wind chill conversion

The wind chill index provides an indicator of how cold temperatures feel based on the actual temperature plus the wind speed.

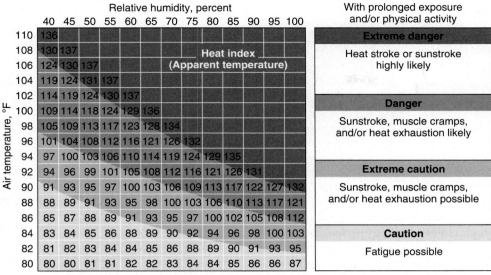

To convert the heat index from °F to °C, use the following equation: °C = (°F − 32) × 5/9

3.13 Heat index conversion

The heat index provides an indicator of how hot temperatures feel based on the actual temperature plus relative humidity.

Temperature Structure of the Atmosphere

In general, the air is cooler at higher altitudes. Remember from Chapter 2 that most incoming solar radiation passes through the atmosphere and is absorbed by the Earth's surface. The atmosphere is then warmed at the surface by latent and sensible heat flows. So it makes sense that, in general, air farther from the Earth's surface will be cooler.

We call the decrease in air temperature with increasing altitude the **lapse rate**. We measure the temperature drop in degrees Celsius per 1000 m (or degrees Fahrenheit per 1000 ft). Figure 3.14 shows how temperature varies with altitude on a typical summer day in the midlatitudes. Temperature drops at an average rate of 6.49°C/1000 m (3.56°F/1000 ft). This average value is known as the **environmental temperature lapse rate**. Looking at the graph, we see that when the air temperature near the surface is a pleasant 20°C (68°F), whereas the air at an altitude of 12 km (40,000 ft) will be a bone-chilling −58°C (−72°F). Keep in mind that the environmental temperature lapse rate is an average value, and that on any given day the observed lapse rate might be quite different.

Figure 3.14 shows another important feature. For the first 12 km (7 mi) or so, temperature falls with increasing elevation. But between about 13 and 15 km (about 8 and 9 mi) elevation, the temperature stops decreasing. In fact, above that height, temperature slowly rises with elevation. Atmospheric scientists use this feature to define two different layers in the lower atmosphere: the troposphere and the stratosphere.

> In the lower atmosphere, air temperatures decrease with increasing altitude. The average rate of temperature decrease with height is termed the environmental temperature lapse rate and is 6.49°C/1000 m (3.56°F/1000 ft).

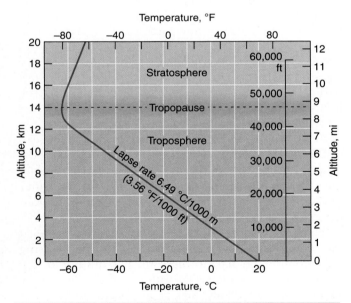

3.14 A typical atmospheric temperature curve for a summer day in the midlatitudes

Temperature decreases with altitude in the troposphere. The rate of temperature decrease with elevation, or lapse rate, is shown at the average value of 6.49°C/1000 m (3.56°F/1000 ft), which is known as the environmental temperature lapse rate. At the tropopause, the decreasing trend stops. In the stratosphere above, temperature is constant or increases slightly with altitude.

TROPOSPHERE

The **troposphere** is the lowest atmospheric layer. All human activity takes place here. Everyday weather phenomena, such as clouds and storms, mainly happen in the troposphere. Here temperature decreases with increasing elevation. The troposphere is thickest in the equatorial and tropical regions, where it stretches from sea level to about 16 km (10 mi). It thins toward the poles, where it is only about 6 km (4 mi) thick.

The troposphere contains significant amounts of water vapor. When the water vapor content is high, vapor can condense into water droplets, forming low clouds and fog, or the vapor can be deposited as ice crystals, forming high clouds. Rain, snow, hail, or sleet—collectively termed *precipitation*—are produced when these condensation or deposition processes happen rapidly. Places where water vapor content is high throughout the year have moist climates. In desert regions water vapor is low, so there is little precipitation. As we saw in Chapter 2, water vapor is an important gas that contributes to greenhouse warming of our planet.

The troposphere contains countless tiny particles that are so small and light that the slightest movements of the air keep them aloft. These are called **aerosols**. They are swept into the air from dry desert plains, lakebeds, and beaches, or released by exploding volcanoes. Oceans are also a source of aerosols. Strong winds blowing over the ocean lift droplets of spray into the air. These droplets of spray lose most of their moisture by evaporation, leaving tiny particles of watery salt that are carried high into the air. Forest fires and brushfires also generate particles of soot as smoke. And meteors, vaporizing in the upper atmosphere, contribute dust particles to the upper layers of air. Closer to the ground, industrial processes that incompletely burn coal or fuel oil release aerosols into the air, as well.

Aerosols are important because water vapor can condense on them to form tiny droplets. When these droplets grow large and occur in high concentration, they are visible as clouds or fog. Aerosol particles scatter sunlight, brightening the whole sky while slightly lowering the intensity of the solar beam.

The troposphere gives way to the stratosphere at the *tropopause*. Here, temperatures stop decreasing with altitude and start to increase. The altitude of the tropopause varies somewhat with season, so the troposphere is not uniformly thick at any location.

STRATOSPHERE AND UPPER LAYERS

The **stratosphere** lies above the tropopause. Air in the stratosphere becomes slightly warmer as altitude increases. The stratosphere reaches up to roughly 50 km (about 30 mi) above the Earth's surface. It is the home of strong, persistent winds that blow from west to east. Air doesn't really mix between the troposphere and stratosphere,

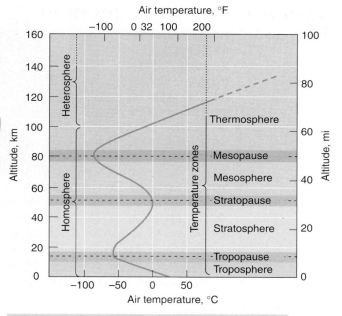

3.15 Temperature structure of the upper atmosphere

Above the troposphere and stratosphere are the mesosphere and thermosphere. The homosphere, in which the air's molecules are well mixed, ranges from the surface to nearly 100 km (about 60 mi) altitude.

so the stratosphere normally holds very little water vapor or dust.

The stratosphere contains the *ozone layer*, which shields Earthly life from intense, harmful ultraviolet energy. It is the ozone molecules that warm the stratosphere, causing temperature to increase with altitude, as they absorb solar energy.

> The troposphere and stratosphere are the two lowermost atmospheric layers. Clouds and weather phenomena occur in the troposphere. The stratosphere is the home of strong, persistent winds that flow from west to east.

Temperatures stop increasing with altitude at the *stratopause*. Above the stratopause we find the *mesosphere*, shown in Figure 3.15. In the mesosphere, temperature falls with elevation. This layer ends at the mesopause, the level at which temperature stops falling with altitude. The next layer is the *thermosphere*. Here, temperature again increases with altitude, but because the density of air is very thin in this layer, the air holds little heat energy.

The gas composition of the atmosphere is uniform for about the first 100 km (about 60 mi) of altitude, which includes the troposphere, stratosphere, mesosphere, and the lower portion of the thermosphere. We call this region the *homosphere*. Above about 100 km (about 60 mi), gas molecules tend to be sorted into layers by molecular weight and electric charge. This region is called the *heterosphere*.

Daily and Annual Cycles of Air Temperature

Let's turn now to how, and why, air temperatures vary around the world. Insolation from the Sun varies across the globe, depending on latitude and season. Net radiation at a given place is positive during the day, as the surface gains heat from the Sun's rays. At night, the flow of incoming shortwave radiation stops, but the Earth continues to radiate longwave radiation. As a result, net radiation becomes negative. Because the air next to the surface is warmed or cooled as well, we get a daily cycle of air temperatures (Figure 3.16).

Why does the temperature peak in the midafternoon? We might expect it to continue rising as long as the net radiation is positive. But on sunny days in the early afternoon, large convection currents develop within several hundred meters of the surface, complicating the pattern. They carry hot air near the surface upward, and they bring cooler air downward. So the temperature typically peaks between 2:00 and 4:00 P.M. By sunset, air temperature is falling rapidly. It continues to fall more slowly throughout the night.

As shown in Figure 3.16, the height of the temperature curves varies with the seasons. In the summer, temperatures are warm and the daily curve is high. In winter, the temperatures are colder. The September equinox is

---

3.16 Daily cycles of insolation, net radiation, and air temperature

These three graphs, which show idealized daily cycles for a midlatitude station at a continental interior location, illustrate how insolation, net radiation, and air temperature are linked.

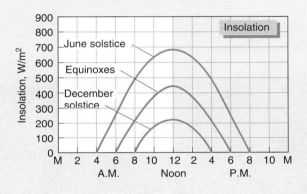

INSOLATION

At the equinox (middle curve), insolation begins at about sunrise (6:00 A.M.), peaks at noon, and falls to zero at sunset (6:00 P.M.). At the June solstice, insolation begins about two hours earlier (4:00 A.M.) and ends about two hours later (8:00 P.M.). The June peak is much greater than at equinox, and there is much more insolation. At the December solstice, insolation begins about two hours later than at equinox (8:00 A.M.) and ends about two hours earlier (4:00 P.M.). The daily total insolation is greatly reduced in December.

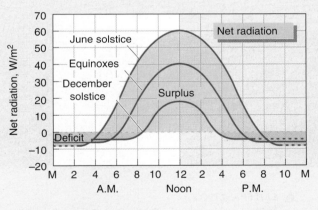

NET RADIATION

Net radiation curves follow closely the insolation curves in (A). At midnight, net radiation is negative. Shortly after sunrise, it becomes positive, rising sharply to a peak at noon. In the afternoon, net radiation decreases as insolation decreases. Shortly before sunset, net radiation is zero—incoming and outgoing radiation are balanced. Net radiation then becomes negative.

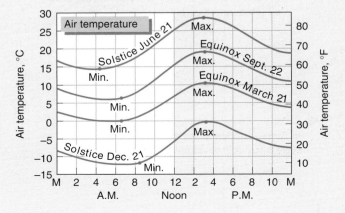

AIR TEMPERATURES

All three curves show that the minimum daily temperature occurs about a half hour after sunrise. Since net radiation has been negative during the night, heat has flowed from the ground surface, and the ground has cooled the surface air layer to its lowest temperature. As net radiation becomes positive, the surface warms quickly and transfers heat to the air above. Air temperature rises sharply in the morning hours and continues to rise long after the noon peak of net radiation.

considerably warmer than the March equinox even though net radiation is the same. This is because the temperature curves lag behind net radiation, reflecting earlier conditions.

LAND AND WATER CONTRASTS

There is another factor that influences the annual temperature cycle. If you have visited San Francisco, you probably noticed that this magnificent city has a unique climate. It's often foggy and cool, and the weather is damp for most of the year. The city's cool climate is due to its location on the tip of a peninsula, with the Pacific Ocean on one side and San Francisco Bay on the other. A southward-flowing ocean current sweeps cold water from Alaska down along the northern California coast, and winds from the west move cool, moist ocean air, as well as clouds and fog, across the peninsula. This airflow keeps summer temperatures low, and winter temperatures above freezing.

Figure 3.17 shows a typical temperature record for San Francisco for a week in the summer. Temperatures hover around 13°C (55°F) and change only a little from day to night. The story is very different at locations far from the water, like Yuma, Arizona. Yuma is in the Sonoran Desert, and average air temperatures there are much warmer on average—about 28°C (82°F). Clearly, no ocean cooling is felt in Yuma! The daily range is also much greater—the daytime temperature in the hot desert drops by nearly 20°C (36°F), producing cool desert nights. The clear, dry air also helps the ground lose heat rapidly.

> Air temperatures at coastal locations tend to be more constant than at interior continental locations because water bodies heat and cool more slowly than the land surface.

3.17 Maritime and continental temperatures

The daily temperature cycle at continental locations is more pronounced than at maritime locations.

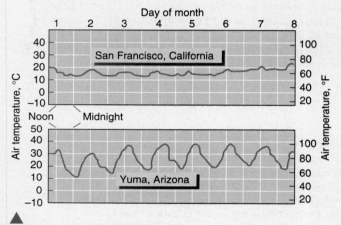

▲ DAILY TEMPERATURE CYCLES
A recording thermometer made these continuous records of the rise and fall of air temperature for a week in summer in San Francisco, California, and Yuma, Arizona.

MELISSA FARLOW/NG Image Collection

▲ SAN FRANCISCO
In San Francisco, on the Pacific Ocean, the daily air temperature cycle is very weak.

George F. Mobley/NG Image Collection

◀ YUMA
In Yuma, a station in the desert, the daily cycle is strongly developed.

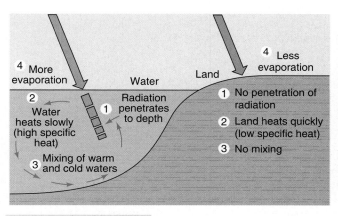

3.18 Land-water contrasts

These four labels describe why a land surface heats more rapidly and more intensely than the surface of a deep water body. As a result, locations near the ocean have more uniform air temperatures—cooler in summer and warmer in winter.

What is behind these differences? The important principle is this: The surface layer of any extensive, deep body of water heats more slowly and cools more slowly than the surface layer of a large body of land when both are subjected to the same intensity of insolation (Figure 3.18). Because of this principle, daily and annual air temperature cycles will be quite different at coastal locations than at interior locations. Together, they make air temperatures above water less variable than those over land. Places located well inland and far from oceans will tend to have stronger temperature contrasts from winter to summer and night to day.

How does the land-water contrast affect the annual air temperature cycle? Let's look in detail at the annual cycle of another pair of stations: Winnipeg, Manitoba, located in the interior of the North American continent, and the Scilly Islands, surrounded by the waters of the Atlantic Ocean off the southwestern tip of England (Figure 3.19).

3.19 Maritime and continental annual air temperature cycles

The annual cycle of monthly temperatures is stronger at continental locations and weaker at marine locations.

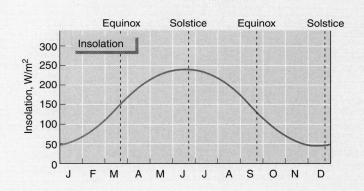

LOCATIONS

Annual cycles of insolation and mean monthly air temperature are shown for two stations at latitude 50° N: Winnipeg, Canada, and Scilly Islands, England. Winnipeg is located in a continental interior, while the Scilly Islands are located in the Celtic Sea off the southwestern tip of England.

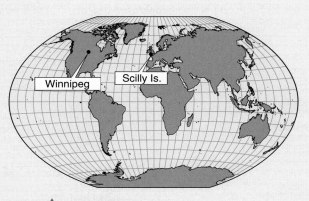

INSOLATION

Insolation is identical for the two stations.

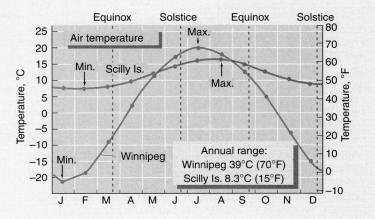

TEMPERATURES

Winnipeg temperatures clearly show the wide annual range and earlier maximum and minimum that are characteristic of this continental location. Scilly Islands temperatures show a maritime location with a narrow annual range and delayed maximum and minimum.

3.20 The relationship between net radiation and temperature

The annual cycle of net radiation drives the annual cycle of temperatures experienced at a particular location.

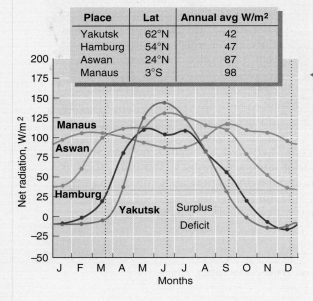

Place	Lat	Annual avg W/m²
Yakutsk	62°N	42
Hamburg	54°N	47
Aswan	24°N	87
Manaus	3°S	98

◄ NET RADIATION

- At Manaus, the average net radiation rate is strongly positive every month. But there are two minor peaks, coinciding roughly with the equinoxes, when the Sun is nearly straight overhead at noon.
- The curve for Aswan shows a large surplus of positive net radiation every month. The net radiation rate curve has a strong annual cycle—values for June and July that are triple those of December and January.
- The net radiation rate cycle for Hamburg, Germany, is strongly developed. There is a radiation surplus for nine months, and a deficit for three winter months.
- During the long, dark winters in Yakutsk, Siberia, the net radiation rate is negative, and there is a radiation deficit that lasts about six months.

GLOBAL LOCATIONS ▶

- Our four stations are selected at latitudes from the Equator to the arctic.
- Manaus, in central Brazil, is in the midst of the equatorial rainforest of the Amazon basin.
- Aswan is near the Tropic of Cancer in the Egyptian desert.
- Hamburg, Germany, is in the midlatitudes, not far from the North Sea.
- Yakutsk lies in central Siberia, only a few degrees of latitude from the Arctic Circle.

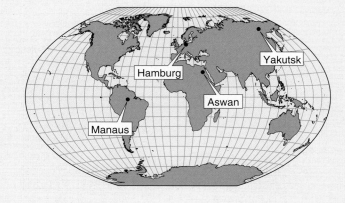

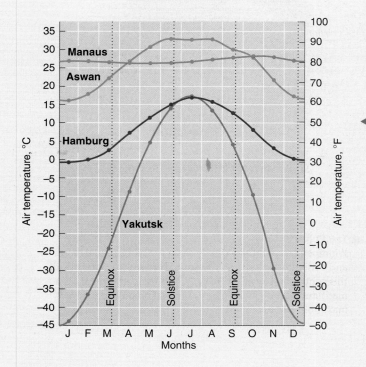

◄ MONTHLY MEAN AIR TEMPERATURE

- Manaus has uniform air temperatures, averaging about 27°C (81°F) for the year. The mean monthly temperature has only a narrow range—about 1.7°C (3.0°F).
- The Aswan temperature follows the cycle of the net radiation rate curve quite consistently, and shows an annual range in monthly temperatures of about 17°C (31°F). June, July, and August are terribly hot, averaging over 32°C (90°F).
- Hamburg's temperature cycle reflects the reduced total insolation at this latitude. Summer months reach a maximum of just over 16°C (61°F), while winter months reach a minimum of just about freezing (0°C or 32°F). The annual range is about 17°C (31°F), the same as at Aswan.
- Yakutsk, Siberia, temperature shows the effect of the extreme net radiation cycle. Temperatures in the three winter months are between −35 and −45°C (about −30 and −50°F), while the July temperature is about the same as Hamburg. Yakutsk's annual temperature range is extremely wide—over 60°C (108°F).

Because these two stations have the same latitude, 50° N, they have the same insolation cycle and receive the same potential amount of solar energy for surface warming.

The temperature graphs for the Scilly Isles and Winnipeg show that the annual range in temperature is much greater for the interior station (39°C; 70°F) than for the coastal station (8°C; 14°F). The nearby ocean waters keep the air temperature at the Scilly Isles well above freezing in the winter, while January temperatures at Winnipeg fall to near −20°C (−4°F).

Another important effect of the land-water contrast concerns the timing of maximum and minimum temperatures. Insolation reaches a maximum at the summer solstice, but it is still strong for a long period afterward, so that net radiation is positive well after the solstice. Therefore, the hottest month of the year for interior regions is July, the month following the summer solstice; the coldest month is January, the month after the winter solstice. The continued cooling after the winter solstice takes place because the net radiation is still negative even though daily insolation has begun to increase.

Over the oceans and at coastal locations, maximum and minimum air temperatures are reached a month later than at inland locations—in August and February, respectively. Because water bodies heat or cool more slowly than land areas, the air temperature changes more slowly. This effect is clearly seen in the Scilly Isles graph, which shows that February is slightly colder than January.

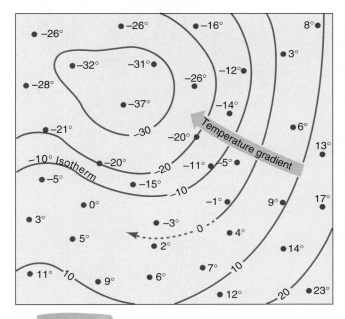

3.21 Isotherms

Isotherms are used to make temperature maps. Each line connects points having the same temperature. Where temperature changes along one direction, a temperature gradient exists. Where isotherms close in a tight circle, a center exists. This example shows a center of low temperature.

ANNUAL NET RADIATION AND TEMPERATURE CYCLES

We have seen that location—maritime or continental—has an important influence on annual temperature cycles. But the greatest effect is caused by the annual cycle of net radiation. Daily insolation varies over the seasons of the year, owing to the Earth's motion around the Sun and the tilt of the Earth's axis. That rhythm produces a net radiation cycle, which, in turn, causes an annual cycle in mean monthly air temperatures. Figure 3.20 examines the cycles of net radiation and temperature at four sites spanning the Equator to the Arctic Circle.

> The annual cycle of net radiation, which results from the variation of insolation with the seasons, drives the annual cycle of air temperature. Where net radiation varies strongly with the seasons, so does surface air temperature.

World Patterns of Air Temperature

So far in this chapter we have learned some important principles about air temperatures. Surface type (urban or rural), elevation, latitude, daily and annual insolation cycles, and location (maritime or continental) all can influence air temperatures. Now let's put all these factors together and see how they affect world air temperature patterns.

> Isotherms are lines of equal temperature drawn on a map. Maps of isotherms show centers of high and low temperatures, as well as temperature gradients.

First, we need a quick explanation of air temperature maps. Figure 3.21 shows a set of **isotherms**, lines connecting locations that have the same temperature. Usually, we choose isotherms that are separated by 5 or 10 degrees, but they can be drawn at any convenient temperature interval.

Isothermal maps clearly show centers of high or low temperatures. They also illustrate the directions along which temperature changes, which are known as *temperature gradients*. In the winter, isotherms dip toward the Equator, while in the summer, they arch poleward (Figure 3.22). Figure 3.23 is a map of the annual range of temperature, which is greatest in northern latitudes between 60 and 70 degrees.

FACTORS CONTROLLING AIR TEMPERATURE PATTERNS

We have already introduced the three main factors that explain world isotherm patterns. The first is latitude. As

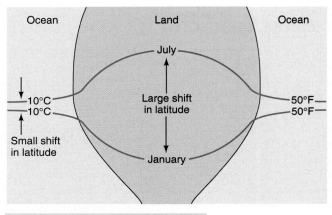

3.22 Seasonal migration of isotherms

Continental air temperature isotherms shift over a much wider latitude range from summer to winter than do oceanic air temperature isotherms. This difference occurs because oceans heat and cool much more slowly than continents.

latitude increases, average annual insolation decreases, and so temperatures decrease as well, making the poles colder than the Equator. Latitude also affects the annual range of monthly insolation, creating a stronger temperature cycle with increasing latitude.

The second factor is the maritime-continental contrast. As we've noted, coastal stations have more uniform temperatures, and are cooler in summer and warmer in winter. Interior stations, on the other hand, have much larger annual temperature variations. Ocean currents can also have an effect because they can keep coastal waters warmer or cooler than you might expect.

Elevation is the third important factor. At higher elevations, temperatures will be cooler, so we expect to see lower temperatures near mountain ranges.

WORLD AIR TEMPERATURE PATTERNS FOR JANUARY AND JULY

World air temperatures for the months of January and July are shown in Figures 3.24 and 3.25. The polar maps (Figure 3.24) show higher latitudes well, while the Mercator maps (Figure 3.25) are best for illustrating trends from the Equator to the midlatitude zones. Following the principles of how air temperatures are related to latitude, maritime-continental contrast, and elevation, it is easy to understand the six important features that are described in these two figures.

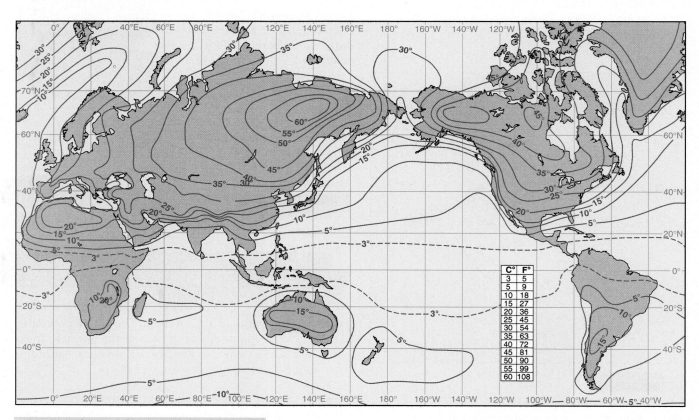

3.23 Annual range of air temperature in Celsius

The annual range of air temperature is defined as the difference between January and July means. Near the Equator, the annual range is quite narrow. In continental interiors, however, the range is much wider—as much as 60°C (108°F) in eastern Siberia.

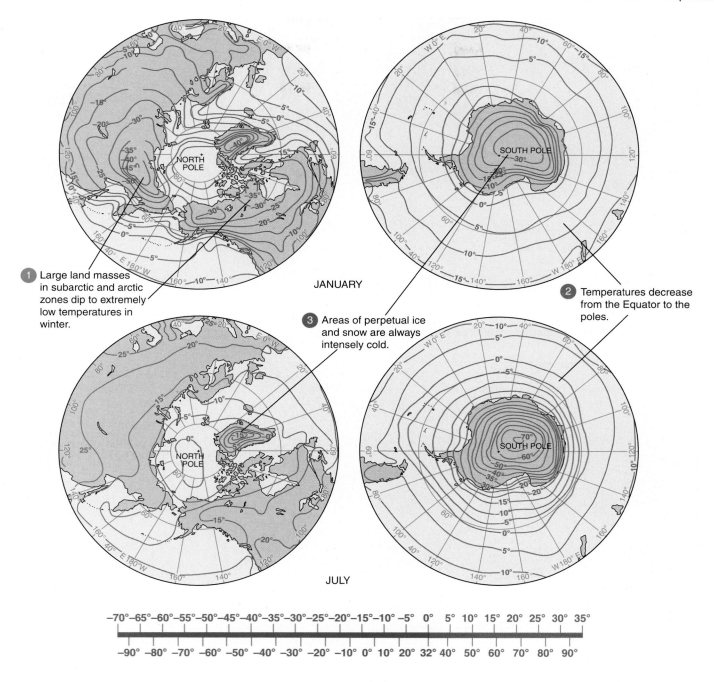

① Large land masses in subarctic and arctic zones dip to extremely low temperatures in winter.

JANUARY

③ Areas of perpetual ice and snow are always intensely cold.

② Temperatures decrease from the Equator to the poles.

JULY

−70° −65° −60° −55° −50° −45° −40° −35° −30° −25° −20° −15° −10° −5° 0° 5° 10° 15° 20° 25° 30° 35°

−90° −80° −70° −60° −50° −40° −30° −20° −10° 0° 10° 20° 32° 40° 50° 60° 70° 80° 90°

3.24 Mean monthly air temperatures for January and July, polar projections

These polar projections show north and south poles in January and July.

① **Large land masses located in the subarctic and arctic zones dip to extremely low temperatures in winter.** Look at North America and Eurasia on the north polar map for January. These low-temperature centers are very clear. The cold center in Siberia, reaches −50°C (−58°F); northern Canada is also quite cold (−35°C; −32°F). The high albedo of the snow helps keep winter temperatures low by reflecting much of the winter insolation back to space.

② **Temperatures decrease from the Equator to the poles.** This is shown on the polar maps by the general pattern of the isotherms as nested circles, with the coldest temperatures at the center, near the poles. The temperature decrease is driven by the difference in annual insolation from the Equator to the poles.

③ **Areas of perpetual ice and snow are always intensely cold.** Our planet's two great ice sheets are located in Greenland and Antarctica. Notice how they stand out on the polar maps as cold centers in both January and July. They are cold for two reasons. First, their surfaces are high in elevation, rising to over 3000 m (about 10,000 ft) in their centers. Second, their white snow surfaces reflect most of the incoming solar radiation.

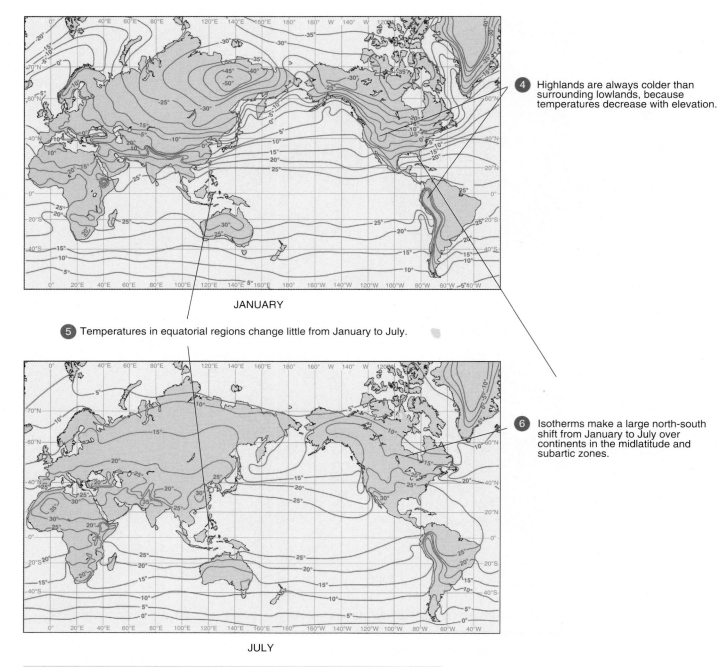

JANUARY

4 Highlands are always colder than surrounding lowlands, because temperatures decrease with elevation.

5 Temperatures in equatorial regions change little from January to July.

6 Isotherms make a large north-south shift from January to July over continents in the midlatitude and subartic zones.

JULY

3.25 Mean monthly air temperatures for January and July—Mercator projections

These Mercator maps show January and July temperatures from 80°N lat. to 60°S lat.

4 **Highlands are always colder than surrounding lowlands because temperatures decrease with elevation.** Look at the pattern of isotherms around the Rocky Mountain chain in western North America. In both summer and winter, the isotherms dip down around the mountains. The Andes Mountains in South America show the effect even more clearly.

5 **Temperatures in equatorial regions change little from January to July.** This is because insolation at the Equator doesn't vary greatly with the seasons. Look at the broad space between the 25°C (77°F) isotherms on both the January and July Mercator maps. In this region, the temperature is greater than 25°C (77°F) but less than 30°C (86°F). Although the two isotherms move a bit from winter to summer, the Equator always falls between them, showing how uniform the temperature is over the year.

6 **Isotherms make a large north-south shift from January to July over continents in the midlatitude and subarctic zones.** The 15°C (59°F) isotherm lies over central Florida in January, but by July it has moved far north, cutting the southern shore of Hudson Bay and then looping far up into northwestern Canada. In contrast, isotherms over oceans shift much less. This is because continents heat and cool more rapidly than oceans.

The Temperature Record and Global Warming

For the past 10,000 years or so, the Earth has experienced a period of moderate climate in which human civilization has developed and thrived. We understand how temperature varies based on the time of day or year, and that our planet can undergo short- and longer-term changes in temperature patterns as a result of natural events. But human activities are now affecting global climate as never before. Today, the science community largely understands the effects of human activities on our climate, and has identified the preventable causes of global warming trends.

THE TEMPERATURE RECORD

Today, satellite technology allows scientists to monitor the surface temperature of the Earth on a daily basis. Establishing a baseline for the Earth's temperature in different regions, and monitoring changes to temperature patterns, requires taking the sea surface temperature and the land surface temperature from a network of sea-based and ground-based meteorological stations, in combination with remote-sensing thermal temperature sensors on satellites and airplanes (Figure 3.26). These quantitative measurements are used for a range of applications, from ocean fisheries to crop forecasting to climate change assessments.

3.26 Measuring the Earth's temperature

NASA and the National Oceanic and Atmospheric Administration (NOAA) employ a constellation of weather and special instrument satellites to provide daily temperature profiles of Earth's surface, the air mass directly above, and the profiles along the atmospheric layers.

TEMPERATURE PREDICTION ▶
Weather satellites provide data for weather prediction; here, forecasting land surface temperatures for urban heat islands affecting the East Coast from Virginia to New York. This image shows a temperature prediction for a day in June.

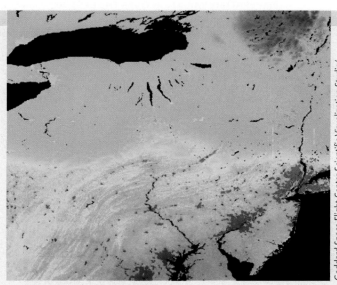

Goddard Space Flight Center Scientific Visualization Studio/ Courtesy NASA

SEA SURFACE TEMPERATURE
Sea surface temperature averages for the month of January 2011 are measured for the globe and recorded by the thermal sensors of the Moderate Resolution Imaging Spectroradiometer (MODIS) aboard NASA's Aqua satellite.

MODIS/Courtesy NASA

3.27 Reconstructing temperature records

Ice cores, coral growth, and tree ring widths provide indirect methods to reconstruct past temperatures.

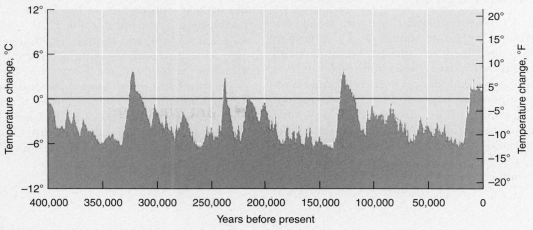

TEMPERATURE VARIATION OVER THE LAST 400,000 YEARS ▲

In ice cores from the Russian Antarctic Vostok research station, oxygen isotope ratios have been used to calculate global temperatures for the past 400,000 years, using known laws of physics and water evaporation rates.

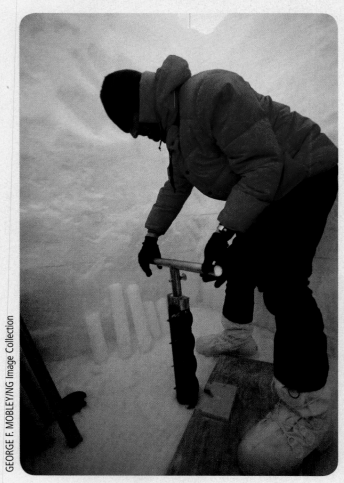

GEORGE F. MOBLEY/NG Image Collection

▲ **ICE CORES**

Scientists drill down into deep ice to extract a column of ice, known as an ice core. Annual cycles of precipitation form distinct layers in the ice core, enabling scientists to determine the age of different portions of the sample. Tiny bubbles in the ice pre-serve samples of the air from the period in which it froze. Here, a scientist is extracting an ice core from a glacier in Kluane National Park, Canada.

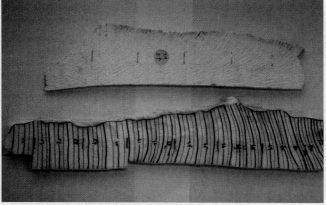

Courtesy Rob Dunbar, Department of Geological and Environmental Sciences, Stanford University

▲ **CORAL GROWTH**

The constant growth of some corals develops structures similar to annual rings in trees. Sampling the chemical composition of the coral from ring to ring can provide a temperature record. Black lines drawn on the lower section mark years; red and blue lines mark quarters.

Taylor S. Kennedy/NG Image Collection

▲ **TREE RINGS**

The width of annual tree rings varies with growing conditions, and in some locations can indicate air temperature.

We have direct records of air temperature measurements dating to the middle of the nineteenth century. If we want to know about temperatures in earlier times, we need to use indirect methods, often called *proxies*. In climates with distinct seasons, many tree species grow annual rings. For trees along the timberline in North America, the ring width is related to temperature—the trees grow better when temperatures are warmer. Because only one ring is formed each year, we can determine the date of each ring by counting backward from the present. Since these trees live a long time, we can extend the temperature record and correlate the tree ring measurements with other proxy measures going back several centuries. The slow growth of some types of corals also provides annual rings of varying chemical composition that indicate water temperature.

Ice cores from glaciers and polar regions also can show layers that are related to annual precipitation cycles. The crystalline structure of the ice, the concentrations of salts and acids, dust and pollen trapped in the layers, and amounts of trapped atmospheric gases, such as oxygen, carbon dioxide, and methane, all reveal information about long-term climate change patterns (Figure 3.27).

Reconstruction of past temperature records shows cycles of higher and lower temperatures. There have also been wide swings in the mean annual surface temperature. Some of this variation is caused by volcanic eruptions. Volcanic activity propels particles and gases—especially sulfur dioxide (SO_2)—into the stratosphere, forming stratospheric aerosols. Strong winds spread the aerosols quickly throughout the entire layer, where they reflect incoming solar radiation and have a cooling effect.

The eruption of Mount Pinatubo in the Philippines lofted 15 to 20 million tons of sulfuric acid aerosols into the stratosphere in the spring of 1991 (Figure 3.28). The aerosol layer produced by the eruption reduced solar radiation reaching the Earth's surface between 2 and 3 percent for the year or so following the blast. In response, global temperatures fell about 0.3°C (0.5°F) in 1992 and 1993.

GLOBAL WARMING

The Earth has been getting warmer, especially in the past 50 years (Figure 3.29). In February 2007, the Intergovernmental Panel on Climate Change (IPCC), a United Nations–sponsored group of more than 2000 scientists, issued a consensus report stating that global warming is "unequivocal." The National Academy of Sciences, the premier science advisors for the U.S. government, has recently amplified the IPCC findings by stating, "Climate change is occurring, is caused largely by human activities, and poses significant risks for— and in many cases is already affecting—a broad range of human and natural systems."

Where we live in the world has a lot to do with our view of climate change. Small island nations have expressed alarm at the rising sea levels they witness. People living in the Arctic are experiencing a greater

> The Earth has been getting warmer, especially in the last 50 years. Human activity, primarily through the enhanced generation of greenhouse gases, is very likely responsible for the warming trend.

Durieux/Sipa Press

3.28 Eruption of Mount Pinatubo, Philippine Islands, April 1991

Volcanic eruptions like this can inject particles and gases into the stratosphere, influencing climate for several years afterward.

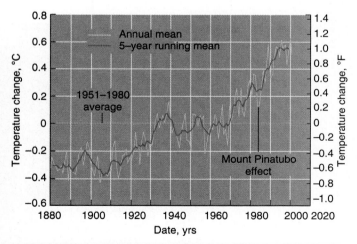

3.29 Mean annual surface temperature of the Earth, 1880–2011

The year 2010 was the hottest on record; and the first decade of the twenty-first century is now the warmest on record. Of the top 11 warmest years between the start of the instrument record to 2011, 10 have occurred since 2000. The yellow line shows the mean for each year; the red line shows a running five-year average. Note the effect of the eruption of Mount Pinatubo.

rise in temperature than people living in lower latitudes (Figure 3.30). Clearly then, latitude plays a significant role in where the effects of global warming are most strongly felt. Currently, the Arctic region is warming at a rate of about 2.5 times the global average.

Causes of Global Warming

In 1995, the IPCC concluded that human activity has probably caused climatic warming by increasing the concentra-

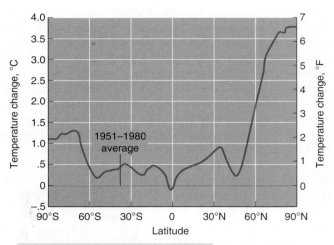

3.30 Temperature trends by latitude

Recent global temperature measurements demonstrate notable warming differences at the highest latitudes, both north and south of the Equator, as compared to average temperatures from the 1951–1980 time period.

tion of **greenhouse gases** in the atmosphere. Recall from Chapter 2 that adding greenhouse gases to the atmosphere enhances the greenhouse effect and increases warming. The IPCC's conclusion, which was judged "likely" in 2001, and upgraded to "very likely" in 2007, was based largely on computer simulations showing that the release of CO_2, CH_4, and other gases from fossil fuel burning and other human activity over the past century has accounted for the pattern of warming recently documented.

Figure 3.31 shows how a number of important factors have influenced global warming since about 1850. Although human activity has both warming and cooling effects on the atmosphere, when we add all these factors together, the warming effects of the greenhouse gases outweigh the cooling effects. The result is a net warming effect of about 1.6 to 2.4 W/m^2, which is about 1 percent of the total solar energy flow absorbed by the Earth and atmosphere.

The year 2010 was the warmest recorded in the instrumental record of thermometer use (since about 1850), according to data compiled by National Aeronautics and Space Administration (NASA) scientists at New York's Goddard Institute of Space Science. It was also the warmest of the past thousand years, according to reconstructions of past temperatures using tree rings and glacial ice cores by University of Massachusetts scientists. In fact, land and ocean temperatures, collected from around the globe, show that by 2011, 9 of the 10 warmest years in the instrumental record had occurred in the twenty-first century. In the past 30 years, the Earth has warmed by 0.6°C (1.1°F). In the past century, it has warmed by 0.8°C (1.4°F).

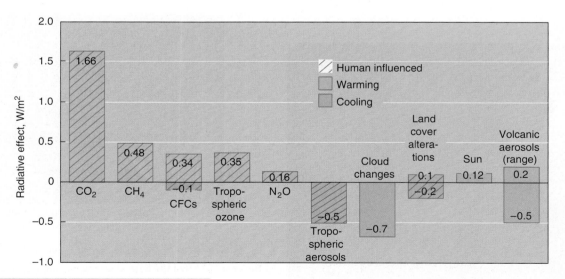

3.31 Factors affecting global warming and cooling

Greenhouse gases act primarily to enhance global warming, while aerosols, cloud changes, and land-cover alterations caused by human activity act to retard global warming. Natural and human-induced factors may be either warming (red) or cooling (blue) or both.

CONSEQUENCES OF GLOBAL WARMING

What will be the future effect of anthropogenic (human-induced) greenhouse warming? Using computer climate models, the scientists of the IPCC have projected that global temperatures will warm between 1.8°C (3.2°F) and 4.0°C (7.2°F) by the year 2100 (Figure 3.32). The range of temperatures predicted by the science models results in part from differences in the technical performance of the models, but it is also driven by different scenarios for economic growth and emission controls. Aerosols from dirty air, for example, could have a cooling effect, as noted in the range of predictions. Since these predictions were developed, observed warming has been on track to hit the high estimate.

A temperature rise of a few degrees may not sound like very much, but consider the fact that the increase of just a few degrees above 37°C (98.6°F) for human body temperature can have serious medical consequences, including brain damage and even death. We are only recently learning about the Earth's temperature norms for a much larger system, and the effects of small temperature changes. The difficulty of this effort is compounded by the realization that many other changes, including melting sea ice and rising sea levels, may accompany a rise in temperature (Figure 3.33).

Climate Change

Global warming is triggering a variety of complex changes in our global climate, which may, conversely, include patterns of cooler weather in some regions. The scientific community is monitoring a variety of ongoing changes in our current climate regime, such as the early onset of spring and the delay of fall. These climate pattern shifts are affecting ecosystems in a wide range of ways, from pine beetle infestations in the Pacific Northwest to bleaching of coral reefs in the Indian Ocean. Climate change could also promote the spread of insect-borne diseases such as malaria. Climate range boundaries may shift positions, making some regions wetter and others drier, leading to greater extremes in the new climate settings. Shifts in agricultural patterns, including desert expansion, could displace large human populations, as well as natural ecosystems.

Global warming is predicted to increase the variability of our climate, along with the amplitude of storms

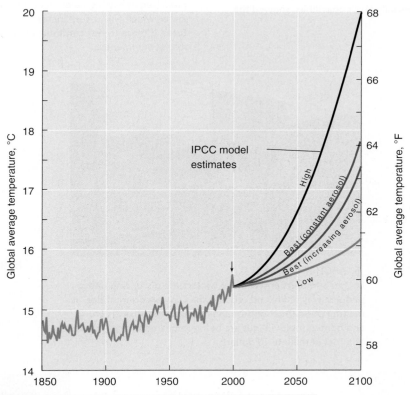

3.32 Computer projections for global temperature increases in the twenty-first century

The IPCC considers numerous computer models in its projections for the next century, based on continued human releases of CO_2, CH_4, and other industry-generated gases. Note that increasing levels of aerosols are predicted to counteract some of the warming trend, by reducing surface insolation.

3.33 Consequences of global warming

Global warming is causing changes to habitats and ecosystems around the globe. The risk of habitat loss is greatest at transitional areas, such as coastlines and the polar regions in North America and Eurasia.

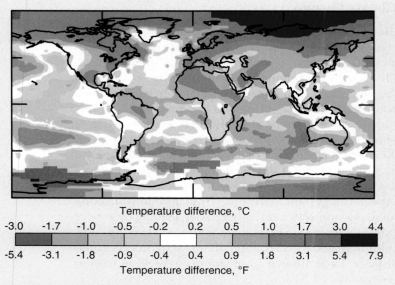

Temperature difference, °C

| -3.0 | -1.7 | -1.0 | -0.5 | -0.2 | 0.2 | 0.5 | 1.0 | 1.7 | 3.0 | 4.4 |

| -5.4 | -3.1 | -1.8 | -0.9 | -0.4 | 0.4 | 0.9 | 1.8 | 3.1 | 5.4 | 7.9 |

Temperature difference, °F

▲ **CHANGE IN GLOBAL SURFACE TEMPERATURES**
This map shows how the 2011 global surface temperatures compare with average surface temperatures from 1951–1980. Although some areas of ocean surface were slightly cooler, nearly all of the Earth's land area was warmer than the temperature base. Arctic regions, and especially arctic coastlines, showed the greatest warming.

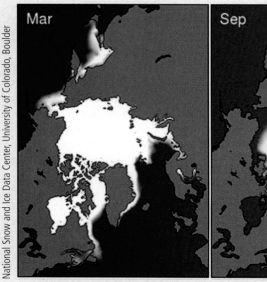
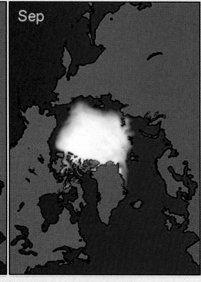

National Snow and Ice Data Center, University of Colorado, Boulder

▲ **POLAR SEA ICE MELTING**
Arctic sea ice reaches its maximum coverage each year in March, and its minimum in September. The extent of September sea ice has shown a decrease of about 10 percent per decade since 1979. Accelerated loss is predicted as the high albedo of the ice is replaced by the lower albedo of water, which will absorb increased solar radiation and raise water temperatures. If these trends continue, the Arctic Ocean will eventually become free of summer ice.

SEA-LEVEL RISE
As glaciers melt and ocean water expands in response to warming, sea levels are predicted to rise 28 to 43 cm (11.0 to 16.9 in.) by the year 2100, overwhelming the ability of both humans and native biodiversity to adapt, and putting at risk as many as 92 million people from annual flooding. The combination of sea-level rise and more frequent and intense storms threatens coastal communities.
▼

imagebroker/Alamy Limited

▲ **ARCTIC THAWING**
Longer summers and warmer winters are disrupting native people's way of life and eroding the coastal communities, as melting permafrost undermines community infrastructures and homes. Inuit villages are being relocated farther inland, at a cost of millions of dollars.

HABITAT LOSS
As a result of rapidly rising sea levels, low-lying islands with their surrounding coral reef ecosystems are subject to increased exposure from storm surges.
▼

© Reg Morrison/Auscape International Pty. Ltd.

Courtesy NOAA

and droughts. Very high 24-hour precipitation—violent snowstorms, rainstorms, sleet, and ice storms—have become more frequent since 1980. Likewise, more frequent and more intense spells of hot and cold weather also may be related to climate warming.

International Response to Global Warming

The combination of rising population numbers, higher living standards, conversion of forests to farmlands, and the ever-increasing consumption of fossil fuel energy are creating a complex turbulence of human affairs affecting our future climate. Worldwide, people have become aware of the problem of global warming, and efforts are underway to address it. For example, at the Rio de Janeiro Earth Summit in 1992, nearly 150 nations signed a treaty limiting emissions of greenhouse gases. And the IPCC, formed in 1988, prepared the First Assessment Report in 1990, which set into motion the creation of the United Nations Framework Convention on Climate Change (UNFCCC) by concluding that:

- Anthropogenic climate change will persist for many centuries.
- Further action is required to address remaining gaps in information and understanding.
- There is a continuing imperative to communicate research advances in terms that are relevant to decision making.

Despite the bold start under the UNFCCC, international consensus has not been reached. Notably, the United States has not been as supportive of international efforts to address climate change as many other nations.

The Kyoto Protocol, initiated in 1997, was the first attempt to codify an agreement whereby industrial nations would reduce their greenhouse gas emissions to about 5 percent below 1990 levels, and would assist developing nations with nonfossil-fuel technologies. Although many lessons have been learned in the ensuing years regarding carbon monitoring and carbon market trading, implementation has, to date, proved too burdensome, leaving individual countries to develop nationally focused programs outside the UN framework.

Carbon taxes, cap-and-trade, and various hybrid policy mechanisms have failed to effectively motivate the collective political will of the UN member nations, including the United States, the world's largest economy.

Individual nations, however, especially China, Germany, and the Scandinavian countries, have demonstrated that significant progress can be made in reducing greenhouse gases through a combination of political leadership, economic incentives, and government support for renewable energy technology development. It seems clear that the ultimate solution to reducing greenhouse gas concentrations and, hence, slowing global climate change will inevitably involve harnessing solar, wind, geothermal, and nuclear energy sources. In the coming decades, energy conservation, in concert with development of new methods of utilizing fossil fuels that produce fewer CO_2 emissions, will probably be the most important factors in reducing greenhouse gas emissions.

WileyPLUS Web Quiz
Take a quick quiz on the key concepts of this chapter.

A Look Ahead

In this chapter we have developed an understanding of both air temperatures and temperature cycles, along with the factors that influence them, including insolation, latitude, surface type, continental-maritime location, and elevation. Air temperatures are a very important aspect of the climates of the Earth, which we will study in Chapter 7.

The other key ingredient of climate is precipitation, the subject we will cover in the next chapter. Here, temperature plays a very important role, since moist air must cool before condensation forms water droplets and, eventually, precipitation.

WileyPLUS Web Links
Visit a NASA site to see today's continental and global temperature maps. Check out global change sites to explore global warming and past climates. See how ice cores and tree rings are used to reconstruct climate. Explore the urban heat island effect.

IN REVIEW **AIR TEMPERATURE**

- Carbon dioxide is one of several greenhouse gases that trap heat radiation leaving the Earth. As a result of fossil fuel burning, the amount of carbon dioxide has increased very significantly since global industrialization, and will probably continue to do so for the next century.
- Future increases in carbon dioxide and other greenhouse gases will depend on economic growth,

efficiency of energy use, and substitution of alternative sources of energy.
- **Air temperature** is influenced by latitude, surface type, coastal or interior location, elevation, and air and ocean circulation patterns.
- Radiation flows into and out of the **surface** of a substance. The energy balance of the ground surface is

determined by **net radiation, conduction** to the soil, and **latent heat transfer** and **convection** to and from the atmosphere.

- Air temperature is measured at 1.2 m (4 ft) above the surface. Weather stations take daily minimum and maximum temperature measurements. The mean daily temperature is the average of minimum and maximum temperatures.
- Temperatures of air and soil at or very close to the ground surface are more variable than air temperature measured at standard height.
- Surface characteristics also affect temperatures. Rural surfaces are generally moist, and slow to heat and cool, while urban surfaces are dry, and readily heat and cool. **Transpiration** of water from plant leaves and evaporation from moist soil surfaces is termed **evapotranspiration.**
- The difference between rural and urban surfaces creates an **urban heat island** effect that makes cities warmer than surrounding areas. Waste heat from urban activities also **significant**ly raises air temperatures of cities.
- Air temperatures observed at mountain loca**tions are lower with** higher elevation; and day-night temperature differences **increase with ele**vation.
- When air temperature increases with altitude, a **temperature inversion** is present. This can develop on clear nights when the surface loses longwave radiation to space.
- The *wind chill index* combines the effects of temperature and wind to describe how cold the air feels to a person outdoors. The *heat index* couples temperature and *relative humidity* to provide a measure of how hot the air feels.
- Air temperatures normally fall with altitude in the lower atmosphere. The decrease in air temperature with increasing altitude is called the **lapse rate**. The average value of decrease with altitude is the **environmental temperature lapse rate**: 6.49°C/1000 m (3.56°F/1000 ft).
- Temperatures decrease with increasing altitude in the lowest layer of the atmosphere, or **troposphere**. This layer includes abundant water vapor that condenses into clouds of water droplets with **aerosol** particles at their centers.
- Above the troposphere is the **stratosphere**, in which temperature stays uniform or increases slightly with elevation. Winds in the stratosphere are strong and persistent. The stratosphere contains the *ozone layer*, which absorbs harmful ultraviolet energy.
- The daily cycle of air temperature is driven largely by net radiation, which is positive during the day and negative at night. Net radiation depends on insolation, which is a function of latitude and season.

- Land and water contrasts affect both daily and annual temperature cycles. Water bodies heat and cool more slowly than land, so maritime locations have less extreme temperatures, while interior continental temperature cycles are more extreme.
- Global temperature patterns for January and July show the effects of latitude, maritime-continental location, and elevation. Equatorial temperatures vary little from season to season. Poleward, temperatures decrease with latitude, and continental surfaces at high latitudes can become very cold in winter. At higher elevations, temperatures are always colder.
- **Isotherms**—lines of equal temperature—over continents swing widely north and south with the seasons, while isotherms over oceans move through a much narrower range of latitude. The annual range in temperature expands with latitude, and is greatest in northern hemisphere continental interiors.
- Within the last few decades, global temperatures have been increasing. CO_2, released by fossil fuel burning, is important in causing warming, but so are the other **greenhouse gases**—CH_4, CFCs, O_3, and N_2O.
- The Earth's temperature record is now measured by satellite instruments and thermometers. Temperatures before about 1850 are estimated from *proxies*, which include tree rings, coral growth rings, and ice cores. Volcanic activity can lower temperatures by injecting aerosols into the stratosphere.
- The global temperature record shows substantial increases, especially in the last 50 years. Recent climate change is caused largely by human activities. The arctic regions are warming especially rapidly.
- Although human activity has both warming and cooling effects on the atmosphere, the result is a net warming equal to about 1 percent of the solar energy absorbed by the Earth and atmosphere.
- By 2011, 9 of the 10 warmest years had occurred in the twenty-first century. In the last 30 years, the Earth has warmed by 0.6°C (1.1°F). Climate variability and the amplitude of extreme weather events have increased. Climate change is presently changing ecosystems and their boundaries.
- International efforts to control global climate change are underway, but no effective consensus has been reached among nations on a plan to curb greenhouse gas emissions.

KEY TERMS

REVIEW QUESTIONS

1. Apart from water vapor, which **greenhouse gas** plays the largest role in warming the atmosphere? Where does it come from? How will its concentration change in the future?
2. Identify five important factors in determining air temperature and air temperature cycles.
3. What factors influence the temperature of a surface?
4. How are *mean daily air temperature* and *mean monthly air temperature* determined?
5. How does the daily temperature cycle measured within a few centimeters or inches of the surface differ from the cycle measured at normal air temperature height?
6. Compare the characteristics of urban and rural surfaces and describe how the differences affect urban and rural air temperatures. Include a discussion of the **urban heat island**.
7. How and why are the temperature cycles of high-mountain stations different from those of lower elevations?
8. What is a **temperature inversion**? When is it most likely to occur?
9. Identify two *temperature indexes* and provide an example of each. Identify a combination of temperature and wind speed that would cause frostbite in 30 minutes. Identify a combination of temperature and humidity that would put a person in extreme danger of sunstroke.
10. Explain how air temperature changes with elevation in the atmosphere, using the terms **lapse rate** and **environmental temperature lapse rate** in your answer.
11. What are the two layers of the lower atmosphere? How are they distinguished? Name the two upper layers. How is the *heterosphere* different from the layers below?
12. Why do large water bodies heat and cool more slowly than land masses? What effect does this have on daily and annual temperature cycles for coastal and interior stations?
13. Explain how latitude affects the annual cycle of air temperature through net radiation by comparing these locations: Manaus, Aswan, Hamburg, and Yakutsk.
14. What three factors are most important in explaining the world pattern of *isotherms*? Explain how and why each factor is important, and what effect it has.
15. Turn to the January and July world temperature maps shown in Figures 3.24 and 3.25. Make six important observations about the patterns, and explain why each occurs.
16. How is the Earth's temperature record established at scales of past decades, past centuries, and past millennia?
17. How is the Earth's temperature changing now? Are there differences by latitude? How is it predicted to change in the future?
18. Identify four important consequences of global warming.
19. What has been the response of the international community to global warming?

VISUALIZING EXERCISES

1. Sketch graphs showing how *insolation*, **net radiation**, and temperature might vary from midnight to midnight during a 24-hour cycle at a midlatitude station such as Chicago.
2. Sketch a graph of air temperature with height showing a low-level **temperature inversion**. Where and when is such an inversion likely to occur?

ESSAY QUESTIONS

1. Portland, Oregon, on the north Pacific coast, and Minneapolis, Minnesota, in the interior of the North American continent, are at about the same latitude. Sketch the annual temperature cycle you would expect for each location. How do they differ, and why? Select one season, summer or winter, and sketch a daily temperature cycle for each location. Again, describe how they differ, and why.
2. Many scientists have concluded that human activities are acting to raise global temperatures. What human processes are involved? How do they relate to natural processes? Are global temperatures increasing now? What are the consequences of global warming?

Chapter 4
Atmospheric Moisture and Precipitation

The island of Moorea, in the Society Islands of French Polynesia, rises from the South Pacific toward the tropical sky, providing a green oasis in a sea of blue. Flanked by a broad coral reef, the island's stunning scenery places it among the most beautiful islands on Earth. The rugged mountains of the island's center, formed by erosion of now-extinct volcanoes, range in elevation to about 1200 m (about 4000 ft). Lying in the path of trade winds, the peaks lift the moist marine air, producing orographic precipitation. Here, an orographic shower bathes the southeast side of the island in a warm rain.

MOOREA, SOCIETY ISLANDS, FRENCH POLYNESIA

©Yann Arthus-Bertrand/Altitude

Atmospheric Moisture and Precipitation

Precipitation is a commonplace event for most people, and as part of our daily lives we don't tend to think too much about it. But without precipitation, our planet would be quite inhospitable. How does water flow between the atmosphere, land surface, and oceans? What factors affect the amount of water vapor present in the atmosphere? What causes water vapor in the atmosphere to condense, forming clouds of water droplets and ice crystals? How do mountains induce precipitation and create rain shadows? How do thunderstorms form, and why are they sometimes severe? What causes tornadoes to form? How have human activities affected air quality? These are questions we will answer in this chapter.

Acid Deposition

Perhaps you've heard about *acid rain* killing fish and poisoning trees. It's made up of raindrops that have been acidified by air pollutants. Fossil fuel burning releases sulfur dioxide (SO_2) and nitric oxide (NO_2) into the air. The SO_2 and NO_2 readily combine with oxygen and water in the presence of sunlight and dust particles to form sulfuric and nitric acid aerosols, which then act as condensation nuclei. The tiny water droplets created around these nuclei are acidic, and when the droplets coalesce in precipitation, the resulting raindrops or ice crystals are also acidic. Sulfuric and nitric acids can also be formed on dust particles, creating dry acid particles. These can be as damaging to plants, soils, and aquatic life as acid rain. Acid rain and acid dust particles, taken together, are termed **acid deposition** (Figure 4.1).

What are the effects of acid deposition? In Europe and in North America, acid deposition has had a severe impact on some ecosystems. In Norway, acidification of stream water has virtually eliminated many salmon runs by inhibiting salmon egg development. In 1990, American scientists estimated that 14 percent of Adirondack lakes were heavily acidic, along with 12 to 14 percent of the streams in the Mid-Atlantic States. Forests, too, have been damaged by acid deposition. In western Germany, the impact has been especially severe in the Harz Mountains and the Black Forest.

Since 1990, the United States has significantly reduced the release of sulfur oxides, nitrogen oxides, and volatile organic compounds, largely by improving and strengthening industrial emission controls. But acid deposition is still a very serious problem in many parts of the world—especially Eastern Europe and the states of the former Soviet Union. There, air pollution controls have been virtually nonexistent for decades. Reducing pollution levels and cleaning up polluted areas will be a major task for these nations over the next decades.

Water in the Environment

Our planet's surface is dominated by water; over 70 percent of the surface of the Earth is covered by liquid water, in oceans and lakes, and solid water, in the ice of glaciers, icecaps, and sea ice. In this chapter, we focus on water in the air, both as vapor and as liquid and solid water. **Precipitation** is the fall of liquid or solid water from the atmosphere that reaches the Earth's land or ocean surface. It forms when moist air is cooled, causing water vapor to form liquid droplets or solid ice particles. If cooling is sufficient, liquid and solid water particles will grow to a size too large to be held aloft by the motion of the atmosphere and will fall toward the Earth.

Before we begin our study of atmospheric moisture and precipitation, however, we will briefly review the properties of water, including its three states (gas, liquid, and solid) and the conversion of one state to another.

4.1 Acid deposition

Deposition of acid droplets and dry acid particles, formed in the atmosphere from gases released by fossil fuel burning, can have severe environmental effects.

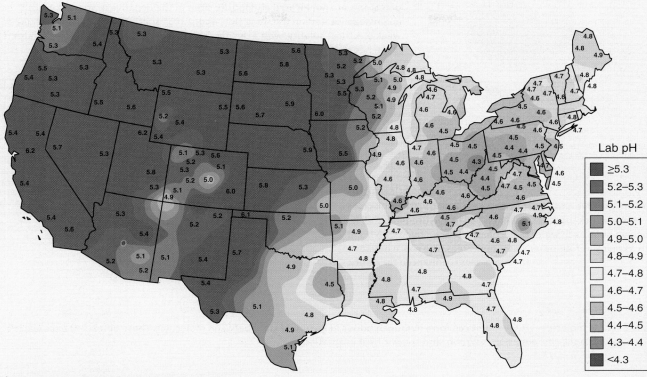

▲ **ACIDITY OF RAINWATER FOR THE UNITED STATES IN 2005**
Values are in pH units, which indicate acidity. The eastern United States, notably Ohio, Pennsylvania, and West Virginia, show the lowest, most acidic values.

Lab pH

	≥5.3
	5.2–5.3
	5.1–5.2
	5.0–5.1
	4.9–5.0
	4.8–4.9
	4.7–4.8
	4.6–4.7
	4.5–4.6
	4.4–4.5
	4.3–4.4
	<4.3

Paal Hermansen/Balance/Photoshot/Newscom

▲ **FORESTS**
This forest in Norilsk, Siberia, Russia, was killed by acid rain and heavy metal pollution.

Al Petteway/NG Image Collection

▲ **BUILDINGS**
Acid rain has eroded and eaten away the face of this stone angel in London, England.

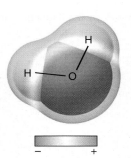

COVALENT BOND
The two hydrogen atoms and one oxygen atom of the water molecule share electrons in what chemists refer to as a covalent bond. The two positive hydrogen atoms are not located at opposite sides of the negative oxygen atom, but are instead located on the same side, about 105° apart. This gives the molecule a weak electrical polarity, with a positive charge on the side of the hydrogen atoms and a negative charge on the oxygen side.

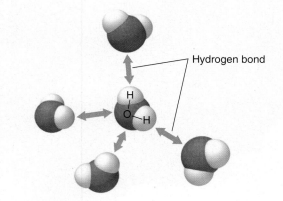

HYDROGEN BOND ▶
Because positive and negative charges attract, water molecules tend to stick together in what is called a hydrogen bond. As shown here, each water molecule can share a hydrogen bond with as many as four other water molecules. The space between the molecules is enhanced for clarity.

4.2 The water molecule

Water is a simple molecule—H_2O—but it can form hydrogen bonds in the liquid state between molecules. These bonds enhance water's surface tension, capillarity, adhesion, solution, and specific heat properties.

PROPERTIES OF WATER

The water molecule (H_2O) is a chemical compound of hydrogen and oxygen, two very abundant elements. The two hydrogen atoms share electrons with the one oxygen atom in what chemists refer to as a covalent bond. As shown in Figure 4.2, the arrangement of the three atoms produces a molecule with a weak positive charge on one side and a weak negative charge on the other. Since positive and negative charges attract, liquid water molecules stick tend to together, negative to positive. This weak attraction is called *hydrogen bonding*. It also helps water molecules form thin films on solid substances, adhering to natural positive and negative charges on substance surfaces.

Hydrogen bonding accounts for number of water's unusual properties. The attraction between water molecules produces *surface tension*, forming a thin skin of surface water molecules that makes a water droplet form a bead. When surface tension is coupled with adherence to a solid surface, thin films of water can be drawn into fine cracks and openings; for example, into rocks and soil openings. This is known as capillary action, or *capillarity*. Water is also a very good solvent. By hydrogen bonding with molecules of other substances, it can dissolve them and carry them away. Thus, water in lakes, rivers, and soils often contains dissolved nutrients, minerals, and small particles that can react with other substances or be carried to new locations.

Another result of hydrogen bonding is the high *specific heat* of water, which makes water heat and cool more slowly than most other substances. In Chapter 3, we noted this property as one reason why locations near water tend to have more uniform climates. Specific heat measures the heat flow required to raise the temperature of 1 gram of a substance by 1°C (1.8°F) at a temperature of 15°C (59°F). Imparting kinetic energy to liquid water molecules requires breaking hydrogen bonds, a process that absorbs much of the heat flow.

THREE STATES OF WATER

Water can exist in three states: as a solid (ice), as a liquid (water), or as an invisible gas (water vapor), as shown in Figure 4.3. If we want to change the state of water from solid to liquid, liquid to gas, or solid to gas, we must add energy. This energy serves to break the hydrogen bonds that keep water either in a crystalline structure as a solid or in a liquid state in which hydrogen bonds keep water molecules in close contact. This energy, called *latent heat*, is drawn in from warmer surroundings. When the change goes the other way, from liquid to solid, gas to liquid, or gas to solid, these bonds are reformed, releasing to the surroundings the energy they absorbed when broken.

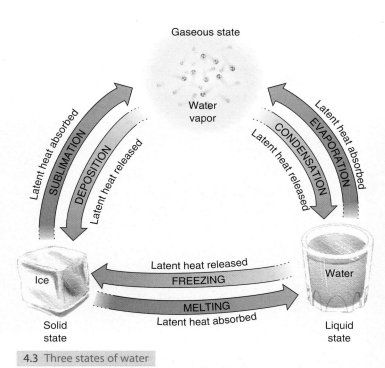

4.3 Three states of water

The arrows show the ways that any one state of water can change into either of the other two states. Heat energy is absorbed or released, depending on the direction of change.

Melting, freezing, evaporation, and *condensation* are familiar terms for describing the changes of state of water. **Sublimation** is the direct transition from solid to vapor. Perhaps you have noticed that ice cubes left in the freezer for a long time shrink away from the sides of the ice cube tray and get smaller. They shrink through sublimation—never melting, but losing mass directly as vapor. In this book, we use the term **deposition** to describe the reverse process, when water vapor crystallizes directly as ice. Frost forming on a cold winter night is a common example of deposition.

> Water exists in solid, liquid, and gaseous states as ice, water, and water vapor. Latent heat energy is released or absorbed as water changes from one state to another.

THE HYDROSPHERE

The realm of water in all its forms, and the flows of water among ocean, land, and atmosphere, are known collectively as the **hydrosphere**, shown in Figure 4.4. About 97.2 percent of the hydrosphere consists of ocean saltwater. The remaining 2.8 percent is freshwater. The next largest reservoir is freshwater stored as ice in the world's ice sheets and mountain glaciers, which accounts for 2.15 percent of total global water.

Fresh liquid water is found above and below the Earth's land surfaces. Subsurface water lurks in openings in soil and rock. Most of it is held in deep storage as groundwater, where plant roots cannot reach. Groundwater makes up 0.63 percent of the hydrosphere.

The small remaining proportion of the Earth's water includes the water available for plants, animals, and human use. Plant roots can access soil water. Surface water is held in streams, lakes, marshes, and swamps. Most of this surface water is about evenly divided between freshwater lakes and saline (salty) lakes. An extremely small proportion makes up the streams and rivers that flow toward the sea or inland lakes.

Only a very small quantity of water is held as vapor and cloud water droplets in the atmosphere—just 0.001 percent of the hydrosphere. However, this small reservoir of water is enormously important. Through precipitation, it supplies water and ice to replenish all freshwater stocks on land. In addition, this water, and its conversion from one form to another in the atmosphere, is an essential part of weather events across the globe. Finally, the flow of water vapor from warm tropical oceans to cooler regions generates a global flow of heat from low to high latitudes.

THE HYDROLOGIC CYCLE

The **hydrologic cycle**, or water cycle, moves water from land and ocean to the atmosphere (Figure 4.5). Water from the oceans and land surfaces evaporates, changing state from liquid to vapor and entering the atmosphere. Total evaporation is about six times greater over oceans than land because oceans cover most of the planet and because land surfaces are not always wet enough to yield much water.

> The hydrologic cycle describes the global flow of water to and from oceans, land, and atmosphere. Water moves by evaporation, precipitation, and runoff.

Water vapor in the atmosphere can condense or deposit to form clouds and precipitation, which falls to Earth as rain, snow, or hail. There is nearly four times more precipitation over oceans than precipitation over land.

When precipitation hits land, it can take three different courses. First, it can evaporate and return to the atmosphere as water vapor. Second, it can sink into the soil and then into the cracks and crevices in rock layers below as groundwater. As we will see in later chapters, this subsurface water emerges from below to feed rivers, lakes, and even ocean margins. Third, precipitation can run off the land, concentrating in streams and rivers that eventually carry it to the ocean or to lakes. This flow is known as *runoff.*

Because our planet contains only a fixed amount of water, a global balance must be maintained among

4.4 Water reservoirs of the hydrosphere

The Earth's reservoirs of water include both salt and freshwater.

▼ ATMOSPHERE
Atmospheric water, although only 0.001 percent of total water, is a vital driver of weather and climate, and sustains life on Earth.

▼ SURFACE WATER
Surface water, including lakes, comprises only a very tiny fraction of Earth's water volume. Here, a bull moose wades along the margin of a lake in search of tasty aquatic plants.

Dr. Maurice G.Hornocker/NG Image Collection

Jodi Cobb/NG Image Collection

▶ DISTRIBUTION OF WATER
Nearly all the Earth's water is contained in its oceans. Fresh surface and soil water make up only a small fraction of the total volume of global water.

Total water

World oceans 97.2%

Fresh water 2.8%

Ice sheets and glaciers 2.15%

Groundwater 0.63%

Surface and atmospheric water 0.02%

Saline lakes and inland seas 0.008%

Fresh-water lakes 0.009%

Soil water 0.005%

Stream channels 0.0001%

Atmosphere 0.001%

Bill Curtsinger/NG Image Collection

▲ OCEANS
Most of the Earth's water is held by its vast oceans. Here, a southern stingray swims along a shallow ocean bottom near Grand Cayman Island.

▶ ICE
Ice sheets and glaciers are the second largest reservoir of water on Earth. Although glaciers are too cold and forbidding for most forms of animal life, Antarctic ice sheets provide a habitat for these Weddell seals.

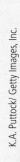
K.A. Puttock/ Getty Images, Inc.

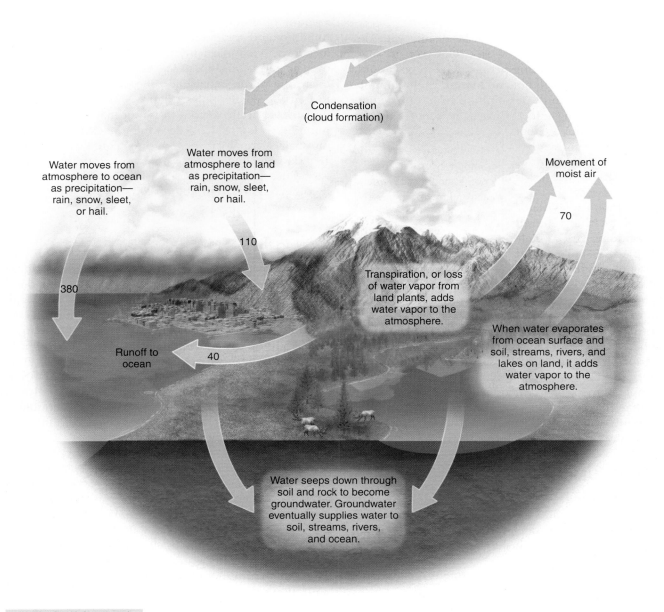

Condensation
(cloud formation)

Water moves from
atmosphere to ocean
as precipitation—
rain, snow, sleet,
or hail.

Water moves from
atmosphere to land
as precipitation—
rain, snow, sleet,
or hail.

Movement of
moist air

70

110

380

Transpiration, or loss
of water vapor from
land plants, adds
water vapor to the
atmosphere.

When water evaporates
from ocean surface and
soil, streams, rivers, and
lakes on land, it adds
water vapor to the
atmosphere.

Runoff to
ocean

40

Water seeps down through
soil and rock to become
groundwater. Groundwater
eventually supplies water to
soil, streams, rivers,
and ocean.

4.5 The hydrologic cycle

In the hydrologic cycle, water moves among the ocean, the atmosphere, the land, and back to the ocean in a continuous process.

flows of water to and from the lands, oceans, and atmosphere. Evaporation leaving the ocean is approximately 420 km³/yr (101 mi³/yr), while the amount entering the ocean via precipitation is 380 km³/yr (91 mi³/yr). Clearly, there is an imbalance between the amount of water lost to evaporation and the amount gained through precipitation. This imbalance is made up by the 40 km³/yr (10 mi³/yr) that flows from the land back to the ocean.

Similarly, there is a balance for the land surfaces of the world. Of the 110 km³/yr (27 mi³/yr) of water

that falls on the land surfaces, 70 km³/yr (17 mi³/yr) is reevaporated back into the atmosphere. The remaining 40 km³/yr (10 mi³/yr) stays in the form of liquid water and eventually flows back into the ocean.

Of all these pathways, we will be most concerned with only one aspect of the hydrologic cycle: the flow of water from the atmosphere to the surface in the form of precipitation. To understand this process, we first need to examine how water vapor in the atmosphere is converted into clouds and, subsequently, into precipitation.

Humidity

Blistering summer heat waves can be deadly, with the elderly and the ill at greatest risk. However, in such weather, even healthy young people need to be careful, especially when it is both hot and humid. High humidity slows the evaporation of perspiration from our bodies, reducing its cooling effect. Clearly, it is not only the temperature of the air that controls how hot weather affects us; the amount of water vapor in the air is important as well.

> Humidity refers to the amount of water vapor present in the air. Warm air can hold much more water vapor than cold air.

The amount of water vapor present in air, referred to as **humidity**, varies widely from place to place and time to time. In the cold, dry air of arctic regions in winter, the humidity is almost zero, while it can reach up to as much as 3 to 4 percent of a given volume of air in the warm wet regions near the Equator.

Consider for a moment an *air parcel*—a cubic meter or kilogram of air, for example. An important principle concerning humidity states that the maximum quantity of water vapor an air parcel can contain is dependent on the air temperature itself. Warm air can contain more water vapor than cold air—a lot more. For example, air at room temperature (20°C; 68°F) can contain about three times as much water vapor as freezing air (0°C; 32°F).

SPECIFIC HUMIDITY

The actual quantity of water vapor contained within a parcel of air is known as its **specific humidity** and is expressed as grams of water vapor per kilogram of air (g/kg). The equation for specific humidity is given as:

$$\text{specific humidity} = \frac{\text{mass of water vapor}}{\text{mass of total air}}$$

Specific humidity is often used to describe the amount of water vapor in a large mass of air. Both humidity and temperature are measured at the same locations in standard thermometer shelters the world over, as well as on ships at sea. Specific humidity is highest in the warm, equatorial zones, and falls off rapidly toward the colder poles (Figure 4.6). Extremely cold, dry air over arctic regions in winter may have a specific humidity as low as 0.2 g/kg, while the extremely warm, moist air of equatorial regions often contains as much as 18 g/kg. The total natural range on a worldwide basis is very wide. In fact, the highest values of specific humidity observed are from 100 to 200 times as great as the lowest values.

DEW-POINT TEMPERATURE

Although the actual amount of water in a given volume of air is called the specific humidity, this is not the same as the maximum quantity of moisture that a given volume of air can contain at any time. Maximum specific humidity, referred to as the *saturation-specific* humidity, is dependent on the air's temperature.

In Figure 4.7 we see, for example, that at 20°C (68°F), the maximum amount of water vapor that the air can contain—the saturation-specific humidity—is about 15 g/kg. At 30°C (86°F), it is nearly double—about 26 g/kg. For cold air, the values are quite low. At –10°C (14°F), the maximum is only about 2 g/kg. Another way of describing the water vapor content of air is by its **dew-point temperature**, also called simply the *dew point*. If air is slowly chilled, its saturation-specific humidity decreases. This can continue until the saturation-specific humidity

> The dew-point temperature of a mass of air is the temperature at which saturation will occur. The more water vapor in the air, the higher the dew-point temperature.

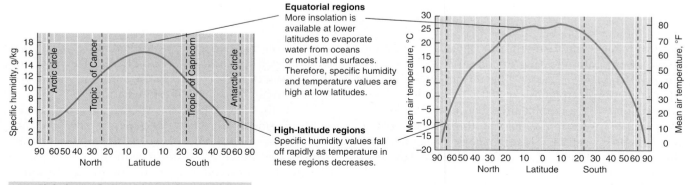

Equatorial regions
More insolation is available at lower latitudes to evaporate water from oceans or moist land surfaces. Therefore, specific humidity and temperature values are high at low latitudes.

High-latitude regions
Specific humidity values fall off rapidly as temperature in these regions decreases.

4.6 Global specific humidity and temperature

Pole-to-pole profiles of specific humidity (left) and temperature (right) show similar trends because the ability of air to hold water vapor (measured by specific humidity) is limited by temperature.

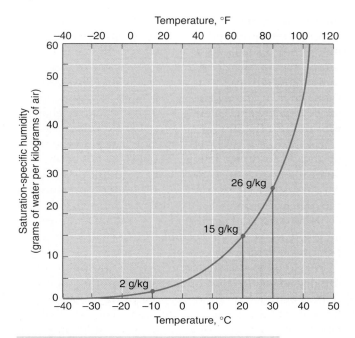

4.7 Saturation-specific humidity and temperature

The maximum specific humidity a mass of air can have—the saturation-specific humidity—increases sharply with rising temperature.

is equal to the specific humidity. When this condition is reached, the air has reached *saturation*, because the air contains the maximum amount of water vapor possible. If further cooling continues, condensation on surfaces will begin. The temperature at which saturation occurs is therefore known as the *dew-point temperature*—that is, the temperature at which dew forms by condensation.

RELATIVE HUMIDITY

When weather forecasters speak of humidity, they are usually referring to **relative humidity**. This measure compares the amount of water vapor present to the maximum amount of vapor that the air can contain at its given temperature. The relative humidity is expressed as a percentage given by:

$$\text{relative humidity} = \frac{100 \times \text{specific humidity}}{\text{saturation-specific humidity}}$$

For example, if the air currently contains half the moisture possible at the present temperature, then the relative humidity is 50 percent. When the humidity is 100 percent, the air contains the maximum amount of moisture possible. The air is saturated, and its temperature is at the dew point. When the specific humidity and saturation-specific humidity are not the same, the air is unsaturated. Generally, when the difference between the two is large, the relative humidity is low, and vice versa.

The relative humidity of the atmosphere can change in one of two ways. First, the atmosphere can directly gain or lose water vapor, thereby changing the specific humidity of the air mass. For example, additional water vapor can enter the air from an exposed water surface or from wet soil. This process is slow because the water vapor molecules must diffuse upward from the surface into the air layer above.

The second way relative humidity changes is through a change of temperature. Even though no water vapor is added, an increase of temperature results in a decrease of relative humidity (Figure 4.8). Recall that the saturation-specific humidity of air is dependent on temperature. When the air is warmed, the saturation-specific humidity increases. The existing amount of water vapor, given by the specific humidity, then represents a smaller fraction of the saturation-specific humidity.

> Relative humidity depends on both the water vapor content and the temperature of air. It compares the amount of water held by air to the maximum amount that can be held at that temperature.

WileyPLUS Weather Station Interactivity
Test your skill at forecasting fog or frost at five locations across the continent based on weather station data. Issue pollution alerts, frostbite, and heat index warnings.

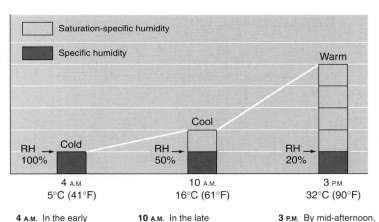

4 A.M. In the early morning hours, the temperature is 5°C (41°F), and the relative humidity of the air is 100 percent. Because the saturation-specific humidity and specific humidity are equal, the air is saturated.

10 A.M. In the late morning hours, the temperature has risen to 16°C (61°F). The relative humidity has dropped to 50 percent, even though the amount of water vapor in the air—the specific humidity—remains the same. Instead, the saturation-specific humidity has increased with temperature.

3 P.M. By mid-afternoon, the air has been warmed by the Sun to 32°C (90°F). The relative humidity has dropped to 20 percent and the air is very dry, because the saturation-specific humidity has greatly increased.

4.8 Relative humidity and air temperature

Relative humidity changes with temperature because warm air can contain more water vapor than cold air. In this example, the amount of water vapor stays the same, and only the saturation-specific humidity of the air mass changes.

Adiabatic Processes

What makes the water vapor in the air turn into liquid or solid particles that can fall to the Earth? The answer is that the air is naturally cooled. When air cools to the dew point, the air is saturated with water. Think about extracting water from a moist sponge. To release the water, you have to squeeze the sponge—that is, reduce its ability to hold water. In the atmosphere, chilling the air beyond the dew point is like squeezing the sponge; it reduces the amount of water vapor the air can contain, forcing some water vapor molecules to change state to form water droplets or ice crystals.

One mechanism for chilling air is nighttime cooling. On a clear night, the ground surface can become quite cold as it loses longwave radiation. If the air is moist, frost can be deposited as water vapor forms ice crystals on ground surfaces. However, this cooling is not enough to form precipitation in the air. Precipitation only forms when a substantial mass of air experiences a steady drop in temperature below the dew point. This happens when an air parcel is lifted to higher and higher levels in the atmosphere.

WileyPLUS The Adiabatic Process
Watch an animation that demonstrates how an air parcel cools as it rises and warms as it descends. The animation shows both dry and moist adiabatic motion.

DRY ADIABATIC LAPSE RATE

If you have ever pumped up a bicycle tire using a hand pump, you might have noticed that the pump gets hot. If so, you have observed the **adiabatic principle**. This important law states that if no energy is added to a gas, its temperature will increase as it is compressed. As you pump vigorously, compressing the air, the metal bicycle pump gets warm. Conversely, when a gas expands, its temperature drops by the same principle. Physicists use the term *adiabatic process* to refer to a heating or cooling process that occurs solely as a result of pressure change, with no heat flowing into or away from a volume of air.

How does the adiabatic principle relate to the uplift of air and to precipitation? The answer is, simply, that atmospheric pressure decreases as altitude increases. As a parcel of air is uplifted, atmospheric pressure on the parcel becomes lower, and the air expands and cools, as shown in Figure 4.9. As a parcel of air descends, atmospheric pressure becomes higher, and the air is compressed and warmed.

We describe this behavior in the atmosphere using the **dry adiabatic lapse rate**, as shown in the lower portion of Figure 4.10. It applies to a rising air parcel that has not yet been cooled to saturation. The dry adiabatic lapse rate has a value of about 10°C per 1000 m (5.5°F per 1000 ft) of vertical rise. That is, if a parcel of air is raised 1 km (3820 ft), its temperature will drop by 10°C (50°F). Conversely, an air parcel that descends will warm by 10°C (50°F) per 1000 m (3820 ft). This is the *dry* rate because no condensation occurs during this process.

There is an important difference to note between the dry adiabatic lapse rate and the environmental temperature lapse rate. The *environmental temperature lapse rate* is simply an expression of how the temperature of still air varies with altitude. This rate will vary from time to time and from place to place, depending on the state of the atmosphere. It is quite different from the dry adiabatic lapse rate. The dry adiabatic lapse rate applies to a mass

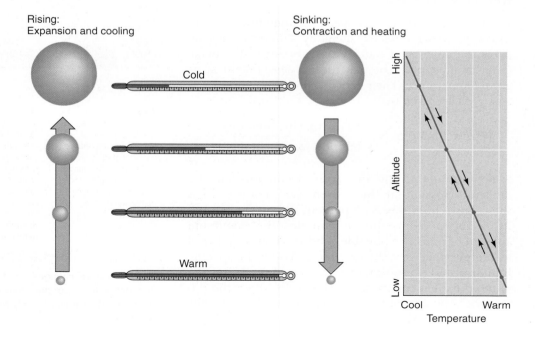

4.9 Adiabatic cooling and heating

When air is forced to rise, it expands and its temperature decreases. When air is forced to descend, its temperature increases.

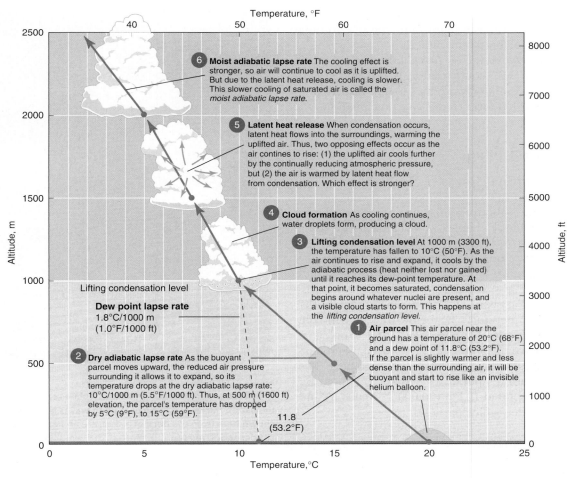

4.10 Adiabatic cooling in a rising parcel of air

Adiabatic decrease of temperature in a rising parcel of air leads to cooling, then to condensation of water vapor into water droplets and the formation of a cloud.

of air moving vertically. It does not vary with time and place, and it is determined by physical laws, not the local atmospheric state.

MOIST ADIABATIC LAPSE RATE

Let's continue examining the fate of a parcel of air that is moving upward in the atmosphere (Figure 4.10). As the parcel moves upward, its temperature drops at the dry adiabatic rate, 10°C/1000 m (5.5°F/1000 ft). Note, however, that the dew-point temperature changes slightly with elevation. Instead of remaining constant, it falls at the *dew-point lapse rate* of 1.8°C/1000 m (1.0°F/1000 ft). As the rising process continues, the air is eventually cooled to its dew-point temperature, and condensation starts to occur. This is shown in Figure 4.10 as the **lifting condensation level**. The lifting condensation level is thus determined by the initial temperature of the air and its initial dew point, and can differ from the example shown here.

If the parcel of saturated air continues to rise, a new principle comes into effect: latent heat energy is released

by the condensing water molecules and warms the surrounding air molecules. In other words, two effects are occurring at once. First, the uplifted air is being cooled by the reduction in atmospheric pressure. Second, it is being warmed by the release of latent heat from condensation.

Which effect is stronger? As it turns out, the cooling effect is stronger, meaning the air will continue to cool as it is uplifted. However, because of the release of latent heat, the cooling will occur at a lesser rate. This cooling rate for saturated air is called the **moist adiabatic lapse rate;** it ranges between 4°C and 9°C per 1000 m (2.2°F and 4.9°F per 1000 ft). The moist adiabatic lapse rate is variable because it depends on the temperature and pressure of the air and its moisture content. For most situations, however, we can use a value

> Rising air cools less rapidly when condensation is occurring, owing to the release of latent heat energy. This explains why the moist adiabatic cooling rate has a lesser value than the dry adiabatic cooling rate.

of 5°C/1000 m (2.7°F/1000 ft). In Figure 4.10, the moist adiabatic rate is shown as a slightly curving line to indicate that its value changes with altitude.

Keep in mind that as the air parcel becomes saturated and continues to rise, condensation is occurring. This condensation can eventually produce liquid droplets and solid ice particles that form clouds and precipitation.

Clouds and Fog

Images of the Earth from space show that about half of our planet is blanketed in **clouds**. Clouds play a complicated role in temperature—both cooling and warming the Earth and atmosphere. In this chapter, we will look at one of the most familiar roles of clouds: producing precipitation.

Clouds are made up of condensed water droplets, ice particles, or a mixture of both, suspended in air. Liquid water turns to ice at 0°C (32°F); but when water is dispersed as tiny droplets in clouds, it can remain in the liquid state at temperatures far below freezing. In fact, clouds consist entirely of supercooled water droplets at temperatures down to about −12°C (10°F). As cloud temperatures grow colder, ice crystals begin to appear. The coldest clouds, with temperatures below −40°C (−40°F), occur at altitudes of 6 to 12 km (20,000 to 40,000 ft) and are made up entirely of ice particles.

CLOUD FORMATION

Cloud particles grow around a tiny center of solid matter. This dust speck of matter, called a **condensation nucleus**, typically has a diameter between 0.1 and 1 micrometers (0.000004 and 0.00004 in.).

The surface of the sea is an important source of condensation nuclei (Figure 4.11). Droplets of spray from the crests of the waves are carried upward by turbulent air. When these droplets evaporate, they leave behind a tiny residue of concentrated salt, suspended in the air, that strongly attracts liquid water molecules, stimulating cloud formation. Nuclei are also thrown into the atmosphere from polluted air over cities, where particulate matter and soot aid condensation and the formation of clouds and fog, thereby increasing rates of precipitation.

> A cloud consists of water droplets, ice particles, or a mixture of both. These form on tiny condensation nuclei, which are normally minute specks of sea salt or dust.

CLOUD FORMS

Anyone who has looked up at the sky knows that clouds come in many shapes and sizes (Figure 4.12). They range from the small, white, puffy clouds often seen in summer to the gray layers that signal a rainy day. Meteorologists name clouds by their vertical structure and the altitudes at which they occur. *Stratiform* clouds are blanketlike and

4.11 Cloud condensation nuclei

Cloud drops condense on small particulates called cloud condensation nuclei. Breaking or spilling waves in the open ocean, shown here from the deck of a ship, are an important source of condensation nuclei.

cover large areas. A common type is *stratus*, a low cloud layer that covers the entire sky. Dense, thick stratus clouds can produce large amounts of rain or snow. Higher stratus clouds are referred to as *altostratus*. *Cirrus* clouds are high, thin clouds that often have a wispy or patchy appearance. When they cover the sky evenly, they form *cirrostratus*.

> There are two major classes of clouds: stratiform (layered) and cumuliform (globular). Cumulonimbus clouds are dense, tall clouds that produce rain or thundershowers.

Cumuliform clouds are clouds with vertical development. The most common cloud of this type is the *cumulus* cloud, which is a globular cloud mass associated with small to large parcels of rising air starting near the surface. There are also *altocumulus*, individual, rounded clouds in the middle layers of the troposphere, and *cirrocumulus*, cloud rolls or ripples in the upper portions of the troposphere. *Nimbus* clouds produce precipitation. Thus, *nimbostratus* is a thick, flat, rain cloud, and *cumulonimbus* is a cumulus rain cloud.

FOG

Fog is simply a cloud layer at or very close to the Earth's surface. For centuries, fog at sea has been a navigational hazard, increasing the danger of ship collisions and

4.12 Cloud gallery

Clouds come in many shapes and sizes. Some common cloud types are shown below.

▼ **CIRRUS**
High, thin, wispy clouds drawn out into streaks are cirrus clouds. Composed of ice crystals, they form when moisture is present high in the air.

John Eastcott and Yva Momatiuk/NG Image Collection

▼ **LENTICULAR CLOUD**
A lenticular, or lens-shaped, cloud forms as moist air flows up and over a mountain peak or range.

Carsten Peter/NG ImageCollection

Classification of clouds according to height and form

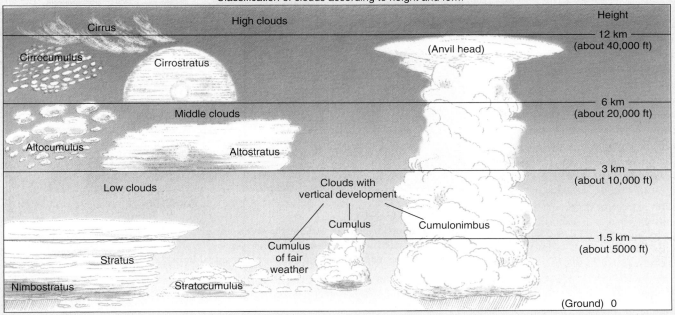

High clouds

Cirrus

Cirrocumulus

Cirrostratus

(Anvil head)

Height

12 km (about 40,000 ft)

Middle clouds

Altocumulus

Altostratus

6 km (about 20,000 ft)

Low clouds

Clouds with vertical development

3 km (about 10,000 ft)

Cumulus

Cumulonimbus

Stratus

1.5 km (about 5000 ft)

Cumulus of fair weather

Nimbostratus

Stratocumulus

(Ground) 0

▲ **CLOUD FAMILIES AND TYPES**
Clouds are grouped into families on the basis of height and vertical development. Individual cloud types are named according to their form.

GORDON WILTSIE/NG Image Collection

▲ **ALTOCUMULUS**
High cumulus clouds grade into a patch of altostratus clouds on the right.

John Eastcott and YvaMomatiuk/NG Image Collection

▲ **CUMULUS**
Puffy, fair-weather cumulus clouds fill the sky above a prairie.

Observing Clouds from GOES

Some of the most familiar images of Earth acquired by satellite instruments are those of the *Geostationary Operational Environmental Satellite* (GOES) system. Images from the GOES series of satellites have been in constant use since 1974. The primary mission of the GOES series is to view cloud patterns and track weather systems by capturing frequent images of the Earth from a consistent viewpoint in space, provided by its geostationary orbit (Figure 4.13).

Altogether, 12 GOES satellites in three major generations have been placed in orbit since 1975. They have provided a nearly continuous stream of images from two points on the Equator that bracket North America at 75° W lat. and 135° W lat. From 135° W lat. (GOES-West), Pacific storm systems can be tracked as they approach the continent and move across the western states and provinces (Figure 4.14). From 75° W lat. (GOES-East), weather systems are observed in the eastern part of the continent as they move in from the west. GOES-East also observes the tropical Atlantic, allowing identification of tropical storms and hurricanes as they form and move eastward toward the Caribbean Sea and the Southeast United States.

Courtesy NASA. Image produced by M. Jentoft-Nilsen, F. Hasler, D. Chesters, and T. Neilsen.

4.13   Earth from GOES-8

In this near-noon image from GOES-8, the entire side of the globe nearest to the satellite is illuminated. Vegetated areas appear green, while semiarid and desert landscapes appear yellow-brown. Clouds are white, and ocean waters are lavender-blue.

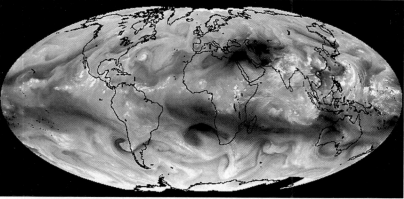

University of Wisconsin - Madison, Space Science and Engineering Center

4.14   Water vapor composite image

Areas of high atmospheric water vapor content are bright in this global image, prepared by merging data from GOES and other geostationary satellites. The brightest areas show regions of active precipitation, while dark areas show low water vapor content. The bright patch in the Atlantic off the Carolina coast is Hurricane Alberto.

groundings. In our industrialized world, fog can be a major environmental hazard. Dense fog on high-speed highways can cause chain-reaction accidents, sometimes involving dozens of vehicles. When airplane flights are shut down or delayed by fog, it is inconvenient to passengers and costs airlines money. Polluted fogs, like London's "pea-soupers" in the early part of the twentieth century, can cause damage to urban dwellers' lungs and, ultimately, take a heavy toll in lives.

One type of fog, known as *radiation fog*, forms at night when the temperature of the air layer at the ground level falls below the dew point. This kind of fog forms in valleys and low-lying areas, particularly on clear winter nights when radiative cooling is very strong.

Another fog type—*advection fog*—results when a warm, moist air layer moves over a cold surface. As the warm air layer loses heat to the surface, its temperature drops below the dew point, and condensation sets in. Advection fog commonly occurs over oceans where warm and cold currents occur side by side. When warm, moist air above the warm current moves over the cold current, condensation occurs. Fogs form in this way off the Grand Banks of Newfoundland, where the cold Labrador current comes in contact with the warmer waters of the Gulf Stream.

Advection fog is also frequently found along the California coast, as seen in Figure 4.15. It forms within a cool marine air layer that is in direct contact with the

ALASKA STOCK IMAGES/NG Image Collection

4.15 Fog

Coastal regions, such as Ketchikan, Alaska, often experience advection fog as moist maritime air moves onshore after crossing a current of cold water.

EYE ON THE LANDSCAPE What else would the geographer see?
From this airborne view, the landscape looks "drowned"—that is, the ocean seems to have filled valleys that extend under the present ocean surface Ⓐ. The valleys were widened and deepened by glaciers during the Ice Age, when sea level was about 120 m (390 ft) lower. Flooded glacial valleys like this are called fiords (see Chapter 17). Ⓑ The strong vertical grain of the vegetation is produced by the tall, thin crowns of conifers. This is the distinctive coastal needleleaf forest of the Pacific Northwest (Chapter 9).

colder water of the California current, and is frequently carried ashore by westerly winds. Similar fogs are also found on continental west coasts in the tropical latitude zones, where cool, Equatorward currents lie parallel to the shoreline.

Precipitation

Precipitation provides the freshwater essential for terrestrial life forms. Precipitation includes not only rain but also snow, sleet, freezing rain, and ice. Annual rates of precipitation vary greatly around the world, as we will see in Chapter 7. Depending on the circumstances, precipitation can be a welcome relief, a minor annoyance, or a life-threatening hazard.

FORMATION OF PRECIPITATION

Clouds are the source of precipitation—the process that provides the freshwater essential for most forms of terrestrial life. Precipitation can form in two ways. In warm clouds, fine water droplets condense, collide, and coalesce into larger and larger droplets that can fall as rain. In colder clouds, ice crystals form and grow in a cloud that contains a mixture of both ice crystals and water droplets.

The first process occurs when saturated air rises rapidly, and cooling forces additional condensation, as shown in Figure 4.16. For raindrops in a warm cloud, the updraft of rising air first lifts tiny suspended cloud droplets upward. By collisions with other droplets, some grow in volume. These larger droplets then collide with other small droplets and continue to grow. Note that a droplet is kept aloft by the force of the updraft on its surface, and as the volume of each droplet increases, so does its weight. Eventually, the downward gravitational force on the drop exceeds the upward force, and the drop begins to fall. Now moving in the opposite direction to the fine cloud droplets, the drop sweeps them up and continues to grow. Collisions with smaller droplets can also split drops, creating more drops that can continue to grow in volume. Eventually, the drop falls out of the cloud, cutting off its source of growth. On its way to the Earth, it may suffer evaporation and decrease in size, or even disappear.

Within cold clouds snow is formed in a different way, known as the *Bergeron process* (Figure 4.17). Cold clouds

> Raindrops form in warm clouds by collision and coalescence. Snow forms in cold clouds as water droplets evaporate and are deposited as ice crystals.

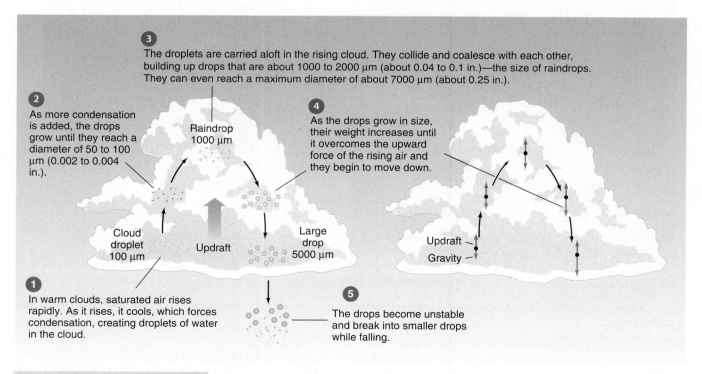

3 The droplets are carried aloft in the rising cloud. They collide and coalesce with each other, building up drops that are about 1000 to 2000 μm (about 0.04 to 0.1 in.)—the size of raindrops. They can even reach a maximum diameter of about 7000 μm (about 0.25 in.).

2 As more condensation is added, the drops grow until they reach a diameter of 50 to 100 μm (0.002 to 0.004 in.).

Raindrop 1000 μm

4 As the drops grow in size, their weight increases until it overcomes the upward force of the rising air and they begin to move down.

Cloud droplet 100 μm

Updraft

Large drop 5000 μm

Updraft

Gravity

1 In warm clouds, saturated air rises rapidly. As it rises, it cools, which forces condensation, creating droplets of water in the cloud.

5 The drops become unstable and break into smaller drops while falling.

4.16 Rain formation in warm clouds

This type of precipitation formation occurs in convectional precipitation in warm clouds typical of the equatorial and tropical zones.

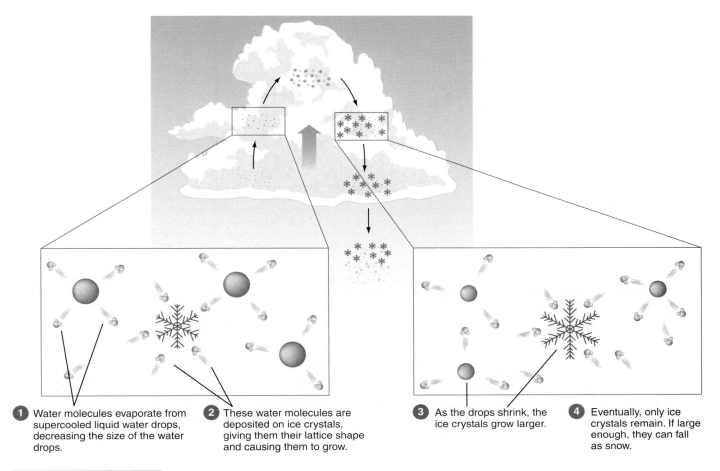

1. Water molecules evaporate from supercooled liquid water drops, decreasing the size of the water drops.

2. These water molecules are deposited on ice crystals, giving them their lattice shape and causing them to grow.

3. As the drops shrink, the ice crystals grow larger.

4. Eventually, only ice crystals remain. If large enough, they can fall as snow.

4.17 The Bergeron process

In cold clouds, precipitation forms as water vapor evaporates from supercooled liquid cloud drops. The water vapor is then deposited on ice crystals, forming snowflakes.

are a mixture of ice crystals and supercooled water droplets. The ice crystals take up water vapor and grow by deposition. At the same time, the supercooled water droplets lose water vapor by evaporation, and shrink. In addition, when an ice crystal collides with a droplet of supercooled water, it freezes the droplet. The ice crystals then coalesce to form ice particles, which can become heavy enough to fall from the cloud.

PRECIPITATION PROCESSES

Air that is moving upward is chilled by the adiabatic process, which leads, eventually, to precipitation. However, one key piece of the precipitation puzzle is still missing: What causes air to move upward in the first place?

Air can move upward in four ways. In this chapter, we discuss the first two: *orographic precipitation* and *convective precipitation*. A third way for air to be forced

> Four types of precipitation are orographic, convective, cyclonic, and convergent.

upward is through the movement of air masses and their interaction with one another. This process occurs in large, spiraling circulations of air, known as cyclones, leading to *cyclonic precipitation*. The fourth way is by *convergence*, in which air currents converge together at a location from different directions, forcing air at the surface upward. We will explore these last two types of motion in Chapters 5 and 6.

OROGRAPHIC PRECIPITATION

Orographic precipitation occurs when a current of moist air flows upward and over a mountainous barrier. The term *orographic* means "related to mountains." To understand the orographic precipitation process, think of what happens to a mass of air moving up and over a mountain range (Figure 4.18). As the

> In orographic precipitation, moist air is forced up and over a mountain barrier, producing cooling, condensation, and precipitation.

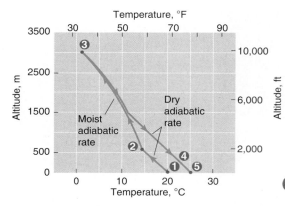

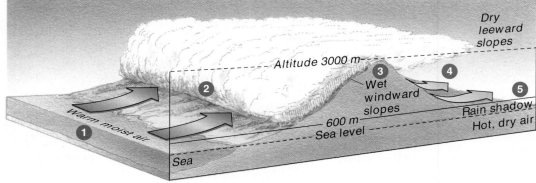

③ After passing over the mountain summit, the air begins to descend down the leeward slopes of the range. As it descends, it is compressed and so, according to the adiabatic principle, it gets warmer. This causes the water droplets and the ice crystals in the cloud to evaporate or sublimate. Eventually the air clears, and it continues to descend, warming at the dry adiabatic rate.

④ At the base of the mountain on the far side, the air is now warmer. It is also drier because much of its moisture has been removed by the precipitation. This creates a rain shadow on the far side of the mountain.

② When the air has cooled sufficiently, water droplets begin to condense, and clouds will start to form. The cloud cools at the moist adiabatic rate until, eventually, precipitation begins. Precipitation continues to fall as air moves up the slope.

⑤ This warm, dry air creates a belt of dry climate extending down the leeward slope and beyond. Several of the Earth's great deserts are formed by rain shadows.

① Air passing over a large ocean surface becomes warm and moist by the time it arrives at the coast. As the air rises on the windward side of the range, it is cooled by the adiabatic process, and its temperature drops according to the dry adiabatic rate.

4.18 Orographic precipitation

Orographic precipitation occurs when warm, moist air flows up and over a mountain barrier.

moist air is lifted, it is cooled, and condensation and rainfall occur. Passing over the mountain summit, the air descends the leeward slopes of the range, where it is compressed and warmed. Because the air is much warmer and much drier than when it started, little precipitation occurs in these regions, producing a **rain shadow** on the far side of the mountain.

California's rainfall patterns provide an excellent example of orographic precipitation and the rain shadow effect. Figure 4.19 contains maps of California's mean annual precipitation that use lines of equal precipitation called *isohyets*. These lines clearly show the orographic effect on air moving across the mountains of California into America's great interior desert zone, which extends from eastern California and across Nevada.

WileyPLUS Orographic Precipitation

View this animation to see how moist air flows over a mountain barrier, creating precipitation and a rain shadow.

CONVECTIVE PRECIPITATION

Air can also be forced upward through convection, leading to **convective precipitation**. In this process, strong updrafts occur within convection cells, vertical columns of rising air that are often found above warm land surfaces. Air rises in a convection cell because it is warmer, and therefore less dense, than the surrounding air.

The convection process begins when a surface is heated unequally. Think of an agricultural field surrounded by a forest, for example. The field surface is largely made up of bare soil with only a low layer of vegetation, so under steady sunshine the field will be

> In convective precipitation, moist air is warmed at the surface, expands, becomes less dense than surrounding, cooler air, and is buoyed upward. At the lifting condensation level, clouds begin to form.

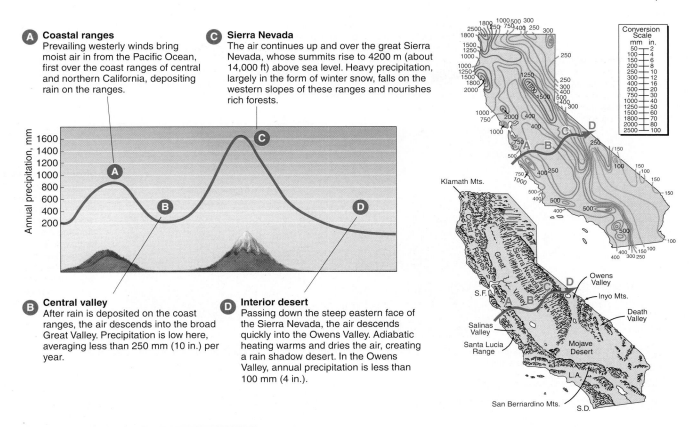

A Coastal ranges
Prevailing westerly winds bring moist air in from the Pacific Ocean, first over the coast ranges of central and northern California, depositing rain on the ranges.

C Sierra Nevada
The air continues up and over the great Sierra Nevada, whose summits rise to 4200 m (about 14,000 ft) above sea level. Heavy precipitation, largely in the form of winter snow, falls on the western slopes of these ranges and nourishes rich forests.

B Central valley
After rain is deposited on the coast ranges, the air descends into the broad Great Valley. Precipitation is low here, averaging less than 250 mm (10 in.) per year.

D Interior desert
Passing down the steep eastern face of the Sierra Nevada, the air descends quickly into the Owens Valley. Adiabatic heating warms and dries the air, creating a rain shadow desert. In the Owens Valley, annual precipitation is less than 100 mm (4 in.).

4.19 Orographic precipitation effects in California

Air moving across California encounters two mountain barriers and a valley between them. The maps on the right show the precipitation and topography of California.

warmer than the adjacent forest. This means that as the day progresses the air above the field will grow warmer than the air above the forest.

The density of air depends on its temperature; warm air is less dense than cooler air. The hot-air balloon operates on this principle. The balloon is open at the bottom, and in the basket below, a large gas burner forces heated air into the balloon. Because the heated air is less dense than the surrounding air, the balloon rises. The same principle will cause a bubble of air to form over the field, rise, and break free from the surface, as in Figure 4.20.

As the bubble of air rises, it is cooled adiabatically; its temperature will decrease as it rises, according to the dry or moist adiabatic lapse rate. The temperature of the surrounding air will normally decrease with altitude as well, but at the environmental lapse rate. For convection to occur, the temperature of the air bubble must be warmer than the temperature of the surrounding atmosphere, even as it rises. When the surrounding air is *stable,* the temperature of the air bubble rapidly reaches the temperature of the surrounding air, and the bubble loses its buoyancy. The uplift stops. This means that the environmental temperature must decrease more rapidly with altitude than the rising air parcel's temperature.

WileyPLUS Convective Precipitation
On this video watch as cumulus clouds form and grow into thunderheads, while you listen to an explanation of the process of convective precipitation.

UNSTABLE AIR

If, however, the air above is relatively cold, the temperature of the bubble will stay warmer and the rising motion will continue. In that situation, the environmental temperature decreases more rapidly with altitude than does the temperature of the bubble. Another way to state this relationship is that the environmental lapse rate must be greater than the dry or moist adiabatic lapse rate. Air with this characteristic is referred to as **unstable air**. Figure 4.21 shows how to determine whether the atmosphere in a given region is able to support convection and precipitation. If the environmental lapse rate is greater than the dry adiabatic rate, the air is unstable. If it is less than the moist adiabatic rate, it is *stable.* Between those two values, the air is *conditionally stable.*

> Unstable air—warm, moist, and heated by the surface—can produce abundant convective precipitation. It is typical of hot summer air masses in the central and southeastern United States.

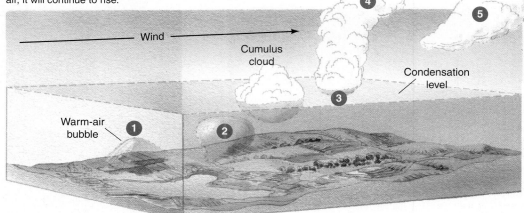

2 Adiabatic cooling As the bubble of air rises, it is cooled adiabatically and its temperature decreases as it rises. However, as long as the bubble is still warmer than the surrounding air, it will continue to rise.

3 Condensation If the bubble remains warmer than the surrounding air and uplift continues, adiabatic cooling chills the bubble to the dew point, and condensation sets in. The rising air column becomes a puffy cumulus cloud. The flat base of the cloud marks the lifting condensation level at which condensation begins.

4 Continued convection The bulging "cauliflower" top of the cloud is the top of the rising warm-air column pushing into higher levels of the atmosphere.

5 Dissipation A small cumulus cloud typically encounters winds aloft that mix it with the local air, reducing the temperature difference and slowing the uplift. After drifting some distance downwind, the cloud evaporates.

1 Surface heating Heated air is less dense than the surrounding air, causing a bubble of warm air to form over the field, then rise and break free from the surface.

4.20 Formation of a cumulus cloud

A bubble of heated air rises above the lifting condensation level to form a cumulus cloud.

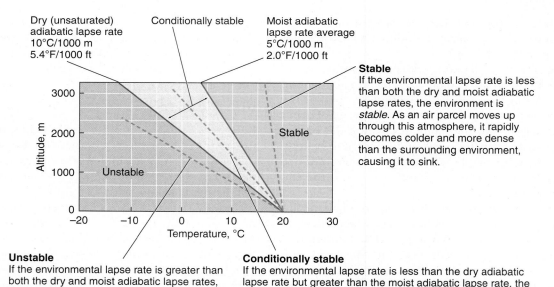

Dry (unsaturated) adiabatic lapse rate
10°C/1000 m
5.4°F/1000 ft

Conditionally stable

Moist adiabatic lapse rate average
5°C/1000 m
2.0°F/1000 ft

Stable
If the environmental lapse rate is less than both the dry and moist adiabatic lapse rates, the environment is *stable*. As an air parcel moves up through this atmosphere, it rapidly becomes colder and more dense than the surrounding environment, causing it to sink.

Unstable
If the environmental lapse rate is greater than both the dry and moist adiabatic lapse rates, the environment is *unstable*. As an air parcel moves up through the atmosphere, it is always warmer and less dense than the surrounding air, causing it to continue to rise.

Conditionally stable
If the environmental lapse rate is less than the dry adiabatic lapse rate but greater than the moist adiabatic lapse rate, the environment is *conditionally stable*. An unsaturated air parcel moving up in this environment becomes colder than the surrounding air and sinks back down. However, a saturated air parcel moving up in this same environment is warmer than the surrounding air and continues to rise.

4.21 Determining stability

Atmospheric stability depends on how air temperature changes with altitude at a particular location at a particular time.

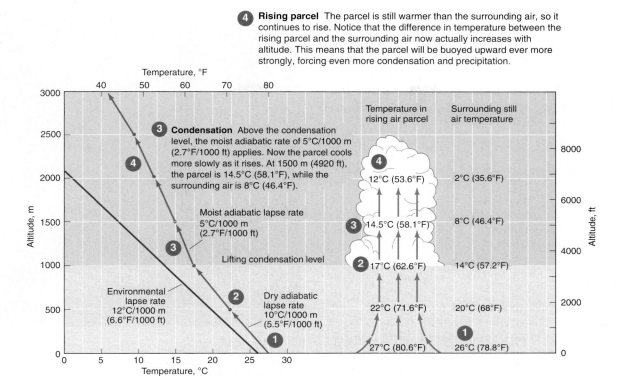

4 **Rising parcel** The parcel is still warmer than the surrounding air, so it continues to rise. Notice that the difference in temperature between the rising parcel and the surrounding air now actually increases with altitude. This means that the parcel will be buoyed upward ever more strongly, forcing even more condensation and precipitation.

3 **Condensation** Above the condensation level, the moist adiabatic rate of 5°C/1000 m (2.7°F/1000 ft) applies. Now the parcel cools more slowly as it rises. At 1500 m (4920 ft), the parcel is 14.5°C (58.1°F), while the surrounding air is 8°C (46.4°F).

Moist adiabatic lapse rate 5°C/1000 m (2.7°F/1000 ft)

Lifting condensation level

Environmental lapse rate 12°C/1000 m (6.6°F/1000 ft)

Dry adiabatic lapse rate 10°C/1000 m (5.5°F/1000 ft)

Temperature in rising air parcel | Surrounding still air temperature

12°C (53.6°F) | 2°C (35.6°F)
14.5°C (58.1°F) | 8°C (46.4°F)
17°C (62.6°F) | 14°C (57.2°F)
22°C (71.6°F) | 20°C (68°F)
27°C (80.6°F) | 26°C (78.8°F)

2 **Uplift** At first, the parcel cools at the dry adiabatic rate. At 500 m (1640 ft), the parcel is at 22°C (71.6°F), while the surrounding air is at 20°C (68°F). Since it is still warmer than the surrounding air, it continues to rise. In this example, it reaches the lifting condensation level at 1000 m (3281 ft). The temperature of the parcel is 17°C (62.6°F).

1 **Initial heating** At ground level, the surrounding air is at 26°C (78.8°F) and has a lapse rate of 12°C/1000 m (6.6°F/1000 ft.). The air parcel is heated by 1°C (1.8°F) to 27°C (80.6°F) and—because it is less dense than the surrounding air—it begins to rise.

4.22 Convection in unstable air

When the air is unstable, a parcel of air that is heated enough to rise will continue to rise to great heights.

Figure 4.22 is a diagram of convection in unstable air. At first, the air parcel rises at the dry adiabatic rate. Above the lifting condensation level, the parcel cools at the moist adiabatic rate. But because the surrounding air is always cooler than the parcel it rises high in the atmosphere, forming a tall cumulus cloud. If uplift and condensation persist, it may grow into a rain-producing cumulonimbus cloud.

The key to the convective precipitation process is the release of latent heat energy. When water vapor condenses into droplets or ice particles, it releases that energy to the surrounding air in the parcel. This heat helps keep the parcel warmer than the surrounding air, fueling the convection process and driving the parcel ever higher.

When the parcel reaches a high altitude, most of its water will have condensed. As adiabatic cooling continues, less latent heat energy will be released, so the uplift weakens. Eventually, uplift stops because the energy source, latent heat release, is gone. The cell dies and dissipates into the surrounding air.

Unstable air masses are often found over the central and southeastern United States in the summer. Summer weather patterns sweep warm, humid air from the Gulf of Mexico over this region, and the intense summer insolation strongly heats the air layer near the ground, producing a steep environmental lapse rate. Thunderstorms are very common in these moist and unstable air masses (Figure 4.23).

4.23 Cumulus clouds over South Dakota

A gathering of storm clouds forms over the Badlands region of South Dakota. Convection within these clouds gives them their vertical cumulus structure.

Annie Griffiths Belt/NG Image Collection

Unstable air is also commonly found in the equatorial and tropical zones. Here, convective showers and thundershowers are frequent. At these low latitudes, much of the orographic rainfall is in the form of heavy showers and thundershowers produced by convection. The forced ascent of unstable air up a mountain slope easily produces rapid condensation, which then triggers the convection process.

Types of Precipitation

Precipitation consists of liquid water drops and solid crystals that fall from the atmosphere and reach the ground. This precipitation can take several different forms.

RAIN

Rain is precipitation that reaches the ground as liquid water. Raindrops can form in warm clouds as liquid water, which through collisions can coalesce with other drops and grow large enough to fall to Earth. They can form through other processes as well. For example, solid ice, in the form of snow or hail, can also produce rain if it falls through a layer of warm air and melts along the way.

> Types of precipitation include rain, snow, sleet and freezing rain, and hail.

To fall from the sky and reach the ground, raindrops usually have to grow larger than 0.2 mm (0.008 in.). At these sizes, we refer to the drops as *mist* or *drizzle*. Once the drops reach 0.5 mm (0.02 in.), they are termed *raindrops*. Typically, raindrops can have a maximum size of only 5 to 8 mm (0.2 to 0.3 in.); drops any larger than that become unstable and break into smaller drops as they fall through the atmosphere.

SNOW

Snow forms as individual water vapor molecules are deposited on existing ice crystals. If these ice crystals are formed by deposition, they take the shape of snowflakes, with their characteristic intricate crystal structure. However, most particles of snow have endured collisions and coalesce with each other and with supercooled water drops. As they do so, they lose their shape and can become simple lumps of ice.

Eventually, whether they are intricate snowflakes or accumulations of ice and supercooled water drops, these ice crystals become heavy enough to fall from the cloud. If the underlying air layer is below freezing, snow produced in cold clouds reaches the ground as a solid form of precipitation; otherwise, the snow melts and arrives as rain.

Ted S. Warren/©AP/Wide World Photos

4.24 Ice storm

When rain falls into a surface layer of below-freezing air, clear ice coats the ground. Driving is particularly hazardous. Shown here is an ice storm that struck the Pacific Northwest in January of 2012. The ice fell into an accumulation of wet snow, adding additional weight to branches and power lines that were already burdened. Many trees were downed and power outages were common.

SLEET AND FREEZING RAIN

When snow falls into a warm air layer at the ground, the melting ice particles are referred to as *sleet*. Perhaps you have experienced an *ice storm*. Ice storms occur when the ground is frozen and the temperature of the lowest air layer is also below freezing. Actually, ice storms are more accurately named "icing" storms because it is not ice that falls but *freezing rain*—supercooled drops that freeze on contact. The freezing rain creates a clear, slippery glaze on roads and sidewalks, making them extremely hazardous. Ice storms cause great damage, especially to telephone and power lines and to tree limbs, which are pulled down by the weight of the ice (Figure 4.24).

HAIL

Hailstones are formed by the accumulation of ice layers on ice pellets that are suspended in the strong updrafts of thunderstorms (Figure 4.25). As these ice pellets—called *graupel*—move through subfreezing regions of the atmosphere, they come into contact with supercooled liquid water droplets, which subsequently freeze to the pellets in a thin sheet. This process, called *accretion*, results in a buildup of concentric layers of ice around each pellet, giving it its typical ball-like shape.

4.25 Hail

Hail is precipitation in the form of frozen pellets or balls of ice.

2 Growth
Kept aloft by updrafts, the ice pellets continue to grow, forming hailstones.

1 Accretion
As small ice pellets, called graupel, move through below-freezing portions of the cloud, they accumulate layers of ice around them.

Updrafts

0°C

◄ **HAIL FORMATION**
Hail forms in thunderstorms with strong updrafts and cold air aloft.

3 Precipitation
Eventually, the hailstones become too big and heavy to be kept aloft by updrafts and hail falls to the Earth.

▼ **HAILSTORM IN PROGRESS**
The white streaks in this photo are marble-sized hailstones, falling into the Henrys Fork River, Idaho, and accumulating on the riverbank in the foreground.

© Nuridsany et Perennou/Photo Researchers, Inc.

▲ **HAILSTONES**
Fresh hailstones, up to about 1 cm (0.4 in.) in diameter.

Peter M. Miller/PhotoResearchers, Inc.

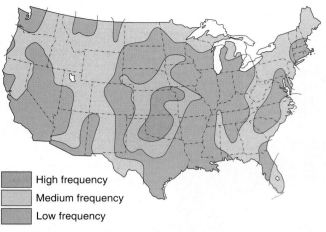

■ High frequency
■ Medium frequency
■ Low frequency

4.26 Frequency of severe hailstorms

As shown in this map of the 48 contiguous United States, severe hailstorms are most frequent in the Midwestern plains states of Oklahoma and Kansas. A severe hailstorm is defined as a local convective storm producing hailstones equal to or greater than 1.9 cm (0.75 in.) in diameter.

With each new layer, the ball of ice—now called *hail*—gets larger and heavier. When it becomes too heavy for the updraft to support, it falls to Earth. When the updrafts are extremely strong, the hail remains aloft, slowly accumulating more mass and getting larger. In that case, hailstones can reach diameters of 3 to 5 cm (1.2 to 2.0 in.).

Hail can occur almost anywhere that severe thunderstorms occur, but serious hailstorms are most frequent in Kansas and Oklahoma (Figure 4.26).

MEASURING PRECIPITATION

Precipitation is recorded as a depth that falls during a certain time—for example, as millimeters or inches per hour or per day. A millimeter of rainfall would cover the ground to a depth of 1 mm if the water did not run off or sink into the soil.

Rainfall is measured with a *rain gauge* (Figure 4.27). This simple meteorological instrument is constructed from a narrow cylinder with a wide funnel at the top.

4.27 Rain gauges

Several types of devices are commonly used to measure rainfall.

RAIN GAUGE TEST ▶
Shown here are five different types of rain gauges installed in Arvada, Colorado. By placing the gauges next to one another, it is possible to compare how their precipitation measurements differ.

PORTABLE RAIN GAUGE ▶
This rain gauge consists of two clear plastic cylinders, here partly filled with red-tinted water. The inner cylinder receives rainwater from the funnel top. When it fills, the water overflows into the larger, outer cylinder.

Arthur N. Strahler

Courtesy USGS, Branch of Quality Systems

The funnel gathers rain from a wider area than the mouth of the cylinder, so the cylinder fills more quickly. The water level gives the amount of precipitation, which is read on a graduated scale.

For meteorological records, snowfall is measured by the amount of liquid water it yields when melted. We can also measure snowfall by depth in millimeters or inches. Ordinarily, a 10-mm (or 10-in.) layer of snow is assumed to be equivalent to 1 mm (or 1 in.) of rainfall, but this ratio may range from 30 to 1 in very loose snow to 2 to 1 in old, partly melted snow.

Precipitation in the form of water droplets or snow particles can be identified using *radar*, a technology in which a beam of radio waves is sent out from a transmitter, strikes the water particles, and then returns to a receiver. By measuring the intensity of the return beam and observing the time it takes for the beam to travel and return it is possible to determine the intensity of precipitation and its location. Images obtained by scanning weather radars are widely used and familiar to anyone who routinely watches weather forecasts on television (Figure 4.28).

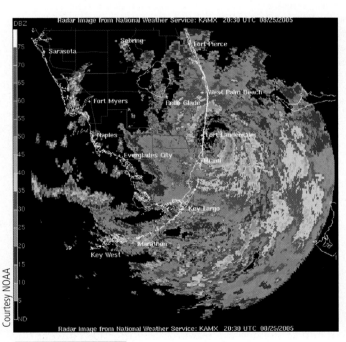

4.28 Weather radar

In this weather radar image, the eye of Hurricane Katrina is about to pass over southern Florida on August 25, 2005. Weather radars scan the horizon in a circular pattern, sending out short pulses of radio waves that are reflected back by raindrops and ice crystals. By measuring the intensity of the return pulse and the delay between the time of sending the pulse and its return, the radar creates a map of precipitation within the radar's view. The radar can also detect the height of storm clouds by broadcasting a series of scans at progressively higher angles above the horizon, creating a three-dimensional picture of precipitation within the area. Using the principle of the Doppler Effect, the weather radar can also measure cloud motion, which can be used to detect tornadoes.

Thunderstorms

A **thunderstorm** is any storm that produces thunder and lightning. At the same time, thunderstorms can also produce high winds, hail, and tornadoes. They are typically associated with cumulus clouds that indicate the presence of rising, unstable air. It is this rising motion that produces the characteristic rainfall and lightning that accompany thunderstorms. Thunderstorms can range from fairly isolated, short-lived storms, sometimes called *air-mass thunderstorms*, to massive, well-organized complexes of storms, called *mesoscale convective systems*.

> A thunderstorm is an intense local storm associated with a tall cumulonimbus cloud in which there are strong updrafts of air. Thunderstorms can produce both hail and cloud-to-ground lightning.

AIR-MASS THUNDERSTORMS

Air-mass thunderstorms are isolated thunderstorms generated by daytime heating of the land surface. They occur in warm, moist air that is often of maritime origin. Triggered by solar heating of the land, they start, mature, and dissipate within an hour or two. Formation stops at night, since surface heating is no longer present.

The typical life cycle of an air-mass thunderstorm involves three stages of development (Figure 4.29). In the *cumulus stage*, unequal surface heating causes air parcels to rise. Isolated cumulus clouds form as the parcels reach and pass through the lifting condensation level. At first, these clouds dissipate as they mix with the surrounding dry air, which evaporates the cloud water droplets. However, this process cools the air aloft and raises its moisture content, causing instability.

As instability increases, convection reaches greater heights and soon a thunderstorm develops. In this *mature stage*, there are both updrafts and downdrafts. The updrafts carry the cloud high into the atmosphere, where upper-level winds draw the cloud downwind to create a characteristic *anvil cloud*.

Downdrafts are created when water droplets become large enough to fall through the cloud and drag the surrounding air downward. Downdrafts can also be caused by the movement of cooler, drier surrounding air into the cloud from the upwind side. This cooler air tends to sink. It is further chilled by the evaporation of cloud water droplets, escalating the sinking motion. At ground level, the downdraft spreads forward, forcing more warm moist air up and into the cloud.

Eventually, the stabilizing effects of the movement of surrounding air into the cloud overcome the destabilizing effects of convection. Widespread downdrafts form and inhibit upward motion and suppress latent heat

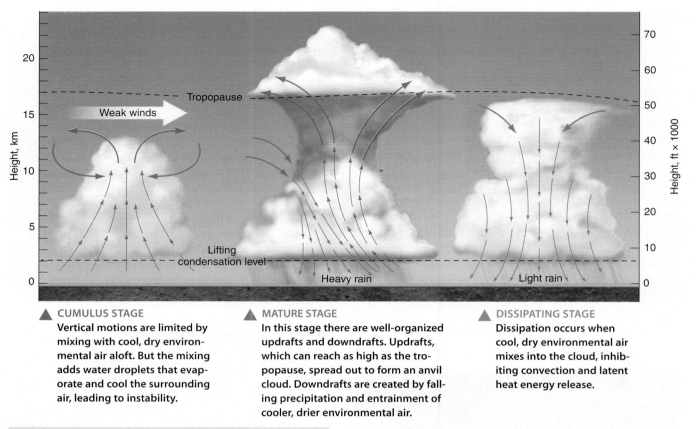

CUMULUS STAGE
Vertical motions are limited by mixing with cool, dry environmental air aloft. But the mixing adds water droplets that evaporate and cool the surrounding air, leading to instability.

MATURE STAGE
In this stage there are well-organized updrafts and downdrafts. Updrafts, which can reach as high as the tropopause, spread out to form an anvil cloud. Downdrafts are created by falling precipitation and entrainment of cooler, drier environmental air.

DISSIPATING STAGE
Dissipation occurs when cool, dry environmental air mixes into the cloud, inhibiting convection and latent heat energy release.

4.29 Stages in the development of an air-mass thunderstorm

There are three development stages of an air-mass thunderstorm: the cumulus, mature, and dissipating. Each stage has characteristic vertical winds and precipitation.

flow, cutting off the power source of the storm. This is the *dissipating stage*.

Air-mass thunderstorms are initiated by surface heating during the day. They start, mature, and dissipate within an hour or two. Formation stops at night, when surface heating is no longer present.

LIGHTNING

Lightning—a giant electric arc passing through the atmosphere—is also generated by convection cell activity. It occurs when updrafts and downdrafts cause positive and negative static charges to build up within different regions of the cloud (Figure 4.30). The exact mechanism is not completely understood, but generally, negative charges are carried downward, and positive charges are carried upward. When the separation of charges reaches a threshold, the molecules and atoms of the atmosphere become conductive, and electric current flows along a narrow path. The atmosphere is heated explosively to produce light and generate a shock wave in the air, which we hear as thunder. Most lightning occurs within the storm cloud. Only about 20 percent of lightning strikes connect to the Earth's surface.

SEVERE THUNDERSTORMS

Severe thunderstorms persist longer than air-mass thunderstorms and have higher winds. They often produce hail or even tornadoes. Although they may start as simple air-mass thunderstorms, they reach a mature stage and then intensify rather than dissipate. In the severe thunderstorm, large amounts of cooler, drier environmental air enter the cloud from the upwind side, creating a strong downdraft (Figure 4.31). As the downdraft spreads out in front of the storm, it creates a *gust front*. The advancing air pushes large volumes of moist surface air upward, feeding the convection. A distinctive *roll cloud* can form that is visible as the storm approaches.

An essential component of the severe thunderstorm is *wind shear*, a change in wind velocity with height that keeps cool, dry air entering from the upwind side while the warm, moist, rising air stays on the downwind side of the cell.

Under unusual circumstances involving a temperature inversion in an upper atmospheric layer, severe thunderstorms can become *supercell thunderstorms*, massive thunderstorms with a single circulation cell comprising very strong updrafts and downdrafts. Because

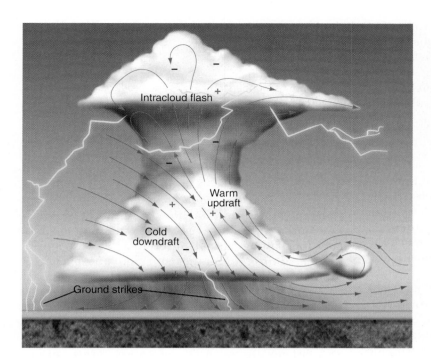

Turbulent upward and downward motions (blue and red arrows) of air, water, and ice pellets in a thunderstorm create areas of positive and negative static charge within the cloud that are discharged by lightning. Most lightning occurs within clouds. In a ground strike, by attraction, a negative charge in the cloud above generates a positive charge in the ground below. When the electrical potential between the cloud and the ground is large enough, a lightning strike occurs.

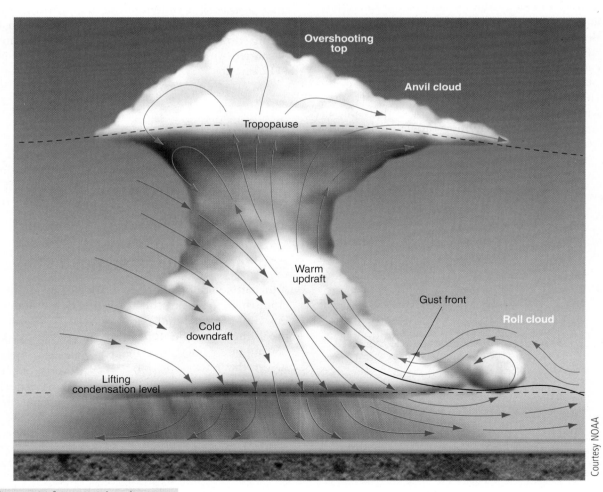

4.31 Anatomy of a severe thunderstorm

A severe thunderstorm can maintain convection and precipitation for long periods of time if it can continuously incorporate warm, moist air. Gust fronts associated with downdrafts extending ahead of the storm system force warm, moist surface air aloft and into the advancing storm.

of their vertical extent, which can be up to 25 km (16 mi), they are affected differently by winds at the surface and winds aloft. If the background wind shear not only involves a change in wind speed with height but also a change in direction—typically in a counterclockwise direction—a rotation of the storm can occur, which is a precursor to the formation of tornadoes.

MICROBURSTS

Another characteristic feature of many severe thunderstorms is the formation of a *microburst,* an intense downdraft or downburst that accompanies the gust front (Figure 4.32). Once the microburst hits the ground, it flows outward in all directions, producing intense, localized winds called *straight-line winds.* A microburst is also often, but not always, accompanied by rain.

The microburst itself can be so intense that it is capable of causing low-flying aircraft to crash. An aircraft flying through the microburst first encounters strong headwinds, which may cause a bumpy ride but does not interfere with the airplane's ability to fly. However, as the airplane passes through the far side of the microburst, it encounters a strong tailwind. The lift of the airplane's wings depends on the speed of the air flowing across them, and the tailwind greatly reduces the air speed, which causes a loss of lift. If the tailwind is strong enough, the airplane cannot hold its altitude and may crash.

Microbursts can be detected by special radar instruments, now installed at many airports, that measure horizontal wind speeds. Training procedures for pilots have also reduced the incidence of aviation accidents in the United States attributed to microbursts and associated wind shear.

WileyPLUS Formation of Thunderstorms
See how a thunderstorm grows and produces hail, lightning, and microbursts.

MESOSCALE CONVECTIVE SYSTEMS

Large, organized masses of severe and supercell thunderstorms sometimes occur under unusual conditions. These are called *mesoscale convective systems.* In one situation, upper-air wind flow patterns cause air to flow into a region and rise. The rising motion stimulates long-lasting clusters of severe and slowly moving thunderstorms. In another situation, the change in wind direction with height caused by the approach of a cold front causes air to rise along a line ahead of the front. The resulting *squall line* of thunderstorms includes storms in different stages of development (Figure 4.33).

A fast-moving squall line of severe storms can produce a *derecho,* a violent, straight-line windstorm that precedes the squall line. Winds are generated by the strong downdrafts of the approaching storm cells. Sustained winds in extreme cases can exceed 50 m/s (112 mi/hr).

Tornadoes

A tornado is a small but intense vortex in which air spirals at tremendous speed. It is associated with thunderstorms spawned by fronts in the midlatitudes of North America. Tornadoes can also occur inside tropical cyclones (hurricanes).

TORNADO CHARACTERISTICS

A **tornado,** seen in Figure 4.34, appears as a dark funnel cloud hanging from the base of a dense cumulonimbus cloud. At its lower end, the funnel may be 100 to 450 m (about 300 to 1500 ft) in diameter. The base of the funnel appears dark because of the density of condensing moisture, dust, and debris swept up by the wind. Wind speeds in a tornado exceed those known

> A tornado is a small but intense spiraling vortex of rising air associated with the strong updraft of an intense thunderstorm.

in any other storm. Estimates of wind speed run as high as 100 m/s (about 225 mph), although generally they are closer to 50 m/s (about 110 mph). As the tornado moves across the countryside, the funnel writhes and

4.32 Anatomy of a microburst

Passing through a microburst, an airplane first experiences a strong headwind, then a strong tailwind that can cause the aircraft to lose lift and crash.

Radar Image from National Weather Service: KBMX   09:51 UTC 04/30/2005

DBZ

+ Lawrenceburg

+ Chattanooga

+ Florence

+ Huntsville

+ Dalton

+ Tupelo

+ Rome

+ Gadsden

Birmingham

+ Columbus

+ Tuscaloosa

+ La Grange

+ Demopolis

+ Columbus

+ Meridian

Montgomery

+ Troy

+ Monroeville

+ Dothan

Radar Image from National Weather Service: KBMX   09:51 UTC 04/30/2005

Courtesy NOAA

4.33 Squall line

This radar image shows convective precipitation in reds and yellows associated with a line of thunderstorms extending across the state of Alabama on April 30, 2005. Gust fronts associated with these storms produced surface winds in excess of 27 m/s (60 mi/hr), knocking down trees and power lines.

4.34 Tornado

A tornado is a rapidly spinning funnel of air that touches the ground. It descends from the cloud base of a severe thunderstorm.

Richard Olsenius/NG Image Collection

twists. Where it touches the ground, it can cause complete destruction of almost anything in its path.

The center of a tornado is characterized by low pressure, which is typically 10 to 15 percent lower than the surrounding air pressure. The result is a very strong-pressure gradient force that generates high wind speeds as the air rushes into the low-pressure center of the tornado. Most tornadoes rotate in a counterclockwise direction, but a few rotate the opposite way.

TORNADO DEVELOPMENT

Tornadoes are usually associated with the presence of severe thunderstorms, which provide one of the key ingredients in the initial development of the tornado—namely, a very strong vertical circulation. The other key ingredient is the presence of significant change in wind speed and direction with height, termed *wind shear* (Figure 4.35). In regions where there is significant wind shear, spinning circulations aligned with the ground—*horizontal vortexes*—can form. Strong convection can then lift portions of the vortex, which results in a vertical tower of slowly rotating air, called a *mesocyclone.*

Initially, the mesocyclone is fairly broad. However, as the storm's vertical convection extends the top of

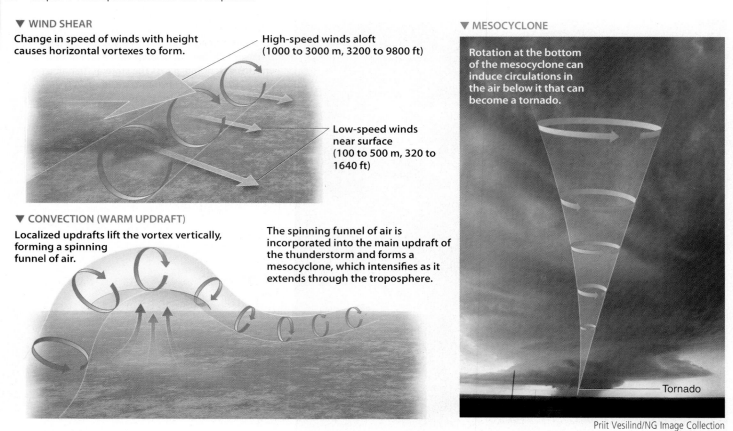

▼ WIND SHEAR

Change in speed of winds with height causes horizontal vortexes to form.

High-speed winds aloft (1000 to 3000 m, 3200 to 9800 ft)

Low-speed winds near surface (100 to 500 m, 320 to 1640 ft)

▼ CONVECTION (WARM UPDRAFT)

Localized updrafts lift the vortex vertically, forming a spinning funnel of air.

The spinning funnel of air is incorporated into the main updraft of the thunderstorm and forms a mesocyclone, which intensifies as it extends through the troposphere.

▼ MESOCYCLONE

Rotation at the bottom of the mesocyclone can induce circulations in the air below it that can become a tornado.

Tornado

Priit Vesilind/NG Image Collection

4.35 Formation of mesocyclones and tornadoes

The combination of vertical wind shear with very strong convection can produce mesocyclones that extend through the troposphere. Rapidly rotating circulations extending from the bottom of the mesocyclones to the surface can then develop into tornadoes.

the mesocyclone high in the atmosphere, the mesocyclone stretches and narrows. As it does so, it begins to spin faster, much as ice skaters spin faster as they pull their arms close to their bodies. The winds associated with the mesocyclone begin to cause the air below it to spin as well. About a fifth of the time, a narrow, rapidly circulating vortex stretches from the base of the mesocyclone down to the ground. When it touches the ground, it officially becomes a tornado.

TORNADO DESTRUCTION

The large majority of tornadoes are relatively weak, lasting only a few minutes and covering only a few hundred meters. On the other hand, large tornadoes, which make up only about 5 percent of all tornadoes, can last for hours and spread destruction over hundreds of kilometers. Devastation from these tornadoes is often complete within the narrow limits of their paths. Only the strongest buildings constructed of concrete and steel can withstand the extremely violent winds.

The most commonly used measure of tornado intensity is the *Enhanced Fujita intensity scale* (Figure 4.36). This scale ranks tornadoes from 0 to 5, weakest to strongest. It is based on the severity of damage found in the tornado's wake by examining 28 types of structures, with up to 12 different "degree of damage" ratings for each. This scale allows scientists to compare tornadoes that have passed through very different regions (for example, industrial and rural areas), enabling them to better estimate the wind speed within each.

Only about 5 percent of tornadoes reach 4 or 5 on the enhanced Fujita intensity scale. However, these tornadoes are responsible for about 70 percent of all tornado-related deaths. On average, about 75 people in the United States die each year from tornadoes, more than from any other natural phenomena except flooding and lightning strikes.

WileyPLUS How Tornados Cause Damage
See how supercell thunderstorms develop and explore how the Enhanced Fujita scale estimates wind speed from hurricane damage.

EF0. Light damage 29–38 m/s (65–85 mi/hr)	Chimneys are damaged, tree branches broken, shallow-rooted trees toppled.

AFP PHOTO/YASSER AL-ZAYYAT/Getty Images, Inc.

EF1. Moderate damage 39–49 m/s (86–110 mi/hr)	Roof surfaces peeled off, some tree trunks broken, garages and trailer homes destroyed.

Peter Carsten/NG Image Collection

EF2. Considerable damage 50–60 m/s (111–135 mi/hr)	Roofs damaged, trailer homes and outbuildings destroyed, damage from airborne debris, large trees uprooted or snapped.

Priit Vesilind/NG Image Collection

EF3. Severe damage 61–73 m/s (136–165 mi/hr)	Roofs and walls torn from structures, small buildings destroyed, non-reinforced masonry buildings destroyed, forest uprooted.

John McCombe/Getty Images, Inc.

EF4. Devastating damage 74–89 m/s (166–200 mi/hr)	Homes destroyed, some lifted distance from foundation, cars blown some distance, debris includes large objects.

Win Henderson/FEMA

EF5. Incredible damage >90 m/s (>200 mi/hr)	Strong homes lifted from foundations, concrete structures damaged, damage from debris the size of automobiles, trees debarked.

Greg Henshal/FEMA

4.36 Enhanced Fujita intensity scale of tornado damage

The enhanced Fujita tornado intensity scale (EF0–EF5) rates the severity of tornadoes, based on the nature and amount of damage done to structures.

Air Quality

Most people living in or near urban areas have experienced air pollution first-hand. Perhaps you've felt your eyes sting or your throat tickle as you drive an urban freeway. Or you've noticed black dust on window sills or window screens and realized that you are breathing in that dust as well.

AIR POLLUTANTS

Air pollution is largely the result of human activity. An **air pollutant** is an unwanted substance injected into the atmosphere from the Earth's surface

> Air pollutants are unwanted substances present in the air. They arise from both human and natural activity.

by either natural or human activities. Air pollutants come as aerosols, gases, and particulates. In earlier chapters we've been introduced to aerosols—small bits of matter in the air, so small that they float freely with normal air movements—and gases. Particulates are larger, heavier particles that sooner or later fall back to Earth.

Most pollutants are generated by the everyday activities of large numbers of people, for example, driving cars, or through industrial activities, such as fossil fuel combustion or the smelting of mineral ores to produce metals. The most common air pollutant is carbon monoxide, followed by particulate matter, volatile organic compounds, nitrogen oxides, sulfur dioxide, and ammonia (Figure 4.37). Volatile organic compounds include evaporated gasoline, dry-cleaning fluids, and incompletely combusted fossil fuels. The buildup of these substances in the air can lead to many types of pollution, including acid deposition and smog.

FALLOUT AND WASHOUT

Pollutants generated by a combustion process are at first carried aloft by convection. However, the larger particulates soon settle under the pull of gravity and return to the surface as *fallout*. Particles too small to settle out are later swept down to Earth by precipitation in a process called *washout*. Through a combination of fallout and washout, the atmosphere tends to be cleaned of pollutants.

Pollutants are also eliminated from the air over their source areas by wind. Strong, through-flowing winds will disperse pollutants into large volumes of cleaner air in the downwind direction. Strong winds can quickly sweep away most pollutants from an urban area; but during periods when winds are light or absent, the concentrations can rise to high values.

SMOG AND HAZE

When aerosols and gaseous pollutants are present in considerable density over an urban area, the resultant mixture is known as **smog** (Figure 4.38). Typically, hazy sunlight can reach the ground through smog, but it can be dense enough to hide aircraft flying overhead from view. Smog also irritates the eyes and throat, and can corrode structures over long periods of time.

Modern urban smog has three main toxic ingredients: nitrogen oxides, volatile organic compounds, and ozone. Nitrogen oxides and volatile organic compounds are largely automobile pollutants. Ozone forms through a photochemical reaction in the air by which nitrogen oxides react with volatile organic compounds in the presence of sunlight. Ozone is harmful to human lungs, and it aggravates bronchitis, emphysema, and asthma.

Haze is a condition of the atmosphere in which aerosols obscure distant objects. Haze builds up naturally in stationary air as a result of human and natural activity. When the air is humid and abundant water vapor is available, water films grow on suspended nuclei. This

Volatile organic compounds (11%)

Sulfur dioxide (8%)

Particulate matter (14%)

Carbon monoxide (54%)

Nitrogen oxides (10%)

Ammonia (3%)

◀ **AIR POLLUTANT EMISSIONS**
Carbon monoxide is the most common air pollutant, accounting for more than half of all emissions. Nitrogen oxides, particulate matter, sulfur dioxide, and volatile organic compounds make up the rest. Ammonia constitutes only a small fraction.

AIR POLLUTANT SOURCES ▶
About half of pollutants by weight are emitted by transportation, mainly in the form of carbon monoxide and volatile organic compounds. Miscellaneous sources make up about a quarter of the emissions, while fuel combustion and industrial processes divide the remainder.

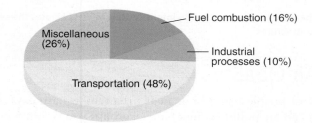

Fuel combustion (16%)

Miscellaneous (26%)

Industrial processes (10%)

Transportation (48%)

4.37 Air pollutant emissions and sources by weight, 2007
Emissions and sources make different contributions to air pollution.

©AP/Wide World Photos

4.38 Smog

Smog, a form of urban fog, is shown here in Beijing, China, on a day in March, 2008. It consists of aerosol haze mixed with toxic gases and hydrocarbon compounds. It is often irritating to the throats and lungs of city dwellers and can cause respiratory distress.

results in aerosol particles large enough to obscure and scatter light, reducing visibility.

INVERSION AND SMOG

The concentration of pollutants over a source area rises to its highest levels when vertical mixing (convection) of the air is inhibited. This happens in an *inversion*, which we have seen as a condition in which the temperature of the air increases with altitude.

According to the adiabatic principle, a heated air parcel emerging from a smokestack or chimney will cool as it rises until it attains the same temperature as the surrounding air. In an inversion, however, the surrounding air gets warmer, not colder, with altitude. So the hot parcel will quickly arrive at the temperature of the surrounding air, and uplift will stop. Pollutants in the air parcel remain trapped below the inversion, keeping concentrations high near the ground.

Two types of inversions are important in causing high air pollutant concentrations: low-level and high-level inversions. When a *low-level temperature inversion* develops over an urban area with many air pollution sources, pollutants are trapped under the "inversion lid." Heavy smog or highly toxic fog can develop. London's poisonous "pea soup" fogs of the past are an example.

Another type of inversion is responsible for the smog problem experienced in the Los Angeles Basin and other California coastal regions, ranging north to San Francisco and south to San Diego. Here, special climatic conditions produce prolonged inversions and smog accumulations (Figure 4.39). Off the California coast is a persistent fair-weather system that is especially strong in the summer. This system produces a layer of warm, dry air at upper elevations. At the same time, a cold current of upwelling ocean bottom water runs along the coast, just offshore. Moist ocean air moves across this cool current and is chilled, creating a cool, marine air layer.

> Persistent inversions can trap pollutants, resulting in low air quality and smog.

The Los Angeles Basin is a low, sloping plain lying between the Pacific Ocean and a massive mountain barrier on the north and east sides. Weak winds from the south and southwest move the cool, marine air inland over the basin. Further landward movement is blocked by the mountain barrier. Since there is a warm layer above this cool marine air layer, the result is a *high-level temperature inversion*. Pollutants accumulate in the cool air layer and produce smog. The upper limit of the smog stands in sharp contrast to the clear air above it,

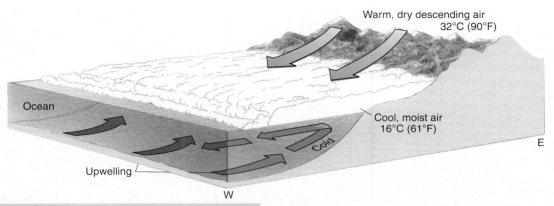

4.39 High-level temperature inversion on the California coast

A layer of warm, dry, descending air from a persistent fair-weather system rides over a cool, moist marine air layer at the surface to create a persistent temperature inversion.

filling the basin like a lake and extending into valleys in the bordering mountains (Figure 4.40).

An actual temperature inversion, in which air temperature increases with altitude, is not essential for building a high concentration of pollutants above a city. Light or calm winds and stable air are all that are required. Stable air has a temperature profile that decreases with altitude, but at a slow rate. Some convective mixing occurs in stable air, but the convective precipitation process is inhibited.

At certain times of the year, slow-moving masses of dry, stable air occupy the central and eastern portions of the North American continent. Under these conditions, a broad *pollution dome* can form over a city or region, and air quality will suffer (Figure 4.41). When there is a regional wind, the pollution from a large city will be carried downwind to form a *pollution plume*.

WileyPLUS Web Quiz
Take a quick quiz on the key concepts of this chapter.

University of Washington Libraries,Special Collections, John Shelton Collection, KC10617

4.40 Smog and haze over Los Angeles

A dense layer of smog and marine haze fills the Los Angeles Basin. The view is from a point over the San Gabriel Mountains looking southwest.

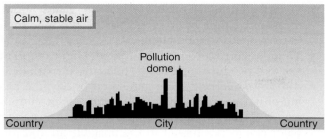

POLLUTION DOME
If calm, stable air overlies a major dity, a pollution dome can form.

POLLUTION PLUME
When wind is present, pollutants are carried away as a pollution plume.

4.41 Pollution dome and pollution plume

Pollution domes and pollution plumes are created when urban activities release pollutants.

A Look Ahead

This chapter has focused primarily on the process of precipitation, which we have examined largely on the scale of the individual cloud. To understand the global pattern of precipitation and how that pattern helps define global climate, we need to examine winds and the global circulation of the atmosphere. These topics are the subjects of the next chapter.

As we will see, wind results from pressure gradients that are created by uneven heating of the air coupled with the turning motion of the Earth's surface. The major role of precipitation in this process is to trap latent heat energy as water evaporates from warm ocean surfaces and release that energy when condensation and precipitation occur at distant locations. The effect is to move heat poleward in a process that creates the distinctive patterns of temperature and precipitation that we term global climates.

WileyPLUS Web Links
Learn more about humidity and how to calculate it.
Observe the different types of clouds and how they form.
View some amazing images of lighting. Monitor global drought.

IN REVIEW ATMOSPHERIC MOISTURE AND PRECIPITATION

- **Acid deposition** of sulfate and nitrate particles, either dry or as acid raindrops, can acidify soils and lakes, causing fish and plant mortality.
- **Precipitation** is the fall of liquid or solid water from the atmosphere to reach the Earth's land or ocean surface.
- Because the water molecule has a weak positive charge on one side and a weak negative charge on the other, liquid water molecules exhibit *hydrogen bonding*, producing strong *surface tension*, adherence to solid substances, *capillarity*, high *specific heat*, and the ability to dissolve many substances.
- Water exists in three states: as a solid (ice), as a liquid (water), or as an invisible gas (water vapor). Evaporating, condensing, melting, freezing, **sublimation**, and **deposition** describe the changes of the state of water.
- The **hydrosphere** is the realm of water in all its forms. Nearly all of the hydrosphere consists of saline ocean water. The remainder is freshwater on the continents and a very small amount of atmospheric water.
- Water moves freely between ocean, atmosphere, and land in the **hydrologic cycle**. The global water balance describes these flows.

- **Humidity** describes the amount of water vapor present in air. The ability of air to hold water vapor depends on temperature. Warm air can hold much more water vapor than cold air.
- **Specific humidity** measures the mass of water vapor in a mass of air in grams of water vapor per kilogram of air. The **dew-point temperature** of a parcel of moist air is the temperature at which the air becomes saturated. The more water vapor in the air, the higher the dew-point temperature.
- **Relative humidity** measures water vapor in the air as the percentage of the maximum amount of water vapor that can be held at the given air temperature.
- The **adiabatic principle** states that when a gas is compressed, it warms, and when a gas expands, it cools. When an air parcel moves upward in the atmosphere, it encounters a lower pressure and so expands and cools. The **dry adiabatic lapse rate** describes the rate of cooling with altitude.
- If the air is cooled below the dew point, condensation or deposition occurs. The altitude at which condensation starts is the **lifting condensation level**. Condensation releases latent heat, which reduces the parcel's rate of cooling with altitude.

When condensation or deposition is occurring, the cooling rate is described as the moist adiabatic lapse rate.

- **Clouds** are composed of droplets of water or crystals of ice that form on *condensation nuclei*. Clouds typically occur in layers, as *stratiform* clouds, or in globular masses, as *cumuliform* clouds.

- **Fog** is a cloud layer at ground level. *Radiation fog* occurs on clear nights in valleys and low-lying areas. *Advection fog* results when a warm, moist air layer moves over a cold surface.

- Weather forecasters use images from the *GOES* satellite system—a pair of geostationary satellites with views of North America and the oceans to its east and west—to monitor clouds and precipitation and track storms.

- Precipitation forms when water droplets in warm clouds condense, collide, and coalesce into droplets that are large enough to fall as rain. In cold clouds, ice crystals grow by deposition, while water droplets shrink by evaporation.

- There are four types of precipitation processes: *orographic, convective, cyclonic,* and *convergent*. In **orographic precipitation**, air moves up and over a mountain barrier. As it moves up, it is cooled adiabatically, and rain forms. As it descends the far side of the mountain, it is warmed, producing a rain shadow effect.

- In **convective precipitation**, unequal heating of the surface causes an air parcel to become warmer and less dense than the surrounding air. Because it is less dense, it rises. As it moves upward, it cools, and condensation with precipitation may occur.

- In *stable air*, the heated parcel rapidly reaches the temperature of the surrounding air.

- In **unstable air**, the surrounding air is always cooler than the parcel, making it rise to great heights with intense condensation. Warm, humid masses of air, heated by the summer Sun, often become unstable, triggering convective precipitation.

- Precipitation from clouds occurs as *rain, snow, sleet, freezing rain,* and *hail*. When *supercooled* rain falls on a surface that has a temperature below freezing, it produces an *ice storm*.

- Precipitation is measured with a *rain gauge*. Snow is measured either by its depth or by the amount of water it yields when melted.

- **Thunderstorms** produce thunder and lightning. They can also produce high winds, hail, and tornadoes.

- *Air-mass thunderstorms* are generated by daytime heating of warm, moist air. In the *cumulus stage*, cumulus clouds form, rise, and evaporate, causing instability. Later clouds rise higher in the unstable air and form *mature stage* thunderstorms, often with *anvil clouds*. Falling precipitation generates downdrafts that eventually lead to the *dissipating stage*.

- *Lightning* occurs in clouds when updrafts and downdrafts cause negative and positive charges to build up in different regions of the cloud. Electric current suddenly passes between the regions, resulting in an arc of lightning. *Severe thunderstorms*, enhanced by wind shear, last longer and have higher winds than air-mass thunderstorms. Massive *supercell thunderstorms* can also form under unusual conditions.

- A *microburst* is an intense downdraft that precedes some severe thunderstorms.

- Upper-air wind flow patterns sometimes produce large masses of severe thunderstorms called *mesoscale convective systems,* or *squall lines* of thunderstorms in different stages of development.

- In the rotating funnel of a **tornado**, wind speeds can exceed those of any other type of storm. Tornadoes are formed when wind shear produces a horizontal vortex that is lifted by convection to create a vertical *mesocyclone*. As the storm develops, rotation of the mesocyclone increases, sometimes generating a tornado.

- Tornado damage is assessed using the *Enhanced Fujita intensity scale*, with ratings from 0 to 5.

- **Air pollutants** are unwanted gases, aerosols, and particulates injected into the air by human and natural activity. Polluting *gases, aerosols,* and *particulates* are generated largely by fuel combustion. Pollutants leave the air by *fallout* from gravity and *washout* from precipitation.

- **Smog**, a common form of air pollution, contains nitrogen oxides, volatile organic compounds, and ozone. *Inversions* can trap smog and other pollutant mixtures in a layer close to the ground, producing unhealthy air.

KEY TERMS

acid deposition, p. 114
precipitation, p. 114
sublimation, p. 117
deposition, p. 117
hydrosphere, p. 117
hydrologic cycle, p. 117
humidity, p. 120
specific humidity, p. 120

dew-point
　temperature, p. 120
relative humidity, p. 121
adiabatic principle, p. 122
dry adiabatic lapse
　rate, p. 122
lifting condensation
　level, p. 123

moist adiabatic lapse
　rate, p. 123
clouds, p. 124
fog, p. 124
orographic
　precipitation, p. 129
rain shadow, p. 130

convective
　precipitation, p. 130
unstable air, p. 130
thunderstorms, p. 137
tornado, p. 140
air pollutant, p. 143
smog, p. 144

REVIEW QUESTIONS

1. To what type of pollution does the term **acid deposition** refer? What are its causes? What are its effects?
2. How does *hydrogen bonding* of water molecules affect the properties of water?
3. Identify the three states of water and the six terms used to describe possible changes of state.
4. What is the **hydrosphere?** Where is water found on our planet? In what amounts? How does water move in the **hydrologic cycle?**
5. Define **specific humidity**. How is the moisture content of air influenced by air temperature?
6. Define **relative humidity**. How is relative humidity measured? Sketch a graph showing relative humidity and temperature through a 24-hour cycle.
7. Use the terms *saturation, dew point*, and *condensation* to describe what happens when an air parcel of moist air is chilled.
8. What is the **adiabatic principle?** Why is it important?
9. Distinguish between **dry** and **moist adiabatic lapse rates**. In a parcel of air moving upward in the atmosphere, when do they apply? Why is the moist adiabatic lapse rate less than the dry adiabatic rate? Why is the moist adiabatic rate variable in amount?
10. How are clouds classified? Name four cloud families, two broad types of cloud forms, and three specific cloud types.
11. What is **fog**? Explain how radiation fog and advection fog form.
12. How is precipitation formed? Describe the process for warm and cold clouds.
13. Identify a satellite system that is used to monitor clouds and track storms. What type of orbit does the system use, and why?
14. Describe the **orographic precipitation** process. What is a **rain shadow?** Provide an example of the rain shadow effect.
15. What is **unstable air?** What are its characteristics?
16. Describe the **convective precipitation** process. What is the energy source that powers this source of precipitation? Explain.
17. Identify and describe five types of precipitation. How is the amount of precipitation measured?
18. What is a **thunderstorm?** Describe the typical life cycle of an *air-mass thunderstorm*.
19. What distinguishes a *severe thunderstorm* from an air-mass thunderstorm? How does a severe thunderstorm produce a microburst? What is a *squall line*?
20. What is a **tornado?** What two ingredients are needed for its formation? How is the intensity of a tornado measured by ground damage?
21. Define **air pollutant** and provide three examples.
22. What is **smog?** What important pollutant forms within smog, and how does this happen?
23. Distinguish between *low-level* and *high-level inversions*. How are they formed? What is their effect on air pollution? Give an example of a pollution situation for each type.

VISUALIZING EXERCISES

1. Lay out a simple diagram showing the main features of the hydrologic cycle. Include flows of water connecting land, ocean, and atmosphere. Label the flow paths.
2. Graph the temperature of a parcel of air as it moves up and over a mountain barrier, producing precipitation.
3. Sketch the anatomy of a thunderstorm cell. Show updraft, downdraft, lifting condensation level, anvil cloud, and other features.

ESSAY QUESTIONS

1. A very important topic for understanding weather and climate is water in the atmosphere. Write an essay or prepare an oral presentation on this topic, focusing on the following questions: What part of the global supply of water is atmospheric? Why is it important? What is its global role? How does the capacity of air to hold water vapor vary? How is the moisture content of air measured? Clouds and fog visibly demonstrate the presence of atmospheric water. What are they? How do they form?
2. Compare and contrast orographic and convective precipitation. Begin with a discussion of the adiabatic process and the generation of precipitation within clouds. Then compare the two processes, paying special attention to the conditions that create uplift. Can convective precipitation occur in an orographic situation? Under what conditions?
3. The thunderstorm is an atmospheric phenomenon that can be violent and dangerous. Tell the story of thunderstorms, including their life cycle and the types of thunderstorms that occur. Describe microbursts and tornadoes as examples of the destructive winds that sometimes occur with thunderstorms.

Chapter 5
Winds and Global Circulation

The winds are fierce atop Antarctica's Mount Discovery, overlooking McMurdo Sound at a height of 2681 m (8796 ft). Cold, dense air drains down from the Transantarctic Mountains during the long Antarctic night and is funneled across the extinct volcano, reaching speeds of up to 300 km/hr (186 mi/hr) on its rush to the sea. Blasting the summit, the wind sculpts the mountain's ice and snow into huge waves and pillows, shown here in the long shadows of the summer Sun. These iceforms are dramatic evidence of the power of the wind.

ICE SCULPTURES AT THE SUMMIT OF MOUNT DISCOVERY, ANTARCTICA

©Yann Arthus-Bertrand/Altitude

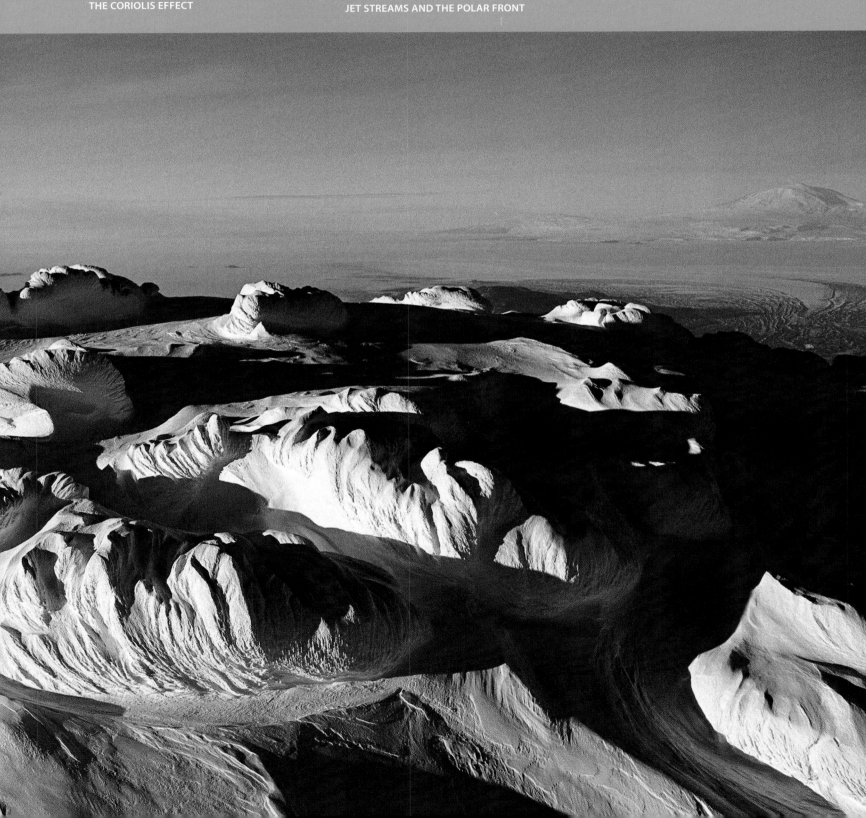

Winds and Global Circulation

T he air around us is always in motion, from a gentle breeze on a summer's day to a cold winter wind. Why does the air move? What are the forces that cause winds to blow? Why do winds blow more often in some directions than others? What is the pattern of wind flow in the upper atmosphere, and how does it affect our weather? How does the global wind pattern produce ocean currents? These are some of the questions we will answer in this chapter.

El Niño

At irregular intervals of about three to eight years, a remarkable disturbance of ocean and atmosphere occurs in the equatorial Pacific region—**El Niño**. Its name comes from Peruvian fishermen, who refer to the Corriente del Niño, or the "Current of the Christ Child," to describe an invasion of warm surface water once every few years around Christmas time that greatly depletes their catch of fish. It lasts more than a year, bringing droughts, heavy rainfalls, severe spells of heat and cold, and a high incidence of cyclonic storms to various parts of the Pacific and its eastern coasts.

Normally, the cool Humboldt (Peru) Current flows northward off the South American coast, and then at about the Equator it turns westward across the Pacific as the south equatorial current. The Humboldt Current is fed by upwelling of cold, deep water, bringing with it nutrients that serve as food for marine life. With the onset of El Niño, upwelling ceases, the cool water is replaced by warm, sterile water from the west, and the abundant marine life disappears. The counterpart to El Niño is *La Niña* (the girl child), in which normal Peruvian coastal upwelling is enhanced, trade winds strengthen, and cool water is carried far westward in an equatorial plume. Figure 5.1 shows two satellite images of sea-surface temperature observed during El Niño and La Niña years.

The major change in sea-surface temperatures that accompanies an El Niño can also shift weather patterns across large regions of the globe. The winter of 1997–1998 experienced one of the strongest El Niños in a century. Torrential rains drenched Peruvian and Ecuadorean coast ranges, producing mudflows, debris avalanches, and extensive river flooding. Conversely, large portions of Australia and the East Indies went rainless for months, and forest fires burned out of control in Sumatra Borneo, and Malaysia. In East Africa, Kenya experienced rainfall 1000 mm (80 in.) above normal.

In North America, a series of powerful winter storms lashed the Pacific coast, doing extensive damage in California. Monstrous tornadoes ripped through Florida, killing over 40 people and destroying more than 800 homes. Meanwhile, mild winter conditions east of the Rockies saved vast amounts of fossil fuel while generating an ice storm that left 4 million people without power in Quebec and the northeastern United States. All told, the deranged weather of the 1997–1998 El Niño was responsible for property damage estimated at $33 billion and the deaths of an estimated 2100 people.

The El Niño of 1997–1998 was rapidly followed by the La Niña of 1998–1999. It resulted in heavier monsoon rains in India and more rain in Australia. In North America, winter conditions were colder than normal in the Northwest and upper Midwest. The eastern Atlantic region endured drought throughout the spring and early summer. The hurricane season of 1998 was the deadliest in the past two centuries. In 2009–2010, yet another strong El Niño occurred, but with ocean warming concentrated in the central equatorial Pacific. It was followed by another strong La Niña in 2011, which produced devastating drought in the American South and Southwest.

What effect will global warming have on El Niño? Because global warming raises sea-surface temperatures and increases equatorial rainfall, climatologists think that global warming will enhance the strength and frequency of El Niño events while reducing the incidence of La Niñas. But the mechanism responsible for the El Niño/La Niña cycle is still not well understood. We'll return to the El Niño/La Niña cycle later in this chapter.

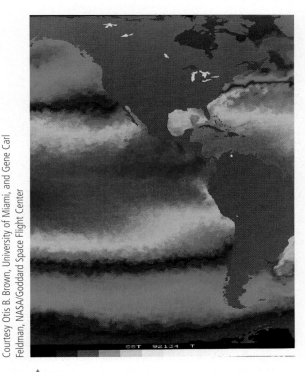

Courtesy Otis B. Brown, University of Miami, and Gene Carl Feldman, NASA/Goddard Space Flight Center

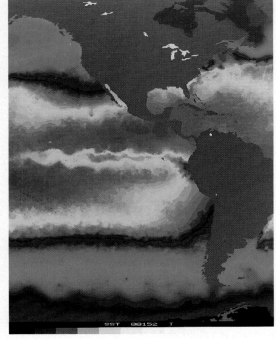

Courtesy Otis B. Brown, University of Miami, and Gene Carl Feldman, NASA/Goddard Space Flight Center

▲ EL NIÑO
Cold water along the South American coast stays to the south. Upwelling off Peru is weak. A region of very warm water develops in the eastern Pacific as the trade winds fail and become weak westerlies.

▲ LA NIÑA
An unusually strong flow of cold water moves up the South American coast toward Peru. Strong upwelling brings more cold water to the surface, which is carried westward in a plume within the south equatorial current. Normal conditions resemble the La Niña, except that current patterns and upwelling are significantly weaker.

5.1 La Niña and El Niño sea-surface temperatures

This striking pair of images shows sea-surface temperatures in the eastern tropical Pacific during El Niño and La Niña years as measured by the NOAA-7 satellite. Green tones indicate cooler temperatures, while red tones indicate warmer temperatures.

Atmospheric Pressure

We live at the bottom of a vast ocean of air—the Earth's *atmosphere* (Figure 5.2). Like the water in the ocean, the air in the atmosphere is constantly pressing on the Earth's surface beneath it and on everything that it surrounds. The atmosphere exerts pressure because *gravity* pulls the gas molecules of the air toward the Earth. Gravity is an attraction among all masses—in this case, between gas molecules and the Earth's vast bulk.

Atmospheric pressure is produced by the weight of a column of air above a unit area of the Earth's surface. At sea level, about 1 kilogram of air presses down on each square centimeter of surface (1 kg/cm²)—about 15 pounds on each square inch of surface (15 lb/in.²). The basic metric unit of pressure is the *pascal* (Pa).

At sea level, the average pressure of air is 101,320 Pa. Many atmospheric pressure measurements are reported in terms of *bars* and *millibars* (mb) (1 bar = 1000 mb 100,000 Pa). In this book we will use the millibar as the metric unit of atmospheric pressure. Standard sea-level atmospheric pressure is 1013.2 mb.

MEASURING AIR PRESSURE

You probably know that a **barometer** measures atmospheric pressure. But do you know how it works? It's based on the same principle as drinking soda through a straw. When using a straw, you create a partial vacuum in your mouth by lowering your jaw and moving your tongue. The pressure of the atmosphere on the liquid in your glass then forces soda up through the straw. The oldest, simplest, and most accurate instrument for measuring

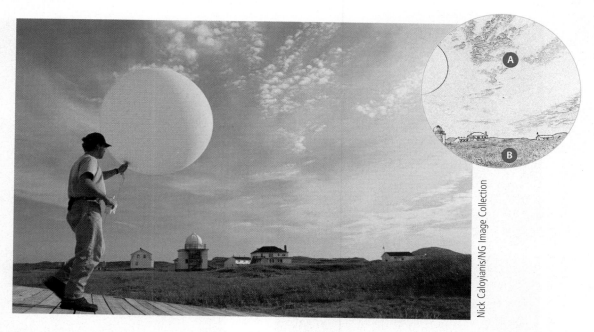

Nick Caloyianis/NG Image Collection

5.2 Ocean of air

The Earth's terrestrial inhabitants live at the bottom of an ocean of air. This buoyant weather balloon will carry upward a package of instruments, known as a radiosonde, that transmits measurements of pressure, altitude, GPS location, temperature, relative humidity, and wind speed and direction. Twice each day, the U.S. National Weather Service launches 100 of the 800 radiosondes worldwide, under agreement with the World Meteorological Organization. Satellite sensors are calibrated with the radiosonde data to provide a snapshot of the atmosphere above us.

EYE ON THE LANDSCAPE What else would the geographer see?
There are at least three types of clouds in this dramatic photo Ⓐ. See if you can identify altocumulus, cirrus, and cirrostratus clouds. The grassy landscape of low, irregular hills Ⓑ suggests a coastal landscape of vegetated dunes. Few plants are adapted to a substrate of pure dune sand.

atmospheric pressure, the *mercury barometer*, works the same way (Figure 5.3).

Because the mercury barometer is so accurate and is used so widely, atmospheric pressure is commonly expressed using the height of the column in centimeters or inches. The chemical symbol for mercury is Hg, and standard sea-level pressure is expressed as 76 cm Hg (29.92 in. Hg). In this book, we will use "in. Hg" as the English unit for atmospheric pressure.

Atmospheric pressure at a single location varies only slightly from day to day. On a cold, clear winter day, the sea-level barometric pressure may be as high as 1030 mb (30.4 in. Hg), while in the center of a storm system it may drop by about 5 percent, to 980 mb (28.9 in. Hg). Changes in atmospheric pressure are associated with traveling weather systems.

In the mercury barometer, air pressure balances the pull of gravity of a column of mercury about 76 cm (30 in.) high.

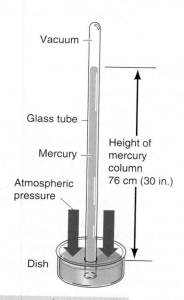

Vacuum

Glass tube

Mercury

Atmospheric pressure

Height of mercury column 76 cm (30 in.)

Dish

5.3 Measuring atmospheric pressure

Mercury barometer Atmospheric pressure pushes the mercury upward into the tube, where a vacuum is present. As atmospheric pressure changes the level of mercury in the tube rises and falls.

AIR PRESSURE AND ALTITUDE

If you have felt your ears "pop" during an elevator ride in a tall building, or on an airliner that is climbing or

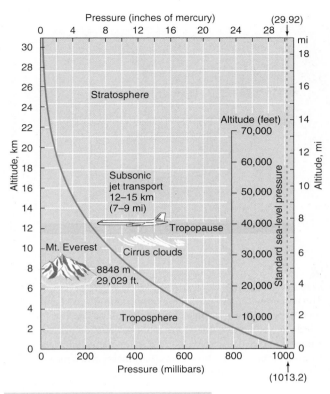

Atmospheric pressure decreases rapidly with altitude near the Earth's surface but much more slowly at higher altitudes.

descending, you've experienced a change in air pressure related to altitude (Figure 5.4). Changes in pressure at higher elevations can have more serious effects on the human body. In the mountains or at high altitudes, with decreased air pressure, less oxygen moves into lung tissues, producing fatigue and shortness of breath (Figure 5.5). These symptoms, sometimes accompanied by headache, nosebleed, or nausea, are known as mountain sickness. They are likely to occur at altitudes of 3000 m (about 10,000 ft) or higher.

Wind

Wind is defined as air moving horizontally over the Earth's surface. Air motions can also be vertical, but these are known by other terms, such as updrafts or downdrafts. Wind direction is identified by the direction from which the wind comes—a west wind blows from west to east, for example. Like all motion, the movement of wind is defined by its direction and velocity. The most common instrument for tracking wind direction is a simple vane with a tail fin that keeps it always pointing into the wind (Figure 5.6).

Anemometers measure wind speed. The most common type consists of three funnel-shaped cups on the ends of the spokes of a horizontal wheel that rotates as the wind strikes the cups. Some anemometers use

Most people who stay in a high-altitude environment for several days adjust to the reduced air pressure. Climbers who are about to ascend Mount Everest usually spend several weeks in a camp partway up the mountain before attempting to reach the summit.

5.6 Measurement of wind

Measuring the wind requires noting both its speed and its direction.

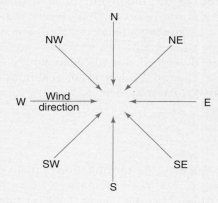

▲ WIND DIRECTION
Winds are designated according to the compass point from which the wind comes. Thus, although a west wind comes from the west, the air is moving eastward.

Courtesy Taylor Instrument Company and Wards Natural Science Establishment, Rochester, New York

▲ COMBINATION CUP ANEMOMETER AND WIND VANE
The anemometer and wind vane observe wind speed and direction, which are displayed on the meter below. The wind vane and anemometer are mounted outside, with a cable from the instruments leading to the meter, which is located indoors. Also shown are a barometer and a maximum–minimum thermometer.

a small electric generator that produces more current when the wheel rotates more rapidly. This is connected to a meter calibrated in meters per second or miles per hour.

What factors affect the strength and direction of wind? Three factors are important. *Pressure gradients*, which are usually associated with air temperature differences, act to push air from high- to low-pressure areas. The *Coriolis effect*, generated by the Earth's rotation, turns the path of moving air sideways, changing the direction of flow. *Friction* with the surface slows the wind in the lower atmosphere and acts in a direction opposite to air motion.

PRESSURE GRADIENTS

Wind is caused by differences in atmospheric pressure from one place to another. Air tends to move from regions of high pressure to regions of low pressure, until the pressure at every level is uniform. On a weather map, lines that connect locations with equal pressure are called **isobars**. A change of pressure, or pressure gradient, occurs at a right angle to the isobars (Figure 5.7). **Pressure**

> In a thermal circulation, heating of the atmosphere creates a pressure gradient that moves air toward the warm region at low levels and away from the warm region at high levels.

gradients develop because of unequal heating in the atmosphere. Figure 5.8 shows how unequal heating creates a **thermal circulation**.

LOCAL WINDS

Local winds are driven by local effects. *Sea* and *land breezes* are simple examples of how uneven heating and cooling of the air can set up thermal circulations and create local winds (Figure 5.9). A similar situation

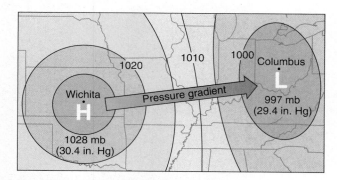

5.7 Isobars and a pressure gradient

This figure shows a pressure gradient. Because atmospheric pressure is higher at Wichita, Kansas, than at Columbus, Ohio, the pressure gradient will push air toward Columbus, producing wind. However, the direction of motion of the wind will also depend on the Coriolis effect and the friction of wind flow with the surface.

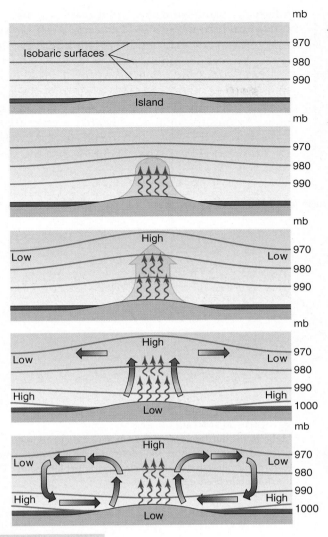

UNIFORM ATMOSPHERE (HEATED EQUALLY)
Imagine a uniform atmosphere above a ground surface. The isobaric levels (or "surfaces") are parallel with the ground surface (isobar = equal pressure).

UNEVEN HEATING
Imagine now that the underlying ground surface, an island, is warmed by the Sun, with cool ocean water surrounding it. The warm surface air rises and mixes with the air above, warming the column of air above the island. Since the warmer air occupies a large volume, the isobaric levels are pushed upward.

PRESSURE GRADIENT
The result is that a pressure gradient is created, and air at higher pressure (above the island) flows toward lower pressure (above the ocean).

SURFACE PRESSURE
Because air is moving away from the island and over the ocean surfaces, the surface pressure changes. There is less air above the island, so the ground pressure there drops. Since more air is now over the ocean surfaces, the pressure there rises.

THERMAL CIRCULATIONS
The new pressure gradient at the surface moves air from the ocean surfaces toward the island, while air moving in the opposite direction at upper levels completes the two loops.

5.8 Thermal circulations

Unequal heating of the Earth's surface produces a thermal circulation.

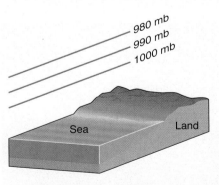

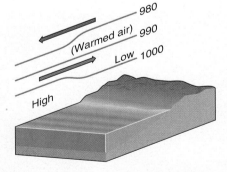

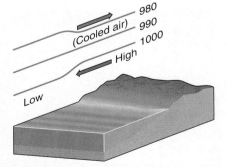

▲ **EARLY MORNING—CALM**
Early in the day, winds are often calm. Pressure decreases uniformly with altitude.

▲ **AFTERNOON—SEA BREEZE**
Later in the day, the Sun has warmed the land and the air above it, creating lower pressure at the surface and higher pressure aloft. The warmed air moves oceanward aloft, while surface winds bring cool marine air landward at the surface to replace it. This is the sea breeze.

▲ **NIGHT—LAND BREEZE**
At night, radiation cooling over land reverses the situation, with higher pressure at the surface and lower pressure aloft, developing a land breeze.

5.9 Sea and land breezes

The contrast between the land surface, which heats and cools rapidly, and the ocean surface, which has a more uniform temperature, induces pressure gradients to create sea and land breezes.

creates *mountain* and *valley winds* (Figure 5.10). Often these winds are moderate and just characteristic of the local environment. But some local winds are more dangerous.

If you're from Southern California, you're probably familiar with the *Santa Ana*,

> Local winds include daily land and sea breezes and mountain and valley winds. The Santa Ana, chinook, and mistral winds also occur in local regions.

a fierce, searing wind that often drives raging wildfire into foothill communities (Figure 5.11). It blows from the interior desert region of Southern California across coastal mountain ranges to reach the Pacific coast. These winds result from the air draining from high-pressure masses created over the Great Basin between the Rockies and the Sierra Nevada mountains. As the dry air flows downhill toward the lower coastal areas, it warms adiabatically. Its relative humidity drops, rapidly drying the moisture in soil and

5.10 Valley and mountain breezes

Daytime heating and nighttime cooling of mountain slopes produces valley and mountain breezes.

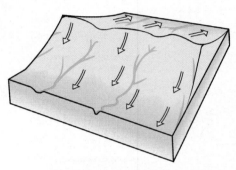

▲ **DAY—VALLEY BREEZE**
During the day, mountain hillslopes are heated intensely by the Sun, causing the air to expand and rise. This draws in air from the valley below, creating a valley breeze.

▲ **NIGHT—MOUNTAIN BREEZE**
At night, the hillslopes are chilled by radiation, setting up a reversed convection loop. The cooler, denser hillslope air moves toward the valley, down the hillslopes, to the plain below. This is the mountain breeze.

5.11 Brush fire

Residents of the Sylmar area of the San Fernando Valley, in Los Angeles, flee from their hillside homes as a brushfire approaches in October 2008. Driven by intense Santa Ana winds, the fires torched mobile homes and industrial buildings and forced the closing a major freeway during rush hour.

©AP/Wide World Photos

5.12 Windmills and wind turbines

Wind turbines vary in size and design. Some are suitable for individual farms, ranches, and homes, whereas others generate power on a commercial scale.

James Stevenson/Science Photo Library/Photo Researchers, Inc.

Arthur N. Strahler

▲
VERTICAL WIND TURBINES

These vertical wind turbines are Darrieus rotors—circular blades turning on a vertical axis. They generate power without needing to face the wind. These rotors are at a "wind farm" east of San Francisco in the Altamont Pass area of Alameda and Contra Costa counties. A group of wind farms here with a total of about 5000 small turbines provides 500 million kilowatt-hours of electricity per year.

▲
WIND FARM

A wind farm in San Gorgonio Pass, near Palm Springs, California. Wind speeds here average 7.5 m/s (17 mi/hr). Wind farms use wind turbines with a generating capacity in the range of 50 to 100 kilowatts.

vegetation, increasing fire danger. It is often funneled through local mountain gaps and across canyon floors with great force. Other local winds include the *chinook*, a warm and dry local wind that results when air passes over a mountain range and descends on the lee side. It is known for rapidly evaporating and melting snow. The *mistral* of the Rhône Valley in southern France is a cold, dry local wind that descends from high plateaus through mountain passes and valleys.

WIND POWER

Wind power is an indirect form of solar power that has been used for centuries. The total supply of wind energy is enormous. The World Meteorological Organization has estimated that about 20 million megawatts could be generated at favorable sites throughout the world, an amount about 100 times greater than the total electrical generating capacity of the United States. Figure 5.12 shows some different types of windmills and wind turbines.

Cyclones and Anticyclones
THE CORIOLIS EFFECT

We have seen that the pressure gradient force pushes air from high pressure to low pressure. For sea and land breezes, which are local in nature, the air moves in about the same direction as the pressure gradient. But on global scales, the direction of air motion is somewhat different. The difference is caused by the Earth's rotation, through the **Coriolis effect** (Figure 5.13).

The Coriolis effect was first identified by the French scientist Gaspard-Gustave de Coriolis in 1835. Because of the Coriolis effect, an object

> The Coriolis effect causes the apparent motion of winds to be deflected away from the direction of the pressure gradient. Deflection is to the right in the northern hemisphere and to the left in the southern hemisphere, as viewed from the starting point.

5.13 The Coriolis effect

The Coriolis effect makes true straight motion on the Earth's surface appear to curve. It can be treated as a force that turns the motion sideward.

▲ ROCKET LAUNCH

Imagine that a rocket is launched from the North Pole toward New York, aimed along the 74° W longitude meridian. As it travels toward New York, the Earth rotates from west to east beneath its straight flight path. If you were standing at the launch point on the rotating Earth below, you would see the rocket's trajectory curve to the right, away from New York and toward Chicago—despite the fact that the rocket has been flying in a straight line from the viewpoint of space. To reach New York, the rocket's flight path would have to be adjusted to allow for the Earth's rotation.

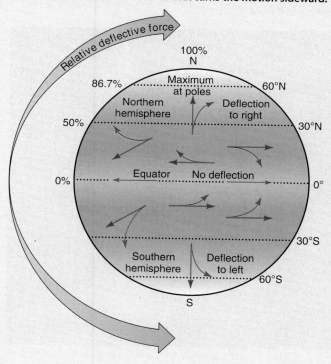

▲ DIRECTION OF DEFLECTION

The Coriolis effect appears to deflect winds and ocean currents to the right in the northern hemisphere and to the left in the southern hemisphere. Blue arrows show the direction of initial motion, and red arrows show the direction of motion apparent to the Earth observer. The Coriolis effect is strongest near the poles and decreases to zero at the Equator.

in the northern hemisphere moves as if a force were pulling it to the right. In the southern hemisphere, objects move as if pulled to the left. This apparent deflection does not depend on direction of motion; it occurs whether the object is moving toward the north, south, east, or west.

Geographers are usually concerned with analyzing the motions of air masses or ocean currents from the viewpoint of an Earth observer on the geographic grid, not from the viewpoint of space. So, as a shortcut, we treat the Coriolis effect as a sideward-turning force that always acts at right angles to the direction of motion. The strength of this Coriolis "force" increases with the speed of motion but decreases with latitude. This trick allows us to describe motion properly within the geographic grid.

WileyPLUS Coriolis Effect

Still confused about the Coriolis effect? View an animation of a game of catch between two players on a rotating turntable to see how an object moving in a straight line appears to follow a curving path.

THE FRICTIONAL FORCE

Closer to the Earth's surface, another force also affects the speed and direction of wind. As wind in the lower troposphere moves over the Earth's surface, a *frictional force* opposes the motion of the wind. In general terms, a rougher surface, with mountains, trees, or buildings, has a greater frictional drag on wind than a smooth surface. The frictional force is greatest close to the surface, and it decreases with altitude. The frictional force always acts in the opposite direction to the air motion (Figure 5.14).

Changes in land cover, including new buildings, deforestation, or new forest growth, can increase the effect of friction on surface winds. For example, in the northern hemisphere, surface wind velocity has been reduced by as much as 15 percent since the 1980s due to changes in land cover, especially forest regrowth in abandoned agricultural regions of Eastern Europe and Western Asia. Planners and developers of wind energy resources are carefully monitoring these changes to

PRESSURE GRADIENT FORCE
The pressure gradient force pushes the parcel toward low pressure

CORIOLIS FORCE
The Coriolis force always acts at right angles to the direction of motion.

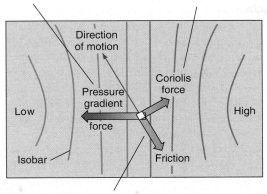

FRICTIONAL FORCE
There is a frictional force exerted by the ground surface that is proportional to the wind speed and always acts in the opposite direction to the direction of motion.

5.14 Balance of forces on a parcel of surface air

A parcel of air in motion near the surface is subjected to three forces. The sum of these three forces produces motion toward low pressure but at an angle to the pressure gradient. This example is for the northern hemisphere.

determine optimal locations for renewable energy development.

CYCLONES AND ANTICYCLONES

When the frictional force is added to the pressure gradient force and the apparent force of the Coriolis effect, it causes winds near the surface to move across the isobars at an angle (Figure 5.14). As a result, the air moves toward low pressure in a spiraling, converging motion. This inward spiral toward a central low pressure forms a **cyclone**, also called a low-pressure system (Figure 5.15). This inward spiraling motion, called *convergence*, also causes the air at the center to rise. In a high-pressure area, air spirals downward and outward to form an **anticyclone**, also called a high-pressure system. This outward spiraling motion is called *divergence*. In the northern hemisphere, cyclones rotate counterclockwise and anticyclones rotate clockwise. In the southern hemisphere, these directions are reversed.

Low-pressure centers (cyclones) are often associated with cloudy or rainy weather, whereas high-pressure centers (anticyclones) are often associated with fair weather. Why is this? When air is forced upward, it is cooled according to

> In a cyclone, winds converge, spiraling inward and upward. In an anticyclone, winds diverge, spiraling downward and outward.

the *adiabatic principle*, allowing condensation and precipitation to begin. So, cloudy and rainy weather often accompanies the inward and upward air motion of cyclones. In contrast, in anticyclones the air sinks and spirals outward. When air descends, it is warmed by the adiabatic process, so condensation can't occur. That is why anticyclones are often associated with fair weather. Cyclones and anticyclones can be a 1000 km (about 600 mi) across, or more. A fair-weather system—an anticyclone—may stretch from the Rockies to the Appalachians. Cyclones and anticyclones can remain more or less stationary, or they can move, sometimes rapidly, to cause weather disturbances.

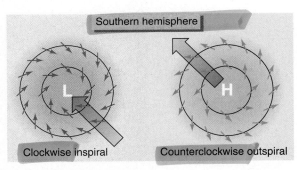

▲ **NORTHERN HEMISPHERE**
In a cyclone, low pressure is at the center, so the pressure gradient is straight inward. In an anticyclone, high pressure is at the center, so the gradient is straight outward. But because of the rightward Coriolis force and friction with the surface, the surface air moves at an angle across the gradient, creating a counterclockwise in-spiraling motion and a clockwise out-spiraling motion.

▲ **SOUTHERN HEMISPHERE**
In the southern hemisphere, the cyclonic spiral will be clockwise because the Coriolis force acts to the left. For anticyclones, the situation is reversed.

5.15 Air motion in surface level cyclones and anticyclones

Air spirals inward in a cyclone and outward in an anticyclone, but the direction of spiraling, clockwise or counterclockwise, depends on the hemisphere.

Global Wind and Pressure Patterns

For simplicity, let's begin by looking at surface winds and pressure patterns on an ideal Earth, one that does not have oceans and continents, or seasons (Figure 5.16). **Hadley cells** are key to understanding the wind patterns on our ideal Earth. These cells form because the Equator is heated more strongly by the Sun than other places, creating thermal circulations. Air rises over the Equator and is drawn poleward by the pressure gradient. But as the air moves poleward, the Coriolis force turns it westward, so it eventually descends at about a 30° latitude, completing the thermal circulation loop. This is a Hadley cell.

The rising air at the Equator produces a zone of surface low pressure known as the *equatorial trough*. Air in both hemispheres moves toward this trough, and there it converges and rises as part of the Hadley circulation. The narrow zone where the air converges is called the **intertropical convergence zone (ITCZ)**. Because the air is largely moving upward, surface winds are light and variable. This region is known as the *doldrums*.

Air descends on the poleward side of the Hadley cell circulation, so there surface pressures are high. This produces two **subtropical high-pressure belts**, each centered at about 30° latitude. Two, three, or four very large and stable anticyclones form within these belts. At the centers of these anticyclones, air descends

> In the Hadley cell thermal circulation, air rises at the intertropical convergence zone and descends in the subtropical high-pressure cells.

and winds are weak. The air is calm as much as one-quarter of the time.

Winds around the subtropical high-pressure centers spiral outward and move toward equatorial as well as middle latitudes. The winds moving toward the Equator are the dependable trade winds that propelled the sailing ships of merchant traders. North of the Equator, they blow from the northeast and are called the *northeast trade winds*. South of the Equator, they blow from the southeast and are the *southeast trades*. Poleward of the subtropical highs, air spirals outward, producing southwesterly winds in the northern hemisphere and northwesterly winds in the southern hemisphere.

Between about 30° and 60° latitude, the pressure and wind patterns become more complex. This is a zone of conflict between air bodies with different characteristics; masses of cool, dry air—*polar outbreaks*—move into the region, heading eastward and toward the Equator, along a border known as the **polar front**. This results in highly variable pressures and winds. On average, however, winds are more often from the west, so the region is said to have *prevailing westerlies*.

At the poles, the air is intensely cold, and high pressure occurs. At the South Pole, out-spiraling of winds around this polar anticyclone creates surface winds from a generally easterly direction, known as *polar easterlies*. In the north polar region, polar easterlies are weaker, and other wind directions are common.

WileyPLUS General Atmospheric Circulation
Start your mastery of global surface and upper-level winds with an overview animation showing the principal features of global atmospheric circulation.

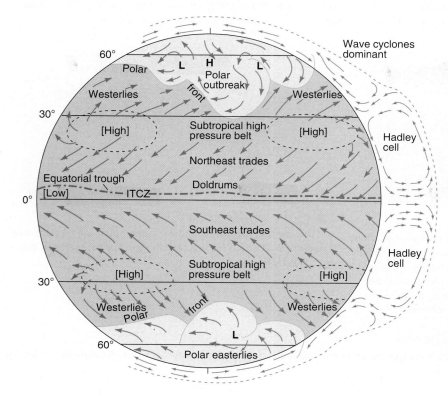

5.16 Global surface winds on an ideal Earth

We can see what global surface winds and pressures would be like on an ideal Earth, without the disrupting effect of oceans and continents and seasonal variations. Surface winds are shown on the disk of the Earth; the cross section at the right shows winds aloft.

SUBTROPICAL HIGH-PRESSURE BELTS

So far we've looked at wind and pressure patterns for a seasonless, featureless Earth. Let's turn now to actual global surface wind and pressure patterns, shown in Figure 5.17. The subtropical high-pressure belts, created by the Hadley cell circulation, are important features. In the northern hemisphere, they include two large high-pressure cells that flank North America—the *Hawaiian* and *Azores highs.* In the southern hemisphere, similar high-pressure cells flank South America and southern Africa.

We've seen that insolation is most intense when the Sun is directly overhead, and that the latitude at which the Sun is directly overhead changes with the seasons. This cycle causes the Hadley cells to intensify and move poleward during their high-Sun season, which also moves the subtropical high-pressure belts.

When the Hawaiian and Azores highs intensify and move northward in the summer, the East and West Coasts of North America feel their effects. On the West Coast, dry, descending air from the Hawaiian high dominates, so fair weather and rainless conditions prevail. On the East Coast, air from the Azores high flows across the continent from the southeast. These winds travel long distances across warm, tropical ocean surfaces before reaching land, picking up moisture and energy along the way. So they generally produce hot, humid

> The intertropical convergence zone and subtropical high-pressure belts are prominent features of the global wind and pressure pattern. They shift northward and southward with the seasons.

5.17 Global winds and pressures, January and July—Mercator projections

These global maps show average sea-level pressures and winds for daily observations in January and July. Values greater than average (1013 mb; 29.2 in. Hg) are shown in red, while values lower than average are shown in green. Pressure units are millibars reduced to sea level. Arrows indicate prevailing wind direction.

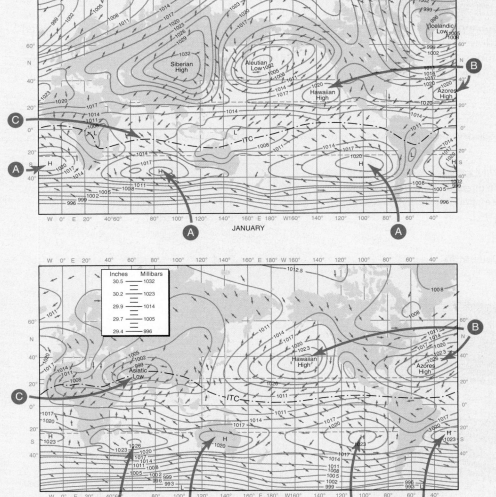

A SOUTHERN HEMISPHERE SUBTROPICAL HIGH-PRESSURE BELT

The most prominent features of the maps are the subtropical high-pressure belts, created by the Hadley cell circulation. The southern hemisphere high-pressure belt has three large high-pressure cells, each developed over oceans, that persist year round. A fourth, weaker, high-pressure cell forms over Australia in July, as the continent cools during the southern hemisphere winter.

B NORTHERN HEMISPHERE SUBTROPICAL HIGH-PRESSURE BELT

The situation is different in the northern hemisphere. The subtropical high-pressure belt shows two large anticyclones centered over oceans, the Hawaiian High in the Pacific and the Azores High in the Atlantic. From January to July, these intensify and move northward.

C SEASONAL SHIFT IN THE ITCZ

There is a huge shift in Africa and Asia, which you can see by comparing the two maps. In January, the ITCZ runs south across eastern Africa, and crosses the Indian Ocean to northern Australia at a latitude of about 15° S. In July, it swings north across Africa along the south rim of the Himalayas, in India, at a latitude of about 25° N—a shift of about 40 degrees of latitude!

weather for the central and eastern United States. In winter, these two anticyclones weaken and move to the south, leaving North America's weather at the mercy of colder winds and air masses from the north and west.

ITCZ AND THE MONSOON CIRCULATION

The ITCZ also shifts along with the Hadley cell circulation. The shift in the ITCZ is moderate in the western hemisphere, with the ITCZ moving a few degrees north from January to July over the oceans (Figure 5.17). In South America, the ITCZ lies across the Amazon in January and swings northward by about 20°.

However, there is a huge shift of the ITCZ in Asia—about 40° of latitude—that is readily visible on the two Mercator maps of Figure 5.17. Why does such a large shift occur? In winter (January map), the snow cover and low insolation in eastern Siberia create an intense and very cold surface high-pressure center—the *Siberian high*. In summer (July map), this high-pressure center has disappeared and is replaced by a surface low centered over the Middle Eastern region. This Asiatic low is produced by the intense summer heating of the desert landscape. The movement of the ITCZ and the change in these surface pressure patterns with the seasons create a reversing wind pattern in Asia known as the **monsoon** (Figure 5.18). In the winter, there is a strong outflow of dry, continental air from the north across China, Southeast Asia, India, and the Middle East. During this *winter monsoon*, dry conditions prevail. In the *summer monsoon*, warm, humid air from the Indian Ocean and the southwestern Pacific moves northward and northwestward into Asia. The cool, dry winter weather is replaced by steamy summer showers.

North America also experiences a monsoon effect, though considerably weaker. During summer, warm, moist air from the Gulf of Mexico tends to move northward across the central and eastern United States. Southwestern states, such as Nevada, Arizona, and New Mexico, receive the major portion of their rainfall in dramatic summer deluges from this monsoon effect. In winter, the airflow across North America generally reverses, and dry, continental air from Canada moves south and eastward.

> The Asian monsoon wind pattern consists of a cool, dry air flow from the northeast during the low-Sun season, and a warm, moist airflow from the southwest during the high-Sun season.

WIND AND PRESSURE FEATURES OF HIGHER LATITUDES

The northern hemisphere has two large continental masses separated by oceans and an ocean at the pole. The southern hemisphere has a large ocean with a cold, ice-covered continent at the center. These differing land–water patterns strongly influence the development of high- and low-pressure centers with the seasons (Figure 5.19).

5.18 Asian monsoon

The Asian monsoon is an alternation of cool, dry northeasterly airflow with warm, moist southwesterly airflow experienced in south Asia.

MONSOON WIND AND PRESSURE PATTERNS ▶
The Asiatic monsoon winds alternate in direction from January to July, responding to reversals of barometric pressure over the large continent.

MONSOON RAINS
Heavy monsoon rains turn streets into canals in Delhi, India.
▼

Steve Raymer/NG Image Collection

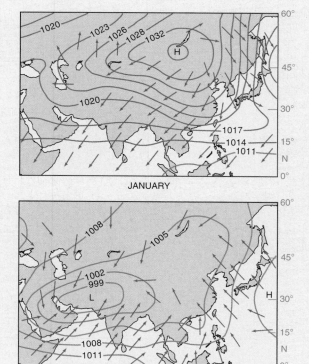

JANUARY

JULY

5.19 Atmospheric pressure—polar projections

These global maps show average sea-level pressure and wind for daily observations in January and July. Values greater than average barometric pressure are shown in red; lower values are in green.

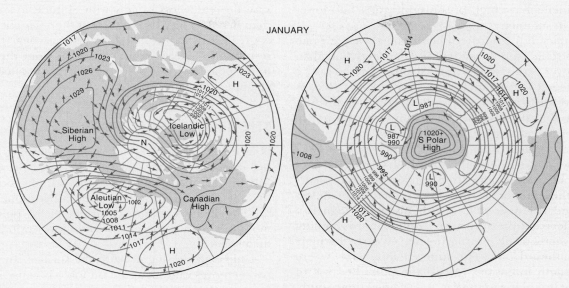

JANIARY

▲ JANUARY

In the northern hemisphere winter, air spirals outward from the strong Siberian high and its weaker cousin, the Canadian high. The Icelandic low and Aleutian low spawn winter storm systems that move southward and eastward onto the continents. In the southern hemisphere summer, air converges toward a narrow low-pressure belt around the south polar high.

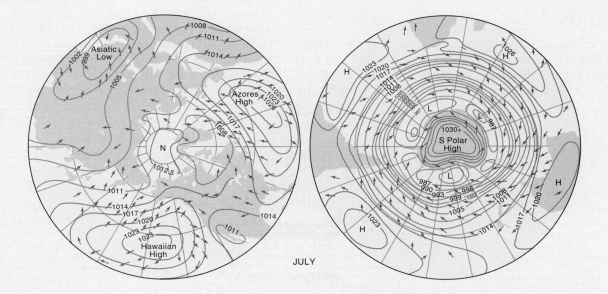

JULY

▲ JULY

In the northern hemisphere summer, the continents show generally low-surface pressure, while high pressure builds over the oceans. The strong Asiatic low brings warm, moist Indian Ocean air over India and Southeast Asia. A weaker low forms over the deserts of the southwestern United States and northwestern Mexico. Out-spiraling winds from the Hawaiian and Azores highs keep the west coasts of North America and Europe warm and dry, and the east coasts of North America and Asia warm and moist. In the southern hemisphere winter, the pattern is little different from winter, although pressure gradients are stronger.

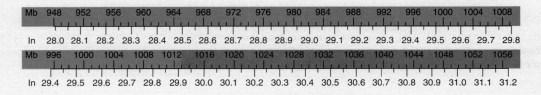

In Chapter 4, we saw that continents are colder in winter and warmer in summer than oceans at the same latitude. We know that cold air is associated with surface high pressure, and warm air with surface low pressure. So, continents have high pressure in winter and lower pressure in summer.

In January in the northern hemisphere, we see a pattern of strong high pressure (Siberian and Canadian highs) over continents and strong low pressure (Icelandic low and Aleutian low) over northern oceans that is driven by winter temperature contrasts. In July, this pattern nearly disappears as continents and northern oceans warm to similar temperatures.

In the southern hemisphere, there is a permanent anticyclone—the South Polar high—centered over Antarctica that varies little from January to July. It occurs because the continent is covered by thousands of meters of ice, and is always cold. Surrounding the high is a band of deep low pressure, with strong, inward-spiraling westerly winds on its northern side. As early mariners sailed southward, they encountered this band, where wind strength intensifies toward the pole. Because of the strong prevailing westerlies, they named these southern latitudes the "roaring forties," "flying fifties," and "screaming sixties."

Winds Aloft

We've looked at airflows at or near the Earth's surface, including both local and global wind patterns. But how does air move at the higher levels of the troposphere? Like air near the surface, winds at upper levels of the atmosphere move in response to pressure gradients and are influenced by the Coriolis effect.

How do pressure gradients arise at upper levels? There is a simple physical principle that states that pressure decreases less rapidly with altitude in warmer air than in colder air. Also recall that the solar energy input is greatest near the Equator and least near the poles, resulting in a temperature gradient from the Equator to the poles. This gives rise to a global pressure gradient (Figure 5.20).

THE GEOSTROPHIC WIND

How does a pressure gradient force pushing poleward produce wind, and what will the wind direction be? Any wind motion is subject to the Coriolis force, which turns it to the right in the northern hemisphere and to the left in the southern hemisphere. So, poleward air motion is toward the east, creating west winds in both hemispheres.

Unlike air moving close to the surface, an upper-air parcel moves without a friction force because it is so far from the source of friction—the surface. So, there are only two forces on the air parcel, the pressure gradient force and the Coriolis force.

Imagine a parcel of air, as shown in Figure 5.21. The air parcel begins to move poleward in response to the pressure gradient force. As it accelerates, the Coriolis force pulls it increasingly toward the right. As its velocity increases, the parcel turns progressively rightward until the Coriolis force just balances the gradient force. At that point, the sum of forces on the parcel is zero, so its speed and direction remain constant. We call this type

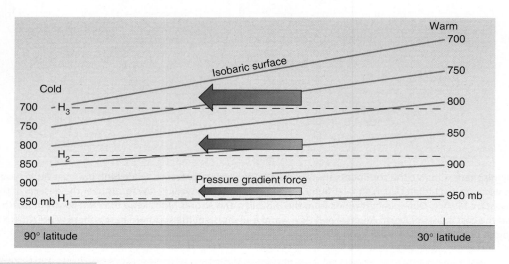

5.20 Upper-air pressure gradient

The isobaric surfaces in this upper-air pressure gradient map slope downward from the low latitudes to the pole, creating a pressure gradient force. Because the atmosphere is warmer near the Equator than at the poles, a pressure gradient force pushes air poleward. The pressure gradient force increases with altitude, bringing strong winds at high altitudes.

5.21 Geostrophic wind

The geostrophic wind blows parallel to the isobars at high altitudes. In this northern hemisphere example, the Coriolis force pulls the motion to the right.

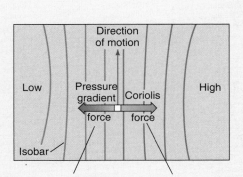

▲ **FORCES ON AN AIR PARCEL**

At upper levels in the atmosphere, a parcel of air is subjected to a pressure gradient force and a Coriolis force.

MOTION OF AN AIR PARCEL ▶

The parcel of air moves in response to a pressure gradient. As it moves, it is turned progressively sideward until the pressure gradient and Coriolis forces balance. This produces the geostrophic wind, which flows parallel to the isobars.

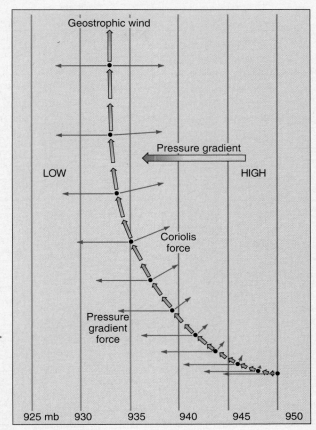

of airflow the **geostrophic wind**. It occurs at upper levels in the atmosphere, and we can see from the diagram that it flows parallel to the isobars.

Figure 5.22 shows an upper-air map for North America on a late June day. A large, upper-air low is centered over the Great Lakes. Since low pressure aloft indicates cold air (Figure 5.20), this mass of cool air, from Canada, has moved southward to dominate the eastern part of the continent. A large, but weaker, upper-air high is centered over the southwestern desert. This will be a mass of warm air, heated by intense surface insolation on the desert below.

The geostrophic wind flow, shown by the wind arrows, follows the contours closely. It is strongest where the contours are closest together, because there the pressure gradient force is strongest. The wind around

> At high altitudes, a moving parcel of air is subjected to two forces, a pressure gradient force and the Coriolis force. When the forces balance, air moves at right angles to the pressure gradient, parallel to the isobars, as the geostrophic wind.

the upper-air low also tends to spiral inward and converge on the low. The converging air descends toward the surface, where surface high pressure is present, and then spirals outward. Similarly, the wind around the upper-air high spirals outward and diverges away from the center. At the surface, low pressure causes air to spiral inward and upward.

GLOBAL CIRCULATION AT UPPER LEVELS

Figure 5.23 sketches the general airflow pattern at higher levels in the troposphere. It has four major features: a polar low, upper-air westerlies, tropical high-pressure belts, and weak equatorial easterlies. We've seen that the general temperature gradient from the tropics to the poles creates a pressure gradient force that generates westerly winds in the upper atmosphere. These *upper-air westerlies* blow in a

> The general pattern of winds aloft is a band of equatorial easterly winds, flanked by tropical high-pressure belts, and a zone of westerly winds at higher latitudes.

5.22 Upper-air wind and pressure map

The upper-air winds blow along the elevation contours of the 500-mb pressure level.

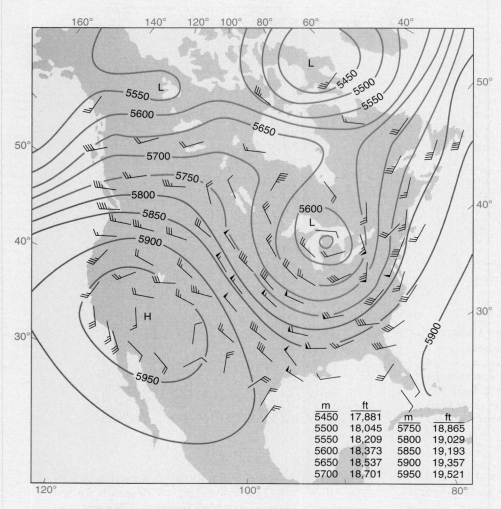

m	ft		m	ft
5450	17,881			
5500	18,045		5750	18,865
5550	18,209		5800	19,029
5600	18,373		5850	19,193
5650	18,537		5900	19,357
5700	18,701		5950	19,521

▲ **AN UPPER-AIR MAP FOR A DAY IN LATE JUNE**
Lines are height contours for the 500-mb surface. This surface dips downward where the air column is colder (low pressure aloft) and domes upward where the air column is warmer (high pressure aloft). So the height contours follow the pressure gradient, and the geostrophic wind follows the height contours.

EXPLANATION OF WIND ARROWS ▶
1 knot (nautical mile per hour) = 0.514 m/s
= 1.15 mi/hr

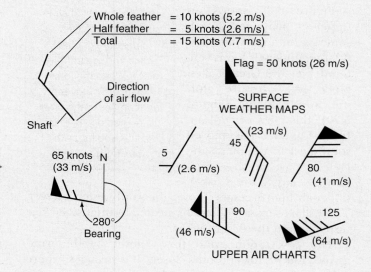

Whole feather = 10 knots (5.2 m/s)
Half feather = 5 knots (2.6 m/s)
Total = 15 knots (7.7 m/s)

Direction of air flow
Shaft

Flag = 50 knots (26 m/s)
SURFACE WEATHER MAPS

65 knots (33 m/s) N
280° Bearing

5 (2.6 m/s)
45 (23 m/s)
80 (41 m/s)
90 (46 m/s)
125 (64 m/s)

UPPER AIR CHARTS

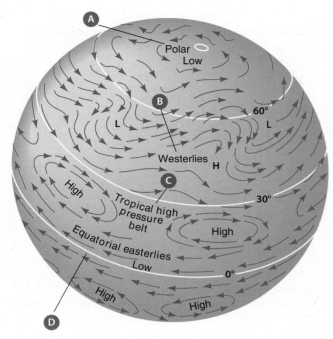

A **POLAR LOW** At high latitudes the westerlies form a huge circumpolar spiral, circling a great polar low-pressure center.

B **UPPER-AIR WESTERLIES** In this generalized plan of global winds high in the troposphere, strong west winds dominate the mid- and high-latitude circulation. They often sweep to the north or south around centers of high and low pressure aloft.

C **TROPICAL HIGH-PRESSURE BELT** Toward lower latitudes, atmospheric pressure rises steadily, forming a tropical high-pressure belt at 15° to 20° N and S lat. This high-altitude part of the surface subtropical high-pressure belt is shifted toward the Equator.

D **EQUATORIAL EASTERLIES** In the equatorial region, there is a zone of low pressure between the high-pressure ridges in which the winds are light but generally easterly. These winds are called the equatorial easterlies.

5.23 Global upper-level winds

complete circuit about the Earth, from about 25° latitude almost to the poles, often undulating from their westerly track.

So, the overall picture of upper-air wind patterns is really quite simple: a band of weak easterly winds in the equatorial zone, belts of high pressure near the Tropics of Cancer and Capricorn, and westerly winds, with some variation in direction, spiraling around polar lows.

JET STREAMS AND THE POLAR FRONT

Jet streams are wind streams that reach great speeds in narrow zones at a high altitude. They occur where atmospheric pressure gradients are strong. Along a jet stream, the air moves in pulses along broadly curving tracks. The greatest wind speeds occur in the center of the jet stream, with velocities decreasing away from it.

There are three kinds of jet streams. Two are westerly streams, and the third is a weaker jet with easterly winds that develops in Asia as part of the summer monsoon circulation. These are shown in Figure 5.24. The most poleward type of jet stream is located along the polar front. It is called the *polar-front jet stream* (or, simply, the "polar jet").

> Jet streams are streams of fast-moving air aloft that occur where atmospheric temperature gradients are strong. Each hemisphere normally exhibits westerly polar and subtropical jet streams. An easterly jet occurs in summer over Asia and Africa.

The polar jet is generally located between 35° and 65° latitude in both hemispheres. It follows the boundary between cold polar air and warm subtropical air. It is typically found at altitudes of 10 to 12 km (about 30,000 to 40,000 ft), and wind speeds in the jet range from 75 to as much as 125 m/s (about 170 to 280 mi/hr).

The subtropical jet stream occurs at the tropopause, just above subtropical high-pressure cells in the northern and southern hemispheres (Figure 5.25). There, westerly wind speeds can reach 100 to 110 m/s (about 215 to 240 mph), associated with the increase in velocity that occurs as an air parcel moves poleward from the Equator.

The tropical easterly jet stream occurs at even lower latitudes. It runs from east to west—opposite in direction to that of the polar-front and subtropical jet streams. The tropical easterly jet occurs only in summer and is limited to a northern hemisphere location over Southeast Asia, India, and Africa.

WileyPLUS Remote Sensing and Climate Interactivity
View satellite images to test your knowledge of global circulation and global cloud patterns. Check out pictures of planet Earth.

DISTURBANCES IN THE JET STREAM

The jet stream is typically a region of confined, high-velocity westerly winds found in the midlatitudes, but it can actually contain broad wavelike undulations called **jet stream disturbances**. These are sometimes also called *Rossby waves*, after Carl-Gustaf Rossby, a Swedish-American meteorologist who first observed and studied them.

Many processes can produce disturbances in the eastward flow of the jet stream. One of the most important is related to *baroclinic instability*, which makes the atmosphere unstable to small disturbances in the jet stream and allows these disturbances to grow over time. (The word "baroclinic" combines the prefix *baro-*, meaning pressure, and the suffix *-cline*, meaning gradient.)

The development of these disturbances is shown in Figure 5.26. For a period of several days or weeks, the jet stream flow may be fairly smooth. Then, an undulation

5.24 Jet Streams

The polar and subtropical jet streams are found in both hemispheres, while the tropical easterly jet is found above Asia in the summer.

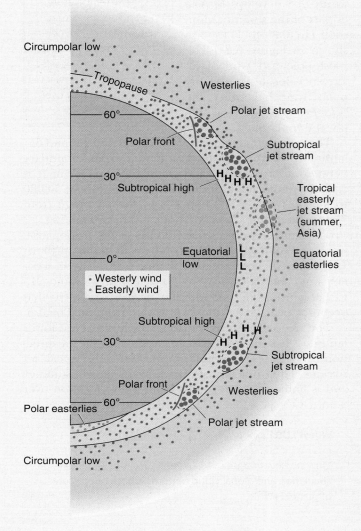

▲ **UPPER-LEVEL CIRCULATION CROSS SECTION**

As shown in this schematic diagram of wind directions and jet streams along a meridian from pole to pole, the four polar and subtropical jets are westerly in direction, in contrast to the single tropical easterly jet.

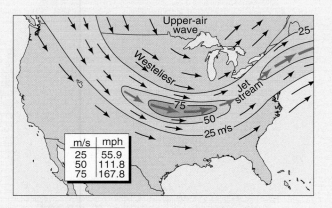

▲ **POLAR JET STREAM**

The polar jet stream is illustrated on this map by lines of equal wind speed.

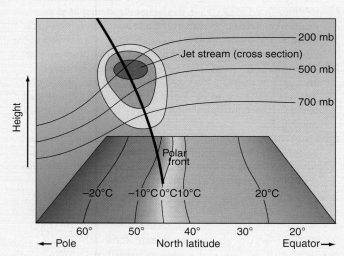

▲ **POLAR JET STREAM AND THE POLAR FRONT**

The polar jet stream is normally located over the polar front. Here, the strong temperature gradient between warm air to the south and cold air to the north produces a strong pressure gradient that can support the high-velocity jet stream.

develops. As the undulation grows, warm air pushes poleward, while a "tongue" of cold air is brought to the south. Eventually, the cold tongue is pinched off, leaving a pool of cold air at a latitude far south of its original location. This cold pool may persist for some days or weeks, slowly warming with time. Because of its cold center, it will contain low pressures aloft. In addition, the cold air in the core will descend and diverge at the surface, creating surface high pressure. Similarly, a

warm air pool will be pinched off far to the north of its original location. Within the core of the warm pool will be rising air, with convergence and low pressure at the surface, as well as high pressure aloft.

What causes a jet stream disturbance to grow over time? This growth is actually a complicated process and involves the interaction of the upper-air circulations—formed by the high- and low-pressure centers—with the global temperature gradient between the warmer,

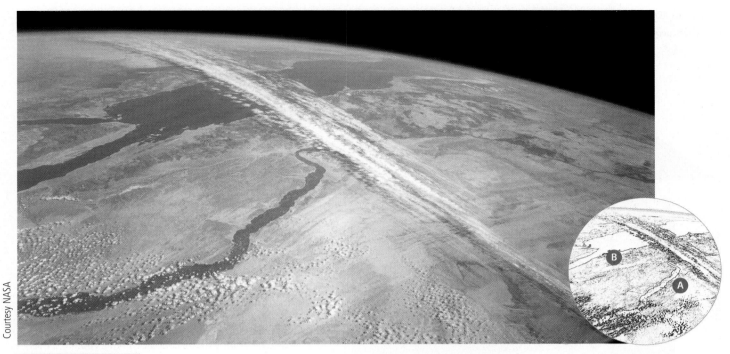

Courtesy NASA

5.25 Jet stream clouds

A strong subtropical jet stream is marked in this space photo by a narrow band of cirrus clouds that occurs on the Equatorward side of the jet. The jet stream is moving from west to east at an altitude of about 12 km (40,000 ft). The cloud band lies at about 25° N lat. In this view, astronauts aimed their camera toward the southeast, taking in the Nile River Valley and the Red Sea. At the left is the tip of the Sinai Peninsula.

EYE ON THE LANDSCAPE What else would the geographer see?
Large expanses of the Middle East are essentially devoid of vegetation, and so a space photo such as this shows variation in soil and surface color. Many of the white areas **A** are wind-blown sand sheets. Ranges of hills and low mountains **B** appear darker.

5.26 Jet stream disturbances

Disturbances form in the upper-air westerlies of the northern hemisphere, marking the boundary between cold polar air and warm tropical air.

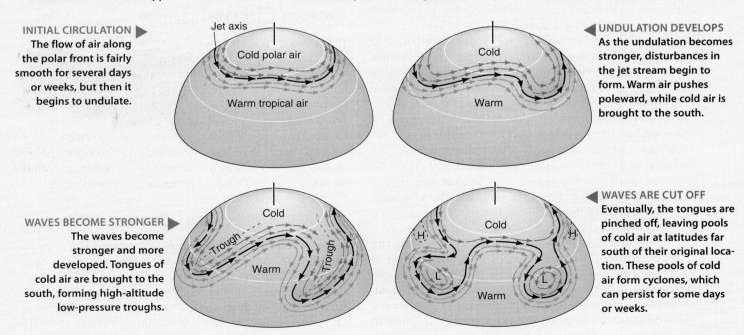

INITIAL CIRCULATION ▶
The flow of air along the polar front is fairly smooth for several days or weeks, but then it begins to undulate.

Jet axis
Cold polar air
Warm tropical air

◀ **UNDULATION DEVELOPS**
As the undulation becomes stronger, disturbances in the jet stream begin to form. Warm air pushes poleward, while cold air is brought to the south.

Cold
Warm

WAVES BECOME STRONGER ▶
The waves become stronger and more developed. Tongues of cold air are brought to the south, forming high-altitude low-pressure troughs.

Cold
Trough
Warm
Trough

◀ **WAVES ARE CUT OFF**
Eventually, the tongues are pinched off, leaving pools of cold air at latitudes far south of their original location. These pools of cold air form cyclones, which can persist for some days or weeks.

Cold
H
H
L
L
Warm

5.27 Growth of disturbances in the jet stream

Once a disturbance begins, positive feedbacks cause it to grow in intensity.

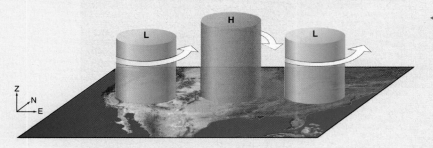

◀ STEP 1

Pressure disturbances in the upper atmosphere are associated with gradients in temperature. Here, the warmer air column (red) produces high pressures aloft, while the cooler air columns (blue) produce lower pressures. The pressure patterns produce disturbances in the jet stream, which circulates geostrophically around the pressure centers.

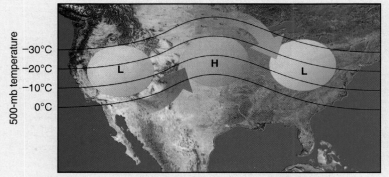

◀ STEP 2

Looking down from above, we see that the change in temperature from one location to another results in a disturbance in the global-scale north–south temperature gradient between low and high latitudes, as shown by the isotherms. The counterclockwise flow around the low-pressure center aloft brings warm, tropical air from the south into the vicinity of the warm-air column, which tends to heat this region even further. Conversely, the high-pressure center aloft brings cold air from the north into the vicinity of the cold-air column, cooling it even further.

◀ STEP 3

As the warm-air column continues to warm, it expands, producing even higher pressures aloft. In addition, the cold-air column continues to cool, thereby decreasing pressures further. The pressure gradient between the cold-air region and the warm-air region intensifies, and so the wind speeds also intensify.

low-latitude air and colder, high-latitude air. This growth process is described in Figure 5.27.

Consider first what happens to a slight disturbance in the jet stream. These disturbances arise from pressure differences between one location and another, resulting in geostrophic wind variations. In addition, the pressure differences are accompanied by temperature differences, with high pressures aloft found over regions of warm air, and low pressures aloft over regions of cool air.

As air circulates around these disturbances, the counterclockwise circulations associated with the low-pressure system bring warm, low-latitude air into a region that is already warm. In contrast, the clockwise circulation around the high-pressure region brings cold, high-latitude air into a region that is already cold. As a result, the warm regions tend to get warmer and the cold regions tend to get colder. In turn, the high pressures associated with the warm-air region tend to get higher and the low pressures associated with the cold-air region tend to get lower. This process intensifies the pressure gradient between the two regions, resulting in stronger winds and an even larger disturbance.

Ocean Circulation

Just as there is a circulation pattern to the atmosphere, there is also a circulation pattern to the oceans that it is driven by differences in density and pressure acting along with the Coriolis force. Pressure differences are created in the water when the ocean is heated unequally, because warm water is less dense than cold water. These pressure differences induce the water to flow. Similarly, saltier water is denser than less salty water, so differences in salinity can also cause pressure differences. The force of wind on the surface water also creates oceanic circulation.

TEMPERATURE LAYERS OF THE OCEAN

The ocean's layered temperature structure is shown in Figure 5.28. At low latitudes throughout the year and in middle latitudes in the summer, a warm surface layer develops, heated by the Sun. Wave action mixes heated surface water with the water below it to form the *warm layer*, which may be as thick as 500 m (about 1600 ft) with a temperature of 20°C to 25°C (68°F to 77°F) in

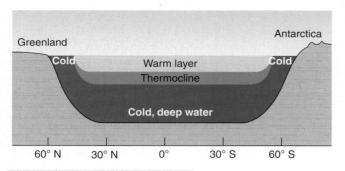

5.28 Ocean temperature structure

A schematic north–south cross section of the world ocean shows that the warm surface water layer disappears in arctic and antarctic latitudes, where very cold water lies at the surface. The thickness of the warm layer and thermocline is greatly exaggerated.

oceans of the equatorial belt. Below the warm layer, temperatures drop rapidly in a zone known as the *thermocline*. Below the thermocline is a layer of very cold water extending to the deep ocean floor. Temperatures near the base of the deep layer range from 0°C to 5°C

(32°F to 41°F). This layering is quite stable because the warm layer is less dense than the cold water and rests on top. In arctic and antarctic regions, the warm layer and thermocline are absent.

SURFACE CURRENTS

An **ocean current** is any persistent, predominantly horizontal flow of ocean water. Current systems exchange heat between low and high latitudes and are essential in sustaining the global energy balance. Surface currents are driven by prevailing winds, while deep currents are powered by changes in temperature and density in surface waters that cause them to sink.

In surface currents, energy is transferred from the prevailing surface wind to water by the friction of the air blowing over the water surface. Because of the Coriolis effect the actual direction of water drift is deflected about 45° from the direction of the driving wind.

The general features of the circulation of the ocean bounded by two continental masses (Figure 5.29) include two large circular movements, called **gyres**,

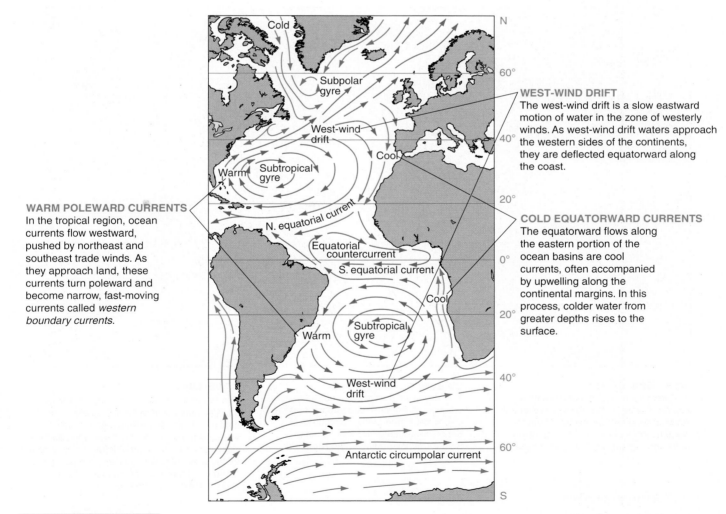

WEST-WIND DRIFT
The west-wind drift is a slow eastward motion of water in the zone of westerly winds. As west-wind drift waters approach the western sides of the continents, they are deflected equatorward along the coast.

WARM POLEWARD CURRENTS
In the tropical region, ocean currents flow westward, pushed by northeast and southeast trade winds. As they approach land, these currents turn poleward and become narrow, fast-moving currents called *western boundary currents*.

COLD EQUATORWARD CURRENTS
The equatorward flows along the eastern portion of the ocean basins are cool currents, often accompanied by upwelling along the continental margins. In this process, colder water from greater depths rises to the surface.

5.29 Ocean current gyres

Two great gyres, one in each hemisphere, dominate the circulation of ocean waters from the surface through 500 to 1000 m (1600 to 3300 ft).

that are centered at latitudes of 20° to 30°. These gyres track the movements of air around the subtropical high-pressure cells. An equatorial current with westward flow marks the belt of the trade winds. Although the trades blow to the southwest and northwest at an angle across the parallels of latitude, the surface water movement follows the parallels. The equatorial currents are separated by an equatorial countercurrent. A slow, eastward movement of surface water over the zone of the westerlies is named the west-wind drift. It covers a broad belt between latitudes 35° and 45° in the northern hemisphere and between latitudes 30° and 60° in the southern hemisphere. These features, as well as others, appear on our map of January ocean currents (Figure 5.30).

Because ocean currents move warm waters poleward, and cold waters toward the Equator, they are important regulators of air temperatures. Warm surface currents keep winter temperatures in the British Isles from falling much below freezing in winter.

COLD WESTERN BOUNDARY CURRENTS
In the northern hemisphere, where the polar sea is largely land-locked, cold currents flow Equatorward along the east sides of continents. Two examples are the Kamchatka Current, which flows southward along the Asian coast across from Alaska, and the Labrador Current, which flows between Labrador and Greenland to reach the coasts of Newfoundland, Nova Scotia, and New England.

WARM POLEWARD CURRENTS
In the equatorial region, ocean currents flow westward, pushed by northeast and southeast trade winds. As they approach land, these currents are turned poleward. Examples are the Gulf Stream of eastern North America and the Kuroshio Current of Japan.

NORTH ATLANTIC DRIFT
West-wind drift water also moves poleward to join arctic and antarctic circulations. In the northeastern Atlantic Ocean, the west-wind drift forms a relatively warm current. This is the North Atlantic drift, which spreads around the British Isles, into the North Sea, and along the Norwegian coast. The Russian port of Murmansk, on the Arctic circle, remains ice-free year round because of the warm drift current.

WEST-WIND DRIFT
The west-wind drift is a slow eastward motion of water in the zone of westerly winds. As west-wind drift waters approach the western sides of the continents, they are deflected equatorward along the coast.

CIRCUMPOLAR CURRENT
The strong west winds around Antarctica produce an Antarctic circumpolar current of cold water. Some of this flow branches Equatorward along the west coast of South America, adding to the Humboldt Current.

UPWELLING
The Equatorward flows are cool currents, often accompanied by upwelling along the continental margins. In this process, colder water from greater depths rises to the surface. Examples are the Humboldt (or Peru) Current, off the coast of Chile and Peru, the Benguela Current, off the coast of southern Africa, and the California Current.

5.30 January ocean currents

Surface drifts and currents of the oceans in January.

Courtesy Otis B. Brown, Robert Evans, and M. Carle, University of Miami, Rosenstiel School of Marine and Atmospheric Science, Florida, and NOAA/ Satellite Data Services Division

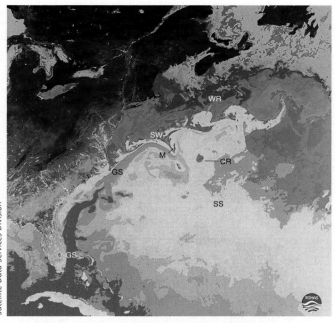

5.31 Sea-surface temperatures in the western Atlantic

In this satellite image showing sea-surface temperature for a week in April from data acquired by the NOAA-7 orbiting satellite, cold water appears in green and blue tones, warm water in red and yellow tones. The image shows the Gulf Stream (GS) and its interactions with cold water of the Continental Slope (SW), brought down by the Labrador Current, and the warm water of the Sargasso Sea (SS). Other features include a meander (M), a warm-core ring (WR), and a cold-core ring (CR).

Cold surface currents keep weather on the California coast cool, even in the height of summer.

Figure 5.31 shows a satellite image of ocean temperature along the east coast of North America for a week in April. The Gulf Stream stands out as a tongue of warm water, moving northward along the southeastern coast. Cooler water from the Labrador Current moves southward along the northern Atlantic coast. Instead of mixing, the two flows remain quite distinct. The boundary between them shows a wavelike flow, much like jet stream disturbances in the atmosphere.

> Global surface ocean currents are dominated by huge, wind-driven circular gyres centered near the subtropical high-pressure cells. Upwelling of cold bottom water often occurs on the west coasts of continents in the subtropical zones.

WileyPLUS Remote Sensing and Climate Interactivity

Check out global satellite images of sea surface temperatures and quiz yourself on ocean surface current patterns.

EL NIÑO AND ENSO

Another phenomenon of ocean surface currents is *El Niño*, described in our opening feature, "Eye on Global Change—El Niño." In an El Niño year, a major change in barometric pressure occurs across the entire stretch of the equatorial zone, as far west as southeastern Asia. Normally, low pressure prevails over northern Australia, the East Indies, and New Guinea, where the largest and warmest body of ocean water is located (Figure 5.32). Abundant rainfall occurs in this area during December, which is the high-Sun period in the southern hemisphere. During an El Niño event, the low-pressure system is replaced by a weak high-pressure zone, and local drought ensues. Pressures drop in the equatorial zone of the eastern Pacific, strengthening the equatorial trough. Rainfall is abundant in this new low-pressure region (right part of figure). The shift in barometric pressure patterns is known as the *Southern Oscillation*, and the two phenomena together are often referred to as *ENSO*.

Surface winds and currents also shift with this change in pressure. During normal conditions, the strong, prevailing trade winds blow westward, causing very warm ocean water to move to the western Pacific and to "pile up" near the western equatorial low. This westward motion causes the normal upwelling along the South American coast, as bottom water is carried up to replace the water dragged to the west. During an El Niño event, the easterly trade winds weaken with the change in atmospheric pressure. Without the westward pressure of the trade winds, warm waters surge eastward. Sea-surface temperatures and actual sea levels rise off the tropical western coasts of the Americas. Through teleconnections that are not completely understood, these changes in Pacific sea-surface temperature and wind fields can impact climate in faraway regions (Figure 5.33).

A related phenomenon, also capable of altering global weather patterns, is *La Niña*, a condition roughly opposite to El Niño. During a La Niña period, sea-surface temperatures in the central and western Pacific Ocean fall to lower than average levels. This happens because the South Pacific subtropical high becomes very strongly developed during the high-Sun season. The result is abnormally strong easterly trade winds. The force of these winds drags a higher-than-normal amount of warm surface water westward, which enhances upwelling along western continental coasts.

El Niño and La Niña occur at irregular intervals and with varying degrees of intensity. When severe, these events can have disastrous effects on weather around the world, as we noted in the "Eye on Global Change" feature beginning this chapter. Rainfall and temperature cycles can change significantly in seemingly unrelated parts of the globe, producing floods, droughts, and temperature extremes. Notable recent El Niño events occurred in 1982–1983, 1989–1990, 1991–1992, 1994–1995, 1997–1998, and 2002–2003, and 2009–2010.

5.32 El Niño

Maps of pressures in the tropical Pacific and eastern Indian Ocean in November during normal and El Niño years.

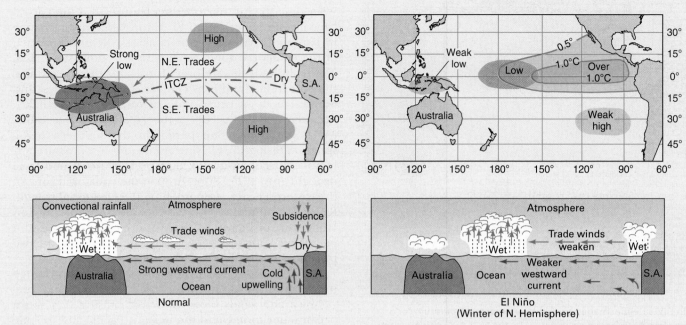

▲ In a normal year, low pressure dominates in Malaysia and northern Australia.

▲ In an El Niño year, low pressure moves eastward to the central part of the western Pacific, and sea-surface temperatures become warmer in the eastern Central Pacific.

5.33 Global weather changes during El Niño

Ocean surface temperature changes cause shifts in weather patterns during El Niño events.

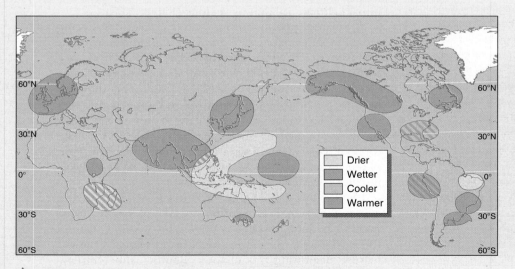

Legend:
- Drier
- Wetter
- Cooler
- Warmer

▲ El Niño events alter global weather, even in areas far from the Pacific Ocean. As a result, some areas are drier, some wetter, some cooler, and some warmer than usual. Typically, northern areas of the contiguous United States are warmer during the winter, whereas southern areas are cooler and wetter.

Image by R.B. Husar, Washington University; land layer by the SeaWiFS Project, fire maps from ESA, sea surface temperature layer from JPL and cloud layer by SSEC, U. Wisconsin.

▲ This image of sea surface temperature shows the buildup of warm waters (red) along the coast of South America and eastern equatorial Pacific.

What causes the ENSO phenomenon? One view is that the cycle is a natural oscillation caused by the way in which the atmosphere and oceans are coupled through temperature and pressure changes. Each ocean–atmosphere state has positive feedback loops that tend to make that state stronger. Thus, the ocean and atmosphere tend to stay in one state or the other until something occurs to reverse the state. In any event, scientists now have good computer models that accept sea-surface temperature along with air temperature and pressure data and can predict El Niño events with reasonable accuracy some months before they occur.

El Niño and its alter ego La Niña show how dynamic our planet really is. As a grand-scale, global phenomenon, El Niño/La Niña shows how the circulation patterns of the ocean and atmosphere are linked and interact to provide teleconnections capable of producing extreme events affecting millions of people throughout the world,

WileyPLUS El Niño
Watch an animation showing how an El Niño develops in detail, including sea surface height changes, atmospheric pressure changes, and movement of the oceanic thermocline.

PACIFIC DECADAL OSCILLATION

Another oceanic phenomenon that influences climate is marked by changes in sea-surface temperature in the northern Pacific Ocean that can produce climate changes across parts of Eurasia, Alaska, and the western United States. These changes interact with ENSO-related changes, either strengthening or weakening them. The changes in the North Pacific pressure pattern can last from weeks to decades. The decadal changes, (Figure 5.34) are called the *Pacific decadal oscillation*. They can persist on a scale of 20 to 30 years. The causes of the oscillation are not well understood.

5.34 Pacific decadal oscillation

The Pacific decadal oscillation is a long-lived climate pattern of Pacific Ocean temperatures. The extreme warm or cool phases can last as long as two decades. During warm conditions, ocean temperatures are higher than normal along the tropical Pacific and off the coast of North America, while ocean temperatures are lower than normal over the central North Pacific. Ocean temperatures suggest that a warm phase began in the late 1970s and has persisted until the present.

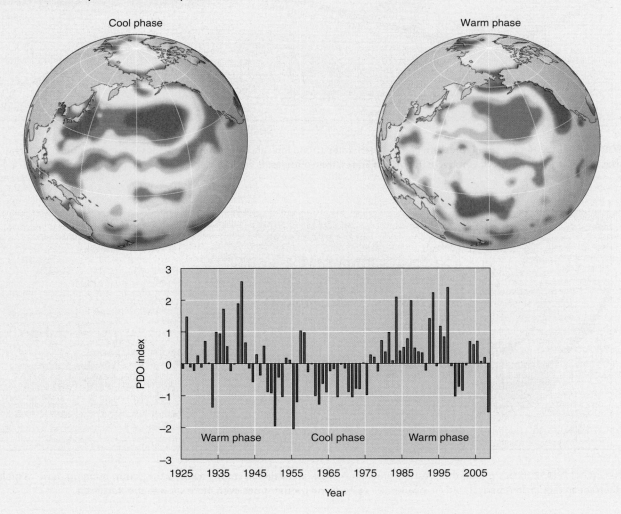

The Pacific decadal oscillation has shown two full cycles in the past century, with cool phases from 1890–1924 and 1947–1976 and warm phases from 1925–1946 and 1977–present. In the warm phase, eastward-moving Pacific storm systems track to the south, leaving the northwestern portion of the United States warm and dry, while the arid Southwest receives more rainfall than normal. In the cool phase, the northwestern United States becomes cooler and wetter, while drought comes to California and the Southwest. The phases also have strong effects on marine coastal fisheries, with the most productive regions shifted northward to Alaskan waters during the warm phase and southward in the cool phase.

NORTH ATLANTIC OSCILLATION

Another ocean–atmosphere phenomenon affecting climate is the *North Atlantic oscillation*, which is associated with changes in atmospheric pressures and sea-surface temperatures over the mid- and high latitudes of the Atlantic (Figure 5.35). The North Atlantic oscillation

5.35 North Atlantic oscillation

The North Atlantic oscillation has two contrasting wintertime phases that are related to the strength of the pressure gradient between the north polar region and the tropical Atlantic.

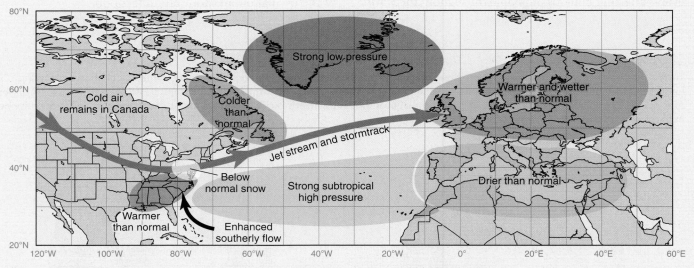

▲ POSITIVE PHASE Here, a strong pressure gradient generates strong low- and high-pressure fields that steer the polar jet stream and Atlantic storms toward northern Europe. In the United States, the Southeast is warmer, and cold air remains in Canada.

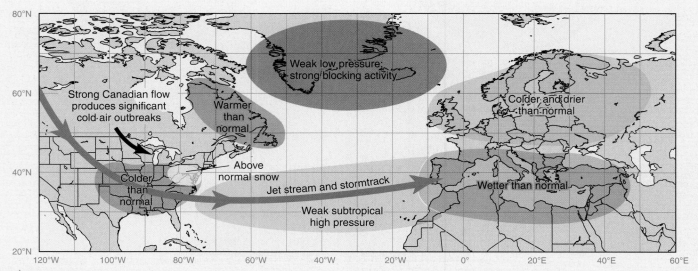

▲ NEGATIVE PHASE In this phase, the pressure gradient is weak and Atlantic storms track to the south, bringing more rainfall to the Mediterranean region. In America, cold air invades the eastern and central states, with more snow in the Northeast.

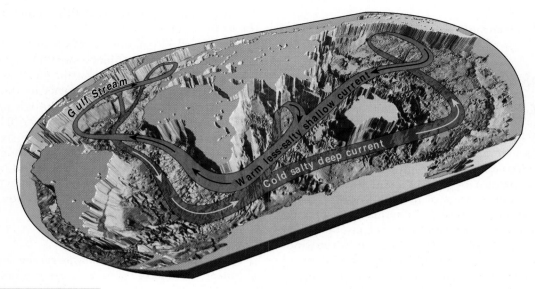

5.36 Thermohaline circulation

This three-dimensional figure shows the "conveyor belt" of deep ocean currents. Warm surface waters in the tropics move poleward, losing heat to the atmosphere en route. The cooler waters sink at higher latitudes, flow toward the Equator, and eventually upwell in oceans far away.

is predominantly an atmospheric phenomenon that is partly related to variations in the surface pressure gradient between the polar sea ice cap and the midlatitudes in both the Atlantic and Pacific Ocean basins. It may also be linked to changes in the ocean surface temperature and, possibly, snow and ice cover over Europe and Greenland.

Shifts in the North American oscillation can occur over the course of weeks, seasons, and even decades. Since about 1975, the positive phase has dominated, bringing wetter but milder conditions to northern Europe, and dry conditions to southern Europe.

DEEP CURRENTS AND THERMOHALINE CIRCULATION

Deep currents move ocean waters in a slow circuit across the floors of the world's oceans. They are generated when surface waters become denser and slowly sink downward. Coupled with these deep currents are very broad and slow surface currents on which the more rapid surface currents, described above, are superimposed. Figure 5.36 diagrams this slow flow pattern, which links all of the world's oceans. It is referred to as *thermohaline circulation*, because it depends on the sinking of cold, salty water along the northern edge of the Atlantic. The sinking leads eventually to upwelling at far distant locations, as described in the figure.

Thermohaline circulation plays an important role in the carbon cycle by moving CO_2-rich surface waters into the ocean depths. As noted in Chapter 3 in "Eye on Global Change • Carbon Dioxide—On the Increase," deep ocean circulation provides a conveyor belt for storage and release of CO_2 in a cycle of about 1500 years' duration. This allows the ocean to moderate rapid changes in atmospheric CO_2 concentration, such as those produced by human activity through fossil fuel burning.

> In the thermohaline circulation, dense, cold, salty surface water sinks in the northern Atlantic Ocean, generating slow bottom currents that in turn cause slow upwelling in the Indian Ocean, western Pacific Ocean, and along the coast of Antarctica.

Some scientists have observed that thermohaline circulation could be slowed or stopped by inputs of freshwater into the North Atlantic. Such freshwater inputs could come from the sudden drainage of large lakes formed by melting ice at the close of the last Ice Age. The freshwater would decrease the density of the ocean water, keeping the water from becoming dense enough to sink. Without sinking, circulation would stop. In turn, this would interrupt a major flow pathway for the transfer of heat from equatorial regions to the northern midlatitudes. This mechanism could result in relatively rapid climatic change and is one explanation for the periodic cycles of warm and cold temperatures experienced since the melting of continental ice sheets about 12,000 years ago.

A Look Ahead

This chapter has examined atmospheric and oceanic circulation processes at several scales, including local winds at the local scale, cyclones and anticyclones at the regional scale, and whole-Earth surface and upper atmosphere circulations at the global scale. We have seen how the pattern of pressure and wind is determined by the basic process of unequal heating of the air, which is largely the result of unequal solar heating of the land and ocean surfaces. In the oceans, the global pattern of surface currents is largely driven by the wind, while deep currents are caused by sinking and rising motions resulting from differences in water temperature and density. In both cases, the flows of water and air are strongly affected by the Earth's rotation as it is felt in the Coriolis force.

The global circulation of winds and currents paves the way for our next subject—weather systems. Recall from Chapter 4 that when warm, moist air rises, precipitation can occur. This happens in the centers of cyclones, where air converges. Although cyclones and anticyclones are generally large surface features, they move from day to day and are steered by the global pattern of winds. Thus, your knowledge of global winds and pressures will help you to understand how weather systems and storms develop and migrate. Your knowledge will also be very useful for the study of climate, which we take up in Chapter 7.

IN REVIEW   WINDS AND GLOBAL CIRCULATION

- **El Niño** is a major disturbance of the equatorial Pacific Ocean and atmosphere. It occurs when cold, upwelling surface water along the Peruvian coast is replaced by warm surface water flowing eastward from the equatorial Pacific. Its counterpart is *La Niña*, in which upwelling is enhanced and cold surface water is carried westward along the Equator.

- El Niño/La Niña can shift weather patterns around the world, bringing unexpected drought, floods, heat, and cold. The El Niño of 1997–1998 was particularly severe, resulting in climate changes across the globe.

- The term **atmospheric pressure** describes the weight of air pressing on a unit of surface area. Atmospheric pressure is measured by a **barometer**. Atmospheric pressure decreases rapidly as altitude increases.

- **Wind** occurs when air moves with respect to the Earth's surface. Wind speed is measured by an *anemometer*. Wind direction indicates the compass direction the wind is coming from.

- Air motion is induced by **pressure gradients** that are formed when air in one location is heated to a temperature that is warmer than another. Heating creates high pressure aloft, which moves high-level air away from the area of heating. This motion induces low pressure at the surface, pulling surface air toward the area of heating, and a **thermal circulation** loop is formed.

- *Local winds* are generated by local pressure gradients. *Sea* and *land breezes*, as well as *mountain* and *valley breezes*, are examples caused by local surface heating. Other local winds include the *Santa Ana, chinook,* and *mistral winds.*

- *Wind power* is a widely available power source with enormous potential.

- The Earth's rotation strongly influences atmospheric circulation through the **Coriolis effect**, which is the result of the Earth's rotation. The *Coriolis force* deflects wind motion, producing circular or spiraling flow paths around **cyclones** (centers of low pressure and convergence) and **anticyclones** (centers of high pressure and divergence).

- Air movement in surface cyclones and anticyclones is also affected by a *frictional force* that opposes the movement. Combined with pressure gradient and Coriolis forces, friction causes winds to spiral inward around surface cyclones, and outward around surface anticyclones.

- Because the equatorial and tropical regions are heated more intensely than the higher latitudes, a vast thermal circulation develops—the **Hadley cells**. This circulation drives the *northeast* and *southeast trade winds*, the convergence and lifting of air at the **intertropical convergence zone (ITCZ)**, and the sinking and divergence of air in the **subtropical high-pressure belts**.

- The **polar front,** lying between about 30° and 60° latitude, marks the boundary between warm, moist tropical air and cool, dry, polar air. At the South Pole, out-spiraling winds create the *polar easterlies*.

- The most persistent features of the global pattern of atmospheric pressure are the **subtropical high-pressure belts**, which are generated by the Hadley cell circulation. They intensify and move poleward during their high-Sun season, affecting the climate of adjacent coasts and continents. The northern belt includes the Hawaiian high, to the west of North America, and the Azores high, to the east.

- The **monsoon** circulation of Asia responds to a reversal of atmospheric pressure over the continent with the seasons. A *winter monsoon* flow of cool, dry air from the northeast alternates with a *summer monsoon* flow of warm, moist air from the southwest.

- In the midlatitudes and poleward, westerly winds prevail. In winter, continents develop high pressure, and intense oceanic low-pressure centers are found off the Aleutian Islands and near Iceland in the northern hemisphere. In the summer, the continents develop low pressure as oceanic subtropical high-pressure cells intensify and move poleward.

- Winds aloft are dominated by a global pressure gradient force between the tropics and pole in each hemisphere that is generated by the hemispheric temperature gradient from warm to cold. Coupled with the Coriolis force, the gradient generates strong westerly **geostrophic winds** in the upper air. In the equatorial region, weak easterlies dominate the upper-level wind pattern. Poleward of the Hadley cells, *upper-air westerlies* circle the Earth in an undulating pattern.

- The *polar-front* and *subtropical* **jet streams** are concentrated westerly wind streams with high wind speeds. The *tropical easterly jet stream* is weaker and limited to Southeast Asia, India, and Africa in summer.

- **Jet-stream disturbances** are large undulating patterns that develop in jet streams. Once formed, they tend to grow larger because of *baroclinic instability*. The undulations bring tongues of cold polar air Equator-ward and warm tropical air poleward that can persist for a number of days or weeks.

- Oceans show a surface warm layer, a *thermocline*, and a deep cold layer. Near the poles, the water is uniformly cold, and the warm layer and thermocline are absent.

- **Ocean currents** are dominated by huge surface **gyres** that are driven by the global surface wind pattern. *Equatorial currents* move warm water westward and then poleward along the east coasts of continents. Return flows bring cold water toward the Equator, along the west coasts of continents.

- *El Niño* events are associated with changes in barometric pressure across the equatorial Pacific, known as the *Southern Oscillation*, giving rise to the term *El Niño-Southern Oscillation,* or *ENSO.*

- The *Pacific decadal oscillation* in sea-surface temperature steers the polar jet stream over the Pacific, which alters winter storm tracks to create 20- to 30-year cycles of wet and dry years along the Pacific coast and arid interior western states.

- The *North Atlantic oscillation* affects weather on both sides of the Atlantic. In the positive phase, the polar jet keeps to the north, bringing more storms to northern Europe. In the negative phase, the polar jet is farther south, bringing cold conditions in the eastern United States and more precipitation in the Mediterranean region.

- Slow, deep, ocean currents are driven by the sinking of cold, salty water in the northern Atlantic that ultimately causes upwelling in the Pacific and Indian oceans. This *thermohaline circulation* pattern involves nearly all the Earth's ocean basins, and acts to moderate the buildup of atmospheric CO_2 by moving CO_2-rich surface waters to ocean depths.

KEY TERMS

REVIEW QUESTIONS

1. Describe the change in sea-surface temperatures that occurs with **El Niño**. What were some of the effects of the severe El Niño of 1997–1998?
2. Explain **atmospheric pressure**. Why does it occur? How is atmospheric pressure measured, and in what units? What is the normal value of atmospheric pressure at sea level?
3. How does atmospheric pressure change with altitude?
4. What is **wind**? How do we define wind direction? What instrument is commonly used to measure wind speed and direction?
5. What is a **pressure gradient**? How is it related to atmospheric heating?
6. Describe a simple **thermal circulation** system, explaining how air motion arises from a pressure gradient force induced by heating.
7. Describe *land* and *sea breezes* and *mountain* and *valley winds*. Identify three other types of local winds.
8. Briefly discuss wind power as a source of energy.
9. What is the **Coriolis effect**, and why is it important? What produces it? How does it influence the motion of wind and ocean currents in the northern hemisphere? In the southern hemisphere?
10. Define **cyclone** and **anticyclone**. How does air move within each? What is the direction of circulation of each in the northern and southern hemispheres? What type of weather is associated with each, and why?
11. What is the *frictional force* and how does it affect wind direction?
12. What is a **Hadley cell**? Describe the circulation in a Hadley cell using the terms **intertropical convergence zone (ITCZ)** and **subtropical high-pressure belt**.
13. What is meant by the term **polar front**?
14. Identify two atmospheric features of the subtropical high-pressure belt in the northern hemisphere. How do these features change with the seasons?
15. What is the Asian **monsoon**? Describe the features of this circulation in summer and winter. How is the ITCZ involved? How is the monsoon circulation related to the high- and low-pressure centers that develop seasonally in Asia?
16. Compare the winter and summer patterns of high and low pressure that develop in the northern hemisphere.
17. Describe the surface wind pattern of the south polar region and Southern Ocean. Does it change with the seasons?
18. How does global-scale heating of the atmosphere create a pressure gradient force that increases with altitude?
19. What is the **geostrophic wind**, and what is its direction with respect to the pressure gradient force?
20. Describe the basic pattern of global atmospheric circulation at upper levels.
21. Identify five **jet streams**. Where do they occur? From which direction do they come?
22. What are **jet stream disturbances**? Describe how *baroclinic instability* causes jet streams to grow. Why are they important?
23. Describe the layered temperature structure of the ocean.
24. What is meant by the term **gyre**?
25. What is the general pattern of ocean surface current circulation? How is it related to global wind patterns?
26. How does the normal pattern of wind, pressure, and ocean currents in the equatorial Pacific change during an El Niño event? a *La Niña* event?
27. What is the *Pacific decadal oscillation*? On what time scale does it operate? What are its effects?
28. Describe the *North Atlantic oscillation*. How does it affect North American weather?
29. How does the *thermohaline circulation* induce deep ocean currents?

VISUALIZING EXERCISES

1. Sketch an ideal Earth (without seasons or ocean–continent features) and its global wind system. Label the following on your sketch: doldrums, equatorial trough, Hadley cell, ITCZ, northeast trades, polar easterlies, polar front, polar outbreak, southeast trades, subtropical high-pressure belts, and westerlies.
2. Draw four spiral patterns showing outward and inward flow in clockwise and counterclockwise directions. Label each as appropriate to cyclonic or anticyclonic circulation in the northern or southern hemisphere.
3. Draw a simplified sketch of the global map of ocean currents showing the larger gyres. Identify at least five currents by name.

ESSAY QUESTIONS

1. An airline pilot is planning a nonstop flight from Los Angeles to Sydney, Australia. What general wind conditions can the pilot expect to find in the upper atmosphere as the airplane travels? What jet streams will be encountered? Will they slow or speed the aircraft on its way?

2. You are planning to take a round-the-world cruise, leaving New York in October. Your vessel's route will take you through the Mediterranean Sea to Cairo, Egypt, in early December. Then you will pass through the Suez Canal and Red Sea to the Indian Ocean, calling at Mumbai, India, in January. From Mumbai, you will sail to Djakarta, Indonesia, and then go directly to Perth, Australia, arriving in March. Rounding the southern coast of Australia, your next port of call is Auckland, New Zealand, which you will reach in April. From Auckland, you head directly to San Francisco, your final destination, arriving in June. Describe the general wind and weather conditions you will experience on each leg of your journey.

Chapter 6
Weather Systems

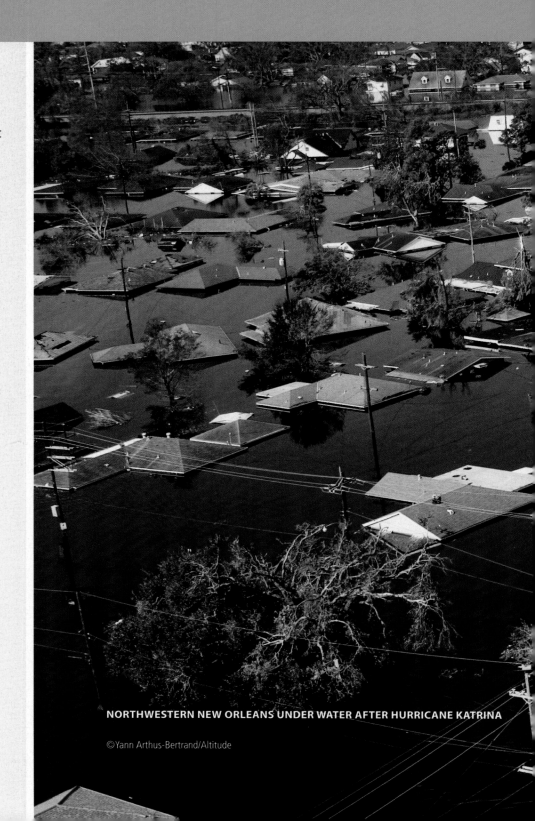

With a death toll exceeding 1300 and property damage in excess of $100 billion, Hurricane Katrina dealt New Orleans and the surrounding coasts a savage blow in 2005. The hurricane's huge storm surge converged on the city from three directions, overtopping and breaking levees, undermining and toppling canal walls, and flooding pumping stations. Eighty percent of the city was flooded, with some parts as much as 6 m (20 ft) underwater. When it was all over, much of the city lay in ruins. Rarely is the awesome power of the tropical cyclone felt more dramatically and at such a cost.

NORTHWESTERN NEW ORLEANS UNDER WATER AFTER HURRICANE KATRINA

©Yann Arthus-Bertrand/Altitude

Weather Systems

Many locations experience changeable weather brought about by weather systems moving through a region. What is a weather system? How do air masses interact in weather system? Why and where in a weather system is precipitation likely to occur? Weather systems include hurricanes and typhoons. How do these destructive storms develop? How and why do they move? How is their intensity measured, and what kinds of damage can they do? These are some of the questions we will answer in this chapter.

Cloud Cover, Precipitation, and Global Warming

Clouds and precipitation represent the most visible and dynamic elements of the hydrologic cycle. Climate change scientists are attempting to address the question of how cloud cover and precipitation patterns will interact with global warming, using highly complex and sophisticated climate models. Recall that global temperatures have been rising over the past few decades, along with human-induced higher atmospheric levels of CO_2. Satellite data show a rise in temperature of the global ocean surface of about 1°C (1.8°F) over the past century. Any rise in sea-surface temperature will raise the rate of evaporation, which will then raise the average atmospheric content of water vapor. What effect will this have on climate?

Water plays several roles in global climate systems. First, in its vapor state, it is one of the greenhouse gases that absorbs and emits longwave radiation, enhancing the warming effect of the atmosphere above the Earth's surface. In fact, water is a more powerful greenhouse gas than CO_2. So, we can predict that an increase in water vapor in the atmosphere should intensify warming. This is what climate scientists refer to as a *positive feedback*—increased CO_2 warms the air, which causes more evaporation, which causes still more warming.

Second, increased water vapor forms clouds. Will more clouds raise or lower global temperatures? Clouds can have different effects on the surface radiation balance, depending on the type of cloud, its thickness, and its moisture content. Large areas of low, white clouds, with their high albedo, can reflect a significant proportion of incoming shortwave radiation back to space, thus acting to lower global temperatures. But cloud droplets and ice particles also absorb longwave radiation from the ground, returning some of that energy as counterradiation. This absorption is an important part of the greenhouse effect, and it is much stronger for water as cloud droplets or ice particles than as water vapor. So, clouds also act to warm global temperatures by intensifying longwave reradiation from the atmosphere to the surface.

Which effect—longwave warming or shortwave cooling—will dominate? This question is at the forefront of climate science. At present, shortwave cooling is stronger than longwave warming. Cloud formation is, therefore, a *negative feedback*; more warming means more clouds, which means more net cooling by the cloud cover (Figure 6.1). That said, global climate models predict that the shortwave cooling effect will become weaker in the future.

Precipitation is also linked to global climate. With more water vapor and more clouds in the air on a warming Earth, more precipitation should result. If precipitation increases in arctic and subarctic zones, snow cover and snow depth could likewise increase. Because snow is a good reflector of solar energy, this would raise the Earth's albedo, thus tending to reduce global temperatures and provide negative feedback to the warming. However, studies to date show that global warming has caused snow to fall later and melt earlier, and sea ice to have a reduced extent. This reduces the average albedo over the year and increases the absorption of solar radiation, so a positive feedback actually results.

The effects of climate change are complex. At this time, global climate models predict that global surface temperatures will be 1.8°C (3.2°F) to 4.0°C (7.2°F) warmer by the year 2100. This warming will be induced primarily by the increase in CO_2 levels predicted for the twenty-first century. But model predictions are always uncertain to some degree. As time goes by, our understanding of global climate, and our ability to accurately predict its changes, are sure to improve.

> Low, thick clouds tend to cool the Earth, and high, thin clouds tend to warm it. On balance, presently, clouds act to cool the Earth.

6.1 Clouds and global climate

If higher surface temperatures evaporate more ocean water, more clouds should result.

Courtesy EUMETSAT, Germany

◀ **EARTH MANTLED BY CLOUDS**
This image, acquired by the European Space Agency's Meteosat–5, shows the Earth's full disk, centered over the Indian Ocean. Cloud height and cloud cover are important factors in determining global climate.

John Eastcott and Yva Momatiuk/NG Image Collection

Raul Touzon/NG Image Collection

▲ **LOW CLOUDS**
Low clouds, like these over Miami, Florida, act to reflect sunlight back to space, cooling the planet.

▲ **HIGH CLOUDS**
High clouds tend to absorb more solar radiation, warming the planet. These wispy cirrus clouds decorate the sky above Denali National Park, Alaska.

Air Masses and Fronts

We know that the Earth's atmosphere is in constant motion, driven by the planet's rotation and its uneven heating by the Sun. The horizontal motion of the wind moves air from one place to another, allowing air to acquire characteristics of temperature and humidity in one region and then carry those characteristics into another region. In addition, as winds at the surface converge and diverge, they generate vertical air motions that affect clouds and precipitation. When air is lifted, it is cooled, enabling clouds and precipitation to form. When air descends, it is warmed, inhibiting the formation of clouds and precipitation. In this way, the Earth's wind systems influence the weather we experience from day to day—the temperature and humidity of the air, cloudiness, and the amount of precipitation.

Some patterns of wind circulation occur commonly and so present recurring patterns of weather. For example, traveling low-pressure centers (*cyclones*) of converging, in-spiraling air often bring warm, moist air in contact with cooler, drier air; clouds and precipitation are typically the result. We recognize these recurring circulation patterns and their associated weather as **weather systems**.

In the midlatitudes, weather systems are often associated with the motion of **air masses**, large bodies of air with fairly uniform temperature and moisture characteristics. An air mass can be several thousand kilometers (or miles) across and can extend upward to the top of the troposphere. We characterize each air mass by its surface temperature, environmental lapse rate, and surface specific humidity. Air masses can be searing hot, icy cold, or any temperature in between. Moisture content can also vary widely between different air masses.

Air masses acquire their characteristics in *source regions*. In a source region, air moves slowly or not at all, which allows the air to acquire temperature and moisture characteristics from the region's land or ocean surface (Figure 6.2). For example, a warm, moist

> Air masses have typical temperature and moisture characteristics that are acquired from a source region. They are classified by latitudinal position (e.g., polar, tropical), and by surface type (i.e., maritime, continental).

6.2 Source regions

Air masses form when large bodies of air acquire the temperature and moisture characteristics of the underlying surface.

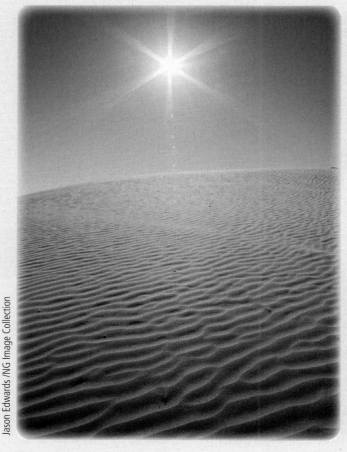

Jason Edwards /NG Image Collection

▲ SUBTROPICAL DESERTS
Over a large subtropical desert, the intense solar radiation and dry surface materials produce a hot air mass with low humidity.

David Doubilet / NG Image Collection

▲ TROPICAL OCEANS
An air mass with warm temperatures and high water vapor content develops over a warm equatorial ocean.

John Dunn / Arctic Light/NG Image Collection

▲ ARCTIC LAND MASSES
A very cold air mass with low water vapor content is generated over cold, snow-covered land surfaces in the arctic zone in winter.

air mass develops over warm equatorial oceans. In contrast, a hot, dry air mass forms over a large subtropical desert. Over cold, snow-covered land surfaces in the arctic zone in winter, a very cold air mass with very low water vapor content occurs.

Pressure gradients and upper-level wind patterns drive air masses from one region to another. When an air mass moves to a new area, it can retain its initial temperature and moisture characteristics for weeks before it comes to fully reflect the new surrounding environment. This is one of the most important properties of air masses: they have the temperature and moisture characteristics of their original source regions, even as they move away from those regions.

We classify air masses by the latitude zone (arctic, polar, tropical, equatorial) and surface type (maritime, continental) of their source regions. Combining these labels produces a list of six important types of air masses (Figure 6.3). Latitudinal position is important because it determines the surface temperature and the environmental temperature lapse rate of the air mass. For example, air-mass temperature can range from $-46°C$ ($-51°F$) for arctic air masses to $27°C$ ($81°F$) for equatorial air masses. The nature of the underlying surface—maritime or continental—usually determines the moisture content of an air mass, given a latitudinal zone. Specific humidity of an air mass can range from 0.1 g/kg (0.0016 oz/lb) over the frozen ground of the arctic to as much as 19 g/kg (0.304 oz/lb) over a warm ocean. In other words, maritime equatorial air can contain about 200 times as much moisture as continental arctic air.

The air masses that form near North America and their source regions have a strong influence on the

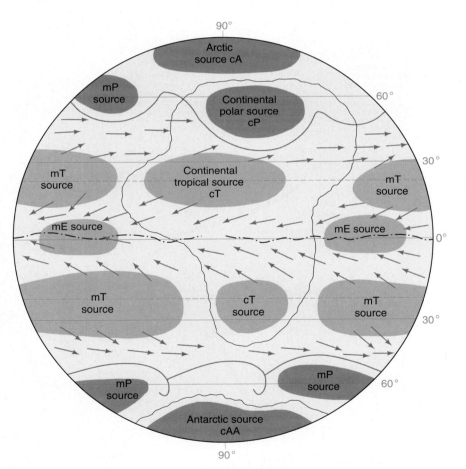

Source regions

Air mass	Symbol	Source region
Arctic	A	Arctic Ocean and fringing lands
Antarctic	AA	Antarctica
Polar	P	Continents and oceans, lat. 50–60° N and S
Tropical	T	Continents and oceans, lat. 20–35° N and S
Equatorial	E	Oceans close to Equator

Surface types

Air mass	Symbol	Surface type
Maritime	m	Oceans
Continental	c	Continents

6.3 **Global air masses and source regions**

In the center of the figure is an idealized continent, which produces continental (c) air masses. It is surrounded by oceans, producing maritime air masses (m). Tropical (T) and equatorial (E) source regions provide warm- or hot-air masses, while polar (P), arctic (A), and antarctic (AA) source regions provide colder air masses of low specific humidity. Polar air masses (mP, cP) originate in the subarctic latitude zone, not in the polar latitude zone. Meteorologists use the word "polar" to describe air masses from the subarctic and subantarctic zones, and we will follow this convention when referring to air masses.

weather. Figure 6.4 shows the air masses and source regions that influence North American weather.

TYPES OF FRONTS

A given air mass usually has a sharply defined boundary between itself and a neighboring air mass. This boundary is termed a **front**. An example is the *polar front*,

described in Chapter 5, where polar and tropical air masses come in contact.

Cold Front

In addition to this polar front, we can also have situations in which a cold-air mass temporarily invades a zone occupied by a warm air mass during the passage of a weather

6.4 North American air-mass source regions and trajectories

Air masses acquire temperature and moisture characteristics in their source regions, then move across the North American continent.

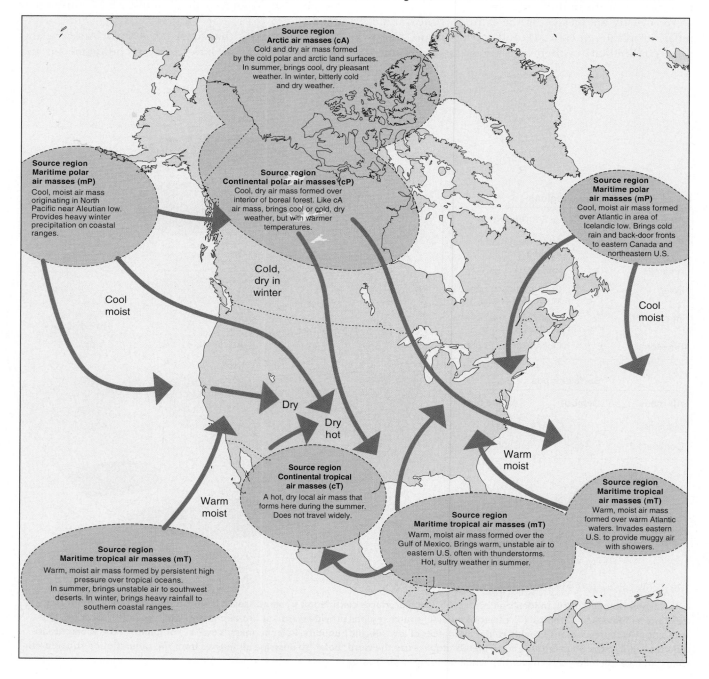

Source region
Arctic air masses (cA)
Cold and dry air mass formed by the cold polar and arctic land surfaces. In summer, brings cool, dry pleasant weather. In winter, bitterly cold and dry weather.

Source region
Maritime polar air masses (mP)
Cool, moist air mass originating in North Pacific near Aleutian low. Provides heavy winter precipitation on coastal ranges.

Source region
Continental polar air masses (cP)
Cool, dry air mass formed over interior of boreal forest. Like cA air mass, brings cool or cold, dry weather, but with warmer temperatures.

Source region
Maritime polar air masses (mP)
Cool, moist air mass formed over Atlantic in area of Icelandic low. Brings cold rain and back-door fronts to eastern Canada and northeastern U.S.

Cold, dry in winter

Cool moist

Cool moist

Dry

Dry hot

Warm moist

Source region
Continental tropical air masses (cT)
A hot, dry local air mass that forms here during the summer. Does not travel widely.

Warm moist

Source region
Maritime tropical air masses (mT)
Warm, moist air mass formed over the Gulf of Mexico. Brings warm, unstable air to eastern U.S. often with thunderstorms. Hot, sultry weather in summer.

Source region
Maritime tropical air masses (mT)
Warm, moist air mass formed over warm Atlantic waters. Invades eastern U.S. to provide muggy air with showers.

Source region
Maritime tropical air masses (mT)
Warm, moist air mass formed by persistent high pressure over tropical oceans. In summer, brings unstable air to southwest deserts. In winter, brings heavy rainfall to southern coastal ranges.

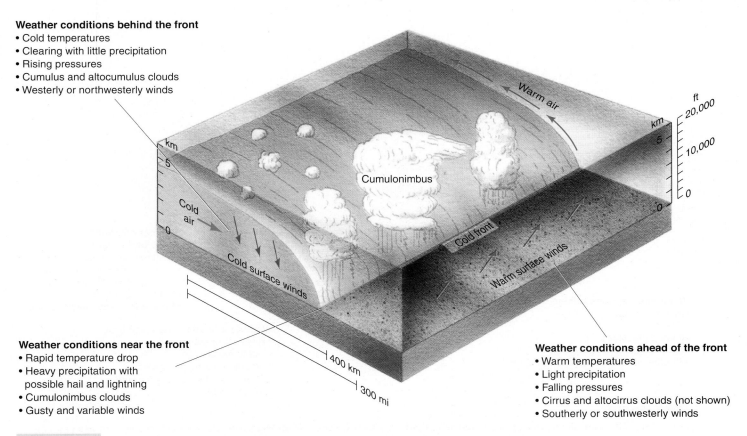

Weather conditions behind the front
• Cold temperatures
• Clearing with little precipitation
• Rising pressures
• Cumulus and altocumulus clouds
• Westerly or northwesterly winds

Weather conditions near the front
• Rapid temperature drop
• Heavy precipitation with
  possible hail and lightning
• Cumulonimbus clouds
• Gusty and variable winds

Weather conditions ahead of the front
• Warm temperatures
• Light precipitation
• Falling pressures
• Cirrus and altocirrus clouds (not shown)
• Southerly or southwesterly winds

6.5 Cold front

In a cold front, a cold air mass lifts a warm air mass aloft. The upward motion can set off a line of showers or thunderstorms. The frontal boundary is actually much less steep than is shown in this schematic drawing.

system. The result is a **cold front**, shown in Figure 6.5. Because the cold air mass is colder, and therefore denser than the warmer air mass, it remains in contact with the ground. As it moves forward, it forces the warmer air mass to rise above it. As the warm air mass rises, it cools adiabatically, water vapor condenses, and clouds form. If the warm, moist air is unstable, severe convection may develop. A cold front often forms a long line of massive cumulus clouds stretching for tens of kilometers (or miles).

As the cold front advances, characteristic weather patterns occur ahead of, behind, and at the front. Before the front arrives, temperatures are still warm, but high cirrus clouds form in the sky, winds shift, and barometric pressure drops, indicating that a front is arriving. As the front arrives, there is a sharp rise in pressure, and temperatures drop rapidly. Winds

> Fronts are boundaries between air masses. When cold air invades warmer air, the boundary is a cold front. When warm air invades colder air, the boundary is a warm front.

are gusty and variable. A dense wall of cumulonimbus clouds produces heavy precipitation, sometimes including hail and lightning. Once the front has passed through the area, the colder air mass brings cooler temperatures and higher pressures. The rain lets up, and clouds begin to clear.

Warm Front

In contrast to a cold front, a **warm front** is one in which warm air moves into a region of colder air, as the cold air retreats (Figure 6.6). Again, the cold air mass remains in contact with the ground because it is denser. As before, the warm-air mass is forced aloft, but this time it rises up on a long ramp over the cold air below. This rising motion, called *overrunning*, creates stratus—large, dense, blanket-like clouds—that often produce precipitation ahead of the warm front. If the warm air is stable, the precipitation will be steady. If the warm air is unstable, convection cells can develop, producing cumulonimbus clouds that provide heavy showers or thunderstorms.

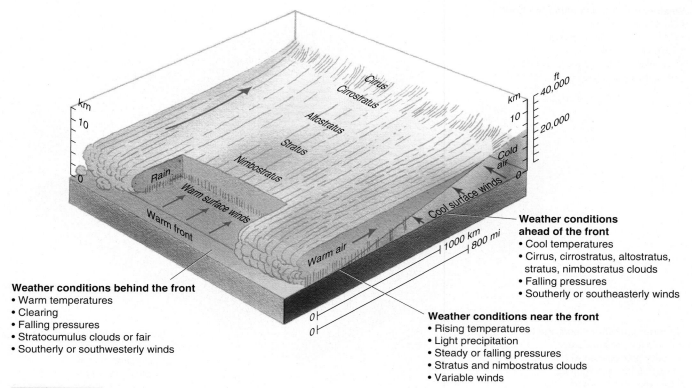

Weather conditions behind the front
• Warm temperatures
• Clearing
• Falling pressures
• Stratocumulus clouds or fair
• Southerly or southwesterly winds

Weather conditions ahead of the front
• Cool temperatures
• Cirrus, cirrostratus, altostratus, stratus, nimbostratus clouds
• Falling pressures
• Southerly or southeasterly winds

Weather conditions near the front
• Rising temperatures
• Light precipitation
• Steady or falling pressures
• Stratus and nimbostratus clouds
• Variable winds

6.6 Warm front

In a warm front, warm air advances toward cold air and rides up and over the cold air. Here, a notch of cloud is cut away to show rain falling from the dense stratus cloud layer.

Warm fronts also bring characteristic weather patterns ahead of, behind, and directly at the front. As a warm front approaches, temperatures are still cool, but pressures begin to drop, and cirrus, cirrostratus, and altostratus clouds form in the sky. As the front arrives, temperatures rise, stratus clouds form, and light precipitation begins. Pressures are steady or falling, and winds are variable.

Occluded and Stationary Fronts

Cold fronts normally move along the ground at a faster rate than warm fronts because the cold, dense air behind the cold front can more easily push through the warm, less dense air ahead of it. The motion of the warm front, which depends on the retreat of the cooler air ahead of it, is slower. Thus, when a cold front and a warm front are in the same region, the cold front can eventually overtake the warm front. The result is an **occluded front**. ("Occluded" means closed or shut off.) The colder air of the fast-moving cold

> In an occluded front, a cold front overtakes a warm front. The warm air is pushed aloft, so that it no longer touches the ground. This abrupt lifting by the denser cold air produces precipitation.

front remains next to the ground, forcing both the warm air and the less cold air ahead to rise over it, as shown in Figure 6.7. The warm air mass is lifted completely free of the ground.

As an occluded front approaches, temperatures are cool, with light to moderate precipitation and falling pressures. As the front arrives, nimbostratus or cumulonimbus clouds form, temperatures fall, and precipitation increases. Pressure remains low but steady. Once the front has passed, temperatures are cold and pressures rise. Cumulus clouds begin to clear and light precipitation clears.

A fourth type of front is known as a *stationary front*, in which two air masses are in contact but there is little or no relative motion between them. Stationary fronts often arise when a cold or warm front stalls and stops moving forward. Clouds and precipitation that were caused by earlier motion will often remain in the vicinity of the now stationary front.

Dry Line

A final type of front—called a *dry line*—can form along the boundary between hot, dry, continental tropical (cT) air and warm, moist, marine tropical (mT) air. Dry lines are usually found out ahead of cold fronts, where southerly and southwesterly winds

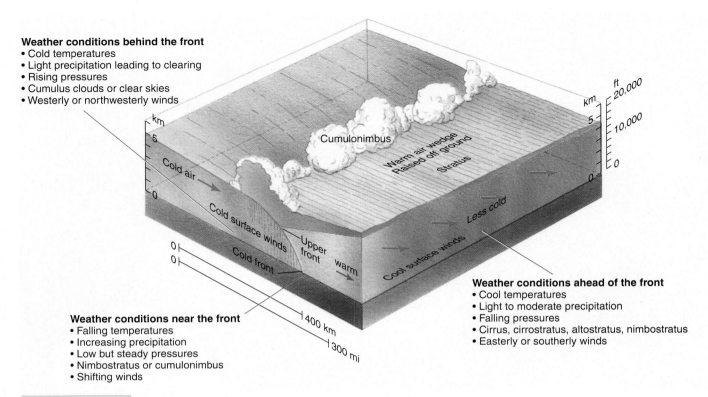

Weather conditions behind the front
- Cold temperatures
- Light precipitation leading to clearing
- Rising pressures
- Cumulus clouds or clear skies
- Westerly or northwesterly winds

Cumulonimbus

Warm air wedge
Raised off ground

Stratus

Cold air

Less cold

Cold surface winds

Upper
front

Cold front

warm

Cool surface winds

Weather conditions near the front
- Falling temperatures
- Increasing precipitation
- Low but steady pressures
- Nimbostratus or cumulonimbus
- Shifting winds

400 km

300 mi

Weather conditions ahead of the front
- Cool temperatures
- Light to moderate precipitation
- Falling pressures
- Cirrus, cirrostratus, altostratus, nimbostratus
- Easterly or southerly winds

6.7 Occluded front

In an occluded front, a cold front overtakes a warm front, lifting a pool of warm, moist air upward. The result is a region of clouds and precipitation.

bring subtropical air from the continental regions and marine regions into contact with one another. Very strong thunderstorms can form along these dry lines as the hot, dry air mass mixes with the warm, moist air mass, raising its temperature and making it very unstable.

Midlatitude Anticyclones and Cyclones

As we saw in Chapter 5, when the Coriolis force, the pressure gradient force, and friction interact, air spirals inward and converges in a *cyclone*, while air spirals outward and diverges in an *anticyclone*. Most types of cyclones and anticyclones are large features that move slowly across the Earth's surface, bringing changes in the weather as they move. These are referred to as *traveling cyclones* and *anticyclones*.

ANTICYCLONES

In an anticyclone, the air diverges and descends. As the air descends, it encounters higher surrounding

pressures, causing the air to warm adiabatically. Warming air does not lead to condensation, so skies are fair, except for occasional puffy cumulus clouds that sometimes develop in a moist surface air layer. For this reason, we often call anticyclones *fair-weather systems*. Toward the center of an anticyclone, winds are light and variable. We find anticyclones in the midlatitudes, typically associated with ridges or domes of clear, dry air that move eastward and toward the Equator. Figure 6.8 shows an example of a large anticyclone centered over eastern North America, bringing fair weather and cloudless skies to the region.

CYCLONES

In a cyclone, the air converges and rises, cooling adiabatically as it does so. If the cooling air reaches saturation, this can cause condensation, leading to precipitation. Many cyclones are weak and pass overhead with little more than a period of cloud cover and light precipitation. However, some cyclones have very intense pressure gradients associated with them that generate strong, intense winds. In addition, the in-spiraling motion associated with these winds can result in significant convergence, and heavy rain or snow can

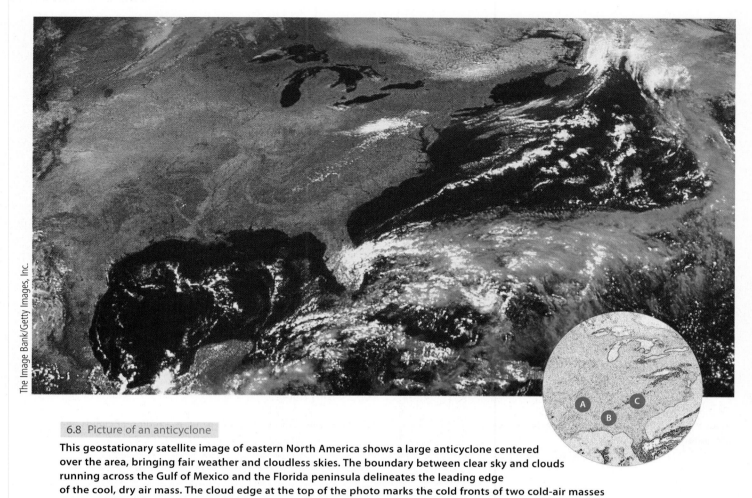

The Image Bank/Getty Images, Inc.

6.8 Picture of an anticyclone

This geostationary satellite image of eastern North America shows a large anticyclone centered over the area, bringing fair weather and cloudless skies. The boundary between clear sky and clouds running across the Gulf of Mexico and the Florida peninsula delineates the leading edge of the cool, dry air mass. The cloud edge at the top of the photo marks the cold fronts of two cold-air masses advancing eastward and southward.

EYE ON THE LANDSCAPE **What else would the geographer see?**

On this clear satellite image, some of the major features of the eastern North American landscape stand out. At **A**, note the difference in soil color along the Mississippi River. These are alluvial soils, deposited over millennia by the Mississippi River during the Ice Ages. At **B**, note the curving arc that marks the boundary between soft coastal plain sediments to the south (light color) and the weathered rocks and soils of the piedmont (dark color), to the north. This boundary can be traced all the way to New York City. Farther inland lie the Appalachians **C**, showing puffy orographic cumulus clouds.

accompany the cyclone. In that case, we call the disturbance a **cyclonic storm**.

There are three types of traveling cyclones. First is the *midlatitude cyclone* of the midlatitude, subarctic, and subantarctic zones, sometimes also called an *extratropical cyclone*. These cyclones range from weak disturbances to powerful storms. Second is the *tropical cyclone* found in the tropical and subtropical zones. Tropical cyclones range from mild disturbances to highly destructive hurricanes, or *typhoons*. A third type is the **tornado**, described in Chapter 5, a small, intense cyclone of enormously powerful winds. The tornado is much smaller in size than other cyclones and is related to strong, localized convective activity.

Midlatitude cyclones can sometimes merge with hurricanes to form very large and dangerous storms. Superstorm Sandy of 2012 is an example (Figure 6.9).

> Traveling cyclones and anticyclones bring changing weather systems. Convergence in cyclones causes condensation and precipitation. Subsidence in anticyclones keeps air clear.

6.9 Superstorm Sandy

On October 22, 2012, a tropical storm formed in the western Caribbean Sea. Intensifying into a Category 1, then 2, hurricane, the storm, now named Sandy, moved northward, wreaking havoc in the Antilles from Jamaica to Cuba. Leaving the Bahamas, heading northeast off the Florida and Carolina coasts, it passed over an unusually warm Gulf stream, gaining energy and new strength. A cold-core midlatitude cyclone, approaching from the west, began to merge with Sandy, providing more energy from the contrast between warm and cold air within the low. Superstorm Sandy was born. It grew to become the largest Atlantic storm in history, with its wind field spanning about 1800 km (1100 mi).

Well to the north, a huge pool of cold air had settled in over Canada's Maritime provinces. Blocking Sandy's path, its clockwise circulation pushed Sandy toward the midatlantic coastline. Sandy's east winds and the east winds around the maritime high, powered by the huge pressure gradient between the two circulation centers, set up an extreme wind field that that pushed water relentlessly toward the shore. As the storm center approached, a full moon pulled the ocean waters ever higher. Sandy's lowest pressures and the spring high tide arrived together between 8:00 and 9:00 PM on the night of October 29. Sustained winds were about 35 m/s (80 mi/hr).

Fed by the wind and tide, a massive storm surge suddenly peaked, flooding vast areas of the New York and New Jersey shoreline. In New York City, the surge measured nearly 14 ft (4.2 m) at the southern tip of Manhattan, flooding tunnels, subways, and electrical infrastructure. In many locations, hurricane-driven breaking waves pushed the water even higher. Between the wind and the water, devastation was complete in many shoreline communities. Much of the region was without power for days or weeks. Damage estimates exceeded $65 billion.

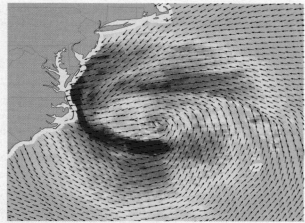

Data courtesy of the Jet Propulsion Laboratory's QuikSCAT and the Indian Space Research Organization OceanSat-2 missions. Courtesy NASA.

▲ SANDY'S WIND FIELD
This NASA image, produced using satellite radar, shows the wind direction and strength of the storm on October 28, just before it curved to the north and west. Landfall was just south of Atlantic City, New Jersey, on the night of October 29..

▼ SEASIDE HEIGHTS AFTER SANDY
On the shoreline of Seaside Heights, New Jersey, houses were wrecked and thrown from their foundations by the wind and the waves of the storm surge. Of the beach boardwalk, only the pilings and crossbeams remain.

Mike Groll / © AP / Wide World Photos

6.10 Life history of a midlatitude cyclone

Midlatitude cyclones are large features, spanning 1000 km (about 600 mi) or more. These are the "lows" that meteorologists show on weather maps. They typically last three to six days. In the midlatitudes, a cyclone normally moves eastward as it develops, propelled by prevailing westerly winds aloft. Circled numbers show the key stages in the life of a cyclone.

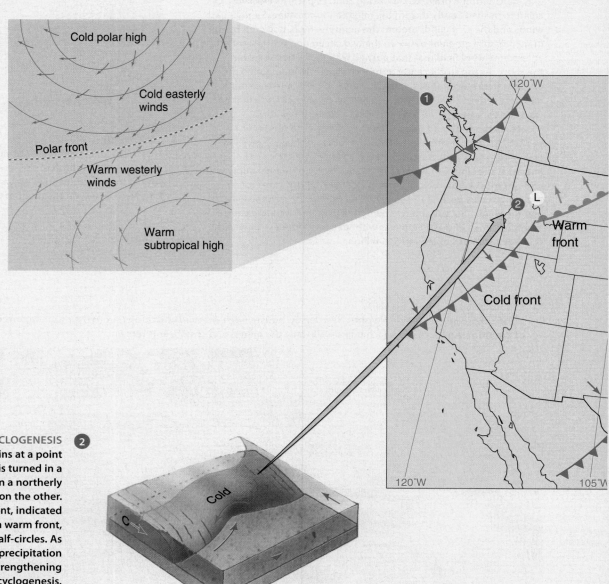

1 INITIAL CONDITIONS FOR A MIDLATITUDE CYCLONE
Along the polar front, two very different air masses are in contact. One air mass is dense, cold polar air, and the other is warm, humid subtropical air. The airflows converge from opposite directions on the two sides of the front, setting up an unstable situation.

Cold polar high

Cold easterly winds

Polar front

Warm westerly winds

Warm subtropical high

Warm front

Cold front

CYCLOGENESIS 2
An undulation or disturbance begins at a point along the polar front. Cold air is turned in a southerly direction, and warm air in a northerly direction, so that each advances on the other. This creates two fronts: a cold front, indicated by a line with blue triangles, and a warm front, indicated by a line with red half-circles. As the fronts begin to move, the precipitation process begins. The formation and strengthening of a cyclone is known as cyclogenesis.

Cold

C

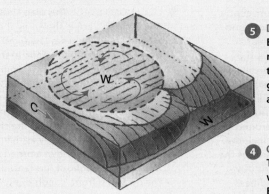

DISSOLVING STAGE
Eventually, the polar front is reestablished, but a pool of warm, moist air remains aloft and north of the polar front. As its moisture content reduces, precipitation dries out, and the clouds gradually dissolve. Soon, another midlatitude cyclone will form along the polar front and move across the continent.

4 **OCCLUDED STAGE**
The faster-moving cold front overtakes the warm front, lifting the warm, moist air mass at the center completely off the ground. Because the warm air is shut off from the ground, this is called an occluded front; it is indicated here by a line with alternating triangles and half-circles. Precipitation continues to occur as warm air is lifted ahead of and behind the occluded front.

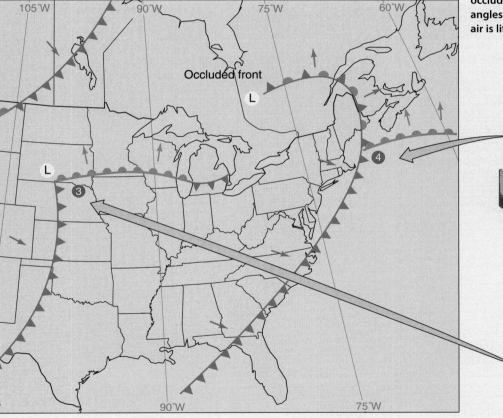

Occluded front

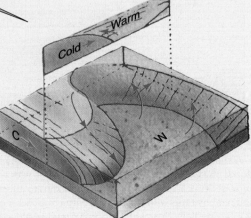

OPEN STAGE **3**
The disturbance along the cold and warm fronts deepens and intensifies. Cold air actively pushes southeastward along the cold front, and warm air actively moves northeastward along the warm front. Precipitation zones along the two fronts are now strongly developed. The precipitation zone along the warm front is wider than the zone along the cold front.

6.11 Simplified surface weather maps and cross sections through a midlatitude cyclone

These maps show weather conditions on two successive days in the eastern United States. The three kinds of fronts are shown by special line symbols, shown in the lower right-hand corner of the Open Stage graphic. Areas of precipitation are shown in gray. We can understand the movement of the respective air masses that generate the fronts by looking at the circulation around the surface low-pressure center, indicated by the isobars. The white line indicates the 0°C (32°F) isotherm.

▼ OPEN STAGE The isobars show a surface low-pressure center with in-spiraling winds. The cold front is pushing south and east, supported by a flow of cold, dry continental polar air circulating around the low-pressure center. Note that the wind direction changes abruptly ahead of the cold front, shifting from southerly to northwesterly. The temperature behind the cold front drops sharply as cP air moves into the region. The warm front is moving north and somewhat east, with warm, moist maritime tropical air circulating around the low pressure, as well. The precipitation pattern includes a broad zone near the warm front and the central area of the cyclone. A thin band of intense precipitation extends down the length of the cold front. Generally, there is cloudiness over much of the cyclone.

▼ OCCLUDED STAGE This map shows conditions 24 hours later. The cyclone track is shown by the red line. The center has moved about 1600 km (1000 mi) in 24 hours—a speed of just over 65 km (40 mi) per hour. In this time, the cold front has overtaken the warm front, forming an occluded front in the central part of the disturbance. A high-pressure area, or tongue of cold polar air, has moved into the area west and south of the cyclone, and the cold front has pushed far south and east. Within the cold-air tongue, the skies are clear. Winds shift from southeasterly to westerly as the occluded front passes. Now precipitation is found in a broad region across the occluded front and throughout the central area of the cyclone. Behind the occluded front, conditions are clear and cold.

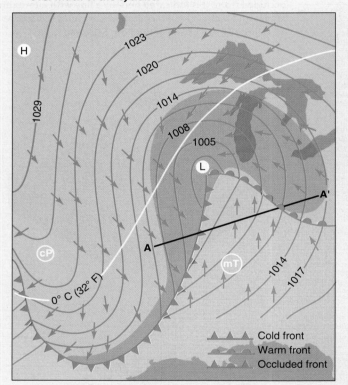

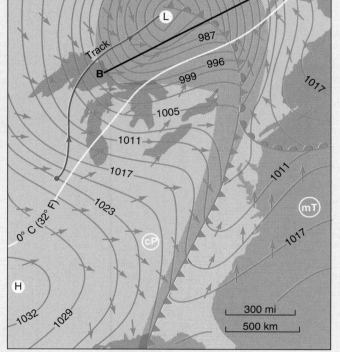

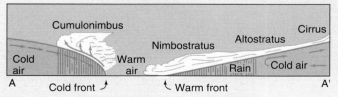

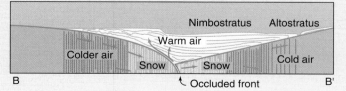

▲ CROSS SECTION OF OPEN STAGE A cross section along the line A–A' shows how the fronts and clouds are related. A broad layer of clouds ahead of the warm front forms a wedge with a thin leading edge of cirrus. Westward, this wedge thickens to altostratus, stratus, and nimbostratus with steady rain. Within the sector of warm air, the sky may partially clear with scattered cumulus. Along the cold front is a narrow belt of cumulonimbus clouds associated with thunderstorms.

▲ CROSS SECTION OF OCCLUDED STAGE A cross section shows conditions along the line B—B', cutting through the occluded part of the storm. Note that the warm air mass is lifted well off the ground and yields heavy precipitation both ahead of and behind the occluded front.

MIDLATITUDE CYCLONES

The **midlatitude cyclone** is the dominant weather system in middle and high latitudes. It is a large in-spiraling of air that repeatedly forms, intensifies, and dissolves along the polar front. Figure 6.10 shows the life history of a midlatitude cyclone, and associated warm and cold fronts, and explains how a midlatitude cyclone forms, grows, and eventually dissolves.

Midlatitude cyclones develop along the *polar front*, which sits between two large anticyclones: the polar high to the north, with its cold, dry air mass, and the subtropical high, with its warm, moist air mass, to the south. At the polar front, the airflow converges from opposite directions, with northeasterly winds to the north of the polar front and southwesterly winds to the south of the polar front. These wind motions lead to a counterclockwise, or cyclonic, circulation that creates a *low-pressure trough* between the two high-pressure cells. Midlatitude cyclones are local intensifications of cyclonic circulation that move along this low-pressure trough. The circular motion

of air around the cyclone generates warm and cold fronts that sweep across large regions, prompting weather changes.

How does weather change as a midlatitude cyclone passes through a region? As the midlatitude cyclone and its accompanying fronts move eastward, a fixed location—like a point south of the Great Lakes—can experience weather ranging from warm, mild conditions with light winds to periods of heavy precipitation with gusty winds to cold, dry, breezy conditions with clear skies overhead. All of these can take place in a span of 24 to 36 hours as the fronts pass through. We can see these changes in Figure 6.11.

MIDLATITUDE CYCLONES AND UPPER-AIR DISTURBANCES

What causes midlatitude cyclones to grow over time? This growth is actually related to the growth of *jet stream disturbances,* which we explained in Chapter 5. To understand this process, look at the upper-level wind and pressure map in Figure 6.12. As the jet stream moves

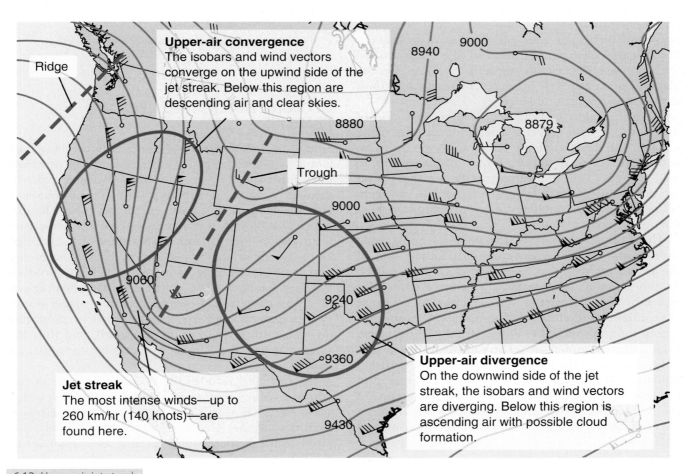

6.12 Upper-air jet streak

This upper-air pressure and wind map shows a jet streak over the southwestern portion of the United States.

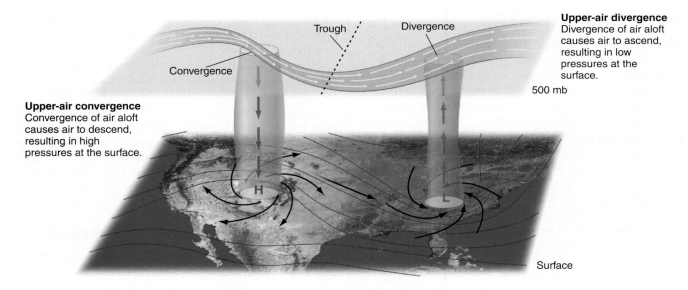

Upper-air divergence
Divergence of air aloft causes air to ascend, resulting in low pressures at the surface.

Trough

Divergence

500 mb

Convergence

Upper-air convergence
Convergence of air aloft causes air to descend, resulting in high pressures at the surface.

H

Surface

6.13 Upper-air and surface pressure patterns

As air diverges and converges around an upper-air low-pressure trough, surface highs and lows are generated.

southward and eastward between the high-pressure disturbance—or *ridge*—and low-pressure disturbance—or *trough*—in the upper atmosphere, it tends to squeeze together or converge. As it does so, the winds around the low-pressure accelerate, forming a *jet streak*. On the eastward side of the low pressure, the winds slow down as they spread apart or diverge.

As the air aloft converges on the upwind side of the jet streak, it produces a descent of air toward the surface. The descending air subsequently produces an anticyclone (or high-pressure center) at the surface, as seen in Figure 6.13. Conversely, as the air diverges on the downwind side of the jet streak, air ascends from below, resulting in a cyclone (or low-pressure center) at the surface.

Thus, the life history of the low-level midlatitude cyclone and its accompanying fronts follows that of the jet stream disturbance, as shown in Figure 6.14. As the upper-air circulation disturbance intensifies and moves eastward, the vertical circulations cause the midlatitude cyclones and anticyclones at the surface to intensify and move along with the disturbance. Eventually, the upper-air jet stream disturbance gets so large it causes the jet stream to circle back on itself, pinching off the upper-air low-pressure center and reestablishing the east-west flow of the jet stream. At the surface, we recognize this as the point at which the midlatitude cyclone occludes and reestablishes the polar front.

WileyPLUS Midlatitude Cyclones
Watch an animation showing the life cycle of a midlatitude cyclone with cold, warm, and occluded fronts.

CYCLONE TRACKS AND CYCLONE FAMILIES

Several important processes can initiate traveling midlatitude cyclones, as seen in Figure 6.15. These processes link upper-air pressure patterns to surface patterns through vertical circulations induced by convergence and divergence, as explained above. The Aleutian and Icelandic lows are regions of frequent jet stream disturbances that spawn eastward-moving cyclones. Disturbances in the upper airflow caused by mountain chains produce *lee-side troughs*, which also produce cyclones. Land-sea coastal boundaries are often regions of strong temperature contrast that can trigger disturbances and cyclones. As the disturbances propagate downwind, their cyclones move as well, and thus the surface storms are "dragged" along by the upper-air *steering winds*.

Because midlatitude cyclones tend to form in certain areas, they tend to travel common paths, called *storm tracks*, as they develop, mature, and dissolve. Cyclones of the Aleutian and Icelandic lows commonly form in succession, traveling as a chain across

6.14 Life history of an upper-air disturbance and accompanying midlatitude cyclone

As an upper-air disturbance in the jet stream develops, vertical circulations in the regions of upper-air convergence and divergence produce changes in surface pressures. As the upper-air disturbance moves and intensifies, so do the accompanying midlatitude cyclones and anticyclones.

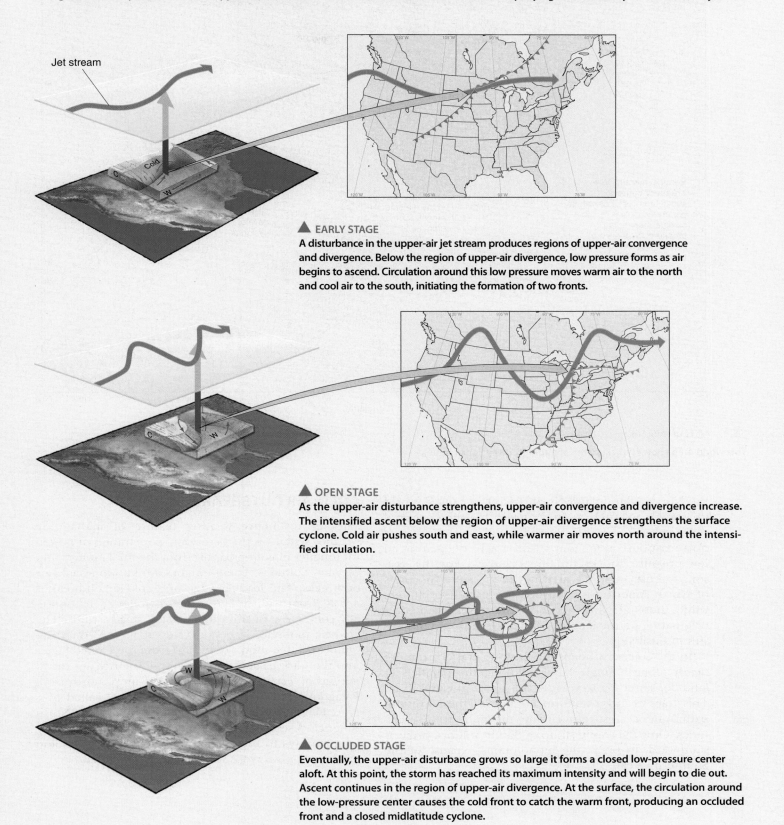

▲ EARLY STAGE

A disturbance in the upper-air jet stream produces regions of upper-air convergence and divergence. Below the region of upper-air divergence, low pressure forms as air begins to ascend. Circulation around this low pressure moves warm air to the north and cool air to the south, initiating the formation of two fronts.

▲ OPEN STAGE

As the upper-air disturbance strengthens, upper-air convergence and divergence increase. The intensified ascent below the region of upper-air divergence strengthens the surface cyclone. Cold air pushes south and east, while warmer air moves north around the intensified circulation.

▲ OCCLUDED STAGE

Eventually, the upper-air disturbance grows so large it forms a closed low-pressure center aloft. At this point, the storm has reached its maximum intensity and will begin to die out. Ascent continues in the region of upper-air divergence. At the surface, the circulation around the low-pressure center causes the cold front to catch the warm front, producing an occluded front and a closed midlatitude cyclone.

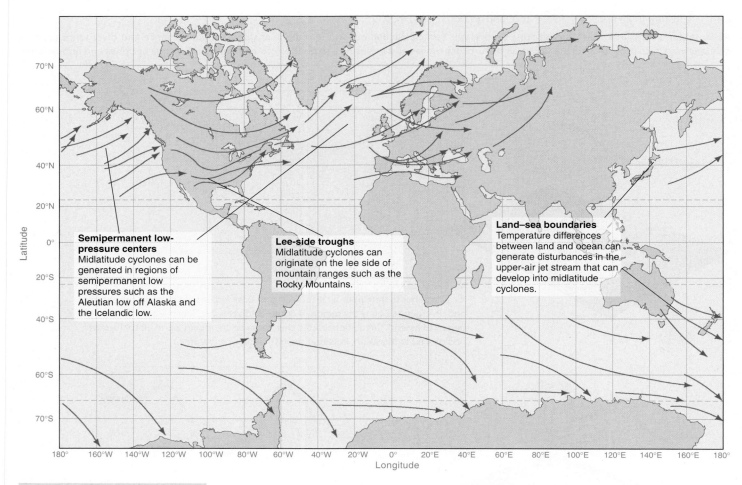

6.15 Paths of midlatitude cyclones

This world map shows typical paths of midlatitude cyclones (blue).

the North Atlantic and North Pacific oceans. A world weather map, such as in Figure 6.16, shows several such *cyclone families*. Each midlatitude cyclone moves northeastward along the storm track, deepening in low pressure and eventually occluding. For this reason, intense cyclones arriving at the western coasts of North America and Europe are usually occluded, while those arriving on the eastern seaboard after originating on the lee side of the Rocky Mountains are still intensifying.

In the southern hemisphere, storm tracks are more nearly along a single lane, following the parallels of latitude. Three such cyclones are shown in Figure 6.16. This track is more uniform because of the consistent pattern of ocean surface circling the globe at these latitudes. Only the southern tip of South America projects southward, to break the monotonous expanse of the Southern Ocean.

COLD-AIR OUTBREAKS

Another distinctive weather feature of midlatitude weather systems is the occasional penetration of powerful tongues of cold polar air from the midlatitudes into very low latitudes. These tongues are known as **cold-air outbreaks**. The leading edge of a cold-air outbreak is a cold front with squalls, which is followed by unusually cool, clear weather and strong, steady winds. The cold-air outbreak is best developed in the Americas. Outbreaks that move southward from the United States into the Caribbean Sea and Central America are called *northers* or *nortes*, whereas those that move north from Patagonia into tropical South America are called *pamperos*. Figure 6.17 shows one such outbreak over North America. A severe polar outbreak may bring subfreezing temperatures to the low latitudes of both regions and damage tropical crops such as citrus and coffee.

6.16 Daily world weather map

A daily weather map of the world for a given day during July or August might look like this map, which is a composite of typical weather conditions.

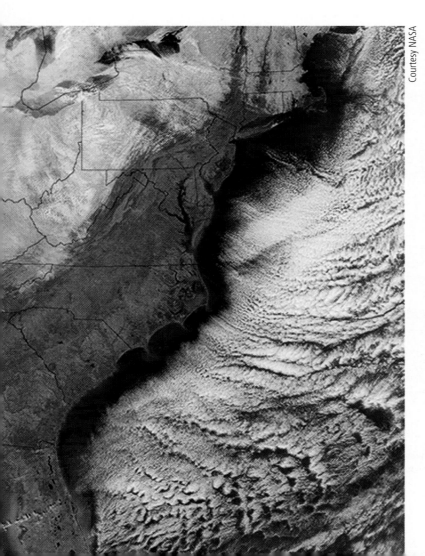

Courtesy NASA

6.17 Cold-air outbreak

This satellite image shows the eastern seaboard during a cold-air outbreak on February 28, 2002. White regions over the central and northeastern portion of the United States are associated with snow cover. To the south, clear, cold conditions are found. This cold-air outbreak brought subfreezing cP air from Canada down into Florida.

Tropical and Equatorial Weather Systems

So far, we have discussed weather systems of the midlatitudes and regions toward the poles. Weather systems of the tropical and equatorial zones show some basic differences from those of the midlatitudes. Upper-air winds are often weak, so air-mass movement is slow and gradual. Air masses are warm and moist, and different air masses tend to have similar characteristics, so fronts are not as clearly defined. Without strong temperature gradients across contrasting air masses, there are no large, intense upper-air disturbances. On the other hand, the high moisture content leads to intense convective activity in low-latitude maritime air masses. Because these air masses are very

> Tropical weather systems include easterly waves and weak equatorial lows. Precipitation results when moist air converges in these systems, triggering convective showers.

moist, only slight convergence and uplift are needed to trigger precipitation.

One of the simplest forms of tropical weather systems is an **easterly wave**, a slow-moving trough of low pressure within the belt of tropical easterlies (the trade winds). These waves occur in latitudes 5° to 30° N and S over oceans, but not over the Equator itself. Figure 6.18 shows circulations and weather features associated with an easterly wave.

Another related weather system is the *weak equatorial low*, a disturbance that forms near the center of the equatorial trough. Moist equatorial air masses converge on the center of the low, causing rainfall from many individual convectional storms. Several such weak lows usually lie along the ITCZ.

TROPICAL CYCLONES

The **tropical cyclone** is the most powerful and destructive type of cyclonic storm (Figure 6.19). It is known as a *hurricane* in the western hemisphere, a *typhoon* in the western Pacific off the coast of Asia, and a *cyclone* in the Indian Ocean. This type of storm typically develops over oceans between 10° and 20° N and S latitudes and no closer than 5° from the Equator.

A tropical cyclone can originate as an easterly wave or weak low that intensifies and grows into a deep, circular low. It can also form as an upper-air disturbance in the subtropical jet stream moves south into the tropical regions. Once formed, the storm moves westward through the trade-wind belt, often intensifying as it travels. It can then curve poleward and eastward, steered by winds aloft. Tropical cyclones can penetrate well into the midlatitudes, as many residents of the southern and eastern coasts of the United States have experienced.

Tropical cyclones grow from *tropical depressions*, which are cyclones with winds below 17 m/s (39 mph). They intensify into *tropical storms* with winds of 18 to 33 m/s (40 to 74 mph). When winds exceed 33 m/s (74 mph), a tropical storm becomes a tropical cyclone.

An intense tropical cyclone is an almost-circular storm center of extremely low pressure (Figure 6.20). Because of the very strong pressure gradient, winds

> Tropical cyclones (also known as hurricanes or typhoons, depending on location) are often powerful and destructive storms, with wind speeds of 30 to 50 m/s (about 65 to 135 mi/hr) and above.

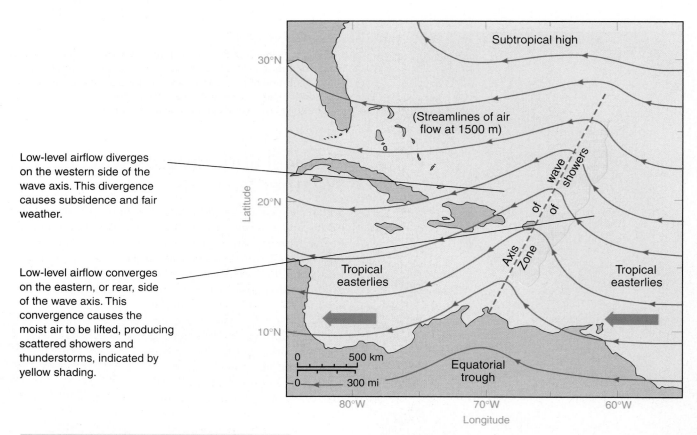

Low-level airflow diverges on the western side of the wave axis. This divergence causes subsidence and fair weather.

Low-level airflow converges on the eastern, or rear, side of the wave axis. This convergence causes the moist air to be lifted, producing scattered showers and thunderstorms, indicated by yellow shading.

6.18 An easterly wave passing over the West Indies

A zone of weak low pressure is at the surface, under the axis of the wave. The wave travels westward at a rate of 300 to 500 km (about 200 to 300 mi) per day. Rainy weather associated with the passage of the wave may last a day or two.

Courtesy NOAA

6.19 Typhoons Aere and Chaba

This satellite image shows Typhoon Aere, situated over Taiwan (upper left), and Typhoon Chaba, situated over the western Pacific (right center), on August 24, 2004. Typhoon Chaba became one of the 15 most intense tropical cyclones on record, surpassing any hurricane in the Atlantic.

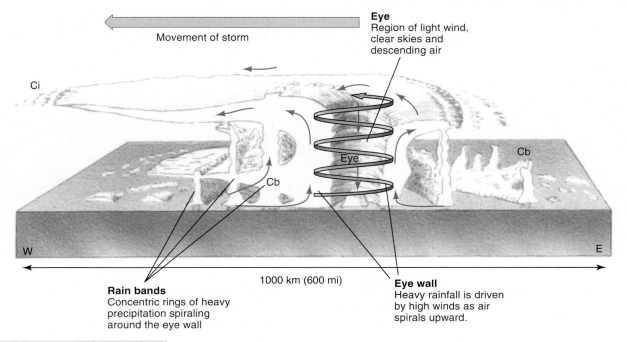

6.20 Structure of a tropical cyclone

In this schematic diagram, cumulonimbus (Cb) clouds in concentric rings rise through dense stratiform clouds. Cirrus clouds (Ci) fringe out ahead of the storm.

6.21 Development and intensification of tropical cyclones

Tropical cyclones are intense wind and rain events. The intensification of these tropical cyclones involves positive feedback loops between the ocean and the atmosphere.

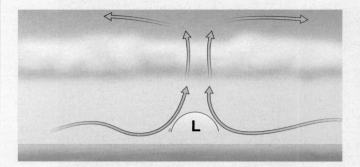

◄ **STARTING THE ENGINE**
Tropical cyclones begin when low-level air flow is disturbed—by an easterly wave or the Equator-ward intrusion of an upper-air disturbance. Either can initiate the convection needed to start a hurricane. Once convection begins, a low-pressure center forms near the surface.

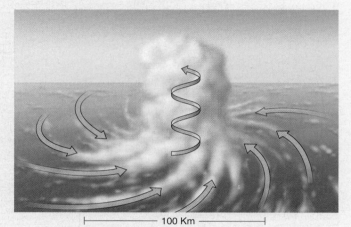

FEEDING IT SOME FUEL ►
The low-pressure center produces in-spiraling air from the tropical ocean. This warm, moist air converges.

◄ **FEEDING IT MORE**
As warm, moist air rises, it expands and cools adiabatically. Once the air cools to the dew-point temperature, condensation begins, releasing tremendous latent heat into the surrounding air. This heating accelerates the upward flow of air.

RUNNING IT WIDE OPEN ►
Convection grows "explosively," accelerating airflow vertically and lowering surface pressures even more. The lowering pressure induces stronger in-spiraling of warm, moist air. As this air rises, its water vapor condenses, releasing more latent heat. This enhances convection further, leading to even lower pressures. Around the center of the hurricane, convection and winds are most intense. However, because the air is spinning so fast, it never reaches the center. Here, calm prevails, with descending air producing a clearing of clouds characteristic of the hurricane eye.

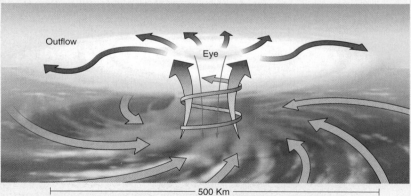

spiral inward at high speed. Convergence and uplift are intense, producing very heavy rainfall. The storm gains its power through the release of latent heat energy as the intense precipitation condenses from the moist air. The storm's diameter may be 150 to 500 km (about 100 to 300 mi). Wind speeds can range from 30 to 50 m/s (about 65 to 135 mi/hr), and sometimes much higher.

Barometric pressure in the storm center commonly falls to 950 mb (28.1 in. Hg) or lower.

Another characteristic feature of a well-developed tropical cyclone is its *eye*, in which clear skies and calm winds prevail. The eye is a cloud-free vortex produced by intense inward spiraling at the center of the storm. Around the eye, the wind is spinning so fast that it cannot converge into the center. Here in the eye, air descends from high altitudes and is adiabatically warmed, causing reevaporation of cloud droplets. As the eye passes over a site, calm prevails, and the sky is clear. It may take about half an hour for the eye to pass, after which the storm strikes with renewed ferocity but with winds in the opposite direction. Wind speeds and precipitation are highest along the cloud wall surrounding the eye. However, precipitation can also be intense in the rain bands that extend in concentric circles away from the hurricane.

TROPICAL CYCLONE DEVELOPMENT

Why do tropical cyclones become so intense? The intensification is due to positive feedback between the ocean and atmosphere, described in Figure 6.21, that allows the atmosphere to draw in massive amounts of energy stored in the ocean.

As an easterly wave moves over the warm ocean waters of the low latitudes, the convergence at the surface results in convective lifting. This lifting subsequently results in condensation, which releases a flow of latent heat energy that warms the surrounding air and produces even greater convection. As the convection intensifies, it further lowers the pressure at the surface. The result is enhanced convergence of warm, moist air, which supplies even more latent heat energy to the system as the air is lifted. As the low pressure at the surface continues to drop, the in-spiraling winds intensify, producing the ferocious winds described in the chapter opening.

At the latitude of tropical cyclone development, the Coriolis force is weaker than in the midlatitudes. Recall from Chapter 5 (see Figure 5.21) that at upper levels in the atmosphere the Coriolis force balances the pressure gradient around a cyclone, and the magnitude of this "balancing act" is proportional to the wind speed. With a weaker Coriolis force, the winds must move faster to balance the pressure gradient. So, given a midlatitude cyclone and a tropical cyclone with equal central low pressures, winds around the tropical cyclone will be faster. As the latitude approaches the Equator, the Coriolis force continues to weaken and finally disappears. This explains why tropical cyclones do not form at latitudes between about 5° N and 5° S. Without the Coriolis force, there is no mechanism to cause winds to revolve around a low-pressure center, so they do not converge to form cyclones.

WileyPLUS Hurricanes
Animations show how a hurricane develops, fueled by the latent heat flow released by condensation as warm, moist ocean air spirals inward and upward.

TROPICAL CYCLONE TRACKS

Tropical cyclones occur only during certain seasons. For hurricanes of the North Atlantic, the season typically runs from June through November, with maximum frequency in late summer or early autumn. In the southern hemisphere, the season is roughly the opposite. These periods follow the annual migrations of the ITCZ to the north and south, with the seasons, and correspond to periods when ocean temperatures are warmest.

Most of the storms originate at 10° to 20° N and S latitude and tend to follow known *tropical cyclone tracks* (Figure 6.22). In the northern hemisphere, they most often travel westward and northwestward through the

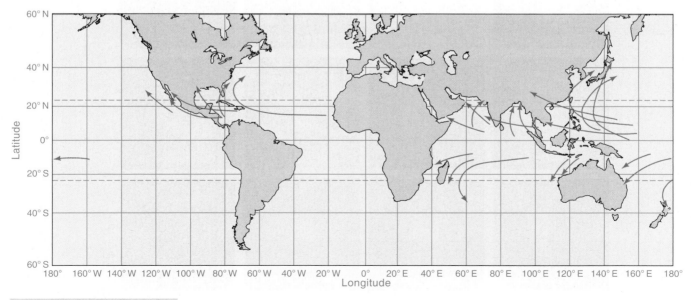

6.22 Paths of tropical cyclones

This world map shows typical paths of tropical cyclones. They develop over warm oceans to the north and south of the Equator, are steered westward on tropical easterlies, and then often turn poleward and eastward into the zone of prevailing westerlies.

6.23 Atlantic tropical cyclone gallery for 2005

The year 2005 was the most active on record for Atlantic hurricanes. Shown here are the 27 named storms that occurred during the official season, from June 1 to November 30. Katrina was the costliest Atlantic storm on record. Wilma was the most intense, at one point observed with a central pressure of 882 mb (26.05 in. Hg) and winds of 83 m/s (185 mph).

trade winds, and then turn northeast at about 30° to 35° N latitude into the zone of the westerlies. Here their intensity lessens, especially if they move over land. In the trade-wind belt, the cyclones typically travel at 10 to 20 km (6 to 12 mi) per hour. In the zone of the westerlies, their speed is more variable.

What factors are required for tropical cyclone development? As shown in Figure 6.22, tropical cyclones form over ocean regions in the low latitudes, but not within 5° of the Equator. Here, we find the ocean waters with temperatures greater than 26.5°C (80°F) that can provide the flow of latent and sensible heat energy needed to drive the tropical cyclone. Also required is an unstable environmental lapse rate, so that convection can easily be initiated by passing easterly waves or upper-air disturbances.

Weak winds with little change in speed or direction with height are also needed, so that the large uniform vertical structure of the tropical cyclone can develop. If this uniform structure is disrupted, the tropical cyclone will quickly die out. Finally, the atmosphere must have a high water content through the bottom 5 km (about 16,000 ft) of the atmosphere. If warm, moist surface air mixes with dry air above, the relative humidity of the air will decrease and the amount of latent heat energy release by condensation will not be sufficient to sustain the needed uplift.

Conversely, what factors prevent the formation of tropical cyclones? Because weak winds are needed at the start of the process, moderate to strong winds above the surface will tend to disrupt the uniform structure of the tropical cyclone. Also, descent of air from above will inhibit convection and the vertical circulations necessary to release latent heat. Finally, if the low is too close to the Equator, the Coriolis force will be too weak to deflect the air motion, and it will rapidly converge into the low-pressure center, evening out the pressure at the surface.

In the western hemisphere, hurricanes originate in the Atlantic off the west coast of Africa, in the Caribbean Sea, or off the west coast of Mexico. In the Indian Ocean, hurricanes are simply called cyclones. They originate both north and south of the Equator, moving north and west, to strike India, Pakistan, and Bangladesh, as well as south and west, to strike the eastern coasts of Africa and Madagascar. *Typhoons* of the western Pacific also form both north and south of the Equator, moving into northern Australia, Southeast Asia, China, and Japan. Curiously, tropical cyclones almost never form in the South Atlantic or southeast Pacific regions. As a result, South America is not threatened by these severe storms.

Once formed, tropical cyclones are given names, to make it easier for weather forecasters to describe and track them. Male and female names are alternated in an alphabetical sequence that is renewed each season. Different sets of names are used within distinct regions, such as the western Atlantic, western Pacific, or Australian regions. In general, names are reused, except for those that identify storms that cause significant damage or destruction; those names are retired from further use.

To track tropical cyclones, we now use satellite images. Within these images, tropical cyclones are often easy to identify by their distinctive pattern of in-spiraling bands of clouds and a clear eye at the storm's center. Figure 6.23 shows a gallery of satellite images of the Atlantic tropical cyclones from 2005, the most active Atlantic hurricane season in recorded history.

WileyPLUS Remote Sensing and Climate Interactivity
Examine satellite images of Hurricanes Hugo and Andrew and watch the storms develop. Learn how cloud types are organized in the hurricane structure.

IMPACTS OF TROPICAL CYCLONES

Tropical cyclones can be tremendously destructive storms, with intense rainfall and very strong winds. The effects of wind, sea-level rise, and rain can cause devastation across very large areas. This destruction can occur along the coasts and farther inland along waterways and over mountain ranges, affecting thousands as the tropical cyclone moves along its track.

The intensity of a tropical cyclone is based on the central pressure of the storm, mean wind speed, and height of the accompanying sea-level rise. Storms are ranked from category 1 (weak) to category 5 (devastating) on the Saffir–Simpson scale, shown in Figure 6.24.

To be classified as a hurricane, a tropical cyclone must have sustained winds of over 33 m/s (74 mph). However, sustained winds in the strongest storms can exceed 70 m/s (about 150 mph), with wind gusts of 90 m/s (about 200 mph) at times.

In addition, rainfall rates can be extremely high. During the passage of some tropical cyclones, 600 mm (2 ft) of rain or more can fall at a location. In some coastal regions, these storms provide much of the summer rainfall. Although this rainfall is a valuable water resource, it can also produce freshwater flooding, elevating levels of rivers and streams above their banks. On steep slopes, soil saturation and high winds can topple trees and produce disastrous earthflows.

Usually, however, the most serious effect of tropical cyclones is coastal destruction, caused by storm waves and very high tides, as shown in Figure 6.25. Since atmospheric pressure at the center of the cyclone is so low, the sea level rises within the center of the storm. In addition, high winds generate damaging surf and push water toward the coast on the side of the storm that has onshore winds, raising the sea level even higher. Waves

The strong onshore winds of a tropical cyclone, coupled with high tides and a favorable offshore sea-bottom configuration, can produce a devastating storm surge that inundates wide areas near the coast. Heavy rains, while recharging water supplies, can also produce damaging floods.

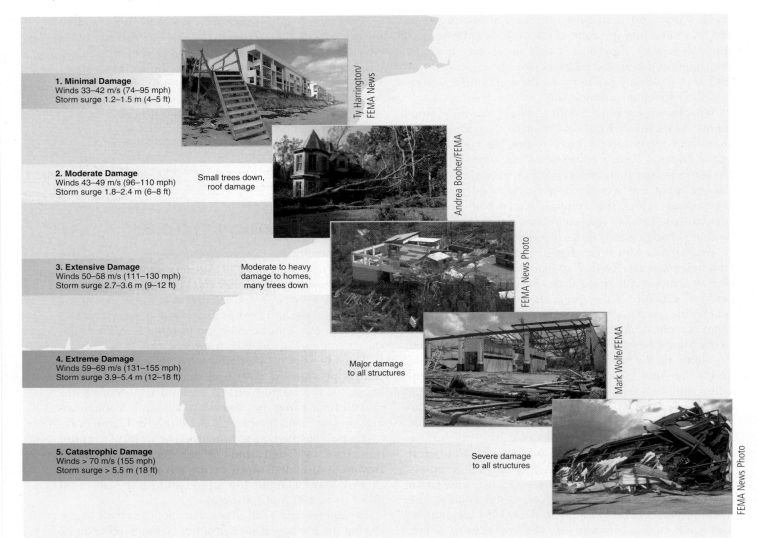

1. Minimal Damage
Winds 33–42 m/s (74–95 mph)
Storm surge 1.2–1.5 m (4–5 ft)

2. Moderate Damage
Winds 43–49 m/s (96–110 mph)
Storm surge 1.8–2.4 m (6–8 ft)

Small trees down, roof damage

3. Extensive Damage
Winds 50–58 m/s (111–130 mph)
Storm surge 2.7–3.6 m (9–12 ft)

Moderate to heavy damage to homes, many trees down

4. Extreme Damage
Winds 59–69 m/s (131–155 mph)
Storm surge 3.9–5.4 m (12–18 ft)

Major damage to all structures

5. Catastrophic Damage
Winds > 70 m/s (155 mph)
Storm surge > 5.5 m (18 ft)

Severe damage to all structures

Ty Harrington/FEMA News

Andrea Booher/FEMA

FEMA News Photo

Mark Wolfe/FEMA

FEMA News Photo

6.24 Saffir-Simpson scale of tropical cyclone intensity

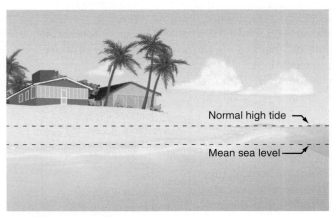

Normal conditions
Structures above normal high tide are not subject to damage from continuously breaking waves.

Normal high tide

Mean sea level

Height of storm surge

Normal high tide

Storm surge
A storm surge combined with a high tide can lift the sea level so that structures are subjected to continuous pounding of heavy surf.

6.25 Storm surge and its effect on coastal areas

As a tropical cyclone moves onshore, it can bring with it a devastating storm surge. By raising the sea level through a combination of high winds and low surface pressures, the storm surge can inundate low-lying areas and subject them to heavy surf.

The Tropical Rainfall Monitoring Mission

In November of 1997, NASA and the Japanese Space Agency launched the Tropical Rainfall Monitoring Mission (TRMM), a joint satellite mission with the goal of monitoring rainfall in the tropical and equatorial regions of the world. Its operational objective is to provide more information about where and when intertropical convective precipitation occurs, particularly over oceans, which are not monitored by weather stations and rain gauges. TRMM is also designed to document how rainfall formation occurs within rain clouds in the intertropical region. Still in operation in 2012, the mission has, over its long lifetime, successfully provided large volumes of information on rainfall amounts and locations, as well as the structure of storms and cyclones. Figure 6.26 shows an example of how TRMM measures precipitation.

The TRMM satellite has a unique orbit, covering only the region from 35° N to 35° S latitudes. With a low altitude and a short orbital period, the platform's instruments sample the complete day/night precipitation cycle over land and ocean surfaces in this region. In this way, scientists have been able to accumulate accurate statistics about rainfall frequency, intensity, duration, and calculate the latent heat energy released in the precipitation process. This capability has, in turn, made it possible for scientists to develop mathematical models of atmospheric circulation, enabling them to predict with much greater accuracy global energy transport, winds, and precipitation.

The TRMM satellite platform has three principal instruments: the precipitation radar, the passive microwave imager, and the visible and infrared scanner. The precipitation radar provides three-dimensional images of storm clouds, including the intensity and distribution of rain, precipitation type, cloud height, and height at which snow melts into rain. The passive microwave imager monitors water vapor, cloud water, and rainfall intensity by measuring the intensity of microwave radiation emitted by liquid water droplets. The visible and infrared scanner tracks clouds as bright objects at visible wavelengths, while its thermal bands measure cloud-top temperature, which indicates cloud height.

Figure 6.27 shows data obtained as TRMM passed near the center of Hurricane Ike on September 4, 2008. The inward-spiraling cloud pattern of this tropical cyclone is clearly shown on the wide track of the instrument, which combines the two imagers (left). The radar transect (right) shows the three-dimensional structure of the storm along a slice through one of its rain bands. Hurricane Ike later crossed Cuba as a category 3 storm, and then hit the Texas coast near Galveston as a very large category 2 storm. Its winds brought storm surges of 1.5 to 4 m (5 to 13 ft) to the coast that leveled the towns of Crystal Beach, Caplen, and Gilchrist, Texas, and destroyed about 13,000 homes in Terrebone Parish, Louisiana.

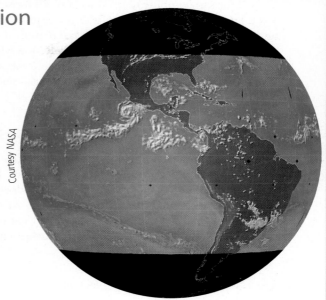

Courtesy NASA

6.26 Intertropical rainfall measured by TRMM

This image shows precipitation as observed over a two-day period by TRMM instruments. A tropical cyclone lies off the Pacific coast of Mexico. Also visible at the top of the image is a line of frontal precipitation across the United States and a band of rainfall across the southern Pacific from the Equator southeast to southern Chile.

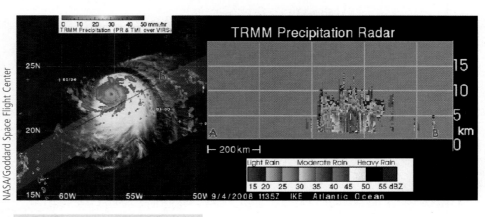

NASA/Goddard Space Flight Center

6.27 TRMM observes Hurricane Ike

This data swath from TRMM instruments shows Hurricane Ike in the Atlantic on September 4, 2008, located about 1000 km (620 mi) east-northeast of San Juan, Puerto Rico. The category 4 storm had sustained winds of 60 m/s (132 mi/hr), with gusts of 72 m/s (161 mi/hr). The image on the left shows the cloud pattern, with a rainfall rate map superimposed. The eye is clearly visible. It was obtained from the TRMM visible/infrared scanner and the microwave imager. The transect A–B, acquired by the platform's precipitation radar, shows a cross section through a rain band to the southeast of the storm's center.

strike the shore at points far inland of the normal tidal range. Low pressure, winds, and the underwater shape of a bay floor can combine to produce a sudden rise of water level, known as a **storm surge**, which carries ocean water and surf far inland. If high tide accompanies the storm, waters will be even higher. For example, low-lying coral atolls of the western Pacific may be entirely swept over by wind-driven seawater, washing away palm trees and houses and drowning the inhabitants.

IMPACTS ON COASTAL COMMUNITIES

Coastal residents of South Florida are particularly familiar with of the damage hurricanes can cause. In 1992, Hurricane Andrew struck the east coast of Florida, near Miami. The second most damaging storm to occur in the United States, it claimed 26 lives and caused more than $35 billion in property damage, measured in today's

dollars. In 2004, four major hurricanes crossed Florida—Charley, Frances, Ivan, and Jeanne. Taken together, these storms destroyed over 25,000 homes in Florida, with another 40,000 homes sustaining major damage.

South Florida is not the only coastal region in the United States to suffer the devastating effects of hurricanes. In 2005, Hurricane Katrina laid waste to the city of New Orleans and much of the Louisiana and Mississippi Gulf coasts (Figure 6.28). Originating southeast of the Bahamas, the hurricane first crossed the South Florida peninsula as a category 1 storm then moved into the Gulf of Mexico, where it intensified to a category 5 storm. Weakening somewhat as it approached the Gulf Coast, its eye came ashore early on August 29 at Grand Isle, Louisiana, with sustained winds of 56 m/s (125 mph).

The city of New Orleans was devastated. Total losses were estimated at more than $100 billion; the official death toll exceeded 1300. The Gulf coasts of Mississippi

6.28 Hurricane Katrina devastates New Orleans

New Orleans is particularly vulnerable to hurricane flooding. Most of its land area has slowly sunk below sea level as underlying Mississippi River sediments have compacted through time. Levees protect the city from Mississippi River flooding on the south, as well as from ocean waters entering from saline Lakes Borgne on the east, and Pontchartrain on the north. Canals and shipping channels connect the river with the lakes and the Gulf.

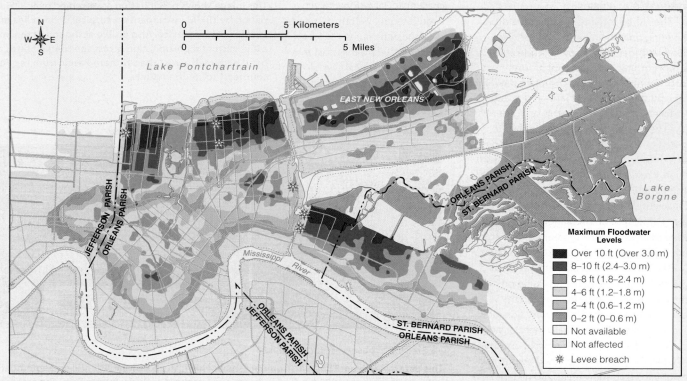

▲   MAXIMUM FLOODWATER LEVELS
This sketch map, based on data compiled by the *New Orleans Times–Picayune* newspaper, shows the extent of flooding during and following the passage of Hurricane Katrina. The storm surge first swept westward from Lake Borgne along a shipping channel, overtopping and eroding levees and flooding east New Orleans and St. Barnard Parish. Penetrating deep into the city, the surge overtopped and breached floodwalls and levees along the main canal connecting the Mississippi River with Lake Pontchartrain. Water levels rose in Lake Pontchartrain, overtopping dikes and levees. Canal walls failed, allowing water to pour into the central portion of the city. Eighty percent of the city was covered with water at depths of up to 6 m (20 ft).

and Louisiana were also hard-hit, with a coastal storm surge as high as 8 m (25 ft) penetrating from 10 to 20 km (6 to 12 mi) inland. Adding insult to injury, much of New Orleans was flooded again just three weeks later by Hurricane Rita, a category 3 storm that made landfall on September 24 at the Louisiana–Texas border.

WileyPLUS Migration of Hurricane Katrina
Explore the development and movement of Hurricane Katrina and view the sequence of the flooding that devastated New Orleans.

Poleward Transport of Energy and Moisture

As we saw in Chapter 5, the general circulation of the atmosphere and oceans is driven by the difference in solar heating between low and high latitudes. This general

▼ DEVASTATION
In the aftermath of Hurricane Katrina, entire neighborhoods in New Orleans lay in ruins for months. Two weeks after the hurricane, the devastation was nearly complete.

©AP/Wide World Photos

circulation, combined with jet stream disturbances in the midlatitudes, serves to redistribute energy and moisture from the Equator to the poles in the process of *poleward energy transport*. Figure 6.29 shows the various mechanisms by which this redistribution of heat energy and moisture takes place.

> Mechanisms of poleward energy transport include Hadley cells, jet stream disturbances, and the Atlantic thermohaline circulation.

An important feature of this redistribution is the Hadley cell circulation, a global convection loop in which moist air converges and rises in the intertropical convergence zone (ITCZ) while subsiding and diverging in the subtropical high-pressure belts.

The Hadley cell convection loop acts to pump heat energy from warm equatorial oceans poleward to the subtropical zone. Near the surface, air flowing toward the ITCZ picks up water vapor evaporated by warm ocean surfaces, increasing the moisture and latent heat energy content of the air. With convergence and uplift of the air at the ITCZ, the energy is released to warm the surrounding air, as condensation occurs. Air traveling poleward in the return circulation aloft retains much of this heat energy, although some is lost to space by radiant cooling. When the air descends in the subtropical high-pressure belts, the energy becomes available at the surface. The net effect is to gather energy from tropical and equatorial zones and release it in the subtropical zone, where it can be conveyed farther poleward by jet stream disturbances into the midlatitudes.

In the mid- and high latitudes, poleward energy transport is produced almost exclusively by jet stream disturbances (Figure 6.30). Because the jet streams flow west to east, there can be no direct poleward transport of energy, as in the Hadley cells. Instead, lobes of cold, dry polar or arctic air associated with growing jet stream disturbances plunge toward the Equator, while tongues of warmer, moister air originating in the subtropics flow toward the poles. Within these disturbances, cyclones also develop along the polar front, producing convergence of air at the surface. The subsequent uplift releases latent heat energy by condensation. Both of these processes result in a heat flow that warms the mid- and higher latitudes well beyond the insolation they receive.

Just as atmospheric circulation plays a role in moving energy from one region of the globe to another, so do oceanic circulations. How is this energy transported? Near the surface, the water in the low latitudes is exposed to intense insolation in the equatorial and tropical zones and is subsequently warmed. It is then transported by the western boundary currents, which bring warm, tropical waters to higher latitudes. As the water travels northward, it cools, losing energy and warming the atmosphere above it. The cooler waters then return south as part of the gyre circulation, where they cool the overlying atmosphere in the tropics.

6.29 Global atmospheric transport of energy and moisture

Radiant solar energy, absorbed by tropical and equatorial land and oceans, is carried toward the poles as sensible and latent heat by global circulation.

◀ HIGH LATITUDES
Disturbances along the polar front allow warm, moist subtropical air to reach far into the high latitudes while also moving cold, dry air from these regions toward lower latitudes. This warms the polar regions.

◀ MIDLATITUDES
Warm, moist air flows north along the surface. Disturbances along the polar front cause this air to lift over the cooler polar air, producing condensation and latent heat energy release.

◀ LOW LATITUDES
Trade winds, converging toward the Equator, evaporate warm ocean surface water, storing latent heat energy in the humid air. As the air rises and condenses in the inter-tropical convergence zone, latent heat energy is released, warming the air. This air flows poleward, eventually subsiding and warming the subtropics.

Diagram labels: Cyclonic-frontal precipitation; 60° Cold, dry; Condensation; 30°; H; Evaporation; Warm, moist; 0° Condensation ITCZ; Warm, moist; Evaporation; H; 30°; Condensation; 60° Cold, dry; Cyclonic-frontal precipitation; Subsidence in highs; Dry; Hadley cell; Convectional precipitation; Hadley cell; Dry; Subsidence in highs.

6.30 Transport of moisture by storm systems

This satellite image tracks water vapor (in shades of white) as it moves through the atmosphere. Trails of water vapor stretch from the tropical Pacific across the eastern United States and into the high latitudes of the Atlantic, marking the poleward transport of water and energy. The release of latent heat energy as this water vapor condenses drives tropical cyclones, like the one seen here over the Gulf of Mexico, and midlatitude cyclones, pictured here as swirling bands of white over the North Atlantic and North Pacific.

Courtesy NASA

0

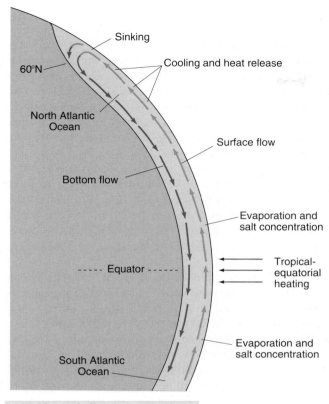

6.31 Thermohaline convection loop

Warm surface waters flow into the North Atlantic, then cool and sink to the deep Atlantic Basin, creating a circulation that warms westerly winds moving onto the European continent.

Energy is also transported by the oceanic thermohaline circulation, shown in Chapter 5. By carrying warm surface water poleward, this loop acts like a heat pump in which energy is acquired in tropical and equatorial regions and is moved northward into the North Atlantic, where it is transferred to the air (Figure 6.31). Since wind patterns move air eastward at higher latitudes, this

energy ultimately warms Europe. The amount of energy transferred to the atmosphere is quite large. A recent calculation shows that it is equal to about 35 percent of the total solar energy received by the Atlantic Ocean north of 40° latitude. This type of circulation does not occur in the Pacific or Indian Ocean, which is why Europe is significantly warmer than the high-latitude regions of the North Pacific and eastern North America at the same latitude.

WileyPLUS Web Quiz
Take a quick quiz on the key concepts of this chapter.

A Look Ahead

This chapter has focused on weather systems and the processes by which they develop. Spawned and steered by upper-level winds, midlatitude weather systems arise from contrasts between air masses that are in contact along the polar front. Tropical weather systems draw their energy from warm tropical ocean waters and are fueled by the release of latent heat energy as moist oceanic air is uplifted and cooled to form clouds and rain.

With a knowledge of weather systems and how they produce precipitation, coupled with an understanding of the global circulation patterns of the atmosphere and oceans, the stage is set for global climate, the topic of Chapter 7. As we will see, the annual cycles of temperature and precipitation that most regions experience are quite predictable, given the changes in wind patterns, air-mass flows, and weather systems that occur with the seasons. Our description of the world's climates grows easily and naturally from the principles you have mastered thus far in your study of physical geography.

WileyPLUS Web Links
Explore this chapter's web links for hurricane information, including the latest Weather Service data and images.

IN REVIEW WEATHER SYSTEMS

- Because global warming, produced by increasing CO_2 levels in the atmosphere, will increase the evaporation of surface water, atmospheric moisture levels will increase. This will tend to enhance the greenhouse effect. But more clouds are likely to form, and this should cool the planet. Increased moisture could also reduce temperatures by increasing the amount and duration of snow cover; but so far, snow cover is

decreasing. Further research on the effect of global warming on climate is needed.
- A **weather system** is a recurring atmospheric circulation pattern with its associated weather. A traveling cyclone (low-pressure center) is an example.
- **Air masses** are distinguished by the latitudinal location (arctic, polar, tropical, equatorial) and type of surface (maritime, continental) of their *source regions*.

- Air masses influencing North America include those of continental and maritime source regions, and of arctic, polar, and tropical latitudes. The United States is affected by the movement of four types of air masses: continental polar (cP), maritime polar (mP), continental tropical (cT), and maritime tropical (mT).
- The boundaries between air masses are termed **fronts**. These include **cold** and **warm fronts**, where cold or warm air masses are advancing. In the **occluded front**, a cold front overtakes a warm front, pushing a pool of warm, moist air mass above the surface. A *stationary front* separates two air masses, with little or no relative motion between them. A *dry line* marks the boundary between cT and mT air masses.
- The traveling anticyclone is typically a fair-weather system. At the center of the anticyclone, pressure is high and air descends.
- **Midlatitude cyclones** form at the boundary between cool, dry air masses and warm, moist air masses. In a cyclone, a vast in-spiraling motion produces cold and warm fronts, and eventually an occluded front. Precipitation normally occurs with each type of front.
- *Jet stream disturbances* cause areas of convergence and divergence in upper-air flows. These create anticyclonic and cyclonic circulations at the surface that move eastward along with the disturbances.
- Midlatitude cyclones tend to form in common areas and follow common paths, called *storm tracks*. They often occur in cyclone families, traveling in sequence.
- **Cold-air outbreaks** occur when tongues of cold polar air invade the low latitudes.
- **Easterly waves** and *weak equatorial lows* are two types of low-latitude weather systems that bring convective showers to unstable tropical and equatorial air. In these areas of low pressure, convergence triggers abundant convective precipitation.
- **Tropical cyclones** can be the most powerful of all storms, growing as large as 500 km (about 300 mi) with sustained winds as high 50 m/s (135 mi/hr) or more.
- Tropical cyclones develop over very warm tropical oceans, where they intensify to become vast in-spiraling systems of very high winds with very low central pressures. They are fueled by latent heat energy, released in condensation, which generates warming, greater uplift, and more condensation. The high winds of the tropical cyclone are partly due to the weakness of the Coriolis force at low latitudes, which requires higher winds to balance pressure gradient forces across the storm.
- Tropical cyclones form in the tropics in the summer and fall, when the waters are warmest over the tropical oceans. They move westward at first, and then often turn poleward and eastward, driven by prevailing westerlies. They are termed *hurricanes* in the Atlantic region, *typhoons* in the western Pacific, and *cyclones* in the Indian Ocean.
- In addition to high winds, tropical cyclones wreak destruction through heavy rainfall, causing debris flows and floods, and by **storm surges,** which bring breaking waves far inland from coastal beaches.
- The Tropical Rainfall Monitoring Mission (*TRMM*) satellite measures and monitors rainfall in tropical and equatorial regions. It has helped scientists model tropical and equatorial weather and climate more accurately.
- Global air and ocean circulation provides the mechanism for poleward energy transport, by which energy derived from solar heating of equatorial and tropical oceans moves toward the poles. In the atmosphere, the energy is carried poleward by the Hadley cell circulation and by movements of warm and cold-air masses in jet stream disturbances.
- In the oceans, a global circulation moves warm surface water northward through the Atlantic Ocean. Heated in the equatorial and tropical regions, the surface water releases energy to the air in the North Atlantic, becomes colder and denser, and sinks to the bottom. These heat flows help make northern and southern climates warmer than we might expect based on solar heating alone.

KEY TERMS

weather systems, p. 188
air masses, p. 188
front, p. 190
cold front, p. 191
warm front, p. 191
occluded front, p. 192
cyclonic storm, p. 194
tornado, p. 194
midlatitude cyclone, p. 199
cold-air outbreaks, p. 202
easterly wave, p. 203
tropical cyclone, p. 204
storm surge, p. 212

REVIEW QUESTIONS

1. How does water, as vapor, clouds, and precipitation, influence global climate? How might water in these forms act to enhance or retard climatic warming?
2. What is an **air mass**? What two features are used to classify air masses? Compare the characteristics and source regions for mP and cT air-mass types.
3. Describe how and where **midlatitude cyclones** form.
4. How is the development of midlatitude cyclones linked to jet stream disturbances? What is the role of converging and diverging winds at the surface and aloft, and how are they linked? Use the term *jet streak* in your answer.
5. What is a **cold-air outbreak**? What other names are used to describe this phenomenon?
6. Identify three **weather systems** that bring rain in equatorial and tropical regions. Describe each system briefly.
7. Describe the structure of a **tropical cyclone**.
8. Once formed, how does a tropical cyclone develop and intensify? How does the weaker Coriolis force of the low latitudes affect the wind speeds of the tropical cyclone?
9. What conditions are necessary for the development of a tropical cyclone? Give a typical track for the movement of a tropical cyclone in the northern hemisphere.
10. Why are tropical cyclones so dangerous? Explain the different types of damage they can cause. Overall, which type of damage is the most severe?
11. What is the goal of the TRMM satellite and its instruments?
12. How does the global circulation of the atmosphere and oceans provide *poleward energy transport?*

VISUALIZING EXERCISES

1. Identify three types of fronts and draw a cross section through each. Show the air masses involved, the contacts between them, and the direction of air-mass motion.
2. Sketch two weather maps, showing a wave cyclone in open and occluded stages. Include isobars on your sketch. Identify the center of the cyclone as a low. Lightly shade areas where precipitation is likely to occur.

ESSAY QUESTIONS

1. Compare and contrast midlatitude and tropical weather systems. Be sure to include the following terms or concepts in your discussion: air mass, convective precipitation, cyclonic precipitation, easterly wave, polar front, stable air, traveling anticyclone, tropical cyclone, unstable air, midlatitude cyclone, and weak equatorial low.
2. Prepare a description of the annual weather patterns that are experienced through the year at your location. Refer to the general temperature and precipitation pattern, as well as the types of weather systems that occur in each season.

Chapter 7
Global Climates and Climate Change

At about 54½° S lat., Ushuaia, Tierra del Fuego, Argentina, is the world's southernmost city. Situated on the shore of the Beagle Channel, sailed by Charles Darwin in HMS Beagle in 1833, Ushuaia's commercial port is the major jumping-off-point for tourist and scientific expeditions to the Antarctic. Fishing, shipping, tourism, and manufacturing are the city's economic strengths. Its 57,000 inhabitants experience a cold marine west-coast climate, with average temperatures of 1.6°C (35°F) in July, the coldest month, and 10.4°C (51°F) in January, the warmest. Light rain or snow falls on about 200 days of the year.

SUMMER IN USHUAIA, TIERRA DEL FUEGO, THE WORLD'S SOUTHERNMOST CITY

©Yann Arthus-Bertrand/Altitude

Global Climates and Climate Change

T he climates of our planet are reflected in a wide range of environments, from hot, wet, equatorial forests to dry, barren deserts to bitterly cold expanses of snow and ice. What factors generate the Earth's climates? How do the global patterns of circulation of the atmosphere and ocean affect climates? How does location—coastal or continental—affect climate, and why? What is the role of elevation? How will global temperatures, precipitation patterns, and weather variability be affected by climate change? These are some of the questions we will answer in this chapter.

Drought in the African Sahel

One of the Earth's more distinctive climates is the wet-dry tropical climate ③. Here, the weather is largely dry for much of the year, though there is a rainy season of a few months duration, during which most of the annual rainfall occurs. Precipitation during the wet season is quite variable, ranging from little or none to torrential rainfalls that produce severe local flooding.

The wet-dry tropical climate ③ is subject to large annual differences in rainfall. Climate records show that two or three successive years of abnormally low rainfall typically alternate with several successive years of average or higher than average rainfall. Variability is a permanent feature of this climate, and both the human inhabitants and natural ecosystems are well adjusted to it.

There is one region, however, where the alternation between wet and dry years takes place on a much longer time scale. Africa's *Sahel region* (Figure 7.1), sandwiched between the

© Alain Nogues/Corbis

▲ SEVERE DROUGHT OF THE 1980s
At the height of the Sahelian drought, vast numbers of cattle perished, and even the goats were hard-pressed to survive.

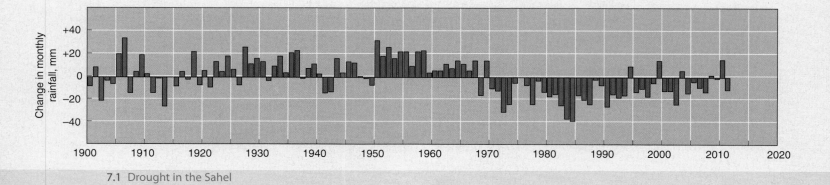

7.1 Drought in the Sahel

continent's monsoon forests and the Sahara Desert, is a band of wet-dry tropical climate ③ with a history of wet and dry periods that last decades. Because the wet periods are long enough for natural ecosystems and human activities to expand their productive range, the lengthy droughts experienced in this region can be particularly devastating.

Although there has been least one severe Sahelian drought in each of the last four centuries, the drought of the late twentieth century, which began in 1968 and ran through the late-1980s, with only one brief respite, was an environmental disaster of the first magnitude. Famine killed about 100,000 people, left three-quarters of a million people dependent on food aid, and severely impacted the agriculture, livestock, and human populations of the region.

As scientists sought to explain the duration and severity of this drought, they first thought that it might have been caused by human activity, caused by overgrazing and conversion of

woodland to agriculture—a process called *desertification* or *land degradation*. These activities reduce precipitation by decreasing atmospheric moisture and raising the surface albedo, which lowers local convective circulation and convective rainfall.

Recent studies now suggest, however, that drought in the Sahel is linked to large-scale atmospheric circulation changes that are produced by multidecadal variations in global sea-surface temperature. A pattern of sea-surface temperatures develops in the Pacific, Atlantic, and Indian Oceans that weakens the African monsoon circulation, shortening or eliminating the Sahel's brief rainy season. Since the pattern can persist for a number of years, the drought produced can be quite long-lasting.

But sea-surface temperature alone is not sufficient to explain the magnitude of the late-twentieth-century Sahelian drought. It now appears that this severe drought in the Sahel was produced by a combination of factors acting synergistically,

THE SAHEL REGION ▶
The Sahel region is a transition between dry Sahara Desert and more moist climates to the south. The green color shows an index that measures the cover of green vegetation. The image uses satellite data from two weeks in June 2005, near the end of the dry season. At this time, most of the Sahel is dry with little or no green vegetation. Patches of green show local areas in the Sahel, typically irrigated, where vegetation persists.

◀ **RAINFALL FLUCTUATIONS FOR THE SAHEL REGION, 1900–2012**
Values shown are the average change in rainfall, in millimeters (mm) per month, for the months June to October (rainy season), expressed as a difference from the average for all years.

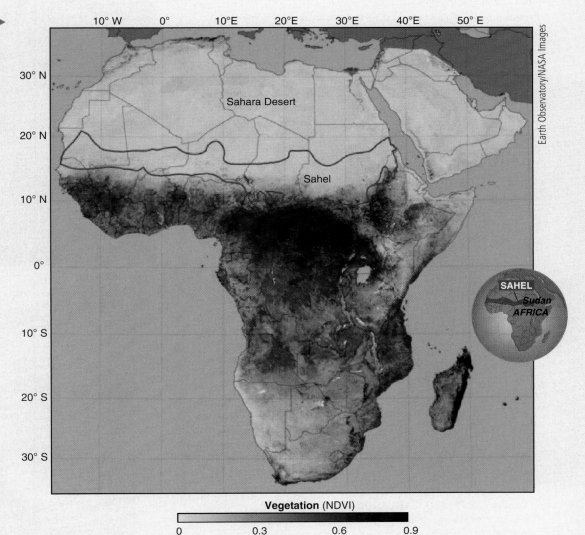

Earth Observatory/NASA Images

Vegetation (NDVI)

0   0.3   0.6   0.9

including sea-surface temperature change, the retreat of natural vegetation cover in the face of drought, and land cover modification by humans.

At present, rainfall in the Sahel is somewhat below historical levels, although drought is not severe. Will the devastating droughts of the 1970s and 1980s return? Some global climate models predict that global warming will lead to the persistence of sea-surface temperature patterns that trigger another Sahelian drought of this magnitude. Coupled with the fact that the Sahelian population is doubling every 20 years, the outlook for the Sahel is not favorable.

Keys to Climate

In Chapter 6 we talked about weather and weather systems. Whereas the term *weather* is used to describe precipitation, winds, and temperatures at a particular time and place, **climate** refers to the average weather of a region. International conventions now designate 30 years as the minimum period for defining a climate. Temperature and precipitation are two of the key measures of climate. To establish the climate of a region, climatologists begin by answering three basic questions: (1) What is the mean annual temperature? (2) What is the mean annual precipitation? (3) What is the seasonal variation in temperature and precipitation?

As we saw in Chapter 3, three primary factors control the magnitude and range of temperatures experienced at a location throughout the year, and influence its annual cycle of air temperature: latitude, coastal versus continental location, and elevation (Figure 7.2). These same three factors are important to precipitation, but precipitation is also affected by annual and monthly

7.2 Climate controls

Factors influencing air temperature include latitude, elevation, and coastal versus continental location.

Jim Richardson/NG Image Collection

George F. Mobley/NG Image Collection

ELEVATION ▶
High-elevation stations show cooler temperatures than sea-level stations, because the atmosphere cools with elevation at the average environmental temperature lapse rate. Kluane National Park, Canada.

LATITUDE ▶
Insolation varies with latitude. The annual cycle of temperature at any place depends on its latitude. Near the Equator, temperatures are warmer and their annual range is low. Toward the poles, temperatures are colder and the annual range is greater. Ellesmere Island, Nunavut, Canada.

John Dunn/NG Image Collection

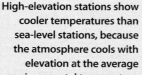

▲ COASTAL-CONTINENTAL LOCATION
Ocean surface temperatures vary less with the seasons than land surface temperatures, so coastal regions experience a smaller annual variation in temperature. Aerial view of Cornwall, England.

Additional factors that also affect precipitation include air masses, annual and monthly air temperatures, mountain barriers, persistent high- and low-pressure centers, and prevailing wind and ocean currents.

▼ AIR MASSES

Air masses that come from continental regions are drier, while air masses from marine locations are moister and can support more precipitation. In regions where two prevailing air masses collide, fronts will form, which can lead to precipitation. Baffin Island, Canada.

▲ ANNUAL AND MONTHLY AIR TEMPERATURE

Warm air can contain more moisture than cold air, so colder regions generally have lower precipitation than warmer regions. Also, precipitation will tend to be greater during the warmer months of the temperature cycle. Mount Des Voeux, Tavenui Island, Fiji Islands.

▼ MOUNTAIN BARRIERS

Location on a mountain can greatly affect the amount of precipitation. On the windward side, the forced uplift of air over the mountains produces condensation and precipitation. On the leeward side, adiabatic warming of the air produces hot, dry conditions. Winery in Paso Robles, California.

▲ PERSISTENT HIGH- AND LOW-PRESSURE CENTERS

Precipitation is affected by surface pressure patterns. In the tropics and the midlatitudes, low pressures at the surface produce convergence and lifting of air that support precipitation. Over the subtropics and polar regions, persistent high-pressure patterns produce divergence of air and descent that inhibit the formation of precipitation. Sahara Desert, Africa.

PREVAILING WIND AND OCEAN CURRENTS ▶

On the western portion of midlatitude continents, the prevailing westerly winds bring warm, moist air off the ocean and onto the continent, resulting in higher precipitation. On the eastern portion, these same winds bring dry air from the continental locations without enough moisture to produce significant precipitation. Redwood National Park, California.

7.3 Global monthly temperature patterns

In general, average annual temperatures are highest at the Equator, and progressively colder toward the poles. The location of a region—coastal or continental—plays a role in determining how temperature varies from month to month throughout the year.

CONTINENTAL INFLUENCE ▶

In continental locations, temperatures generally follow monthly insolation with latitude. Seasonal variation is least at the Equator and more pronounced at higher latitudes.

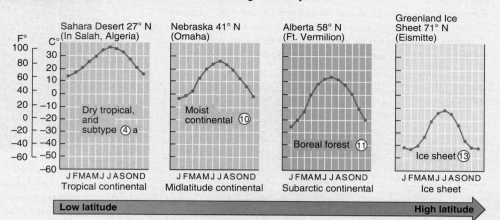

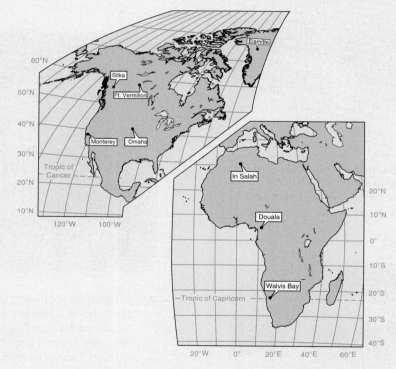

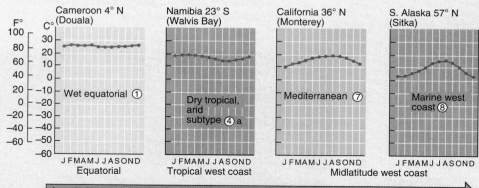

MARINE (COASTAL) INFLUENCE ◀

In coastal locations, temperatures are more moderate, and seasonal variations are curtailed, with the least at the Equator and the greatest toward the poles.

air temperatures, prevailing air masses, and relation to mountain barriers, position of persistent high- and low-pressure centers, and prevailing wind and ocean currents. Keep these key ideas in mind as you read this chapter, as doing so will help make the climates we discuss easier to explain and, thus, understand.

GLOBAL MONTHLY TEMPERATURE PATTERNS

Figure 7.3 shows annual cycles of monthly air temperature for different temperature regimes. Overall, we find that (1) annual variation in insolation, determined by latitude, provides the basic control on temperature; and (2) the effect of location—maritime or continental—moderates that variation.

Another factor affecting air temperature is elevation, as we will see in a discussion of highland climates on later pages. Temperatures are cooler with elevation because the atmosphere cools at the average environmental temperature lapse rate of 6.4°C/1000 m (3.5°F/1000 ft).

> The month-to-month variation in temperature at a location depends on latitude, location, and elevation. At higher latitudes, in continental interiors, and at higher elevations, the variation is greater.

GLOBAL PRECIPITATION PATTERNS

Global precipitation patterns are largely determined by air-mass characteristics and their movements, which in turn are produced by global air circulation patterns. Our map of mean annual precipitation (Figure 7.4) shows seven important features:

1. *Wet equatorial belt.* Moist mE air masses converge in this equatorial zone of warm temperatures to generate abundant convective rainfall.
2. *Trade-wind coasts.* Here, low-latitude easterly trade winds move moist mT air masses from warm oceans to land, where hills and mountains generate orographic rainfall.
3. *Tropical deserts.* Stationary subtropical high-pressure cells, driven by Hadley cell circulation, provide hot and dry subsiding cT air masses that produce barren deserts.
4. *Midlatitude deserts and steppes.* In midlatitude continental interiors, far from oceanic moisture sources, dry air masses dominate, and precipitation is low.
5. *Moist subtropical regions.* These midlatitude regions lie on the western sides of subtropical high-pressure cells, where they receive moist mT air masses from

tropical oceans that provide ample cyclonic and convective precipitation.
6. *Midlatitude west coasts.* Prevailing westerly winds bring cool and moist mP air masses to mountainous west coasts, producing abundant orographic precipitation.
7. *Arctic and polar deserts.* These regions are dominated by cold cP and cA air masses, which are too cold to hold much moisture, and precipitation is low.

GLOBAL MONTHLY PRECIPITATION PATTERNS

Total annual precipitation is a useful factor for establishing the character of a climate type, but it does not account for the seasonality of precipitation. If there is a pattern of alternating dry and wet seasons, instead of a uniform distribution of precipitation throughout the year, we can expect that the natural vegetation, soils, crops, and human use of the land will all be different. Figure 7.5 presents eight examples of different types of monthly precipitation patterns.

Monthly precipitation patterns can be grouped into three types:

1. *Uniformly distributed precipitation.* This includes a wide range of possibilities, from little or no precipitation in any month to abundant precipitation in all months.
2. *Precipitation maximum during the summer (or season of high Sun).* Plants grow best during the season with highest insolation, and if the warm season is also wet, growth of both native plants and crops is enhanced.
3. *Precipitation maximum during the winter or cooler season (season of low Sun).* This means that the warm season is also dry. The stress on growing plants can be intense, and crops will most likely require irrigation.

> Monthly patterns of precipitation are typically of three types: uniformly distributed, summer (high-Sun) maximum, and winter (low-Sun) maximum.

Climate Classification

Mean monthly values of air temperature and precipitation, which are widely available, are good descriptors of how we experience the climate at a location. A **climograph** combines graphs of monthly temperature and

7.4 World precipitation

This global map of mean annual precipitation uses *isohyets*—lines drawn through all points having the same annual precipitation—labeled in mm (in.). Mean annual precipitation varies widely across the globe, ranging from nearly zero to more than 5000 mm (200 in.).

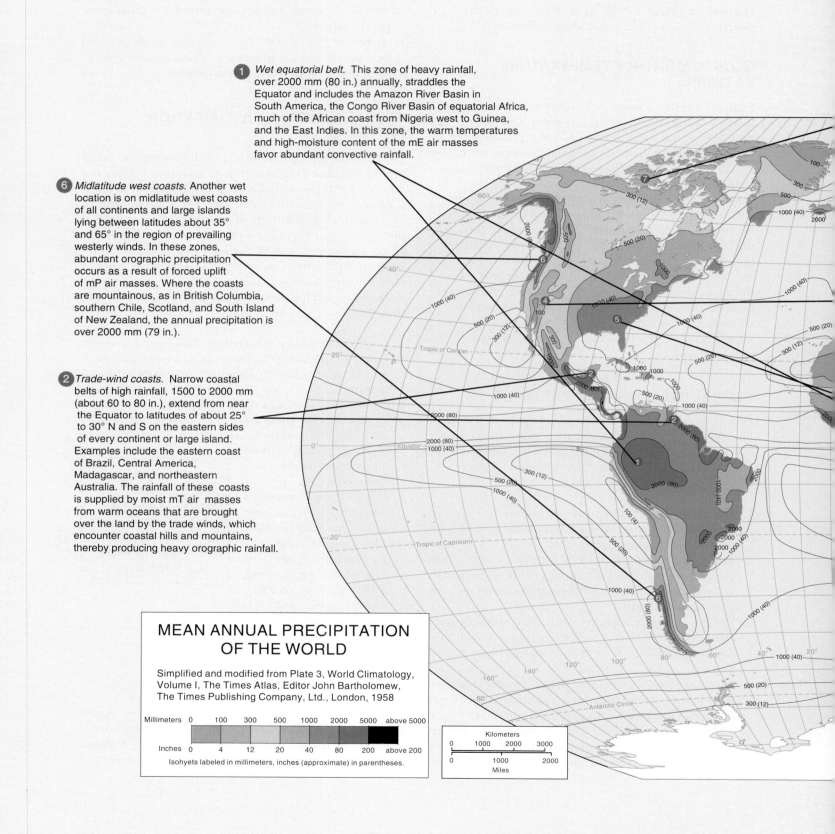

1 *Wet equatorial belt.* This zone of heavy rainfall, over 2000 mm (80 in.) annually, straddles the Equator and includes the Amazon River Basin in South America, the Congo River Basin of equatorial Africa, much of the African coast from Nigeria west to Guinea, and the East Indies. In this zone, the warm temperatures and high-moisture content of the mE air masses favor abundant convective rainfall.

6 *Midlatitude west coasts.* Another wet location is on midlatitude west coasts of all continents and large islands lying between latitudes about 35° and 65° in the region of prevailing westerly winds. In these zones, abundant orographic precipitation occurs as a result of forced uplift of mP air masses. Where the coasts are mountainous, as in British Columbia, southern Chile, Scotland, and South Island of New Zealand, the annual precipitation is over 2000 mm (79 in.).

2 *Trade-wind coasts.* Narrow coastal belts of high rainfall, 1500 to 2000 mm (about 60 to 80 in.), extend from near the Equator to latitudes of about 25° to 30° N and S on the eastern sides of every continent or large island. Examples include the eastern coast of Brazil, Central America, Madagascar, and northeastern Australia. The rainfall of these coasts is supplied by moist mT air masses from warm oceans that are brought over the land by the trade winds, which encounter coastal hills and mountains, thereby producing heavy orographic rainfall.

MEAN ANNUAL PRECIPITATION OF THE WORLD

Simplified and modified from Plate 3, World Climatology, Volume I, The Times Atlas, Editor John Bartholomew, The Times Publishing Company, Ltd., London, 1958

| Millimeters | 0 | 100 | 300 | 500 | 1000 | 2000 | 5000 | above 5000 |
| Inches | 0 | 4 | 12 | 20 | 40 | 80 | 200 | above 200 |

Isohyets labeled in millimeters, inches (approximate) in parentheses.

Kilometers 0 1000 2000 3000

Miles 0 1000 2000

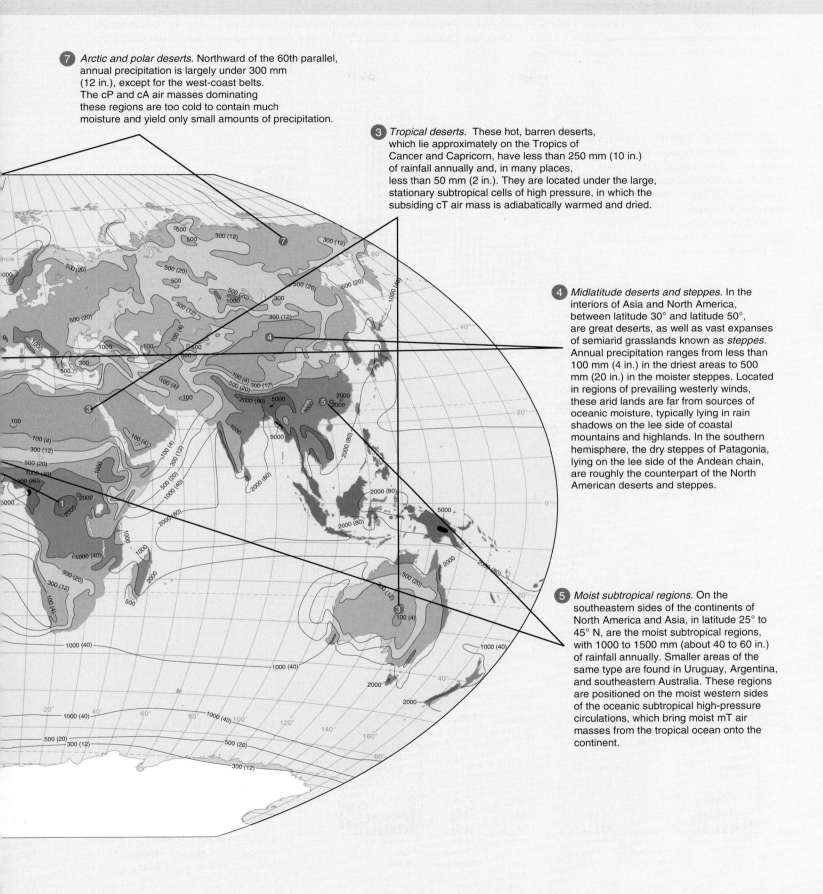

7 *Arctic and polar deserts.* Northward of the 60th parallel, annual precipitation is largely under 300 mm (12 in.), except for the west-coast belts. The cP and cA air masses dominating these regions are too cold to contain much moisture and yield only small amounts of precipitation.

3 *Tropical deserts.* These hot, barren deserts, which lie approximately on the Tropics of Cancer and Capricorn, have less than 250 mm (10 in.) of rainfall annually and, in many places, less than 50 mm (2 in.). They are located under the large, stationary subtropical cells of high pressure, in which the subsiding cT air mass is adiabatically warmed and dried.

4 *Midlatitude deserts and steppes.* In the interiors of Asia and North America, between latitude 30° and latitude 50°, are great deserts, as well as vast expanses of semiarid grasslands known as *steppes.* Annual precipitation ranges from less than 100 mm (4 in.) in the driest areas to 500 mm (20 in.) in the moister steppes. Located in regions of prevailing westerly winds, these arid lands are far from sources of oceanic moisture, typically lying in rain shadows on the lee side of coastal mountains and highlands. In the southern hemisphere, the dry steppes of Patagonia, lying on the lee side of the Andean chain, are roughly the counterpart of the North American deserts and steppes.

5 *Moist subtropical regions.* On the southeastern sides of the continents of North America and Asia, in latitude 25° to 45° N, are the moist subtropical regions, with 1000 to 1500 mm (about 40 to 60 in.) of rainfall annually. Smaller areas of the same type are found in Uruguay, Argentina, and southeastern Australia. These regions are positioned on the moist western sides of the oceanic subtropical high-pressure circulations, which bring moist mT air masses from the tropical ocean onto the continent.

7.5 Global monthly precipitation patterns

Seasonal patterns of precipitation can have a relatively uniform monthly pattern, ranging from very wet (Singapore) to very dry (Tamanrasset), or a precipitation maximum in summer during the high-Sun period (Chittagong, Kaduna, Shanghai, Harbin), or a precipitation maximum in the winter during the low-Sun period (Palermo, Shannon).

EQUATORIAL AND TROPICAL ZONES ▶
Singapore, near the Equator, is uniformly wet, while Tamanrasset, in the Sahara Desert, is uniformly dry. Kaduna has a wet high-Sun season, when the ITCZ is nearby, and Chittigong has a strong high-Sun season precipitation maximum produced by the Asian monsoon circulation. Both have very dry low-Sun seasons.

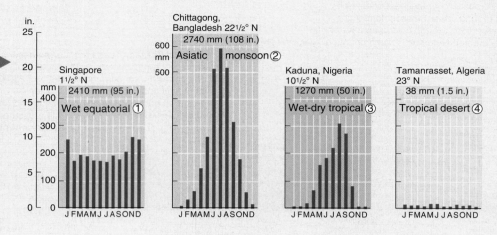

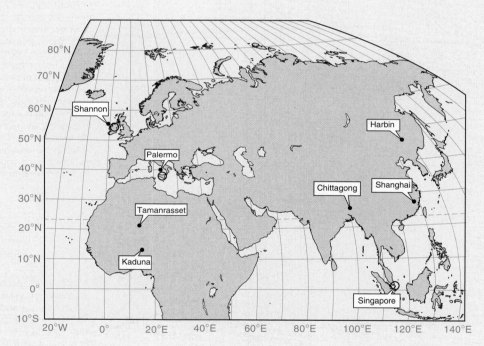

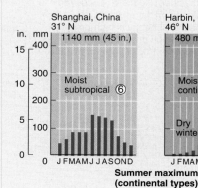

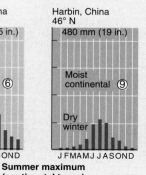

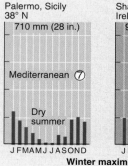

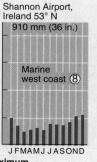

Summer maximum
(continental types)

Winter maximum
(west coast types)

◀ SUBTROPICAL AND MIDLATITUDE ZONES
Shanghai and Harbin show strong summer precipitation maxima. With their continental east coast locations, they receive moist mT air masses from the nearby Pacific in summer and dry cP air masses in the winter. Palermo and Shannon have dry summer patterns, caused by the northward summer migration of subtropical high pressure that blocks or reduces exposure to moist mT and mP air masses.

precipitation for an observing weather station, yielding a quick and effective picture of variation through the year (Figure 7.6).

CLIMATE CLASSIFICATION SYSTEMS

Climatology is the science of analyzing climate—weather over the long term—as it varies over time and around the globe. Climatologists are often computer modelers with strong backgrounds in physics and math, and they use computer programs to simulate the Earth's weather at various time scales. They study past and present climate systems to predict future trends. Based on our current understanding of global warming trends, climate change will be a major challenge for society as a whole over the next century.

Geographers and applied climatologists have devised a number of classification systems to group the climates of individual locations into distinctive climate types. Like all other scientific models, these scientific classification systems are not exact replicas of the real-world climates they represent, but they do provide a useful method to summarize, communicate, and exchange information about them. For example, the Swedish botanist Carolus Linnaeus created a classification system for describing plants and animals by genus and species (for example, *Homo sapiens*), which biologists still use as the universal framework for their discussion of species. Because classification systems reflect the methods and values of the scientists who create them, they can be more or less appropriate for different purposes.

One climate classification system still in wide use today was developed by the Austrian climatologist Vladimir Köppen in 1918 and modified by Rudolf Geiger and Wolfgang Pohl in 1953. Primarily designed to capture variability of vegetation around the globe, it features a system of letters to label and define climates by mean annual precipitation and temperature, as well as precipitation in the driest month. The Köppen climate system is easy and convenient to use, given monthly weather data, but it is not directly related to the underlying processes that differentiate the Earth's climates. (For more details on this system, refer to our

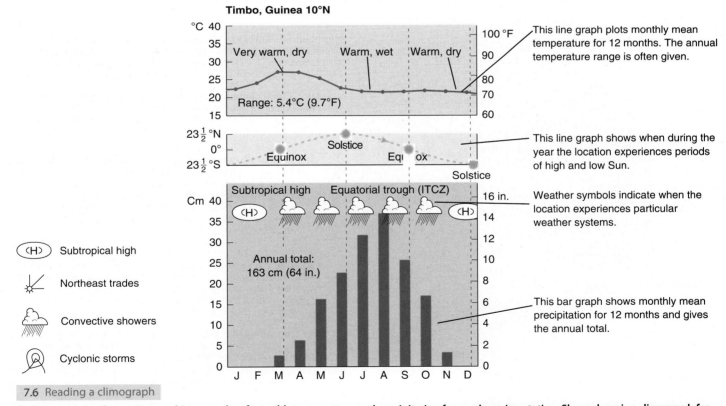

7.6 Reading a climograph

A climograph is a figure that combines graphs of monthly temperature and precipitation for an observing station. Shown here is a climograph for Timbo, Guinea, not far from the Equator in Africa. We will see this climograph again as an example of the tropical wet-dry climate.

Special Supplement • The Köppen Climate System, located at the end of this chapter.)

In comparison, the classification system we use in this chapter is designed to be explained, and understood, according to the characteristics of air masses, their movements, and their interactions along frontal boundaries—that is, by the typical weather patterns various regions experience throughout the year. This system is based on monthly variation in soil moisture (see Appendix 1), and it captures these causes of climate variation quite well. In this book we will discuss 13 distinctive climate types, all of which follow quite naturally from the principles governing temperature and precipitation described in prior chapters.

CLIMATE GROUPS

Recall from Chapter 6 that air masses are classified by a two- or three-letter code, according to the general latitude of their source regions and the surface type—land or ocean—within that region. The latitude determines the temperature of the air mass, which can also depend on the season. The type of surface, land or ocean, controls the moisture content of the air mass. Because the air-mass characteristics control the two most important climate variables—temperature and precipitation—we can use air masses as a guide for explaining and understanding climates around the globe.

We also know that frontal zones are regions in which air masses come in contact. When unlike air masses are in contact with one another, cyclonic precipitation is likely to develop. Because the position of frontal zones changes with the seasons, seasonal movements of frontal zones also influence annual cycles of temperature and precipitation.

By combining what we know about air-mass source regions and frontal zones, we can subdivide the globe into bands, according to latitude, to define three broad groups of climates: low-latitude (Group I), midlatitude (Group II), and high-latitude (Group III), shown in Figure 7.7. Within each of these three climate groups are several climate types (or, simply, climates)—four low-latitude climates, six midlatitude climates, and three high-latitude climates—for a total of 13 climate

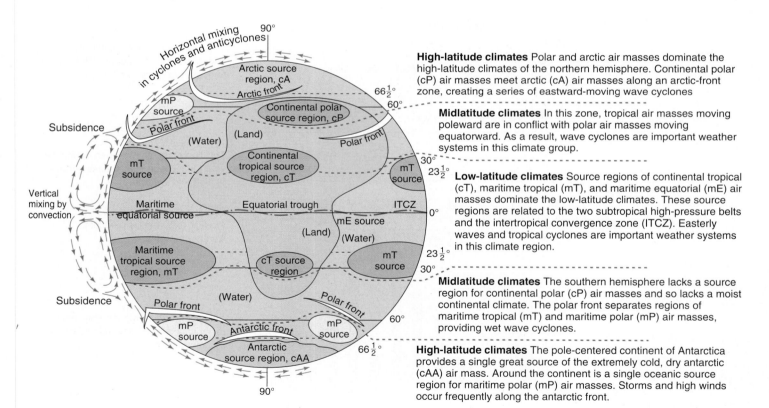

High-latitude climates Polar and arctic air masses dominate the high-latitude climates of the northern hemisphere. Continental polar (cP) air masses meet arctic (cA) air masses along an arctic-front zone, creating a series of eastward-moving wave cyclones

Midlatitude climates In this zone, tropical air masses moving poleward are in conflict with polar air masses moving equatorward. As a result, wave cyclones are important weather systems in this climate group.

Low-latitude climates Source regions of continental tropical (cT), maritime tropical (mT), and maritime equatorial (mE) air masses dominate the low-latitude climates. These source regions are related to the two subtropical high-pressure belts and the intertropical convergence zone (ITCZ). Easterly waves and tropical cyclones are important weather systems in this climate region.

Midlatitude climates The southern hemisphere lacks a source region for continental polar (cP) air masses and so lacks a moist continental climate. The polar front separates regions of maritime tropical (mT) and maritime polar (mP) air masses, providing wet wave cyclones.

High-latitude climates The pole-centered continent of Antarctica provides a single great source of the extremely cold, dry antarctic (cAA) air mass. Around the continent is a single oceanic source region for maritime polar (mP) air masses. Storms and high winds occur frequently along the antarctic front.

7.7 Climate groups, air masses, and frontal zones

Using a map of air-mass source regions for a simplified hypothetical continent, we can identify five global bands associated with three major climate groups: low latitude, midlatitude, and high latitude. Each group has a set of distinctive climates with unique characteristics that are explained by the movements of air masses and frontal zones.

types. Each climate has a name and a number. The name describes the general nature of the climate and suggests its global location. The number helps to identify the climate on maps and diagrams. In this text, for convenience, we will include both the climate name and number. The world map of climates, given in Figure 7.8, shows the actual distribution of climate types on the continents.

DRY, MOIST, AND WET-DRY CLIMATES

All but 2 of our 13 climate types are classified as either **dry climates** or **moist climates**. In dry climates, total annual evaporation of moisture from the soil and from plant foliage exceeds annual precipitation by a wide margin. The soil is dry much of the year, and plant cover is sparse or lacking. Generally speaking, the dry climates do not support permanently flowing streams. In moist climates, rainfall keeps the soil moist through much of the year. The vegetation cover is forest (of many types) or prairies of dense, tall grasses. Rainfall generally provides a year-round flow in the larger streams.

Within the dry climates, there is a wide range of aridity, from very dry deserts nearly devoid of plant life to moister regions that support a partial cover of grasses or shrubs. We will refer to two dry climate subtypes. The *semiarid* (steppe) subtype, designated by the letter *s*, is found next to moist climates. It has enough precipitation to support sparse grasses and shrubs. The *arid* subtype, indicated by the letter *a*, ranges from extremely dry climates to climates that are almost semiarid.

> In dry climates, evaporation and transpiration exceed precipitation by a wide margin. In moist climates, precipitation maintains soil moisture and active stream flow for most or all of the year. Wet-dry climates alternate between wet and dry states.

In addition, 2 of our 13 climates cannot be accurately described as either dry or moist climates. These are the wet-dry tropical ③ and Mediterranean ⑦ climate types, which show a seasonal alteration between a very wet season and a very dry season. This striking contrast gives a special character to the two climates, so they are identified as **wet-dry climates**.

HIGHLAND CLIMATES

Highland climates occupy mountains and high plateaus at any latitude. They tend to be cool to cold because air temperatures in the atmosphere normally decrease with altitude. They are also often moist from orographic precipitation, becoming wetter with elevation. However, rain shadows can occur on sheltered sides of mountains and high plateaus. Highland areas usually derive their annual temperature cycle and the times of their wet and dry seasons from the climate of the surrounding lowland. The Indian cities of New Delhi (lowland) and Shimla (highland), shown in Figure 7.9, provide an example.

> **WileyPLUS** Interaction of Climate, Vegetation, and Soil
> View climate, vegetation, and soils maps, superimposed to display common patterns—animations for North America and Africa.

Low-Latitude Climates (Group I)

The low-latitude climates lie, for the most part, between the Tropics of Cancer and Capricorn, occupying all of the equatorial zone (10° N to 10° S), most of the tropical zone (10° to 15° N and S), and part of the subtropical zone. The low-latitude climate regions include the equatorial trough of the intertropical convergence zone (ITCZ), the belt of tropical easterlies (northeast and southeast trades), and large portions of the oceanic subtropical high-pressure belt. There are four low-latitude climates: wet equatorial ①, monsoon and trade-wind coastal ②, wet-dry tropical ③, and dry tropical ④. We will now look at each one in detail, beginning with the wet equatorial ① and monsoon and trade-wind coastal ② climates (Figure 7.10).

WET EQUATORIAL CLIMATE ①

(KÖPPEN: Af)

The **wet equatorial climate** ① is controlled by the intertropical convergence zone (ITCZ) and is dominated by warm, moist maritime equatorial (mE) and maritime tropical (mT) air masses that yield heavy convectional rainfall (Figure 7.11). This climate has uniformly warm temperatures and abundant precipitation in all months, although there may be a period with less rainfall when the ITCZ shifts toward the opposite tropic. Rainforest is the natural vegetation cover. Streams flow abundantly throughout most of the year, and the riverbanks are lined with dense forest vegetation.

> The wet equatorial climate ① is dominated by warm, moist tropical and equatorial maritime air masses, yielding abundant rainfall year-round. The monsoon and trade-wind coastal climate ② has a strong wet season that occurs when the ITCZ is nearby.

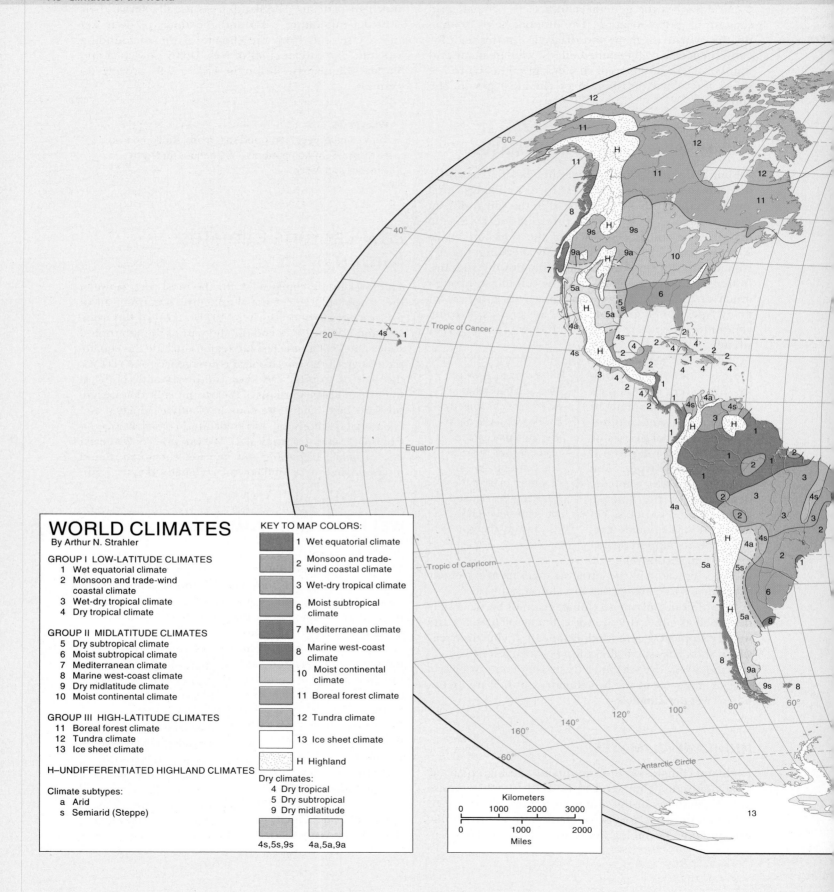

WORLD CLIMATES
By Arthur N. Strahler

GROUP I LOW-LATITUDE CLIMATES
  1 Wet equatorial climate
  2 Monsoon and trade-wind
     coastal climate
  3 Wet-dry tropical climate
  4 Dry tropical climate

GROUP II MIDLATITUDE CLIMATES
  5 Dry subtropical climate
  6 Moist subtropical climate
  7 Mediterranean climate
  8 Marine west-coast climate
  9 Dry midlatitude climate
  10 Moist continental climate

GROUP III HIGH-LATITUDE CLIMATES
  11 Boreal forest climate
  12 Tundra climate
  13 Ice sheet climate

H–UNDIFFERENTIATED HIGHLAND CLIMATES

Climate subtypes:
  a Arid
  s Semiarid (Steppe)

KEY TO MAP COLORS:

1 Wet equatorial climate

2 Monsoon and trade-
   wind coastal climate

3 Wet-dry tropical climate

6 Moist subtropical
   climate

7 Mediterranean climate

8 Marine west-coast
   climate

10 Moist continental
    climate

11 Boreal forest climate

12 Tundra climate

13 Ice sheet climate

H Highland

Dry climates:
  4 Dry tropical
  5 Dry subtropical
  9 Dry midlatitude

4s,5s,9s   4a,5a,9a

Kilometers
0   1000   2000   3000

0   1000   2000
Miles

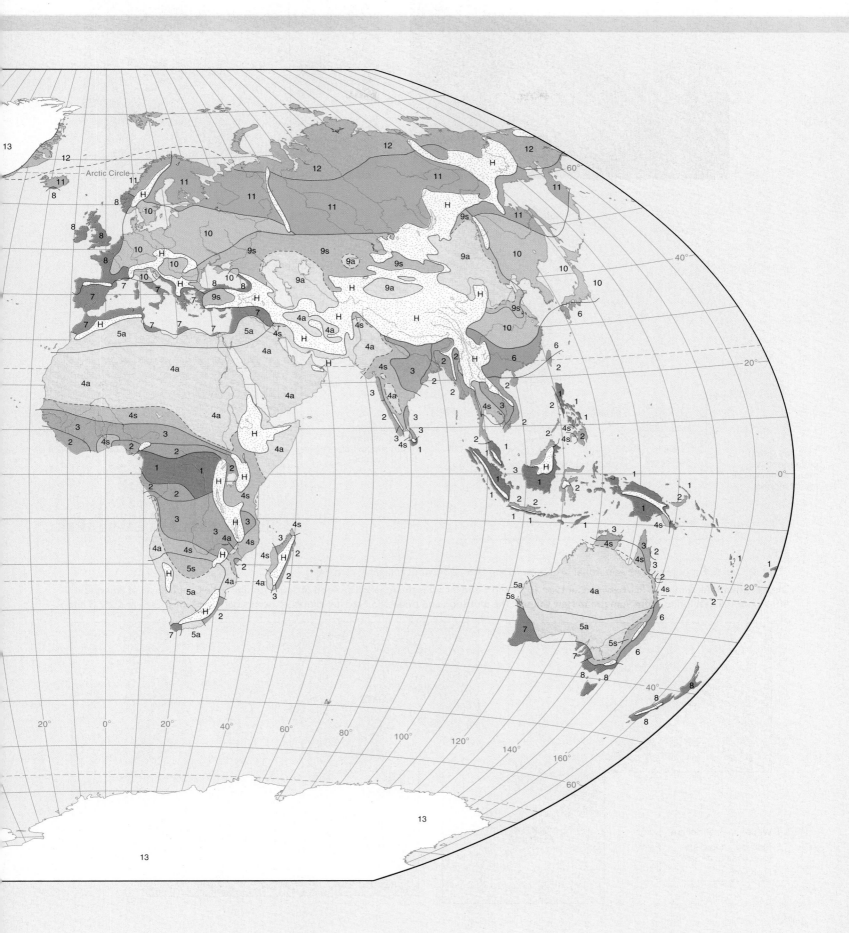

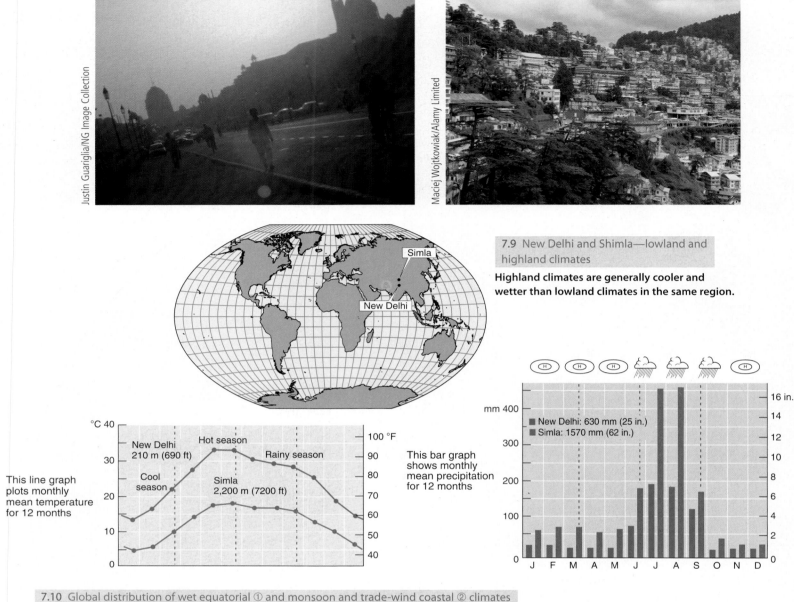

Justin Guariglia/NG Image Collection

Maciej Wojtkowiak/Alamy Limited

7.9 New Delhi and Shimla—lowland and highland climates

Highland climates are generally cooler and wetter than lowland climates in the same region.

This line graph plots monthly mean temperature for 12 months

°C 40 / 100 °F

New Delhi
210 m (690 ft)

Hot season

Rainy season

Cool season

Simla
2,200 m (7200 ft)

This bar graph shows monthly mean precipitation for 12 months

■ New Delhi: 630 mm (25 in.)
■ Simla: 1570 mm (62 in.)

mm 400 / 16 in.

J F M A M J J A S O N D

7.10 Global distribution of wet equatorial ① and monsoon and trade-wind coastal ② climates

The wet equatorial climate ① lies between latitude 10° N and S and includes the Amazon lowland of South America, the Congo Basin of equatorial Africa, and the East Indies, from Sumatra to New Guinea. The monsoon and trade-wind coastal climate ② occurs between latitudes 5° and 25° N and S.

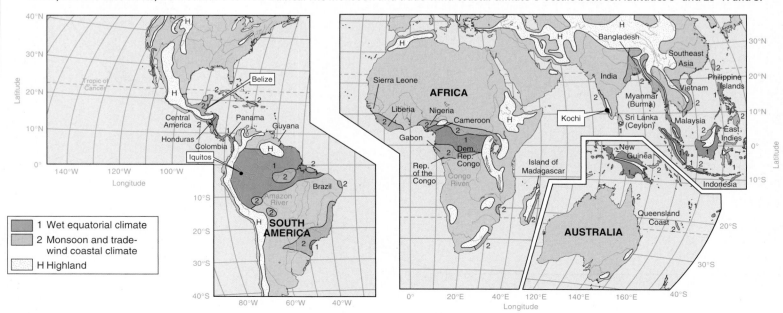

■ 1 Wet equatorial climate
■ 2 Monsoon and trade-wind coastal climate
▨ H Highland

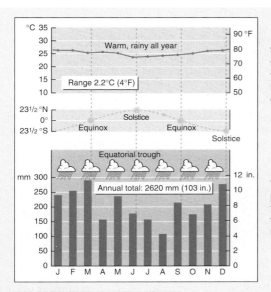

There are uniformly warm temperatures throughout the year, with mean monthly and mean annual temperatures close to 27°C (81°F).

There's a large amount of precipitation every month, but there is a seasonal rainfall pattern, with heavier rain when the ITCZ migrates into the region.

William Albert Allard/NG Image Collection

7.11 Wet equatorial climate ①

Iquitos, Peru (lat. 3° S), located in the upper Amazon lowland, shows mean monthly temperature and precipitation patterns that are characteristic of the warm, rainy, wet equatorial climate ①.

MONSOON AND TRADE-WIND COASTAL CLIMATE ②

(KÖPPEN: Af, Am)

Like the wet equatorial climate ①, the **monsoon and trade-wind coastal climate** ② experiences high annual rainfall and warm temperatures throughout the year. Rainfall has a distinctly seasonal rhythm. In the high-Sun season, the ITCZ is nearby, so monthly rainfall is greater. In the low-Sun season, when the ITCZ has migrated to the other hemisphere, the region is dominated by subtropical high pressure, so there is less monthly rainfall. The warmest temperatures occur during clear skies in the high-Sun season, just before the clouds and rain of the wet season appear. Minimum temperatures occur at the time of low Sun.

As its name suggests, the monsoon and trade-wind coastal climate ② is produced by two different situations. On trade-wind coasts, rain falls more heavily during the high-Sun season as a result of the seasonal migration of the ITCZ (Figure 7.12). During the high-Sun season, moisture-laden maritime tropical (mT) and maritime equatorial (mE) air masses move onshore, onto narrow coastal zones, propelled by trade winds or monsoon circulation patterns. As the warm, moist air passes over

Raymond Gehman/NG Image Collection

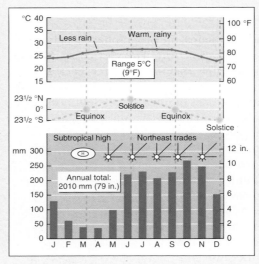

Temperatures are warm throughout the year.

Rainfall is abundant from June through November, when the ITCZ is nearby. Following the December solstice, rainfall is greatly reduced, with minimum values in March and April, when the ITCZ is farthest away.

7.12 Trade-wind coastal climate ②

The climograph for Belize City, a Central American east-coast city (lat. 17° N), demonstrates the warm temperatures and seasonally heavy rainfall that are characteristic of a trade-wind coastal climate.

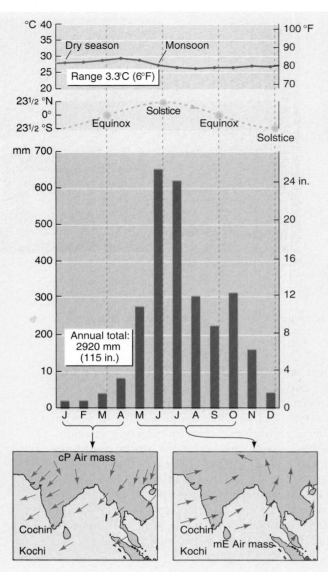

Air temperatures show only a very weak annual cycle, cooling a bit during the rains, so the annual range is small.

There is an extreme peak of rainfall during the rainy monsoon, and there is a short dry season at time of low Sun.

During the summer, monsoon circulation brings moist mE air to the western coasts of the continent.

7.13 Monsoon coastal climate ②

Kochi, on the southwestern tip of India (lat. 10° N), has the weak temperature cycle but strong precipitation cycle of the monsoon climate. The photo shows downtown Kochi in the rainy season. Warm mE air masses moving in from the ocean provide frequent showers at this time. (Kochi is positioned on the map in Figure 7.10.)

© SAJEEV KRISHNAN/Alamy Limited

coastal hills and mountains, the orographic effect touches off convectional shower activity. The east coasts of land masses experience this trade-wind effect because the trade winds blow from east to west. Trade-wind coasts are found along the east sides of Central and South America, the Caribbean Islands, Madagascar (Malagasy), Southeast Asia, the Philippines, and northeast Australia.

A similar seasonal pattern of heavy rainfall occurs in monsoon coasts. (Figure 7.13). In the Asiatic summer monsoon, the monsoon circulation brings mE air onshore. Because the onshore monsoon winds blow from southwest to northeast, the western coasts of land masses are exposed

In the wet periods of the monsoon and trade-wind coastal climate ②, equatorial east coasts receive warm, moist air masses from easterly trades, while tropical south and west coasts receive moist air from southwesterly monsoon winds.

to this moist airflow. Western India and Myanmar (formerly Burma) are examples of regions with this type of climate. Moist air also penetrates well inland in Bangladesh, bringing with it very heavy monsoon rains.

In the wet equatorial climate ① and the wetter areas of the monsoon and trade-wind coastal region ②, a special environment occurs: the low-latitude rainforest (Figure 7.14). In the rainforest, temperatures are uniformly warm throughout the year, and rainfall is high. Streams flow abundantly most of the year, and the riverbanks are lined with dense forest vegetation.

WET-DRY TROPICAL CLIMATE ③

(KÖPPEN: Aw, Cwa)

As we move farther poleward, the seasonal cycles of rainfall and temperature become stronger in the **wet-dry tropical climate** ③ (Figure 7.15). This climate is produced

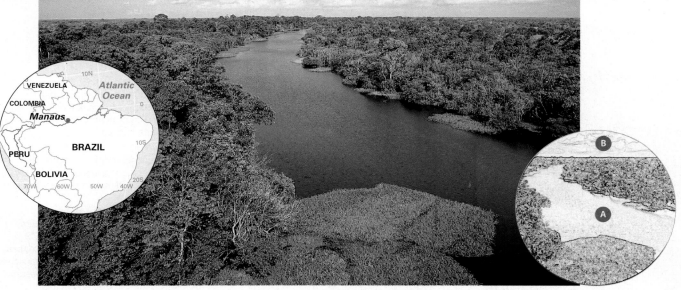

© Will & Deni McIntyre/Photo Researchers, Inc.

7.14 Rainforest of the western Amazon lowland

This aerial view, taken near Manaus, Brazil, shows a tributary of the Amazon lined with forests. The dense canopy includes many different tree species.

EYE ON THE LANDSCAPE What else would the geographer see?

The river is blue **A**, rather than brown, indicating that it doesn't carry a lot of sediment. The vegetation cover binds the soil, keeping it from running into the river during the frequent rains. The warm, moist air equatorial air displays fleecy cumulus clouds **B**.

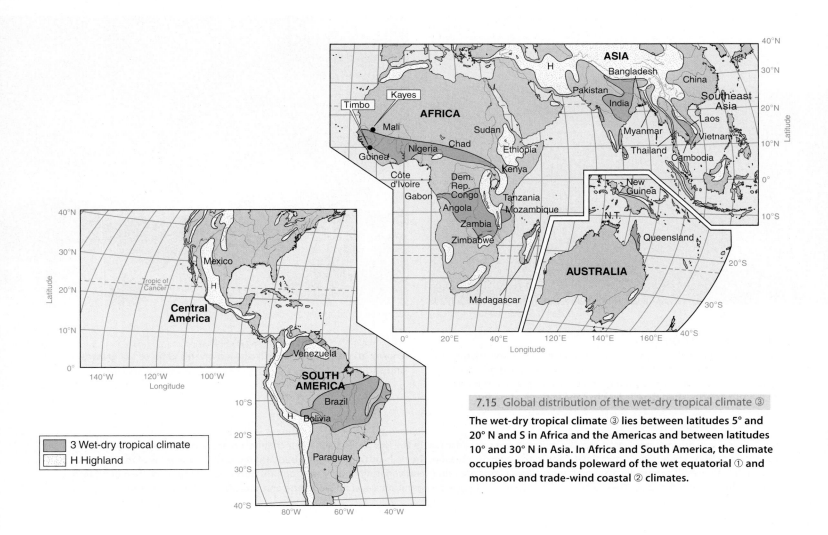

7.15 Global distribution of the wet-dry tropical climate ③

The wet-dry tropical climate ③ lies between latitudes 5° and 20° N and S in Africa and the Americas and between latitudes 10° and 30° N in Asia. In Africa and South America, the climate occupies broad bands poleward of the wet equatorial ① and monsoon and trade-wind coastal ② climates.

Legend:
3 Wet-dry tropical climate
H Highland

In February and March, insolation increases and air temperature rises sharply. When the rains set in, the cloud cover and evaporation of rain cause temperatures to drop.

The rainy season begins just after the March equinox and peaks when the ITCZ has migrated to its most northerly position.

Monthly rainfall peaks in August, but as the low-Sun season arrives, the ITCZ moves south and precipitation decreases. December through February are practically rainless, when subtropical high pressure dominates the climate, and stable, subsiding continental tropical (cT) air pervades the region.

AP/Wide World Photos

7.16 Wet-dry tropical climate ③

Timbo, Guinea (lat. 10° N), is typical of the wet-dry tropical climate ③ region of western Africa. Shown in the photo is the marketplace in Kindia, Guinea, which is about 150 km (about 100 mi) from Timbo.

Kari Niemelainen/Alamy Limited

bildagentur-online.com/th-foto/Alamy Limited

▲ SAVANNA WOODLAND

Here, coarse grasses occupy the open space between the rough-barked and thorny trees. There may also be large expanses of grassland. In the dry season, the grasses turn to straw, and many of the tree species shed their leaves to cope with the drought.

▲ THORNTREE-TALL GRASS SAVANNA

Thorny acacia trees are widely spaced on this plain of tall grasses and shrubs.

7.17 Savanna environments

The tropical wet-dry climate ③ produces the unique savanna landscape. In the low-Sun season, river channels that are not fed by nearby moist mountain regions are nearly or completely dry. In the high-Sun season, they fill with runoff from abundant rains. When the rains of the high-Sun season fail, the result can be devastating famine. Most plants are rain-green vegetation; they are dormant during the dry period, then burst into leaf and bloom with the rains. There are two basic types of rain-green vegetation: savanna woodland and thorntree-tall grass-savanna.

by an alternating dominance of the wet ITCZ and dry subtropical high pressure (Figure 7.16). During the high-Sun season, the ITCZ is nearby, and moist maritime tropical (mT) and maritime equatorial (mE) air masses dominate, creating the wet season. As the ITCZ migrates toward the opposite tropic, subtropical high pressure moves into the region, and dry continental tropical (cT) air masses prevail, creating the dry season. Cooler temperatures in the dry season give way to a very hot period before the rains begin.

In central India and Vietnam, Laos, and Cambodia, mountain barriers tend to block some flows of warm,

> The wet-dry tropical climate ③ has a very dry season alternating with a very wet season. A typical vegetation cover in this climate is savanna woodland, a sparse cover of trees over grassland.

moist mE and mT air provided by trade and monsoon winds. This reduces overall rainfall and enhances the dry season, yielding a wet-dry tropical climate ③.

The native vegetation of the wet-dry tropical climate ③ must survive alternating seasons of very dry and very wet weather. This situation produces a *savanna environment* of sparse vegetation (Figure 7.17).

DRY TROPICAL CLIMATE ④

(KÖPPEN: BWh, BSh)

The **dry tropical climate** ④ is found in the center and east sides of subtropical high-pressure cells (Figure 7.18). As we saw in Chapter 5, air descends and warms adiabatically in these regions, inhibiting condensation, so rainfall is very rare, occurring only when unusual weather conditions move moist air into the region. Skies are clear most of the time, so the Sun heats the surface intensely, keeping air

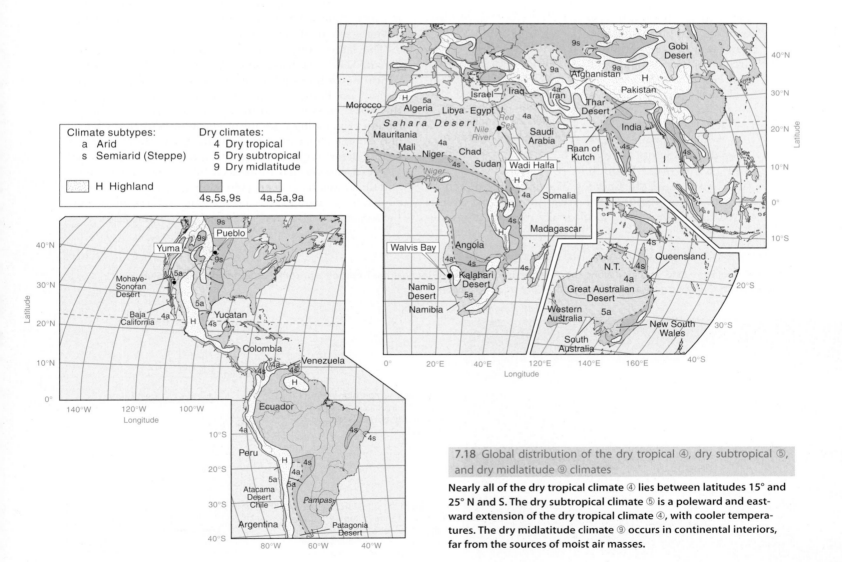

7.18 Global distribution of the dry tropical ④, dry subtropical ⑤, and dry midlatitude ⑨ climates

Nearly all of the dry tropical climate ④ lies between latitudes 15° and 25° N and S. The dry subtropical climate ⑤ is a poleward and eastward extension of the dry tropical climate ④, with cooler temperatures. The dry midlatitude climate ⑨ occurs in continental interiors, far from the sources of moist air masses.

There is a strong annual temperature cycle with a very hot period at the time of high Sun. Daytime maximum air temperatures are frequently between 43º and 48ºC (about 110º to 120ºF) in the warmer months. There is a comparatively cool season at the time of low Sun.

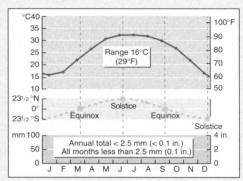

There is too little rainfall to show on the climograph. Over a 39-year period, the maximum rainfall recorded in a 24-hour period at Wadi Halfa was only 0.75 cm (0.3 in.)

Andrew McConnell/Alamy Limited

7.19 Dry tropical climate ④, dry desert subtype

Wadi Halfa, on the Nile River in Sudan (lat. 22° N), almost on the Tropic of Cancer, has a climograph that is typical of the high temperatures and low rainfall of the dry tropical climate ④, dry desert subtype.

temperatures high. During the high-Sun period, heat is extreme, and during the low-Sun period, temperatures are cooler. Given the dry air and lack of cloud cover, the daily temperature range is very wide (Figure 7.19).

The driest areas of the dry tropical climate ④ are near the Tropics of Cancer and Capricorn. Rainfall increases toward the Equator. Traveling this direction, we encounter regions that have short rainy seasons when the ITCZ is near, until finally the climate grades

> The dry tropical climate ④ lies in the belt of persistent subtropical high pressure, so rainfall is rare. Temperatures are very hot during the high-Sun season but are significantly cooler during the low-Sun season.

into the wet-dry tropical ③ type. The boundary line between climate types can shift as a result of even small changes in climate or land use.

The largest region of dry tropical climate ④ is the Sahara–Saudi Arabia–Iran–Thar desert belt of North Africa and southern Asia; it includes some of the driest regions on Earth. Another large region of this type is the desert of central Australia. The west coast of South America, including portions of Ecuador, Peru, and Chile, also exhibits the dry tropical climate ④, but these western coastal desert regions, along with those on the west coast of Africa, are strongly influenced by cold ocean currents and the upwelling of deep, cold water, just offshore. There, cool water moderates coastal zone temperatures, reducing the seasonality of the temperature cycle (Figure 7.20).

7.20 Dry tropical climate ④, western coastal desert subtype

Walvis Bay, Namibia (lat. 23° S), situated along the desert coast of western Africa near the Tropic of Capricorn, has a climograph typical of the west-coast deserts that constitute a subcategory of the dry tropical climate.

Karsten Wrobel/Alamy Limited

The monthly temperatures are remarkably cool for a location that is nearly on the Tropic of Capricorn.

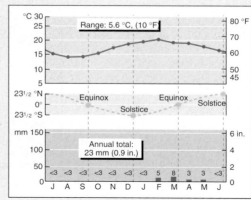

Because of the coastal location, the annual range of temperatures is also small—only 5ºC (9ºF). Coastal fog is a persistent feature of this climate, providing most of the environmental moisture.

The Earth's desert landscapes are actually quite varied. Although these dry landscapes are largely extremely arid, broad zones at the margins are semiarid. These steppes have a short wet season that supports the growth of grasses, on which animals (both wild and domestic) graze (Figure 7.21).

Midlatitude Climates (Group II)

The midlatitude climates almost fully occupy the land areas of the midlatitude zone and a large proportion of the subtropical latitude zone. They also extend into the subarctic latitude zone along the western edge of Europe, reaching to latitude 60°. Unlike the low-latitude climates, which are about equally distributed between northern and southern hemispheres, nearly all of the midlatitude climate area is in the northern hemisphere. In the southern hemisphere, the land area poleward of latitude 40° is so small that the climates are dominated by a great southern ocean.

> In midlatitude climate regions, tropical and polar air masses interact, producing traveling cyclones and anticyclones and frontal boundaries. Midlatitude climates range from very wet to very dry and usually show a strong variation in temperature and/or precipitation throughout the year.

In the northern hemisphere, the midlatitude climates lie between two groups of very dissimilar air masses that interact intensely. Tongues of maritime tropical (mT) air masses enter the midlatitude zone from the subtropical zone, where they meet and conflict with tongues of maritime polar (mP) and continental polar (cP) air masses along the polar-front zone.

The midlatitude climates include the poleward halves of the great subtropical high-pressure systems and much of the belt of prevailing westerly winds. As a result, weather systems, such as traveling cyclones and their fronts, characteristically move from west to east. This global airflow influences the distribution of climates from west to east across the North American and Eurasian continents.

There are six midlatitude climate types, ranging from those with strong wet and dry seasons to those with uniform precipitation. Temperature cycles for these climate types are also quite varied. We will now examine each of these climate types in more detail.

DRY SUBTROPICAL CLIMATE ⑤

(KÖPPEN: BWh, BWk, BSh, BSk)

The **dry subtropical climate** ⑤ is simply a poleward extension of the dry tropical climate ④, characterized by the warm, dry weather of the poleward side of the

James P. Blair/NG Image Collection

7.21 Steppes

The semiarid steppes bordering many of the world's deserts often support nomadic grazing cultures. Here, Shahsavan tribespeople near Tabriz, Iran, pack their possessions in preparation for a move to summer pastures.

subtropical high-pressure cells. As latitude increases, insolation decreases somewhat and becomes more seasonal. This tends to lower temperatures overall and heighten the contrast between summer and winter temperatures (Figure 7.22). As a result, the lower-latitude portions of this climate have a distinct cool season, and the higher-latitude portions have a cold season. The cold season, which occurs at a time of low Sun, is caused in part by the invasion of cold continental polar (cP) air masses from higher latitudes. Wave cyclones occasionally move into the subtropical zone in the low-Sun season, producing precipitation. There are both arid (a) and semiarid (s) subtypes in this climate.

> The dry subtropical climate ⑤ resembles the dry tropical climate ④, but has cooler temperatures during the low-Sun season, when continental polar air masses can invade the region.

The dry subtropical climate ⑤ stretches in a broad band across North Africa, connecting with the Near East. This climate is also part of the environment of Southern Africa and southern Australia; and a band of it occurs in Patagonia, in South America.

In North America, the Mojave and Sonoran deserts of the American Southwest and northwest Mexico are regions of dry subtropical climate ⑤. At the northern margin, in the interior Mojave Desert of southeastern California, at about latitude 34° N, the intense summer heat is comparable to that experienced in the Sahara Desert, but low Sun brings a winter season not experienced in the tropical deserts. Here, cyclonic precipitation can occur in most months, including the cool low-Sun months. Desert plants and animals have adapted to this dry environment (Figure 7.23).

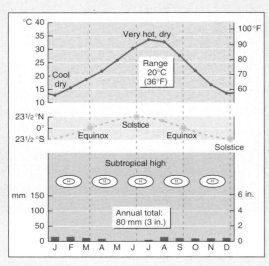

There is a strong seasonal temperature cycle, with a dry hot summer and freezing temperatures in December and January.

Precipitation is small in all months but has peaks in late winter and late summer. The August maximum is caused by the invasion of maritime tropical (mT) air masses, which bring thunderstorms to the region. Higher rainfalls from December through March are produced by midlatitude wave cyclones following a southerly path. Two months, May and June, are nearly rainless.

7.22 Dry subtropical climate ⑤

Yuma, Arizona (lat. 33° N), has a climograph that is typical of an arid region in the dry subtropical climate ⑤.

Richard Cummins/SuperStock

Jeff Foott/Discovery Channel Images/Getty Images, Inc.

7.23 Mojave Desert

These odd-looking plants are Joshua trees, which are abundant in the Mojave Desert. Most areas of this desert have fewer plants.

MOIST SUBTROPICAL CLIMATE ⑥

(KÖPPEN: Cfa)

On the eastern side of continents that lie at latitudes of about 20° to 35° N and S, the subtropical high-pressure cells provide an eastward flow of maritime tropical (mT) air to the **moist subtropical climate** ⑥ (Figure 7.24). The flow of warm, moist air dominates this climate, resulting in warm temperatures, high humidity, and high precipitation (Figure 7.25). Nearby warm ocean currents, like the Gulf Stream, add moisture and warmth.

In summer, this mT air often becomes unstable, providing abundant convective precipitation. Occasional tropical cyclones add to the summer rainfall total. There is also plenty of winter precipitation, produced by midlatitude cyclones. Continental polar

> The moist subtropical climate ⑥, found on the eastern sides of continents in the midlatitudes, has abundant precipitation. In summer, maritime tropical air masses provide convective showers and tropical cyclones, while in winter, midlatitude cyclones yield rain and occasional snow.

(cP) air masses frequently invade this climate region in winter, bringing spells of subfreezing weather. In Southeast Asia, this climate is characterized by a strong monsoon effect, with much more rainfall in the summer than in the winter. The native vegetation of the moist subtropical climate ⑥ is forest of several different types (Figure 7.26).

THE MEDITERRANEAN CLIMATE ⑦

(KÖPPEN: Csa, Csb)

The **Mediterranean climate** ⑦, found between latitudes 30° and 45° N and S (Figure 7.27), is unique because it has a wet winter and a very dry summer. This climate is located along the west coasts of continents, just poleward of the dry eastern side of the subtropical high-pressure cells. When the subtropical high-pressure cells move poleward in summer, they enter the Mediterranean climate region. Dry continental tropical (cT) air dominates, producing the dry summer season. In winter, the subtropical high-pressure cells weaken and move toward the Equator. Moist mP air mass invades, bringing cyclonic storms that generate ample rainfall.

7.24 Global distribution of the moist subtropical climate ⑥

Regions with a moist subtropical climate ⑥ include most of the southeastern United States, from the Carolinas to eastern Texas. In Asia, this climate is found in southern China, Taiwan, and southernmost Japan. In South America, it is experienced in parts of Uruguay, Brazil, and Argentina. And in Australia, the moist subtropical climate consists of a narrow band between the eastern coastline and the eastern interior ranges.

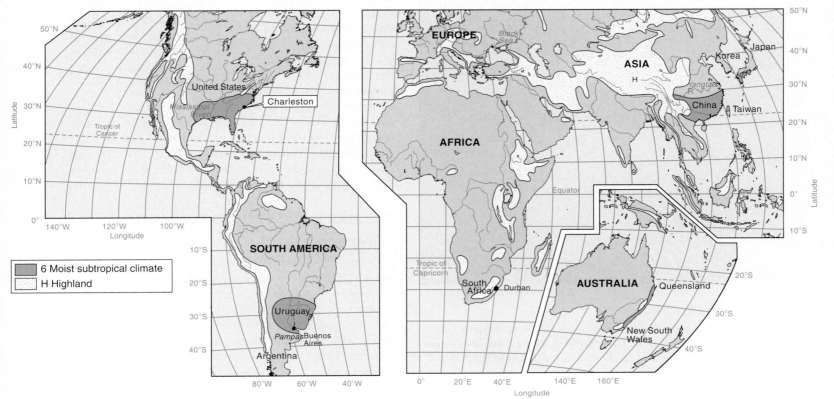

There is a strongly developed annual temperature cycle, with a large annual range. Winters are mild, with the January mean temperature well above the freezing mark.

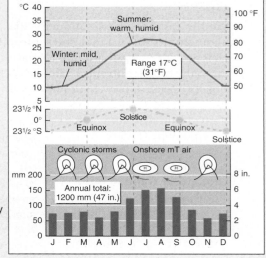

Ample precipitation falls in every month, with a definite summer maximum.

Raymond Gehman/NG Image Collection

7.25 Moist subtropical climate ⑥

Charleston, South Carolina (lat. 33° N), located on the eastern seaboard, has a climograph that is typical of the moist subtropical climate ⑥, with a mild winter and a warm summer and year-round precipitation.

The Mediterranean climate ⑦ spans arid to humid climates, depending on location. Generally, the closer an area is to the tropics the stronger the influence of subtropical high pressure will be, and thus the drier the climate. The temperature range is moderate, with warm to hot summers and mild winters. Coastal zones between latitudes 30° and 35° N and S, such as Southern California, experience a narrower annual range, with very mild winters (Figure 7.28).

The native vegetation of the Mediterranean climate environment is adapted to survive through the long summer drought. Shrubs and trees typically produce small, hard, or thick leaves that resist water loss through transpiration. These plants are called *sclerophylls*, a word that combines the prefix *scler*, from the Greek for "hard," with *phyllo*, Greek for "leaf."

> The Mediterranean climate ⑦, found along midlatitude west coasts, is distinguished by its dry summer and wet winter. In summer, dry subtropical high pressure blocks rainfall, while in winter, wave cyclones produce ample precipitation.

THE MARINE WEST-COAST CLIMATE ⑧

(KÖPPEN: Cfb, Cfc)

Like the Mediterranean climate, the **marine west-coast climate** ⑧ has a winter precipitation maximum. Located on midlatitude west coasts poleward of the Mediterranean climate ⑦, it is similarly influenced by subtropical high pressure in the summer, but not as strongly. This makes the summers drier than the winters. These locations receive the prevailing westerlies from a large ocean, and experience frequent

Stephen Alvarez/NG Image Collection

7.26 Moist subtropical forest

A mix of broadleaf deciduous trees and shrubs (e.g., oaks, hickories, poplars), and occasionally pines, are quite typical of forests in this climate zone. Broadleaf evergreen trees and shrubs, such as the mountain laurel shown here on the left, can also appear.

At Monterey, the annual temperature cycle is weak because of the coastal location.

The summer is very dry. Rainfall drops to nearly zero for four consecutive summer months but rises to substantial amounts in the rainy winter season.

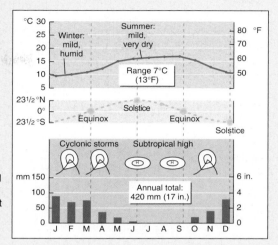

Sisse Brimberg/NG Image Collection

7.27 Mediterranean climate ⑦

Monterey, California, has a typical Mediterranean precipitation pattern, with an almost rainless summer. The temperature range here is moderated by the coastal location, with westerly winds moving across cool currents just offshore. Coastal upwelling of bottom waters brings nutrients to surface waters, supporting extensive fisheries.

7.28 Global distribution of the Mediterranean ⑦ and marine west-coast ⑧ climates

In North America, the Mediterranean climate ⑦ is found in central and southern California. In the southern hemisphere, it occurs along the coast of Chile, in the Cape Town region of South Africa, and along the southern and western coasts of Australia. In Europe, this climate type surrounds the Mediterranean Sea, giving the climate its name. The general latitude range of marine west-coast climate ⑧ is 35° to 60° N and S. It occupies west coasts poleward of the Mediterranean climate ⑦.

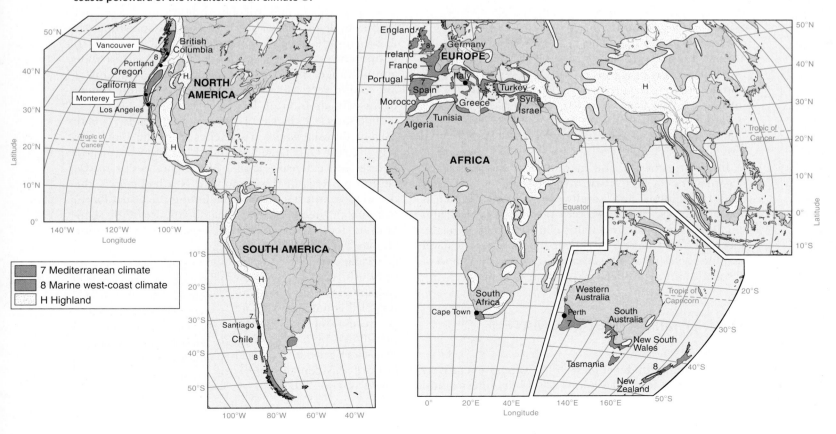

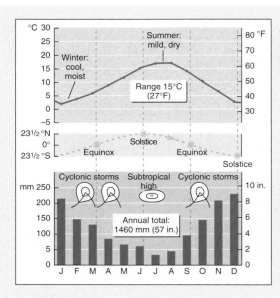

7.29 Marine west-coast climate ⑧

Vancouver, British Columbia (lat. 49° N), has a climograph typical of the marine west-coast climate ⑧. The conifer forests seen lining the nearby mountains thrive on the abundant orographic precipitation provided by the moist marine air.

The annual temperature range is small, and winters are very mild for this latitude.

The annual precipitation total is large, with most precipitation falling in winter.

David Alan Harvey/NG Image Collection

cyclonic storms involving cool, moist mP air masses that provide ample precipitation in all months. Where the coast is mountainous, the orographic effect causes large amounts of precipitation annually. The annual temperature range is comparatively narrow for midlatitudes. The marine influence keeps winter temperatures milder than at inland locations at equivalent latitudes (Figure 7.29).

> The marine west-coast climate ⑧ features frequent cyclonic storms that provide abundant precipitation, especially when enhanced by an orographic effect. Summers are drier, due to subtropical high pressure that moves poleward, blocking storm tracks.

In North America, the marine west-coast climate ⑧ occupies the western coast, from Oregon to northern British Columbia and southwestern most Alaska. In Western Europe, the British Isles, northern Portugal, and much of France fall under this climate category. In the southern hemisphere, this climate includes New Zealand and the southern tip of Australia, as well as the island of Tasmania and the Chilean coast south of latitude 35° S.

THE DRY MIDLATITUDE CLIMATE ⑨

(Köppen: Bwk, Bsk)

The **dry midlatitude climate** ⑨ is almost exclusively limited to the interior regions of North America and

Eurasia, where sources of moist maritime air are far away. Much of the area lies within the rain shadow of mountain ranges on the west or south. These mountain ranges effectively block the eastward flow of maritime air masses. In winter, continental polar (cP) air masses dominate. In summer, solar heating creates a local dry continental air mass; but occasionally, maritime air masses invade, causing convective rainfall. The annual temperature cycle is strongly developed, with a wide annual range. Summers are warm to hot, and winters are cold to very cold (Figure 7.30). The latitude range of this climate is 35° to 55° N.

The largest expanse of the dry midlatitude climate ⑨ is in Eurasia, where it stretches from the Black Sea region to the Gobi Desert and northern China. True arid deserts and extensive areas of highlands occur in the central portions of this region. In North America, the dry western interior regions, including the Great Basin, Columbia Plateau, and the Great Plains, are of the semi-arid, or steppe, subtype. A small area of dry midlatitude climate ⑨ is found in southern Patagonia, near the tip of South America. The low precipitation and cold winters of this semiarid climate produce a steppe landscape dominated by hardy perennial short grasses. A typical crop of this climate type is wheat (Figure 7.31).

> The dry midlatitude climate ⑨ occupies continental interiors that are in rain shadows or far from oceanic moisture sources. Precipitation is low, and the annual temperature variation is great.

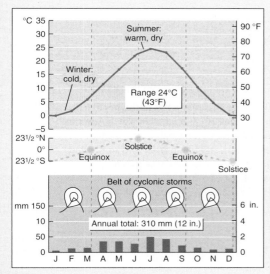

The temperature cycle has a large annual range, with warm summers and cold winters. January, the coldest winter month, has a mean temperature just below freezing.

Most of the annual precipitation is convective summer rainfall, which occurs when moist maritime tropical (mT) air masses invade from the south and produce thunderstorms. In the winter, snowfall is light.

Andre Jenny/The Image Works

7.30 Dry midlatitude climate ⑨

Pueblo, Colorado (lat. 38° N), just east of the Rocky Mountains, has a climograph typical of the dry midlatitude climate, with a peaked summer maximum of rainfall in the summer months.

Jim Richardson/NG Image Collection

7.31 Wheat harvest

Wheat is a major crop of the semiarid, dry midlatitude steppe lands, but harvests are at the mercy of rainfall variations from year to year. Good spring rains lead to an ample harvest; but if spring rains fail, so does the wheat crop. Shown here is an aerial view of combines bringing in a Kansas wheat harvest.

MOIST CONTINENTAL CLIMATE ⑩

(KÖPPEN: Dfa, Dfb, Dwa, Dwb)

The **moist continental climate** ⑩ lies in the midlatitudes at the polar-front zone, where polar and tropical air masses meet. As a result, day-to-day weather is highly variable. Located in the midlatitudes in the central and eastern parts of North America and Eurasia (Figure 7.32), this climate can experience wave cyclones at any time of year. Precipitation is ample throughout the year, increasing in summer when maritime tropical (mT) air masses invade. Seasonal temperature contrasts are strong. Cold winters are dominated by continental polar (cP) and continental arctic (cA) air masses from subarctic source regions (Figure 7.33).

Most of the eastern half of the United States, from Tennessee to the north, as well as the southernmost strip of eastern Canada, lies in the moist continental climate zone. The moist continental climate ⑩ region in China, Korea, and Japan has more summer rainfall and drier winters than North America. This is an effect of the monsoon circulation, which moves moist maritime tropical (mT) air across the eastern side of Asia in summer, and dry continental polar (cP) air southward through the region in winter. In Europe, the moist continental climate ⑩ lies in a higher latitude belt (45° to 60° N) and receives precipitation from mP air masses coming from the North Atlantic. This climate does not occur in the southern hemisphere, since there are no large land

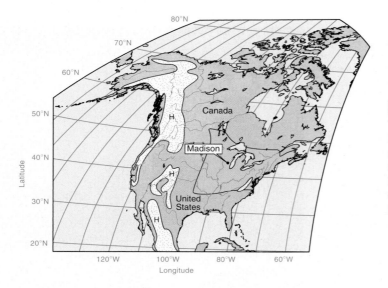

7.32 Global distribution of the moist continental climate ⑩

The moist continental climate ⑩ is restricted to the northern hemisphere, between latitudes 30° and 55° N in North America and Asia. Most of the north and eastern half of the United States, from Tennessee to the Canadian border and somewhat beyond, lies in this climate region. In Asia, it occurs in northern China, Korea, and Japan. Most of central and eastern Europe in latitudes 45° to 60° N has a moist continental climate ⑩.

| | 10 Moist continental climate |
| | H Highland |

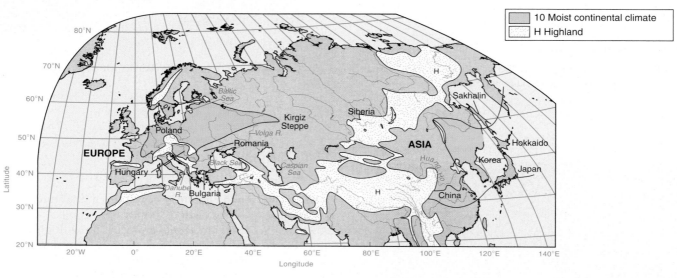

7.33 Moist continental climate ⑩

Madison, Wisconsin (lat. 43° N), has a moist continental climate ⑩. Much of the winter precipitation is in the form of snow, as shown in this photo, and it remains on the ground for long periods.

Exactostock/SuperStock

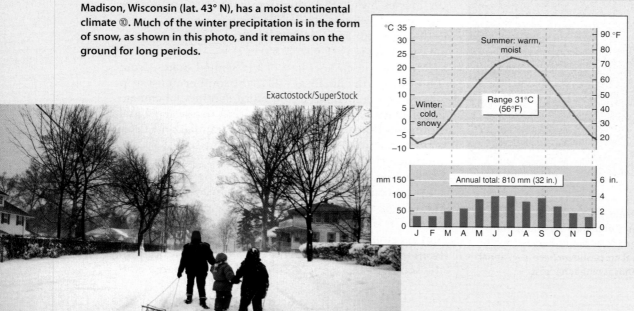

Winters are cold—with three consecutive monthly means well below freezing—and summers are warm, making the annual temperature range very large.

There is ample precipitation in all months, and the annual total is large. There is a summer maximum of precipitation when the maritime tropical (mT) air mass invades, and convective precipitation forms along moving cold fronts and squall lines.

masses or source regions to sustain cP air masses.

Forests are the dominant natural vegetation cover throughout most of this climate, although tall, dense grasses are the natural cover where the climate grades into drier climates, such as the dry midlatitude climate ⑨.

> The moist continental climate ⑩ lies in the polar-front zone, where day-to-day weather is highly variable. Ample frontal precipitation is enhanced in summer by maritime tropical air masses. Winter temperatures fall well below freezing.

High-Latitude Climates (Group III)

By and large, the high-latitude climates occupy the northern subarctic and arctic latitude zones of the northern hemisphere. They also extend southward into the midlatitude zone as far south as about latitude 47° N in eastern North America and eastern Asia. The boreal forest climate and tundra climate are absent in the southern hemisphere, which lacks a high-latitude land mass between 50° and 70° S. However, the ice sheet climate is present in both hemispheres, in Greenland and Antarctica.

The high-latitude climates coincide closely with the belt of prevailing westerly winds that circles each pole. In the northern hemisphere, this circulation sweeps maritime polar (mP) air masses, formed over the northern oceans, into conflict with continental polar (cP) and continental arctic (cA) air masses on the continents. Jet-stream disturbances form in the westerly flow, bringing lobes of warmer, moister air poleward into the region in exchange for colder, drier air that is pushed toward the Equator. As a result of these processes, wave cyclones are frequently produced along the arctic-front zone.

BOREAL FOREST CLIMATE ⑪

(KÖPPEN: Dfc, Dfd, Dwc, Dwd)

The **boreal forest climate** ⑪ occupies the source region for cP air masses, which are cool or cold, dry, and stable. As a result, temperatures are cool or cold, particularly in winter, and precipitation is low. In North America, the boreal forest climate stretches from central and western Alaska across the Yukon and Northwest Territories to Labrador on the Atlantic coast. In Europe and Asia, it reaches from the Scandinavian Peninsula eastward across all of Siberia to the Pacific Ocean (Figure 7.34).

The boreal forest climate ⑪ is a continental climate, with long, bitterly cold winters and short, cool summers (Figure 7.35). Colder cA air masses very commonly invade the region. The annual range of temperature is greater here than in any of the other climates, and is greatest in Siberia, in Russia.

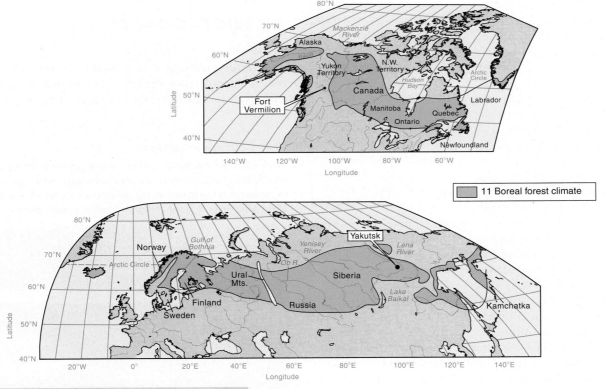

7.34 Global distribution of the boreal forest climate ⑪

The boreal forest climate ⑪ ranges from latitude 50° to 70° N.

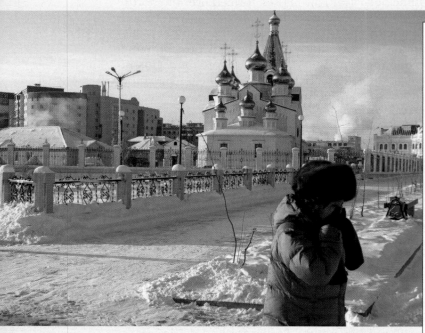

Bjoern Steinz/Panos Pictures

Temperatures are below freezing for more than half the year. The summers are short and cool.

Precipitation has a marked annual cycle with a summer maximum, but the total annual precipitation is small.

7.35 Boreal forest climate ⓵

Fort Vermilion, Alberta 11 (lat. 58° N), and Yakutsk, Siberia (lat. 62° N), both have climographs characteristic of the boreal forest climate (11), with extreme winter cold and a very wide annual range in temperature. A snow cover remains over solidly frozen ground, shown in this photo from Yakutsk, throughout the entire winter.

Precipitation increases substantially in summer, when maritime air masses penetrate the continent with wave cyclones, but the total annual precipitation is low. Although much of the boreal forest climate ⓵ is moist, large areas in western Canada and Siberia have low annual precipitation, and are therefore cold and dry.

The land surface features of much of the boreal forest climate region were shaped beneath the great ice sheets of the last Ice Age. Erosion by the moving ice exposed hard bedrock over vast areas and created numerous shallow rock basins, which are now lakes (Figure 7.36).

The dominant upland vegetation of the boreal forest climate region is boreal forest, consisting largely of needleleaf trees. Although the growing season in the boreal forest climate is short, cultivating crops there is possible. Farming is, however, largely limited to lands surrounding the Baltic Sea, bordering Finland and Sweden. Crops grown in this area include barley, oats, rye, and wheat. The needleleaf forests provide paper, pulp, cellulose, and construction lumber.

> The boreal forest climate ⓵ has long, bitterly cold winters and short, cool summers. For most of the year, cold, dry continental arctic and polar air masses dominate. In summer, occasional maritime air masses provide moisture for precipitation.

THE TUNDRA CLIMATE ⓬

(KÖPPEN: ET)

The **tundra climate** ⓬ is a cold maritime climate of the polar and arctic regions. It occupies arctic coastal fringes, ringing the Arctic Ocean and extending across

7.36 Lakes in a boreal forest

Much of the boreal forest consists of low but irregular topography, formed by continental ice sheets during the Ice Age. Low depressions scraped out by the moving ice are now lakes. Alaska's Mulchatna River is in the foreground of this aerial photo.

Doug Cheeseman/Alamy Limited

the island region of northern Canada (Figure 7.37). Areas with this climate include the Alaska North Slope, the Hudson Bay region, and the Greenland coast in North America. In Eurasia, this climate type occupies the northernmost fringe of the Scandinavian Peninsula and the Siberian coast. The Antarctic Peninsula (not shown in Figure 7.37) also belongs to this climate.

This climate is dominated by polar (cP and mP) and arctic (cA) air masses. Winters are long and severe, but the nearby ocean water moderates temperatures so they don't fall to the extreme lows found in the continental interior. There is a very short mild season, but many climatologists do not recognize it as a true summer (Figure 7.38).

The term *tundra* describes both an environmental region and a major class of vegetation. Hardy perennial grasses and sedges, accompanied by low, frost-resistant shrubs, are arranged sparsely on rocky soils.

Peat bogs are numerous. In some places, a distinct tree line—roughly along the 10°C (50°F) isotherm of the warmest month—separates the boreal forest and tundra.

Because of the cold temperatures experienced in the tundra and northern boreal forest climate zones, the ground is typically frozen to great depths. This perennially frozen ground, or *permafrost*, prevails over the tundra region. That said, temperature increases in the Arctic associated with global warming have thawed the upper permafrost layer at some locations, leaving unstable soils that are unable to support roads and buildings.

The tundra climate ⑫ occupies arctic coastal fringes. Although the climate is very cold, the maritime influence keeps winter temperatures from falling to the levels of the ice sheet climate (13) mild season provides a few months of thaw.

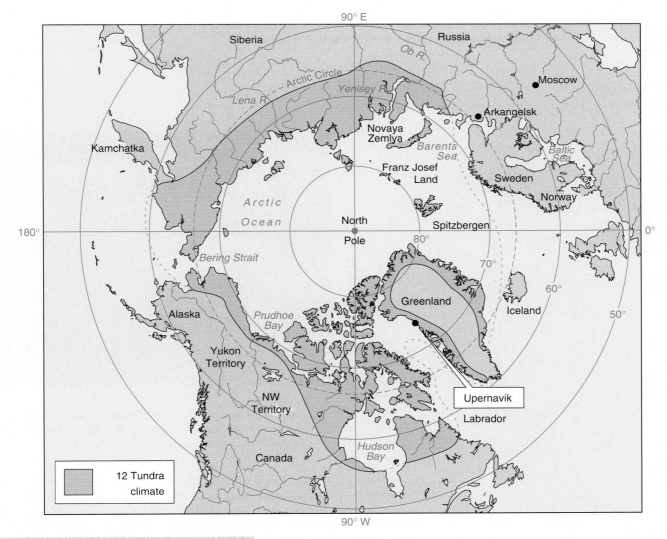

7.37 Global distribution of the tundra climate ⑫

The latitude range for the tundra climate ⑫ is 60° to 75° N and S, except for the northern coast of Greenland, where tundra occurs at latitudes above 80° N.

The long winter is very cold, but the annual temperature range is not large, given the high latitude.

Total annual precipitation is low. In July, the sea-ice cover melts and the ocean water warms, raising the moisture content of the local air mass, increasing precipitation.

Hinrich Baesemann/Landov LLC

7.38 Tundra climate ⑫

Upernavik, located on the west coast of Greenland (lat. 73° N), has a climograph typical of the tundra climate ⑫, with a short, mild period of above-freezing temperatures and little annual precipitation.

7.39 Ice sheet climate ⑬

It is always below freezing in the ice sheet climate, and bitterly cold in winter.

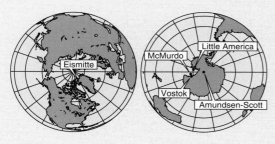

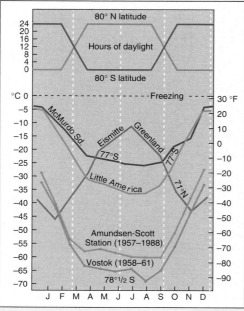

Maria Stenzel/NG Image Collection

▲ **ANTARCTICA**
Snow and ice accumulate at higher elevations here in the Dry Valleys region of Victoria Land, Antarctica. Most of Antarctica is completely covered by ice sheets of the polar ice cap.

◄ **TEMPERATURE GRAPHS FOR FIVE ICE SHEET STATIONS**
Estimitte is on the Greenland ice cap; the other stations are in Antarctica. Temperatures in the interior of Antarctica are far lower than at any other place on Earth. A low of −89.2°C (−128.6°F) was observed in 1983, at Vostok, about 1300 km (about 800 ml) from the South Pole at an altitude of about 3500 m (11,500 ft). At the pole (Amundsen-Scott Station) July, August, and September have averages of about −60°C (−76°F). Temperatures are considerably higher, month for month, at the Little America research base (now abandoned) in Antarctica because it is located close to the Ross Sea and is at a low altitude.

THE ICE SHEET CLIMATE ⑬

(KÖPPEN: EF)

The **ice sheet climate** ⑬ coincides with the source regions of arctic (A) and antarctic (AA) air masses, situated on the vast, high ice sheets of Greenland and Antarctica and over the polar sea ice of the Arctic Ocean (Figure 7.39). The mean annual temperature in the ice sheet climate is much lower than that of any other climate, with no monthly mean above freezing. Strong temperature inversions, caused by radiative loss from the surface, develop over the ice sheets. In Antarctica and Greenland, the high surface altitude of the ice sheets intensifies the cold. Strong cyclones with blizzard winds are frequent. There is very little precipitation, almost all occurring as snow, and the snow accumulates because of the continuous cold. The latitude range for this climate is 65° to 90° N and S.

> The ice sheet climate ⑬ has the lowest temperatures found on Earth. No month shows mean temperatures above freezing, and winter mean monthly temperatures can fall to 40°C (−40°F) and below.

Climate Change

So far in this chapter, we have focused on the characteristics of the Earth's major climates as we experience them today. But it's important to realize that these climate characteristics have changed many times over the course of the Earth's history. Based on a variety of field data, scientists now agree that our planet has experienced a series of climate shifts over the course of its recent geologic history, alternating between colder periods of glaciation and milder interglacial periods, such as the present (Figure 7.40).

Over time, plant and animal species adjust to their individual environments, and when those environments change, to survive, they must either adapt to, or move to keep pace with, the evolving areas of their older environments. During the last glaciation, many common plant species moved toward the Equator and then migrated back during the first few millennia of the interglacial climate.

Since the last glaciation ended, approximately 10,000 years ago, the human species has thrived in the current distribution of relatively mild climates. However, the Earth is entering a new era of climate change, triggered to a significant degree by increased levels of carbon dioxide (CO_2) and other greenhouse gases in our atmosphere, as discussed in detail in prior chapters. Like other species, we, too, will have to adapt to our altered environments.

SHIFTING CLIMATE CHARACTERISTICS

As we established at the beginning of this chapter, climates are characterized by their annual patterns of temperature and precipitation, both of which are expected to change as a result of global warming. However, climate change will not bring a simple or uniform poleward shift of boundaries among the different climate zones discussed in this chapter. Instead, there will be great variability in the effects experienced in different regions. Nearly all land regions will get warmer, but with greater variability in temperature and precipitation.

Temperature increases are expected to be highest at the poles, due to the circulation patterns that move energy away from the Equator. Based on current trends, scientists expect late-summer arctic sea

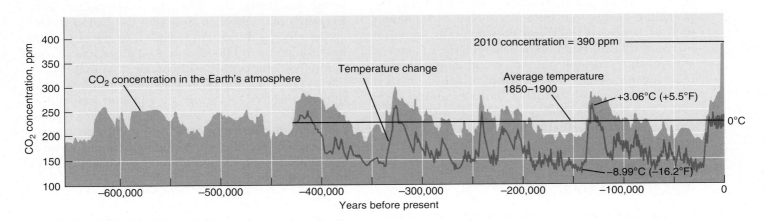

7.40 Past temperature and CO_2 records

Scientific data from ice cores and other field measurements portray a cycle of climate change that includes ice ages and interglacial periods over the past few hundred thousand years. This graph, which draws the correlation between past CO_2 levels and temperature cycles, shows that today's levels of CO_2 are much higher than anything observed in the last 700,000 years.

Climate change is having its greatest effect in the climates of high latitudes. Measurable impacts in the Arctic and in Antarctica include perma-frost melting, ice loss, and seasonal shifts. These images show the loss of perennial, or year-round, sea ice in the Arctic Ocean between 1979 and 2003. The ice is receding at a rate of 9 percent per decade.

1979   2003

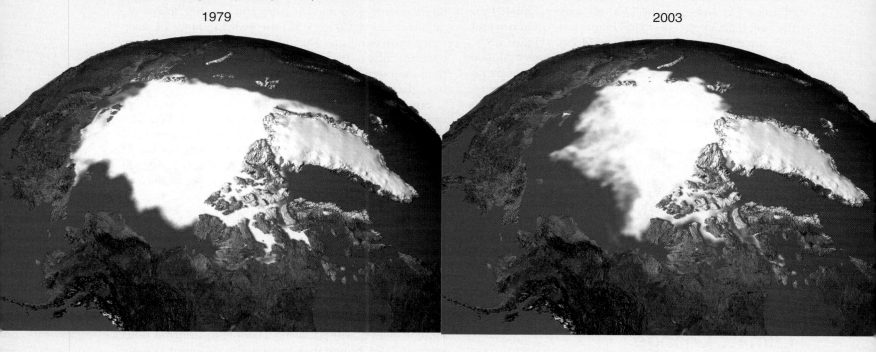

ice to nearly disappear by the middle of the century (Figure 7.41). Warming at the poles influences the pressure zones that push air masses into the mid-latitudes and control weather patterns. These shifting weather patterns will affect every aspect of our modern lives, from agricultural output to human health and natural disasters.

Global warming will trigger changes in average annual precipitation around the world. Scientists generally expect increased precipitation and evaporation in some areas and drier conditions in others. The prevailing factors that influence global precipitation, such as ocean currents and pressure cells, are shifting, and will shift farther. For example, the subtropical high-pressure cells will strengthen and move poleward, reducing precipitation in subtropical zones. In turn, rainfall will increase at higher latitudes and in some areas of the tropics. Still, a great deal of uncertainty remains about just where and how precipitation patterns will change.

WEATHER VARIABILITY

Climate change will affect more than the temperature and precipitation patterns of different climates around the world. The frequency of extreme weather events is also predicted to change. For example, the number

of extreme temperature events is forecast to rise in California. In the western mountain ranges, it is expected there will be less precipitation overall, but that winter snowfall extremes will be more likely.

Sound evidence also points to an increase in extreme weather events. Variations in storm paths and intensities reflect variations in major features of the atmospheric circulation patterns. The number of tropical storms and hurricanes varies from year to year, but data show substantial increases in the intensity and duration of these severe storms since the 1970s. The recent increase in the frequency of the El Niño–Southern Oscillation (ENSO) cycles, which is thought to be linked to global warming, has brought record floods and droughts across all continents. The number of weather-related natural disasters has also gone up, with cyclone activity and above-average rainfall triggered by warmer sea and land temperatures (Figure 7.42). In 2010, insurance companies reported that natural disasters caused $130 billion in property damage. This trend of more, and more intense, weather events is one that will require international attention to mitigate disasters.

FUTURE CHALLENGES AND ADAPTATIONS

As our climate changes and global temperatures rise, humans, other animals, and plants will all have to face

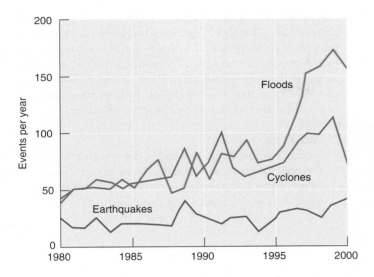

7.42 Frequency of natural disasters

This graph shows that natural disasters have been occurring more frequently in recent years. Some of this increase may be attributed to population growth, as well as improved information gathering and reporting. There can be no doubt, however, that weather-related natural disasters, such as floods and cyclones, have been occurring much more frequently; at the same time, the incidence of earthquakes has remained about the same, suggesting that some of the increase is related to climate change.

new challenges and develop new strategies to adapt. Over the past few years, climate research scientists have used **global climate models** to investigate the Earth's past and future climates. These are complex computer programs that re-create and model the physical, chemical, and biological conditions in the Earth's atmosphere, oceans, land, and ice. Global climate models are calibrated to match the Earth's existing climate by using records from past decades. The capability of these models to re-create our known weather patterns establishes their reliability for future projections.

Once a global climate model has been proven to be reliable scientists can project what the Earth's future climate might be. Global climate models are run under different sets of assumptions for greenhouse gas emissions—for example, whether emissions stay the same, decrease as called for by international goals, or increase according to current trends (Figure 7.43). About a dozen major climate change models have been included in the studies summarized by the Intergovernmental Panel on Climate Change (IPCC). No single model gives a definitive picture of the Earth's future climate, but together these climate projections show a range of possibilities and suggest how policy decisions made now might affect future conditions.

Climate scientists are careful to point out that the Earth is much more complicated than the models and mathematical equations they are using to simulate the Earth–ocean–atmosphere system. It is beyond even today's state-of-the-science to predict our future climate with great accuracy. That said, the convergence of global climate models does make it possible to make predictions that are more likely to be correct in some areas than in others.

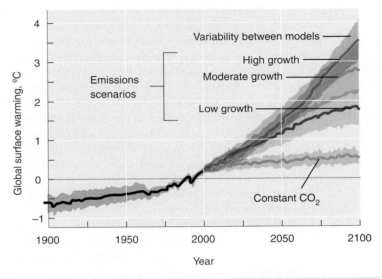

7.43 Predictions of global surface warming from global climate models

Different scenarios for the Earth's temperature increase are generated based on various assumptions about the growth of future greenhouse gas emissions, such as low, moderate, or high.

The big questions facing communities and governments around the world now is how human societies will be affected by climate change, and how they will adapt to that change. For physical geographers and other scientists, the coming decades will prove both professionally challenging and rewarding, as they work to understand and help us all to mitigate the negative effects of environmental change.

WileyPLUS Web Quiz
Take a quick quiz on the key concepts of this chapter.

A Look Ahead

This chapter has focused on the Earth's climates at the global scale. The worldwide pattern of climates is created by global- and continental-scale atmospheric circulation processes interacting with land and sea surfaces of the Earth's continents and oceans. As we have seen repeatedly through the first seven chapters in this book, these processes are driven by solar energy, coupled with the Earth's rotation.

The climates of the Earth are remarkably diverse, ranging from the hot, humid wet equatorial climate ① at the Equator to the bitterly cold and dry ice sheet climate ⑬ at the poles. The environments associated with the Earth's climates are also highly varied, from the lush, equatorial rainforest to the stunted willows of the tundra. Climate exerts strong controls on vegetation and soils, especially at the global level. With an understanding of climate, we can now turn to a more detailed look at biogeography, vegetation, and soils in the next three chapters.

WileyPLUS Web Links

This chapter's web links lead you to lots of basic information about climates, as well as to sources on such climate-related topics as the monsoon, desertification, tundra and taiga, and the dust bowl. Check them out!

IN REVIEW GLOBAL CLIMATES

- Africa's *Sahel region* has a wet-dry climate with multi-year cycles of low and high rainfall. During wet years, population and land use expand, leading to human stress and famine when the dry years arrive. These cycles depend on slow changes in global sea-surface temperature patterns and are enhanced by human alteration of the land surface.

- **Climate** is the average weather of a region. It is usually characterized by average temperature and precipitation, as well as their variation throughout the year. Various climate factors control these characteristics, including a location's latitude, elevation, relation to the coast, prevailing air masses, and nearby persistent high- and low-pressure centers.

- The variation in monthly temperature at a particular location depends largely on two factors: latitude, which determines the month-to-month pattern of insolation, and location—continental or maritime—which enhances or moderates the annual insolation cycle. At higher elevations, temperatures are lower.

- Global precipitation patterns are largely determined by air masses and their movements, which in turn are produced by global air circulation patterns. Precipitation regimes are also influenced by the location of high- and low-pressure patterns associated with these global circulations. The main features of the global pattern of rainfall are:
  - A wet equatorial belt, produced by convectional precipitation around the ITCZ.
  - Trade-wind coasts that receive moist flows of mT air from trade winds, as well as tropical cyclones.
  - Tropical deserts, which are located under subtropical high-pressure cells of hot and dry subsiding air.

  - Midlatitude deserts and steppes, which are dry because they are far from maritime moisture sources.
  - Moist subtropical regions, which receive westward flows of moist mT air in the summer and eastward-moving wave cyclones in winter.
  - Midlatitude west coasts, which are subjected to eastward flows of mP air and occluded wave cyclones by prevailing westerly winds.
  - Polar and arctic deserts, where little precipitation falls because the air is too cold to hold much moisture.

- Global monthly precipitation falls into three patterns: uniform, ranging from abundant to scarce; high-Sun (summer) maximum; and low-Sun (winter) maximum.

- Climate classification systems provide a useful framework to summarize climates of particular locations. The system used in this book interprets climate classes using air-mass characteristics and frontal zones.

- In *dry climates*, precipitation is largely evaporated from soil surfaces and transpired by vegetation. There are two subtypes: *arid* (driest) and *semiarid* or *steppe* (a little wetter). In *moist climates*, precipitation exceeds evaporation and transpiration during most months. In *wet-dry climates*, strong wet and dry seasons alternate.

- *Highland climates* are generally colder and wetter than the climates of surrounding lowlands.

- There are three groups of climate types, arranged by latitude: low-latitude climates (Group I), midlatitude climates (Group II), and high-latitude climates (Group III).

- The **wet equatorial climate** ① is warm to hot with abundant rainfall. This is the steamy climate of the Amazon and Congo basins. The ITCZ is always nearby, so rainfall is abundant throughout the year. The annual temperature cycle is very weak.

- The **monsoon and trade-wind coastal climate** ② is warm to hot, with very wet rainy seasons. Easterly trade winds or southwesterly monsoon winds in Asia bring high rainfall in the high-Sun season, when the ITCZ is nearby. Rainfall is lower when the ITCZ is in the opposite tropic. Temperatures are highest in the dry weather, before the onset of the wet season. The climates of western India, Myanmar, Vietnam, and Bangladesh are good examples.

- The **wet-dry tropical climate** ③ is warm to hot with very distinct wet and dry seasons. It has a very dry period at the time of low Sun, and a wet season at the time of high Sun, when the ITCZ is near. Temperatures peak strongly just before the onset of the wet season. The Sahel region of Africa is a good example.

- The **dry tropical climate** ④ describes the world's hottest deserts—extremely hot in the high-Sun season, a little cooler in the low-Sun season, with little or no rainfall. Here, subtropical high pressure dominates year-round. The Sahara Desert, Saudi Arabia, and the central Australian desert are examples. This climate also includes cooler deserts along subtropical west coasts.

- The **dry subtropical climate** ⑤ also includes desert climates, but is found farther poleward than the dry tropical climate. It is also dominated by subtropical high pressure, but has a wider annual temperature range and a distinct cool season. Wave cyclones and polar air masses occasionally reach this climate in winter. This type includes the hottest part of the American Southwest desert.

- The **moist subtropical climate** ⑥ is a climate of cool winters and warm, humid summers with abundant rainfall. It is dominated by maritime tropical (mT) air masses with both midlatitude and tropical cyclones. The southeastern United States and southern China are examples.

- The **Mediterranean climate** ⑦ is marked by hot, dry summers and rainy winters. Continental tropical (cT) air from subtropical high pressure dominates in summer, to produce hot, dry conditions, while moist mP air invades in winter, generating ample rainfall. Southern and central California, Spain, southern Italy, Greece, and the coastal regions of Lebanon and Israel are regions of Mediterranean climate.

- The **marine west-coast climate** ⑧ has warm summers and cool winters, with more rainfall in winter than the Mediterranean climate ⑦. Subtropical high pressure invades in summer, blocking precipitation.

In winter, occluded cyclones provide abundant precipitation. Regions of this climate include the Pacific Northwest—coastal Oregon, Washington, and British Columbia.

- The **dry midlatitude climate** ⑨ prevails in midlatitude continental interiors. A local, dry air mass dominates, with occasional invasions of moist maritime air. The steppes of central Asia and the Great Plains of North America are familiar locales of this climate—warm to hot in summer, cold in winter, and with low annual precipitation.

- The **moist continental climate** ⑩ is found in the eastern United States and lower Canada—cold in winter, warm in summer, with ample precipitation through the year. Summer is wetter, with frequent invasions of maritime tropical (mT) air. In winter, continental polar (cP) and arctic (cA) air masses bring cold, dry weather. The American Northeast and upper Midwest are examples of this type.

- The **boreal forest climate** ⑪ is a snowy climate with short, cool summers and long, bitterly cold winters. It is the source region of continental polar (cP) air masses. Northern Canada, Siberia, and central Alaska are regions of the boreal forest climate ⑪.

- The **tundra climate** ⑫ of arctic coastal fringes has a long, severe winter, although cold temperatures are somewhat moderated by the nearby Arctic Ocean. Polar and arctic air masses (cP, mP, and cA) dominate. This is the climate of the coastal arctic regions of Canada, Alaska, Siberia, and Scandinavia.

- The **ice sheet climate** ⑬ is bitterly cold, with temperatures always below freezing. It is the climate of arctic (A) and antarctic (AA) air masses. It is restricted to Greenland and Antarctica.

- The Earth's climates have shifted many times in the past. Cold periods of glaciation have alternated with milder interglacial periods, such as the present. However, the Earth is entering a new climate era, one marked by global warming caused by the buildup of greenhouse gases.

- Future climates will be generally warmer, with greater variability in temperature and precipitation. Polar and arctic warming will be pronounced. Some climates will be wetter, others drier.

- Weather will be more variable, with an increased frequency of rare, extreme events, such as floods, droughts, and severe storms.

- **Global climate models** are complex computer programs that simulate weather and climate over land, ice, water, and in the atmosphere. Because individual model predictions vary, they provide a range of climate projections in response to scenarios of global change. Adapting to climate change and mitigating its effects will be a major challenge in our future.

KEY TERMS

climate, p. 222
climograph, p. 225
dry climate, p. 231
moist climate, p. 231
wet equatorial climate ①, p. 231
monsoon and trade-wind coastal
    climate ②, p. 235

wet-dry tropical climate ③, p. 236
dry tropical climate ④, p. 239
dry subtropical climate ⑤, p. 241
moist subtropical climate ⑥, p. 243
Mediterranean climate ⑦, p. 243
marine west-coast climate ⑧, p. 244
dry midlatitude climate ⑨, p. 246

moist continental climate ⑩, p. 247
boreal forest climate ⑪, p. 249
tundra climate ⑫, p. 251
ice sheet climate ⑬, p. 253
global climate model, p. 256

REVIEW QUESTIONS

1. Discuss the use of monthly records of average temperature and precipitation to characterize the **climate** of a region. Why are these measures useful?

2. Why are latitude and location (maritime or continental) important factors in determining the annual temperature cycle of a station?

3. List and describe the important climate control factors that influence temperature and precipitation in given locations.

4. Identify seven important features of the global map of precipitation. Which factors produce them?

5. The seasonality of precipitation at a station can be described as following one of three patterns. What are they, and how do they arise? Give examples.

6. What is a **climate classification system**? How does the Köppen climate system compare to the system used in this text?

7. Identify three groups of global climates based on latitude. How is each group influenced by air masses and global circulation patterns?

8. Why is the annual temperature cycle of the **wet equatorial climate** ① so uniform?

9. The **wet-dry tropical climate** ③ has two distinct seasons. Which factors produce the dry season? The wet season?

10. Why is the **dry tropical climate** ④ dry? How do the arid and semiarid subtypes of this climate differ? How does the **dry subtropical climate** ⑤ differ from the **dry tropical climate** ④?

11. Both the **moist subtropical** ⑥ and **moist continental** ⑩ climates are found on eastern sides of continents in the midlatitudes. What are the major factors that determine their temperature and precipitation cycles? How do these two climates differ?

12. Both the **Mediterranean** ⑦ and **marine west-coast** ⑧ climates are found on the west coasts of continents. Why do they experience more precipitation in winter than in summer? How do the two climates differ?

13. The **boreal forest climate** ⑪ and **tundra climate** ⑫ are both climates of the northern regions, but the tundra is found fringing the Arctic Ocean, whereas the boreal forest is located farther inland. Compare these two climates from the viewpoint of coastal-continental effects.

14. What is the coldest climate on Earth? How is the annual temperature cycle of this climate related to the cycle of insolation?

15. Have global climates been stable over geologic time? What changes in weather and climate can we expect in the future, based on the projections of global climate models?

VISUALIZING EXERCISES

1. Sketch the temperature and rainfall cycles for a typical station in the monsoon and trade-wind coastal climate ②. What factors contribute to the seasonality of the two cycles?

2. Sketch climographs for the Mediterranean climate ⑦ and the dry midlatitude climate ⑨. What are the essential differences between them? Explain why they occur.

3. Suppose South America were upside down. That is, imagine that the continent was cut out and flipped over end-for-end so that the southern tip was at about 10° N latitude and the northern end (Venezuela) was positioned at about 55° S. The Andean chain would still be on the west side, but the shape of the land mass would now be quite different. Sketch this continent and draw possible climate boundaries, using your knowledge of global air circulation patterns, frontal zones, and air-mass movements.

ESSAY QUESTIONS

1. The intertropical convergence zone (ITCZ) moves north and south with the seasons. Describe how this movement affects the four low-latitude climates.

2. Discuss the role of the polar front and the air masses that come in conflict in the polar-front zone in the temperature and precipitation cycles of the midlatitude and high-latitude climates.

3. Which climate is your home in? Compare it to the climate of another location with which you are familiar. In your comparison, stress the factors that determine the monthly patterns of temperature and precipitation, including the global circulation patterns, air masses, and frontal zones that influence each climate. Construct a possible climograph for each.

SPECIAL SUPPLEMENT • THE KÖPPEN CLIMATE SYSTEM

Air temperature and precipitation data comprise the basis for several climate classifications. One of the most important of these is the Köppen climate system, devised in 1918 by Dr. Vladimir Köppen of the University of Graz in Austria. For several decades, this system, with various revisions, was the most widely used climate classification among geographers. Köppen was both a climatologist and plant geographer, so that his main interest lay in finding climate boundaries that coincided approximately with boundaries between major vegetation types.

Under the Köppen system, each climate is defined according to assigned values of temperature and precipitation, computed in terms of annual or monthly values. Any given station can be assigned to its particular climate group and subgroup solely on the basis of the records of temperature and precipitation at that place.

Note that "mean annual temperature" refers to the average of 12 monthly temperatures for the year, and "mean annual precipitation" refers to the average of the entire year's precipitation as observed over many years.

The Köppen system features a shorthand code of letters assigned to designate major climate groups, subgroups within the major groups, and further subdivisions to distinguish particular seasonal characteristics of temperature and precipitation. Five major climate groups are designated by capital letters as follows:

A   Tropical rainy climates
The average temperature of every month is above 18°C (64.4°F). These climates have no winter season. Annual amount of rainfall is large and exceeds annual evaporation.

B   Dry climates
Annual evaporation exceeds annual precipitation. There is no water surplus; hence, no permanent streams originate in B climate zones.

C   Mild, humid (mesothermal) climates
The coldest month has an average temperature of under 18°C (64.4°F), but above −3°C (26.6°F); at least one month has an average temperature above 10°C (50°F). The C climates thus have both a summer and a winter.

D   Snowy-forest (microthermal) climates
The coldest month has an average temperature of under −3°C (26.6°F). The average temperature of the warmest month is above 10°C (50°F). (Forest is not generally found where the warmest month is colder than 10°C (50°F).)

E   Polar climates
The average temperature of the warmest month is below 10°C (50°F). These climates have no true summer.

Note that four of these five groups (A, C, D, and E) are defined by temperature averages, whereas one (B) is defined by precipitation-to-evaporation ratios. Groups A, C, and D have temperatures and precipitation amounts that can support the growth of forest and woodland vegetation. Figure S7.1 shows the boundaries of the five major climate groups, and Figure S7.2 is a world map of Köppen climates. Subgroups within the five major groups are designated by a second letter according to the following codes:

S   Semiarid (steppe)
W   Arid (desert)

The capital letters S and W are applied only to the dry B climates.

f   Moist, adequate precipitation in all months; no dry season. This modifier is applied to A, C, and D groups.
w   Dry season in the winter of the respective hemisphere (low-Sun season).
s   Dry season in the summer of the respective hemisphere (high-Sun season).
m   Rainforest climate, despite short, dry season in monsoon-type precipitation cycle. Applies only to A climates.

From combinations of the two letter groups, 12 distinct climates emerge:

A   Tropical rainforest climate
The rainfall of the driest month is 6 cm (2.4 in.) or more.

Am   Monsoon variety of Af
The rainfall of the driest month is less than 6 cm (2.4 in.). The dry season is strongly developed.

Aw   Tropical savanna climate
At least one month has rainfall less than 6 cm (2.4 in.). The dry season is strongly developed.

Figure S7.3 shows the boundaries between Af, Am, and Aw climates as determined by both annual rainfall and rainfall of the driest month.

BS   *Steppe climate*
Characterized by grasslands, this semiarid climate occupies an intermediate position between the desert climate (BW) and the more humid climates of the A, C, and D groups. Boundaries are determined by formulas given in Figure S7.4.

BW   *Desert climate*
Desert has an arid climate with annual precipitation of usually less than 40 cm (15.7 in.). The boundary with the adjacent steppe climate (BS) is determined by formulas given in Figure S7.4.

S7.1 Generalized Köppen climate map

Highly generalized world map of major climate regions according to the Köppen classification. Highland areas are in black.

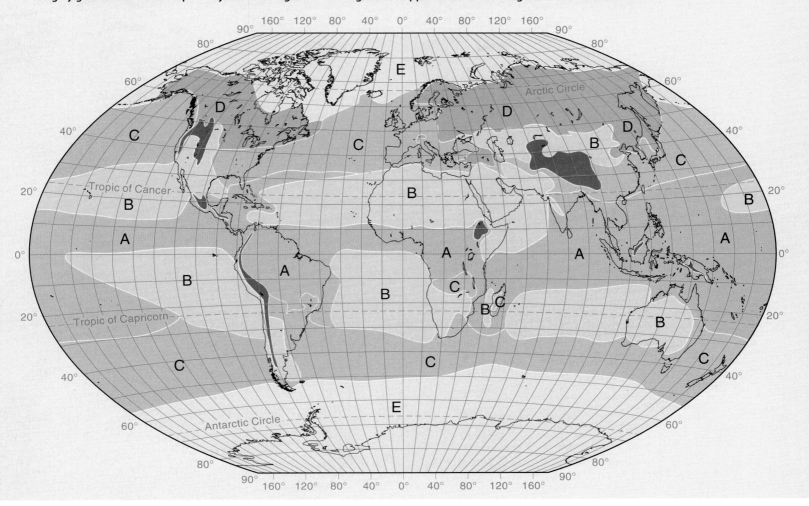

C   Mild humid climate with no dry season
Precipitation of the driest month averages more than 3 cm (1.2 in.).

Cw   Mild humid climate with a dry winter
The wettest month of summer has at least 10 times the precipitation of the driest month of winter. (Alternative definition: 70 percent or more of the mean annual precipitation falls in the warmer six months.)

Cs   Mild humid climate with a dry summer
Precipitation of the driest month of summer is less than 3 cm (1.2 in.). Precipitation is at least three times as much as the driest month of summer. (Alternative definition: 70 percent or more of the mean annual precipitation falls in the six months of winter.)

Df   Snowy-forest climate with a moist winter
No dry season.

Dw   Snowy-forest climate with a dry winter

ET   Tundra climate
The mean temperature of the warmest month is above 0°C (32°F) but below 10°C (50°F).

EF   Perpetual frost climate
In this ice sheet climate, the mean monthly temperatures of all months are below 0°C (32°F).

To denote further variations in climate, Köppen added a third letter to the code group. The meanings are as follows:

a   With hot summer; warmest month is over 22°C (71.6°F); C and D climates.

b   With warm summer; warmest month is below 22°C (71.6°F); C and D climates.

c   With cool, short summer; less than four months are over 10°C (50°F); C and D climates.

d   With very cold winter; coldest month is below −38°C (−36.4°F); D climates only.

h   Dry-hot; mean annual temperature is over 18°C (64.4°F); B climates only.

k   Dry-cold; mean annual temperature is under 18°C (64.4°F); B climates only.

As an example of a complete Köppen climate code, BWk refers to a cool desert climate, and Dfc to a cold, snowy-forest climate with a cool, short summer.

S7.2 Köppen climates of the world

World map of climates, drawn according to the Köppen–Geiger–Pohl system.

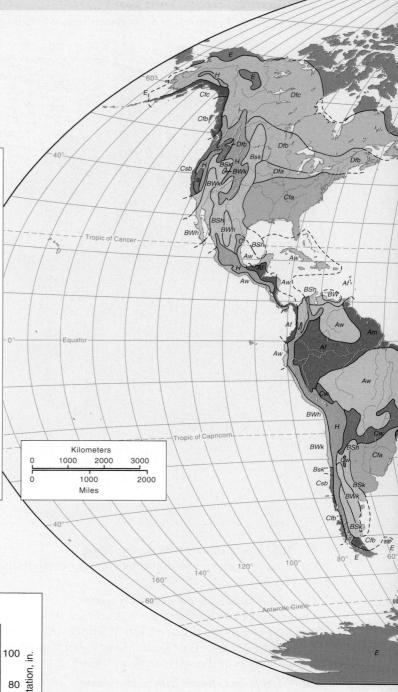

KÖPPEN-GEIGER SYSTEM OF CLIMATE CLASSIFICATION

After R. Geiger and W. Pohl (1953)

Key to letter code designating climate regions:

FIRST LETTER

A C D Sufficient heat and precipitation for growth of high-trunked trees.
A *Tropical climates.* All monthly mean temperatures over 18°C (64.4°F).
B *Dry climates.* Boundaries determined by formula using mean annual temperature and mean annual precipitation (see graphs).
C *Warm temperature climates.* Mean temperature of coldest month: 18°C (64.4°F) down to –3°C (26.6°F).
D *Snow climates.* Warmest month mean over 10°C (50°F). Coldest month mean under –3°C (26.6°F).
E *Ice climates.* Warmest month mean under 10°C (50°F).

SECOND LETTER

S *Steppe climate.* } Boundaries determined by formulas (See graphs).
W *Desert climate.*
f Sufficient precipitation in all months.
m Rainforest despite a dry season (i.e., monsoon cycle).
s Dry season in summer of the respective hemisphere.
w Dry season in winter of the respective hemisphere.

THIRD LETTER

a Warmest month mean over 22°C (71.6°F).
b Warmest month mean under 22°C (71.6°F). At least 4 months have means over 10°C (50°F).
c Fewer than 4 months with means over 10°C (50°F).
d Same as c, but coldest month mean under –38°C (–36.4°F).
h Dry and hot. Mean annual temperature over 18°C (64.4°F).
k Dry and cold. Mean annual temperature under 18°C (64.4°F).

H Highland climates.

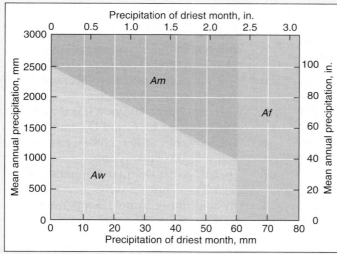

S7.3 Boundaries of A climates

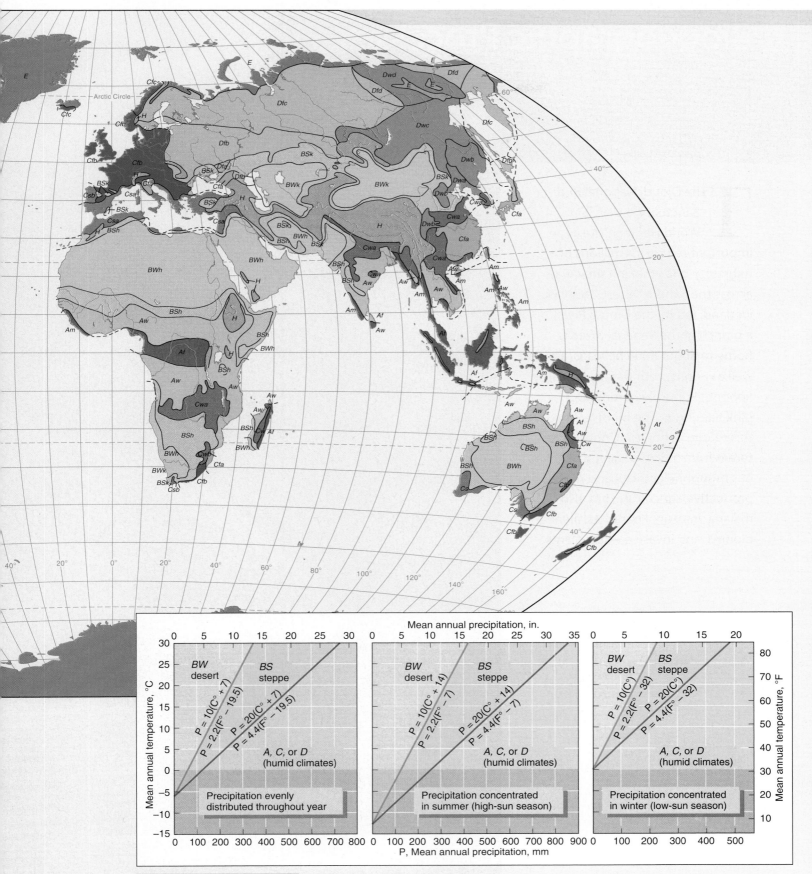

S7.4 Boundaries of B climates

Upper formulas: metric system; lower formulas: English system.

Chapter 8
Biogeographic Processes

The Djoudj National Bird Sanctuary, a UNESCO World Heritage site, is an important refuge for millions of migratory birds making their way across the Sahara Desert. This vast wetland, fed by the Djoudj River, a branch of the Senegal River, is home to about 5000 pelicans, as well as egrets, cormorants, African Spoonbills, purple herons, and pink flamingos. The biodiversity of this refuge is, however, now under threat from a dam project that has brought electricity and more productive agriculture, but also malaria, intestinal parasites, algae blooms, and invasive water plants.

PELICANS IN THE DJOUDJ NATIONAL BIRD SANCTUARY, SENEGAL

Biogeographic Processes

This chapter is about the processes that determine where and when organisms are found on the Earth's varied land surface. How do organisms use the Sun to power their life activities? What environmental factors limit the distribution of organisms? How do organisms interact with one another? How do plant and animal communities change with time, if left undisturbed? Where do species come from? How do they change and evolve? How do they find their way from place to place across the globe? These are some of the questions we will answer in this chapter.

Human Impact on the Carbon Cycle

Carbon, an element that is abundant at the Earth's surface, is essential for life. Carbon cycles continuously among the land surface, atmosphere, and ocean in many complex pathways. However, these flows are now strongly influenced by human activity. The most important human impact on the carbon cycle is the burning of fossil fuels, which releases carbon dioxide (CO_2) into the atmosphere and enhances global warming. The impact of human activity is also felt in alterations to the Earth's land covers—for example, by clearing forests or abandoning agricultural areas—which can cause the release or absorption of atmospheric CO_2. Let's look at these impacts in more detail.

Figure 8.1 shows a simple diagram of the global atmospheric carbon budget for the period 2000–2005. The magnitudes of the annual flows are shown in gigatons (Gt) of carbon per year (1 gigaton = 10^9 metric tons = 10^{12} kg = 1.1×10^9 English tons = 1.1 English gigatons). Note that these flows are estimates, and a second value after the first indicates the degree of uncertainty.

Fossil fuel burning contributes about 7.2 ± 0.3 Gt of carbon per year, nearly all in the form of carbon dioxide. About 2.2 ± 0.5 Gt of carbon per year is taken up by the oceans, reducing the atmospheric content by that amount. In addition, yearly uptake of carbon dioxide by land ecosystems is estimated at about 0.9 ± 0.6 Gt of carbon. Taken

> Fossil fuel burning contributes more than 7 Gt of carbon to the atmosphere per year. Oceanic processes remove about 2 Gt, and land ecosystems about 1 Gt, leaving about 4 Gt of carbon accumulation per year.

together, these two flows out of the atmosphere leave about 4.1 ± 0.1 Gt of carbon remaining in the atmosphere each year.

Two processes are responsible for the uptake of carbon by the oceans. First, carbon dioxide dissolves in seawater, which removes carbon from the atmosphere. Second, phytoplankton—microscopic plants living in the ocean—take up carbon dioxide in photosynthesis. When they die, they produce organic matter that sinks to the ocean floor, removing it from short-term circulation. The removal of carbon dioxide is moderated somewhat by another process by which the formation of calcium carbonate by diatoms and other marine organisms releases CO_2. Taken together, these oceanic processes remove about 30 percent of the carbon released to the atmosphere each year by fossil fuel burning.

Land ecosystems cycle carbon by photosynthesis, respiration, decomposition, and combustion. Photosynthesis and respiration are basic physiological processes of plants that absorb CO_2 from the atmosphere and release it, respectively. Decomposition is the process by which bacteria and fungi digest dead organic matter; it is actually a form of respiration. Combustion refers to uncontrolled combustion, as when an ecosystem burns in a forest fire.

These processes, taken as a whole, remove about 0.9 Gt of carbon per year from the atmosphere—about 12 percent of the contribution by fossil fuel burning. This uptake of atmospheric CO_2 by land ecosystems means that plant biomass—the amount of carbon-bearing material contained in living and dead plant matter—is increasing at that rate. However, forests are presently diminishing in area as they are logged or converted to farmland or grazing land. This conversion, which is primarily occurring in tropical and equatorial regions, is estimated to release about 1.6 Gt of carbon per year to the atmosphere. Assuming this large amount is being released, then the remaining area of land ecosystems must be taking up at least that much carbon, and more—totalling about 2.5 Gt per year—to provide a net uptake of 0.9 Gt/yr.

8.1 Human impact on the carbon cycle

Human induced changes in vegetation and soils are affecting the carbon cycle.

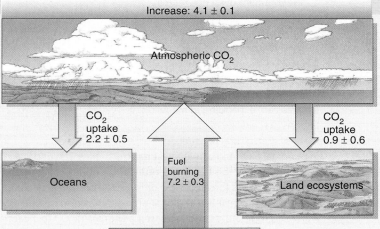

Increase: 4.1 ± 0.1

Atmospheric CO_2

THE GLOBAL CARBON CYCLE ▶
Values are in gigatons of carbon per year.

CO_2 uptake 2.2 ± 0.5

Oceans

Fuel burning 7.2 ± 0.3

CO_2 uptake 0.9 ± 0.6

Land ecosystems

Fossil fuel

James P. Blair/NG Image Collection

◀ **FOREST CLEARING**
Clearing forests and shrublands for agriculture and grazing lands releases carbon through burning or enhanced decay of new biomass. Here, fires burn in the Amazon River Basin as cattle ranchers convert forest to grassland.

▼ **FOREST GROWTH**
Growth of forests, planted or natural, builds terrestrial biomass and removes carbon from the atmosphere.

Jim Zipp/Photo Researchers, Inc.

Bob Gibbons/Photo Researchers, Inc.

▲ **CARBON IN SOILS**
Soil organic matter releases carbon dioxide when digested by microorganisms. When soil temperature rises in response to global warming, the rates of digestion and CO_2 release will also increase. Shown here is a peat bog near Zakopane, Poland. Grass covers the peat in the foreground. Slabs of peat, cut from the bog's surface, are drying and will later be burned as fuel.

Independent evidence seems to confirm this conclusion. In Europe, for example, forest statistics show an increase of growing stock—the volume of living trees—of at least 25 percent since 1970. In North America, forest areas are expanding in many regions in the wake of abandonment of agricultural production in marginal areas to natural forest regrowth. New England is a good example of this trend. A century ago, only a small portion of New England was forested. Now only a small portion is cleared.

> The clearing of forests or shrublands for agriculture or grazing releases carbon to the atmosphere; conversely, vegetation regrowth and expansion into abandoned lands takes up carbon.

Some of the increase in global biomass may also be due to warmer temperatures and increased atmospheric CO_2 concentrations, which enhance photosynthesis and so make plants more productive. Another stimulating factor is nitrogen fertilization of soils caused by the washout of nitrogen pollutant gases in the atmosphere.

Clearly, the dynamics of forests are important in the global carbon cycle, but soils may be even more important. Recent inventories estimate that about four times as much carbon resides in soils than in aboveground plant biomass. The largest reservoir of soil carbon is in the boreal forest. In fact, there is about as much carbon in boreal forest soils as in all aboveground vegetation.

This soil carbon has accumulated over thousands of years under cold conditions, which have retarded its decay. However, there is now great concern that global warming, which is acting more strongly at high latitudes, will raise the rate of decay of this vast carbon pool, releasing CO_2 as microorganisms digest the organic matter. Boreal forests, which are still taking up CO_2, may soon start releasing it, intensifying the warming trend. Figure 8.1 shows some terrestrial sources and sinks of carbon, including soil organic matter.

Reducing the rate of carbon dioxide buildup in the atmosphere is a matter of heightened concern worldwide. As we noted in Chapter 3, the nations of the world have been struggling to implement an effective plan to control these emissions. While much good progress has been made, more work is needed. Global commitment to the reduction of CO_2 releases and control of global warming still eludes us.

Energy and Matter Flow in Ecosystems

This chapter is the first of two that discuss biogeography. **Biogeography** focuses on the distribution of plants and animals—the *biota*—over the Earth. It identifies and describes the processes that influence plant and animal distribution patterns. *Ecological biogeography* looks at how the distribution patterns of organisms are affected by the environment. *Historical biogeography* focuses on how spatial distribution patterns of organisms arise over time and space.

But before we turn to those processes we must first consider a number of ideas from the domain of **ecology**, which is the science of the interactions among life-forms and their environments. These ideas concern how organisms live and interact as ecosystems, and how energy and matter are cycled by these ecosystems. The term **ecosystem** refers to a group of organisms and their environments (Figure 8.2). Ecosystems take up matter and energy as plants and animals grow, reproduce, and maintain life. That matter and energy can be recycled within the ecosystem or exported out of it. Ecosystems balance the various processes and activities within them. Many of these balances are robust and self-regulating, but some are highly sensitive and can be easily upset or destroyed.

THE FOOD WEB

Energy is transferred through an ecosystem in steps, making up a *food chain* or a **food web**. At the bottom of the chain are the *primary producers*, which absorb sunlight and use the light energy to convert carbon dioxide and water into *carbohydrates* (long chains of sugar molecules) and, eventually, into other biochemical molecules, by *photosynthesis* (Figure 8.3). The primary producers support the *consumers*—organisms that ingest other organisms as their food sources. Finally, *decomposers* feed on decaying organic matter, from all levels of the web. Decomposers are largely microscopic organisms (microorganisms) and bacteria.

> The food web traces how food energy flows from organism to organism within an ecosystem. Primary producers support primary, secondary, and higher-level consumers. Decomposers feed on dead plant and animal matter from all levels.

The food web is really an energy flow system, tracing the path of solar energy through the ecosystem. Solar energy is absorbed by the primary producers and stored in the chemical products of photosynthesis. As these organisms are eaten and digested by consumers, chemical energy is released. This chemical energy is used to power new biochemical reactions, which again produce stored chemical energy in the consumers' bodies.

Energy is lost at each level in the food web through *respiration*. You can think of this lost energy as fuel burned to keep the organism operating. Energy expended in respiration is ultimately lost as waste heat and cannot be stored for use by other organisms higher up in the food chain. This means that, generally, both the numbers of organisms and their total amount of living tissue must decrease substantially up the food chain. In general, only 10 to 50 percent of the energy stored in organic matter at one level can be passed up the chain to the next level. Normally, there are about four levels of consumers.

The number of individuals of any species present in an ecosystem depends on the resources available to support them. If these resources provide a steady supply of energy, the population size will normally stay steady. But resources can vary with time; for example, in an annual cycle. In those cases, the population size of a species depending on these resources may fluctuate in a corresponding cycle.

8.2 Ecosystem gallery

Tundra, marsh, and savanna are just three of the Earth's many distinctive types of ecosystems.

Ted Kerasote/Photo Researchers, Inc.

▲ **CARIBOU IN THE FOOTHILLS OF THE BROOKS RANGE**

The caribou, a large grazing mammal, is one of the more important primary consumers of the tundra ecosystem.

EYE ON THE LANDSCAPE What else would the geographer see?

The snow patches near the ridgeline Ⓐ may be shallow depressions, called cirques, that once held the heads of alpine glaciers. Notice the "blobs" of soil on the slope at Ⓑ, distinguished by color and shape. These are most likely solifluction lobes, produced in the short arctic summer when the snow melts and the surface layer of soil thaws. The saturated soil creeps downhill in lobes until it dries out and its motion stops.

Raymond Gehman/NG Image Collection

▲ **FRESHWATER MARSH**

The marsh or swamp ecosystem supports a wide variety of life-forms, both plant and animal. Here, a group of white ibises forages for food in the shallow waters of Okefenokee Swamp, Georgia.

Beverly Joubert/NG Image Collection

SAVANNA ▶

The savanna ecosystem, with its abundance of grazing mammals and predators, has a rich and complex food web. Here, a line of African elephants travels single-file across the savanna plains of the Savuti region, Botswana.

8.3 The food web of a salt marsh

A salt marsh is a good example of an ecosystem. It contains a variety of organisms—algae and aquatic plants, microorganisms, insects, snails, and crayfish, fishes, birds, shrews, mice, and rats. There are also inorganic components—water, air, clay particles and organic sediment, inorganic nutrients, trace elements, and light energy.

David Lyons/Alamy Limited

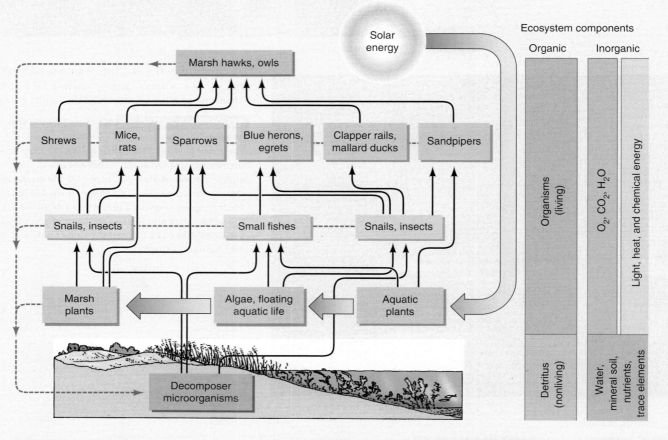

Ecosystem components

Organic   Inorganic

Solar energy

Marsh hawks, owls

Shrews | Mice, rats | Sparrows | Blue herons, egrets | Clapper rails, mallard ducks | Sandpipers

Snails, insects | Small fishes | Snails, insects

Marsh plants | Algae, floating aquatic life | Aquatic plants

Decomposer microorganisms

Organisms (living)

O_2, CO_2, H_2O

Light, heat, and chemical energy

Detritus (nonliving)

Water, mineral soil, nutrients, trace elements

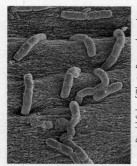

David Scharf/Photo Researchers, Inc.

Decomposers Microscopic organisms and bacteria feed on detritus, or decaying organic matter, from all levels of the web.

Florida Images/Alamy Limited

Primary producers The plants and algae in the food web are the primary producers. They use light energy to convert carbon dioxide and water into carbohydrates. These organisms, engaged in photosynthesis, form the base of the food web.

Paul Whitten/Photo Researchers, Inc.

Primary consumers The lowest level of consumers are the primary consumers (the snails, insects and fishes).

Mike Baylan/U.S. Fish &Wildlife Service

Secondary consumers At the next level are the secondary consumers (the mammals, birds, and larger fishes), which feed on the primary consumers.

Alan Carey/©Corbis

Higher level consumers Still higher levels of feeding occur in the salt-marsh ecosystem as marsh hawks and owls consume the smaller animals below them in the food web. In most ecosystems there are about four levels of consumers.

PHOTOSYNTHESIS AND RESPIRATION

Simply put, photosynthesis is the production of carbohydrate (Figure 8.4). *Carbohydrate* is a general term for a class of organic compounds that are made from the elements carbon, hydrogen, and oxygen. Carbohydrate molecules are composed of short chains of carbon bonded to one another. Hydrogen (H) atoms and hydroxyl (OH) molecules are also attached to the carbon atoms. We can symbolize a single carbon atom with its attached hydrogen atom and hydroxyl molecule as –CHOH–. The leading and trailing dashes indicate that the unit is just one portion of a longer chain of connected carbon atoms.

Photosynthesis of carbohydrate requires a series of complex biochemical reactions using water (H_2O) and carbon dioxide (CO_2), plus light energy. This process requires *chlorophyll*, a complex organic molecule that absorbs light energy for use by the plant cell. A simplified chemical reaction for photosynthesis can be written as:

> Photosynthesis is the process by which plants combine water, carbon dioxide, and solar energy to form carbohydrate. Respiration is the reverse process: carbohydrate is oxidized in living tissues to yield the energy that sustains life.

$$H_2O + CO_2 + \text{light energy} \rightarrow -CHOH- + O_2$$

Oxygen gas molecules (O_2) are a by-product of photosynthesis. Because gaseous carbon as CO_2 is "fixed" to a solid form in carbohydrate, we also call photosynthesis a *carbon fixation* process.

Respiration is the opposite of photosynthesis. In this process, carbohydrate is broken down and combines with oxygen to yield carbon dioxide and water. The overall reaction is:

$$-CHOH- + O_2 \rightarrow CO_2 + H_2O + \text{chemical energy}$$

As with photosynthesis, the actual reactions involved are not this simple. The chemical energy released is stored in several types of energy-carrying molecules in living cells and used later to synthesize all the biological molecules necessary to sustain life.

Although much of the chemical energy released in respiration is taken up in the new chemical bonds of these molecules, a certain portion raises the temperature of the new molecules. As the molecules cool, this internal energy is lost to the environment as heat flow. In a similar way, each subsequent biochemical reaction releases internal energy and adds to the heat flow. Eventually, all the chemical energy gained by respiration is lost to the surroundings in these transformations.

Because respiration consumes carbohydrate, we have to take respiration into account when talking about the amount of new carbohydrate placed in storage. *Gross photosynthesis* is the total amount of carbohydrate produced by photosynthesis. *Net photosynthesis* is the amount of

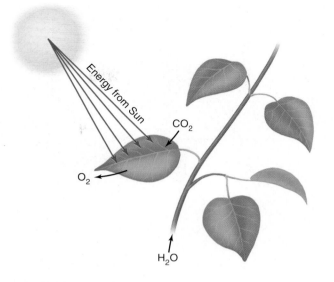

Leaves take in CO_2 from the air, and H_2O from their roots, using solar energy absorbed by chlorophyll to combine them, forming carbohydrate. In the process, O_2 is released. Photosynthesis takes place in chloroplasts, tiny grains in plant cells that have layers of chlorophyll, enzymes, and other molecules in close contact.

synthesized carbohydrate remaining after respiration has broken down sufficient carbohydrate to power the plant:

Net photosynthesis = Gross photosynthesis − Respiration

The rate of net photosynthesis depends on the intensity of light energy available, up to a limit. Most green plants only need about 10 to 30 percent of full summer sunlight for maximum net photosynthesis. Once the intensity of light is high enough for maximum net photosynthesis, the duration of daylight becomes an important factor in determining the rate at which the products of photosynthesis build up in plant tissues (Figure 8.5). The rate of photosynthesis also increases as air temperature increases, up to a limit (Figure 8.6).

> Day length, air and soil temperature, and water availability are the most important climatic factors that control net primary productivity.

NET PRIMARY PRODUCTION

Plant ecologists measure the accumulated net production by photosynthesis in terms of **biomass**, which is the dry weight of organic matter. This quantity could, of course, be stated for a single plant or animal, but a more useful measurement is the biomass per unit of surface area within the ecosystem—that is, grams of biomass per square meter or (metric) tons of biomass per hectare (1 hectare = 10^4 m^2). Of all ecosystems, forests have the greatest biomass because of the large amount of wood that the trees accumulate through time. The biomass of grasslands and croplands is much smaller in comparison; and the biomass of

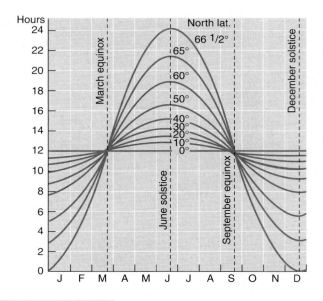

8.5 Day-length variation

The graph shows the duration of the daylight period at various latitudes in the northern hemisphere throughout the year. The angle of the Sun's rays also changes with latitude and the seasons. The vertical scale gives the number of hours the Sun is above the horizon, with changing seasons. At low latitudes, days are not far from the average 12-hour length throughout the year. At high latitudes, days are short in winter but long in summer. In subarctic latitudes, photosynthesis can go on in summer during most of the 24-hour day, compensating for the short growing season.

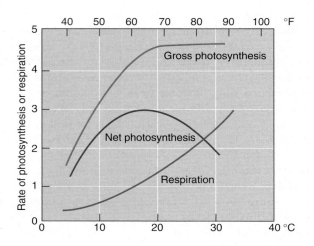

8.6 Temperature and energy flow

The figure shows the results of a laboratory experiment in which sphagnum moss was grown under constant illumination and increasing temperature. Gross photosynthesis increased rapidly, to a maximum at about 20°C (68°F), then leveled off. But net photosynthesis—the difference between gross photosynthesis and respiration—peaked at about 18°C (64°F), then fell off rapidly because respiration continued to increase with temperature.

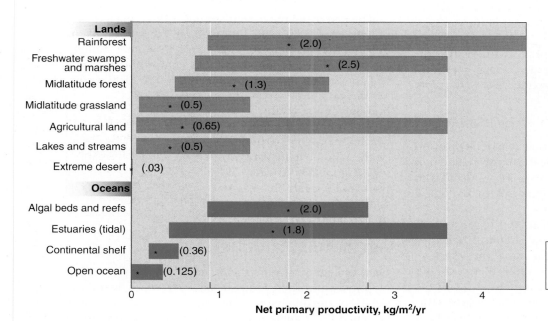

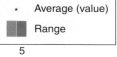

8.7 Net primary production of ecosystems

Net primary production of ecosystems ranges widely across lands and oceans. On land, freshwater swamps and marshes are most productive, on average. In marine environments, coral reefs and algal beds head the list.

freshwater bodies and the oceans is about one-hundredth that of the grasslands and croplands.

The amount of biomass per unit area tells us about the amount of photosynthetic activity, but it can be misleading. In some ecosystems, biomass is broken down very quickly by consumers and decomposers. So if we want to know how productive the ecosystem is, it's better to work out the annual yield of useful energy

produced by the ecosystem, or the *net primary production*. Net primary production represents a source of renewable energy derived from the Sun that can be exploited to fill human energy needs. The use of biomass as an energy source involves releasing solar energy that has been fixed in plant tissues through photosynthesis. It can take place in a number of ways—by burning wood for fires, for example.

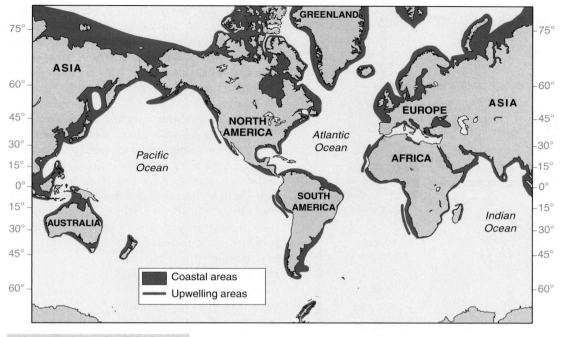

8.8 Distribution of world fisheries

Coastal areas and upwelling areas together supply most of the world production of fish.

Figure 8.7 shows the net primary production of various ecosystems in units of kilograms of dry organic matter produced annually from 1 square meter of surface. The highest values are in two quite unlike environments: forests and wetlands (swamps, marshes, and estuaries). Agricultural land compares favorably with grassland, but the range is very large in agricultural land, reflecting many factors such as availability of soil water, soil fertility, and use of fertilizers and machinery.

> Net primary production measures the rate of accumulation of carbohydrate by primary producers. Equatorial rainforests and freshwater swamps and marshes are among the most productive ecosystems, while deserts are least productive.

Open oceans aren't generally very productive. Continental shelf areas are more so; in fact, they support much of the world's fishing industry (Figure 8.8). Upwelling zones are also highly productive.

WileyPLUS Remote Sensing and Biosphere Interactivity
Explore satellite images from local to global scales to examine land and ocean productivity. Identify ocean algae blooms.

BIOGEOCHEMICAL CYCLES

We've seen how energy from the Sun flows through ecosystems, passing from one part of the food chain to the next. Ultimately, that energy is radiated to space and lost from the biosphere. Matter also moves through ecosystems, but because gravity keeps surface material earthbound, matter can't be lost in the global ecosystem. As

molecules are formed and reformed by chemical and biochemical reactions within an ecosystem, the atoms that compose them are not changed or lost. In this way, matter is conserved, and atoms and molecules are used and reused, or cycled, within ecosystems.

Atoms and molecules move through ecosystems under the influence of both physical and biological processes. We call the pathways that a particular type of matter takes through the Earth's ecosystem a **biogeochemical cycle** (sometimes referred to as a *material cycle* or *nutrient cycle*).

The major features of a biogeochemical cycle are diagrammed in Figure 8.9. Any area or location of concentration of a material is a **pool**. There are two types of pools: *active pools*, where materials are in forms and places easily accessible to life processes, and *storage pools*, where materials are more or less inaccessible to life. A system of pathways of material flows connects the various active and storage pools within the cycle. Pathways can involve the movement of material in all three states of matter: gas, liquid, and solid. For example, carbon moves freely in the atmosphere as carbon dioxide gas, and freely in water as dissolved CO_2 and carbonate ion ($CO_3^=$). It also takes the form of a solid in deposits of limestone and dolomite (calcium and magnesium carbonate).

The Carbon Cycle

Of all the biogeochemical cycles, the **carbon cycle** is the most important. All life is composed of carbon compounds of one form or another. That is why it is of such grave concern today that human activities since the Industrial Revolution have modified the carbon cycle in significant ways.

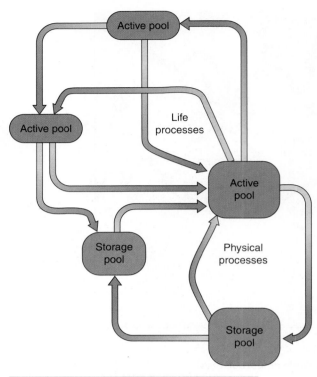

8.9 General features of a biogeochemical cycle

dissolved gas in fresh- and saltwater. Atmospheric carbon dioxide makes up less than 2 percent of all the carbon, excluding carbonate rocks and sediments. The atmospheric pool of carbon is supplied by plant and animal respiration in the oceans and on land, by outgassing volcanoes, and by fossil fuel combustion in industry.

In the sedimentary portion of its cycle, we find carbon in carbohydrate molecules in organic matter, as hydrocarbon compounds in rock (petroleum and coal), and as mineral carbonate compounds such as calcium carbonate ($CaCO_3$). A lot of carbon is incorporated into shells of marine organisms, large and small. When they die, their shells settle to the ocean floor, where they dissolve or accumulate as layers of sediment. This provides an enormous carbon storage pool, but it is not available to organisms until it is later released by rock weathering. Organic compounds synthesized by phytoplankton also settle to the ocean floor where they are eventually transformed into the hydrocarbon compounds that make up petroleum and natural gas.

As noted earlier, humans are affecting the carbon cycle by burning fossil fuels.

> The carbon cycle is a biogeochemical cycle in which carbon flows among storage pools in the atmosphere, ocean, and on the land. Human activity has affected the carbon cycle, causing carbon dioxide concentrations in the atmospheric storage pool to increase.

Carbon moves through the cycle as a gas, as a liquid, and as a solid (Figure 8.10). In the gaseous portion of the cycle, carbon moves largely as carbon dioxide (CO_2), a free gas in the atmosphere and a

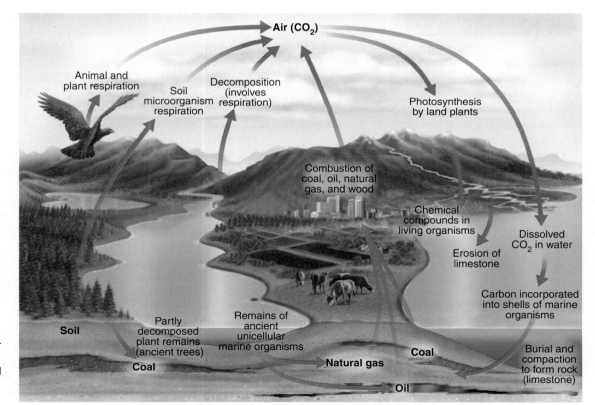

8.10 The carbon cycle

Ecosystems cycle carbon through photosynthesis, respiration, decomposition, and combustion. The movement of carbon between the atmosphere, ocean, and living organisms is known as the carbon cycle.

CO_2 is being released into the atmosphere at a rate far beyond that of any natural process, causing global warming. Human impact is also felt through the changes we make to Earth's land covers—for example, by clearing forests, abandoning agricultural areas, or letting agricultural areas grow back to forests or rangelands.

The Nitrogen Cycle

The *nitrogen cycle* is another important biogeochemical cycle. Nitrogen makes up 78 percent of the atmosphere by volume, so the atmosphere is a vast storage pool in this cycle. Nitrogen, as N_2 in the atmosphere cannot be assimilated directly by plants or animals, but certain microorganisms supply nitrogen to plants through *nitrogen fixation*. Nitrogen-fixing bacteria belonging to the genus *Rhizobium* are associated with the roots of certain plants, including legumes such as beans and peas. Once these bacteria have fixed the nitrogen into ammonia (NH_3), it can be assimilated by plants. Animals then assimilate the nitrogen when they eat the plants. Nitrogen is returned to the soil in the waste of the animals. Other soil bacteria convert nitrogen from usable forms back to N_2, in a process called *denitrification*, thereby completing the organic portion of the nitrogen cycle.

Human activities such as agriculture and industry add more nitrogen to the system, disrupting the nitrogen cycle (Figure 8.11). At present rates, nitrogen fixation

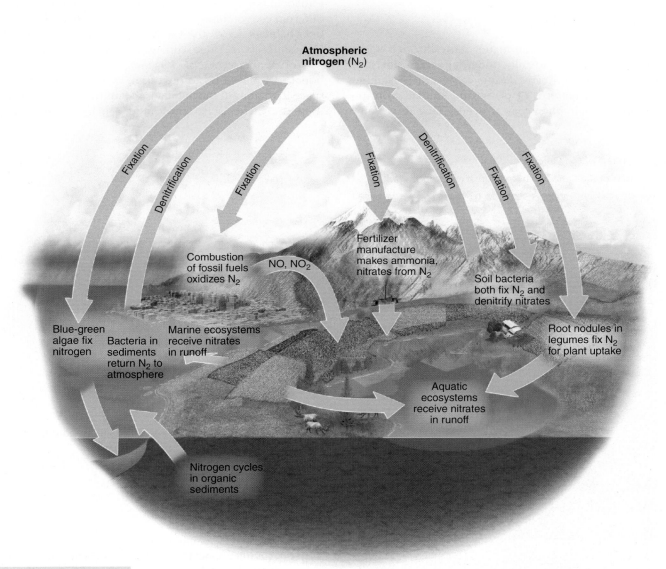

8.11 The nitrogen cycle

At the present time, due to human activity, nitrogen fixation far exceeds natural levels of denitrification. Nitrogen is fixed naturally by oceanic blue-green algae, soil bacteria, and wild legumes. Human-caused nitrogen fixation is caused by activities such as fertilizer manufacture, fuel combustion, and growth of legumes as crops. Bacteria in soil and sediments return N_2 to the atmosphere through denitrification. Nitrates from fertilization run off into freshwater and marine aquatic ecosystems, causing pollution problems.

from human activity nearly equals all natural biological fixation; and usable nitrogen is accumulating, resulting in a number of negative effects. Nitrous oxides in the atmosphere (NO and NO_2), formed by fuel combustion, contribute to climate change. Industrially emitted nitrogen may fall back to the surface as acid rain, harming plants, corroding buildings, and acidifying soils, lakes, and streams. Reactive nitrogen is also transferred from the land into rivers, where it reduces biodiversity, and then to the seas and oceans, where it can create zones devoid of life. These problems will worsen in years to come due to the fact that industrial fixation of nitrogen in fertilizer manufacture is currently doubling about every six years. The global impact of such large amounts of nitrogen reaching rivers, lakes, and oceans remains unknown.

Ecological Biogeography

We've seen how energy and matter move through ecosystems. But if we want to fully understand ecosystems, we also need to look at *ecological biogeography*, which examines the distribution patterns of plants and animals from the viewpoint of their physiological needs. That is, we must

understand how the individual organisms of an ecosystem interact with their environment. From fungi digesting organic matter on a forest floor to ospreys fishing in a coastal estuary, each organism has a range of environmental conditions that limits its survival; each also has a set of characteristic adaptations that it exploits to obtain the energy it needs to live.

Let's start by considering the relationship between organisms and their physical environment. Figure 8.12 shows how living conditions can change across the Canadian boreal forest such that different regions support different ecosystems. In this way, we can distinguish six distinct **habitats** across the Canadian boreal forest: upland, bog, bottomland, ridge, cliff, and active sand dune.

We use the term *ecological niche* to describe the functional role played by an organism, as well as the physical space it inhabits. If the habitat is the individual's "address," then the niche is its "profession," including how and

> The habitat of a species describes the physical environment that harbors its activities. The ecological niche describes how it obtains its energy and how it influences other species and its own environment.

8.12 Habitats of the Canadian boreal forest

The habitats of the Canadian boreal forest are quite varied.

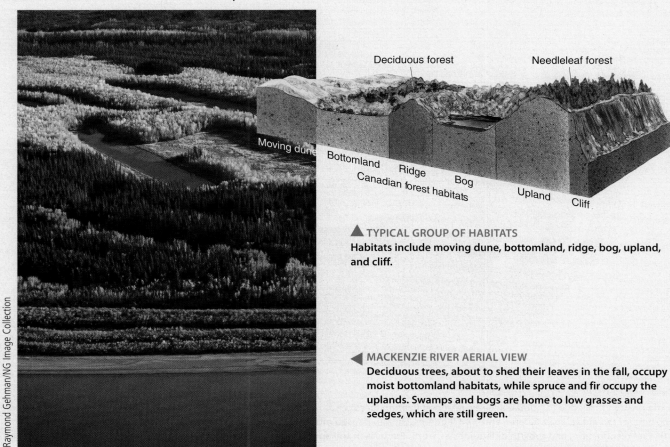

Raymond Gehman/NG Image Collection

▲ TYPICAL GROUP OF HABITATS
Habitats include moving dune, bottomland, ridge, bog, upland, and cliff.

◄ MACKENZIE RIVER AERIAL VIEW
Deciduous trees, about to shed their leaves in the fall, occupy moist bottomland habitats, while spruce and fir occupy the uplands. Swamps and bogs are home to low grasses and sedges, which are still green.

where it obtains its energy and how it influences other species and the environment around it.

When describing the ecological niche, we talk about the organism's tolerance of and responses to changes in moisture, temperature, soil chemistry, illumination, and other factors. Although many different species may occupy the same habitat, only a few of them will ever share the same ecological niche, for, as we'll see shortly, evolution will tend to separate those that do.

As we move from habitat to habitat, we find that each is home to a group of organisms that occupy different but interrelated ecological niches. We can define a *community* as an assemblage of organisms that live in a particular habitat and interact with one another. Although every organism must adjust to variations in the environment on its own, we find that similar habitats often contain similar communities. Biogeographers and ecologists recognize specific types of communities, called *associations*, in which typical organisms are likely to be found together. These associations are usually defined by species, as in the beech-birch-maple forest that is found from the Great Lakes region to New England in suitable habitats.

WATER NEED

Let's now turn to the environmental factors that help determine where organisms, as individuals and species, are found. The first of these is the availability of water.

Plants and animals have adapted to cope with the abundance and/or scarcity of water in a variety of ways (Figure 8.13). Plants that are adapted to drought conditions are called **xerophytes** (from the Greek roots *xero-*, meaning "dry," and *phyton*, meaning "plant").

Some xerophytes have a thick layer of wax or wax-like material on their leaves and stems, helping them to seal water inside. Others adapt to a desert environment by greatly reducing their leaf area or by bearing no leaves at all. Needlelike leaves, or spines in place of leaves, also conserve water.

Plants in water-scarce environments are also better at obtaining and storing water. For example, their roots may extend deeply into the soil to reach moisture far from the surface—that is, groundwater. Plants drawing from groundwater are called *phreatophytes*. Other desert plants produce a widespread, but shallow, root system, enabling them to absorb water from short desert downpours that saturate only the uppermost soil layer. Leaves and stems of desert plants known as *succulents* are often thickened by a spongy tissue that stores water. The common prickly pear cactus is an example (Figure 8.13).

> Xerophytes are plants that are adapted to a dry and sometimes hot environment. Examples are phreatophytes, which have deep roots, and succulents, which store water internally in spongy tissues.

8.13 Organisms adapted to water scarcity

Plants in water-scarce environments have developed effective ways to obtain and store water.

Annie Griffiths Belt/NG Image Collection

▲ SUCCULENT
The prickly pear cactus (*Opuntia*), a desert succulent, retains water in its thick, fleshy stems (cactus pads) for use during long periods without rainfall. San Pedro Valley, Arizona.

SCLEROPHYLL ▼
This California live oak holds most of its tough, waxy leaves through the dry season. Such hard-leaved evergreen trees and woody shrubs are called sclerophylls.

Dennis Flaherty/Photo Researchers, Inc.

Many small desert plants have a very short life cycle—germinating from seed, leafing out, bearing flowers, and producing seed in the few weeks immediately following a heavy rain shower. They survive the dry period as dormant seeds that require no moisture.

Certain climates, such as the wet-dry tropical climate, have a yearly cycle with one season in which water is unavailable to plants because of lack of precipitation. In these climates, some species of trees, termed *tropophytes*, are deciduous; they shed their leaves at the onset of the dry season and grow new ones with the arrival of the wet season. The Mediterranean climate also has a strong seasonal wet-dry alternation, with dry summers and wet winters. Plants in this climate often have hard, thick, leathery, evergreen leaves and are referred to as *sclerophylls* (Figure 8.13).

Xeric animals have evolved methods that are somewhat similar to those used by the plants just described. Many of the invertebrates stay dormant during the dry period. When rain falls, they emerge to take advantage of the new and short-lived vegetation that often results. Many species of birds only nest when the rains occur, when food for their offspring is most abundant. The tiny brine shrimp of the Great Basin may lie dormant for many years until normally dry lakebeds fill with water, an event that occurs perhaps only three or four times a century. The shrimp then emerge and complete their life cycles before the lake evaporates again. Other animals have evolved more unique adaptations, such as changing their body color to absorb or reflect solar energy, depending on their internal temperature. Mammals are by nature poorly adapted to desert environments, but many survive through a variety of mechanisms that enable them to prevent water loss. Similar to plants that reduce transpiration to conserve water, many desert mammals do not sweat through skin glands; instead they rely on other methods of cooling, such as avoiding the Sun and

8.14 Temperature adaptations

Animals that live in extreme temperatures have adapted by regulating their body temperature.

Jeff Lepore/Photo Researchers, Inc.

BATS

These little brown bats are hibernating together in a cluster. Their body temperature falls, and their heartbeat slows. They can survive for almost half the year in this state.

▼ **BROWN BEAR**

A heavy coat and a thick layer of body fat insulate this Alaskan brown bear, allowing it to maintain a constant body temperature.

Roy Toft/NG Image Collection

W. Phillip Kahl/Photo Researchers, Inc.

▲ **CHAMELEON**

The namaqua chameleon lives in the Kalahari Desert of southern Africa. It changes its skin color to regulate its body temperature, turning black in the morning to absorb solar rays and then light gray during the day to reflect them.

becoming active only at night. In this respect, they are joined by most of the rest of the desert fauna, which spend their days in cool burrows in the soil, coming out only at night to forage for food.

TEMPERATURE

The temperature of the air and soil directly influences the rates of physiological processes in plant and animal tissues. In general, each plant species has an optimum temperature associated with each of its functions, such as photosynthesis, flowering, fruit-

> Temperature affects physiological processes occurring in plant and animal tissues. In general, colder climates have fewer plant and animal species.

ing, or seed germination. There are limiting lower and upper temperatures for these individual functions, as well, and for the all-around survival of the plant itself.

Temperature can also act indirectly on plants and animals. Higher air temperatures lower the relative humidity of the air, enhancing transpiration from plant leaves, as well as increasing direct evaporation of soil water.

In general, the colder the climate the fewer the species capable of surviving. We only find a few plants and animals in the severely cold arctic and alpine environments of high latitudes and high altitudes. In plants, ice crystals can grow inside cells in freezing weather, disrupting cellular structures. Cold-tolerant plant species are able to expel excess water from cells to spaces between cells, where freezing does no damage.

Most animals can't regulate their temperature internally. These animals, which include reptiles, invertebrates, fish, and amphibians, are referred to as *cold-blooded*; their body temperature passively follows that of the environment. With a few exceptions (notably, fish and some social insects), these animals are active only during the warmer parts of the year. They survive the cold weather of the midlatitude zone winter by becoming dormant.

Some vertebrates enter a state called *hibernation*, in which their metabolic processes virtually stop and their body temperatures closely parallel those of their surroundings (Figure 8.14). Most hibernators seek out burrows, nests, or other environments where winter temperatures do not reach extremes or fluctuate rapidly. Soil burrows are particularly suited to hibernation because below the uppermost layers, soil temperatures don't vary a great deal.

Warm-blooded animals, like us, maintain their body tissue at a constant temperature, by internal metabolism. This group includes the birds and mammals. Fur, hair, and feathers insulate the animals by trapping dead airspaces next to the skin surface. A thick layer of fat will also provide excellent insulation (Figure 8.14). Other adaptations are for cooling; for example, sweating or panting uses the high latent heat of vaporization of water to remove heat. The seal's flippers and bird's feet expose blood-circulating tissues to the cooler surroundings, promoting heat loss (Figure 8.14).

OTHER CLIMATIC FACTORS

Light also helps determine local plant distribution patterns. Some plants are adapted to bright sunlight, whereas others require shade (Figure 8.15). The amount of light available to a plant will depend in large part on the plant's position. Tree crowns in the upper layer of a forest receive maximum light but correspondingly reduce the amount available to lower layers. In extreme cases, forest trees so effectively cut off light that the forest floor is almost free of shrubs and smaller plants.

In certain deciduous forests of midlatitudes, the period of early spring, before the trees are in leaf, is one of high light intensity at ground level, permitting the smaller plants to go through a rapid growth cycle. In summer, these plants largely disappear as the tree leaf canopy is completed. Other low plants in the same

8.15 Sun-loving and shade-loving plants

California poppies and desert dandelions thrive in bright sun (left). In contrast, cow parsnip prefers the deep shade in Mount Hood National Forest, Oregon (right).

Marc Moritsch/NG Image Collection   Phil Shermeister/NG Image Collection

habitat require shade and do not appear until the leaf canopy is well developed.

In addition to temperature and moisture, ecological factors of light intensity, length of the daylight period and growing season, and wind duration and intensity act to influence plant and animal distribution patterns.

The light available for plant growth varies by latitude and season. As we saw earlier, the number of daylight hours in summer increases rapidly with higher latitude and reaches its maximum poleward of the Arctic and Antarctic Circles, where the Sun may be above the horizon for 24 hours. The rate of plant growth in the short frost-free summer is greatly accelerated by the prolonged daylight.

In midlatitudes, where many species are deciduous, the annual rhythm of increasing and decreasing periods of daylight determines the timing of budding, flowering, fruiting, and leaf shedding. Even on overcast days there is usually enough light for most plants to carry out photosynthesis at their maximum rates.

Light also influences animal behavior. The day/night cycle controls the activity patterns of many animals. Birds, for example, are generally active during the day, whereas small foraging mammals, such as weasels, skunks, and chipmunks, are more active at night. In midlatitudes, as autumn days grow shorter and shorter, squirrels and other rodents hoard food for the coming winter season. Later, increasing hours of daylight in the spring trigger such activities as mating and reproduction.

Wind is also an important environmental factor in the structure of vegetation in highly exposed positions. Wind causes excessive drying, desiccating the exposed side of the plant and killing its leaves and shoots. Trees of high-mountain summits are often distorted in shape, with trunks and branches bent to near-horizontal, facing away from the prevailing wind direction.

Taken separately or together, moisture, temperature, light, and wind can limit the distribution of plant and animal species. Biogeographers recognize that there is a critical level of climatic stress beyond which a species cannot survive. This means that we can mark out a *bioclimatic frontier*, a geographic boundary showing the limits of the potential distribution of a species.

GEOMORPHIC FACTORS

Geomorphic, or landform, factors help differentiate habitats for ecosystems. Among these factors are slope steepness, slope aspect (the orientation of a sloping ground surface with respect to geographic north), and relief (the difference in elevation of divides and adjacent valley bottoms).

Slope steepness affects the rate at which precipitation drains from a surface, which indirectly influences plants and animals. On steep slopes, surface runoff is rapid; in contrast, on gentle slopes, more precipitation penetrates into the soil, providing a moister habitat. Steep slopes often have thin soil because they erode more rapidly while soil on gentler slopes is thicker.

Geomorphic (landform) factors influencing plant and animal distributions include slope angle, slope aspect, and relief. Edaphic (soil) factors include soil particle size and the amount and nature of organic matter in the soil.

Slope aspect controls plant exposure to sunlight and prevailing winds. Slopes facing the Sun have a warmer, drier environment than slopes that face away from the Sun. In midlatitudes, these slope-aspect contrasts may be strong enough to produce quite different biotic communities on north-facing and south-facing slopes (Figure 8.16).

On peaks and ridge crests, rapid drainage dries the soil, which is also more exposed to sunlight and drying winds. By contrast, the valley floors are wetter because water converges there. In humid climates, the groundwater table on valley floors may lie close to or at the ground surface, producing marshes, swamps, ponds, and bogs.

EDAPHIC FACTORS

Soils can vary widely from one small area to the next, influencing the local distribution of plants and animals. *Edaphic factors* are connected to the soil. For example, sandy soils store less water than soils with abundant silt and clay, so they are often home to xerophytes. If there's a high amount of organic matter in the soil, then the soil will be rich in nutrients and will harbor more plant species. The relationship can work in the opposite direction, too: biota can change soil conditions, as when prairie grassland develops a rich, fertile soil beneath it.

DISTURBANCE

Disturbance includes fire, flood, volcanic eruption, storm waves, high winds, and other infrequent catastrophic events that damage or destroy ecosystems and modify habitats. Although disturbance can greatly alter the nature of an ecosystem, it is often part of a natural cycle of regeneration that gives short-lived or specialized species the opportunity to grow and reproduce.

For example, fire will strike most forests sooner or later (Figure 8.17). In many cases, the fire is beneficial. It cleans out the understory and consumes dead and

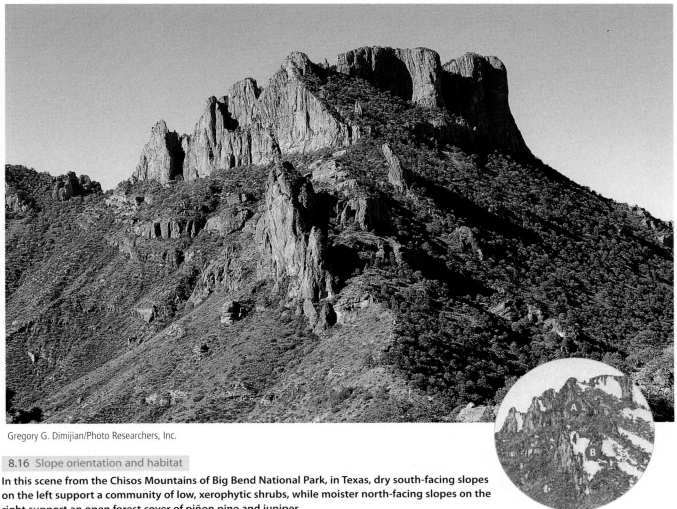

Gregory G. Dimijian/Photo Researchers, Inc.

8.16 Slope orientation and habitat

In this scene from the Chisos Mountains of Big Bend National Park, in Texas, dry south-facing slopes on the left support a community of low, xerophytic shrubs, while moister north-facing slopes on the right support an open forest cover of piñon pine and juniper.

EYE ON THE LANDSCAPE What else would the geographer see?
The vertical faces of these cliffs **A** and pinnacles **B** are most likely joint planes—planes of fractures in the rock resulting from cooling or from release of the pressure of overlying rock removed by erosion.

Kent Dannen/Photo Researchers, Inc.

8.17 Forest fire

Fire sweeps through a forest of pines. In some types of forests, frequent fires are beneficial to maintaining community habitats.

decaying organic matter while leaving most of the overstory trees untouched. Fire also helps expose mineral soil on the forest floor, and fertilizes it with new ash, providing a productive environment for dormant seeds. In addition, shrubs and low plants no longer shade the soil from sunlight. Among tree species, pines are typically well adapted to germinating under such conditions. In fact, the jack pine of eastern North America and the lodgepole pine of the intermountain West have cones that remain tightly closed until the heat of a fire opens them, allowing the seeds to be released.

Fires also preserve grasslands. Grasses are fire-resistant because they have extensive root systems belowground and germinal buds located at or just below the surface. But woody plants that might otherwise invade grassland areas are not so resistant and are usually killed by grass fires.

In many regions, active fire suppression has reduced the frequency of burning to well below natural levels. That may sound like a good thing, but in forests, this causes dead wood to build up on the forest floor. So, when a fire does start, it's destructive rather than beneficial, burning hotter and more rapidly and consuming the crowns of many overstory trees.

Flooding is another important disturbance. It displaces animal communities and deprives plant roots of oxygen. Where flooding brings a swift current, mechanical damage rips limbs from trees and scours out roots. High winds

Disturbance by factors such as fire, flooding, or high winds is a natural process to which many ecosystems are adapted. In semiarid regions, fires act to maintain grasslands and open forests.

Paul Collis/Alamy Limited

8.18 Tree throw

When strong winds blow down a healthy tree, the root mat lifts off the ground, leaving a pit. The lifted soil eventually falls back next to the pit, forming a mound, which is a favored spot for the germination of young trees.

are another significant disturbance factor; they can topple individual trees as well as whole forest stands (Figure 8.18).

INTERACTIONS AMONG SPECIES

Species don't react with just their physical surroundings. They also interact with each other. That interaction may benefit at least one of the species, have a negative effect on one or both species, or have no effect on either species.

Competition is a negative interaction. It happens whenever two species need a common resource that is in short supply (Figure 8.19). Both populations suffer

8.19 Competition

A pride of lions and a herd of elephants peaceably share a water hole in Chobe National Park, Botswana. Other animals must wait their turns.

Beverly Joubert/NG Image Collection

All Canada Photos/Alamy Limited

8.20 Predation

This South American giant anteater enjoys a lunch of Brazilian termites.

from lowered growth rates than they would have had if only one species were present. Sometimes one species will win the competition and crowd out its competitor. At other times, the two species may remain in competition indefinitely.

Competition is an unstable situation. If a genetic strain within one of the populations emerges that can use a substitute resource, its survival rate will be higher than that of the remaining strain, which still competes. The original strain may become extinct. In this way, evolutionary mechanisms tend to reduce competition among species.

Predation and *parasitism* are other negative interactions between species. Predation occurs when one species feeds on another (Figure 8.20). There are obvious benefits for the predator species, which obtains energy for survival; but of course, the interaction has a negative outcome for the prey species. Parasitism occurs when one species gains nutrition from another, typically when the parasite organism invades or attaches to the body of the host in some way.

Although we tend to think that predation and parasitism are always negative—benefiting one species at the expense of the other—in some cases it works out well for the prey or host populations, too, in the long run. A classic example is the rise and fall of the deer herd on the Kaibab Plateau north of the Grand Canyon in Arizona (Figure 8.21). Predation and parasitism will also remove the

> Negative interactions among species include competition, predation, parasitism, herbivory, and allelopathy. Positive interactions include commensalism, proto-cooperation, and mutualism, which are three forms of symbiosis.

8.21 Rise and fall of the Kaibab deer herd

The population of the Kaibab deer herd exploded without predators, then crashed when the food ran out.

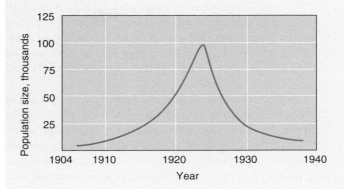

▲ POPULATION SIZE WITH TIME

This graph plots the population size of the deer herd in the Kaibab National Forest, Arizona. The herd grew from about 4000 to nearly 100,000 between 1907 and 1924, when the government began controlling predatory wolves, coyotes, and mountain lions, and protecting game. But confined in an area of 283,000 hectares (700,000 acres), the huge deer population proved too much for the land, and overgrazing led to a population crash. In one year, half the animals starved to death; by the late 1930s, the population had declined to a stable level, near 10,000. Previously, predation had maintained the deer population at levels that were in harmony with the supportive capability of the environment.

▼ KAIBAB MULE DEER

This buck is a member of today's Kaibab deer herd, a population of mule deer. Named for their large ears, which resemble those of a mule, they are common throughout western North America.

Mira/Alamy Limited

weaker individuals, improving the genetic makeup of the attacked species.

Another type of negative interaction between species is *herbivory*. When animals graze, they can reduce the viability of the plant species population. Although some plants can maintain themselves well in the face of grazing pressure, others are quite sensitive to overgrazing. *Allelopathy*, also a negative interaction, occurs when one plant species produces chemical toxins that inhibit other species.

When we looked at the nitrogen cycle earlier in the chapter, we mentioned the symbiotic relationship between legumes and nitrogen-fixing *Rhizobium* bacteria, which benefits both types of organisms. Symbiosis includes three types of positive interactions: commensalism, protocooperation, and mutualism. In *commensalism*, one of the species benefits and the other is unaffected. Sometimes the relationship benefits both parties but isn't essential for their survival. This type of relationship is called *protocooperation*. If the relationship reaches a point where one or both species cannot survive alone, it's called *mutualism*. The relationship between nitrogen-fixing bacteria and legumes is a classic example of mutualism because the bacteria need the plants for their own survival.

WileyPLUS Remote Sensing and Biosphere Interactivity
See disturbance at work by analyzing satellite images of fires and deforestation from Los Alamos to Madagascar.

Ecological Succession

Plant and animal communities change through time. Walk through the country and you'll see patches of vegetation in many stages of development, from open, cultivated fields through grassy shrublands to forests. Clear lakes gradually fill with sediment and become bogs. We call these changes—by which biotic communities succeed one another on the way to a stable end point—**ecological succession**.

In general, succession leads to the most complex community of organisms possible, given its physical conditions of the area. The series of communities that follow one another is called a *sere*, and each of the temporary communities is referred to as a *seral stage*. The stable community, which is the end point of succession, is the *climax*. If succession begins on a

> In ecological succession, an ecosystem proceeds through seral stages to reach a climax. Primary succession occurs on new soil, while secondary succession occurs where disturbance has removed or altered existing communities.

newly constructed deposit of mineral sediment, it is called *primary succession*. If, on the other hand, succession occurs on a previously vegetated area that has been recently disturbed, perhaps by fire, flood, windstorm, or human activity, it is referred to as *secondary succession*.

Primary succession could happen on a sand dune, a sand beach, the surface of a new lava flow or freshly fallen layer of volcanic ash, or the deposits of silt on the inside of a river bend that is gradually shifting, for example. Such sites are often little more than deposits of coarse mineral fragments. In other cases—floodplain silt deposits, for example—the surface layer is made of redeposited soil, containing substantial amounts of organic matter and nutrients.

Succession begins with the *pioneer stage*. It includes a few plant and animal pioneers that are unusually well adapted to otherwise inhospitable conditions, which may be caused by rapid water drainage, dry soil, excessive sunlight exposure, wind, or extreme ground and lower air temperatures. As these pioneering plants grow, their roots penetrate the soil. When the plants decay, their roots add organic matter directly to the soil, while their fallen leaves and stems add an organic layer to the ground surface. Large numbers of bacteria and invertebrates begin to live in the soil. Grazing mammals feed on the small plants, and birds forage the newly vegetated area for seeds and grubs.

The pioneers soon transform conditions, making them favorable for other species to invade the area and displace the pioneers. The new arrivals may be larger plants with foliage that covers the ground more extensively. If this happens, the climate near the ground will have less extreme air and soil temperatures, higher humidity, and less intense insolation. These changes allow still other species to invade and thrive. When the succession has finally run its course, a climax community of plant and animal species in a more or less stable composition will have been established.

Sand dune colonization is a good example of primary succession (Figure 8.22). Animal species also change as succession proceeds. This is especially noticeable among the insects and invertebrates, which go from sand spiders and grasshoppers on the open dunes to sowbugs and earthworms in the dune forest.

Secondary succession can occur after a disturbance alters an existing community. *Old-field succession*, taking place on abandoned farmland, is a good example of secondary succession (Figure 8.23).

WileyPLUS Succession
View this animation to see the sequence of successional changes that occur when a beaver dam turns a low valley in the boreal forest into a bog.

8.22 Dune succession

PINES AND HOLLIES ▶

Once the dunes are stabilized, low tough shrubs take over, paving the way for drought-resistant tree species, such as pines and hollies. This coastal dune forest is on Dauphin Island, Alabama.

▼ **DUNE GRASS**

In the earliest stages of succession on coastal dunes, beach grass colonizes the barren habitat, as shown here in Cape Cod, Massachusetts. It propagates via underground stems that creep beneath the surface of the sand and send up shoots and leaves.

SUCCESSION, CHANGE, AND EQUILIBRIUM

So far, we've been describing successional changes caused by the actions of the plants and animals themselves; one group of inhabitants paves the way for the next. As long as nearby populations of species provide colonizers, the changes lead automatically from bare soil or fallow field to climax forest. This type is called *autogenic* (self-producing) *succession*.

But in many cases, autogenic succession does not run its full course. Environmental disturbances, such as wind, fire, flood, or land clearing for agriculture interrupt succession temporarily, or even permanently. For example, winds and waves can disturb autogenic succession on seaside dunes, or a mature forest may be destroyed by fire. In addition, inhospitable habitat conditions such as site exposure, unusual bedrock, or impeded drainage can hold back or divert the course

We can view the pattern of plant and animal communities on the landscape as a balance between succession, in which the community modifies its own habitat and composition, and environmental disturbance, such as wind, flood, fire, or logging.

of succession so successfully that the climax is never reached.

The Introduction of an *invasive species* can also greatly alter existing ecosystems and successional pathways. For example, after the parasitic chestnut blight fungus was introduced from Asia to New York City in 1904, it spread across the eastern states, decimating the entire population of the American chestnut tree within a period of about 40 years. This tree species, which may have accounted for as many as one-fourth of the mature trees in eastern forests, is now found only as small blighted stems sprouting from old root systems.

While succession is a reasonable model to explain many of the changes that we see in ecosystems over time, we must also take into account other effects. External forces can reverse or rechannel autogenic change, temporarily or permanently. The landscape is a mosaic of distinctive biotic communities with different biological potentials and histories.

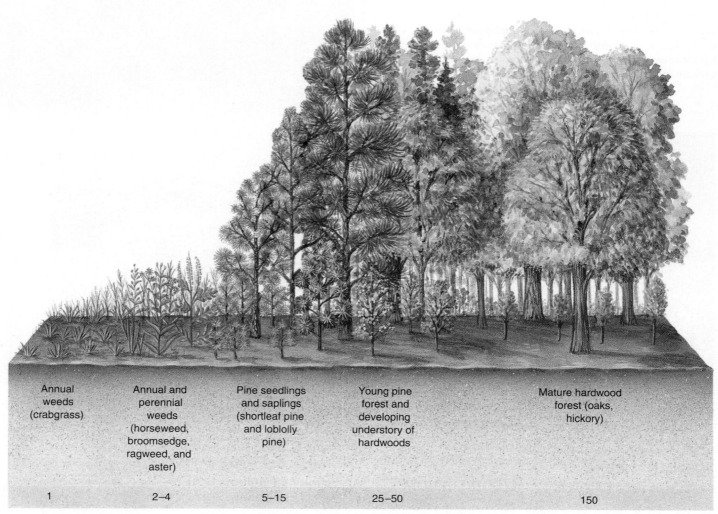

Annual weeds (crabgrass)	Annual and perennial weeds (horseweed, broomsedge, ragweed, and aster)	Pine seedlings and saplings (shortleaf pine and loblolly pine)	Young pine forest and developing understory of hardwoods	Mature hardwood forest (oaks, hickory)
1	2–4	5–15	25–50	150

Years after cultivation

8.23 Old-field succession in the Southeast United States

When cultivation ceases, grasses and forbs colonize the bare soil. The first stages depend on the last use of the land. If row crops were the last to be cultivated, the pioneers will be annuals and biennials. If small grain crops were last, the pioneers are often perennial herbs and grasses. If pasture has been abandoned, those pioneers that were not grazed will have a head start. Where mineral soil was freshly exposed by plowing, pines often follow the first stages of succession because pine seeds favor disturbed soil and strong sunlight for germination. The pines eventually shade out the other plants and become dominant. Pine dominance is only temporary, though, because their seeds cannot germinate in shade and litter on the forest floor. Hardwoods such as hickories and oaks can, however, germinate and so their seedlings grow quickly to fill holes in the canopy. After several more decades, the deciduous hardwoods shade out the pines, resulting in the oak-hickory climax forest.

FOCUS ON REMOTE SENSING

Remote Sensing of Fires

Wildfires occur frequently on the Earth's land surface, and biomass burning has important effects on both local and global ecosystems. Biomass burns inefficiently, releasing not only carbon dioxide and water, but also a number of other greenhouse gases, which absorb outgoing longwave radiation and enhance the greenhouse effect. Aerosols are another by-product of inefficient combustion that can affect atmospheric processes. Burning mobilizes such nutrients as nitrogen, phosphorus, and sulfur in ash so that they become available for a new generation of plants; but it also carries them upward in smoke and as gases.

Fire affects ecosystems by changing species composition and creating a patchy structure of diversity on the landscape. It can also stimulate runoff and soil erosion causing the loss of a significant layer of vegetation to fire. For these reasons and more, monitoring of fires by remote sensing is a topic of intense interest to global change researchers.

Fires can be remotely sensed in several ways. Thermal imagers detect active fires as bright spots because they emit more heat energy than normal surfaces. However, when fires are obscured under clouds, smoke plumes can also identify the location of fires; they are, however, hard to distinguish from clouds in some images. Burn scars can also be detected after the fire.

Figure 8.24, from May 8, 2009, shows the Jesusita fire, in the hills above Santa Barbara, California, as imaged by NASA's MODIS instrument. Although not a large fire by California standards, this intense blaze destroyed 77 homes, damaged another 22, and caused the evacuation of about 30,000 people.

The burn scar image was acquired on May 10 by the Advanced Land Imager aboard NASA's Earth Observing-1 satellite platform. This imager uses a state-of-the-art technology to produce multispectral images that are similar to those of Landsat. In the burned area, little vegetation remains outside of canyon bottoms.

8.24 Santa Barbara's Jesusita fire

The fire is imaged here by MODIS, aircraft camera, and Landsat.

Jeff Schmaltz MODIS Land Rapid Response Team, NASA GSFC

▲ MODIS VIEWS THE FIRE
This MODIS image, acquired on May 8, 2009, shows the smoke plume from the Jesusita fire blowing southward, across the city of Santa Barbara and the Channel Islands. The large urban area in the right-center of the image is Los Angeles.

JESUSITA FIRE INVADES THE CITY ▶
Late afternoon "sundowner" winds, strong and hot, drove the fire into the city, burning many homes.

NASA image created by Jesse Allen, using data provided courtesy of the NASA EO-1 Team

▲ BURN SCAR FROM THE JESUSITA FIRE
Five days after the fire, NASA's Advanced Land Imager captured this true-color image of the huge burned area, shown in reddish-purple tones. The road at the top of the photo marks the summit of the mountain ridge above the city. The street grid of Santa Barbara is visible in the lower right.

Lamberts Photography/WENN.com, via NewsCom

Historical Biogeography

Thus far, we've looked at ecological processes that produce biogeographic patterns at local and regional spatial scales. We now turn to patterns at continental and global scales that develop over longer time periods. *Historical biogeography* focuses on how these spatial distribution patterns arise over space and time. It happens through four key processes: evolution, speciation, extinction, and dispersal.

EVOLUTION

An astonishing number of organisms exist on Earth, each adapted to the ecosystem in which it carries out its life cycle (Figure 8.25). Scientists have described and identified about 40,000 species of microorganisms, 350,000 species of plants, and 2.2 million species of animals, including some 800,000 insect species—probably only a fraction of the actual number of species found on Earth.

8.25 Diversity of life-forms on Earth

The forms of life on Earth are amazingly varied.

▼ **MICROORGANISM**

Diatoms, like this one of the genus *Corethron,* are a class of algae with over 70,000 known species. They contain silicified skeletons and are extremely abundant in fresh and ocean water.

Bill Curtsinger/NG Image Collection

▼ **FUNGUS**

Fungi, like this stinkhorn on the floor of a Costa Rican rainforest, are plants without chlorophyll; they take their nutrition from dead and decaying organic matter.

Roy Toft/NG Image Collection

Jason Edwards/NG Image Collection

▲ **REPTILE**

An Australian thorny devil lizard makes its way across an arid landscape. Inhabiting the sand plains of South Australia, this small reptile lives on ants. It gets fluid from the dew that condenses on its body and is carried to its mouth in fine grooves between its spines.

Darlyne A. Murawski/NG Image Collection

▲ **INSECT**

These placid blue nose caterpillars on a leaf are herbivores. They are well protected from predation by defensive spines that sting and detachable burrs that work their way into the skin of a predator.

How have life forms become this astonishingly diverse? Through the process of **evolution**, the environment itself has acted on organisms to create this diversity. You've probably heard of Sir Charles Darwin, whose monumental biological work, *The Origin of Species by Means of Natural Selection*, was published in 1859. Through exhaustive studies, Darwin showed that all life possesses *variation*, the differences that arise between parent and offspring. He proposed that the environment acts on variation in organisms in much the same way that a plant or an animal breeder does, picking out the individuals with qualities that are best suited to their environment. These individuals are more likely to live longer, propagate, and pass on their useful qualities.

Darwin termed this survival and reproduction of the fittest **natural selection**. He saw that, through time, when acted upon by natural selection, variation could bring about the formation of new species, whose individuals differed greatly from their ancestors.

> Natural selection acts on the variation that occurs within populations of species to bring about evolution. Variation is produced by mutation, in which genetic material is altered, and recombination, from which new combinations of existing genetic material arise.

But how and why does this variation occur in the first place? Although Darwin couldn't provide an explanation, we now know the answer. Variation comes from two interacting processes: *mutation* and *recombination*. A reproductive cell's genetic material (DNA, or deoxyribonucleic acid) can mutate when the cell is exposed to heat, ionizing radiation, or certain types of chemical agents. Chemical bonds in the DNA are broken and reassembled. Most mutations either have no effect or are harmful. But a small proportion of mutations have a positive effect on the individual's genetic makeup. If that positive effect makes the individual organism more likely to survive and reproduce, then the altered gene is likely to survive as well and thus be passed on to offspring.

Recombination describes the process by which an offspring receives two slightly different copies, or *alleles*, of each gene from its parents. One allele may be dominant and suppress the other, or the two alleles may act simultaneously. Because each individual receives two alleles of each gene, and there are typically tens of thousands of genes in an organism, the possible number of genetic combinations is very large. Thus, recombination serves as a constant source of variation that acts to make every offspring slightly different from the next.

SPECIATION

Mutations change the nature of species through time. But just what is a species? For our purposes, we can define a **species** (plural, also *species*) as a collection of individuals capable of interbreeding to produce fertile offspring. A *genus* (plural, *genera*) is a collection of closely related species that share a similar genetic evolutionary history (Figure 8.26).

Speciation is the process by which species are differentiated and maintained. Actually, speciation is not a single process. It arises from a number of component processes acting together through time. We've already looked at two of these: mutation and natural selection.

Richard Parker/Photo Researchers, Inc.

Blue Line Pictures/Getty Images, Inc.

▲ RED OAK
The acorns of red oak (*Quercus rubra*) have a flat cap and stubby nut, with pointed bristle tips on the leaf lobes.

▲ WHITE OAK
The acorns of white oak (*Quercus alba*) have a deeper cap and a longer nut, with rounded leaf-lobe tips.

8.26 Red and white oaks

Although similar in general appearance, these two species of the genus *Quercus* are easy to distinguish.

A third speciation process is *genetic drift*. Chance mutations that don't have any particular benefit can still change the genetic composition of a breeding population until it diverges from other populations. Genetic drift is a weak factor in large populations. But in small populations, such as a colony of a few pioneers in a new habitat, random mutations are more likely to be preserved. *Gene flow* is the opposite process. Evolving populations exchange alleles as individuals move among populations, keeping the gene pool uniform.

Speciation often occurs when populations become isolated from one another, so there's no gene flow between them. This *geographic isolation* can happen in several ways. For example, geologic forces may uplift a mountain range that separates a population into two different subpopulations by a climatic barrier. Or a chance long-distance dispersal may establish a new population far from the main one. These are examples of *allopatric speciation*. As genetic drift and natural selection proceed, the populations gradually diverge and eventually lose the ability to interbreed.

The evolution of finch species on the Galápagos Islands is a classic example of allopatric speciation (Figure 8.27). Charles Darwin visited this cluster of five

> Speciation is the process that differentiates and maintains species. Component processes affecting speciation include mutation, natural selection, genetic drift, and gene flow. Isolation, in which breeding populations are separated, enhances speciation.

8.27 Allopatric speciation of Galápagos finches

Five genera and 14 species of finch evolved from a single ancestral population. As the story has been reconstructed, the Galápagos Islands were first colonized by a single original finch species, the blue-black grassquit. Over time, through natural selection, individual populations became adapted to conditions on particular islands; these adaptations were enhanced by their isolation on different islands, and eventually they evolved into different species. Later, some of these species successfully reinvaded other islands, thereby continuing the speciation and evolution process. The finches' beak shapes are adapted to their primary food sources: seeds, buds, or insects.

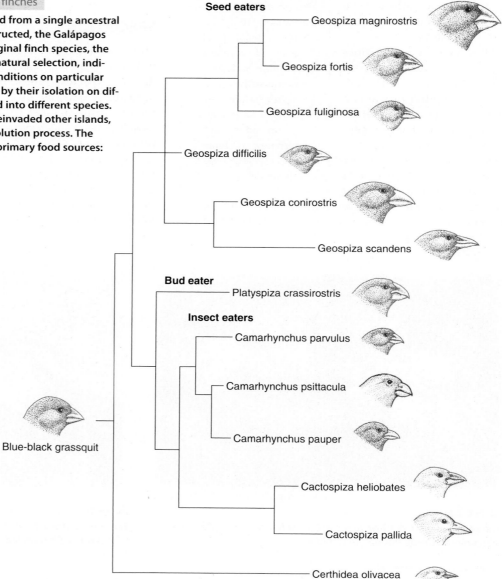

Seed eaters
Geospiza magnirostris
Geospiza fortis
Geospiza fuliginosa
Geospiza difficilis
Geospiza conirostris
Geospiza scandens

Bud eater
Platyspiza crassirostris

Insect eaters
Camarhynchus parvulus
Camarhynchus psittacula
Camarhynchus pauper
Cactospiza heliobates
Cactospiza pallida

Certhidea olivacea

Blue-black grassquit

8.28 Galápagos tortoises

These land dwellers are the largest living tortoises, with a life expectancy of 100 to 150 years. On the Galápagos Islands they diverged by allopatric speciation into 12 subspecies, of which 10 still exist. The subspecies generally have shells of two distinctive shapes: dome-back and saddle-back.

Eric Rorer/Aurora Photos Inc.

Brandon Cole/Alamy Limited

▲ DOME-BACK SHELL
On the wetter islands with abundant ground vegetation, the tortoises developed dome-shaped shells, along with shorter necks and legs suited to ground grazing.

▲ SADDLE-BACK SHELL
On the drier islands with less ground vegetation, the tortoises evolved saddle-shaped shells with wide openings, and longer necks and legs, to allow grazing on low trees and shrubs.

major volcanic islands and nine lesser ones, located about 800 km (500 mi) from the coast of Ecuador, and they inspired his ideas about evolution. The famous giant tortoises of the Galápagos are another example (Figure 8.28). Each of the larger islands bears at least one distinctly different population of these reptiles. Like the finches, they are believed to have evolved from a single ancestral stock that colonized the island chain and then diverged into unique types.

Sympatric speciation, by contrast, occurs only within a larger population. Imagine a species that has two different primary food sources. Eventually, mutations will emerge that favor one food source over the other. For example, birds could develop two different lengths or shapes of beak, with one beak type better adapted for eating fruit and the other better suited to seeds. As these mutations are exposed to natural selection, they will, over time, produce two different populations, each adapted to its own food source. Eventually, the populations may become separate species.

Another mechanism of sympatric speciation, one that is quite important in plants, is *polyploidy*. Normal organisms have two sets of genes and chromosomes—that is, they are *diploid*. Through accidents in the reproduction process, two closely related species can cross in such a way that the offspring carries both sets of genes from both parents. Although these *tetraploids* are fertile, they can't reproduce with the populations from which they arose, and so they are instantly isolated as new species. About 70 to 80 percent of higher plant species probably arose in this fashion.

EXTINCTION

Over geologic time, all species are doomed to **extinction**. When conditions change more quickly than populations can evolve new adaptations to cope with the changes, population size falls. When that happens, the population becomes more vulnerable to chance occurrences, such as a fire, a rare climatic event, or an outbreak of disease. Ultimately, the population is wiped out.

Some extinctions occur very rapidly, particularly those induced by human activity, such as in the classic example of the passenger pigeon (Figure 8.29).

> Extinction occurs when all individuals of a species die in response to ecological or environmental change. Extreme events, such as the collision of a meteorite with the Earth about 65 million years ago, can cause mass extinctions.

Joel Sartore/NG Image Collection

8.29 Passenger pigeon

The passenger pigeon was a dominant bird of eastern North America in the late nineteenth century. But because they were easily captured in nets, and then shipped to markets for food, they were virtually extinct by 1890. The last known passenger pigeon died in the Cincinnati Zoo in 1914.

Rare but extreme events can also cause extinctions. Strong evidence suggests that the Earth was struck by a large meteorite about 65 million years ago, wiping out the dinosaurs and many other groups of terrestrial and marine organisms (Figure 8.30). The impact sent huge volumes of dust into the atmosphere, choking

8.30 Chicxulub crater

When a large meteorite hit the Earth about 65 million years ago, it created a huge crater centered near Chicxulub, Mexico, on the Yucatan Peninsula. The curving shoreline seen here, created by a jutting shelf of limestone, is thought to be a remnant of the crater. Many scientists now believe that the impact of this meteorite was responsible for the extinction of dinosaurs and many other species.

Wes C. Skiles/NG Image Collection

off sunlight globally for several years; as a result, many plants and animals froze to death.

DISPERSAL

Nearly all types of organisms can move from their location of origin to new sites. Often this *dispersal* is confined to one life stage, as in the dispersal of higher plants as seeds. Even for animals, however, there is often a single developmental stage when they are more likely to move from one site to the next.

Normally, dispersal doesn't change the species' geographic range. Seeds fall near their sources, and animals seek out nearby habitats to which they are adjusted. Dispersal is thus largely a method for gene flow that helps to encourage the cross-breeding of organisms throughout a population. When land is cleared or new land is formed, dispersal moves colonists into the new environment, as explained in the discussion of succession. Species also disperse by *diffusion*, the slow extension of their range from year to year. An example of diffusion is the northward colonization of the British Isles by oaks at the end of the Ice Age (Figure 8.31).

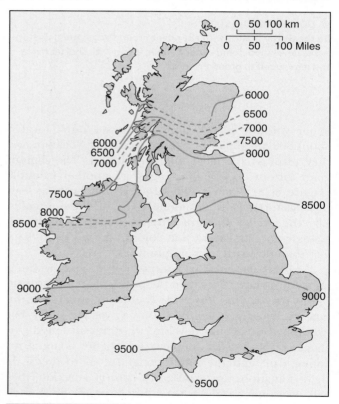

8.31 Diffusion of oaks

Following the retreat and melting of continental glaciers at the close of the Ice Age, oak species diffused northward across the British Isles. Contoured lines on the map indicate northern borders known with some certainty. Dashed lines show boundaries known less accurately. The oaks took about 3500 years, from about 9500 to 6000 years before the present day, to reach their northern limit.

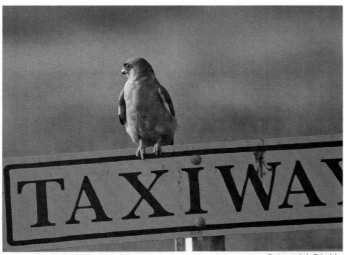

© Jeremiah Trimble

8.32 Dispersal

This small red-footed falcon was sighted for the first time in the western hemisphere in August 2004 at a grassy meadow airstrip on Martha's Vineyard Island, Massachusetts. The species normally winters in Africa and summers in eastern Europe. Munching happily on butterflies, grasshoppers, and small voles, it remained at the airstrip for about two weeks before heading for parts unknown.

this case, it's beyond the species' physiological limits to cross the barrier. But there may be ecological barriers as well—for example, a zone heavily populated with predators, or a region occupied by strong and successful competitor species.

There are also corridors that help dispersal. For example, Central America forms a present-day land bridge, connecting North and South America, that has been in place for about 3.5 million years. Other corridors existed in the recent past. The Bering Strait region between Alaska and easternmost Siberia was dry land during the early Cenozoic Era (about 60 million years ago) and during the Ice Age, when sea level dropped by more than 100 m (325 ft). Many plant and animal species of Asia are known to have crossed this bridge and then spread southward into the Americas. One notable migrant species of the last continental glaciation was the aboriginal human, and evidence suggests that these skilled hunters caused the extinction of many of the large animals of the era, including wooly mammoths and ground sloths, which disappeared from the Americas about 10,000 years ago (Figure 8.33).

A rare, long-distance dispersal event can be very significant, as we saw with the Galápagos finches. Some species, such as the coconut, are especially well adapted to long-distance dispersal. Among the animals, birds, bats, and insects are frequent long-distance travelers (Figure 8.32). Generally, nonflying mammals, freshwater fishes, and amphibians are less likely to make long journeys, with rats and tortoises the exceptions.

Dispersal often requires surmounting barriers. That might mean bridging an ocean or an ice sheet by an unlikely accident, such as transport in a windstorm or on a raft of floating debris. But other barriers are not so obvious. For example, the basin and range country of Utah, Nevada, and California is a sea of desert interspersed with islands of forest. Whereas birds and bats can move easily from one island to the next, a small mammal would not be likely to cross this desert sea under its own power. In

> Dispersal is the capacity to move from a location of origin to new sites. In diffusion, species extend their range slowly from year to year. In long-distance dispersal, unlikely events establish breeding populations at remote locations.

Kenneth W. Fink/Photo Researchers, Inc.

8.33 Wooly mammoth

A reconstruction of the wooly mammoth, a huge tusked mammal that inhabited North America throughout the Ice Age; it became extinct about 10,000 years ago, most likely from hunting by prehistoric humans.

8.34 Endemic and cosmopolitan species

Endemic species are restricted in range, whereas cosmopolitan species are found in many habitats worldwide.

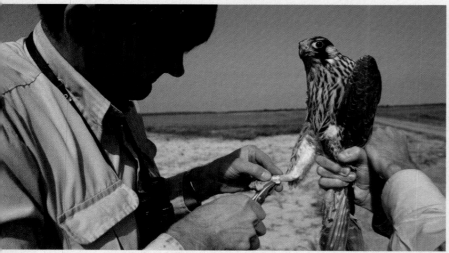

James P. Blair/NG Image Collection

▲ **COSMOPOLITAN SPECIES**

Shown here are two cosmopolitan species: the peregrine falcon (*Falco peregrinus*) and the human (*Homo sapiens*). Both are found widely over the globe. The human, a scientist, is attaching a band to the leg of the bird as part of a study to learn how the falcon is affected by chemical contamination.

ENDEMIC SPECIES ▶

The gingko tree was widespread throughout the Mesozoic Era (245 to 60 million years ago) but until recently was endemic to a small region in eastern China. Owing to human activity, it is now much more widely distributed around the world. It is widely planted in North America as an urban street tree.

S.W. Carter/Photo Researchers, Inc.

DISTRIBUTION PATTERNS

Over time, evolution, speciation, extinction, and dispersal have distributed many species across the Earth, creating a number of spatial distribution patterns. An **endemic** species is found in one region or location and nowhere else. An endemic distribution can arise in two ways: the species simply stays within a small range of its original location, or it contracts from a broader range. Some endemic species are ancient relics of biological strains that have otherwise gone extinct (Figure 8.34). In contrast to endemics are *cosmopolitan species*, which are distributed very widely (Figure 8.34). Very small organisms, or organisms with very small propagating forms, are often cosmopolitan because they can be distributed widely by atmospheric and oceanic circulations. *Disjunction* is another interesting pattern, in which one or more closely related species are found in widely separated regions.

BIOGEOGRAPHIC REGIONS

When we examine the spatial distributions of species on a global scale, we find common patterns. Closely related species tend to live near one another or to occupy similar regions. But larger groups of organisms, such as families and orders, often have disjunct distribution patterns. For example, the South America–Africa–Australia–New Zealand pattern for the ratite birds, described in Figure 8.35, fits the distribution pattern of many other ancient families of plants and animals. This reflects the bird's common ancestry on the supercontinent of Gondwana, which existed about 210 million years ago and then gradually split apart

> Cosmopolitan species are found very widely, whereas endemic species are restricted to a single region or location. Disjunctions occur when closely related species are found in widely separated regions.

8.35 Disjunct distribution

These three bird species share a common ancestry in the ratite group, which scientists theorize originated on the southern supercontinent Gondwana before it split apart. As Gondwana split into South America, Africa, Australia, and New Zealand, ratite bird populations became isolated, allowing separate but related species to evolve into their present-day distributions.

200 million years ago

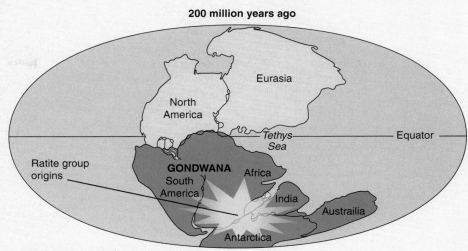

Eurasia

North America

Tethys Sea

Equator

Ratite group origins

GONDWANA
South America

Africa

India

Austrailia

Antarctica

Present-day distribution of ratite bird group

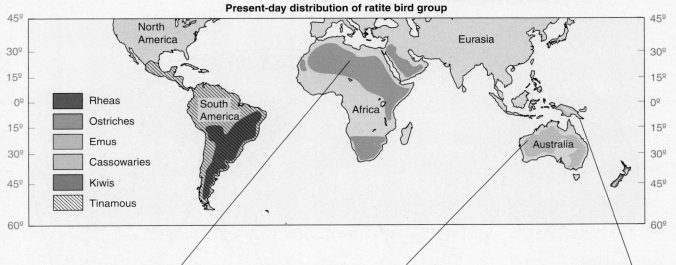

North America

Eurasia

South America

Africa

Australia

Rheas
Ostriches
Emus
Cassowaries
Kiwis
Tinamous

Bates Littlehales/NG Image Collection

▲ OSTRICH

The ostrich is restricted to Africa and the Middle East, although it was formerly found in Asia.

Joel Sartore/NG Image Collection

▲ EMU

The emu inhabits most of Australia, where it is commonly encountered in the wild.

Tim Laman/NG Image Collection

▲ CASSOWARY

The cassowary hails from New Guinea and northeastern Australia. It is a rainforest dweller.

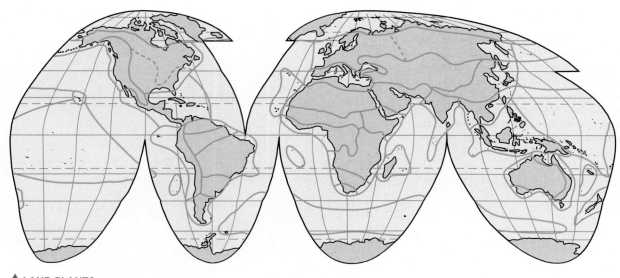

▲ LAND PLANTS
Biogeographic regions of land plants.

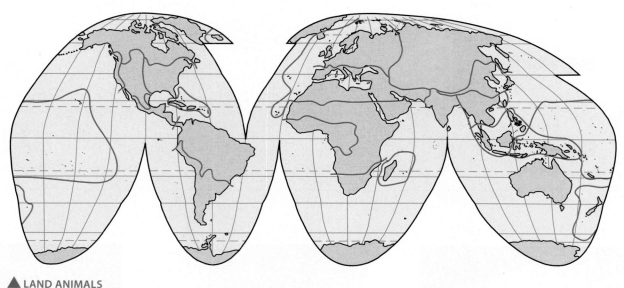

▲ LAND ANIMALS
Biogeographic regions of land animals.

8.36 Biogeographic regions

Note that many of the boundaries on these two maps are fairly close, indicating that at the global scale, plants and animals have similar and related histories of evolution and environmental affinity.

(see Chapter 11). Global climate also plays an important role in distribution. Often, members of the same lineage have made similar adaptations to the environment, and so they are found in similar climatic regions.

We can define **biogeographic regions** as areas in which the same or closely related plants and animals tend to be found together. When we cross the boundary between two biogeographic regions, we pass from one group of distinctive plants and animals to another. Figure 8.36 shows the major biogeographic regions for plants and animals.

Biodiversity

Today, global **biodiversity**—the variety of biological life on Earth—is rapidly decreasing. Two out of every five species on the planet that have been assessed by scientists face extinction, according to the International Union for Conservation of Nature and Natural Resources (Figure 8.37).

Our species, *Homo sapiens*, has ushered in a wave of extinctions unlike any known in recent geologic history. In the last 40 years, several hundred land-animal

8.37 Endangered species

Human activities are endangering many species. Here are two examples.

▲ BLACK-FOOTED FERRET

This small mammal, related to otters and badgers, was driven nearly to extinction as its prey, the prairie dog, was hunted and poisoned throughout the western United States.

▲ MANATEE

The West Indian manatee is a large aquatic mammal found year-round in the West Indies and southern Florida, and in the summer as far north as coastal Virginia. Loss of habitat and collisions with boats and barges reduced the population until it reached endangered-species status.

species have disappeared. Aquatic species have also been severely affected, with 40 species or subspecies of freshwater fish lost in North America alone in the last few decades. In the plant kingdom, botanists estimate that over 600 species have become extinct in the past four centuries. These documented extinctions may be only the tip of the iceberg. Many species have yet to be discovered and may become extinct before we ever learn about them. Figure 8.38 shows the conservation status of some important groups of plants and animals. Many species are already extinct or imperiled.

> Biodiversity expresses the variety of biological life. Human activity has reduced biodiversity by modifying natural habitats and causing extinctions.

How has human activity caused extinctions? Over our history, we've dispersed new organisms to regions where they outcompete or prey on existing organisms. Many islands have been subjected to waves of invading species, ranging from rats to weeds, brought first by prehistoric humans, and later by explorers and conquerors, and still later by colonists. Hunting by prehistoric humans alone was sufficient to exterminate many species. Then, as humans learned to use fire, large areas became subject

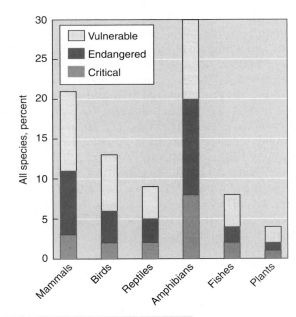

8.38 Degree of endangerment of species

The number of plant and animal species is decreasing at a rate not equaled since about 65 million years ago. Amphibians are especially sensitive to subtle changes in the environment and so may be considered indicator species that point to further trends in extinction rates. (Data from the Red List of the International Union for the Conservation of Nature.)

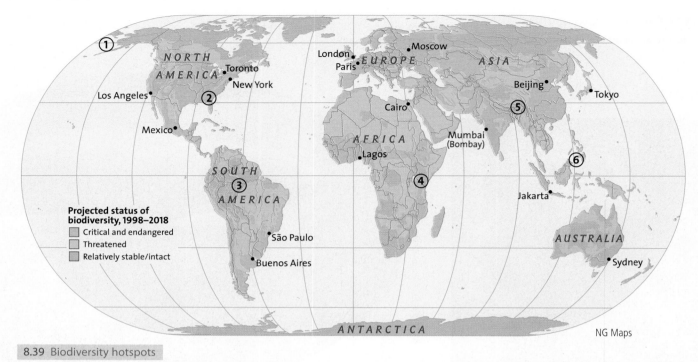

8.39 Biodiversity hotspots

Conservation International identified 34 hotspots of biodiversity, defined as habitats supporting at least 1500 endemic plant species and having lost 70 percent of their original extent. Many of these are in the six major regions identified on the map.

to periodic burning. And, over time, human alteration and fragmentation of habitats isolated many plant and animal populations, causing populations to shrink and, sometimes, become extinct.

Biodiversity is not uniform over the Earth's surface. In general, tropical and equatorial regions have more species and greater variation in species composition among different habitats. Geographic areas in which biodiversity is especially high, and also threatened, are referred to as *hotspots* (Figure 8.39). An important strategy for the preservation of global biodiversity is to first identify hotspots and then take conservation measures to protect them. In this way, such efforts can be most effective.

Why is biodiversity important? Nature has provided an incredibly rich array of organisms that interact with each other in a seamless web of organic life. When we cause the extinction of a species, we break a link in that web. Ultimately, the web begins to fray, with unknown consequences for both the human species and all other forms of life on Earth.

WileyPLUS Web Quiz

Take a quick quiz on the key concepts of this chapter.

A Look Ahead

In this chapter we've focused on the processes that determine the spatial patterns of biota at scales ranging from local to global. We've examined how energy and matter flow in local ecosystems and how different ecosystems have different levels of productivity. We have seen how organisms adjust to their individual environments and how natural selection works in response to environmental pressures. We have also explored how the processes of evolution, dispersal, and extinction generate patterns of species distribution and determine biodiversity at continental and global scales. Our next chapter takes a more functional view of the life layer by inventorying the global *biomes*, major divisions of ecosystems that are based largely on the dominant life-form of their vegetation covers, and where they occur on Earth.

WileyPLUS Web Links

Learn more about food webs, carbon dioxide cycling, wildfires, and more at the web sites on this chapter's web links list. Explore evolution and extinction, and pay a virtual visit to the Galápagos islands.

IN REVIEW BIOGEOGRAPHIC PROCESSES

- Fossil fuel burning releases large amounts of carbon to the atmosphere as CO_2. Less than half of that carbon is removed by oceanic and land processes, leaving the remainder to accumulate and increase global warming.

- Oceans take up atmospheric CO_2 by solution and in the photosynthesis of phytoplankton. Land ecosystems release CO_2 by land clearing, biomass burning, and oxidation of organic matter in soils, but take up

CO_2 as forests grow and forest area expands to cover abandoned lands.

- **Biogeography** focuses on the distribution of plants and animals over the Earth. *Ecological biogeography* examines how relationships between organisms and the environment help determine when and where organisms are found. *Historical biogeography* examines how, where, and when species have evolved and how they have been distributed over longer periods of time and at broader scales.

- **Ecology** is the science of interactions among organisms and their environments. Its focus is the **ecosystem**, which by interaction among components provides pathways for flows of energy and cycles of matter. The **food web** of an ecosystem details how food energy flows from *primary producers* through *consumers* and on to *decomposers*. Because energy is lost at each level, only a relatively few top-level consumers are normally present at any one time.

- *Photosynthesis* is the production of carbohydrate from water, carbon dioxide, and light energy by primary producers. *Respiration* is the opposite process, by which carbohydrate is broken down into carbon dioxide and water to yield chemical energy and, thus, power organisms. *Net photosynthesis* is the amount of carbohydrate remaining after respiration has reduced *gross photosynthesis*. Net photosynthesis increases with more light and higher temperature, up to a point.

- **Biomass** is the accumulated net production of an ecosystem. Forests and wetlands are ecosystems with high rates of *net primary production*, while grasslands and agricultural lands generally have lower rates. Oceans are most productive in coastal and upwelling zones near continents. Among climate types, those with abundant rainfall and warm temperatures are most productive.

- **Biogeochemical cycles** consist of *active* and *storage* pools linked by flow paths. The **carbon cycle** includes an active pool of biospheric carbon and atmospheric CO_2, with a large storage pool of carbonate, in sediments, and carbon, in fossil fuels. Human activities have paved a pathway from storage to active pools by burning fossil fuels.

- The *nitrogen cycle* also has an atmospheric pool, but the nitrogen is largely held in the form of N_2, which cannot be used directly by most organisms. *Nitrogen fixation* occurs when N_2 is converted to more useful forms, by bacteria or blue-green algae, often in symbiosis with higher plants. Human activity has doubled the rate of nitrogen fixation, largely through fertilizer manufacture.

- The physical environment that harbors a species is termed its **habitat**. A *community* of organisms shares a particular habitat. The functional role of each organism in the community is its *ecological niche. Associations* are recurring types of communities that are often defined by characteristic species. Environmental factors influencing the distribution patterns of organisms include moisture, temperature, light, and wind.

- Organisms require water to live, and so they are limited by the availability of water. **Xerophytes** are adapted to arid habitats. They reduce water loss by having waxy leaves, or spines instead of leaves, or no leaves at all. *Phreatophytes* have deep roots; *succulents* store water in spongy tissues. *Tropophytes* are deciduous during the dry season. *Sclerophylls* have thick, leathery leaves that resist drying during the Mediterranean summer climate.

- *Xeric animals* include vertebrates that are nocturnal or have other adaptations to conserve water. Invertebrates such as brine shrimp can adjust their life cycles to survive prolonged drought.

- Temperature acts on plants to trigger and control stages of their growth, as well as to limit growth at temperature extremes. Survival below freezing requires special adaptations, and so only a small proportion of plants are frost-tolerant. *Cold-blooded animals* have body temperatures that follow the environment, but they can moderate these temperatures by seeking out warm or cool places. Mammals and birds maintain constant internal temperatures through a variety of adaptation mechanisms.

- The light available to a plant depends on its position in the structure of the community. Duration and intensity of light vary with latitude and season and serve as a cue to initiate growth stages in many plants. The day/night cycle regulates much of animal behavior. Wind deforms plant growth by desiccating buds and young growth on the windward side of the plant.

- A *bioclimatic frontier* marks the potential distribution boundary of a species. Other factors may also limit the distribution of a species.

- *Geomorphic factors* of slope steepness and orientation affect both the moisture and temperature environment of the habitat and serve to differentiate the microclimate of each community.

- Soil, or *edaphic,* factors can also limit the distribution patterns of organisms, or affect community composition.

- *Disturbance* is set in motion by catastrophic events that damage or destroy ecosystems. Fire is a very common type of disturbance that influences forests, grasslands, and shrublands. Floods, high winds, and storm waves are others. Many ecosystems include specialized species that are well adapted to disturbance.

- Species interact in a number of ways, including **competition**, *predation* and *parasitism*, and *herbivory*. In *allelopathy*, plant species literally poison the soil environment against competing species. Positive (beneficial) interaction between species is termed **symbiosis**.

- **Ecological succession** comes about as ecosystems change in predictable ways through time. A series of stable communities follows a *sere* to a *climax. Primary succession* occurs on new soil substrate, while *secondary succession* occurs on disturbed habitats. Succession on coastal sand dunes follows a series of stages from dune grass to an oak and holly forest. *Old-field succession,* which occurs on abandoned farmland, leads to a forest climax.

- Although succession is a natural tendency for ecosystems to change with time, it is opposed by natural disturbances and limited by local environmental conditions. *Invasive species* can alter ecosystems and successional pathways.
- Fire contributes greenhouse gases to the atmosphere, as well as changing species composition, stimulating runoff, and increasing soil erosion.
- *Historical biogeography* focuses on evolution, speciation, extinction, and dispersal, and how they influence the distribution patterns of species.
- Life has attained its astonishing diversity through **evolution**. In this process, **natural selection** acts on *variation* to produce populations that are progressively better adjusted to their environments. Variation arises from *mutation* and *recombination*.
- A **species** is best defined as a population of organisms that are capable of interbreeding successfully. **Speciation** is the process by which species are differentiated and maintained. It includes mutation, natural selection, *genetic drift*, and *gene flow*.
- *Geographic isolation* acts to segregate subpopulations of a species, allowing genetic divergence and speciation to occur. The finches of the Galápagos Islands provide an example of *allopatric speciation* by geographic isolation. In *sympatric speciation*, adaptive pressures force a breeding population to separate into different subpopulations that may become species. Sympatric speciation of plants has included *polyploidy*, which is an important mechanism of evolution for higher plants.
- **Extinction** occurs when populations become very small and thus vulnerable to chance occurrences of fire, disease, or climate anomaly. Rare but very extreme events can cause mass extinctions. An example is the meteorite

impact that the Earth suffered about 65 million years ago. The global dust cloud that lingered for several years caused extremely cold temperatures, which wiped out many important lines of plants and animals.

- Species change their ranges by *dispersal*. Plants are generally dispersed by seeds, whereas animals often disperse under their own power. Since most dispersal happens within the range of a species, it acts primarily to encourage gene flow between subpopulations. Long-distance dispersal, though very rare, may still be very important in establishing biogeographic patterns. *Barriers*, often climatic or topographic, inhibit dispersal and induce geographic isolation. Geographic corridors serve as pathways that facilitate dispersal.
- **Endemic** species are found in one region or location and nowhere else. They arise either by a contraction of the range of a species or by a recent speciation event. *Cosmopolitan* species are widely dispersed and nearly universal. *Disjunction* occurs when one or more closely related species appear in widely separated regions.
- **Biogeographic regions** capture patterns of occurrence in which the same or closely related plants and animals tend to be found together. They result because their species have common histories and similar environmental preferences.
- **Biodiversity** is rapidly decreasing as human activity progressively affects the Earth. Extinction rates for many groups of plants and animals are as high or higher today than they have been at any time in the past. Humans act to disperse predators, parasites, and competitors widely, disrupting long-established evolutionary adjustments of species to their environments. Hunting and burning have exterminated many species. Habitat alteration and fragmentation also lead to extinctions. Preservation of global biodiversity includes a strategy of protecting *hotspots* where diversity is greatest.

KEY TERMS

biogeography, p. 268
ecology, p. 268
ecosystem, p. 268
food web, p. 268
biomass, p. 271
biogeochemical cycle, p. 273

pool, p. 273
carbon cycle, p. 273
habitats, p. 276
xerophytes, p. 277
competition, p. 282
ecological succession, p. 284

evolution, p. 289
natural selection, p. 289
species, p. 289
speciation, p. 289
extinction, p. 291
endemic, p. 294

biogeographic regions, p. 296
biodiversity, p. 296
symbiosis, p. 299

REVIEW QUESTIONS

1. What is the fate of carbon released by fossil fuel burning to the atmosphere?
2. Describe the processes that move CO_2 between the ocean and atmosphere and between land ecosystems and the atmosphere.
3. How do changes in land cover affect the global carbon budget? What is the role of soil carbon?
4. Define the terms **biogeography**, **ecology**, and **ecosystem**.
5. What is a **food web** or *food chain*? What are its essential components? How does energy flow through the food web of an ecosystem?
6. Compare and contrast the processes of *photosynthesis* and *respiration*. What classes of organisms are associated with each?

7. How is *net primary production* related to biomass? Identify some types of terrestrial ecosystems that have a high rate of net primary production and some with a low rate.

8. Which areas of oceans and land are associated with high net primary productivity? How is net primary production on land related to climate?

9. What is a **biogeochemical cycle?** What are its essential features?

10. What are the essential features and flow pathways of the **carbon cycle?** How have human activities impacted the carbon cycle?

11. What are the essential features and flow pathways of the *nitrogen cycle?* What role do bacteria play? How has human activity modified the nitrogen cycle?

12. What is a **habitat?** What are some of the characteristics that differentiate habitats? Compare habitat with *ecological niche.*

13. Contrast the terms *ecosystem, community,* and *association.*

14. Although water is a necessity for terrestrial life, many organisms have adapted to arid environments. Describe some of the adaptations that plants and animals have evolved to cope with the desert.

15. Terrestrial temperatures vary widely. How does the annual variation in temperature influence plant growth, development, and distribution? How do animals cope with variation in temperature?

16. How does the ecological factor of light affect plants and animals? How does wind affect plants?

17. How do *geomorphic* and *edaphic factors* influence the habitat of a community?

18. Identify several types of *disturbance* experienced by ecosystems. How does fire affect forests and grasslands?

19. Contrast the terms used to describe interactions among species. Provide an example of beneficial predation.

20. Describe ecological succession using the terms *sere, seral stage, pioneer,* and *climax.* Use dune succession as an example.

21. How do *primary succession* and *secondary succession* differ? Describe old-field succession as an example of secondary succession.

22. How does the pattern of ecosystems on the landscape reflect a balance between succession and disturbance?

23. What are the effects of fire on the atmosphere and on ecosystems?

24. Explain Darwin's theory of **evolution** by means of **natural selection.** What key point was Darwin unable to explain?

25. What two sources of variation act to differentiate offspring from parents?

26. What is **speciation?** Identify and describe four component processes of speciation.

27. What is the effect of *geographic isolation* on speciation? Provide an example of *allopatric speciation.*

28. How does *sympatric speciation* differ from allopatric speciation? Provide an example of sympatric speciation.

29. What is **extinction?** Provide some examples of extinctions of species.

30. Describe the process of *dispersal.* Provide a few examples of plants and animals suited to long-distance dispersal.

31. Contrast barriers and corridors in the dispersal process.

32. How does an **endemic** distribution pattern differ from a *cosmopolitan* pattern? What is *disjunction?*

33. How are biogeographic regions differentiated?

34. What is **biodiversity?** How has human activity impacted biodiversity?

VISUALIZING EXERCISES

1. Sketch a graph showing the relationship among gross photosynthesis, net photosynthesis, respiration, and temperature. How is net photosynthesis obtained from gross photosynthesis and respiration for each temperature?

2. Diagram the general features of a biogeochemical cycle in which storage pools and active pools are linked by life processes and physical processes.

3. Draw a timeline illustrating old-field succession. Between the stages, indicate the environmental changes that occur.

4. Carefully compare the two maps of Figure 8.36. Which boundaries are similar and which are different? Speculate on possible reasons for the similarities and differences.

ESSAY QUESTIONS

1. Suppose atmospheric carbon dioxide concentration doubles. What will be the effect on the carbon cycle? How will flows change? Which pools will increase? Decrease?

2. Select three distinctive nearby habitats for plants and animals with which you are familiar. Organize

a field trip (real or virtual) to visit them. Compare their physical environments and describe the basic characteristics of the ecosystems found there.

3. Invent a biological history of the Galápagos Islands, describing how and when organisms such as finches and tortoises evolved using the processes of speciation.

Chapter 9
Global Biogeography

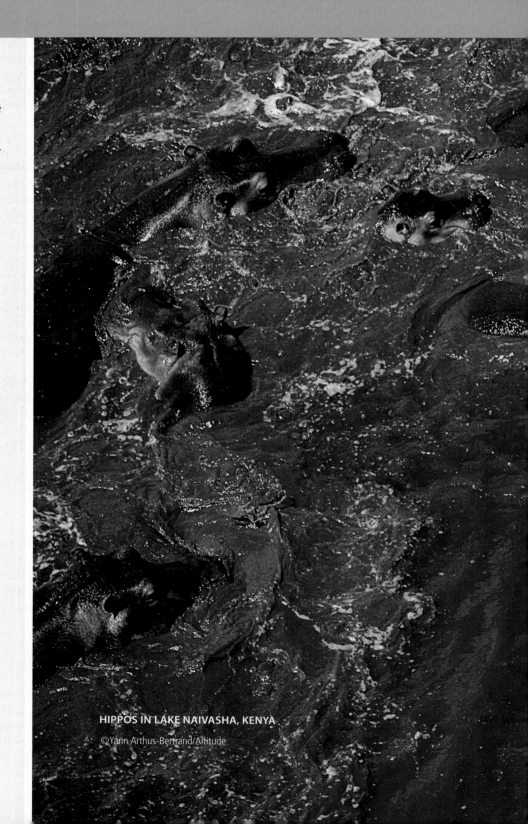

Among the roughly 40,000 threatened species on the International Union for the Conservation of Nature's "Red List of Threatened Species" is the hippopotamus. Cantankerous, unpredictable, and dangerous, the hippo's notorious bad behavior made it a target of fisherman, due to its habit of attacking them and their boats. But it's the poachers and ivory hunters who are more to blame for decimating the populations of this once-ubiquitous lake, river, and marsh dweller. Now classified as "vulnerable," the hippopotamus is losing more ground, particularly in West Africa, to its human predators.

HIPPOS IN LAKE NAIVASHA, KENYA
©Yann Arthus-Bertrand/Altitude

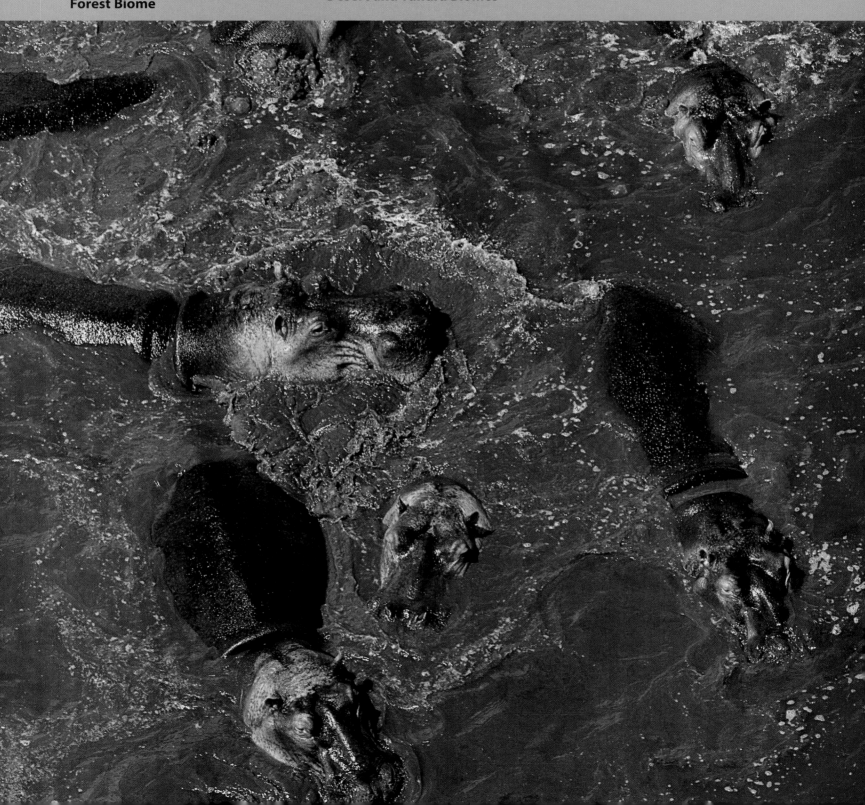

Global Biogeography

T he Earth's land environments are home to a vast diversity of plants and animals. How do biogeographers and ecologists classify the ecological realms of the land into biomes and vegetation formation classes? What are the five biomes of the world? How do they compare? Where are they found? What important formation classes do they contain? How is the global and continental pattern of biomes and formation classes related to climate? These are some of the questions we will answer in this chapter.

Exploitation of the Low-Latitude Rainforest Ecosystem

Many of the world's equatorial and tropical regions are home to the *rainforest ecosystem*. This ecosystem is perhaps the most diverse on Earth, supporting more species of plants and animals than any other. Very large tracts of rainforest still exist in South America, South Asia, and some parts of Africa. Ecologists regard this ecosystem as a genetic reservoir of countless species of plants and animals. But as human populations continue to expand, in concert with the quest for agricultural land, low-latitude rainforests are being threatened, by land clearing, logging, cultivation of cash crops, and domestic animal grazing.

In the past, low-latitude rainforests were farmed by native peoples using the *slash-and-burn* method—cutting down all the vegetation in a small area and then burning it (Figure 9.1). They do this because in a rainforest ecosystem, most of the nutrients are held within living plants rather than in the soil, and burning the vegetation on the site releases the trapped nutrients, returning a portion of them to the soil, where they then become available to growing crops.

The supply of nutrients derived from the original vegetation cover is small, however, and the harvesting of crops rapidly depletes the nutrients. After a few seasons of cultivation, the old field has to be abandoned and a new field cleared. Nearby rainforest plants soon reestablish their hold on the abandoned area and, eventually, the rainforest returns to its original state.

> The rainforest ecosystem, home of the world's most diverse collection of plants and animals, is threatened by deforestation and conversion to cropland and rangeland.

©Jacques Jangoux/Getty Images, Inc.

9.1 Slash-and-burn clearing

This rainforest in Maranhão, Brazil, has been felled and burned in preparation for cultivation.

In contrast to these practices, modern intensive agriculture requires large areas of land, and is not compatible with the rainforest ecosystem. When large areas are abandoned, seed sources are so far away that the original forest species cannot regain their hold. Instead, secondary species dominate, often accompanied by successful invaders from other vegetation types. Once these invaders enter an area, they tend to stay, and their dominance is permanent, at least on the human time scale. The rainforest ecosystem is thus a resource that, once cleared, will never return in quite the same way. Over time, the destruction of low-latitude rainforest will result in the disappearance of thousands of species of organisms from the rainforest environment—representing millions of years of evolution—together with the loss of the most complex ecosystem on Earth.

In Amazonia, to transform large areas of rainforest into agricultural land, heavy machinery is put in motion carving out major highways and innumerable secondary roads and trails. Large fields for cattle pasture or commercial crops are shaped by cutting, bulldozing, clearing, and burning the vegetation. In some regions, the great broad-leafed rainforest trees are cut down for commercial lumber.

According to a recent report issued by the United Nations Food and Agriculture Organization, about 0.6 percent of the world's rainforest is lost annually by conversion to other uses. More rainforest land—2.2 million hectares (about 8500 mi²)—is lost annually in Asia than in Latin America and the Caribbean, where 1.9 million hectares (about 7300 mi²) are converted every year. Africa's loss of rainforest was estimated at about 470,000 hectares per year (1800 mi²). Among individual countries, Brazil and Indonesia are the loss leaders, accounting for nearly half of the rainforest area converted to other uses. And these numbers do not even include larger losses of moist deciduous forests in these regions. Deforestation in low-latitude dry deciduous forests and hill and montane forests is also very serious.

To combat very rapid deforestation rates in some regions, many nations are now working to stem the loss of their rainforest environments. But they are up against powerful adversaries, because the rainforest can provide valuable agricultural land, minerals, and timber; so the pressure to allow deforestation remains intense.

Natural Vegetation

Over the last few thousand years, human societies have come to dominate much of the land area of our planet. We've changed the natural vegetation—sometimes drastically—of many regions. What exactly do we mean by **natural vegetation**? Natural vegetation is a plant cover that develops with little or no human interference. It responds to natural forces, storms, or fires that can modify or even destroy it. Nevertheless, natural vegetation can still be seen over vast areas of the wet equatorial climate ①, although the rainforests there are slowly being cleared. And much of the arctic tundra and the boreal forest of the subarctic zones remain in a natural state.

In contrast to natural vegetation is *human-influenced vegetation*. Much of the midlatitude land surface is totally under human control, through intensive agriculture, grazing, or urbanization. Other areas that appear to be untouched may actually be dominated by human activity in a more subtle manner (Figure 9.2). For example, most national parks and national forests in the United States have been protected from fire for many decades. As a result, however, dead branches and debris have accumulated on the forest floor, generating fuel loads that encourage hot, damaging, crown fires, rather than cooler, sparser, understory fires that leave the larger, healthier, trees alive.

9.2 The great Yellowstone fire

Yellowstone National Park has been little disturbed by natural catastrophe or human interference for at least the past two centuries. But through August and September of 1988, 45 forest fires—mostly started by lightning strikes—burned out of control in the park. They were most destructive in long-unburned areas where dead wood and branches had accumulated for decades.

▼ **AFTER THE BURN**
This stand of lodgepole pines in Yellowstone was killed by an intensely hot, crown fire.

▼ **TEN YEARS LATER**
After 10 years, regeneration had started a new pine forest.

Jonathan Blair/©Corbis

M.P. Kahl/Photo Researchers, Inc.

Humans have also moved plant species from their original habitats to foreign lands and environments. Some of these exported plants thrive like weeds, forcing out natural species and becoming a major nuisance. Other human activities such as clear-cutting,

> Most of the Earth's land surface is influenced subtly or strongly by human activities, including clearing for agriculture and grazing, fire suppression, and introduction of alien plants and animals.

slash-and-burn agriculture, overgrazing, and wood gathering have had profound effects on the plant species and the productivity of the land (Figure 9.3).

STRUCTURE AND LIFE-FORM OF PLANTS

Plants come in many types, shapes, and sizes. Botanists recognize and classify plants by species. The biogeographer, in contrast, is less concerned with individual species and more interested in plant cover as a whole.

9.3 Deforestation and desertification

Forest clearing, if carried out unsustainably, reduces global biomass and biodiversity while contributing to global warming. Desertification, or land degradation, lowers productivity by overusing the land for grazing, wood gathering, and other consumptive activities.

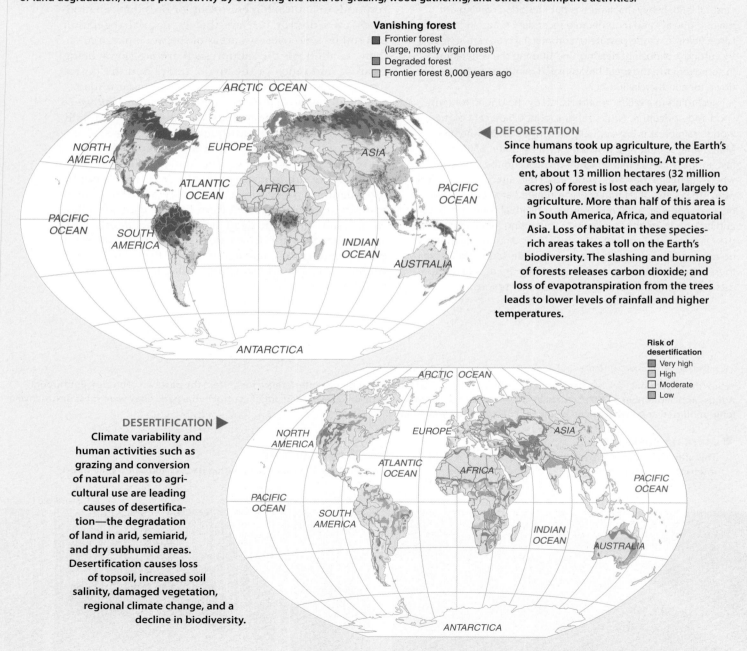

Vanishing forest
- Frontier forest (large, mostly virgin forest)
- Degraded forest
- Frontier forest 8,000 years ago

◀ **DEFORESTATION**
Since humans took up agriculture, the Earth's forests have been diminishing. At present, about 13 million hectares (32 million acres) of forest is lost each year, largely to agriculture. More than half of this area is in South America, Africa, and equatorial Asia. Loss of habitat in these species-rich areas takes a toll on the Earth's biodiversity. The slashing and burning of forests releases carbon dioxide; and loss of evapotranspiration from the trees leads to lower levels of rainfall and higher temperatures.

Risk of desertification
- Very high
- High
- Moderate
- Low

DESERTIFICATION ▶
Climate variability and human activities such as grazing and conversion of natural areas to agricultural use are leading causes of desertification—the degradation of land in arid, semiarid, and dry subhumid areas. Desertification causes loss of topsoil, increased soil salinity, damaged vegetation, regional climate change, and a decline in biodiversity.

So, when talking about plant cover, plant geographers discuss the **life-form** of the plant—its physical structure, size, and shape. Most life-form names are in common use, and though you're probably already familiar with them, we'll review them quickly here.

Figure 9.4 illustrates various plant life-forms. Trees and shrubs are erect, woody plants. They are *perennial*, meaning that their woody tissues endure from year to year. Most have life spans of many years. *Trees* are large plants with a single upright main trunk, often with few branches in the lower part but branching in the upper part to form a crown. *Shrubs* have several stems branching from a base near the soil surface, creating a mass of foliage close to ground level.

Lianas are also woody plants, but they take the form of vines, which are supported on trees and shrubs. Lianas include tall, heavy vines in the wet equatorial and tropical rainforests, as well as some woody vines of midlatitude forests. English ivy, poison ivy or oak, and Virginia creeper are familiar North American examples of lianas.

Digital Vision/Getty Images, Inc.

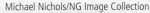
CLEAR-CUTTING
Clear-cutting of large tracts of timber, without sustainable replanting, contributes to deforestation, erosion, and loss of habitat.

SLASH-AND-BURN
Large areas of equatorial rainforest in the Amazon Basin are now being converted to grazing land and agriculture. The first step in this process is to cut the forest for timber and burn the debris to release nutrients to soil. The nutrients are soon exhausted, leaving the land impoverished and unproductive.

Michael Nichols/NG Image Collection

© Dieter Telemans/Panos Pictures

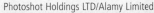
WOOD GATHERING
Where fuel is in short supply, firewood is stripped from living trees, diminishing the vegetation cover. When trees are gone, dung is burned, further impoverishing the soil.

OVERGRAZING
Overgrazing strips vegetation from the land, reducing evapotranspiration, raising temperatures, and leaving the soil without cover.

Photoshot Holdings LTD/Alamy Limited

Sugar maples   Hemlock   Beech   Ash   Sugar maples

Liana

Tree layer

Dogwood

Shrub layer

Herb layer

Moss layer

9.4 Layers of a beech–maple–hemlock forest

Tree crowns form the uppermost layer, shrubs an intermediate layer, and herbs a lower layer. Mosses and lichen grow very close to the ground. In this schematic diagram, the vertical dimensions of the lower layers are greatly exaggerated.

Herbs make up a major class of plant life-forms. They lack woody stems and so are usually small, tender plants. They occur in a wide range of shapes and leaf types. Some are *annuals*, living only for a single season; others are broad-leafed; and still others are narrow-leafed, such as grasses. Herbs share few characteristics with each other, though all usually form a lower layer than shrubs and trees. *Lichens*, which also grow close to the ground, are life-forms in which algae and fungi live together, forming a single plant structure (Figure 9.5). Lichens dominate the vegetation in some alpine and arctic environments.

Forest is a vegetation structure in which trees grow close together. The crowns of forest trees often touch, so

9.5 Lichens

Lichens are plant forms that combine algae and fungi in a single symbiotic organism. They are abundant in some high-latitude habitats, and they also range to the tropics. Pictured here is a mountain tundra landscape, Regebufjellet, Norway, with a carpet of reindeer lichen in white in the foreground. Darker crustose lichens grow in patches on the rocks.

Kevin Prönnecke/ © imagebroker/Alamy Limited

that their foliage largely shades the ground. Many forests in moist climates have at least three layers of life-forms: the tree, shrub, and herb layers. There is sometimes a fourth, as well: the lowermost layer of mosses and related very small plants. In *woodland*, tree crowns are separated by open areas that usually have a low herb or shrub layer.

Terrestrial Ecosystems—The Biomes

For humans, ecosystems may be regarded as huge natural factories producing food, fiber, fuel, and structural materials. These useful products are manufactured by organisms taking energy from the Sun, and we harvest that energy by using these ecosystem products. The products and productivity of ecosystems depend on their climate. Where temperature and rainfall cycles permit, ecosystems provide a rich bounty. Where temperature or rainfall cycles restrict ecosystems, human activities also may be limited. Of course, humans too are part of the ecosystems that we modify for our own benefit.

Ecosystems fall into two major groups, aquatic and terrestrial. *Aquatic ecosystems* include marine environments and the freshwater environments of the lands. Marine ecosystems include the open ocean, coastal estuaries, and coral reefs. Freshwater ecosystems include lakes, ponds, streams, marshes, and bogs. In this book, we'll focus on the **terrestrial ecosystems**, which are dominated by land plants spread widely over the upland surfaces of the continents. Because these ecosystems are directly influenced by climate, they are closely woven into the fabric of physical geography.

We divide terrestrial ecosystems into **biomes**. Although the biome includes both plant and animal life, green plants dominate simply because of their enormous biomass. Plant geographers concentrate on the characteristic life-form of the green plants within the biome—principally trees, shrubs, lianas, and herbs—but also other life-forms in certain biomes.

There are five principal biomes:

- The **forest biome** is dominated by trees, which form a closed or nearly closed canopy. Forests require an abundance of soil water, so they are found in moist climates. Temperatures must also be suitable, requiring at least a warm season, if not warm temperatures year-round.
- The **savanna biome** is transitional between forest and grassland. It exhibits an open cover of trees with grasses and herbs underneath.
- The **grassland biome** develops in regions with moderate shortages of soil water. The semiarid regions of the dry tropical, dry subtropical, and dry midlatitude climates are the home of the grassland biome. Temperatures must also provide adequate warmth during the growing season.
- The **desert biome** includes organisms that can survive a moderate to severe water shortage for most, if not all, of the year. Temperatures can range from very

> Biogeographers recognize five principal biomes: forest, grassland, savanna, desert, and tundra. Formation classes are subdivisions of biomes based on vegetation structure and life-form.

hot to cool. Plants of the desert biome are often xerophytes, which have adapted to the dry environment.
- Plant life in the **tundra biome** is limited by cold temperatures. Only small plants that can grow quickly when temperatures warm above freezing in the warmest month or two can survive.

Biogeographers break down the five principal biomes into smaller vegetation units, called *formation classes*, using the life-form of the plants. For example, at least four and perhaps as many as six kinds of forests can be distinguished within the forest biome. At least three kinds of grasslands are easily recognizable. Deserts, too, span a wide range in terms of the abundance and life-form of plants. The formation classes described in the remaining portion of this chapter are major, widespread types that are clearly associated with specific climate types.

WileyPLUS Natural Vegetation Regions of the World
Take a second look at the world map of vegetation regions, and click on a formation class or biome to see a characteristic photo. An animation.

BIOMES, FORMATION CLASSES, AND CLIMATE

The pattern of formation classes depends heavily on climate. As climate changes with latitude or longitude, vegetation will also change. Figure 9.6 shows how vegetation formation classes respond to temperature and

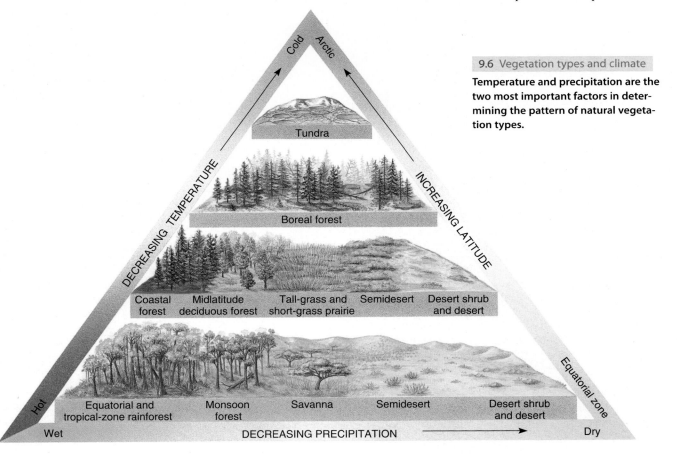

9.6 Vegetation types and climate

Temperature and precipitation are the two most important factors in determining the pattern of natural vegetation types.

precipitation gradients. In both low- and midlatitude environments, strong precipitation gradients produce vegetation types grading from forest to desert. At high latitudes, decreasing temperatures control the transition from forest to tundra.

The diagram does not include seasonality. In low latitudes, savanna and grassland formation classes are found in regions with a distinct dry season. In the midlatitudes, west coasts are marked with a strong summer dry period, providing sclerophyll vegetation

9.7 Natural vegetation regions of the world

The world's regions of natural vegetation range from forests and prairies to deserts and tundra.

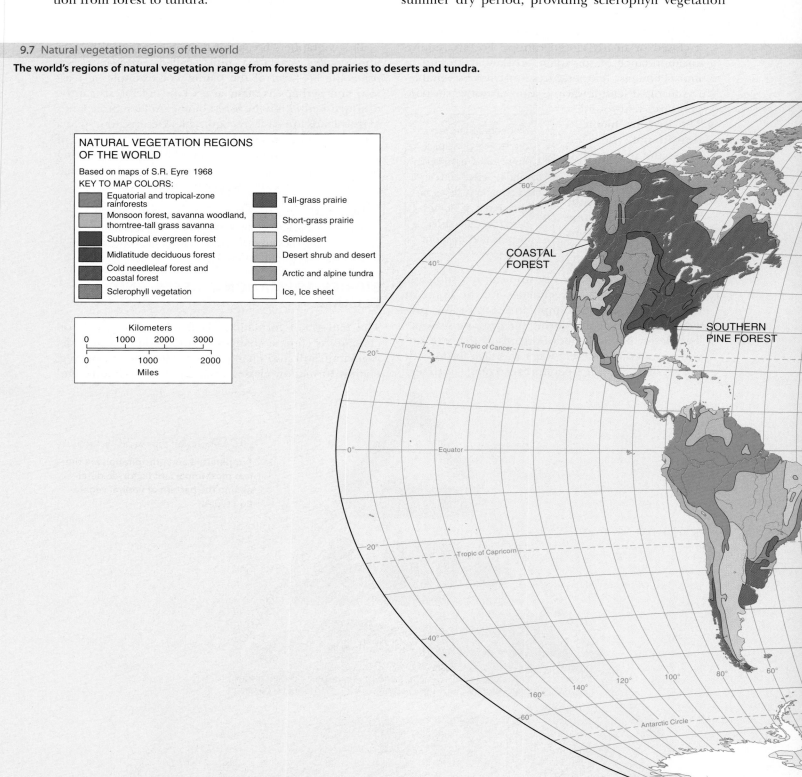

NATURAL VEGETATION REGIONS
OF THE WORLD

Based on maps of S.R. Eyre 1968
KEY TO MAP COLORS:

- Equatorial and tropical-zone rainforests
- Monsoon forest, savanna woodland, thorntree-tall grass savanna
- Subtropical evergreen forest
- Midlatitude deciduous forest
- Cold needleleaf forest and coastal forest
- Sclerophyll vegetation
- Tall-grass prairie
- Short-grass prairie
- Semidesert
- Desert shrub and desert
- Arctic and alpine tundra
- Ice, Ice sheet

Kilometers
0   1000   2000   3000

0   1000   2000
Miles

COASTAL FOREST

SOUTHERN PINE FOREST

(not shown) in coastal regions and, farther poleward, encouraging the growth of lush coastal forests of tall and stately conifers.

Figure 9.7 is a generalized world map of vegetation formation classes. It simplifies the very complex patterns of natural vegetation to show large uniform regions in which a given formation class might be expected to occur. Although the boundaries between vegetation types are shown as distinct lines, many real boundaries are gradational and located approximately.

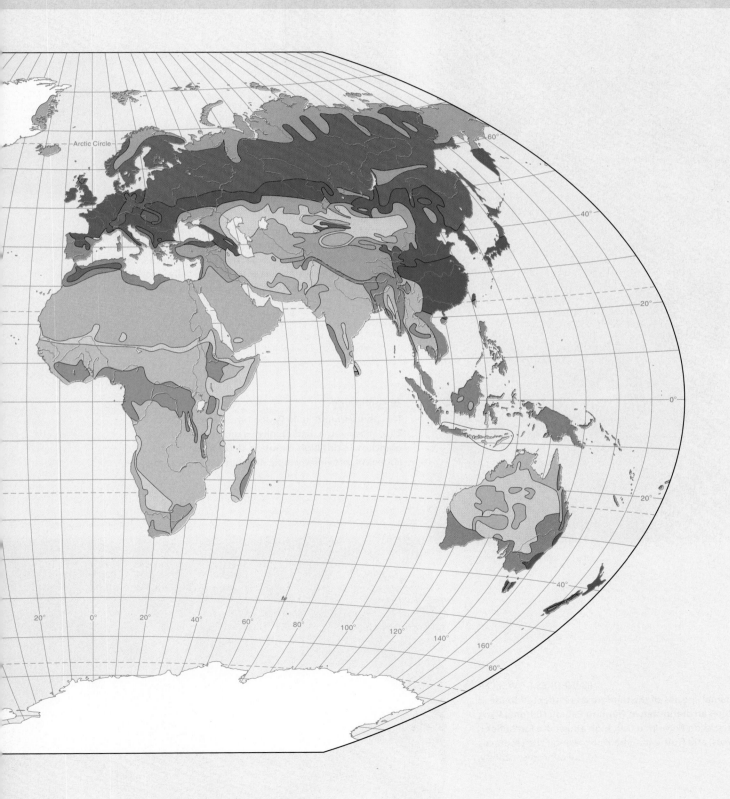

9.8 Rainforest

Rainforest environments include unique plants and animals.

Frans Lanting/NG Image Collection

Tim Laman/NG Image Collection

▲ **RAINFOREST INTERIOR**

Crowns in this rainforest in Kalimantan, Borneo, Indonesia, form a continuous canopy that shades the lower layers. The lower two-thirds of the trees are characteristically smooth-barked and unbranched. Many rainforest species, especially in low or wet areas, have buttress roots extending out from the base of the tree.

◄ **EPIPHYTES**

This photo from Braulio Carillo National Park, Costa Rica, shows red-leafed epiphytes growing on tree branches. In this high-elevation cloud forest, rainfall is abundant, providing ample moisture for these plants.

Tim Laman/NG Image Collection

RAINFOREST DWELLER ▶

Many animal species of the rainforest are adapted to life in the trees. Here, an orangutan in Gunung Palung National Park, Borneo, snacks on *Polyalthia* fruit high above the forest floor. Toucans, parrots, and fruit-eating bats also exploit the resources of the rainforest canopy.

WileyPLUS Interaction of Climate, Vegetation, and Soil
Return to the animations for Chapter 7 to view climate, vegetation, and soils maps superimposed to reveal common patterns. Animations for Africa and North America.

Forest Biome

Within the **forest biome**, we can recognize six major formations: low-latitude rainforest, monsoon forest, subtropical evergreen forest, midlatitude deciduous forest, needleleaf forest, and sclerophyll forest. Ecologists also sometimes recognize three principal types of forest as separate biomes, based on their widespread nature and occurrence in different latitude belts: low-latitude rainforest, midlatitude deciduous and evergreen forest, and boreal forest.

LOW-LATITUDE RAINFOREST

Low-latitude rainforest, found in the equatorial and tropical latitude zones, consists of tall, closely spaced trees (Figure 9.8). Equatorial and tropical rainforests are not jungles of impenetrable plant thickets; rather, the floor of the low-latitude rainforest is usually so densely shaded by a canopy of tree crowns that plant foliage is sparse close to the ground. The trees define a number of distinct rainforest layers (Figure 9.9).

Typical of the low-latitude rainforest are thick, woody lianas supported by the trunks and branches of trees. They climb to the upper canopy, where light is available, and develop numerous branches of their own. *Epiphytes* ("air plants") are also common in low-latitude rainforest (Figure 9.8). These plants attach themselves to the trunks, branches, or foliage of trees and lianas. Their "host" is used solely as a means of physical support. Epiphytes include plants of many different types, among them ferns, orchids, mosses, and lichens.

Figure 9.10 shows the world distribution of low-latitude rainforests. These rainforests develop in a climate that is continuously warm, frost-free, and with abundant precipitation in all months of the year (or, at most, with only one or two dry months). These conditions occur

9.9 Rainforest layers

Crowns of the trees of the low-latitude rainforest tend to form two or three layers. The highest layer consists of scattered "emergent" crowns that protrude from the closed canopy below, often rising to 40 m (130 ft). Some emergent species develop wide buttress roots, which serve to support them. Below the layer of emergents is a second, continuous layer, which is 15 to 30 m (about 50 to 100 ft) high. A third, lower layer consists of small, slender trees 5 to 15 m (about 15 to 50 ft) high with narrow crowns.

9.10 Global distribution of low-latitude rainforest

This world map of low-latitude rainforest shows equatorial and tropical rainforest types. A large area of rainforest lies astride the Equator and extends poleward through the tropical zone (lat. 10° to 25° N and S) along monsoon and trade-wind coasts. Within the low-latitude rainforest are many highland regions where temperatures are cooler and rainfall is increased by the orographic effect. The canopy of this montane forest is more open, with lower trees and smaller plants that luxuriate in the abundant rainfall.

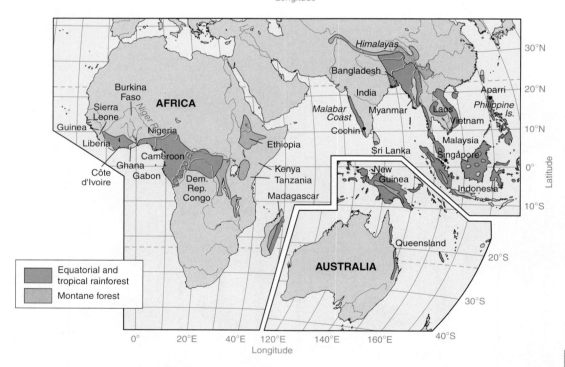

Equatorial and tropical rainforest

Montane forest

in the wet equatorial climate ① and the monsoon and trade-wind coastal climate ②. Plants grow continuously throughout the year.

A particularly important characteristic of the low-latitude rainforest is the large number of species of plants and animals that coexist within it. Equatorial rainforests contain as many as 3000 tree species within a few square kilometers.

Large herbivores are, however, uncommon in the low-latitude rainforest. They include the African okapi and the tapir of South America and Asia. Most herbivores in this environment are climbers, and include many primates—monkeys and apes (Figure 9.8). Tree sloths spend their lifetimes hanging upside down as they browse the forest canopy. There are only a few

Low-latitude rainforests are very diverse, containing large numbers of plant and animal species. Broadleaf evergreen trees dominate the vegetation cover. The rainforest climate is wet all year, or has a short dry season.

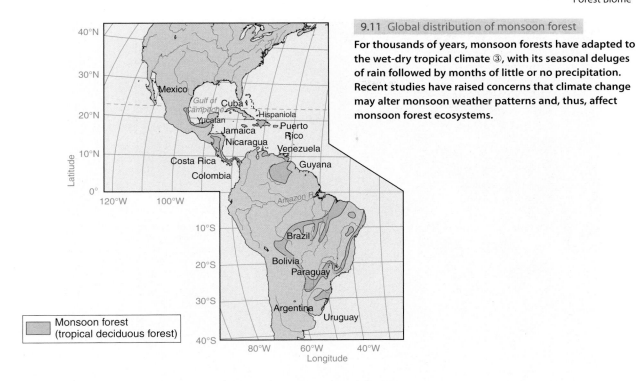

9.11 Global distribution of monsoon forest

For thousands of years, monsoon forests have adapted to the wet-dry tropical climate ③, with its seasonal deluges of rain followed by months of little or no precipitation. Recent studies have raised concerns that climate change may alter monsoon weather patterns and, thus, affect monsoon forest ecosystems.

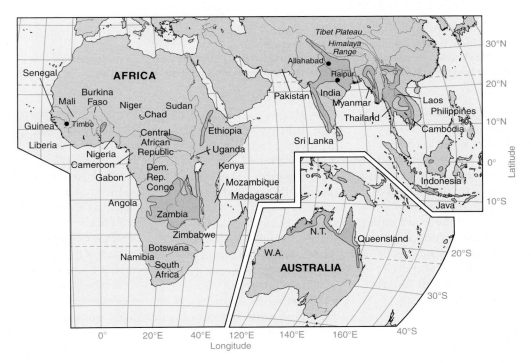

Monsoon forest (tropical deciduous forest)

large predators here as well. Notable among them are the leopards of African and Asian forests and the jaguars and ocelots of the South American forests.

MONSOON FOREST

Figure 9.11 is a world map of the *monsoon forest*. It is typically open, but grades into woodland, with open areas occupied by shrubs and grasses (Figure 9.12). Monsoon forest of the tropical latitude zone differs from tropical rainforest in that it is *deciduous*; that is, most of the trees of the monsoon forest shed their

Monsoon forest is an open cover of deciduous trees that shed their leaves during a pronounced dry season. It occurs in the wet-dry tropical climate ③, ranging from South America to Africa and southern Asia.

Aditya "Dicky" Singh/Alamy Limited

Rajasthan, India

9.12 Monsoon woodland

Trees of the monsoon forest shed their leaves in the dry season. The forest cover is sparser and the trees shorter than in rainforest. Tree trunks are massive, often with thick, rough bark. Branching starts at a comparatively low level and produces large, round crowns. Shown here is the monsoon forest of the Ranthambore Tiger Reserve, Rajasthan, India.

leaves due to stress during the long dry season, which occurs at the time of low Sun and cool temperatures.

This forest develops in the wet–dry tropical climate ③, where a long rainy season alternates with a dry, rather cool season. The typical regions of monsoon forest are found in Myanmar, Thailand, and Cambodia. In the monsoon forest of southern Asia, the teakwood tree was once abundant, but it was cut down and the wood widely exported to the Western world to make furniture, paneling, and decking. Now this magnificent tree has been logged out. Large areas of monsoon forest also occur in south-central Africa and in Central and South America, bordering the equatorial and tropical rainforests.

SUBTROPICAL EVERGREEN FOREST

Subtropical evergreen forest is generally found in regions of moist subtropical climate ⑥ where winters are mild and there is ample rainfall throughout the year (Figure 9.13). This forest occurs in two forms: broadleaf and needleleaf (Figure 9.14).

The *subtropical broadleaf evergreen forest* has fewer tree species than the low-latitude rainforests, which also are home to broadleaf evergreen types. Trees are not as tall here as in the low-latitude rainforests, and their leaves tend to be smaller and more leathery; thus, the leaf canopy is less dense. The subtropical broadleaf evergreen forest often has a well-developed lower layer of vegetation. Depending on the location, this layer may include tree ferns, small palms, bamboos, shrubs, and herbaceous plants. Many lianas and epiphytes also grow here.

The subtropical broadleaf evergreen forest is associated with the moist subtropical climate of the southeastern United States, southern China, and southern Japan. Centuries ago, however, the land in these areas was cleared of natural vegetation and replaced by agriculture.

Figure 9.14 shows the subtropical evergreen forests of the northern hemisphere. The *subtropical needleleaf evergreen forest* occurs only in the southeastern United States. Here it is referred to as the southern pine forest, because it is dominated by species of pine. Much of this area is now contained within commercial pine plantations that produce lumber, kraft paper, cardboard, and related wood products.

> The subtropical evergreen forest includes both broadleaf and needleleaf types and is found in moist subtropical climate ⑥ regions of southeastern North America and Southeast Asia. Most of this formation class has been lost to cultivation.

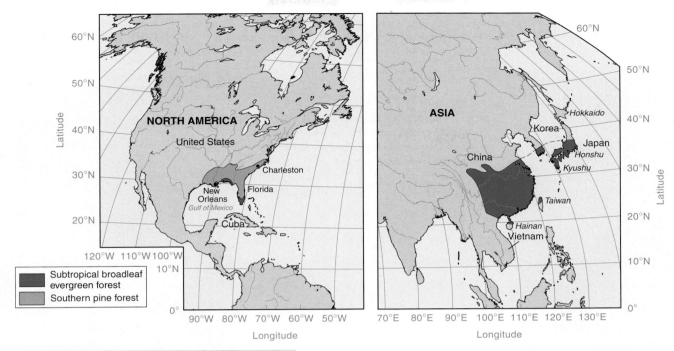

9.13   Global distribution of subtropical evergreen forest

This northern hemisphere map of subtropical evergreen forests includes the needleleaf southern pine forest. Broadleaf evergreen forest includes trees of the laurel and magnolia families, so it is also termed "laurel forest." Because this region is intensely cultivated, little natural laurel forest remains.

9.14   Subtropical evergreen forest

This forest of mild and moist climates includes both broadleaf and needleleaf types.

▼ **BROADLEAF**

Here, broad-leafed tree species dominate over a lower layer of smaller plants. This example, from New South Wales, Australia, includes many species of eucalyptus.

George Gall/NG Image Collection

▼ **NEEDLELEAF**

The subtropical needleleaf evergreen forest inhabits dry, sandy soils, and experiences occasional droughts and fires. Pines are well-adapted to these conditions. A layer of broadleaf shrubs and small trees often is established beneath the pines. This longleaf pine stand is near Aiken, South Carolina.

Raymond Gehman/NG Image Collection

9.15 Deciduous forest

Four distinct seasons are notably displayed in the annual spring-through-winter vegetation growth cycles of these forest ecosystems.

◀ **WOODCHUCK**
Many small mammals burrow in forest soils for shelter or food, including the woodchuck, or groundhog, shown here. They are joined by ground squirrels, mice, and shrews.

FOX SQUIRREL ▶
Scampering freely from the trees to the forest floor, the fox squirrel feeds primarily on nuts and seeds of the deciduous forest. In the canopy, squirrels are joined by many species of birds and insects.

▼ **TREES OF THE FOREST**
Shown here in Monongahela National Forest, West Virginia, is a forest of maple, oak, and hickory in fall colors.

▼ **WHITE-TAILED DEER**
Among the large herbivores that graze in the deciduous forest are the white-tailed deer of North America and the red deer and roe deer of Eurasia.

MIDLATITUDE DECIDUOUS FOREST

Midlatitude deciduous forest is the forest type native to eastern North America and western Europe (Figure 9.15). It is dominated by tall, broadleaf trees that provide a continuous and dense canopy in summer but shed their leaves completely in the winter. Lower layers of small trees and shrubs are weakly developed. In the spring, a lush layer of low herbs quickly develops then soon fades after the trees reach full foliage and shade the ground. The deciduous forest is home to a wide variety of animal life, stratified according to canopy layers.

Figure 9.16 is a map of midlatitude deciduous forests, which are found almost entirely in the northern hemisphere. This forest type is associated with the moist continental climate ⑩, which receives adequate precipitation in all months, normally with a summer maximum. It experiences is a strong annual temperature cycle, with a cold winter season and a warm summer.

Trees common to the deciduous forest of eastern North America, southeastern Europe, and eastern Asia are oak, beech, birch, hickory, walnut, maple, elm, and ash. Where the deciduous forests have been cleared by lumbering, pines readily develop as second-growth forest.

In western Europe, the midlatitude deciduous forest is associated with the marine west-coast climate. Here, the dominant trees are mostly oak and ash, with beech found in cooler and moister areas. In Asia, the midlatitude deciduous forest occurs as a belt between the boreal forest to the north and steppelands to the south. A small area of deciduous forest is found in Patagonia, near the southern tip of South America.

> Midlatitude deciduous forest consists largely of trees that drop their leaves during the cold season. It is characteristic of the marine west-coast ⑧ and moist continental ⑩ climates.

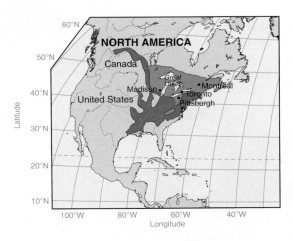

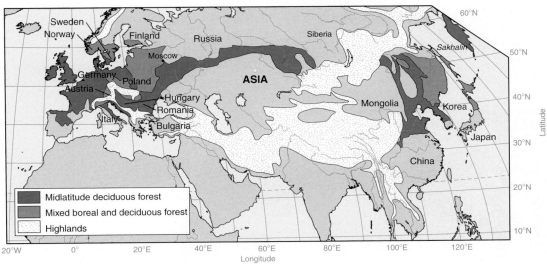

9.16 Northern hemisphere distribution of midlatitude deciduous forest

In Western Europe, the midlatitude deciduous forest is associated with the marine west-coast climate ⑧. Here, the dominant trees are mostly oak and ash, with beech in cooler and moister areas. Regions of deciduous forest are also found in a belt across Asia and near the southern tip of South America.

NEEDLELEAF FOREST

Needleleaf forest is composed largely of *conifers*, straight-trunked, cone-shaped trees with relatively short branches and small, narrow, needlelike leaves. Most conifers are evergreen; they retain their needles for several years before shedding them. At any location, species are usually few in number. In fact, large tracts of needleleaf forest consist almost entirely of only one or two species. Where the needleleaf forest is dense, it provides continuous and deep shade to the ground. Lower layers of vegetation are sparse or absent, except for possibly a thick carpet of mosses.

Boreal forest is the cold-climate needleleaf forest of high latitudes (Figure 9.17). It occurs in two great continental belts, one in North America and one in Eurasia (Figure 9.18). These belts span their land masses from west to east in latitudes 45° N to 75° N, and they closely correspond to the region of boreal forest climate ⑪.

The boreal forest of North America, Europe, and western Siberia is composed of such evergreen conifers as spruce and fir, while the boreal forest of north-central and eastern Siberia is dominated by larch. The larch tree sheds its needles in winter and is thus a deciduous needleleaf tree. Broadleaf deciduous trees, such as

9.17 Boreal forest

A view of the boreal forest of Denali National Park, Alaska, pictured here just after the first snowfall of the season. At this location, near the northern limits of the boreal forest, the tree cover is sparse. The golden leaves of aspen mark the presence of this deciduous species.

©Michael Townsend

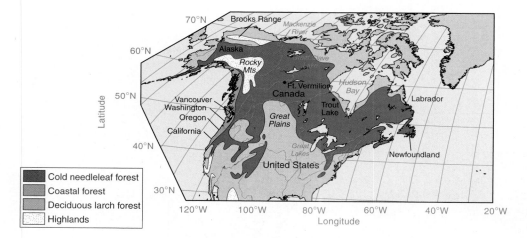

Cold needleleaf forest
Coastal forest
Deciduous larch forest
Highlands

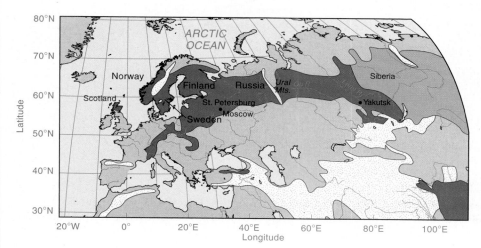

9.18 Northern hemisphere distribution of needleleaf forest

This northern hemisphere map shows cold-climate needleleaf forests, including coastal forest.

aspen, balsam poplar, willow, and birch, tend to take over rapidly in areas of needle-leaf forest that have been burned over. These species can also be found bordering streams and in open places. Between the boreal forest and the midlatitude decidu-ous forest lies a broad transi-tion zone of mixed boreal and deciduous forest.

> Needleleaf forest includes boreal and coastal forest. Boreal forest stretches across the northern reaches of North America and Eurasia. Coastal forest is restricted to the coastal ranges of the Pacific Northwest region.

Coastal forest is a distinctive needleleaf evergreen forest of the Pacific Northwest coastal belt, ranging in latitude from northern California to southern Alaska. Here, in a band of heavy orographic precipitation, mild temperatures, and high humidity, are perhaps the dens-est of all conifer forests, with magnificent specimens of cedar, spruce, and Douglas fir. At the extreme southern end, coastal forest includes the world's largest trees—redwoods (Figure 9.19). Individual redwood trees attain heights of over 115 m (377 ft) and diameters of over 7 m (23 ft) at the base.

9.19 Coastal forest

Along the western coast of North America, from central California to Alaska, is a forest dominated by many unique species of needleleaf trees. Shown here is a stand of coast redwoods, *Sequoia sempervirens*, a spe-cies found at the southern end of this range. It is gener-ally considered the tallest of tree species, attaining a height of 115 m (377 ft) and diameter of 7 m (23 ft) at the base. These redwoods are at Muir Woods National Monument, not far from San Francisco.

Raymond Gehman/NG Image Collection

SCLEROPHYLL FOREST

The native vegetation of the Mediterranean climate ⑦ is adapted to survive through the long summer drought. Shrubs and trees that can withstand such drought are equipped with small, hard, or thick leaves that resist water loss through transpiration. These plants are called *sclerophylls*.

Sclerophyll forest is made up of trees with small, hard, leathery leaves. The trees are often low-branched and gnarled, with thick bark. The formation class includes *sclerophyll woodland*, an open forest in which only 25 to 60 percent of the ground is cov-ered by trees. Also included are extensive areas of *scrub*, a plant formation type consist-ing of shrubs covering some-what less than half of the ground area. The trees and shrubs are evergreen, retain-ing their thickened leaves despite a severe annual drought. Our map of sclero-phyll vegetation, given in Figure 9.20, includes forest, woodland, and scrub types.

> Sclerophyll forest is dominated by low trees with thick, leathery leaves that are well-adapted to the long summer drought of the Mediterranean climate ⑦. Southern California's chaparral, found on coast-range slopes, is a form of sclerophyll scrub.

Sclerophyll forest is lim-ited to west coasts between 30° and 40° to 45° N and S latitude. In the California coastal ranges, the sclerophyll forest or woodland is typically dominated by live oak and white oak. Grassland occupies the

9.20 Global distribution of sclerophyll forest

Sclerophyll forest is closely associated with the Mediterranean climate ⑦ and occurs along west coasts between 30° and 40° to 45° N and S latitude.

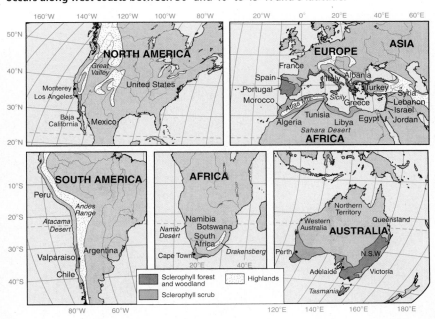

open ground between the scattered oaks. Much of the remaining vegetation is sclerophyll scrub known as *chaparral* (Figure 9.21).

In the Mediterranean lands, the sclerophyll forest forms a narrow coastal belt ringing the Mediterranean Sea. Here, the Mediterranean forest consists of such trees as cork oak, live oak, Aleppo pine, stone pine, and olive.

Important areas of sclerophyll forest, woodland, and scrub are found in southeast, south-central, and southwest Australia, Chile, and the Cape region of South Africa.

Over the centuries, human activity has reduced the sclerophyll forest to woodland, or destroyed it entirely. Today, large areas of this former forest consist of dense scrub.

Earl Scott/Photo Researchers, Inc.

9.21 Chaparral

Chaparral varies in composition with elevation and exposure. It may contain wild lilac, manzanita, mountain mahogany, poison oak, and live oak. Most of the plants are drought-resistant sclerlophylls, with hard, leathery leaves.

Mattias Klum/National Geographic/Getty Images, Inc.

9.22 Clear-Cutting

Clear-cutting of large tracts of timber, without sustainable replanting, contributes to erosion and loss of habitat.

DEFORESTATION

Since humans took up agriculture, the Earth's forests have been diminishing through the process of deforestation (Figure 9.22). At present, about 13 million hectares (32 million acres) of forest are lost each year, largely to agriculture. More than half of this area is in South America, Africa, and equatorial Asia. Much of the deforestation can be linked to economic development policies that promote cultivation of cash crops such as soybeans and palm oil. Loss of habitat in these species-rich areas takes a toll on the Earth's biodiversity. Slashing and burning of forest also releases carbon dioxide, and loss of evapotranspiration from trees leads to less rainfall and higher temperatures.

The island of Borneo, which is divided among three countries—Brunei, Indonesia, and Malaysia—is experiencing rapid deforestation and loss of biodiversity. Figure 9.23 shows how satellite monitoring of the island, combined with models of economic development, can be used to predict the course of future deforestation.

Savanna and Grassland Biomes

The savanna and grassland biomes of African safaris and Argentine gauchos support an important variety of grazing mammals and their predators, as well as the world's great migrating herds. Unfortunately, these lands are easily converted to agriculture when irrigated by water drawn from deep wells, and this practice extends the loss of these unique biomes.

SAVANNA BIOME

The **savanna biome** is usually associated with the tropical wet-dry climate ③ of Africa and South America. Its vegetation ranges from woodland to grassland. In *savanna woodland*, the trees are spaced rather widely apart because there is not enough soil moisture during the dry season to support a full tree cover (Figure 9.24). In the open spacing, a dense lower layer develops, which usually consists of grasses. The woodland has an open, parklike appearance. Savanna woodland usually lies in a broad belt adjacent to equatorial rainforest.

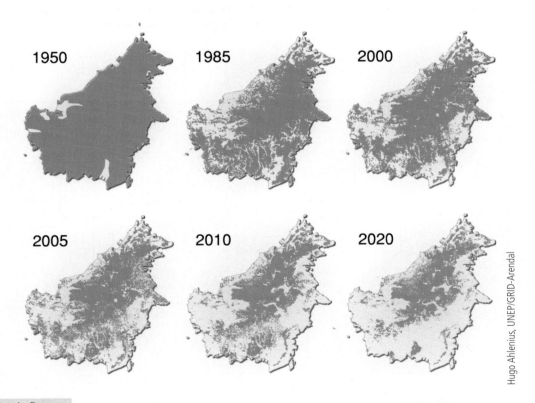

Hugo Ahlenius, UNEP/GRID-Arendal

9.23 Deforestation in Borneo

Both tropical lowland and highland forests of Borneo, containing a rich biodiversity of plants and animals, have decreased rapidly since 1950. Scientists from the United Nations (UN) used a combination of historic maps and aerial photography to create a base map of Borneo's original terrestrial ecosystems (shown in green). Next, they analyzed Landsat data to create a series of GIS maps delineating the extent of deforestation for a 60-year period. Field investigations showed that the burned, logged, and cleared areas, previously habitat for species such as orangutans and elephants, were converted to agricultural land, developed areas, or palm oil plantations (shown in light brown). The UN team made projections to 2020, based on national policies and agricultural land deeds.

9.24 Savanna

Grasses and drought-resistant trees occur in varying proportions throughout the savanna biome.

Peter Johnson/©Corbis

Annie Griffiths Belt/NG Image Collection

▲ **SAVANNA WOODLAND**

Where the trees are more closely spaced, we have savanna wood-land. This example is from the central Kalahari Desert, Botswana. The rich green of the landscape identifies the time of year as just after the rainy season.

▲ **THORNTREE-TALL-GRASS SAVANNA**

The long dry season of the tropical wet-dry climate ③ restricts the vegetation to grasses with an open canopy of drought-resistant trees, such as the acacia shown in this photo. The zebra, here in Serengeti National Park, Tanzania, is one of more than a dozen species of antelope that graze the savanna.

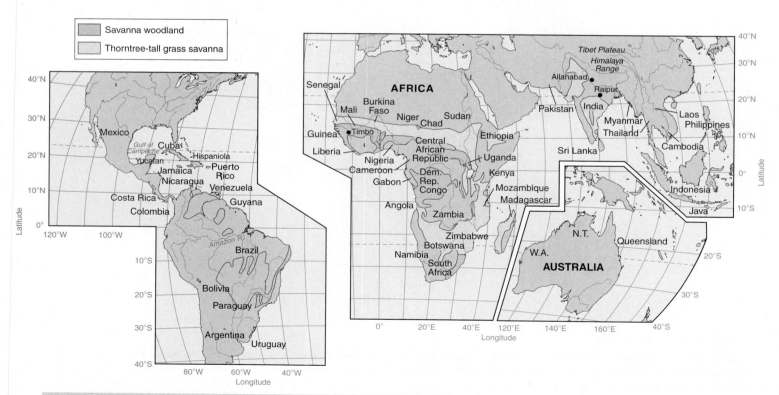

9.25 Global distribution of savanna woodland and thorntree-tall-grass savanna

Savanna and thorntree-tall-grass savanna occurs in South America to the north and south of the Amazon Basin, as well as in an African crescent, stretching from Senegal to Botswana. Thorntree-tall-grass savanna is also found in India.

In the tropical savanna woodland of Africa, the trees are of medium height. Tree crowns are flattened or umbrella-shaped, and the trunks have thick, rough bark. Some species of trees are xerophytic forms, adapted to the dry environment with small leaves and thorns. Others are broad-leafed deciduous species that shed their leaves in the dry season. In this respect, savanna woodland resembles monsoon forest.

Fires occur frequently in the savanna woodland during the dry season, but the tree species are particularly resistant to fire. Many geographers think that periodic burning of the savanna grasses keeps forest from invading the grassland. Fire doesn't kill the underground parts of grass plants; rather, it limits tree growth to fire-resistant species. So, many rainforest tree species that might otherwise grow in the wet-dry climate regime are suppressed by fires. Browsing animals also kill many young trees, helping maintain grassland, at the expense of forest.

The regions of savanna woodland are shown in Figure 9.25. In Africa, the savanna woodland grades into a belt of *thorntree-tall-grass savanna*, a formation class transitional to the desert biome (Figure 9.24). The trees are largely of thorny species, and are more widely scattered. The open grassland is more extensive than in the savanna woodland. One characteristic tree is the flat-topped acacia. Elephant grass, another common species of the woodland, can grow to a height of 5 m (16 ft) to form an impenetrable thicket.

> The savanna biome is adapted to a strong wet-dry annual cycle. Grazing by large mammals, and periodic burning in the dry season, maintain the openness of the savanna by suppressing tree seedlings.

Savanna biome vegetation is described as rain-green. That's because the thorntree-tall-grass savanna is closely identified with the semiarid subtype of the dry tropical ④s and dry subtropical ⑤s climates. In the semiarid climate, soil-water storage is only enough for plants during the brief rainy season. After rains begin, the trees and grasses quickly green-up. Vegetation of the monsoon forest is also rain-green.

The African savanna is widely known for the diversity of its large grazing mammals (Figure 9.26). With these

9.26 Animals of the African savanna

A diverse assortment of mammals populate African savannas.

Norbert Rosing/NG Image Collection

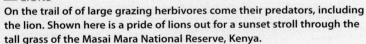
Michael Nichols/NG Image Collection

▲ **LIONS**
On the trail of of large grazing herbivores come their predators, including the lion. Shown here is a pride of lions out for a sunset stroll through the tall grass of the Masai Mara National Reserve, Kenya.

▲ **WILDEBEEST**
These strange-looking animals are actually antelopes. More than a dozen antelope species graze on the savanna. Each one has a particular preference for eating either the blade, sheath, or stem of the grasses. Grazing stimulates the grasses to continue to grow, and so the ecosystem is more productive when grazed than when left alone.

DanitaDelimont.com/NewsCom

SPOTTED HYENA ▶
Another savanna predator is the hyena; unlike the lion, which attacks its prey directly, the hyena runs it to exhaustion. Small packs of 6 to 12 animals do the hunting, employing different strategies for different antelope prey. They fear only the big cats, such as lions. Shown here is a hyena in the Ngorongoro Conservation Area, Tanzania, carrying a dead wildebeest calf in its vise-grip jaws.

9.27 Grasslands

Grasslands range from lush to sparse.

◀ **TALL-GRASS PRAIRIE**
In addition to grasses, tall-grass prairie vegetation includes many forbs, such as the wildflowers shown in this photo. The grasses are deeply rooted and form a thick and continuous turf.

Joel Sartore/NG Image Collection

STEPPE ▶
Buffalo grass and blue grama grass are typical of the American steppe, seen here at the Pawnee National Grassland, near Fort Collins, Colorado. There may also be some scattered shrubs and low trees near watercourses. Bare soil is often exposed between the low grasses and weeds.

James Steinberg/Photo Researchers, Inc.

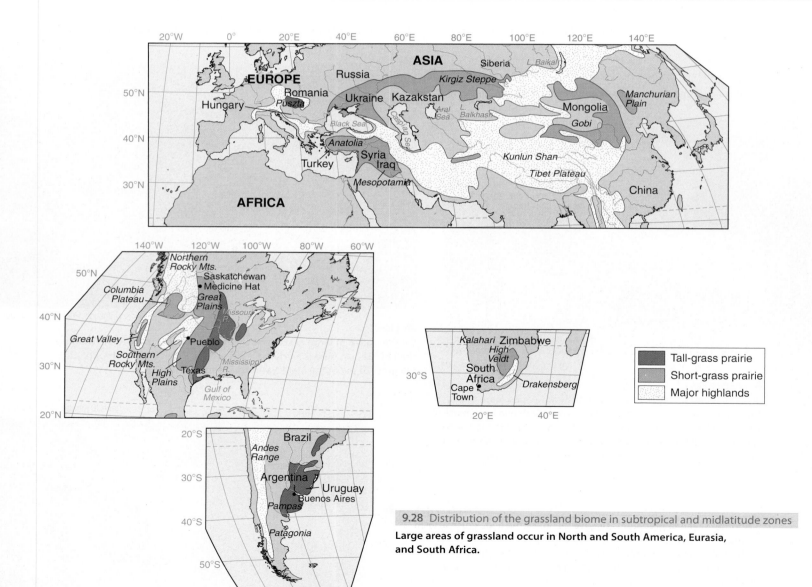

9.28 Distribution of the grassland biome in subtropical and midlatitude zones

Large areas of grassland occur in North and South America, Eurasia, and South Africa.

Legend:
- Tall-grass prairie
- Short-grass prairie
- Major highlands

grazers come a large variety of predators—lions, leopards, cheetahs, hyenas, and jackals. Elephants are the largest animals of the savanna and adjacent woodland regions.

GRASSLAND BIOME

The **grassland biome** includes two major formation classes: tall-grass prairie and steppe (Figure 9.27). *Tall-grass prairie* is a ground cover of tall grasses along with some broad-leafed herbs, named *forbs*. Like the savanna biome, grasslands are maintained by frequent burning, which kills trees and shrubs that might otherwise dominate the grasses. As a result, trees and shrubs are not found on the prairie, but they do occur in narrow bands and patches of forest in and along stream valleys.

Figure 9.28 shows the distribution of grassland around the world. Prairie grasslands are associated with the drier areas of moist continental climate ⑩, and steppe grasslands correspond well with the semiarid subtype of the dry continental climate ⑨s.

North American tall-grass prairies once stretched in a belt from the Texas Gulf coast to southern Saskatchewan, and extended eastward into Illinois. Today, they have been converted almost entirely to agricultural land. Another major area of tall-grass prairie is the Pampa region of South America, which occupies parts of Uruguay and eastern Argentina. The Pampa region falls into the moist subtropical climate ⑥ with mild winters and abundant precipitation.

Steppe, or *short-grass prairie*, consists of sparse clumps of short grasses. Steppe grades into semidesert in dry environments and into prairie where rainfall is higher. Steppe grassland is concentrated largely in the midlatitude areas of North America and Eurasia.

The animals of the grassland are distinctive, and feature many grazing mammals. The grassland ecosystem supports some rather unique adaptations to life (Figure 9.29). Animals such as jackrabbits and jumping mice have learned to jump or leap, to gain an unimpeded view of their surroundings. Another leaper, the

9.29 Grassland animals

Grassland animals are well-adapted to home environments of short and tall grasses.

Raymond Gehman/NG Image Collection

◄ **AMERICAN BISON**
Prairie grasslands are the home of many types of grazing animals, including the American bison, also known as the buffalo. Once extremely widespread throughout the Great Plains, it was hunted to near extinction in the nineteenth century. The animals here are part of a managed herd in Kentucky. Other prairie grazers include the elk and the pronghorn antelope.

Bates Littlehales/NG Image Collection

▲ **PRAIRIE DOG**
Many animals burrow into the prairie soil for shelter, such as the prairie dog. These highly social animals live in colonies, or "dog-towns," home to hundreds of animals. They feed on grasses, forbs, and insects.

Joel Sartore/NG Image Collection

▲ **JACKRABBIT**
The jackrabbit is a common grazer of the prairies and steppes. It has developed a leaping capability, as have the pronghorn and jumping mouse, that enables it to see above the grass as it moves.

pronghorn, combines its leap with great speed, which enables it to avoid predators and fire. Many grassland animals also burrow, because the soil provides the only shelter in the exposed landscape. Rodents, including prairie dogs, gophers, and field mice all burrow; and rabbits exploit old burrows, using them for nesting or shelter. Invertebrates too seek shelter in the soil, and many are adapted to living within the burrows of rodents, where extremes of moisture and temperature are substantially moderated.

> The grassland biome includes tall-grass and short-grass prairie (steppe). Tall-grass prairie provides rich agricultural land suited to cultivation and cropping. Short-grass prairie occupies vast regions of semidesert and is suited to grazing.

Desert and Tundra Biomes

The desert is a highly evolved ecosystem that supports a multitude of plants and animals. Insects, reptiles, mammals, and birds can occasionally be spotted at night when the Sun ceases its unfiltered radiation of the sparse vegetation. Rare and fantastic plants may flower after many decades, when sufficient rain finally falls, triggering the germination of long-dormant seeds. And lucky are those who see them, for these desert blooms may last only a few days, or weeks.

DESERT BIOME

The **desert biome** includes several formation classes that are transitional from grassland and savanna biomes into

9.30 Global distribution of the desert biome

This world map of the desert biome includes desert and semidesert formation classes.

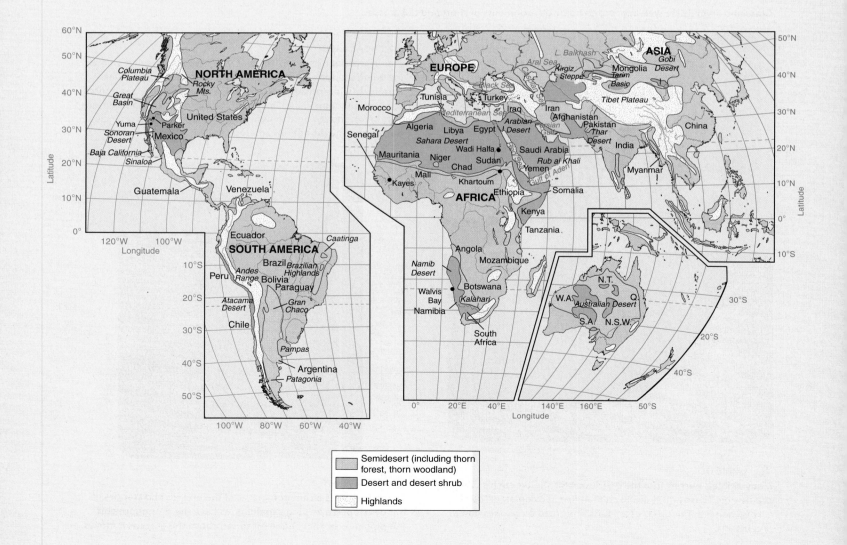

Semidesert (including thorn forest, thorn woodland)

Desert and desert shrub

Highlands

9.31 Life of the desert biome

Desert plants and animals have evolved to make careful use of scarce water resources.

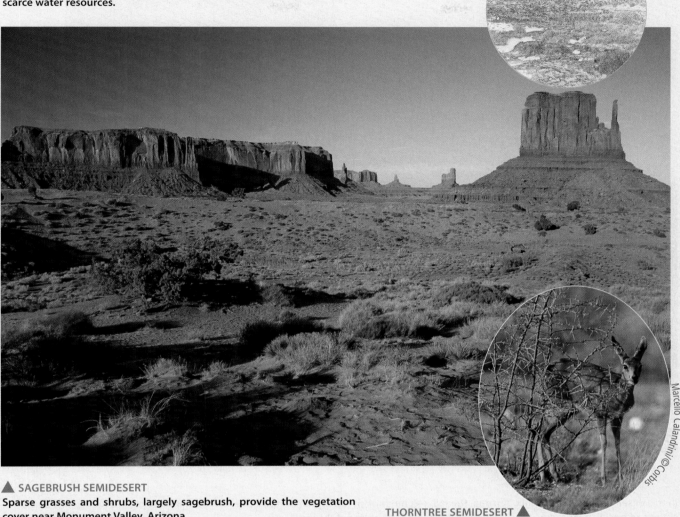

Rich Reid/NG Image Collection

Marcello Calandini/©Corbis

▲ **SAGEBRUSH SEMIDESERT**
Sparse grasses and shrubs, largely sagebrush, provide the vegetation cover near Monument Valley, Arizona.
EYE ON THE LANDSCAPE **What else would the geographer see?**
These tall, columnar landforms are buttes Ⓐ. Mesas are larger, isolated rock platforms Ⓑ. Here in Monument Valley, they are formed by a thick layer of uniform sandstone, which breaks up into rectangular blocks as it weathers. The blocks fall away, leaving a vertical cliff face behind.

THORNTREE SEMIDESERT ▲
The thorntree semidesert formation is found in tropical climates with very long dry seasons and short, but intense, rainy seasons. It consists of a sparse vegetation cover of grasses and thorny shrubs, which are dormant for much of the year. This photo shows a steenbok (a small African antelope) in Etosha National Park, Namibia.

---

vegetation of the arid desert. Our map of the desert biome (Figure 9.30) recognizes two basic formation classes: semidesert and dry desert.

Semidesert is a transitional formation class found in a wide latitude range, from the tropical zone to the midlatitude zone (Figure 9.31). It is identified primarily with

> The desert biome includes semidesert and dry desert and occupies the tropical, subtropical, and midlatitude dry climates (④, ⑤, and ⑨). Desert plants vary widely in appearance and in adaptation to the dry environment.

the arid subtypes of all three dry climates. Semidesert consists primarily of sparse xerophytic shrubs. One example is the sagebrush vegetation of the middle and southern Rocky Mountain region and Colorado Plateau. In recent times, overgrazing and trampling by livestock have caused semidesert shrub vegetation to expand widely into areas of the western United States that used to be steppe grasslands.

Thorntree semidesert of the tropical zone is made up of xerophytic trees and shrubs that are adapted to a climate with a very long, hot dry season and only a very brief, but intense, rainy season. We find these conditions

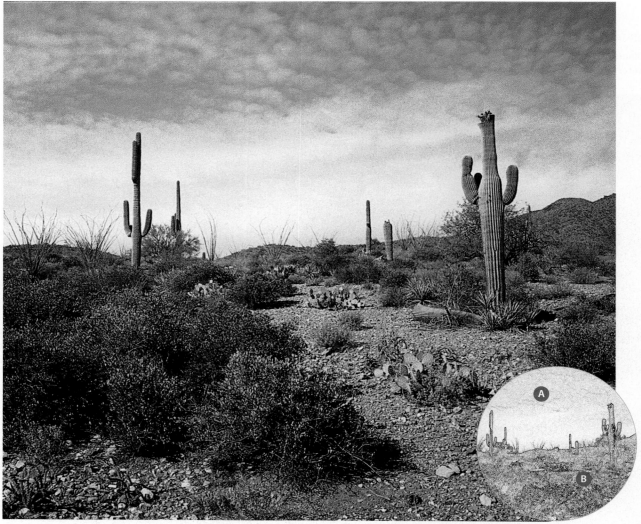

Alan Strahler

9.32 Desert plants

In this desert scene near Phoenix, Arizona, the tall, columnar plant is a saguaro cactus; the delicate wandlike plant is an ocotillo. Small clumps of prickly pear cactus are seen between groups of hard-leafed shrubs.

EYE ON THE LANDSCAPE What else would the geographer see?
This sky of puffy altocumulus clouds and layered altostratus **A** indicates moisture at higher levels in the troposphere. The photo, taken in July, reflects conditions of the local monsoon season, in which moist air from the Gulf of California moves into southern Arizona, often generating intense thunderstorms and flash floods. The gravel-covered ground surface **B** is characteristic of deserts, where wind and storm runoff remove fine particles and leave coarser rock fragments behind.

in the semiarid and arid subtypes of the dry tropical ④ and dry subtropical ⑨ climates. The thorny trees and shrubs are known locally as thorn forest, thornbush, or thornwoods. In some places, cactus plants are abundant.

Dry desert is a formation class of plants that are widely dispersed over the ground. It consists of small, hard-leafed,

or spiny shrubs, succulent plants (such as cactus), and/or hard grasses. Many species of small annual plants appear only after rare and heavy downpours. In fact, many of the areas mapped as desert vegetation have no plant cover at all because the surface consists of shifting dune sands or sterile salt flats.

9.33 Desert animals

Small desert mammals live mostly out of sight in underground burrows, surviving on a low water budget. Many are noctural, avoiding the hot part of the day. Shown in the background is a scene from the Kalahari Desert, featuring the endangered quiver tree, a species of aloe, on a landscape of fractured red rock.

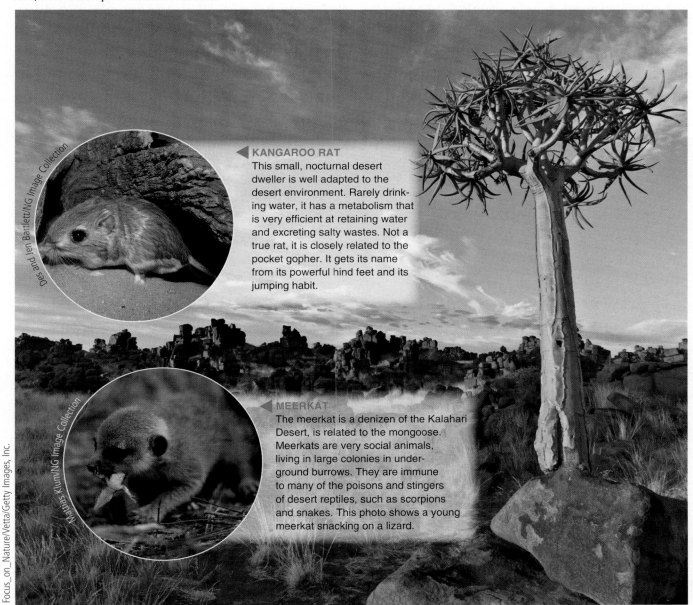

KANGAROO RAT
This small, nocturnal desert dweller is well adapted to the desert environment. Rarely drinking water, it has a metabolism that is very efficient at retaining water and excreting salty wastes. Not a true rat, it is closely related to the pocket gopher. It gets its name from its powerful hind feet and its jumping habit.

MEERKAT
The meerkat is a denizen of the Kalahari Desert, is related to the mongoose. Meerkats are very social animals, living in large colonies in underground burrows. They are immune to many of the poisons and stingers of desert reptiles, such as scorpions and snakes. This photo shows a young meerkat snacking on a lizard.

Desert plants around the world look very different from each other. In the Mojave and Sonoran deserts of the southwestern United States, for example, plants are often large, giving the appearance of a woodland (Figure 9.32).

Desert animals, like the plants, are typically adapted to the dry conditions of the desert (Figure 9.33).

Important herbivores in American deserts include kangaroo rats, jackrabbits, and grasshopper mice. Insects are abundant, as are, not surprisingly, insect-eating bats and birds such as the cactus wren. Reptiles, especially lizards, are also common.

TUNDRA BIOME

Arctic tundra is a formation class of the **tundra biome**, with a tundra climate ⑫. In this climate, plants grow during the brief summer of long days and short (or absent) nights. At this time of year, air temperatures rise above freezing, and a

> The tundra biome includes low plants that are adapted to survival through a harsh, cold winter. They grow, bloom, and set seed during a short summer thaw.

shallow surface layer of ground ice thaws. The permafrost beneath, however, remains frozen, keeping the meltwater at the surface. These conditions produce a marshy environment for a short time over wide areas. Because plant remains decay very slowly within the cold meltwater, layers of organic matter build up in the marshy ground. Frost action in the soil fractures and breaks large roots, keeping tundra plants small (Figure 9.34). In winter, wind-driven snow and extreme cold injure plant parts that project above the snow.

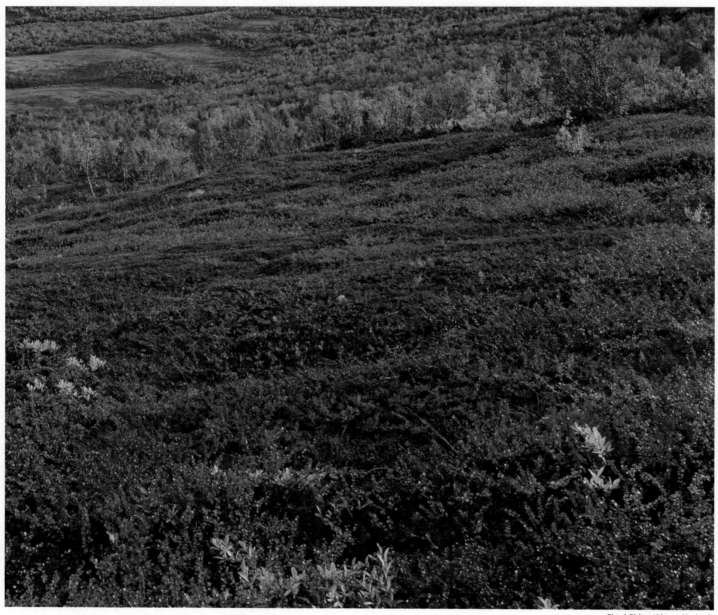

Chad Ehlers/Alamy Limited

9.34 Arctic tundra in Lapland

Found in areas of extreme winter cold, with little or no true summer, arctic tundra plant life consists of low perennial grasses, sedges, herbs, and dwarf shrubs, accompanied by lichens and mosses. This photo shows tundra in fall colors in Lapland, Finland. In the background, a sparse stand of trees grows in a sheltered spot.

Tundra vegetation is also found at high elevations, above the limit of tree growth and below the vegetation-free zone of bare rock and perpetual snow. This *alpine tundra* resembles arctic tundra in many physical respects.

Although only a few plant and animal species are suited to the tundra, they are often represented by a large number of individuals (Figure 9.35). The food web of the tundra ecosystem is simple and direct.

The important producer is reindeer moss, the lichen *Cladonia rangifera*. Caribou, reindeer, lemmings, ptarmigan (arctic grouse), and snowshoe rabbits all graze on this lichen. Foxes, wolves, and lynxes prey on those animals, although they may all feed directly on plants, as well. During the summer, the abundant insect population helps support the migratory waterfowl populations.

9.35 Animals of the tundra

◀ CARIBOU
Barren-ground caribou roam the tundra, constantly grazing the lichens and plants of the tundra and boreal zone. They migrate long distances between calving and feeding grounds.

▼ MUSK OXEN
These wooly tundra-grazers are more closely related to goats than cattle. They feed on grasses, sedges, and other ground plants, scratching their way through the snow to find them in winter. Hunted close to extinction, they are now protected. Originally restricted to Alaska, Canada, and Greenland, they have been introduced to northern Europe.

Michael S. Quinton/NG Image Collection

▼ SANDPIPER
This small migratory bird, a dunlin sandpiper, travels long distances to return to the tundra to nest and fledge its young. It probes the tundra with its sensitive beak, searching for insects. The boggy tundra presents an ideal summer environment for many other migratory birds, such as waterfowl and plovers.

Joel Sartore/NG Image Collection

Joel Sartore/NG Image Collection

Mapping Global Land Cover by Satellite

Imagine yourself as an astronaut living on an orbiting space station, watching the Earth revolve underneath you. One of the first things that would strike you about the land surface is its color, and how it changes from place to place and time to time. Deserts are in shades of brown, dotted with white salty playas and the black spots and streaks of recent volcanic activity. Equatorial forests are green and lush, dissected by branching lines of dark rivers. Shrublands are marked by earth colors, but with a greenish tinge.

Some regions show substantial change throughout the year. In the midlatitude zone, forests and agricultural lands go from intense green in the summer to brown, as leaves drop and crops are harvested. Snow expands toward the Equator in the fall and winter and retreats poleward in the spring and summer. In the tropical zones, grasslands and savannas go from brown to green and to brown again as the rainy season comes and goes. Some features, such as lakes, remain nearly unchanged throughout the year.

Ever since the first satellite images of the Earth were received, scientists have used color—and the change of color with time—as an indicator of land-cover type. For example, there is a 30-year history of producing land-cover maps for local areas using individual Landsat images from cloud-free dates. But global-scale mapping of land cover requires instruments like MODIS that can observe the surface on a daily or near-daily basis to acquire cloud-free images of regions that are frequently cloud-covered.

Figure 9.36 is a map of global land cover produced from MODIS images for the year 2005. The legend recognizes 17 types of land cover, including forests, shrublands, savannas, grasslands, and wetlands. The global pattern of land-cover types is rather similar to that shown in Figure 9.7. Evergreen broadleaf forest dominates the equatorial belt, stretching from South America through Central Africa to south Asia. Adjoining the equatorial forest belt are regions of savanna

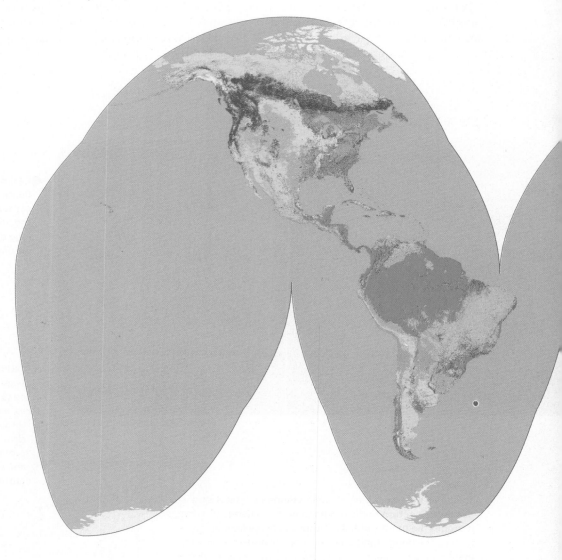

9.36 Global land cover from MODIS

This map of global land-cover types was constructed from MODIS data acquired during 2005. The map has a spatial resolution of 1 km²—that is, each square kilometer of the Earth's land surface is independently assigned a land-cover type label.

and grassland, which have strong wet-dry climates. The vast desert region running from the Sahara to the Gobi is barren or sparsely vegetated. It is flanked by grasslands on the west, north, and east. Broadleaf deciduous forests are prominent in eastern North America, western Europe, and eastern Asia. Evergreen needleleaf forests span the boreal zone from Alaska and northwest Canada to Siberia. Croplands are found throughout most regions of human habitation, except for dry desert regions and cold boreal zones.

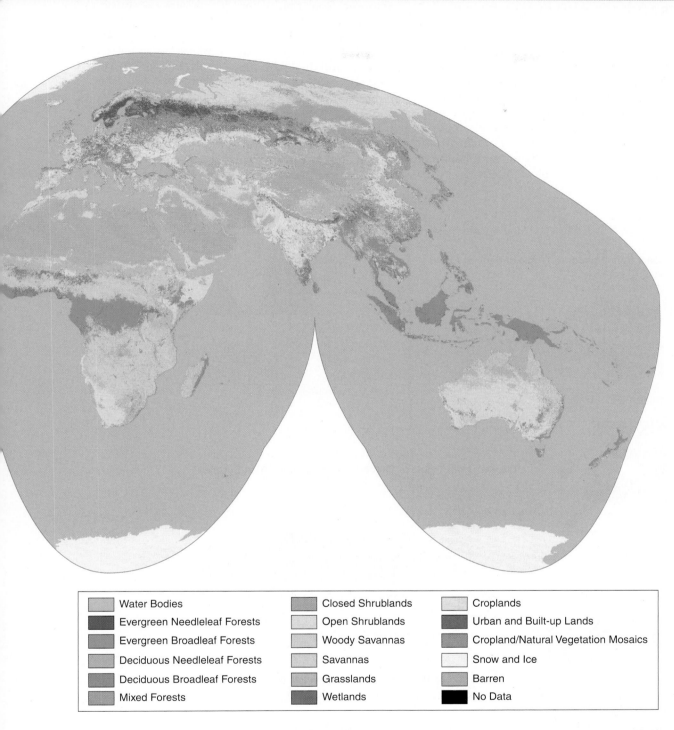

Water Bodies	Closed Shrublands	Croplands
Evergreen Needleleaf Forests	Open Shrublands	Urban and Built-up Lands
Evergreen Broadleaf Forests	Woody Savannas	Cropland/Natural Vegetation Mosaics
Deciduous Needleleaf Forests	Savannas	Snow and Ice
Deciduous Broadleaf Forests	Grasslands	Barren
Mixed Forests	Wetlands	No Data

The MODIS map was constructed from both color (spectral) and time-based (temporal) information using a process called classification. In short, a computer program is presented with many images of each land-cover type. It then "learns" the examples and uses them to classify pixels depending on their spectral and temporal pattern. The MODIS global land-cover map shown was prepared with more than 2500 examples of the 17 land-cover types. It is estimated to be 75 to 80 percent accurate.

Land-cover mapping is a common application of remote sensing. Given the capability of spaceborne instruments to image the Earth consistently and repeatedly, classification of remotely sensed data is a natural way of extending our knowledge from the specific to the general to provide valuable new geographic information.

Climate and Altitude Gradients

CLIMATE GRADIENTS AND BIOME TYPES

As we have seen, biomes and formation classes change along with climate. Figure 9.37 shows three continental transects that illustrate this principle.

The upper transect stretches from the Equator to the Tropic of Cancer in Africa. Across this region, climate ranges through all four low-latitude climates: wet equatorial ①, monsoon and trade-wind coastal ②, wet-dry tropical ③, and dry tropical ④. Vegetation grades from equatorial rainforest, savanna woodland, and savanna grassland to tropical scrub and tropical desert.

The middle transect is a composite from the Tropic of Cancer to the Arctic Circle in Africa and Eurasia. Climates include many of the mid- and high-latitude types: dry subtropical ⑤, Mediterranean ⑦, moist continental ⑩, boreal forest ⑪, and tundra ⑫. The vegetation cover grades from tropical desert through subtropical steppe to sclerophyll forest in the Mediterranean. Farther north is the

> Because climate factors of temperature and precipitation vary with elevation and over space, vegetation patterns often show zonation with altitude and systematic variation on long transects.

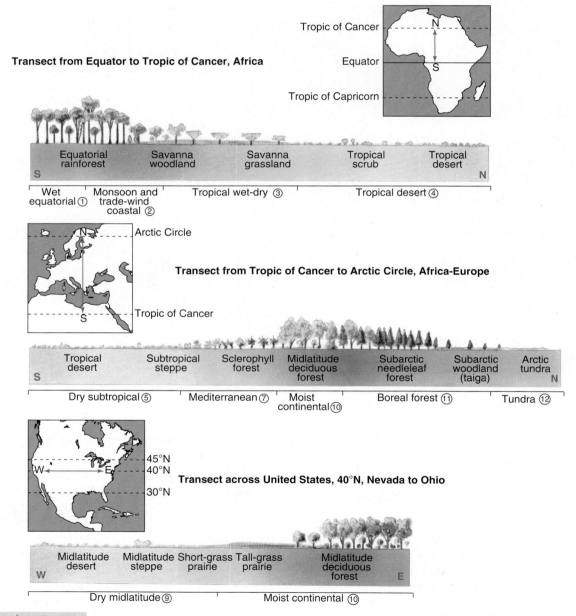

Transect from Equator to Tropic of Cancer, Africa

Transect from Tropic of Cancer to Arctic Circle, Africa-Europe

Transect across United States, 40°N, Nevada to Ohio

9.37 Vegetation transects

Three continental transects show the sequence of plant formation classes across climatic gradients. Effects of mountains or highland regions are not shown.

midlatitude deciduous forest in the region of moist continental climate ⑩, which grades into boreal needleleaf forest, subarctic woodland, and, finally, tundra.

The lower transect ranges across the United States, from Nevada to Ohio. On this transect, the climate begins as dry midlatitude ⑨. Precipitation gradually increases eastward, reaching moist continental ⑩ near the Mississippi River. The vegetation changes from midlatitude desert and steppe to short-grass prairie, tall-grass prairie, and midlatitude deciduous forest.

ALTITUDE GRADIENTS

Climate varies not only with latitude and longitude, but also with elevation. At higher elevations, temperatures are cooler, and precipitation is usually greater. This can produce a zonation of ecosystems with elevation that resembles a poleward transect.

The vegetation zones of the Colorado Plateau region in northern Arizona and adjacent states provide a striking example of altitude zonation. Figure 9.38 shows a

9.38 Altitude zones of vegetation near Grand Canyon, Arizona

With altitude comes cooler temperatures and increased precipitation. Plant species change along with the climate.

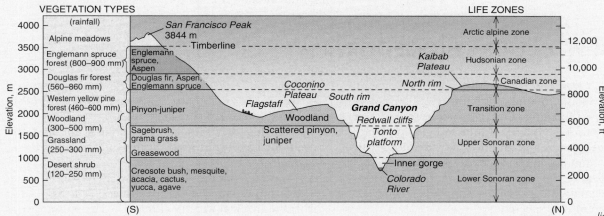

▲ **LIFE ZONES OF THE COLORADO PLATEAU**

A series of life zones identified by biogeographers describes the zonation of ecosystems with altitude in this region. They are named for geographic regions that have similar vegetation, from the Sonoran Desert to Hudson's Bay.

TRANSITION ZONE ▶

At intermediate elevations, we find a woodland of pinyon pine and juniper, with sparse grasses between the trees.

Jim Zipp/Photo Researchers, Inc.

Ethan Welty / Aurora Creative/Getty Images, Inc.

U.S. Forest Service

▲ **ARCTIC-ALPINE LIFE ZONE**

At the highest elevations we find an alpine tundra of sparse low plants and angular rock fragments.

LOWER SONORAN LIFE ZONE ▶

Cactus is often part of desert shrub communities near the floor of the Grand Canyon.

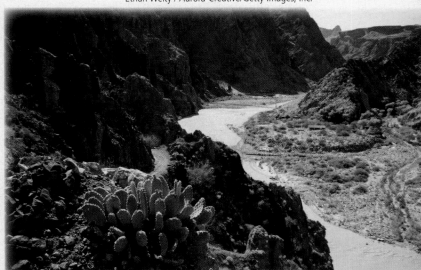

cross section of the land surface in this region, along with photos of typical vegetation types you might see at different elevations. Rainfall ranges from about 120 to 900 mm (about 5 to 35 in.) as elevation increases from about 700 m (about 2300 ft) at the bottom of the Grand Canyon to 3844 m (12,608 ft) at the top of San Francisco Peak. Temperature also decreases substantially with elevation.

Noting the variation in vegetation cover with altitude, biogeographers working in this region developed a series of *life zones* to refer to these cover types. They relate the appearance of the vegetation cover to regions you might encounter on a transect from Mexico to the Arctic Ocean.

A Look Ahead

As we've seen, global and continental patterns of biomes and formation classes are strongly related to corresponding patterns of climate. The key ingredients of climate are temperature, moisture, and the variation of temperature and moisture through the year. These same factors are also important to the formation of soils, which is the subject of the next chapter. Vegetation cover also influences soil formation. For example, soils developed on grasslands are very different from those developed under conifer forests. Other important determinants of soil formation are the nature of the soil's parent material, as it is derived from weathered rock, and the time allowed for soil formation to proceed.

IN REVIEW GLOBAL BIOGEOGRAPHY

- The *rainforest ecosystem* can rapidly rebound from the low-intensity, slash-and-burn agriculture of native peoples, but it is also under threat by deforestation and conversion of forest to agricultural and grazing land.
- **Natural vegetation** is a plant cover that develops with little or no human interference. Although much vegetation appears to be in a natural state, humans influence the vegetation cover by fire suppression and the introduction of new species.
- The **life-form** of a plant refers to its physical structure, size, and shape. Life-forms include *trees, shrubs, lianas, herbs,* and *lichens.*
- The largest unit of terrestrial ecosystems is the **biome**: forest, grassland, savanna, desert, and tundra. Within biomes are vegetation formation classes. At the global and continental scales, the distribution of biomes and formation classes is determined by climate.
- The **forest biome** includes six important forest formation classes. The *low-latitude rainforest* exhibits a dense canopy and open floor, and supports a very large number of species. *Monsoon forest* is largely deciduous, with most species shedding their leaves after the wet season. *Subtropical evergreen forest* occurs in *broadleaf* and *needleleaf* forms in the moist subtropical climate ⑥.
- *Midlatitude deciduous forest* is associated with the moist continental climate ⑩. Its species shed their leaves before the cold season. *Needleleaf forest* consists largely of evergreen conifers. It includes the *boreal forest* of high latitudes and the *coastal forest* of

the Pacific Northwest. *Sclerophyll forest,* composed of trees with small, hard, leathery leaves, is found in the Mediterranean climate ⑦ region.
- The Earth's forests are slowly shrinking due to deforestation, largely through cutting and conversion to agriculture.
- The **savanna biome** consists of widely spaced trees with an understory, often of grasses. It is associated with the tropical wet-dry climate ③. Dry-season fire is frequent in the savanna biome, limiting the number of trees but encouraging the growth of grasses.
- The **grassland biome** of midlatitude regions includes *tall-grass prairie* in moister environments and *short-grass prairie,* or **steppe**, in semiarid areas. Like the savanna biome, it is partly maintained by fire. Most of the tall-grass prairie is now agricultural land.
- Vegetation of the **desert biome** ranges from a *semidesert* of xerophytic or thorny shrubs and small trees to a *dry desert,* home to species adapted to the driest of environments.
- **Tundra biome** vegetation is limited largely to low herbs and a few shrubs that are adapted to the severe drying cold and frost action experienced on the fringes of the Arctic Ocean.
- Continental transects clearly demonstrate how patterns of climate are related to patterns of biomes and formation classes.
- Time series of satellite images of the Earth's land surface can render maps of land cover at global and continental scales.

KEY TERMS

natural vegetation, p. 305
life-form, p. 307
forest, p. 308

terrestrial
    ecosystems, p. 308
biome, p. 308

forest biome, p. 309
savanna biome, p. 309
grassland biome, p. 309

desert biome, p. 309
tundra biome, p. 309
Steppe, p. 327

REVIEW QUESTIONS

1. How do traditional agricultural practices in the low-latitude rainforest compare to present-day practices? What are the implications for the rainforest environment?
2. What global regions of the *rainforest ecosystem* are most threatened by deforestation?
3. What is **natural vegetation?** How do humans influence vegetation?
4. Define and differentiate the following terms: *tree, shrub, herb, liana, perennial, deciduous, evergreen, broadleaf, needleleaf, forest, woodland.*
5. What are the five main **biome** types that ecologists and biogeographers recognize? Describe each briefly.
6. *Low-latitude rainforests* occupy a large region of the Earth's land surface. What are the characteristics of these forests? Include forest structure, types of plants, diversity, and climate in your answer.
7. *Monsoon forest* and *midlatitude deciduous forest* are both deciduous but for different reasons. Compare the characteristics of these two formation classes and their climates.
8. *Subtropical broadleaf evergreen forest* and *tall-grass prairie* are two vegetation formation classes that have been greatly altered by human activities. How has this been done, and why?
9. Distinguish among the types of *needleleaf forest*. What characteristics do they share? How are they different? How do their climates compare?
10. Which type of forest, with related woodland and scrub types, is associated with the Mediterranean climate? What are the features of these vegetation types? How are they adapted to the Mediterranean climate?
11. Describe the formation classes of the **savanna biome**. Where is this biome found, and in what climate types? What role does fire play in the savanna biome?
12. Compare the two formation classes of the **grassland biome**. How do their climates differ?
13. Describe the vegetation types of the **desert biome**.
14. What are the features of *arctic* and *alpine tundra*? How does the cold tundra climate influence the vegetation cover?

VISUALIZING EXERCISES

1. Forests often contain plants of many different life-forms. Sketch a cross section of a forest including typical life-forms, and identify them with labels.
2. Figure 9.6 shows a triangular diagram of the relationships among natural vegetation, temperature, precipitation, and latitude. Make a copy of the figure, and indicate the climate types by name and number associated with the vegetation types in each block. Consult the global maps of climate, vegetation, and transects shown in Figure 9.36.
3. How does elevation influence vegetation? Sketch a hypothetical mountain peak in the southwestern American desert that rises from a plain at about 500 m (about 1600 ft) to a summit at about 4000 m (about 13,000 ft), and label the vegetation zones you might expect to find on its flanks.

ESSAY QUESTIONS

1. Figure 9.36 presents a vegetation transect from Nevada to Ohio. Expand the transect on the west so that it begins in Los Angeles. On the east, extend it northeast from Ohio through Pennsylvania, New York, western Massachusetts, and New Hampshire, to end in Maine. Sketch the vegetation types in your additions, and label them, as in the diagram. Below your vegetation transect, draw a long bar subdivided to show the climate types.
2. Construct a similar transect of climate and vegetation from Miami to St. Louis, Minneapolis, and Winnipeg.

Chapter 10
Global Soils

Long pursued as a folk remedy for its medicinal properties, the ginseng plant is cultivated in loose, well-drained, organic-rich soils that are much like the forest soils of its natural habitat. A low-light environment is required, and soon these rows will be covered with light-filtering canopies. After two to three years of growth, the roots are harvested and preserved for sale and use. High-value crops like ginseng can justify the cost of special soil preparation and growing conditions, but for other crops, the available soils and local climate determine what can be planted and grown.

PLANTING GINSENG IN SOUTH KOREA

©Yann Arthus-Bertrand/Altitude

Global Soils

T he soil layer supports the world's natural vegetation, land animals, and, through agriculture, its human population. How is soil formed? What processes break up and alter the minerals of surface material? How do soils develop distinctive layers, known as horizons? How does climate affect soil development? What are the major types of world soils, and where are they found? What impact will global change have on agriculture? These are some of the questions we will answer in this chapter.

Global Change and Agriculture

For the remainder of the twenty-first century, and probably well beyond, our global climate will change. The Earth will become warmer, especially in mid- and high latitudes. Most areas will have more precipitation, although higher temperatures will often bring more summer drought stress. Extreme events—heavy rainfalls and high winds—will be more frequent. How will global climate change impact agriculture (Figure 10.1)?

In general, higher temperatures will increase the productivity of most mid- and high-latitude crops by lengthening their growing seasons. But once temperatures get too high, the effects will turn negative, as heat stress restricts growth. Global warming is also expected to lower minimum temperatures. This will be beneficial for many mid- and high-latitude crops but detrimental to some tropical and equatorial crops.

A significant CO_2 fertilization effect will be felt, as well, as atmospheric CO_2 concentrations double. Because CO_2 is present in the atmosphere only in low concentrations, CO_2 limits photosynthesis in many situations. At higher CO_2 concentrations, plants become more productive. This fertilization effect is well documented in many studies of plants growing in greenhouses with increased CO_2 concentrations, and in studies of agricultural fields downwind from free-air release of CO_2 gas. What are the combined effects of rising temperatures and higher levels of CO_2 on crop yields? Recent studies show that, in general, if warming is less than about 2.5°C (4.5°F), yields will increase; but if greater, they will decrease. That said, the predicted changes for different crops and regions vary widely.

An important factor affecting yields is adaptation. Adaptation describes actions that respond to climate change, such as altering planting and harvesting dates, selecting different strains of crops, varying the crops that are planted, adding or modifying irrigation, and using more or different fertilizers and pesticides. Studies show that adaptation can mitigate many, but not all, of the negative effects of predicted changes in temperature and rainfall.

Other factors are also important. Addressing soil degradation, which includes erosion of productive soil layers and chemical depletion of nutrients, is a major challenge. Although the amount of irrigated land is increasing, so is irrigated land that is degraded by salt accumulation and waterlogging. Soil erosion will become more widespread as more frequent high-rainfall events wash soil from fields into nearby creeks and rivers. Wind deflation of productive soils will also become more common if global warming increases wind speeds. Insect pests will migrate as climate changes, infesting new regions but possibly deserting old ones. Weeds are stimulated by warm temperatures and higher levels of CO_2, and in some cases, will compete more effectively with crops. For livestock, climate change will affect the nature and availability of forage and grain, as well as animal diseases and parasites.

What are the likely effects of global climate change on food availability and prices? So far, agricultural production has been able to ramp up to meet the needs of the world's expanding population. But will it be able to keep pace with demand? Most studies predict that with global warming greater than 2.5°C (4.5°F), demand will exceed supply and food prices will rise. Vulnerable populations, such as marginal farmers and poor urban dwellers, will be at greater risk of hunger. On the other hand, if adaptation to climate change is effective, these risks will be minimized.

Agriculture is a human activity that is essential to our

> Initially, global warming will generally enhance crop production. But as temperatures continue to rise, the effects on agriculture will be decidedly negative—though adaptation will mitigate some of the effects.

10.1 Agriculture and climate change

Climate change will ultimately affect crop yields and cause crop-growing regions to shift to new locations.

Russ Munn/AGStockUSA/Alamy Limited

▲ YOUNG CORN FIELD, IOWA

Climate change will impact how crops grow. Earlier seasonal warming will allow a longer growing season, but hotter summers with higher drought potential may limit later growth. However, increasing atmospheric CO_2 will tend to increase growth potential.

Leonard Lee Rue III/Bruce Coleman, Inc./Photoshot Holdings Ltd.

▲ SOIL EROSION

Soil erosion, which strips off the most fertile part of the soil, makes it challenging to increase crop production and yield. More frequent and intense rainfall events, which are predicted for midlatitude summers, will escalate soil loss by rilling and gullying unprotected surfaces.

▼ VEGETABLE VENDOR, CAIRO, EGYPT

Climate change will affect the availability, and hence the price, of food from region to region. With global warming estimated at greater than 2.5°C (4.5°F), most studies predict that, at present levels, future supply will not meet future demand, and thus prices will rise.

Charlie Waite/Getty Images, Inc.

Andrey Rudakov/ Bloomberg via Getty Images, Inc.

▲ CEREAL CROP HARVEST, KRASNODAR REGION, RUSSIA

As climates change, so will crop yields; exactly how will depend on the crop, the region, and the degree of adaptation employed by growers.

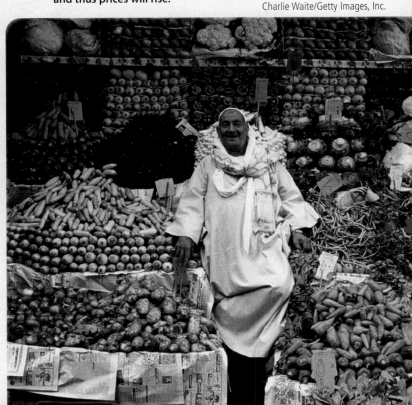

species, and as the climate heats up and CO_2 levels rise, the effects of global warming on agriculture will be undeniable. Let's hope that we are able to adapt to the changes, and smart enough to figure out how to provide enough food for us all.

The Nature of the Soil

This chapter is devoted to soil systems. **Soil** is the uppermost layer of the land surface that plants use and depend on for nutrients, water, and physical support. Geographers are keenly interested in the differences in soils from place to place around the globe, especially agricultural soils (Figure 10.2). The capability of the soils and climate of a region to produce food largely determines the size of the population it will support. In spite of the growth of cities, most of the world's inhabitants still live close to the soil that furnishes their food. And many of those same inhabitants die prematurely

when the soil cannot furnish enough food for all, either through loss of soil productivity or by damaging effects of climate change.

Soils are influenced by factors and processes that can be very different from place to place. As a result, the soils themselves can vary greatly from continent to continent, region to region, and even field to field. Soil scientists recognize five important factors that influence soil development:

- *Climate:* Temperature and precipitation
- *Organisms:* Plants and animals that live in and on the soil, including humans
- *Relief:* The slope of the land surface and the direction that the slope faces toward the Sun
- *Parent material:* The type of rock or sediment from which the soil develops
- *Time:* The length of time that soil-forming processes have been at work

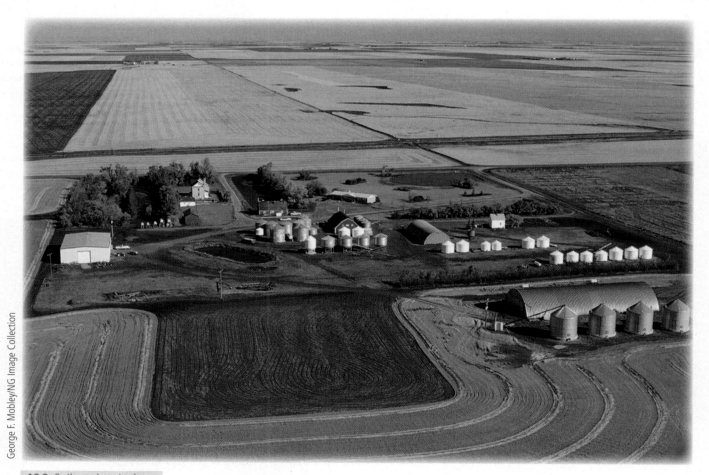

George F. Mobley/NG Image Collection

10.2 Soils and agriculture

North American prairies are famous for their fertile soils. This farm in Grand Coulee, near Regina, Saskatchewan, specializes in growing crop seeds.

In this chapter we'll look at these factors and others in more detail in order to understand the many different types of soil found over the globe.

INTRODUCING THE SOIL

Soil includes matter in all three states—solid, liquid, and gas. Because the matter in these states is constantly changing and interacting through chemical and physical processes, soil is a very dynamic layer.

Soil characteristics develop over a long period of time through a combination of many processes acting together. Physical processes break down rock fragments into smaller and smaller pieces. Chemical processes alter the mineral composition of the original rock, producing new minerals. Taken together, these physical and chemical processes are referred to as *weathering*.

Over time, weathering processes soften, disintegrate, and break apart bedrock, forming a layer of *regolith* (Figure 10.3). Other kinds of regolith are mineral particles transported by streams, glaciers, waves, and water currents, or winds. For example, dunes formed of sand transported by wind are a type of regolith on which soil can form.

Soil development starts with the **parent material**, the mineral matter available to make up the soil. Normally this matter consists of regolith formed from underlying rock or sediments transported to the location. Because different minerals weather at different rates to form different products, the composition of the parent material

can be quite important in determining the nature of the soil that develops at a site.

In addition to mineral matter, the soil contains *organic matter*, which includes both live and dead plants, animals, microorganisms, and organic materials that remain after weathering. Soil scientists use the term *humus* (pronounced "hew-muss") to describe the finely divided, partially decomposed, organic matter in soils. Some humus rests on the soil surface and some is carried into soil layers by downward-percolating rainfall. Humus particles give the soil a brown or black color.

We also find air and water in soil. Water contains high levels of dissolved substances, such as nutrients. Because of the activities of plant roots and microorganisms, air in soils often has high levels of carbon dioxide or methane and low levels of oxygen. Soil is a complex mixture of solids, liquids, and gases. Solid matter includes both mineral and organic matter. Watery solutions and atmospheric gases are also found in soils.

Although we may think that soil is found all over the world, large expanses of continents don't have soil. For example, dunes of moving sand, bare rock surfaces of deserts and high mountains, and surfaces of fresh lava near active volcanoes do not have a soil layer.

SOIL COLOR AND TEXTURE

Color is the most obvious feature of a soil (Figure 10.4). Some color relationships are quite simple. For example, Midwest prairie soils are black or dark brown in color because they contain many humus particles. And the red or yellow soils of the Southeast are created by the presence of iron-containing oxides.

In some areas, soil color is inherited from the mineral parent material, but more generally the color is generated during soil formation. For example, dry climates often have soils with a white surface layer of mineral salts that have been brought upward by evaporation. In the cold, moist climate of the boreal forest, a pale, ash-gray layer near the top of the soil is created when organic matter and colored minerals are washed downward, leaving only pure, light-colored mineral matter behind.

The mineral matter of the soil consists of individual mineral particles that vary widely in size. The term **soil texture** refers to the proportion of particles that fall into each of three size grades: *sand*, *silt*, and *clay* (Figure 10.5). The finest soil particles, which are included in the clay size grade, are called *colloids*. We ignore the coarsest particles—*gravel*—in discussing soil texture because they don't play an important role in soil processes.

> Soil texture refers to the proportions of sand, silt, and clay found in a soil. Soil color, which is usually determined by soil-forming processes, varies widely.

Vegetation

Soil

Regolith

Bedrock

Forest litter

10.3 A cross section through the land surface

The soil is a surface layer of mineral and organic matter that supports vegetation. The soil is derived largely from the regolith, a zone of weathered rock or sediment underlying the soil layer.

10.4 Soil color

The color of soil provides vital clues to its type and condition.

B. Anthony Stewart/NG Image Collection

◄ **BLACK SOIL**
Dark soil colors normally indicate the abundance of organic matter in the form of humus. Here, students are weeding rows of onions in a field near Rome, New York.

© Eduardo Pucheta Photo/Alamy Limited

▲ **WHITE SOIL**
Salts form a white surface deposit on a plateau in western Argentina. Salts in the soil are carried upward and accumulate as water from infrequent rainstorms evaporates at the surface. Salt-tolerant plants form a spotty vegetation cover.

◄ **RED SOILS OF THE SOUTHEAST**
The red-brown color of this soil, near Cedar Mountain, Culpeper County, Virginia, is caused by iron oxides. With proper treatment, the ancient soils of this region can be highly productive.

Sam Abell/NG Image Collection

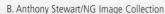

mm   in.

Gravel — 0.1

2
1 — 2.0 — 0.01

0.1   Sand

— 0.05 — 0.001

0.01   Silt

0.001 — 0.002 — 0.0001

Clay

0.0001 — 0.00001

Colloids — 0.000001

0.00001 — 0.0000001

10.5 Mineral particle sizes

Size grades are named sand, silt, and clay (which includes colloids). They are defined using the metric system, and each unit on the scale represents a power of 10.

Soils can be characterized by their proportions of sand, silt, and clay (Figure 10.6). Soil texture determines the capability of the soil to hold water and support microbial and animal life. A loam is a soil that contains a proportion of each of the three grades that are most effective at storing moisture. This mixture is often added to gardens because it is particularly suitable for plant growth. Loams can be further classified as sandy, silty, or clay-rich, depending on which grade is dominant.

Why is soil texture important? Soil texture determines the capability of the soil to hold water. Coarse-textured (sandy) soils have many small passages between touching mineral grains, which serve to quickly conduct water through to deeper layers. But soils of fine particles have far smaller passages and spaces, so water will penetrate down more slowly and tend to be held in the upper layers. We will return to the water-holding capability of soils shortly.

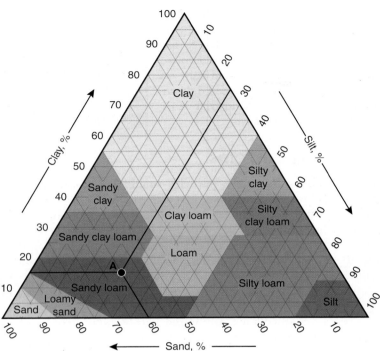

10.6 Soil texture classification

Soils can be classified based on their proportion of sand, silt, and clay. Loams contain a proportion of sand, silt, and clay that is particularly effective for holding water. On the diagram, follow the three grid lines at a point to the sides of the triangle and read the proportion. For example, the texture at point A is 60 percent sand, 15 percent clay, and 25 percent silt.

SOIL STRUCTURE

Soil structure refers to the way soil grains are clumped together into larger masses, called **peds**, ranging from small grains to large blocks (Figure 10.7). Peds form when clay minerals shrink as they dry out. Cracks form in the soil, which define the surfaces of the peds. Small peds, shaped roughly like spheres, give the soil a granular or crumb structure. Larger peds form an angular, blocky structure. Platy structure tends to indicate compaction of the soil by animals or human

10.7 Soil structure

Soil structure is described by the shape and size of soil aggregates, called peds.

▲ GRANULAR STRUCTURE
Looking similar to cookie crumbs, granular peds are rounded little pellets the size of BBs. They are loosely packed and found in the surface layers, along with organic material and roots.

▲ BLOCKY STRUCTURE
These cube-shaped peds form in the subsoil and are 5 to 50 mm (0.2 to 2 in.). These are typical of B horizons with high clay content.

▲ PLATY STRUCTURE
Larger than granular peds, these thin, flat soil masses, looking much like a stack of pancakes, occur near the surface and are associated with soil compaction.

▲ COLUMNAR STRUCTURE
Typical of dry regions and sandy soils, these peds are vertical columns found in the lower soil horizons. Often several centimeters in length, they frequently occur with a salt cap from arid illuviation.

activity. A columnar structure can occur in arid or sandy soils.

Soils with a well-developed granular or blocky structure are easy to cultivate—a factor that is especially important for nonmechanized farming using animal-drawn plows. Soils with a high clay content can lack peds. They are sticky and heavy when wet and difficult to cultivate. When dry, they become too hard to be worked.

Soil Chemistry

In addition to physical properties such a soil texture and structure, soil chemistry and soil moisture are important factors that determine the suitability of soil to support vegetation. The chemistry of a soil determines how well it can provide nutrients to plants, and the availability of soil moisture determines the capability of the vegetation to grow and thrive.

ACIDITY AND ALKALINITY

One important element in soil chemistry is pH, which is a measure of how acidic or alkaline the soil is. Soil pH is an important indicator of soil fertility (Figure 10.8). It affects how minerals dissolve in water solutions and their availability for uptake by plant roots. More specifically, pH is a measure of the free hydrogen ions in water or soil. A pH of 7 indicates a neutral balance between positively charged hydrogen (H^+) and negatively charged hydroxide (OH^-). Acidic soils (with more H^+ ions and fewer OH^- ions) can be formed when an excess of organic decomposition occurs because of insufficient leaching or drainage. Limestone parent material, on the other hand, can be a factor in development of alkaline soils (with fewer H^+ ions and more OH^- ions). Most plants prefer a pH level between 6 and 8 for maximum growth.

Soils in cold, humid climates are often highly acidic, while in arid climates, soils are typically alkaline. Acidic soils can be modified by applying lime, a compound of calcium, carbon, and oxygen ($CaCO_3$), which removes acid ions and replaces them with the alkaline cation, calcium

(cations are described in the next section). Such soil modification strategies are sometimes used to combat the effects of acid rain on lakes and soils located downwind from industrial emissions sources.

> Soil acidity varies widely. Soils of cool, moist regions are generally acidic, while soils of arid climates are alkaline. Acidic soils are often low in base ions.

SOIL COLLOIDS

Soil colloids are particles smaller than one ten-thousandth of a millimeter (0.0001 mm; 0.000,004 in.). Like other soil particles, some colloids are mineral, whereas others are organic. Mineral colloids are usually very fine clay particles. Under a microscope, they display thin, plate-like bodies. When these particles are well mixed in water, they remain suspended indefinitely, turning the water murky. Organic colloids are tiny bits of organic matter that are resistant to decay.

Soil colloids are important because their surfaces attract soil nutrients dissolved in soil water as positively charged mineral ions, or *cations* (Figure 10.9). Some cations are needed for plant growth, including calcium (Ca^{++}), magnesium (Mg^{++}), potassium (K^+), and sodium (Na^+). These are referred to as **base cations,** or simply, **bases.** They need to be dissolved in a soil-water solution to be available for plant uptake. Colloids hold these ions but also exchange them with plants when they are in close contact with root membranes. The fertility of the soil-water solution for plants is based on the capability of the soil to hold and exchange cations; this is referred to as the **cation-exchange capacity**. Without soil colloids, most vital nutrients would be leached out of the soil by percolating gravitational water and carried away in streams.

The relative amount of bases held by a soil determines the *base status* of the soil. A high base status means that the soil holds an abundant supply of base cations necessary for plant growth. If soil colloids hold a small supply of bases, the soil is of low base status and is, therefore, less fertile. Humus colloids have a high capacity to hold bases, so they are associated with high soil fertility.

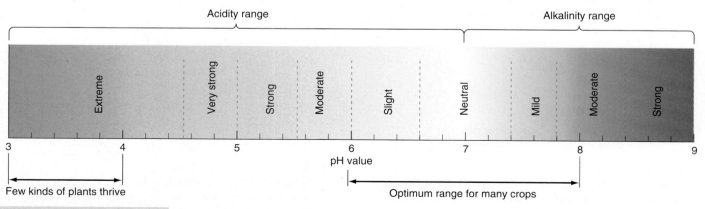

10.8 The pH scale and soil fertility

Acidity and alkalinity ranges are used to define optimal soil pH for various agricultural and native plants. The optimal pH for most farm crops is 6.5.

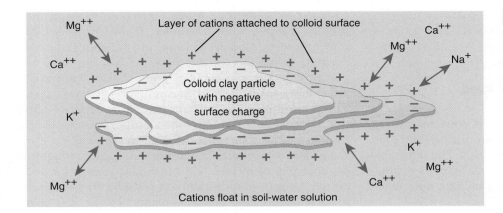

Cation-exchange capacity is defined by the quantity of positively charged mineral ions (nutrients) that move between the negatively charged colloid particle and the soil-water solution. This capacity determines soil fertility.

Acid ions have the power to replace the nutrient bases clinging to the surfaces of the soil colloids. As the acid ions displace the bases and build up, the bases are released to the soil solution. The bases are then gradually washed downward below rooting level, weakening soil fertility. When this happens, the soil acidity is increased. Aluminum ions (Al^{+++}), which are not plant nutrients, also have the capability to displace base cations, reducing base status and soil fertility.

MINERAL ALTERATION

There are two classes of soil minerals: primary and secondary. *Primary minerals* are compounds found in unaltered rock. These are mostly silicon- and oxygen-containing minerals with varying proportions of aluminum, calcium, sodium, iron, and magnesium. Although primary minerals make up a large fraction of the solid matter of many kinds of soils, they don't play an important role in sustaining plant or animal life.

When primary minerals are exposed to air and water at or near the Earth's surface, their chemical composition is slowly altered. *Mineral alteration* is a chemical weathering process that is explained in more detail in Chapter 13. The primary minerals are altered into **secondary minerals**, which are essential to soil development and to soil fertility.

> Primary minerals in soils are those that remain from unaltered rock. Secondary minerals are formed by mineral alteration. Clay minerals and sesquioxides are important secondary minerals.

The most important secondary minerals in soils are *clay minerals;* they make up the majority of mineral colloids. Some clay minerals hold bases tightly, and others loosely, which can affect the soil's base status.

Mineral oxides are also significant secondary minerals. We find them in many kinds of soils, particularly those that have developed in warm, moist climates over very long periods of time (hundreds of thousands of years). Under these conditions, minerals are chemically broken down into simple oxides, compounds in which a single element is combined with oxygen.

Oxides of aluminum and iron are the most important oxides in soils. Two atoms of aluminum are combined with three atoms of oxygen to form the *sesquioxide* of aluminum (Al_2O_3). (The prefix *sesqui* means "one and a half" and refers to the chemical composition of one and one-half atoms of oxygen for every atom of aluminum.) In soils, aluminum sesquioxide and water molecules bind together to make the mineral bauxite. Where bauxite layers are thick and uniform, they are sometimes strip-mined as aluminum ore (Figure 10.10).

Iron sesquioxide (Fe_2O_3) combines with water molecules to produce *limonite*, a yellowish to reddish mineral

At this location on the island of Jamaica, ancient soils have weathered to provide a deep layer of aluminum oxide. Stained red by iron oxide, the layer is strip-mined by power shovels to unearth aluminum ore.

James L. Amos/NG Image Collection

that makes the color of soils and rocks reddish to chocolate-brown. At one time, shallow accumulations of limonite were mined for iron.

Soil Moisture

The soil layer is also a reservoir of moisture for plants. Soil moisture is a key factor in determining how the soils of a region support vegetation and crops. The interrelationships among soil moisture, climate, and vegetation help us understand how changes in climate may affect agriculture and native ecosystems.

SOIL-WATER STORAGE

After a rainfall or snowmelt event, the soil becomes partially or completely filled with water. Where does this water go? The excess water, called *gravitational water*, slowly drains downward into the regolith underneath. However, some water clings to soil particles by cohesion. This water resists the pull of gravity because of the force of microscopic *capillary tension*.

To understand this force, think about a droplet of condensation that has formed on the cold surface of a glass of ice water. The water droplet seems to be enclosed in a film of surface molecules, drawing the droplet together into a rounded shape. That surface is produced by capillary tension, which keeps the drop clinging to the side of the glass, defying the force of gravity. Similarly, tiny films of water stick to individual soil grains, particularly at the points of grain contact. The portion of water held in the soil by capillary tension is available for evaporation or absorption by plant rootlets, so it is sometimes called *available water*.

Soil also stores water in another way, holding it tightly by molecular attraction. This *hygroscopic water* is located on the surfaces of individual soil particles. Hygroscopic water is largely unavailable to plants because it is held firmly by molecular attractions so strong that they cannot be broken by plants or through evaporation.

When a soil has been saturated by water and then allowed to drain under gravity until no more water moves downward, the soil is said to be holding its **storage capacity** of water. Maintaining soil moisture at or near storage capacity is important for agricultural productivity. In times of drought, when all available water has been removed from the soil, the soil is said to be at its *wilting point*. Although there is still hygroscopic water present in the soil, it is inaccessible to plants, and they begin to wilt. Figure 10.11 shows the three forms of soil water in relation to storage capacity and wilting point.

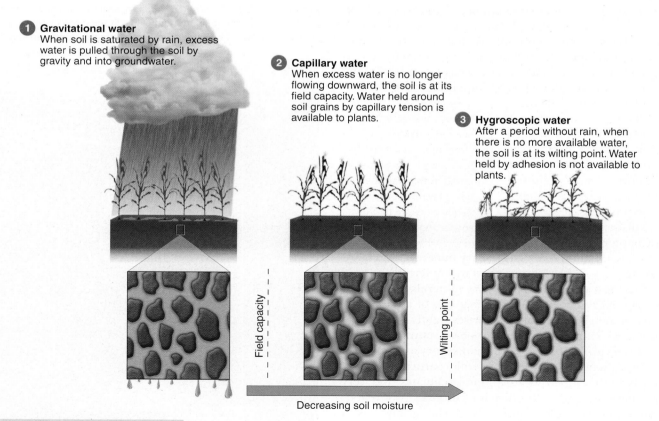

1 Gravitational water
When soil is saturated by rain, excess water is pulled through the soil by gravity and into groundwater.

2 Capillary water
When excess water is no longer flowing downward, the soil is at its field capacity. Water held around soil grains by capillary tension is available to plants.

3 Hygroscopic water
After a period without rain, when there is no more available water, the soil is at its wilting point. Water held by adhesion is not available to plants.

Field capacity

Wilting point

Decreasing soil moisture

10.11 Three forms of soil moisture

Soil holds water in three forms: gravitational, capillary, and hygroscopic. Capillary water is the most important source of water for plants.

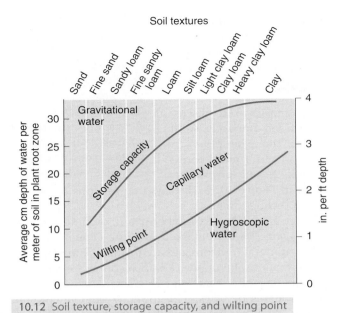

Soil textures

10.12 Soil texture, storage capacity, and wilting point

Finer-textured soils hold more water. They also hold water more tightly, raising the wilting point. The soils with the most available water are loamy.

given time is determined by the *soil-water balance*, which includes the gain and loss of water stored in the soil. The diagram in Figure 10.13 illustrates the components of the soil-water balance.

Precipitation reaching the surface of the soil takes two paths. One part enters the soil and recharges the reservoir of soil-water storage; the other part runs off the surface as overland flow into streams and rivers. Water leaves the soil-water storage reservoir through evapotranspiration, which depends on the air temperature and the amount of vegetation cover. Gravity also pulls water from the soil layer into groundwater below. As we will see in Chapter 14, the extra groundwater eventually emerges from the bottoms of streambeds and lakes, joining overland flow to constitute total runoff.

During warm weather, evapotranspiration increases, and water flows out of the soil-water reservoir. If precipitation does not increase as well, soil water decreases. Eventually, soil-water storage can be drawn down below the wilting point, and plants and crops suffer. When precipitation increases or temperatures fall, reducing transpiration, the soil-water reservoir is recharged.

> The soil-water balance controls the availability of soil water for agriculture or natural vegetation. Depending on the balance, soil moisture may be lowered or recharged.

The availability and flow of water in the soil depends largely on its texture (Figure 10.12). Coarse-textured (sandy) soils store less water because they are more porous, allowing water to drain through to deeper layers. Because fine-textured (silty) soils have far smaller passages and spaces, they are less porous, allowing water to penetrate down more slowly, with some water held longer in the upper layers. As a result, storage capacity increases as grain size decreases.

Finer-textured soils hold more water than coarser-textured soils because fine particles have a much larger surface area in a unit of volume than do coarse particles. Although this increases hygroscopic water, such

> Storage capacity, which depends on soil texture, measures the amount of water held by a soil after excess water has drained away. Available water remains, held by capillary action.

water remains unavailable to plants. Capillary water has significantly more spaces to occupy in finer-textured soils. However, because fine particles also hold water more tightly, it is more difficult for plants to extract moisture from fine soils. As a result, the wilting point also increases as grain size decreases. This means that plants can wilt in fine-textured soils even though they hold more soil water than coarse-textured soils. Loamy soils are best for plant growth because they offer the most available water.

SOIL-WATER BALANCE

It's clear that soil water is a critical resource needed for plant growth. The amount of water available at any

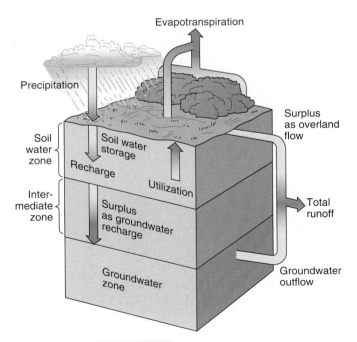

10.13 The soil-water balance

The soil-water storage reservoir loses water through evapotranspiration and by drainage to groundwater. The soil-water storage reservoir gains water from precipitation.

The soil-water balance is an important determiner of agricultural productivity, and it depends largely on climate—the variation in temperature and precipitation through the year. In a climate with a dry summer growing season, like the Mediterranean climate ⑦, irrigation is required for crops. In a climate with a wet summer or abundant year-round precipitation, such as the moist subtropical climate ⑥, soil water can maintain lush vegetation or crops through the growing season.

Soil Development

How do soils develop their distinctive characteristics? Let's turn to the processes that form soils and soil layers.

SOIL HORIZONS

Most soils have distinctive horizontal layers that differ in physical and chemical composition, organic content, or structure (Figure 10.14). We call these layers **soil horizons**. They develop through interactions among climate, living organisms, and the land surface, over time. Horizons usually develop either by selective removal or accumulation of certain ions, colloids, and chemical compounds. This removal or accumulation is normally produced by water seeping through the soil profile from the surface to deeper layers. Horizons often have different soil textures and colors.

To simplify our discussion of soil horizons, we'll look at the example of a soil found in a moist forest climate. A **soil profile**, as shown in Figure 10.15, displays the horizons on a cross section through the soil.

> Soil horizons are distinctive layers found in soils that differ in physical or chemical composition, organic content, or structure. The soil profile refers to the display of horizons on a cross section through the soil.

There are two types of soil horizons: organic and mineral. *Organic horizons*, marked with the capital letter O, lie over mineral horizons and are formed from plant and animal matter. The upper O_i *horizon* contains decomposing organic matter that you can easily recognize by eye, such as leaves or twigs. The lower O_a *horizon* contains humus, which has broken down beyond recognition.

There are four main mineral horizons. The *A horizon* is enriched with organic matter, washed downward from the organic horizons. Next is the *E horizon*. Clay particles and oxides of aluminum and iron are removed from the E horizon by downward-percolating water, leaving behind pure grains of sand or coarse silt.

The *B horizon* receives the clay particles, aluminum, and iron oxides, as well as organic matter washed down from the A and E horizons. It's dense and tough because its natural spaces are filled with clays and oxides.

Beneath the B horizon is the *C horizon*. It consists of the parent mineral matter of the soil. Below this regolith lies bedrock or sediments of much older age than the soil. Soil scientists limit the term *soil* to the A, E, and B horizons, which plant roots can readily reach.

SOIL-FORMING PROCESSES

There are four classes of soil-forming processes: soil enrichment, removal, translocation, and transformation. Figure 10.16 diagrams these processes and shows how they work together on the landscape.

In *soil enrichment*, matter—organic or inorganic—is added to the soil. Surface mineral enrichment of silt by river floods or as wind-blown dust is an example. Organic enrichment occurs as water carries humus from the O horizon into the A horizon below.

In *removal* processes, material is removed from the soil body. This occurs when erosion carries soil particles into streams and rivers. *Leaching*, the loss of soil compounds and minerals by solution in water flowing to lower levels, is another important removal process.

Translocation describes the movement of materials upward or downward within the soil. Fine particles—particularly clays and colloids—are translocated downward, a process called **eluviation**. This leaves behind grains

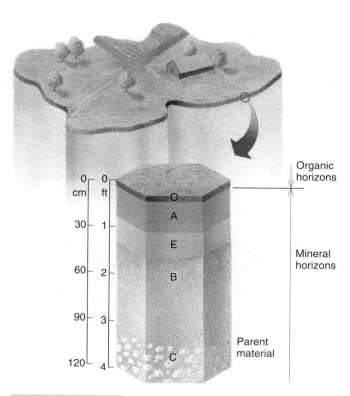

10.14 Soil horizons

A column of soil will normally show a series of horizons, which are horizontal layers with different properties.

▼ **HORIZON SEQUENCE**

A sequence of horizons that might appear in a forest soil developed under a cool, moist climate.

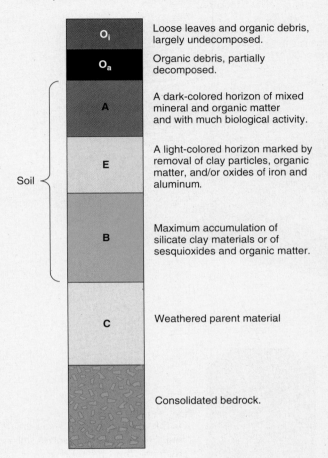

Oᵢ	Loose leaves and organic debris, largely undecomposed.
Oₐ	Organic debris, partially decomposed.
A	A dark-colored horizon of mixed mineral and organic matter and with much biological activity.
E	A light-colored horizon marked by removal of clay particles, organic matter, and/or oxides of iron and aluminum.
B	Maximum accumulation of silicate clay materials or of sesquioxides and organic matter.
C	Weathered parent material
	Consolidated bedrock.

Soil

FOREST SOIL PROFILE ▶

This photo shows an actual forest soil profile on outer Cape Cod, Massachusetts. The brown horizon in the upper middle of the photo is the A horizon, which is followed by the pale grayish E horizon and the reddish B horizon. A thin layer of wind-deposited silt and dune sand (pale brown layer) has been deposited on top.

of sand or coarse silt, forming the E horizon. Material brought downward from the E horizon—clay particles, humus, or sesquioxides of iron and aluminum—accumulates in the B horizon, a process called **illuviation**.

The soil profile shown in the photo in Figure 10.15 displays the effects of both soil enrichment and translocation. The topmost layer

The four classes of soil-forming processes are enrichment, removal, translocation, and transformation. In translocation, fine particles are transported downward by eluviation and accumulate in lower horizons by illuviation.

of the soil is a thin deposit of wind-blown silt and dune sand, which has augmented the soil profile. Humus, moving downward from decaying organic matter in the O horizon, has enriched the A horizon, giving it a brownish color. Eluviation has removed colloids and sesquioxides from the whitened E horizon, and illuviation has added them to the B horizon, which displays the orange-red colors of iron sesquioxide.

The translocation of calcium carbonate is another important process. In moist climates, a large amount of surplus soil water moves downward to the groundwater zone. This water movement leaches calcium carbonate from the entire soil in a process called *decalcification*.

10.16 Soil formation processes

Enrichment, removal, translocation, and transformation are four important processes of soil formation.

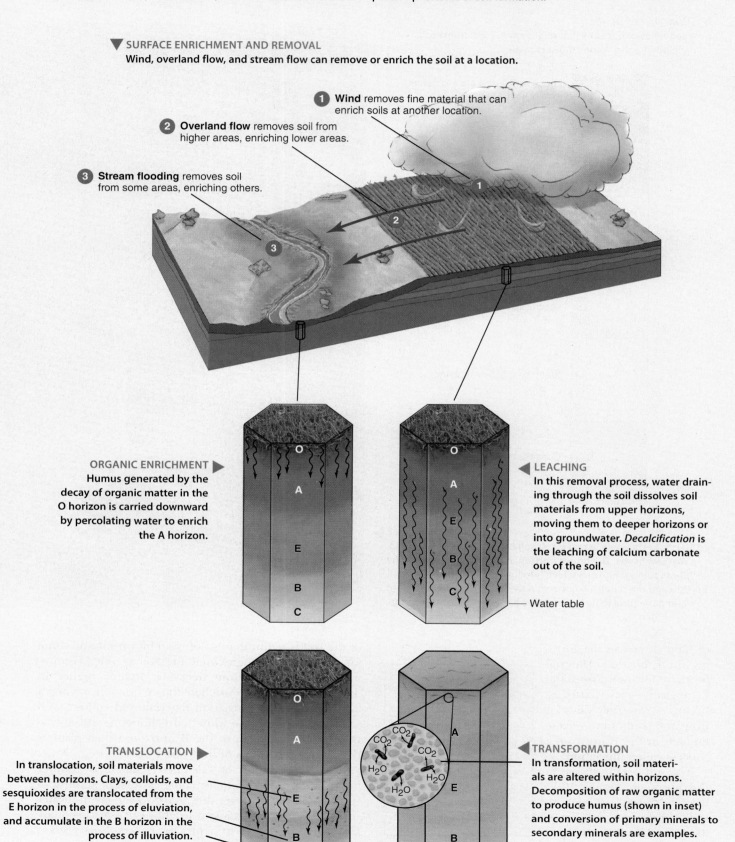

▼ SURFACE ENRICHMENT AND REMOVAL
Wind, overland flow, and stream flow can remove or enrich the soil at a location.

1 **Wind** removes fine material that can enrich soils at another location.

2 **Overland flow** removes soil from higher areas, enriching lower areas.

3 **Stream flooding** removes soil from some areas, enriching others.

ORGANIC ENRICHMENT ▶
Humus generated by the decay of organic matter in the O horizon is carried downward by percolating water to enrich the A horizon.

◀ LEACHING
In this removal process, water draining through the soil dissolves soil materials from upper horizons, moving them to deeper horizons or into groundwater. *Decalcification* is the leaching of calcium carbonate out of the soil.

Water table

TRANSLOCATION ▶
In translocation, soil materials move between horizons. Clays, colloids, and sesquioxides are translocated from the E horizon in the process of eluviation, and accumulate in the B horizon in the process of illuviation.

◀ TRANSFORMATION
In transformation, soil materials are altered within horizons. Decomposition of raw organic matter to produce humus (shown in inset) and conversion of primary minerals to secondary minerals are examples.

Soils that have lost most of their calcium are also usually acidic, and so they are low in bases. Adding lime or pulverized limestone will not only correct the acid condition, but will also restore the missing calcium, an important plant nutrient.

In dry climates, annual precipitation is not sufficient to leach the carbonate out of the soil and into the groundwater below. Instead, it is carried down to the B horizon, where it is deposited as white grains, plates, or nodules, in a process called *calcification*. Calcification can produce a cemented layer, known as a *hard pan*, that interferes with both eluviation and illuviation. This renders the soil less fertile by preventing the exchange of nutrients.

In colder climates, a pan can also form from the accumulation of oxides of iron and aluminum by illuviation. This type of pan can block drainage and keep the soil saturated for long periods, resulting in chemical-reducing conditions. This chemical environment can affect iron and manganese compounds, causing spots or streaks of blue and green colors in the soil called *mottles*.

Upward translocation can also occur in desert climates. In some low areas, a layer of groundwater lies close to the surface, producing a flat, poorly drained area. As water at or near the soil surface evaporates, groundwater is drawn upward to replace it by capillary tension, much like a cotton wick draws oil upward in an oil lamp. This groundwater is often rich in dissolved salts. When the salt-rich water evaporates, the salts are deposited and build up. This process is called *salinization*. Large amounts of these salts are toxic to many kinds of plants. When salinization occurs in irrigated lands in a desert climate, the soil can be ruined, with little hope of revival.

The last class of soil-forming processes involves the *transformation* of material within the soil body. An example is the conversion of minerals from primary to secondary types, which we have already described. Another example is decomposition of organic matter by microorganisms to produce humus, a process termed *humification*. In warm moist climates, transformation of organic matter to carbon dioxide and water can be nearly complete, leaving virtually no organic matter in the soil.

WileyPLUS Soil Horizons
Watch an animation demonstrating eluviation and illuviation through the soil profile.

FACTORS OF SOIL FORMATION

Let's revisit the factors of soil development we listed at the beginning of this chapter in more detail: climate, organisms, relief, parent material, and time.

Climate

Climate, measured by precipitation and temperature, is an important determinant of soil properties. As we have seen, precipitation controls the downward movement of nutrients and other chemical compounds in soils by translocation. If precipitation is high, water will wash nutrients deeper into the soil and out of reach of plant roots. If precipitation is low, salts will build up in the soil and restrict fertility.

Soil temperature affects the chemical development of soils and the formation of horizons. Below 10°C (50°F), biological activity is slowed; and at or below the freezing point (0°C; 32°F), biological activity stops and chemical processes affecting minerals are inactive. Thus, decomposition is slow in cold climates, and so organic matter accumulates to form a thick O horizon. This material becomes humus, which is carried downward to enrich the A horizon. In contrast, bacteria rapidly decompose plant material in the warm, moist climates of low latitudes. O horizons are lacking, and the entire soil profile will contain very little organic matter.

> Temperature affects soil development. In cold climates, decomposition of organic matter is slow, and organic matter accumulates. In warm, wet climates, where organic matter decomposes rapidly, soil organic matter is scarce.

Organisms

Living plants and animals, as well as their nonliving organic products, have an important effect on soil. We have already noted the role that organic matter as humus plays in soil fertility. It also helps bind the soil into crumbs and clumps, which allows water and air to penetrate the soil freely. Plant roots, by their growth, mix and disturb the soil and provide organic material directly to upper soil horizons.

Vegetation is an important factor in determining soil qualities. For example, some of America's richest soils developed in the prairies of the Middle West under a cover of thick grass. The deep roots of the grass, in a cycle of growth and decay, deposited nutrients and organic matter throughout the thick soil layer. In the Northeast, conifer forests provide a surface layer of decaying needles that keeps the soil quite acidic. Acid ions replace base cations, which are leached below root depth, out of the reach of plants. When farmed today, these soils need applications of lime to reduce their acidity and enhance their fertility.

Organisms living in the soil include many species, from bacteria to burrowing mammals (Figure 10.17). Earthworms continually rework the soil not only by burrowing, but also by passing soil through their intestinal tracts. They ingest large amounts of decaying leaf matter, carry it down from the surface, and incorporate it into the mineral soil horizons. Many forms of insect larvae perform a similar function. And moles, gophers, rabbits, badgers, prairie dogs, and other burrowing animals make larger, tubelike openings.

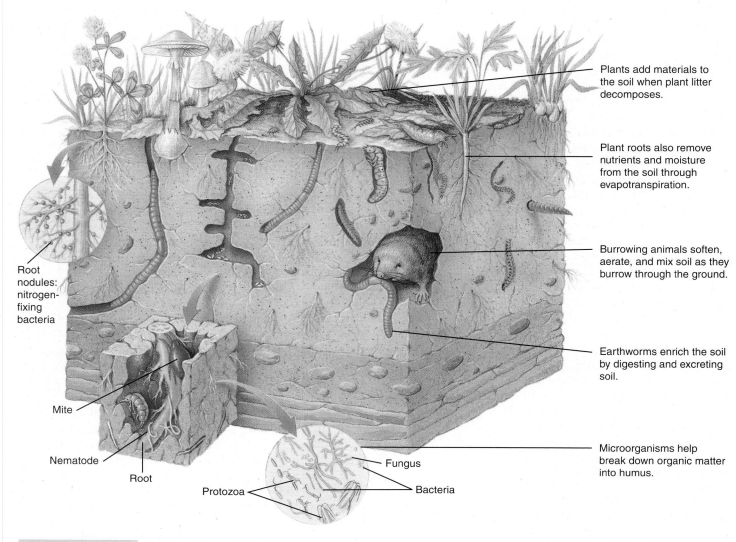

Root nodules: nitrogen-fixing bacteria

Plants add materials to the soil when plant litter decomposes.

Plant roots also remove nutrients and moisture from the soil through evapotranspiration.

Burrowing animals soften, aerate, and mix soil as they burrow through the ground.

Earthworms enrich the soil by digesting and excreting soil.

Microorganisms help break down organic matter into humus.

Mite

Nematode

Root

Protozoa

Fungus

Bacteria

10.17 Soil organisms

The diversity of life in fertile soil includes plants, algae, fungi, earthworms, flatworms, roundworms, insects, spiders and mites, bacteria, and burrowing animals such as moles and groundhogs. Soil horizons are not drawn to scale.

Relief

The configuration, or shape, of the ground surface, known as *relief*, also influences soil formation. Generally speaking, soil horizons are thick on gentle slopes and thin on steep slopes. This is because the soil is more rapidly removed by erosion on the steeper slopes. In addition, slopes facing away from the Sun are sheltered from direct insolation and so tend to have cooler, moister soils. Slopes facing toward the Sun are exposed to direct solar rays, raising soil temperatures and increasing evapotranspiration.

Parent Material

Soil chemistry is influenced by the original source of parent material. For example, iron-rich bedrock produces soils rich in iron oxides, whereas limestone forms calcium-rich soils. Some types of secondary minerals, weathered from particular primary minerals, can produce soils with unique properties. Also, soil texture is largely determined by the size of mineral grains within the parent material.

Time

The characteristics and properties of soils require time for development. For example, a fresh deposit of mineral matter, like the clean, sorted sand of a dune, may require hundreds to thousands of years to acquire the structure and properties of a sandy soil. Soil formation is based on processes that occur under a set of climatic conditions for moisture and temperature regimes—regimes that may continue for centuries.

A soil scientist's rule of thumb is that it takes about 500 years to form 2.5 cm (1 in.) of topsoil.

Human Activity

Human activity also influences the physical and chemical nature of the soil. Clearing of native vegetation for crops can induce erosion, removing upper layers that are rich in organic matter. Large areas of agricultural soils have been plowed and planted for centuries. As a result, both the structure and composition of these agricultural soils have undergone great changes. These altered soils are often recognized as distinct soil classes that are just as important as natural soils.

The Global Scope of Soils

How soils are distributed around the world helps to determine the quality of environments of the globe.

Soil fertility, along with the availability of freshwater, is a basic measure of the capability of an environmental region to produce food for human consumption.

We classify soils according to a system developed by scientists of the U.S. Natural Resources Conservation Service, in cooperation with soil scientists of many other nations. In this book, we'll discuss the two highest levels of this classification system. The top level contains 12 **soil orders**. We'll also mention a few important *suborders* that make up the second classification level. For this analysis, we will group the soil orders based on four factors that can characterize a particular order: maturity, climate, parent material, and high organic matter (Table 10.1). Figure 10.18 is a map of world soil orders. Let's look at the soil orders in more detail.

> Soils are classified by soil order and suborder. We group these divisions by dominant factors of maturity, climate, parent material, and high organic content.

Table 10.1 Soil Orders

Soil Order	Distribution	Characteristics
Soils Characterized by Maturity		
Entisols	Recently deposited materials or materials that don't form horizons	■ No distinct horizons
Inceptisols	Young soils with minerals capable of further weathering and alteration	■ Weakly developed horizons
Alfisols	Humid and subhumid climates, usually with forest cover	■ Subsurface accumulation of clay ■ High base status
Spodosols	Cold, moist climates, often with boreal forest cover	■ Low in humus ■ Low base status
Ultisols	Equatorial to subtropical zones, and warm, moist climates with a weak to strong dry season	■ Subsurface accumulation of clay ■ Low base status
Oxisols	Moist climates in equatorial to subtropical zones, often with rainforest vegetation	■ Subsurface accumulation of mineral oxides ■ Very low base status
Soils Characterized by Climate		
Mollisols	Semiarid to subhumid midlatitude grasslands	■ Dark, organic-rich, upper horizon ■ Loose texture ■ Very high base status
Aridisols	Dry climates of deserts and semideserts	■ Low content of organic matter ■ Accumulation of salts
Gelisols	Cold, frozen climates	■ Underlain by permafrost ■ Subject to churning by freeze/thaw
Soils Characterized by Parent Material		
Vertisols	Clay parent material in climates with alternating wet and dry seasons	■ Clay-rich ■ High base status ■ Subject to churning and cracking
Andisols	Soils developed on volcanic ash	■ Dark, fertile soils with high amounts of carbon ■ Found near active volcanoes
Soils High in Organic Matter		
Histosols	Cool, poorly drained areas	■ Very high content of organic matter

10.18 Soils of the world

The map shows the general distribution of 11 soil types, including 7 soil orders and 4 suborders of Alfisols that correspond well to 4 basic climate zones. Soils found on floodplains, glacial landforms, sand dunes, marshlands, bogs, and volcanic ash deposits are not represented because they are of local occurrence. In highlands, the soil patterns are too complex to show at a global scale.

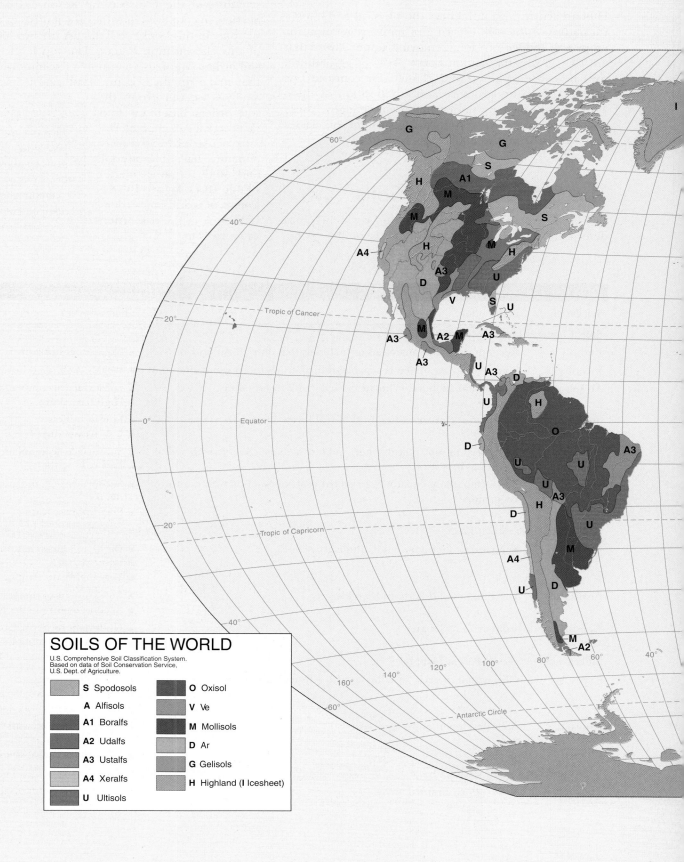

SOILS OF THE WORLD

U.S. Comprehensive Soil Classification System.
Based on data of Soil Conservation Service,
U.S. Dept. of Agriculture.

S   Spodosols
A   Alfisols
A1 Boralfs
A2 Udalfs
A3 Ustalfs
A4 Xeralfs
U   Ultisols

O   Oxisol
V   Ve
M   Mollisols
D   Ar
G   Gelisols
H   Highland (I Icesheet)

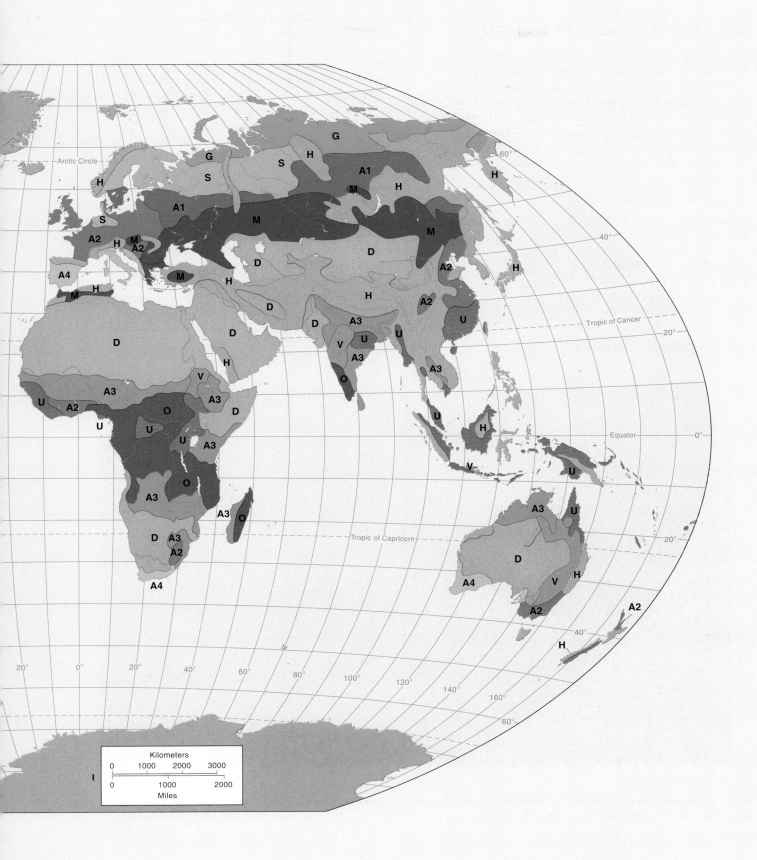

WileyPLUS Soils of the World
Take a second look at the world map of soils and click on a soil type to see a photo of the soil profile. An animation.

SOILS CHARACTERIZED BY MATURITY

Where materials have been recently deposited soils have no horizons or poorly developed horizons and are capable of further mineral alteration. This situation can also occur when parent materials are not appropriate for horizon formation. In contrast, soils that have experienced long-term adjustment to prevailing soil temperature and water conditions have well-developed horizons or fully weathered soil materials.

Entisols and Inceptisols

Entisols are mineral soils without distinct horizons. They are soils in the sense that they support plants, and they may be found in any climate and under any vegetation. They don't have distinct horizons for

> Entisols lack horizons, often because they are only recently deposited. They may occur in any climate or region.

two reasons: either the parent material—for example, quartz sand—is not appropriate for horizon formation, or not enough time has passed for horizons to form in recent deposits of alluvium or on actively eroding slopes. **Inceptisols** are soils with weakly developed horizons, usually because the soil is quite young. Entisols and Inceptisols can be found anywhere from equatorial to arctic latitude zones. Often, they occur as patches too small to show on the global soils map (Figure 10.18).

Entisols and Inceptisols of floodplains and delta plains in warm and moist climates are among the most highly productive agricultural soils in the world because of their favorable texture, ample nutrient content, and large soil-water storage. In Southeast Asia, Inceptisols support dense populations of rice farmers (Figure 10.19). Annual river floods cover low-lying plains and deposit layers of fine silt. This sediment is rich in primary minerals that yield bases as they weather chemically over time. The constant enrichment of the soil explains its high fertility in a region where uplands develop Ultisols with low fertility.

> Inceptisols have only weakly developed horizons. Inceptisols of river floodplains and deltas are often very productive.

Steve Raymer/NG Image Collection

10.19 Fertile Inceptisols

Broad river floodplains are often the sites of fertile Inceptisols, given their abundant water and a frequent flooding cycle that adds nutrients to the soil. Shown here are rice paddies in Vietnam.

10.20 Alfisols

Alfisols have a gray, brownish, or reddish surface horizon. Unlike the clay-rich horizon of the Ultisols, the B horizon of Alfisols is enriched by silicate clay minerals that can hold bases such as calcium and magnesium, giving Alfisols a high base status. Below the A horizon is a pale E horizon, which has lost some of the original bases, clay minerals, and sesquioxides by eluviation. These materials have become concentrated in the B horizon.

Estate of Henry D. Foth

▲ UDALF FROM MICHIGAN
Udalfs are a suborder of Alfisols closely associated with the moist continental climate in North America, Europe, and eastern Asia. The E horizon is the lighter layer below the top darker plow layer.

Estate of Henry D. Foth

▲ USTALF FROM TEXAS
Ustalfs, another suborder of Alfisols, range from the subtropical zone to the Equator, and are often associated with the wet-dry tropical climate in Southeast Asia, Africa, Australia, and South America. This Ustalf, however, is from Texas.

Alfisols and Spodosols

The **Alfisols** are soils characterized by a clay-rich horizon produced by illuviation and a high base status (Figure 10.20). The world distribution of Alfisols is extremely wide in latitude, ranging from as high as 60° N in North America and Eurasia to the equatorial zone in South America and Africa.

Because the Alfisols span an enormous range in climate

> Alfisols have horizons of eluviation and illuviation of clays. They also have a high base status and can be very productive.

types, four important suborders of Alfisols, each with its own climate affiliation, are shown on our global soils map.

- *Boralfs* are Alfisols of cold (boreal) forest lands of North America and Eurasia. They have a gray surface horizon and a brownish subsoil.
- *Udalfs* are brownish Alfisols of the midlatitude zone.
- *Ustalfs* are brownish to reddish Alfisols of the warmer climates.
- *Xeralfs* are Alfisols of the Mediterranean climate ⑦, with its cool moist winter and dry summer. The Xeralfs are typically brownish or reddish in color.

10.21 Spodosol profile

▼ SCHEMATIC DIAGRAM OF A MOLLISOL PROFILE

The Spodosol profile is marked by a light gray or white albic horizon, an E horizon that is bleached by organic acids percolating downward from a slowly decomposing O horizon. Also diagnostic is the spodic horizon, a dense mixture of organic matter and compounds of aluminum and iron, all brought downward by eluviation from the overlying E horizon. Spodosols are strongly acidic and are low in plant nutrients such as calcium and magnesium. They are also low in humus. Although the base status of the Spodosols is low, forests of pine and spruce are supported through the process of recycling of the bases.

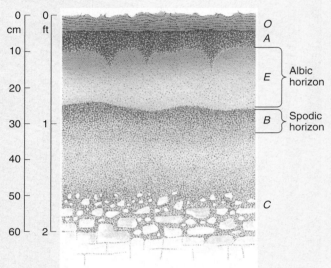

◀ SPODOSOL PROFILE
This Spodosol from France, of suborder Orthod, clearly shows the albic E horizon above the yellowish spodic B horizon. The O and A horizons have been largely removed by erosion.

Poleward of the Alfisols in North America and Eurasia extends a great belt of soils of the order **Spodosols**, formed in the cold boreal forest climate beneath a needleleaf forest. They are distinguished by a *spodic horizon*, a dense accumulation of iron and aluminum oxides and organic matter, and a gray or white *albic E horizon* that lies above the spodic horizon (Figure 10.21).

Spodosols are closely associated with regions recently covered by the great ice sheets of the Late-Cenozoic Ice Age. These soils are, therefore, very young. Typically, the parent material is coarse sand, consisting largely of the mineral quartz. This mineral cannot weather to form clay minerals, so Spodosols are naturally poor soils in terms of agricultural productivity. Because they are acidic, lime application is essential. They also need heavy applications of fertilizers. With proper management, Spodosols can be highly productive, if the soil texture is favorable;

> Spodosols have a light-colored albic horizon of eluviation, and a dense spodic horizon of illuviation. They develop under cold needleleaf forests and are quite acidic.

for example, they produce high yields of potatoes in Maine and New Brunswick.

Oxisols and Ultisols

Oxisols and Ultisols have developed over long time spans in an environment with warm soil temperatures and plentiful soil water, either in a wet season or throughout the year.

Oxisols have developed in the moist climates of the equatorial, tropical, and subtropical zones on land surfaces that have been stable over long periods of time. We find these soils over vast areas of South America and Africa in the wet equatorial climate ①, where the native vegetation is rainforest. The wet-dry tropical climate ③ in South America and Africa, with its large seasonal water surplus, is also associated with Oxisols. Figure 10.22 shows a soil profile for an Oxisol in Hawaii.

Ultisols are similar to the Oxisols, but have a subsurface clay horizon (Figure 10.23). They originate in closely related environments. In a few areas, the Ultisol profile contains a subsurface horizon of sesquioxides called *laterite* (Figure 10.24). This horizon

10.22 Oxisols

Oxisols usually lack distinct horizons, except for darkened surface layers. Soil minerals are weathered to an extreme degree and are dominated by stable sesquioxides of aluminum and iron, giving them red, yellow, and yellowish-brown colors. The soil has a very low base status because nearly all the bases have been removed. There's a small store of nutrient bases very close to the soil surface. The soil is quite easily broken apart, so rainwater and plant roots can easily penetrate.

▲ AN OXISOL IN HAWAII
Sugarcane is being cultivated here on the island of Oahu.

Alan H. Strahler

SOIL PROFILE FOR AN OXISOL ▶
The intense red color is produced by iron sesquioxides. This profile shows an Oxisol of the suborder Torrox in Hawaii.

Estate of Henry D. Foth

can harden into bricklike blocks when it is exposed to the air.

We find Ultisols throughout Southeast Asia and the East Indies. Other important areas are in eastern Australia, Central America, South America, and the southeastern United States. Ultisols extend into the lower midlatitude zone in the United States, where they correspond quite closely with areas of moist subtropical climate ⑥. In lower latitudes, Ultisols are identified with the wet-dry tropical climate ③ and the monsoon and trade-wind coastal climate ②. Note that all these climates have a dry season, even though it may be short.

> Oxisols and Ultisols develop over long time periods in warm, moist climates. Oxisols have substantial accumulations of iron and aluminum sesquioxides. Ultisols have a horizon of clay accumulation.

Before the advent of modern agricultural technology, both Oxisols and Ultisols of low latitudes were cultivated for centuries using a primitive slash-and-burn agriculture. Without fertilizers, these soils can sustain crops on freshly cleared areas for only two or three years at most, before the nutrient bases are exhausted and the garden plot must be abandoned. To sustain high-crop yields, substantial amounts of lime and fertilizers are required. Ultisols are also vulnerable to devastating soil erosion, particularly on steep hill slopes.

SOILS CHARACTERIZED BY CLIMATE

Three soil orders include soils that develop under special climatic conditions. *Mollisols* are soils of grasslands in climates with a distinct dry period. *Aridisols* are desert soils, where temperatures are high and precipitation is

Estate of Henry D. Foth

10.23 Soil profile for an Ultisol

Ultisols are reddish to yellowish in color. They have a subsurface horizon of clay, which is not found in the Oxisols. It is a B horizon and has developed by illuviation. Although the characteristic native vegetation is forest, Ultisols have a low base status. As in the Oxisols, most of the bases are found in a shallow surface layer where they are released by the decay of plant matter. They are quickly taken up and recycled by the shallow roots of trees and shrubs. This profile is for an Ultisol from North Carolina. The pale layer is the E horizon, and the thin, dark top layer is all that remains of the A horizon after erosion.

James P. Blair/NG Image Collection

10.24 Laterite

Soil erosion began when this slope in Borneo was stripped of vegetation. Natural accumulations of laterite, formed below the soil surface, were soon exposed. They now form tough, solid caps on columns of weaker soil.

low and sporadic. *Gelisols* are soils of cold climates where the soil may thaw for only a few weeks of the year.

Mollisols

Mollisols are grassland soils that occupy vast areas of semiarid and subhumid climates in midlatitudes. They are unique in having a very thick, dark brown to black surface horizon called a *mollic epipedon* (Figure 10.25).

Most areas of Mollisols are closely associated with the semiarid subtype of the dry midlatitude climate ⑨ and the adjacent portion of the moist continental climate ⑩. In North America, Mollisols dominate the Great Plains region, the Columbia Plateau, and the northern Great Basin. In South America, a large area of Mollisols covers the Pampa region of Argentina and Uruguay. In Eurasia, a great belt of Mollisols stretches from Romania eastward across the steppes of Russia, Siberia, and Mongolia.

Because of their loose texture and very high base status, Mollisols are among the most naturally fertile soils in the world. They now produce most of the world's commercial grain crop, though yield varies considerably from one year to the next because seasonal rainfall is highly variable. Interestingly, most of these soils have been used for crop production only in the last century. Before that, they were used mainly for grazing by nomadic herds. The Mollisols have favorable properties for growing cereals in large-scale mechanized farming and are relatively easy to manage.

10.25 Mollisols

Mollisols are soils that develop under a cover of grassland vegetation.

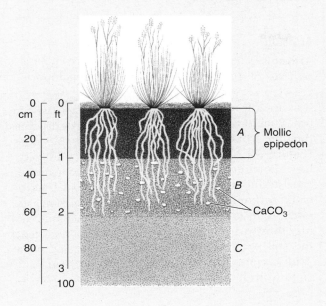

◀ A SCHEMATIC DIAGRAM OF A MOLLISOL PROFILE
Mollisols have a very dark brown to black surface horizon that lies within the A horizon. It is always more than 25 cm (9.8 in.) thick. The soil has a loose, granular structure, or a soft consistency when dry. Mollisols are dominated by calcium among the bases of the A and B horizons, giving them a very high base status.

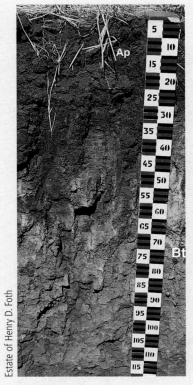

Estate of Henry D. Foth

▲ BOROLL
A Mollisol of cold-climate regions. This example is from Russia.

Prof. Randall Schaetzl

▲ UDOLL
A Mollisol of moist midlatitude climates. This example is from Wisconsin.

Prof. Randall Schaetzl

▲ USTOLL
A Mollisol from a midlatitude semiarid climate. This example is developed in an accumulation of wind-blown silt.

A brief mention of four suborders of the Mollisols commonly found in the United States and Canada will help explain important regional soil differences related to climate (Figure 10.25). *Borolls,* the cold-climate suborder of the Mollisols, are found in a large area extending on both sides of the U.S.– Canadian border east of the Rocky Mountains, as well as in Russia. *Udolls* are Mollisols of a relatively moist climate. They used to support tall-grass prairie, but today they are closely identified with the corn belt in the American Midwest, as well as the Pampa region of Argentina. *Ustolls* are Mollisols of the semiarid subtype of the dry midlatitude climate ⑨, with a substantial soil-water shortage in the summer months. They underlie much of the short-grass prairie region east of

> Mollisols are soils of grasslands in subhumid to semiarid climates. They have a thick, dark brown surface layer, termed a mollic epipedon. Because of their loose texture and high base status, they are highly productive.

the Rockies. *Xerolls* are Mollisols of the Mediterranean climate ⑦, with its tendency to cool, moist winters and rainless summers.

Aridisols

Aridisols, soils of the desert climate, are dry for long periods of time. Because the climate supports only very sparse vegetation, the soils are low in organic matter. They are pale-colored, ranging from gray to red (Figure 10.26). They may have horizons with significant accumulations of calcium carbonate or soluble salts. The salts, containing mostly sodium, make the soil very alkaline. The Aridisols are closely correlated with the arid subtypes

> Aridisols are desert soils with weakly developed horizons. They often exhibit subsurface layers composed of an accumulation of calcium carbonate or soluble salts. With irrigation and proper management, they are quite fertile.

10.26 Aridisols

The soil horizons of Aridisols are often weakly developed, but there may be important subsurface horizons with accumulations of calcium carbonate or soluble salts.

Estate of Henry D. Foth

▲ **ARIDISOL FROM COLORADO**
Carbonate accumulations are visible as white blotches and patches in this Aridisol, subclass Argid, from Colorado.

Courtesy Soil Conservation Service

▲ **SALIC HORIZON**
The white layer close to the surface in this Aridisol profile in the Nevada desert is a salic horizon.

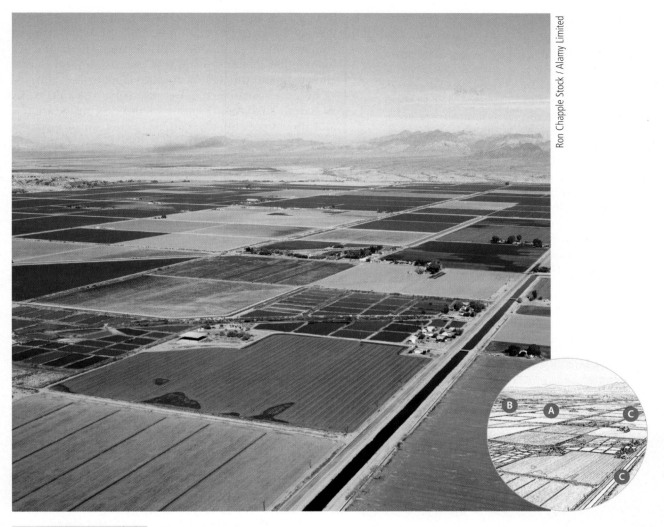

Ron Chapple Stock / Alamy Limited

10.27 Irrigated Aridisols

Cotton farming along the Colorado River, near Parker, Arizona, is shown in this oblique air photo.

EYE ON THE LANDSCAPE **What else would the geographer see?**
The broad floodplain of the Colorado River shows how productive Aridisols can be when properly managed and irrigated. A consistent climate with abundant sunlight, coupled with abundant river water, makes for high productivity and crop yields. Fallow fields **A** show typical Aridisol colors of gray to brown, while cotton fields are vibrant green **B**. Note the network of canals **C** serving the fields.

of the dry tropical climate ④, dry subtropical climate ⑤, and dry midlatitude climate ⑨.

Most Aridisols are still used for nomadic grazing, just as they have been through the ages. With low and sporadic rainfall, it's difficult to cultivate crops. But with irrigation, Aridisols can be highly productive (Figure 10.27).

Gelisols

Gelisols are soils of cold regions that are underlain by permanently frozen ground, or *permafrost*. They usually consist of very recent parent material, left behind by glacial activity during the Ice Age, along with organic matter that decays slowly at low temperatures. They are subjected to frequent freezing and thawing, which causes ice lenses and wedges to grow and melt in the soil. This action churns the soil, mixing parent material and organic matter irregularly and inhibiting development of a soil profile.

> Gelisols are soils of permafrost regions that are churned by freeze/thaw ice action.

SOILS CHARACTERIZED BY PARENT MATERIALS

Two soil orders include soils with parent materials that give the soils special characteristics. *Vertisols* are black, clay-rich soils, and *Andisols* are dark soils that develop on volcanic ash.

Vertisols

Vertisols have a unique set of properties that stand in sharp contrast to the Oxisols and Ultisols (Figure 10.28). They are black in color and have a high content of the clay mineral montmorillonite, which is formed from the weathering of particular volcanic rocks. This mineral swells and shrinks with wetting and drying, producing deep vertical cracks in the soil during the dry season. With onset of the rainy season, the surface materials are washed into the cracks, effectively mixing the soil into a thick, uniform layer. The abundance of clays gives these soils a high base status. Vertisols typically form under grass and savanna vegetation in subtropical and tropical climates with a pronounced dry season. These climates include the semiarid subtype of the dry tropical steppe climate ④ and the wet-dry tropical climate ③.

Because Vertisols require particular volcanic rocks as parent materials, regions of this soil are scattered and show no distinctive pattern on the world map. An important region of Vertisols is the Deccan Plateau of western India, where basalt, a dark variety of igneous rock, supplies the silicate minerals that are altered into the necessary clay minerals.

> Vertisols develop on certain types of volcanic rock in wet-dry climates under grassland and savanna vegetation. They expand and contract with wetting and drying, creating deep cracks in the soil.

Andisols

Andisols are soils in which more than half of the parent mineral matter is volcanic ash, which after spewing high into the air from the craters of active volcanoes, came

10.28 Vertisols

Vertisols have a high base status and are particularly rich in calcium and magnesium nutrients, with a moderate content of organic matter. The soil retains large amounts of water because of its fine texture, but much of this water is held tightly by the clay particles and is not available to plants. The clay minerals that are abundant in Vertisols shrink when they dry out, producing deep cracks in the soil surface. As the dry soil blocks are wetted and softened by rain, some fragments of surface soil drop into the cracks before they close, so that the soil "swallows itself" and is constantly being mixed.

Estate of Henry D. Foth

Dinodia Photos/Alamy Limited

▲ **VERTISOL IN TEXAS**
On drying, Vertisols develop deep vertical cracks.

◄ **SOIL PROFILE FOR A VERTISOL IN INDIA**
The very dark color and shiny surfaces of the clay particles are typical of Vertisols. This profile is for a Vertisol of suborder Ustert in India.

to rest in layers over the surrounding landscape. The fine ash particles are glass-like shards. Andisols have a high proportion of carbon, formed from decaying plant matter, so the soil is very dark. They form over a wide range of latitudes and climates and are generally fertile soils. In moist climates they support a dense natural vegetation cover.

Andisols aren't shown on our world map because they are found in small patches associated with individual volcanoes that are located mostly in the "Ring of Fire," the chain of volcanic mountains and islands that surrounds the great Pacific Ocean. Andisols are also found on the island of Hawaii, where volcanoes are presently active.

> Andisols are unique soils that form on volcanic ash of relatively recent origin. They are dark in color and typically fertile.

SOILS HIGH IN ORGANIC MATTER

Our last group of soil orders has only one member, *Histosols*, which are composed almost entirely of decaying organic matter in thick layers.

Histosols

Throughout the northern regions of Spodosols are countless patches of **Histosols**. This unique soil order has a very high content of organic matter in a thick, dark upper layer (Figure 10.29). Most Histosols go by common names such as peats or mucks. They have formed in shallow lakes and ponds by accumulation of

> Histosols are organic soils, often termed peats or mucks. They are typically formed in cool or cold climates in areas of poor drainage.

10.29 Histosols

Histosols are soils consisting largely or completely of organic matter.

▲ A HISTOSOL FROM MINNESOTA
This soil consists almost entirely of a deep layer of weathered peat.

Estate of Henry D. Foth

▲ PEAT BOG
This peat bog in Galway, Ireland, has been trenched to reveal a Histosol profile. Peat blocks, seen to the sides of the trench, are drying for use as fuel.

Farrell Grehan/Photo Researchers, Inc.

EYE ON THE LANDSCAPE What else would the geographer see?
Peat bogs are typical of glacial terrain. Moving ice sheets remove and take up loose soil and rock, then leave sediment behind as they melt. The low, hummocky terrain Ⓐ at the edge of this bog may be part of a moraine, a landform that occurs at the melting edge of an ice sheet, as wasting ice leaves piles of sediment behind in a chaotic and disorganized fashion.

partially decayed plant matter. In time, the water is replaced by a layer of organic matter, or *peat*, and becomes a *bog*. Peat from bogs is dried and baled for sale as mulch for use on suburban lawns and shrubbery beds. For centuries, Europe has used dried peat from bogs of glacial origin as a low-grade fuel.

Some Histosols are *mucks*, organic soils composed of fine black materials of sticky consistency. These are agriculturally valuable in midlatitudes, where they occur as beds of former lakes in glaciated regions. After appropriate drainage and application of lime and fertilizers, these mucks are remarkably productive for garden vegetables. Histosols are also found in low latitudes where poor drainage has favored thick accumulations of plant matter.

WileyPLUS Web Quiz

Take a quick quiz on the key concepts of this chapter.

A Look Ahead

An important message of the last three chapters is that climate, vegetation, and soils are closely interrelated. The nature of the vegetation cover depends largely on the climate, varying from desert to tundra biomes. Soils depend on both climate and vegetation to develop their unique characteristics. But soils can also influence vegetation development. Similarly, the climate at a location can be influenced by its vegetation cover.

The next two chapters also comprise a set of topics that are closely related. They deal with the lithosphere, the realm of the solid Earth. Our survey of the lithosphere will begin with the nature of Earth materials and then move on to a discussion of how continents and ocean basins are formed, and how they are continually changing, even today. In Chapter 12, we will discuss landforms occurring within continents that are produced by such lithospheric processes as earthquake faulting and volcanic activity.

WileyPLUS Interaction of Climate, Vegetation, and Soil

Return to the animations for Chapter 7 to view climate, vegetation, and soils maps superimposed to illustrate common patterns. Animations for Africa and North America.

IN REVIEW GLOBAL SOILS

- At first, global climate change will increase crop yields as temperatures rise; but, eventually, the added stress on crop plants will cause yields to drop. Food prices will probably rise, but adjustments in farming practices may compensate.
- **Soil** is the uppermost layer of the land surface. The major factors influencing soil and soil development are climate, organisms, relief, parent material, and time.
- The soil layer is a complex mixture of solid, liquid, and gaseous components. The **parent material** of a soil is the mineral matter available to make up the soil. Normally, it is *regolith* that is produced from rock by *weathering*.
- Soil colors range from white to red or yellow to brown and black. Color is usually generated in the soil formation process.
- **Soil texture** refers to the proportions of *sand*, *silt*, and *clay* that are present. A loam is a mixture of sizes. Soil texture determines water-holding characteristics.
- **Peds** are soil grains ranging in size from small to large blocks. *Soil structure* can range from granular to columnar. Granular and blocky structures are the easiest to cultivate.
- Soils range from acidic to alkaline. Soils of cool and cold climates are usually acidic, while arid climates often have alkaline soils.
- **Soil colloids,** the finest particles in soils, are important because they help retain nutrient cations, or **bases**, that are used by plants. If *base status* is high, the soil provides nutrients for plant growth. Acid ions tend to displace bases, making acidic soils generally less fertile.
- In soils, *primary minerals* are chemically altered to become **secondary minerals;** these include *mineral oxides* and *clay minerals*. The nature of the clay minerals determines the soil's base status. *Mineral oxides* are products of long-term weathering.
- The soil receives water from precipitation. It loses water by evapotranspiration and by percolation of water down to groundwater below.
- After a soil is fully wetted by heavy rainfall or snowmelt, *gravitational water* drains from the soil and the soil reaches its **storage capacity** of *available water*. If soil water is lost by evapotranspiration and is not

- recharged, only *hygroscopic water* remains and the *wilting point* is eventually reached.
- The *soil-water balance* describes the gain and loss of soil-water storage. The balance depends on precipitation, evapotranspiration, and runoff.
- Most soils possess distinctive horizontal layers called **soil horizons** that are produced by soil processes acting through time. The **soil profile** describes the layered structure of a particular soil.
- *Organic horizons* (*O horizons*) lie at the top of the soil profile and consist of organic matter in various stages of decay.
- *Mineral horizons* include the *A horizon*, enriched with organic matter; the *E horizon*, from which clay particles and oxides are removed; the *B horizon* in which clays and oxides accumulate; and the *C horizon*, composed of unaltered regolith.
- Mineral horizons are developed by processes of enrichment, removal, translocation, and transformation. In *enrichment*, organic or mineral matter is added to the soil. In *removal*, matter is lost, for example, by erosion.
- In downward *translocation*, humus, clay particles, mineral oxides, and calcium carbonate are removed by **eluviation** from an upper horizon; they accumulate by **illuviation** in a lower one.
- Downward translocation of calcium carbonate can cause *calcification* in dry climates, leading to the formation of a *hard pan*. In *salinization*, salts are translocated upward by evaporating irrigation water; they accumulate near the surface and reduce crop yields.
- In *transformation*, primary minerals are altered to secondary minerals. *Humification* is a transformation process by which organic matter is broken down by bacterial decay.
- Precipitation and temperature are important climate factors in soil development. With high precipitation, nutrients are washed down below rooting depth. With low precipitation, carbonates and salts accumulate in the soil.
- Low temperatures slow the decay of organic matter, allowing humus to accumulate in the soil. High temperatures speed decay, causing soils to lack organic matter.
- Plants can strongly affect soil development. Prairie grasses generate a thick, rich layer of highly productive soil. Conifer forests produce decaying needles that can acidify a soil, diminishing soil nutrients.
- Organisms contribute organic matter to the soil while mixing and reworking it, improving soil texture and structure.
- *Relief* can affect soil development. Soils are thicker on flat and gentle slopes, and thinner on steeper slopes, where runoff carries soil particles downhill.

- Parent material can affect soil development by influencing soil chemistry and providing secondary minerals with unique properties.
- Soil characteristics and properties require time for development—often hundreds to thousands of years.
- Human activity, in the form of plowing and planting crops, can alter soils. Vegetation clearing can induce soil erosion.
- Global soils are classified into 12 **soil orders**, often determined by the presence of one of more diagnostic horizons.
- **Entisols** and **Inceptisols** are soil orders with poorly developed horizons or no horizons. Entisols are composed of fresh parent material and have no horizons. The horizons of Inceptisols are only weakly developed. Entisols and Inceptisols of floodplains and deltas can be very productive.
- **Alfisols** have a horizon of clay accumulation and a high base status. They are very widely distributed, found in moist climates from equatorial to subarctic zones.
- **Spodosols**, found in cold, moist climates supporting needleleaf forests, exhibit an *albic horizon* of eluviation and a *spodic horizon* of illuviation. The base status of Spodosols is low.
- **Oxisols** are old, highly weathered soils of low latitudes. They have a horizon of mineral oxide accumulation and a low base status. They occur in moist climates from equatorial to subtropical zones.
- **Ultisols** are also found in low latitudes with a dry season, extending into the lower midlatitudes. They have a horizon of clay accumulation and are also of low base status.
- **Mollisols** have a thick upper layer rich in humus, with a high base status. They are soils of midlatitude grasslands and are very productive.
- **Aridisols** are soils of desert climates. They are marked by horizons of accumulation of carbonate minerals or salts. With irrigation, and fertilizer, they can be very productive.
- **Gelisols** are soils of permafrost regions that are churned by frost action, inhibiting the development of a soil profile.
- **Vertisols** are rich in a type of clay mineral that expands and contracts with wetting and drying, creating deep cracks that keep the soil well mixed. Many Vertisols are found in subtropical regions with a pronounced dry season. Base status is high.
- **Andisols** are weakly developed soils occurring on young volcanic deposits.
- **Histosols** have a thick upper layer formed almost entirely of organic matter.

KEY TERMS

soil, p. 344
parent material, p. 345
soil texture, p. 345
peds, p. 347
soil colloids, p. 348
base cations or
  bases, p. 348

cation-exchange
  capacity, p. 348
secondary minerals, p. 349
storage capacity, p. 350
soil horizons, p. 352
soil profile, p. 352
eluviation, p. 352

illuviation, p. 353
soil orders, p. 357
Entisols, p. 360
Inceptisols, p. 360
Alfisols, p. 361
Spodosols, p. 362
Oxisols, p. 362

Ultisols, p. 362
Mollisols, p. 364
Aridisols, p. 366
Gelisols, p. 367
Vertisols, p. 368
Andisols, p. 368
Histosols, p. 369

REVIEW QUESTIONS

1. How will increasing temperature affect agricultural crop yields? What will be the effect of increasing atmospheric CO_2 concentrations?

2. What actions can farmers and farm managers take to mitigate the effects of climate change?

3. What is the predicted impact of global climate change on food prices? Who will be most affected?

4. Which important factors condition the nature and development of the soil?

5. Identify important soil materials and components. What is *weathering?* What is *regolith?* What is **parent material**?

6. *Soil color* and *soil texture* are used to describe soils and soil horizons. Identify these terms, showing how they are applied.

7. How does soil structure affect cultivation? What are **peds**? Identify four types of structure.

8. Identify factors that can make a soil *acidic* or *alkaline*.

9. What are **soil colloids**? Why are they important? How is the capability of colloids to hold bases affected by soil acidity?

10. Identify two important classes of **secondary minerals** in soils and provide examples of each class.

11. How does the capability of soils to hold water vary, and how does this capability relate to soil texture? In your answer, define *storage capacity* and *wilting point.*

12. Which processes determine the soil-water balance? How are they related to climate?

13. What is a **soil horizon**? Distinguish between organic and mineral horizons. Generally describe *A, E, B,* and *C horizons.*

14. Identify four classes of soil-forming processes, and briefly describe each.

15. What are *translocation* processes? Explain using the terms **eluviation** and **illuviation**. How is calcium carbonate translocated in soils?

16. Briefly describe the effect of the following factors on soil development: climate, organisms, relief, parent material, time, and human activity.

17. How many **soil orders** are there? Try to name them all.

18. Which soil orders lack distinct horizons? How are they distinguished from each other?

19. Compare **Alfisols** and **Spodosols**. What features do they share? What features distinguish them? Where are they found?

20. Identify two soil orders that are especially associated with low latitudes. For each order, provide at least one distinguishing characteristic and explain it.

21. Describe a **Mollisol** profile. How are the properties of Mollisols related to climate and vegetation cover? Name four suborders within the Mollisols.

22. What are the key features of **Aridisols**? Where are they found?

23. How are **Gelisols** related to climate?

24. Distinguish between **Vertisols** and **Andisols** based on their parent materials.

25. How do **Histosols** differ from other soil orders?

VISUALIZING EXERCISES

1. Sketch the profile of a Spodosol, labeling O, A, E, B, and C horizons. Diagram the movement of materials from the zone of eluviation to the zone of illuviation.

2. Examine the world soils map (Figure 10.18) and identify the soil type nearest to your location.

Also note which soil types of limited occurrence (not shown on the map) might be found nearby. Develop a short list of characteristics that would help you tell these soil types apart.

ESSAY QUESTIONS

1. Document the important role of clay particles and clay mineral colloids in soils. What is meant by the term *clay*? What are *colloids*? What are their properties? How does the type of clay mineral influence soil fertility? How does the amount of clay influence the water-holding capacity of the soil? What is the role of clay minerals in horizon development?

2. Using the world maps of global soils and global climate, compare the pattern of soils on a transect along the 20° E longitude meridian with the patterns of climate encountered along the same meridian. What conclusions can you draw about the relationship between soils and climate? Be specific.

Chapter 11
Earth Materials and Plate Tectonics

Coal is an abundant fossil fuel that provides about one-fourth of the world's energy. Open-pit mines, in which overlying soils and rocks are removed to reveal coal seams, now supply much of the world's coal. Although this form of mining is safer for the miners, the environmental impacts, in the form of spoil banks and toxic wastewater, can be costly. Burning the coal at the power plant releases carbon dioxide, increasing atmospheric concentrations and promoting global warming. Continued releases of CO_2 are not sustainable and eventually must end.

OPEN-PIT COAL MINE, DELMAS, SOUTH AFRICA

©Yann Arthus-Bertrand/Altitude

Earth Materials and Plate Tectonics

T he geography of the Earth—its continents, oceans, mountains, and plains—is determined by geologic processes acting over millions of years. How do these geologic processes work? What is the inner structure of the Earth? What are the outermost layers of the Earth's structure, and why are they important? How are Earth materials formed and transformed? What are the major relief features of our planet? What processes create them? Today's continents are in motion. What happens when they split apart? When they collide? These are some of the questions we will answer in this chapter.

11.1 Geologic time

The geologic timescale encompasses a conceptual framework essential for the study and comprehension of the incremental changes that have shaped today's Earth. Nearly all the landscape features visible to us today were formed within the Paleozoic to Cenozoic eras. Geographers study the evolution of the landscape and biodiversity using the geologic timescale as a universal tool for common reference and understanding.

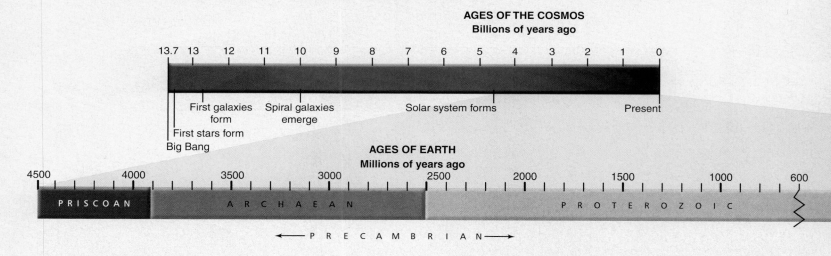

AGES OF THE COSMOS
Billions of years ago

13.7 13 12 11 10 9 8 7 6 5 4 3 2 1 0

First galaxies form
First stars form
Big Bang
Spiral galaxies emerge
Solar system forms
Present

AGES OF EARTH
Millions of years ago

4500 4000 3500 3000 2500 2000 1500 1000 600

PRISCOAN   ARCHAEAN   PROTEROZOIC

← P R E C A M B R I A N →

© David Hutt /Alamy Limited

◀ PRISCOAN
The earliest-known rocks on the Earth date from this period. Scientists used radiometric dating to date these 4.3-billion-year-old rocks in Canada's Northwest Territories.

Courtesy Didier Descouens

PROTEROZOIC ▶
Around 2.8 to 2.4 billion years ago, photosynthetic bacteria, growing as layers and mats in sunlit shallow water, began to add oxygen to the atmosphere. The rock shown here is formed from layers of these fossilized microbes built up over time.

The Changing Earth

This chapter begins the section of the book dedicated to *geomorphology*, the study of the shape of the Earth's features and how they change over time. Geographers build upon the foundations of geomorphology to understand the differences in the distribution of humans, plant and animal species, and natural resources. At broader scales, the shape of the Earth's surface features depends on the underlying rocks, so we begin our study with a close look at the Earth's inner structure, materials, and global topography.

To reconstruct the Earth's history over geologic time, geologists rely on several big ideas. One of these is *uniformitarianism*, the idea that the same geologic processes we can observe today have operated since the beginning of the Earth's history. This means that the same cycles and forces that shape the Earth today can help us understand how it has changed since its earliest history. In other words, the present is the key to the past.

THE TIMESCALE FOR GEOLOGIC CHANGE

In order to talk about the time frame of geologic events, we use the *geologic timescale* (Figure 11.1). Geologists divide the 4.5 billion years since the Earth formed into *eons*, *eras*, and *periods*. Eons are vast chunks of time

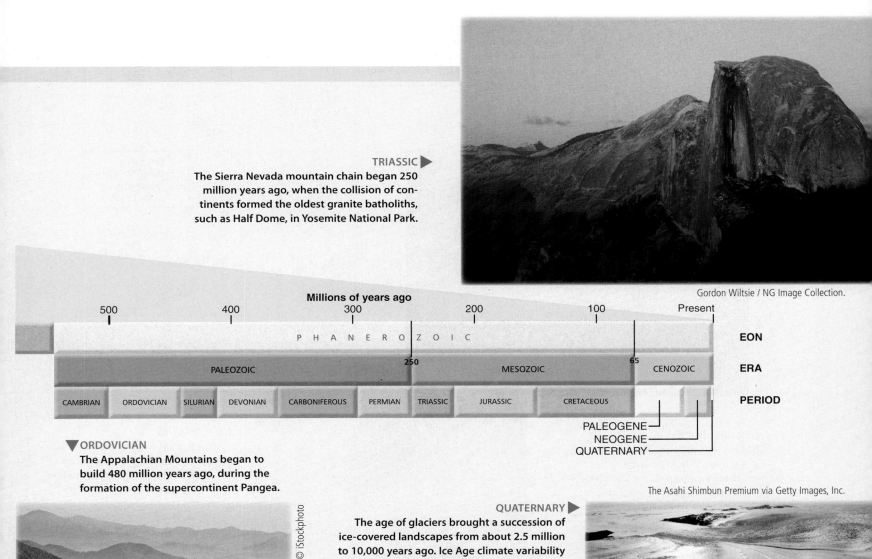

TRIASSIC ▶
The Sierra Nevada mountain chain began 250 million years ago, when the collision of continents formed the oldest granite batholiths, such as Half Dome, in Yosemite National Park.

Gordon Wiltsie / NG Image Collection.

Millions of years ago

500   400   300   200   100   Present

PHANEROZOIC | **EON**

| PALEOZOIC | 250 | MESOZOIC | 65 | CENOZOIC | **ERA** |

| CAMBRIAN | ORDOVICIAN | SILURIAN | DEVONIAN | CARBONIFEROUS | PERMIAN | TRIASSIC | JURASSIC | CRETACEOUS | | | **PERIOD** |

PALEOGENE
NEOGENE
QUATERNARY

▼ORDOVICIAN
The Appalachian Mountains began to build 480 million years ago, during the formation of the supercontinent Pangea.

© iStockphoto

The Asahi Shimbun Premium via Getty Images, Inc.

QUATERNARY ▶
The age of glaciers brought a succession of ice-covered landscapes from about 2.5 million to 10,000 years ago. Ice Age climate variability influenced species distribution and abundance, including modern *Homo sapiens*. The Antarctic's present-day landscape of glaciers portrays the challenging environment of ice-covered landforms.

divided into eras, which in turn are divided into periods. The divisions are broad in the early parts of the Earth's history and much narrower in the past few hundred million years, which we know more about. Many of the names of the divisions describe the world or ecosystems at the time. For example, Paleozoic means "old life" and includes early land plants, insects, reptiles, and amphibians; Mesozoic means the era of "middle life" dominated by the dinosaurs. Period names often come from rock formations, such as the coals of the Carboniferous period.

Geologists use *radiometric dating* to determine the ages of different rocks (Figure 11.2). This technique uses what we know about physics and the rate of radioactive decay of rock elements to determine the ages of rocks. Many geologic timescale divisions were originally drawn based on the dating of rock layers. As scientists use advances in radiometric techniques to date the layers or individual rock formations, a more accurate picture is emerging of the Earth's geologic history,

A very important benchmark in the geologic timescale is the *Cambrian period*, when life on Earth began to flourish. In *Precambrian* time, life had early beginnings but is generally absent from the geologic record. Nearly all the landscape features visible today were formed

11.2 Radiometric dating

Using the principles of radioactive decay and measurements of the ratios of particular isotopes of uranium and lead, geophysicists can date the time of formation of certain types of crystals within a rock.

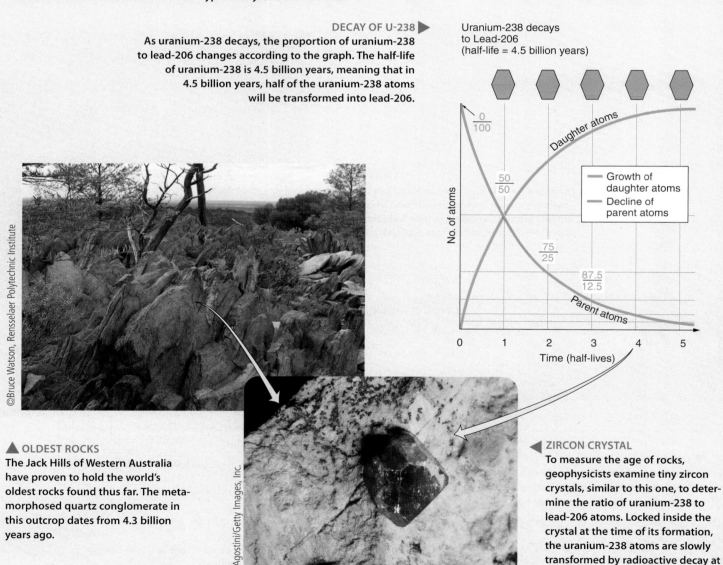

DECAY OF U-238 ▶
As uranium-238 decays, the proportion of uranium-238 to lead-206 changes according to the graph. The half-life of uranium-238 is 4.5 billion years, meaning that in 4.5 billion years, half of the uranium-238 atoms will be transformed into lead-206.

Uranium-238 decays to Lead-206 (half-life = 4.5 billion years)

Growth of daughter atoms
Decline of parent atoms

Daughter atoms
Parent atoms

No. of atoms

Time (half-lives)

©Bruce Watson, Rensselaer Polytechnic Institute

DeAgostini/Getty Images, Inc.

▲ **OLDEST ROCKS**
The Jack Hills of Western Australia have proven to hold the world's oldest rocks found thus far. The metamorphosed quartz conglomerate in this outcrop dates from 4.3 billion years ago.

◀ **ZIRCON CRYSTAL**
To measure the age of rocks, geophysicists examine tiny zircon crystals, similar to this one, to determine the ratio of uranium-238 to lead-206 atoms. Locked inside the crystal at the time of its formation, the uranium-238 atoms are slowly transformed by radioactive decay at a known rate into lead-206.

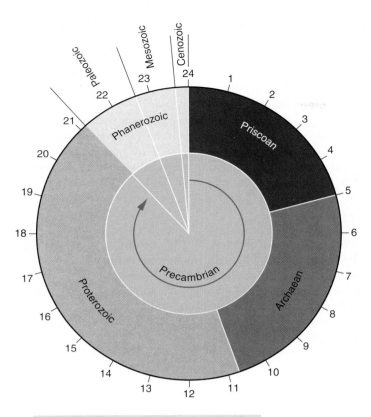

11.3 The Earth's history on a 24-hour clock

If the Earth's history were compressed into a single 24-hour day, the last 2 hours and 50 minutes would delineate the proliferation of life.

within the *Cenozoic era*, which is the most recent.

The vast scale of the Earth's history is hard to grasp. If you think of the history of the Earth since its formation as spanning a single 24-hour day, you can place the age of each geologic time division on a 24-hour timescale (Figure 11.3). Precambrian time ends at about 21:10. That means that life only proliferated on Earth during the last 2 hours and 50 minutes of this day. The human genus itself arises at about 30 seconds before midnight, and the last 5000 years of human civilization occupy about half a second—truly a fleeting moment in our planet's vast history.

> Life on Earth became abundant during the Cambrian period. The Cenozoic era is the most recent, and nearly all the landscape features visible today were formed within that era.

FORCES OF GEOLOGIC CHANGE

The Earth's surface is constantly changing, as old crust is broken down and new crust is formed. Volcanic and tectonic activity brings fresh rock to the planet's surface. We call these internal, or *endogenic, processes*, because they work from within the Earth. External, or *exogenic, processes*, such as weathering by wind and water, work at the Earth's surface. They lower continental surfaces by removing and transporting mineral matter through the action of running water, waves and currents, glacial ice, and wind.

The specific shapes of the Earth's surface are called *landforms*. We distinguish between landforms that have been freshly created and those that have been shaped by weathering, erosion, and mass wasting. *Initial landforms* are newly created by volcanic and tectonic activity. Exogenic processes wear down initial landforms to create *sequential landforms* (Figure 11.4). By examining landforms, we can begin to understand what kinds of Earth processes formed and shaped them. Geomorphology's major processes and landforms are the major focus of this, the last, section of the book.

The Structure of the Earth

Before we begin to discuss how the Earth's surface is changing, we need to establish what we know about the structure of the Earth and how this affects what we see at the surface.

▼ CRUSTAL ACTIVITY
This endogenic process produces an initial landform in the shape of this mountain block.

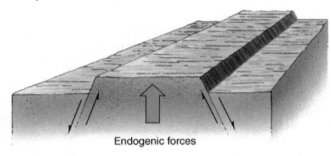

Endogenic forces

Exogenic forces

Erosional:
Canyon
Divide

Depositional (fan)

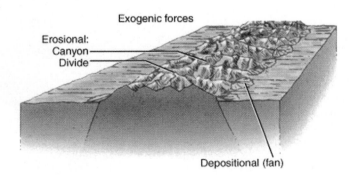

▲ EROSION AND DEPOSITION
These exogenic processes carve the mountain into a sequential landform.

11.4 Initial and sequential landforms

Scientists gain knowledge about the Earth's interior from a variety of observations. Active lava flows from volcanoes show us what lies below the Earth's crust. We can study exposed rock layers in road cuts and canyons to learn how sediments and other rock formations are laid down over tens of thousands, even millions, of years to create a record of past climates and Earth-forming events. With deep drilling, we can study the composition of the Earth near the surface. To explore the Earth at even deeper levels, we can observe the paths of earthquake waves traveling through its center. A combination of field sampling and laboratory testing, in conjunction with scientific publications and vibrant discussions among Earth scientists, has helped to reveal the hidden world of our planet's structure.

Our planet contains a central core, with several layers, or shells, surrounding it (Figure 11.5). The densest matter is at the center, and each layer above it and up to the surface is increasingly less dense and cooler.

THE CORE

The Earth's central **core** is about 3500 km (about 2200 mi) in radius and is very hot—somewhere between 3000°C and 5000°C (about 5400°F to 9000°F). We know from measurements of earthquake waves passing through the Earth that the core consists of two distinct layers. The outer core is liquid, as demonstrated by the

fact that energy waves suddenly change behavior when they reach this boundary. In contrast, the inner core is solid and made mostly of iron, with some nickel. The inner core remains solid despite the high temperatures because of the extreme pressure of all the Earth materials surrounding it.

The liquid iron core creates a magnetic field as the fluid flows around the solid core and interacts with the Earth's existing magnetic field. This process in turn generates a dynamic energy condition that maintains the Earth's perpetual magnetic field.

THE MANTLE

The core is surrounded by the **mantle**, a shell about 2900 km (about 1800 mi) thick, made of mafic (a word formed from "ma," for *ma*gnesium-bearing, and "fic," from *fer*ic, or iron-bearing) silicate minerals. Mantle temperatures range from about 2800°C (about 5100°F) near the core to about 1800°C (about 3300°F) near the crust. The mantle is the largest of the Earth's layers, making up more than 80 percent of the Earth's total volume.

Like the core, the mantle can be subdivided further into zones, characterized by different temperatures and compositions. The lower mantle is hotter than the upper mantle, but it is largely rigid, as a result of intense pressure at this depth.

11.5 The structure of the Earth

Seismic studies and other direct and indirect observations have given scientists a good idea of the internal structure of the Earth.

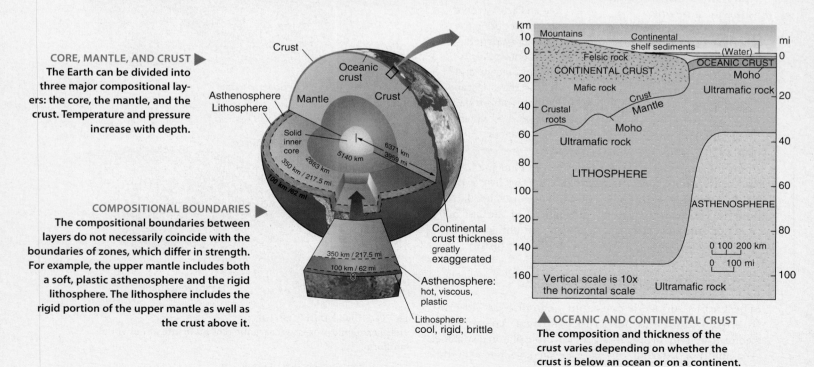

In the upper mantle, although the temperature is lower, the pressure is also lower, so rocks are less rigid. This softer, plastic part of the mantle is the *asthenosphere*. This is the most fluid layer of the mantle, where molten material forms in hotspots as a result of heat generated by endogenic radioactive decay. Processes occurring in this region provide the energy to drive earthquake and volcanic activity at the Earth's surface, along with continuous motions that deform the Earth's crust.

> The layers of the Earth's interior include the crust, mantle, liquid outer core, and solid inner core. Continental crust has both felsic and mafic rock zones, while oceanic crust has only mafic rock.

THE CRUST AND LITHOSPHERE

The thin, outermost layer of our planet is the Earth's **crust**. The Earth's crust is separated from the mantle by the boundary called the *Moho* (short for Mohorovičić discontinuity). At the Moho, seismic waves indicate that a sudden change in the density of materials occurs. The crust, composed of varied rocks and minerals, ranges from about 7 to 40 km (about 4 to 25 mi) thick and contains the continents and ocean basins. It is the source of soil on the lands, salts of the sea, gases of the atmosphere, and all the water of the oceans, atmosphere, and lands.

The crust that lies below ocean floors—*oceanic crust*—consists almost entirely of mafic rocks. The *continental crust* consists of two continuous zones: a lower zone of dense mafic rock and an upper zone of lighter felsic rock. Felsic rock (a word formed from *feld*spar and *s*ilica) is composed of silicates of aluminum, sodium, potassium, and calcium. Another key distinction between continental and oceanic crust is that the crust is much thicker beneath the continents—35 km (22 mi) on average—than it is beneath the ocean floors, where it is typically 7 km (4 mi).

Geologists use the term *lithosphere* to describe an outer Earth shell of rigid, brittle rock, including the crust and the cooler, upper part of the mantle. The lithosphere ranges in thickness from 60 to 150 km (40 to 95 mi). It is thickest under the continents and thinnest under the ocean basins.

The lithosphere floats on the plastic asthenosphere, much like an iceberg floats in water, in an equilibrium state known as *isostasy*. As erosion reduces the height and weight of a mountain block over time, it becomes lighter and floats higher on the asthenosphere by the process of isostasy. This causes the block to rise, making the mountains higher again and subject to more erosion. The process continues until the mountains become a range of low hills with shallow mountain roots.

You can think of the lithosphere on top of the asthenosphere as a hard, brittle shell resting on a soft, plastic underlayer. Because the asthenosphere is soft and plastic, the rigid lithosphere can move easily over it. The lithospheric shell is divided into large pieces called **lithospheric plates**. As we will see later in this chapter, a single plate can be as large as a continent and move independently of the plates that surround it—like a great slab of floating ice on the polar sea. Lithospheric plates can separate from one another at one location, while elsewhere they may collide in crushing impacts that raise great mountains. The major relief features of the Earth—its continents and ocean basins—were created by the continuous movements of these plates in geologic timescales on the surface of the Earth, a phenomenon we will discuss in greater detail throughout the upcoming chapters.

> The lithosphere is the solid, brittle outermost layer of the Earth. It includes the crust and the cooler, brittle upper part of the mantle. The asthenosphere, which lies below the lithosphere, is plastic.

Earth Materials and the Cycle of Rock Change

The most abundant elements in the Earth's crust are oxygen, silicon, aluminum, iron, calcium, sodium, potassium, and magnesium (Figure 11.6). They exist in a variety of rock combinations and are formed through

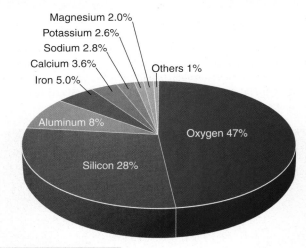

Magnesium 2.0%
Potassium 2.6%
Sodium 2.8%
Calcium 3.6%
Iron 5.0%
Others 1%
Aluminum 8%
Oxygen 47%
Silicon 28%

11.6 Crustal elements

The Earth's crust is composed of the principal elements listed. Oxygen, which is bound with other elements as oxides, represents approximately half of the crustal volume. If we include water (H_2O) located in the crust, groundwater, and aquifers, the amount of oxygen is much higher.

physical and chemical processes, both endogenic and exogenic. Oxygen, the most abundant element, readily combines with many of these elements in the form of oxides. Crustal elements can form chemical compounds that we recognize as minerals (Figure 11.7).

11.7 Examples of minerals

▼ **SALT CRYSTALS**
Minerals are naturally occurring crystalline chemical compounds. Salt, sodium chloride, is an example. These clear salt crystals were deposited near the vent of an underwater volcano. In many rocks, mineral crystals are too small to be seen without magnification.

Carsten Peter/NG Image Collection

▼ **QUARTZ CRYSTALS**
Quartz, or silicon dioxide, is a very common mineral. Under unusual circumstances, quartz is found as regular six-sided crystals, shown here in this sample from Venezuela. Usually, it is found as a clear or light-colored mineral, and it is often present in sediments such as beach sand or river gravel.

Michael Nichols/NG Image Collection

A **mineral** is a naturally formed solid, inorganic substance with a characteristic crystal structure and chemical composition.

Rocks are usually composed of two or more minerals. Often, many different minerals are present, but a few rock varieties are made almost entirely of one mineral. Most rock in the Earth's crust is extremely old, dating back many millions of years, but rock is also being formed at this very hour, as active volcanoes emit lava that solidifies on contact with the atmosphere or ocean.

> Rocks are composed of minerals, naturally occurring inorganic substances. The three classes of rocks are igneous, sedimentary, and metamorphic.

Rocks fall into three major classes: *igneous, sedimentary*, and *metamorphic*. Each class of rock has unique properties and structures that affect how the rocks are shaped into landforms. Different types of rocks are worn down at different rates. Some are easily eroded, whereas others are much more resistant. For example, we usually find weak rocks under valleys, and strong rocks under hills, ridges, and uplands (Figure 11.8).

WileyPLUS Virtual Rock Lab Interactivity
How are minerals and rocks identified? Learn the characteristics of common minerals and rocks and test the knowledge you've gained in the chapter with this interactivity.

IGNEOUS ROCKS

Igneous rocks, are formed when molten material, or **magma**, solidifies. The magma moves upward from pockets a few kilometers below the Earth's surface through fractures in older solid rock. There the magma cools, forming rocks of mineral crystals.

Magma that solidifies below the Earth's surface and remains surrounded by older, preexisting rock is called *intrusive igneous rock*. Because intrusive rocks cool slowly, they develop mineral crystals that are visible to the eye. If the magma reaches the surface and emerges as lava, it forms *extrusive igneous rock* (Figure 11.9). Extrusive igneous rocks cool very rapidly on the land surface or ocean bottom and thus show crystals of only microscopic size. You can see formation of extrusive igneous rock today where volcanic processes are active.

Most igneous rock consists of *silicate minerals*, chemical compounds that contain silicon and oxygen atoms. These rocks also contain mostly metallic elements. The mineral grains in igneous rocks are very tightly interlocked, so the rock is normally very hard. Quartz (see Figure 11.7B), which is made of silicon dioxide (SiO_2), is the most common mineral of all rock classes. It is quite hard and resists

11.8 Landforms and rock resistance

Weaker rock erodes more rapidly than resistant rock, leaving the resistant rock standing as mountains, ridges, or belts of hills. Igneous rock, consisting of crystals formed by cooling of molten rock, is often resistant to erosion, forming mountains and high plateaus. Most metamorphic rocks are also resistant to erosion, typically forming hills and hill belts. Sedimentary rocks can be strong or weak, depending on their composition. Strong sedimentary rocks form ridges or hills, while weak ones form valleys.

Universal Images Group via Getty Images, Inc.

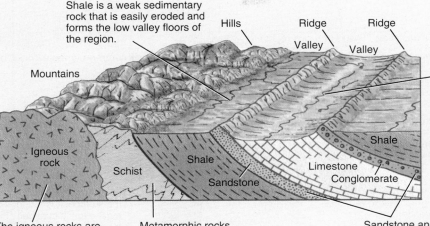

Shale is a weak sedimentary rock that is easily eroded and forms the low valley floors of the region.

Mountains · Hills · Ridge · Valley · Ridge · Valley

Limestone is dissolved by carbonic acid in rain and surface water, also forming valleys in humid climates. In arid climates limestone is a resistant rock and usually forms ridges and cliffs.

Igneous rock · Schist · Shale · Sandstone · Conglomerate · Limestone · Shale

The igneous rocks are resistant—typically forming uplands or mountains rising above adjacent areas of shale and limestone.

Metamorphic rocks vary in resistance. Metamorphic rocks are sometimes found in contact with igneous rock that provides the heat for the metamorphic process.

Sandstone and conglomerate are typically resistant and form ridges or uplands.

▲ GRANITE

This igneous rock is rich in quartz, which is very resistant to erosion and decay. Here, a climber ascends a steep granite slope in the White Mountains of New Hampshire.

SANDSTONE ▶

Sandstone is a resistant rock, and sandstone layers often stand out as ridges along the landscape. In this image of Virginia's Shenandoah Valley, taken from the Space Shuttle, sandstone ridges alternate with softer rocks of limestone and shale, which are extensively farmed.

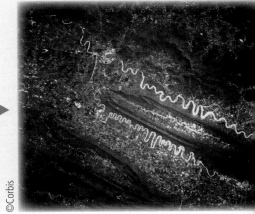

©Corbis

11.9 Lava

Lava is molten rock that reaches the Earth's surface. It can flow out from the volcanic vent as a thick liquid, or be ejected violently as fine volcanic ash and coarser particles, sometimes as large as boulders. These photos are from recent eruptions of Kilauea volcano in Hawaii Volcanoes National Park.

© Douglas Peebles Photography/Alamy Limited

Ralph Lee Hopkins/NG Image Collection

◀ FIRE FOUNTAIN

During a large eruption, lava can run down the flanks of the volcano as a river of fire. At Kilauea, some of these lava flows reach the sea.

LAVA TONGUE ▶

As shown by its red-hot interior, this recent tongue of lava is still cooling. The cooled lava forms the igneous extrusive rock basalt.

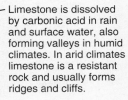

chemical breakdown. Beach sand is largely composed of tiny, tough quartz grains.

Figure 11.10 shows some other common silicate minerals and the intrusive and extrusive rocks that are made from them. Felsic rock contains mostly *felsic minerals*, which are light-colored and less dense. Mafic rock contains mostly *mafic minerals*, which are dark-colored and denser. *Ultramafic* rock is dominated by mafic minerals rich in magnesium and iron and is the densest of the three rock types.

11.10 Silicate minerals and igneous rocks

Only the most common silicate mineral groups are listed, along with six common igneous rocks, both volcanic and plutonic.

Grain size →

Felsic

Extrusive rocks

Rhyolite lies at the felsic, high-silica end of the scale. It is usually pale, ranging from nearly white to shades of gray, yellow, red, or lavender.

Intrusive rocks

Granite the intrusive equivalent of rhyolite, is common because felsic magmas usually crystallize before they reach the surface. It is found most often in continental crust, especially in the cores of mountain ranges.

Courtesy Brian J. Skinner

Silica content

Andesite is an intermediate-silica rock. It is usually light to dark gray, purple, or green.

Diorite is the intrusive equivalent of andesite, an intermediate-silica rock.

Courtesy Brian J. Skinner

Basalt, a mafic rock, is dominant in oceanic crust and the most common igneous rock on Earth. It typically has a dark gray, dark green, or black color.

Gabbro is the intrusive equivalent of basalt, a low-silica rock.

Courtesy Brian J. Skinner

Mafic

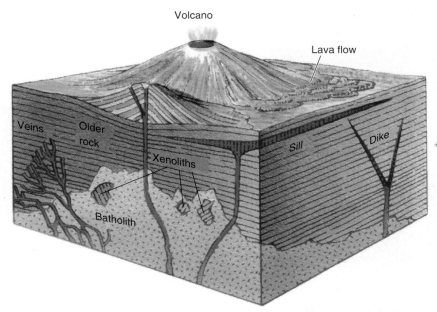

Volcano

Lava flow

Veins
Older rock
Xenoliths
Batholith
Sill
Dike

11.11 Plutons

Igneous intrusions of hot magma cool to form plutons. The largest of these intrusions are *batholiths*. When magma pushes its way vertically between fractures in the rock, it cools to form a *dike*; and when it is horizontal, it forms a *sill*. Other plutons are the channels of extinct volcanoes. Batholiths may contain pieces of surrounding, unmelted rock called *xenoliths*.

When igneous rock is extruded at the surface, it forms volcanoes and lava flows (Figure 11.11). When igneous rock is intruded into surrounding rocks and cools, it forms **plutons**, intrusive rock bodies of any size. The largest plutons are *batholiths*. Smaller plutons include *sills* and *dikes*. Sometimes, batholiths contain *xenoliths*, fragments of surrounding rock that are incorporated into the batholith without melting.

> Intrusive igneous rocks cool slowly below the Earth's surface and develop visible mineral crystals. Extrusive igneous rocks cool rapidly on the land surface or ocean bottom and show microscopic crystals.

WileyPLUS Igneous Rock Animation
Take another look at our diagram of silicate minerals and igneous rocks. Click on the diagram to see photos of the minerals and rocks and learn more about their characteristics.

SEDIMENTS AND SEDIMENTARY ROCKS

Now let's turn to the second rock class, the **sedimentary rocks**. Sedimentary rocks are made from layers of mineral particles found in other rocks (igneous, sedimentary, and metamorphic) that have been released by weathering. They also include rocks made from newly formed organic matter, both plant biomass and invertebrates. Most inorganic minerals in sedimentary rocks are from igneous rocks.

Rocks exposed at the Earth's surface are broken down into fragments of many sizes in a process called *weathering*. Physical weathering divides rock into ever-smaller pieces, and chemical weathering alters the chemical composition of mineral grains through exposure to oxygen and water. When weathered rock or mineral particles are transported by air, water, or glacial ice, we call them *sediment*. Streams and rivers carry sediment to lower levels, where it builds up. Sediment usually accumulates on shallow seafloors bordering continents, but it also collects in inland valleys, lakes, and marshes. Wind can also transport sediment.

Over geologic timescales, the sediment becomes compacted and hardens to form sedimentary rock, with distinctive visible characteristics. Sediment is generally deposited by wind and water in layers, called **strata**. The study of thickness and patterns of these strata can tell us much about the history of an area, including the past presence of oceans and rivers, as well as the history of faulting and deformation. In the Grand Canyon, for example, light-colored layers indicate long drought periods, while dark-colored strata indicate periods of more moist conditions.

There are three major classes of sediment: clastic, chemically precipitated, and organic (Figure 11.12).

> Sedimentary rocks are composed of sediment, which may be clastic, chemically precipitated, or organic. Layers of sediment are termed strata.

Clastic sediment

Sediment that is made up of inorganic rock and mineral fragments (clasts) is called *clastic sediment*. Clastic sediments can come from igneous, sedimentary, or metamorphic rocks, and so they can include a very wide range of minerals. Quartz and feldspar usually dominate clastic sediment.

11.12   Some common sedimentary rock types

Subclass	Rock type		Composition
Clastic (Composed of rock and/ or mineral fragments)	Sandstone		Cemented sand grains
	Siltstone		Cemented silt particles
	Conglomerate		Sandstone contain in pebbles of hard rock
	Mudstone		Silt and clay, with some sand
	Claystone		Clay
	Shale		Clay, breaking easily into flat flakes and plates

Bill Hatcher/NG Image Collection

Walter Meayers Edwards/
NG Image Collection

Subclass	Rock type		Composition
Chemically precipitated (formed by chemical precipitation from sea water, sometimes with the help of micro-or macro organisms)	Limestone		Calcium carbonate, formed by precipitation on sea or lake floors
	Dolomite		Magnesium and calcium carbonates, similar to limestone
	Chert		Silica, a microcrystalline form of quartz
	Evaporites		Minerals formed by evaporation of salty solutions in shallow inland lakes or coastal lagoons

Norbert Rosing/NG Image Collection

Subclass	Rock type		Composition
Organic (formed from organic material)	Coal		Rock formed from peat or other organic deposits; may be burned as a mineral fuel
	Petroleum (mineral fuel)		Liquid hydrocarbon found in sedimentary deposits; not a true rock but a mineral fuel
	Natural gas (mineral fuel)		Gaseous hydrocarbon found in sedimentary deposits; not a true rock but a mineral fuel

Melissa Farlow/NG Image Collection

The size of clastic sediment particles determines how easily and how far they are transported by wind and water currents. Fine particles are easily suspended in slowly moving fluids, while coarser particles are moved only by stronger currents of water or air. When the fluid velocity decreases, coarser particles settle out first. In this way, particles of different sizes are sorted by the fluid motion.

When layers of clastic sediment build up, the lower strata are pushed down by the weight of the sediments above them. This pressure compacts the sediments, squeezing out excess water. Dissolved minerals recrystallize in the spaces between mineral particles in a process

called cementation. Figure 11.13 shows some important varieties of clastic sedimentary rocks.

Chemically precipitated sediment

Chemically precipitated sediment is made of solid inorganic mineral compounds that separate out from saltwater solutions or from the silica and calcium carbonate shells of microorganisms. Rock salt

> Clastic sedimentary rocks are formed when sediments are compressed and cemented. Sandstone and shale are common examples. Limestone is formed by chemical precipitation in a marine environment.

11.13 Some important varieties of clastic sedimentary rocks

Bill Hatcher/NG Image Collection

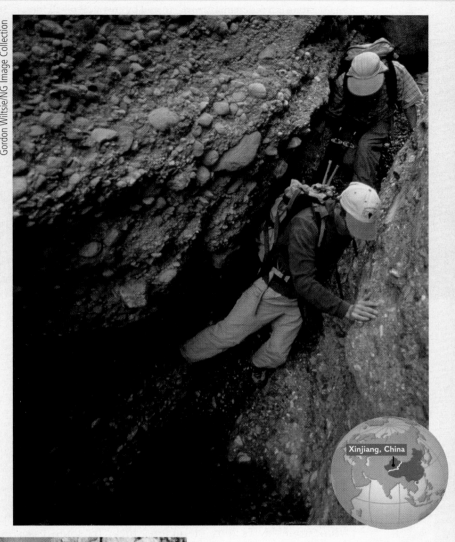

Gordon Wiltsie/NG Image Collection

Xinjiang, China

▲ SANDSTONE

Sandstone is composed of sand-sized particles, normally grains of eroded quartz that are cemented together in the process of rock formation. This example is Navajo sandstone, which is found on the Colorado Plateau in Utah and Arizona. It was originally deposited in layers by moving sand dunes.

CONGLOMERATE ▶

Conglomerate is a sedimentary rock of coarse particles of many different sizes. These climbers are working their way through some beds of conglomerate on their way to the top of Shipton's Arch, Xinjiang, China. On the left, the softer sediment between hard cobbles has eroded away, leaving the rounded rocks sticking out.

◀ SHALE

Shale is a rock formed mostly from very fine-textured particles—silt and clay—deposited in calm water. It is typically gray or black in color and breaks into flat plates, as shown here. Some shale deposits contain fossils, like these ancient trilobites, marine animals of Cambrian age, found near Antelope Springs, Utah.

Walter Meayers Edwards/NG Image Collection

11.14 Chalk cliffs provide a record of climate change

Limestone is a sedimentary rock created by chemical precipitation. Some limestones are rich in microfossils, which can give clues to the ocean environment at the time of precipitation.

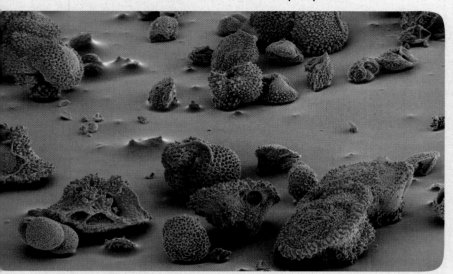

Eric Condliffe, University of Leeds Electron Optics Image Laboratory

▲ DIATOMS

By examining the fossil shells of each layer under a microscope, scientists can observe the shifts in ocean water temperatures over tens of thousands of years, and thus compile a record of the Earth's climate changes in past millennia.

Norbert Rosing/NG Image Collection

▲ CHALK CLIFFS

The white chalk cliffs of the island of Rügen off the coast of Germany, are composed of layers of calcium carbonate from the shells of ancient microorganisms. These microorganisms form their tiny calcium carbonate shells in different shapes, depending on the temperature and acidity of the ocean water.

is an example of chemically precipitated sediment. One of the most common sedimentary rocks formed by chemical precipitation is limestone (Figure 11.14). Sediments of the tiny shells of microscopic organisms (diatoms and foraminifera, for example) can form limestone layers tens of meters thick. In some cases, the composition of the shells can indicate ocean water conditions at the time of precipitation. Limestone can also be formed from coarse fragments of shells or corals.

Organic sediment

Organic sediment is made up of the tissues, or biomass, of plants and animals. Peat is an example of an organic sediment. This soft, fibrous, brown or black substance accumulates in bogs and marshes where the acidic water conditions prevent the decay of plant or animal remains.

Peat is a form of *hydrocarbon*, a compound of hydrogen, carbon, and oxygen. Hydrocarbon compounds are the most important type of organic sediment—one that we increasingly depend on for fossil fuel. They formed from the remains of plants or tiny zooplankton and algae that built up over millions of years and were compacted under thick layers of inorganic clastic sediment. Hydrocarbons can be solid (peat and coal), liquid (petroleum), or gas (natural gas). Coal is the only hydrocarbon that is a rock (Figure 11.15). We often find natural gas and petroleum in open, interconnected

pores in a thick sedimentary rock layer, such as in a porous sandstone (Figure 11.15).

The Earth's **fossil fuels** have accumulated over many millions of years. These fuels are considered *nonrenewable resources*, meaning that, as far as we are concerned, once they are gone, there will be no more; for even in a thousand years, the amount of these fuels created will scarcely be measurable in comparison to the stores produced in the geologic past. And as our industrial society continues to consume them, ever more rapidly, they become scarcer, and more costly, as well.

> Hydrocarbons in sedimentary rocks include coal, petroleum, natural gas, and peat. These mineral fuels power modern industrial society.

The amount of carbon currently being released into the atmosphere from the burning of fossil fuels is causing alarm among climate change scientists. Rapid release of geologically stored carbon is pushing the atmospheric levels of CO_2, a greenhouse gas, to the highest recorded levels in over 600,000 years. The vast majority of these scientists believe that these increases are responsible for raising global temperatures, which in turn are creating weather systems of greater variability and magnitude.

WileyPLUS Clastic Rocks Animation
Learn the terms used for clasts of different sizes, and examine several different types of clastic rocks.

11.15 Fossil fuels

Knowledge of sedimentary rock formations helps geologists search for fossil fuels.

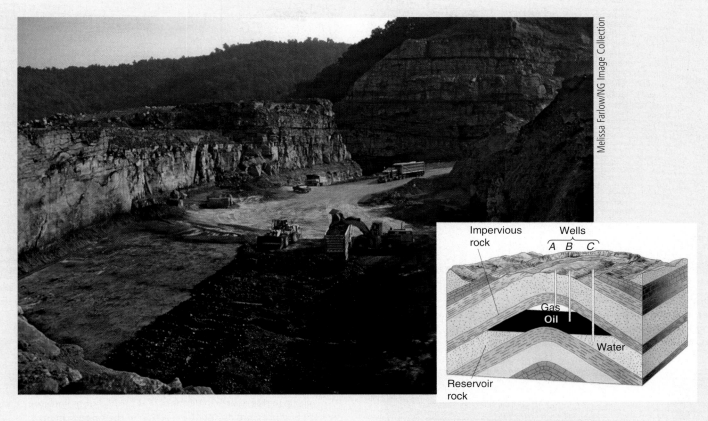

▲ STRIP MINE

In strip mining, including the controversial practice of mountaintop removal, layers of coal are mined by removing overlying rock, allowing direct access to the coal deposit. This strip mine is located in Man, West Virginia.

▲ TRAPPING OIL AND GAS

This idealized cross section shows an oil pool on a dome structure in sedimentary strata. Well A will draw gas, well B will draw oil, and well C will draw water. Here, the impervious rock is shale, and the reservoir rock is sandstone.

METAMORPHIC ROCKS

The mountain-building, or *orogenic*, processes of the Earth's crust involve tremendously high pressures and temperatures. These extreme conditions alter igneous and sedimentary rock, transforming the rock so completely in texture and structure that we have to reclassify it as **metamorphic rock** (Figure 11.16). In many cases, the mineral components of the parent rock are changed into different mineral varieties. In some cases, the original minerals may recrystallize.

Metamorphic rocks tend to be more resistant to weathering than their parent rocks because the heat and pressure welds their mineral grains together and may transform their

> Metamorphic rocks are formed from preexisting rocks by intense heat and pressure, which alter rock structure and chemical composition. Shale is transformed into slate or schist, sandstone into quartzite, limestone into marble, and igneous rocks into gneiss.

minerals into stronger forms. For example, extreme heat and pressure transform shale into *slate* or *schist*, sandstone into *quartzite*, limestone into *marble*, and igneous rocks or clastic sediments into *gneiss*.

Metamorphism can occur in several different situations. When magma intrudes into deep crustal rocks, the heat can transform the adjacent rock, in a process called *contact metamorphism*. For example, marble forms when heat and pressure in the Earth's crust causes limestone to recrystallize and form larger, more uniform crystals of calcite.

Regional metamorphism is associated with large areas of tectonic activity from the collision of lithospheric plates. This form of metamorphism is responsible for the more common rocks such as slate, schist, and gneiss that are associated with major mountain ranges.

THE CYCLE OF ROCK CHANGE

Rocks are constantly being transformed from one class to another in the **cycle of rock change**, which

11.16 Some common metamorphic rock types

Metamorphic rocks are formed from preexisting rocks by heat and pressure.

Raymond Gehman/NG Image Collection

▲ QUARTZITE

Under heat and pressure, sandstone recrystallizes to form a very strong rock of connected quartz grains called quartzite. Seneca Rocks, West Virginia, provides a striking example.

Ralph Lee Hopkins/NG Image Collection

▲ SCHIST

When shale is exposed to heat and pressure for long periods, the minerals in the shale recrystallize and grow together to form a stronger rock called schist. Shown is Granite Gorge, Grand Canyon National Park.

◀ MARBLE

Marble can serve as a useful and beautiful building material. The marble in this quarry in Vermont was formed from limestone by heat and pressure during the creation of the Appalachian mountain chain.

Robert Sisson/NG Image Collection

Rock Type	→	Description
Slate	→	Shale exposed to heat and pressure that splits into hard flat plates
Schist	→	Shale exposed to intense heat and pressure that shows evidence of shearing
Quartzite	→	Sandstone that is "welded" by a silica cement into a very hard rock of solid quartz
Marble	→	Limestone exposed to heat and pressure, resulting in larger, more uniform crystals
Gneiss	→	Rock resulting from the exposure of clastic sedimentary or intrusive igneous rocks to heat and pressure

recycles crustal minerals over many millions of years (Figure 11.17). This cycle, operating over the geologic timescale, involves various physical, chemical, and biological processes that create, transform, and recycle the Earth's crust.

In the surface environment, rocks weather into fragments that form sediments through erosion and deposition. As sediments accumulate, they are buried and compressed by the weight of sediments above and are cemented by mineral-rich solutions to form sedimentary rock. In the deep-Earth environment, igneous or sedimentary rock is transformed by heat and pressure into metamorphic rock, which may ultimately melt to form magma. When the magma cools, either on or near the surface, it forms igneous rock that can be weathered into sediment, completing the cycle.

The cycle of rock change has been active since our planet became solid and internally stable, continuously

In the surface environment, rocks weather into sediment. In the deep environment, heat and pressure transform sediment into rock that is eventually exposed at the surface.

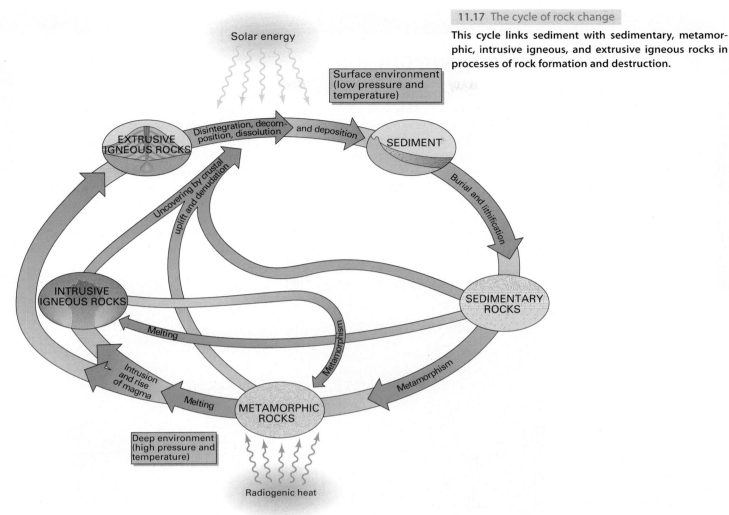

This cycle links sediment with sedimentary, metamorphic, intrusive igneous, and extrusive igneous rocks in processes of rock formation and destruction.

forming and reforming rocks of all three major classes. Not even the oldest igneous and metamorphic rocks found so far are the "original" rocks of the Earth's crust. These were recycled eons ago.

Global Topography

Having outlined the structure of the Earth and the composition of the Earth's crust, we will now explore the way the crust is distributed into its major surface features, its *topography*. Modern technology allows scientists to map much of the Earth's topography through remote sensing (Figure 11.18). The most obvious topographic patterns on the Earth's surface are its relief features—mountain chains, midoceanic ridges, high plateaus, and ocean trenches, for example. In upcoming chapters, we will discuss in greater detail the processes responsible for these features.

RELIEF FEATURES OF THE CONTINENTS

We divide continents into two types of regions: active mountain-making belts and inactive regions of old, stable rock.

Alpine chains

Active mountain-making belts are narrow zones that are usually found along the margins of lithospheric plates. We call these belts *alpine chains* because they are characterized by high, rugged mountains, such as the Alps of Central Europe. Even today, alpine mountain-building continues in many places.

The mountain ranges in the active belts grow through one of two very different geologic processes. First is *volcanism*, in which massive accumulations of volcanic rock are formed by extrusion of magma. Many lofty mountain ranges consist of chains of volcanoes built of extrusive igneous rocks. For example, the Cascade Range in the Pacific Northwest region of the United States comprises a chain of volcanic mountains, including Mount St. Helens, Mount Baker, and Mount Hood.

The second mountain-building process is *tectonic activity*, the breaking and bending of the Earth's crust under internal Earth forces. Tectonic activity usually occurs when lithospheric plates come together in collisions, as we will see in more detail in the next chapter. Crustal masses that are raised by tectonic activity create mountains and plateaus. A well-known example is the Himalaya mountain range, which resulted from

Mapping the Earth's Topography from Space

The Shuttle Radar Topography Mission (SRTM) collected elevation data for most of the planet to create a high-resolution digital topographic database of the Earth. Mounting a specially modified radar system onboard the space shuttle *Endeavour* during an 11-day mission in February 2000, NASA scientists acquired images showing measurements of the local height of the Earth's land surface between 60° N and 60° S latitude. They then assembled these measurements into a global database of elevations on a grid scale of 30 m (98 ft) for the United States and 90 m (295 ft) for the rest of the world.

The simulated image of Mount St. Helens (Figure 11.18) shows how detailed topographic information can be used to display a realistic landscape. Given a viewpoint above the Earth's surface and a direction of view, a computer calculates the distance to the nearest ground point in every desired downward direction. Using those calculations and a direction of solar illumination, the program synthesizes an image that resembles the true view of a sunlit landscape, complete with shadows. To add color to the image, an elevation scale is used, with colors from green to brown to white indicating increasing elevation.

Mt. Adams   Mt. St. Helens   Mt. Hood

Courtesy JPL/NASA Images

11.18 Mount St. Helens

This simulated image of Mount St. Helens was produced from the Shuttle Radar Technology Mission's topographic database. Mount Adams and Mount Hood can also be seen in the distance.

the collision of the subcontinent of India with the continent of Asia. In some instances, volcanism and tectonic activity combine to produce a mountain range. Tectonic activity can also lower crustal masses to form depressions.

> The two basic subdivisions of continental masses are active belts of mountain making, and inactive regions of old, stable rock. Mountains are built by volcanism and tectonic activity.

Continental shields

Belts of recent and active mountain making account for only a small portion of the continental crust. Most of the rest is made up of much older, comparatively inactive rock, which we call continental shields and mountain roots.

Continental shields are regions of low-lying igneous and metamorphic rocks that are resistant to weathering; since ancient times they have formed the foundation for continent building. The core areas of some shields are made up of rock dating back to the Archean eon, 2.5 to 3.9 billion years ago. Shields may be either exposed or covered by layers of sedimentary rock (Figure 11.19).

Remains of older mountain belts lay within the shields in many places. These *mountain roots* are mostly made up of ancient sedimentary rocks that have been intensely bent and folded and, in some locations, changed into metamorphic rocks. Thousands of meters of overlying rocks have been removed from these old tectonic belts so that only the lowermost structures remain. Roots appear as chains of long, narrow ridges, rarely rising over 1000 m (about 3280 ft) above sea level. The Appalachian mountain chain and its extensions to the Taconic, Green, and White Mountains are a familiar example (Figure 11.20).

> Inactive continental regions of stable rocks include continental shields and ancient mountain roots. Continental shields are low-lying areas of old igneous and metamorphic rock.

11.19 Continental shields

Large areas of the continents are underlain by ancient, erosion-resistant igneous and metamorphic rocks.

All Canada Photos/Alamy Limited

CANADIAN SHIELD ▶

During the Ice Age, continental glaciers stripped the Canadian shield of its sediments, leaving a landscape of low hills, rock outcrops, and many lakes, shown here in this view of the Sudbury Basin, near Sudbury, Ontario, Canada, in autumn colors.

▼ MAP OF CONTINENTAL SHIELDS

Shields are areas of ancient rocks that have been eroded to levels of low relief. Mountain roots are shown with a brown line. The areas of oldest rock are circled in red dashed lines.

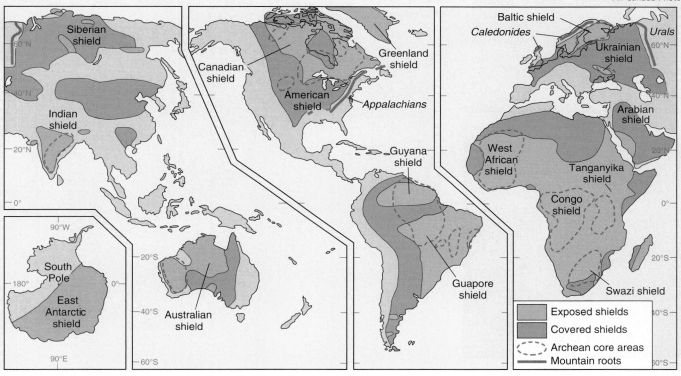

Medford Taylor/National Geographic/Getty Images, Inc.

11.20 Mountain roots

The Appalachians of the eastern United States are a good example of ancient mountain roots. Shown here are the rounded ridges and knobs of the blue Ridge Mountains, underlain by igeneous and metamorphic rocks of Paleozoic age.

Continental landforms

Along with the alpine ranges, the continents feature six major types of landforms: widely spaced mountains, plains, mountains, depressions, high plateaus, hills and low plateaus, and ice sheets (although the latter are not always regarded as landforms). Figure 11.21 shows a map of these global-scale landforms. Their locations on

11.21 Global landforms

At the global scale, there are six major types of continental landforms. Tectonic activity creates mountain belts; crustal uplift shapes high plateaus; and spreading results in widely spaced mountains. Crustal subsidence produces depressions. In stable regions, long-continued erosion yields low plateaus, hills, and plains. Some global regions are covered by ice sheets or ice caps.

Witold Skrypczak/Lonely Planet Images/Getty Images, Inc.

► WIDELY SPACED MOUNTAINS

Ranges of block mountains such as the Great Basin are separated by wide valleys filled with mountain sediment. Such ranges are found in regions of crustal uplift and streching.

ICE SHEETS ►

In some regions, the land surface is covered by a thick sheet of giacial ice. Greenland's vast ice cap is as much as 3 km (1.86 ml) thick as its center.

Courtesy NASA Johnson Space Center Collection

Wolfgang Kaehler/SuperStock

▲ PLAINS

Gently sloping regions of stable rock such as the Great Plains of the United States have been eroded over long geologic time.

Mark Newman / Photo Researchers/Getty Images, Inc.

◄ MOUNTAINS

Tectonic activity lifts mountain ranges such as the Andes above the surrounding land. These landforms are associated with active or former collisions of lithospheric plates.

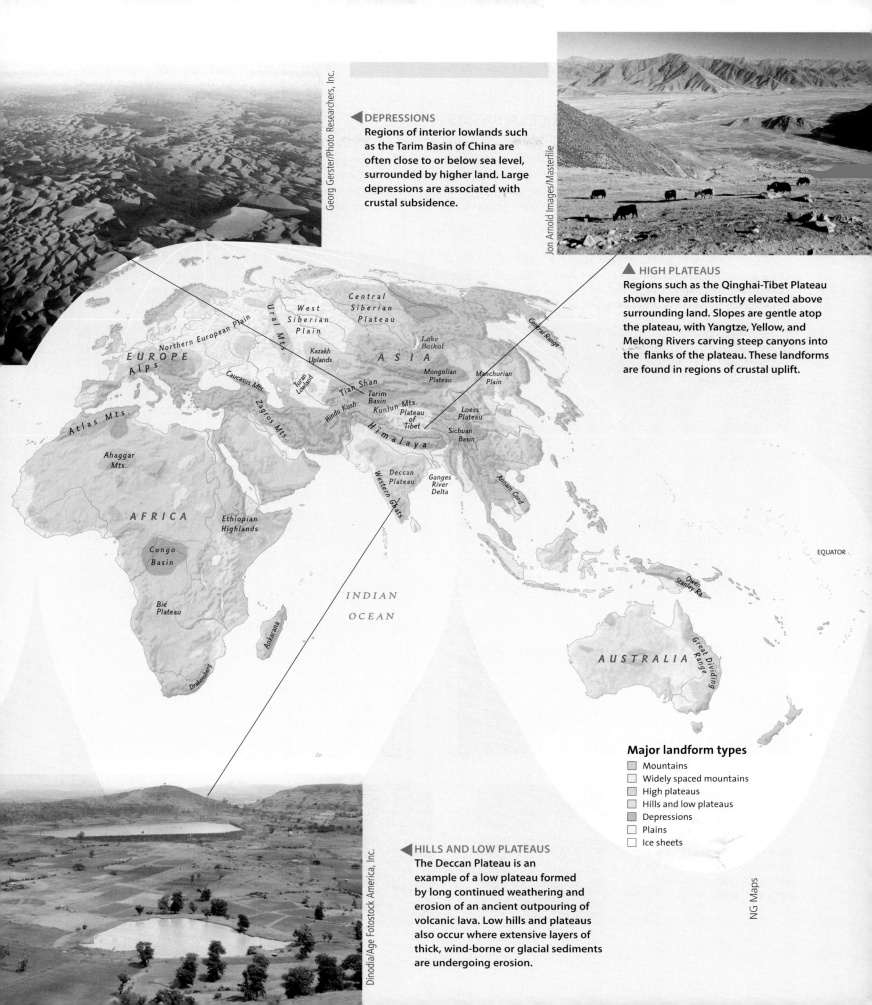

▶ **DEPRESSIONS**

Regions of interior lowlands such as the Tarim Basin of China are often close to or below sea level, surrounded by higher land. Large depressions are associated with crustal subsidence.

▲ **HIGH PLATEAUS**

Regions such as the Qinghai-Tibet Plateau shown here are distinctly elevated above surrounding land. Slopes are gentle atop the plateau, with Yangtze, Yellow, and Mekong Rivers carving steep canyons into the flanks of the plateau. These landforms are found in regions of crustal uplift.

Central Siberian Plateau

West Siberian Plain

Ural Mts.

Northern European Plain

Kazakh Uplands

EUROPE

Alps

ASIA

Lake Baikal

Central Range

Caucasus Mts.

Turan Lowland

Tian Shan

Mongolian Plateau

Manchurian Plain

Atlas Mts.

Zagros Mts.

Hindu Kush

Tarim Basin

Kunlun Mts.

Plateau of Tibet

Loess Plateau

Sichuan Basin

Himalaya

Ahaggar Mts.

Ethiopian Highlands

Deccan Plateau

Western Ghats

Ganges River Delta

Annam Cord.

AFRICA

Congo Basin

Bié Plateau

Ankarana

Drakensberg

INDIAN OCEAN

EQUATOR

Owen Stanley Ra.

AUSTRALIA

Great Dividing Range

Major landform types

- ☐ Mountains
- ☐ Widely spaced mountains
- ☐ High plateaus
- ☐ Hills and low plateaus
- ☐ Depressions
- ☐ Plains
- ☐ Ice sheets

◀ **HILLS AND LOW PLATEAUS**

The Deccan Plateau is an example of a low plateau formed by long continued weathering and erosion of an ancient outpouring of volcanic lava. Low hills and plateaus also occur where extensive layers of thick, wind-borne or glacial sediments are undergoing erosion.

the globe are largely related to plate tectonic activity, as we will see in greater detail in Chapter 12.

RELIEF FEATURES OF THE OCEAN BASINS

Oceans make up 71 percent of the Earth's surface. Relief features of oceans are quite different from those of the continents. Much of the oceanic crust is less than 60 million years old, compared to the great bulk of the continental crust, which is over 1 billion years old. The young age of the oceanic crust is quite remarkable. We will see later that plate tectonic theory explains this age difference.

Just like continental topography, undersea topography is largely determined by tectonic activity (Figure 11.22). In areas where two lithospheric plates collide, the ocean floor shows deep trenches where one plate is being pushed under the other. These areas, such as the Pacific Basin, are characterized by earthquakes and volcanic activity, as we will discuss further in Chapter 12.

11.22 Undersea topography

Using precise radar measurements of the height of the ocean surface, it is possible to infer the depth of the water and, therefore, draw a map of undersea topography. Deeper regions are shown in tones of purple, blue, and green, while shallower regions are in tones of yellow and reddish brown. Data were acquired by the U.S. Navy Geosat satellite altimeter.

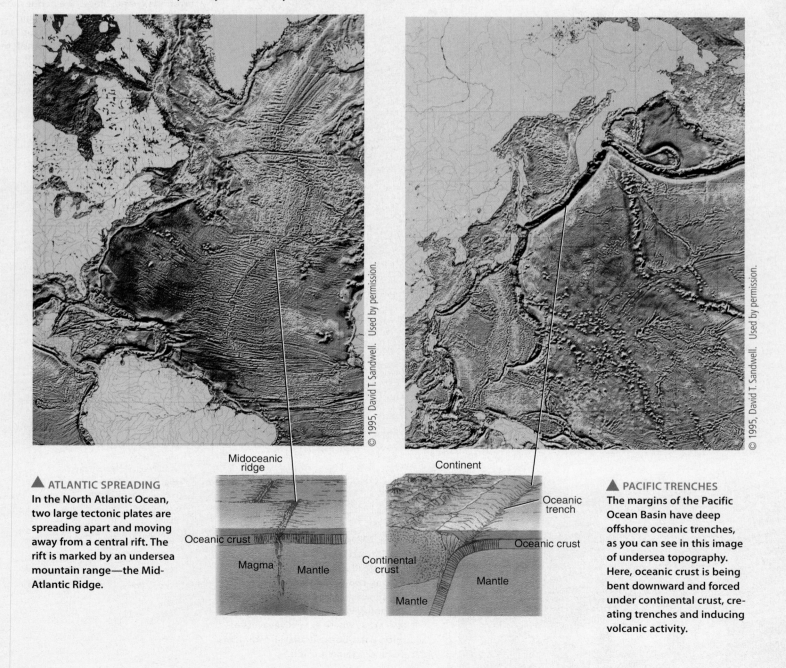

© 1995, David T. Sandwell. Used by permission.

▲ ATLANTIC SPREADING
In the North Atlantic Ocean, two large tectonic plates are spreading apart and moving away from a central rift. The rift is marked by an undersea mountain range—the Mid-Atlantic Ridge.

Midoceanic ridge

Oceanic crust

Magma   Mantle

Continent

Oceanic trench

Oceanic crust

Continental crust

Mantle

Mantle

▲ PACIFIC TRENCHES
The margins of the Pacific Ocean Basin have deep offshore oceanic trenches, as you can see in this image of undersea topography. Here, oceanic crust is being bent downward and forced under continental crust, creating trenches and inducing volcanic activity.

In areas where plates spread, or move apart, a *midoceanic ridge* of submarine hills divides the basin in about half. Precisely in the center of the ridge, at its highest point, is a narrow trenchlike feature, the *axial rift*. The location and shape of this rift suggest that the crust is being pulled apart along the line of the rift. Where the crust pulls apart at the rift, magma wells up and hardens to fill the void, emerging as new oceanic crust. The discovery that the youngest area of the Earth's crust was created along midoceanic ridges was an important clue to our modern understanding that the lithospheric plates are in motion.

> Ocean basins include a midoceanic ridge with a central axial rift where crust is being pulled apart.

CONFIGURATION OF THE CONTINENTS

The long history of how the Earth's surface features formed is driven by the movement of lithospheric plates sliding over the hot viscous asthenosphere. If the plates are in motion, then the configuration of our continents must have changed many times over the history of the Earth. The way the continents look today is but a brief snapshot of the Earth's dynamic form in geologic timescales.

The Theory of Continental Drift

As higher-quality navigational charts became available, geographers were able to see the close correspondence between the outlines of the eastern coast of South America and the western coastline of Africa. In the early twentieth century, German meteorologist and geophysicist Alfred Wegener proposed the first full-scale scientific theory describing the breakup of a single supercontinent, which he named Pangea, into multiple continents that drifted apart, in a process he called *continental drift*. Wegener suggested that Pangea existed intact as early as about 300 million years ago, in the Carboniferous period. He supported his theory by demonstrating that similar fossils and present-day plant species were found on separate continents, suggesting they had once been adjacent (Figure 11.23). Unfortunately, Wegener's explanation of the physical process that caused the continents to separate was incomplete, and most scientists of the time rejected his theory.

> Alfred Wegener proposed that today's continents had broken apart from a single supercontinent named Pangaea. Although many doubted his ideas, he was eventually proven right.

Seafloor spreading

In the 1960s, seismologists discovered the mechanism responsible for the motion of the plates—seafloor spreading, produced by slow-moving currents in the

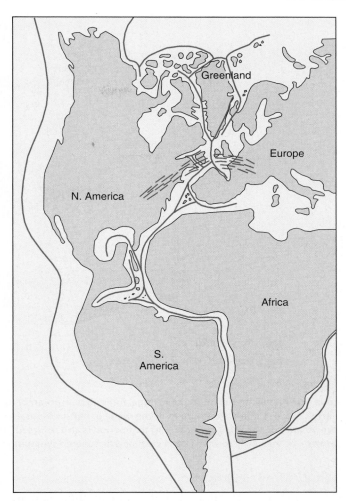

11.23 Wegener's Pangea

Alfred Wegener's 1915 map fits together the continents that today border the Atlantic Ocean Basin. The sets of dashed lines show the fit of Paleozoic tectonic structures between Europe and North America and between southernmost Africa and South America.

plastic asthenosphere (Figure 11.24). Seafloor spreading was recognized when geophysicists studied patterns of magnetism on the seafloor and later obtained sediment samples to determine the relative age of the ocean crust. They found that the crust was youngest along the seafloor spreading zones and grew progressively older as it spread away from the midoceanic ridges. This evidence showed that new crust was formed by magma upwelling along the ocean ridge, while older crust cooled and sank back down into the mantle at ocean trenches. As the tectonic plates come apart at spreading centers, the continents of the plates move apart. The theory that describes our current understanding of how the plates move is known as *plate tectonics*, which we will examine in the next section.

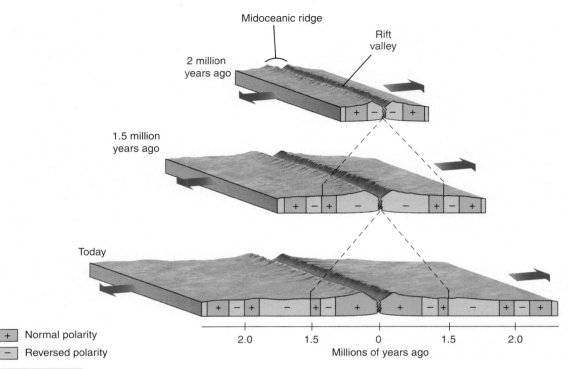

11.24 Seafloor spreading

Lava extrudes along the midoceanic ridge, forming new oceanic crust. When lava cools, it becomes magnetized in polarity with the Earth's magnetic field. Periodically, in the range of tens of thousands of years, the polarity of the Earth's magnetic field is reversed, and the polarity of crust that forms during that period is also reversed. Teams of geoscientists have mapped the successive ages—magnetic bands—of the newly formed ocean crust and discovered symmetry in the age of crust on both sides of the midoceanic ridge.

History of the continents

The continents are moving today, just as they have in the past. Data from orbiting satellites shows that rates of separation between or convergence of two plates are on the order of 5 to 10 cm (about 2 to 4 in.) per year, or 50 to 100 km (about 30 to 60 mi) per million years. At that rate, global geography must have been very different in past geologic eras than it is today (Figure 11.25).

As it turns out, Wegener was largely correct about the supercontinent Pangea; but today we know that there were even earlier supercontinents than Pangea. An earlier supercontinent, Rodinia, was fully formed about 700 million years ago; and some interesting evidence suggests that there may have been yet another supercontinent before Rodinia. In fact, over the billions of years of the Earth's geologic history, the union of the continents, and their subsequent breakup, is a repeating process, one that has occurred half a dozen times or more.

By studying the changing continental arrangements and locations over time, we can recognize that as the continents changed latitude, their climates, soils, and vegetation evolved as well.

WileyPLUS Continents of the Past

Follow today's continents from the Precambrian age to the present as they converge into a supercontinent and then split apart. An animation.

Plate Tectonics

As described earlier in this chapter, the Earth's lithosphere is fractured into more than 50 separate tectonic plates, ranging from very large to very small. The motions of lithospheric plates are responsible for shaping our planet, from the tops of mountains to the trenches of the sea bottom. Over geologic time, changes in the configuration of oceans and continents produced by the movement of lithospheric plates have altered the Earth's climates and influenced the distribution and abundance of plants and animals across the Earth's landscapes. The body of knowledge about lithospheric plates and their motions is referred to as **plate tectonics**.

> Tectonic processes include extension and compression. Extension causes fracturing and faulting of the crust, while compression produces folds and overthrust faults.

EXTENSION AND COMPRESSION

The movement of lithospheric plates generates two types of force on the crust: extension and compression (Figure 11.26). *Extension*, or *rifting*, occurs when the lithosphere is pulled apart. Rocks break and shift along

11.25 History of the continents

The Earth's lithospheric plates have twisted and turned, collided, and then broken apart over the past 600 million years to create the geography of the Earth as we know it today.

1 600 million years ago
A supercontinent, known as Rodinia, split apart, and oceans filled the basins. Fragments collided, thrusting up mountain ranges. Glaciers spread, twice covering the equator. A new polar supercontinent, Pannotia, formed.

2 500 million years ago
A breakaway chunk of Pannotia moved north, splitting into three masses—Laurentia (North America), Baltica (northern Europe), and Siberia. In shallow waters, the first multicellular animals with exoskeletons appeared, and the Cambrian explosion of life began.

3 300 million years ago
Laurentia collided with Baltica and later with Avalonia (Britain and New England). The Appalachian Mountains arose along the edge of the supercontinent Pangea.

4 200 million years ago
Dinosaurs roamed Pangea, which stretched nearly from pole to pole and almost encircled Tethys, the oceanic ancestor of the Mediterranean Sea. The immense Panthalassic Ocean surrounded the supercontinent.

5 100 million years ago
Pangea broke apart. The Atlantic poured in between Africa and the Americas. India split away from Africa, and Antarctica and Australia were stranded near the South Pole.

6 50 million years ago
Moving continental fragments collided—Africa into Eurasia, pushing up the Alps, and India into Asia, raising the Plateau of Tibet. With dinosaurs now extinct, birds and once-tiny mammals began to evolve rapidly.

7 Present day
Formation of the Isthmus of Panama and the split of Australia from Antarctica changed ocean currents, changing the climate. North America and Eurasia encircled the Arctic Ocean, restricting its circulation. Ice sheets waxed and waned in many cycles, and sea levels rose and fell.

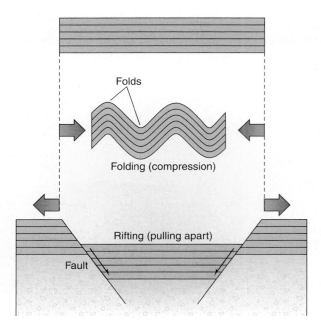

11.26 Two basic forms of tectonic activity

Flat-lying rock layers may be compressed and form folds, or be pulled apart to produce faults by rifting.

planes of breakage, called *faults*, creating valleys flanked by higher blocks. When extension causes a lithospheric plate to fracture and spread apart, magma rises from below to fill the opening. *Compression* occurs when plates are pushed together. Rock strata are tightly compressed into wavelike structures, called *folds*, and may even be pushed on top of one another along fault lines into piles of layered strata called *thrust sheets* (Figure 11.27).

PLATES AND BOUNDARIES

Figure 11.28 shows the major features of plate interactions. As we saw earlier in Chapter 11, there are two types of lithosphere, oceanic and continental. The *oceanic lithosphere* is thinner and denser (about 50 km, or 30 mi thick), whereas the *continental lithosphere* is thicker and lighter (about 150 km, or 95 mi thick). Because both types of lithosphere are "floating" on the plastic asthenosphere below, the surface of the thicker and lighter continental plate rises above the ocean floor.

The figure shows three types of plate boundaries: spreading, converging, and transform. Plates X and Y are pulling apart along a **spreading boundary**, which lies along the axis of a midocean ridge. The gap in the crust is filled by magma rising from the mantle beneath. Near the ocean floor, the extrusive magma cools to create basalt, while at depth, the intrusive magma forms gabbro. Together, the basalt and gabbro continually form new oceanic crust.

At the right is the **converging boundary** between plates Y and Z. Because the oceanic plate is comparatively thin and dense, in contrast to the thick, buoyant continental

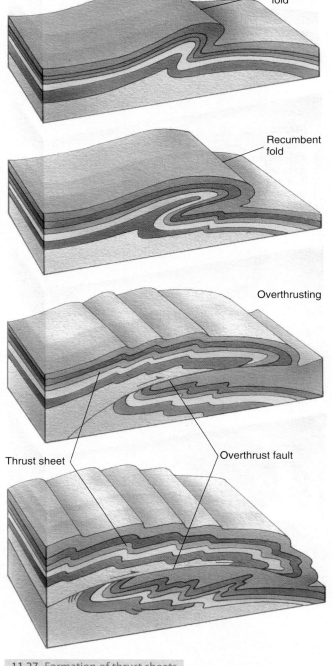

11.27 Formation of thrust sheets

Under the severe compressive forces that can accompany a collision between continents, layered rocks are folded so tightly that they break and move along overthrust faults to produce thrust sheets.

plate, the oceanic lithosphere bends down and plunges into the asthenosphere. The process in which one plate is carried beneath another is called **subduction**.

The leading edge of the descending plate is cooler and therefore denser than the surrounding hot, soft asthenosphere. As a result, the slab sinks under its own

11.28 Plate motions and boundaries

A plate of oceanic lithosphere is moving to the right, away from a spreading boundary at an axial rift at the left. At the converging boundary on the right, the plate is subducting under a plate of continental lithosphere.

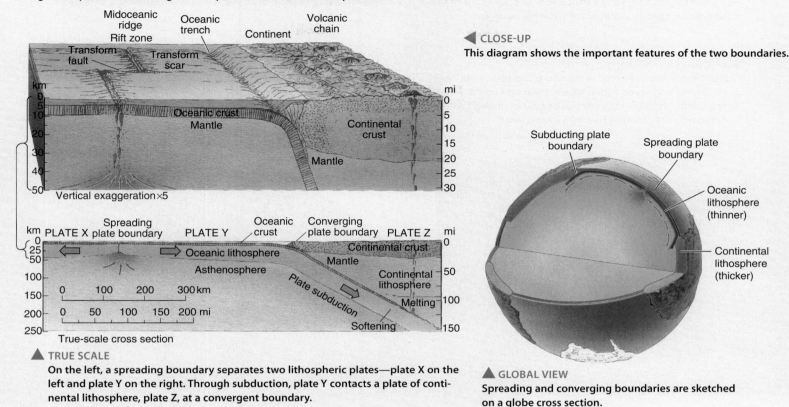

▲ TRUE SCALE

On the left, a spreading boundary separates two lithospheric plates—plate X on the left and plate Y on the right. Through subduction, plate Y contacts a plate of continental lithosphere, plate Z, at a convergent boundary.

◀ CLOSE-UP
This diagram shows the important features of the two boundaries.

▲ GLOBAL VIEW
Spreading and converging boundaries are sketched on a globe cross section.

weight once subduction has begun. However, as the slab descends, it is heated and softened. The underportion, which is mantle rock in composition, simply reverts to mantle rock.

The descending plate is covered by a thin upper layer of less dense mineral matter derived from oceanic and continental sediments. This material can melt and become magma. The magma tends to rise because it is less dense than the surrounding material. The figure shows some magma pockets on the right, formed from the upper edge of the slab. They are shown rising through the overlying continental lithosphere. When they reach the Earth's surface, they form a chain of volcanoes lying about parallel with the deep oceanic trench that marks the line of descent of the oceanic plate. Because the edge of the continent is the site of subduction and volcanoes, it is called an *active continental margin*.

In addition to spreading and converging boundaries, there is also the *transform boundary* (Figure 11.29). Here, one lithospheric plate slides past the other without separating or converging. The two plates are in contact along a vertical fracture, called a *transform fault*.

> At a spreading boundary, crust is being pulled apart. At a converging boundary, one plate is subducted beneath another. At a transform boundary, two plates move past each other.

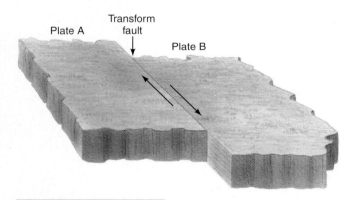

11.29 Transform boundary

At a transform boundary, two lithospheric plates slide past each other along a nearly vertical transform fault.

WileyPLUS Plate Tectonics
See the structure of active and passive plate margins in this animation. Watch as the seafloor spreads, forming oceanic crust that collides with a continent, and is subducted.

Of the Earth's many large and small lithospheric plates, seven are significantly larger than the others: Pacific, North American, Eurasian, Antarctic, Australian, and South American. Figure 11.30 shows these major

11.30 Lithospheric plates and their motions

The ever-shifting lithosphere is cracked into tectonic plates that are in slow but constant motion. As they collide or pull apart from each other, they generate volcanoes and earthquakes and form long chains of mountains. Map arrows point to the direction and relative rate of plate motion, and longer arrows indicate faster movement.

▼ SPREADING CENTERS

New lithosphere is formed in the spreading zone, both on continents and ocean floors, where two plates are moving apart. This rift valley in Iceland is the only example of a land-based rift valley created by a midoceanic ridge.

Emory Kristof/NG Image Collection

▼ EARTHQUAKES

Stress and release of the Earth's crust from the movement of plates create earthquakes. Major devastation and loss of life resulted from the 2010 Haiti earthquake. Major earthquakes can be mapped along the boundaries of the plates.

Alison Wright/NG Image Collection

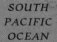

Tectonic feature

Plate boundary
- Divergent
- Convergent
- Transform zone

Plate motion
- Divergent (arrow length proportional to plate motion speed)
- Convergent
- Hot spot

Major tectonic event in the last 100 years
Earthquake
- Ten deadliest
- Ten costliest
- Other

Volcanic eruption
- Notable
- Known during the past 10,000 years

NG Maps

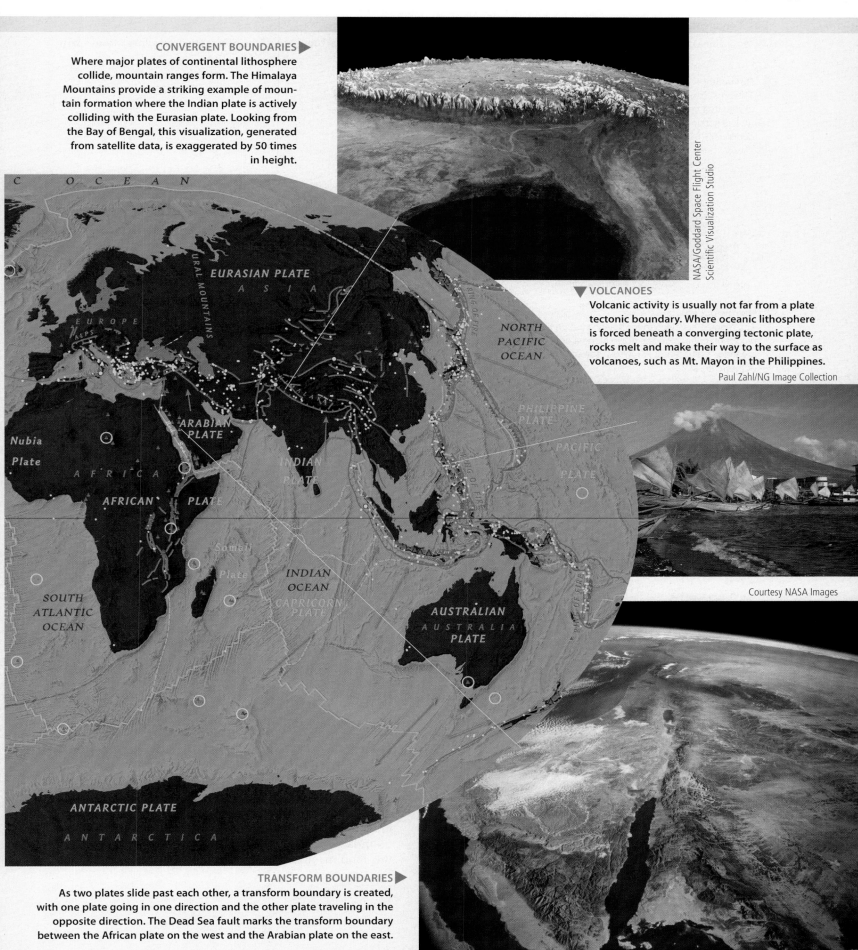

CONVERGENT BOUNDARIES ▶

Where major plates of continental lithosphere collide, mountain ranges form. The Himalaya Mountains provide a striking example of mountain formation where the Indian plate is actively colliding with the Eurasian plate. Looking from the Bay of Bengal, this visualization, generated from satellite data, is exaggerated by 50 times in height.

▼ VOLCANOES

Volcanic activity is usually not far from a plate tectonic boundary. Where oceanic lithosphere is forced beneath a converging tectonic plate, rocks melt and make their way to the surface as volcanoes, such as Mt. Mayon in the Philippines.

TRANSFORM BOUNDARIES ▶

As two plates slide past each other, a transform boundary is created, with one plate going in one direction and the other plate traveling in the opposite direction. The Dead Sea fault marks the transform boundary between the African plate on the west and the Arabian plate on the east.

11.31 Continental rupture and spreading

When extensional tectonic activity occurs beneath the continental lithosphere, the continent ruptures and the new continental edges spread apart.

1 Rift valley

The crust is uplifted and stretched apart, causing it to break into blocks that become tilted on faults. Eventually, a long narrow rift valley appears. Magma rises up from the mantle to continually fill the widening crack at the center.

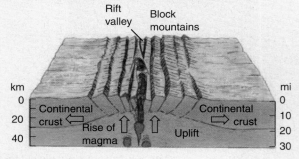

2 Narrow ocean

The magma solidifies to form new crust on the rift valley floor. Crustal blocks on either side slip down along a succession of steep faults, creating mountains. A narrow ocean develops, floored by new oceanic crust.

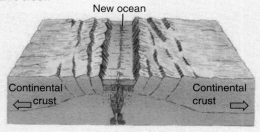

3 Large ocean

The ocean continues to widen until a large ocean has formed and the continents are widely separated. As the ocean basin widens, the passive continental margins subside and receive sediments from the continents. (Note: The vertical scale is greatly exaggerated to emphasize surface features.)

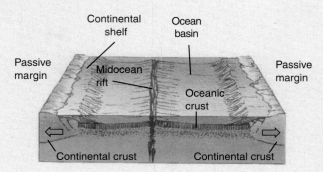

plates and others, as well as the spreading, converging, and transform boundaries that surround them.

WileyPLUS Tectonic Plate Boundary Relationships
Play this movie to watch the landscapes of spreading, converging, and transform boundaries between lithospheric plates. Iceland, the Cascades, Himalayas, Alps, and the San Andreas fault are featured.

WileyPLUS Remote Sensing and Tectonic Landforms Interactivity
Select "tectonics" to review the names and locations of tectonic plates and to examine a famous transform fault—the San Andreas.

CONTINENTAL RUPTURE AND NEW OCEAN BASINS

So far, we have examined spreading boundaries in oceanic lithosphere. What happens when continental lithosphere fractures and splits apart? This is the process of *continental rupture* (Figure 11.31), which occurs when tectonic forces uplift a plate of continental lithosphere and pull it apart. At first, a *rift valley* forms, and as the bottom of the rift valley sinks below sea level, seawater enters. Eventually, a wide ocean forms, with an axial rift down its center and continental edges on either side. These continental edges are called *passive continental margins*. Here, continental lithosphere is joined to oceanic lithosphere, but there is no motion between the two types of lithosphere.

The Red Sea is a good example of a continental rupture in progress. Figure 11.32 is a photo, taken by an astronaut, of the Red Sea where it joins the Gulf of Aden. As clarified in the inset map, this is a triple junction of three spreading boundaries established by the motion of the Arabian

> Continental rupture begins with the formation of a rift valley and tilted block mountains. Ocean soon invades the rift. As the continental crust recedes, oceanic crust fills the gap.

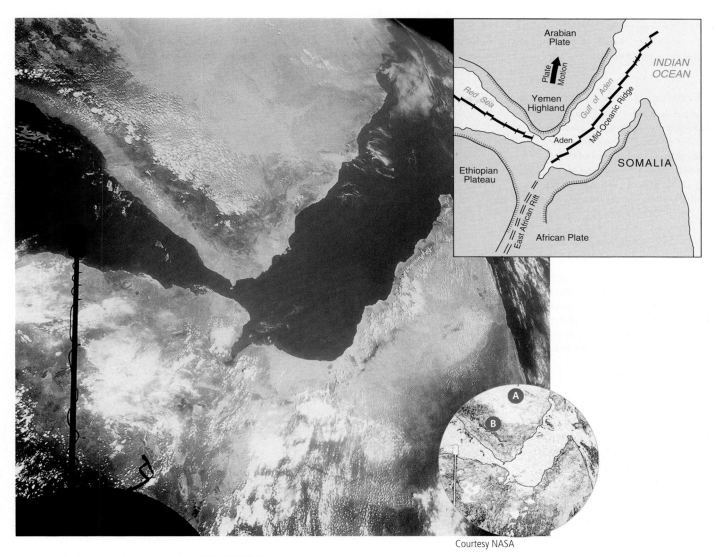

Courtesy NASA

11.32 The Red Sea and Gulf of Aden from orbit

This spectacular photo, taken by astronauts on the Gemini XI mission, shows the southern end of the Red Sea and the southern tip of the Arabian Peninsula.

EYE ON THE LANDSCAPE **What else would the geographer see?**
The pinkish tones of the interior Arabian desert Ⓐ are caused by iron oxides in the desert rocks and soils; note the absence of any dark vegetation. The cumulus clouds Ⓑ are the result of the lifting of moist air over the coastal mountain ranges.

plate pulling away from the African plate. It is easy to visualize how the two plates have split apart, allowing the ocean to enter.

ISLAND ARCS AND COLLISION OF OCEANIC LITHOSPHERIC PLATES

When a continent ruptures to form an ocean basin with axial rift, the two new plates move apart and create a new ocean. But eventually the plate motions may reverse, and the ocean basin may start to close. For this to happen, a plate must fracture and produce a subduction boundary. If the fracture occurs at a passive continental margin, oceanic crust will be subducted below continental crust, as shown in Figure 11.28. But what happens if the fracture occurs in the middle of a

> At a convergent boundary where plates of oceanic lithosphere collide, an arc of volcanic islands rises just beyond the subduction zone.

11.33 Formation of an island arc and arc-continent collision

When an ocean closes, the oceanic lithosphere is subducted under another plate of oceanic lithosphere, building an island chain. Eventually, the continental lithosphere collides with the oceanic lithosphere at the subduction site, acquiring the remains of the island arc.

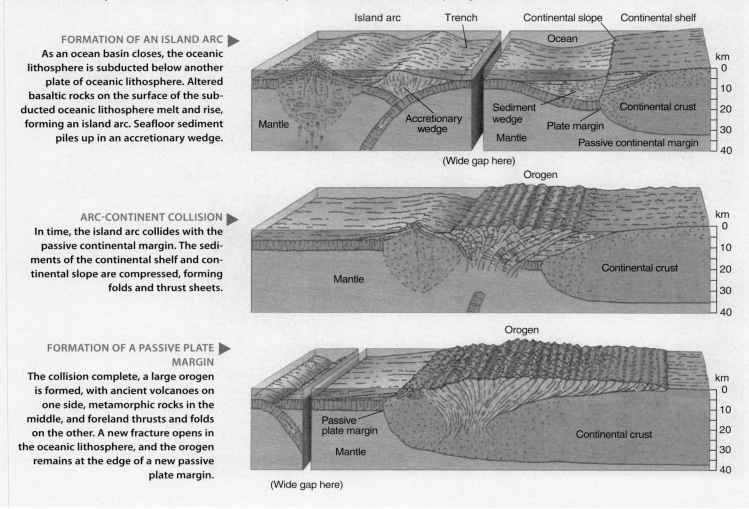

FORMATION OF AN ISLAND ARC ▶
As an ocean basin closes, the oceanic lithosphere is subducted below another plate of oceanic lithosphere. Altered basaltic rocks on the surface of the subducted oceanic lithosphere melt and rise, forming an island arc. Seafloor sediment piles up in an accretionary wedge.

ARC-CONTINENT COLLISION ▶
In time, the island arc collides with the passive continental margin. The sediments of the continental shelf and continental slope are compressed, forming folds and thrust sheets.

FORMATION OF A PASSIVE PLATE ▶
MARGIN
The collision complete, a large orogen is formed, with ancient volcanoes on one side, metamorphic rocks in the middle, and foreland thrusts and folds on the other. A new fracture opens in the oceanic lithosphere, and the orogen remains at the edge of a new passive plate margin.

plate of oceanic lithosphere? This situation is shown in Figure 11.33.

As the subducted oceanic lithosphere plunges downward, oceanic crust is carried into the mantle. Since it originally came from the mantle and is of the same composition, it simply softens and disappears into the mantle rock. However, the minerals in the upper layer of subducting oceanic crust have reacted with water, altering their structure. At depth, they melt more readily than unaltered basalt, forming magma that eventually rises and erupts on the seafloor. The new volcanoes grow into an *island arc;* the Aleutian Islands are a good example.

As the process continues, seafloor sediment piles up in the trench, taking shape as an accretionary wedge of sediments. Meanwhile, the continuing rise of magma fortifies the island arc from below, increasing the height and width of the volcanic mass.

ARC-CONTINENT COLLISION

If ocean-basin closing continues, the island arc eventually collides with a passive continental margin. Since the island arc is thick and buoyant, it is not subducted but, rather, pushed up against the continent. The layers of sediment that have accumulated on the continental shelf and continental slope are crushed and deformed. The sediments are thrust far inland over the older continental rocks. The mass of collided rocks is called an *orogen,* and the process of its formation is described as an *orogeny.* If the collision continues, another oceanic fracture develops and a new subduction boundary is drawn.

> Eventually, the island arc can collide with a passive continental margin producing an arc-continent collision.

11.34 Continental suture

Where plates of continental lithosphere collide, the rock formations along the boundaries may be transformed under extreme pressure into newly formed metamorphic rocks. The resulting continental suture binds the two plates into a single, larger plate.

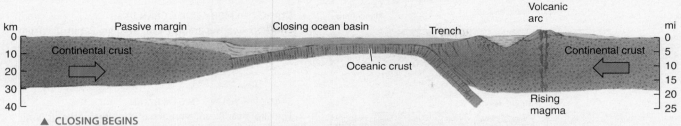

▲ **CLOSING BEGINS**
Subduction is occurring on the right, where oceanic lithosphere collides with the continental lithosphere. The continent on the left has a passive margin.

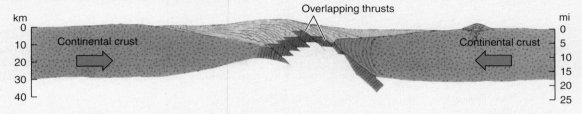

▲ **SUTURING BEGINS**
As the continents move closer, the ocean between the converging continents is eliminated. The oceanic crust is compressed, creating a succession of overlapping thrust faults that ride up, one over the other.

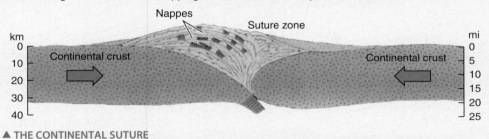

▲ **THE CONTINENTAL SUTURE**
As the slices become more and more tightly squeezed, they are forced upward. A mass of metamorphic rock takes shape between the joined continental plates, welding them together. The new rock mass is a continental suture.

CONTINENT-CONTINENT COLLISION

Where the subduction boundary closing the ocean basin lies at the edge of a continent, ongoing closing results in a *continent-continent collision* (Figure 11.34). The collision permanently unites the two plates, so that there is no further tectonic activity along that collision zone. The collision zone is called a **continental suture**.

Continent-continent collisions have occurred many times since the late Precambrian time, including many with island arcs sandwiched between the colliding landmasses. Ancient sutures marking early collisions include the Ural Mountains, which divide

> When two continental lithospheric plates collide in an orogeny, continental rocks are crumpled and over-thrust. The plates become joined in a continental suture.

Europe from Asia, the Appalachian Mountains of eastern North America, and the Caledonian Mountains of Scotland, Norway, Svalbord, and eastern Greenland. More recently, in the Cenozoic era, continent-continent collisions occurred along a great tectonic line that marks the southern boundary of the Eurasian plate. The line begins with the Atlas Mountains of North Africa and runs, with a few gaps, to the great Himalayan range, where it is still active. Each segment of this collision zone represents the collision of a different north-moving plate against the single and relatively immobile Eurasian plate.

THE WILSON CYCLE AND SUPERCONTINENTS

Geologic evidence has shown that ocean basins have opened and closed many times in the geologic past. The

11.35 The Wilson cycle

The Wilson cycle describes the opening and closing of ocean basins by plate tectonic activity.

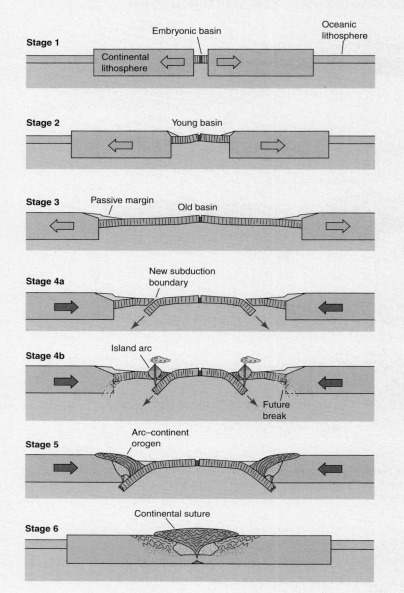

STAGE 1—EMBRYONIC OCEAN BASIN
The Red Sea separating the Arabian Peninsula from Africa is an active example.

STAGE 2—YOUNG OCEAN BASIN
The Labrador Basin, a branch of the North Atlantic lying between Labrador and Greenland, is an example of this stage.

STAGE 3—OLD OCEAN BASIN
This includes all of the vast expanse of the North and South Atlantic oceans and the Antarctic Ocean. Passive margin sedimentary wedges have become wide and thick.

STAGE 4A
The ocean basin begins to close as continental plates move together. New subduction boundaries begin to emerge.

STAGE 4B
Island arcs have risen and grown into great volcanic island chains. These are found surrounding the Pacific plate, with the Aleutian arc as an example.

STAGE 5—CLOSING CONTINUES
Formation of new subduction margins close to the continents is followed by arc-continent collisions. The Japanese Islands represent this stage.

STAGE 6
The ocean basin has finally closed with a collision orogen, forming a continental suture. The Himalayan orogen is a recent example, with activity continuing today.

cycle of opening and closing is called the *Wilson cycle*, named for the Canadian geophysicist J. Tuzo Wilson.

The Wilson cycle begins with continental rupture and the formation of a wide ocean basin with passive margins (Figure 11.35, stages 1–3). As the plates reverse their motion, and the ocean basin begins to close, oceanic lithosphere fractures and new subduction boundaries take shape (stage 4a). Island arcs soon appear and grow (stage 4b). Eventually, fractures occur

> Ocean basins open and close in the Wilson cycle, which describes how continents split and are reunited. As many as 6 to 10 supercontinents have formed throughout Earth's history.

at the continental margins, and the arcs collide with continents, producing arc-continent orogens (stage 5). In the final stage, the orogens collide, producing a continental suture (stage 6).

As noted above, there is strong evidence that all the continents were once joined in a supercontinent, Rodinia, about 600 million years ago. Rodinia then broke apart, with its continents converging again about 400 million years later into the supercontinent Pangaea. There may have been as many as 6 to 10 such cycles of supercontinent formation, followed by breakup, in the Earth's ancient history (Figure 11.36). This time-cycle of supercontinents now is the basic theme of the geologic evolution of our planet.

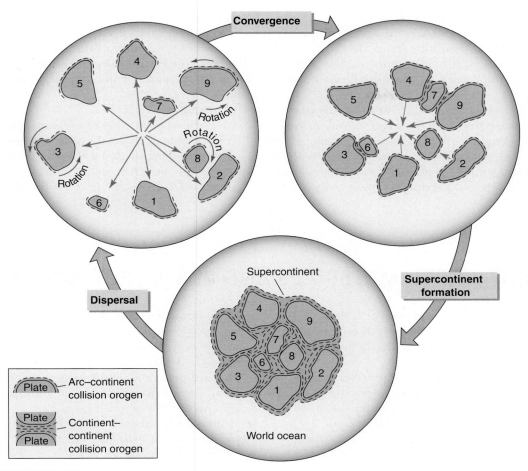

11.36 Supercontinent cycle

Over hundreds of millions of years, supercontinents are formed and reformed in a cycle of formation, dispersal, and convergence. Wiggly dashes show collision orogens that accumulate around and between the continents.

THE POWER SOURCE FOR PLATE MOVEMENTS

Lithospheric plates are huge, so it must take enormous power to drive their motion. Where does this power come from? From *radiogenic heating*, caused by radioactive decay of unstable isotopes that occur naturally in the rock beneath the continents. We don't know exactly how radiogenic heating sets plates in motion, but one theory is that they are generated by *convection currents* in hot, plastic mantle rock. Since hotter rock is less dense than cooler rock, unequal heating could produce streams of upwelling mantle rocks that rise steadily beneath spreading plate boundaries.

> Radiogenic heat is the power source for plate tectonic motions.

Some geologists hypothesize that rising mantle rocks push a portion of a lithospheric plate upward until it fractures. The pieces then move horizontally away from the spreading axis under the influence of gravity. This is called *gravity gliding*. Another theory is that once the plate begins to descend, the descending part pulls the rest of the plate along with it because it is cooler and, therefore, more dense than the mantle rock. Explaining the mechanisms that drive plate motions is a future research goal of many geophysicists.

WileyPLUS Web Quiz
Take a quick quiz on the key concepts of this chapter.

A Look Ahead

This chapter began by focusing on the processes by which the minerals and rocks of the Earth's surface are formed. As we have seen, these processes do not occur everywhere and at all times. Instead, there is a grand plan that organizes the formation and destruction of rocks and distributes the processes of the cycle of rock change in a geographic pattern. That plan also determines the locations of the largest and most obvious features of our planet's surface: the continents and oceans at the global scale, and its mountain ranges and basins at a continental scale. The grand plan is plate tectonics, the scheme for understanding the dynamics of the Earth's crust over millions of years of geologic time.

Plate tectonics also sets the stage for the development of landforms in response to tectonic and volcanic activity. By bending, folding, and breaking rocks, tectonic activity produces rock structures that erosion then sculpts into unique landforms. By generating upwelling magma at spreading centers and subduction boundaries, plate motions also yield volcanic landforms on the surface, as well as intrusive rock configurations that are eventually revealed by erosion of overlying rocks. These landforms, along with the earthquakes and eruptions that accompany their formation, are the subjects of our next chapter.

WileyPLUS Web Links

View more rocks and minerals and examine the formation of petroleum. Dive to the ocean floor and explore undersea tectonic landforms. Watch the continents move through the eons. Chart geologic time and discover ancient environments. It's all in this chapter's web links.

IN REVIEW   EARTH MATERIALS AND PLATE TECTONICS

- Geologists trace the history of the Earth through the *geologic timescale*, which has divisions of *eons, eras,* and *periods*. The Cambrian period marks the beginning of widespread life on Earth.

- *Endogenic processes* of volcanic and tectonic activity shape *initial landforms*, while *exogenic processes*, such as erosion and deposition by running water, waves, wind, and glacial ice, sculpt *sequential landforms*.

- At the center of the Earth lies the **core**, a dense mass of liquid iron and nickel that is solid at the very center. Enclosing the metallic core is the **mantle**, composed of *mafic rock*. The outermost layer is the **crust**. *Continental crust* consists of two zones: a lighter zone of *felsic rock* atop a denser zone of mafic rock. *Oceanic crust* consists only of denser, mafic rock.

- The *lithosphere*, the outermost shell of rigid, brittle rock, includes the crust and an upper layer of the mantle. Below the lithosphere is the *asthenosphere*, a region of the mantle in which mantle rock is soft or plastic. The lithosphere is divided into **lithospheric plates**.

- **Minerals** are naturally occurring, inorganic substances, often with a crystalline structure; they are largely composed of oxygen, silicon, aluminum, iron, calcium, sodium, potassium, and magnesium.

- **Rocks** are naturally occurring assemblages of minerals. They fall into three major classes: igneous, sedimentary, and metamorphic.

- **Igneous rocks** are largely composed of silicate minerals. They are formed when **magma** cools and solidifies. If magma erupts on the surface to cool rapidly as lava, the rocks formed are *extrusive* and have a fine crystal texture. If the magma cools slowly below the surface, the rocks are *intrusive* and the crystals are larger.

- Felsic rocks contain mostly *felsic minerals* and are least dense of the rock types; mafic rocks, containing mostly *mafic minerals*, are denser; and *ultramafic rocks* are most dense. Because felsic rocks are least dense, they are generally found in the upper layers of the Earth's crust, while mafic and ultramafic rocks are generally found in lower layers.

- **Plutons** are bodies of intrusive igneous rocks. Types of plutons include *batholiths, sills,* and *dikes*.

- **Sedimentary rocks** are formed in layers, or **strata**, composed of transported rock fragments called *sediment*. *Clastic sedimentary* rocks are composed of sediments that usually accumulate on ocean floors. As the layers are buried more and more deeply, water is pressed out and particles are cemented together. *Sandstone* and *shale* are common examples.

- *Chemical precipitation* also produces sedimentary rocks, such as *limestone*. Organic sediment is composed of tissues of plants and animals—*peat* is an example. Coal, petroleum, and natural gas are hydrocarbon compounds occurring in sedimentary rocks, commonly known as **fossil fuels**.

- **Metamorphic rocks** are formed when igneous or sedimentary rocks are exposed to heat and pressure. Shale is altered to *slate* or *schist*; sandstones become *quartzite*; limestone becomes *marble*; and intrusive igneous rocks or clastic sediments metamorphose into *gneiss*.

- In the **cycle of rock change**, there are two environments. Rocks are altered, fragmented, and deposited as sediment in the surface environment. In the deep environment, sediment or preexisting rock is altered by heat and pressure or melted to form magma. It reaches the surface environment by extrusion, as lava, or is revealed by erosion.

- The Earth's global topography consists of major features, such as mountain chains, midoceanic ridges, high plateaus, and ocean trenches.
- The Shuttle Radar Technology Mission used a radar system to draw precise maps of detailed surface topography.
- Continental landmasses consist of active belts of mountain making and inactive regions of old, stable rock. Mountain belts, known as *alpine chains*, are built by *volcanism* and *tectonic activity*.
- *Continental shields* are regions of low-lying igneous and metamorphic rocks. They may be exposed or covered by layers of sedimentary rocks. Ancient *mountain roots* lie within some shield regions.
- Continental-scale landforms include widely spaced mountains, plains, mountains, depressions, high plateaus, hills and low plateaus, and ice sheets.
- The ocean basins are marked by a *midoceanic ridge* with its central *axial rift*. This ridge occurs at the site of crustal spreading. Where two lithospheric plates collide, deep trenches form as one plate is pushed under the other.
- Alfred Wegener assembled substantial evidence showing that the major continents were once assembled into a supercontinent called *Pangaea*, which subsequently drifted apart.
- Patterns of magnetism in the rocks of ocean basins illustrate that the younger rocks are nearer to the spreading zones, and the older, farther away.
- The union of continents into supercontinents, and their later breakup, is a repeating process that has probably occurred half a dozen times or more throughout the Earth's geologic history.
- **Plate tectonics** is the body of knowledge about lithospheric plates and their motions.
- *Extension* occurs when the lithosphere is pulled apart, causing *faults*. *Compression* occurs when plates are pushed together, producing *folds* and *thrust sheets*.
- The *continental lithosphere* includes the thicker, lighter continental crust and a rigid layer of mantle rock beneath. The *oceanic lithosphere* comprises the thinner, denser oceanic crust and rigid mantle below.

- The lithosphere is fractured and broken into a set of **lithospheric plates**, large and small, that move with respect to one another.
- Where plates move apart, a **spreading boundary** occurs. At **converging boundaries**, plates collide.
- When the oceanic lithosphere and continental lithosphere collide, the denser oceanic lithosphere plunges beneath the continental lithospheric plate in a process called **subduction**. A trench marks the site of downplunging. Some subducted oceanic crust melts and rises to the surface, producing volcanoes.
- At *transform boundaries*, plates move past one another on a *transform fault*.
- In a *continental rupture*, extensional tectonic forces fracture and move a continental plate in opposite directions, creating a *rift valley*. Eventually, the rift valley widens and opens to the ocean. New oceanic crust forms as spreading continues.
- The closing of an ocean basin can cause two plates of lithospheric crust to collide, and subduction shapes an *island arc* of volcanic islands.
- An *arc-continent collision* occurs when continued subduction draws a passive continental margin up against an island arc, forming an *orogen*.
- Eventually, the closing produces a *continent-continent collision*, in which two continental plates are welded together in a zone of metamorphic rock, called a **continental suture**.
- The *Wilson cycle* of ocean-basin opening and closing has occurred many times in the geologic past. It is part of a cycle of formation of *supercontinents* in which the continents form one large landmass, split apart, and then hundreds of millions of years later rejoin in a new supercontinent.
- Plate movements are thought to be powered by *radiogenic heat*. The exact mechanism is unknown, but may include *convection currents* in the plastic mantle rock of the asthenosphere, *gravity gliding* of plates away from an uplifted axial rift, and the gravitational pull of descending plates into a subduction zone.

KEY TERMS

core, p. 380
mantle, p. 380
crust, p. 381
lithospheric
  plates, p. 381
mineral, p. 382
rocks, p. 382

igneous rock, p. 382
magma, p. 382
plutons, p. 385
sedimentary rock, p. 385
strata, p. 385
fossil fuels, p. 388
metamorphic rock, p. 389

cycle of rock change, p. 389
plate tectonics, p. 398
spreading boundary, p. 400
converging boundary, p. 400
subduction, p. 400
continental suture, p. 407

REVIEW QUESTIONS

1. What are the divisions of the *geologic timescale?* Why is the *Cambrian period* important?
2. Compare and contrast the terms *exogenic processes, endogenic processes, initial landforms,* and *sequential landforms.*
3. Describe the Earth's inner structure, from the center outward. What types of crust are present? How are they different? What is the *Moho?*
4. Explain *isostasy* using the terms *lithosphere,* **lithospheric plate**, and *asthenosphere.*
5. Define the terms **mineral** and **rock**. Name the three major classes of rocks.
6. What are *silicate minerals?* Identify three types of silicate minerals based on color and density.
7. How do igneous rocks differ when *magma* cools (a) at depth and (b) at the surface?
8. What is *sediment?* Define and describe three classes of sediment.
9. How are **sedimentary rocks** formed by *chemical precipitation?*
10. What types of sedimentary deposits consist of *hydrocarbon compounds?* How are they formed?
11. What are **metamorphic rocks?** Identify three types of metamorphic rocks and describe how they are formed.
12. What are the two basic subdivisions of continental masses?
13. What term is attached to belts of active mountain making? What are the two basic processes by which mountain belts are constructed?
14. What is a *continental shield?* How old are continental shields?
15. Describe the undersea topography associated with separating lithospheric plates.
16. What was Wegener's theory of *continental drift?* Why was it opposed at the time?
17. Referring to Figure 11.25, briefly summarize the history of the Earth's continents since about 600 million years ago.
18. Compare the structure and composition of *oceanic* and *continental lithosphere.* What is a *lithospheric plate?* Identify three types of plate boundaries.
19. Describe the process of *subduction* as it occurs at a **converging boundary** of continental and oceanic lithospheric plates. How is subduction related to volcanic activity?
20. What are *transform faults?* Where do they occur?
21. Name the seven largest lithospheric plates. Identify an example of a spreading boundary by general geographic location and the plates involved. Do the same for a converging boundary.
22. How does continental rupture produce *passive continental margins?* Describe the process of rupturing; itemize its various stages.
23. How are *island arcs* formed? What type of plate collision is involved?
24. What is meant by the term *arc-continent collision?* Describe how it occurs.
25. What is a **continental suture?** How does it form?
26. How is the principle of convection thought to be related to plate tectonic motions? What role might gravity play in the motion of lithospheric plates?

VISUALIZING EXERCISES

1. Sketch a block or cross section of the Earth showing the following features: batholith, sill, dike, veins, lava, and volcano.
2. Sketch the cycle of rock change and describe the processes that act within it to form igneous, sedimentary, and metamorphic rocks.
3. Sketch a cross section showing a collision between the oceanic and continental lithospheres at an active continental margin. Label the following features: oceanic crust, continental crust, mantle, oceanic trench, and rising magma. Indicate where subduction is occurring.
4. Sketch a continent–continent collision and describe the formation process of a continental suture. Provide a present-day example where a continental suture is being formed, and give an example of an ancient continental suture.

ESSAY QUESTIONS

1. A granite is exposed at the Earth's surface, high in the Sierra Nevada mountain range. Describe how mineral grains from this granite might be released and altered, to eventually form a sedimentary rock. Trace the route and processes that would incorporate the same grains in a metamorphic rock.

2. Invent a medium-sized continent, name it, and sketch a map of it on which is located one or more of the following features: mountain chains, widely spaced mountains, interior plains, depressions, high plateaus, and hills and low plateaus. Use the locations of these features on the global landform map (Figure 11.21) as a guide. Describe the geologic processes that have produced the topography of your continent.

3. Suppose astronomers discover a new planet that, like Earth, has continents and oceans. They dispatch a reconnaissance satellite to photograph the new planet. What features would you tell them to look for, and why, to detect past and present plate tectonic activity on the new planet?

Chapter 12
Tectonic and Volcanic Landforms

Located near the northern end of the Atlantic rift zone, Iceland was formed from magma upwelling along the spreading boundary between the Eurasian and North American lithospheric plates. In 1783, two fissures, each about 25 km (16 mi) long, opened in southern Iceland, spewing about 15 km³ (about 4 mi³) of molten rock. Volcanic gases and ash blanketed the entire island. Approximately three-fourths of the island's livestock perished; and after a second eruption, some 100,000 people died of famine. Now dormant, the volcanic cones along the rift are covered with a thick blanket of green moss.

VOLCANIC CONES LAKAGÍGAR, ICELAND
©Yann Arthus-Bertrand/Altitude

Tectonic and Volcanic Landforms

T ectonic and volcanic processes create landforms, ranging from block mountains and rift valleys to lofty volcanic cones. What types of rock structures does plate tectonic activity create? How does erosion act on these rock structures to produce distinctive landforms and landscapes? How do earthquakes signal tectonic activity? Why are they dangerous? What kinds of volcanoes are there, and where do they occur? Why are active volcanoes environmental hazards? These are some of the questions we will answer in this chapter.

The Great Tohoku Earthquake of 2011

At 2:46:24 PM on March 11, 2011, off the northeast seacoast of Honshu, Japan, the Earth began to tremble, signaling the start of the largest earthquake in Japan's history. The great Tohoku earthquake registered a magnitude of 9.0; the world's fourth largest since 1900. The quake resulted from a rupture along a 300-km (186-mi) subduction boundary in the Japan Trench, located on the Pacific Ring of Fire. More than 700 aftershocks were felt for several days afterward, as the Pacific plate settled itself under the Eurasian plate.

Thousands perished in the initial moments after the initial quake, despite the fact that Japan has in place the best earthquake preparation protocols in the world, which include regular testing of state-of-the-art early-warning sirens and practicing evacuation drills. Minutes after the quake shook the land, a 9.2-m (30-ft) tsunami swept away tens of thousands of people and literally wiped towns off the map along the coast of Sendai and Fukushima (Figure 12.1). Without the early-warning systems and constant training, hundreds of thousands more undoubtedly would have perished. Escalating the crisis and complicating rescue and recovery operations, the tsunami damaged nuclear reactors located on the coast at Fukushima Dai-ichi, which led to the release of radioactive material into the local environment.

In the wake of this natural disaster, seismic experts humbly acknowledged that no one had predicted such a massive earthquake, and that science still could not foresee the timing, location, or magnitude of the next devastating earthquake.

Tectonic Landforms

Plate tectonics provides a convenient framework for our study of landforms, which begins with this chapter. The motion of lithospheric plates warps, folds, and breaks rock layers apart in the processes occurring at spreading, converging, and transform boundaries. This creates *rock structures*, such as folds, faults, and thrust sheets. Although these structures are often formed deep within the Earth, erosion over many millions of years can remove great thicknesses of overlying rocks to reveal the structures at the surface. The result is a group of landforms that depend on both rock structure and the way particular rock layers and rock bodies—weak or strong—respond to erosion.

ROCK STRUCTURES OF CONVERGING BOUNDARIES

According to the Wilson cycle (Chapter 11), ocean basins open and close over long periods of time. When they close, lithospheric plates come together in a converging boundary. We begin our examination of rock structures of converging boundaries with fold belts.

Fold Belts

When two continental lithospheric plates collide, the plates are squeezed together at the boundary. As the ocean closes, sediments that have accumulated in shallow ocean basins next to the continents are pushed into **folds**. The folds in turn create a set of alternating *anticlines*, or uparching bends, and troughs, called *synclines* (Figure 12.2).

The Asahi Shimbun/Getty Images Inc.

12.1 Fukushima, Japan, shortly after the great Tohoku earthquake of 2011

When fold belts are eroded, they become a **ridge-and-valley landscape** (Figure 12.3), evident, for example, on the eastern side of the Appalachian Mountains from Pennsylvania to Alabama. In this landscape, weaker formations such as shale and limestone are eroded away, leaving hard strata, such as sandstone to stand in bold relief as long, narrow ridges. These folds are continuous and even-crested, producing almost parallel ridges.

Anticlines are not, however, always ridges, even though they are upfolds. If a resistant rock type at the center of the anticline is eroded through, to reveal softer rocks underneath, an *anticlinal valley* can form. A *synclinal mountain*

Upfolds are anticlines, and downfolds are synclines. Fold belts create a ridge-and-valley landscape of alternating ridges of resistant rock and valleys of weak rock.

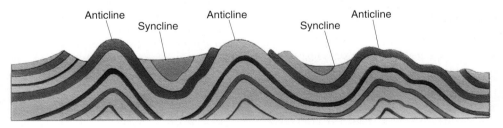

12.2 Anticlines and synclines

Compression where continents collide can produce folded sedimentary rocks. Anticlines are upward folds, and synclines are downward folds. Erosion acts on the rock layers differently, depending on their resistance, etching a pattern of hills and valleys related to the rock structure underneath. This example is a cross section of sedimentary rocks in the Jura Mountains of France and Switzerland.

12.3 Ridge-and-valley landscape

Deep erosion of simple, open folds produces a ridge-and-valley landscape. Following erosion, folds that dip downward or rise upward produce zigzag ridges.

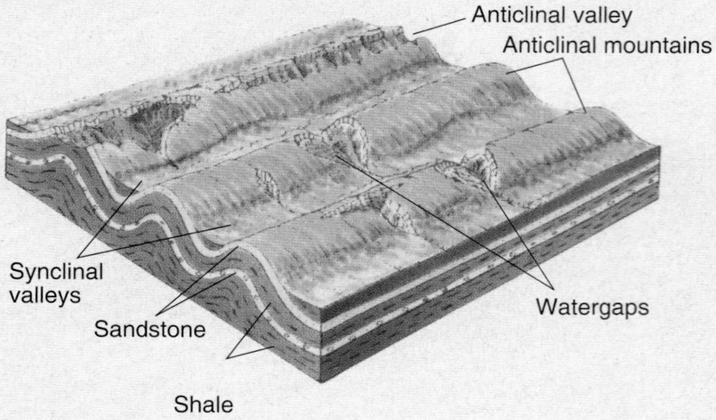

◀ EROSION OF WEAK ROCKS
Weaker formations such as shale and limestone are eroded away, leaving long, narrow ridges of hard strata, such as sandstone or quartzite. At some locations, the resistant rock at the center of the anticline is eroded through to reveal softer rocks underneath, creating an anticlinal valley.

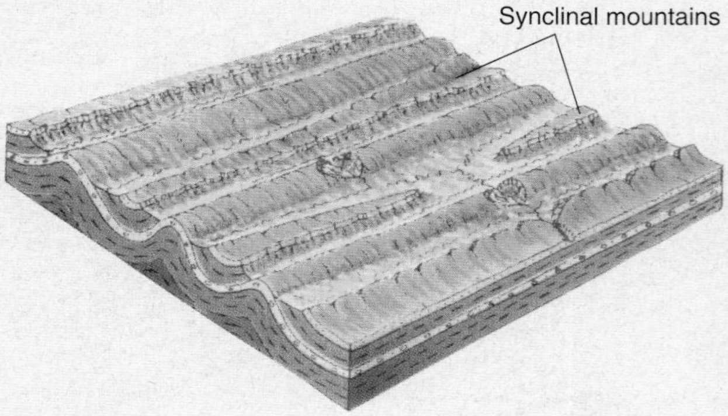

▲ CONTINUED EROSION
Synclinal mountains also can be formed after continued erosion, when resistant rock at the center of a syncline is exposed. The resistant rock stands up as a ridge.

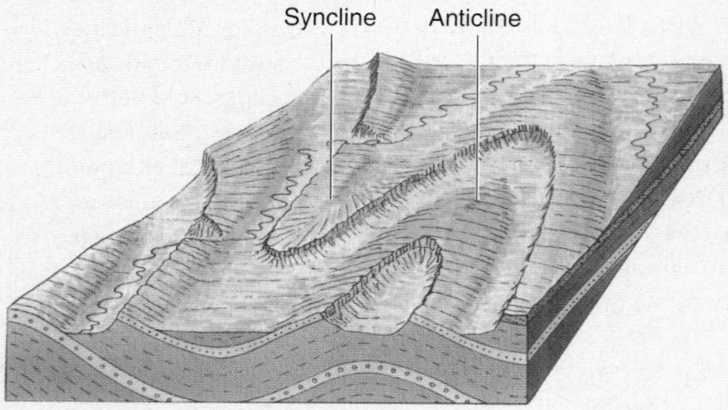

▲ PLUNGING FOLDS
When folds dip down, or plunge, their erosion creates zigzag ridges.

Science Source/Photo Researchers, Inc.

▲ RIDGE-AND-VALLEY LANDSCAPE IN PENNSYLVANIA
This Landsat image displays the ridge-and-valley country of south-central Pennsylvania in color-infrared. The zigzag ridges are formed by bands of hard quartzite. Plunging folds are readily visible. The strata were crumpled during a continental collision that took place over 200 million years ago.

is also possible. It occurs when a resistant rock type is exposed at the center of a syncline, and the rock forms a ridge. In some regions, the fold crests plunge, rising up in places and dipping down in others. This results in a pattern of curving mountain crests and valleys.

WileyPLUS Folding
In a video focusing on the Appalachian Mountains, watch how a landscape of ridges and valleys is created by folding. In an animation, view the development of anticlinal and synclinal mountains and valleys. See how plunging folds produce zigzag and hairpin ridges.

Metamorphic Belts

Closer to the center of two colliding plates of continental lithosphere, folding can be more intense and accompanied by heat and pressure, altering the strata into metamorphic rocks. The result is still a landscape of ridges and valleys, but the ridges and valleys won't be as sharp or as straight as in belts of open folds. Even so, resistant rocks, such as quartzite, gneiss, and schist, will form highlands and ridges, while weak rocks, such as slate and marble, will shape lowlands and valleys (Figure 12.4). Parts of New England, particularly the Taconic and Green Mountains of eastern New York, western Massachusetts, and Vermont, illustrate the landforms eroded on an ancient metamorphic belt. The larger valleys trend north and south and are underlain by marble. They are flanked by ridges of gneiss, schist, slate, or quartzite.

> Slate and marble are weak metamorphic rocks that underlie valleys. Schist, gneiss, and quartzite are more resistant and underlie uplands and ridges.

Warped Rock Layers

Sometimes sedimentary beds are warped into a steplike bend, without folding. This type of structure is called a *monocline*. When erosion chips away at the layers, stripping out the weaker ones, the stronger layers form *hogback ridges* (Figure 12.5). If the front of the warped section is carved into short valleys by stream action, the result is a series of triangular landforms known as *flatirons*.

Another distinctive set of landforms is created when strata are warped upward into a **sedimentary dome**. Igneous intrusions of rocks at great depths may be responsible for some domes. For others, upward motion

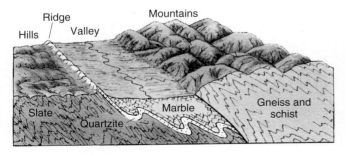

12.4 Metamorphic belts

Marble is shaped into valleys in this humid environment, while slate and schist emerge as hill belts. Quartzite stands out boldly and may produce conspicuous narrow hogback ridges. Areas of gneiss form highlands.

12.5 Erosion of monoclines

Where sedimentary rock beds are warped by tectonic forces, erosion can reveal hogback ridges or flatirons.

University of Washington Libraries, Special Collections, John Shelton Collection, KC12017

▲ HOGBACK RIDGES

Sandstone formations of the Colorado front range are tilted upward and eroded to form hogback ridges.

▼ FLATIRONS

Visible in this upward bending of rock layers in Comb Ridge, Utah, are tilted beds cut into flatiron shapes by erosion of streams draining the steep front of the landform.

AirPhoto-Jim Wark

on deep faults may have been the cause. Figure 12.6 shows how domes are eroded over time to emerge as hogback ridges and *circular valleys*. An example of a large and rather complex dome is the Black Hills dome of western South Dakota and eastern Wyoming. Its central core is rich in mineral deposits, including the famous Homestake Mine, one of the world's richest gold producers.

> Erosion of monoclines and domes produces hogback ridges and flatirons, and in the case of domes, circular valleys.

PLATE INTERIORS

Far from spreading and converging boundaries, the centers of continental lithospheric plates exhibit rock structures, sometimes quite ancient, that also produce distinctive landforms. Passive continental margins have unique features as well.

Exposed Batholiths and Monadnocks

Where the stable continental interior of a lithospheric plate has been stripped of sedimentary cover

12.6 Erosion of domes

Sedimentary domes erode to form hogback ridges and circular valleys.

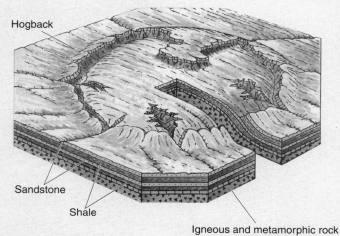

▲ **EROSION OF SEDIMENTARY LAYERS**
Erosion has partially revealed the summit region of the dome, exposing older strata beneath. Eroded edges of steeply dipping strata form sharp-crested hogback ridges.

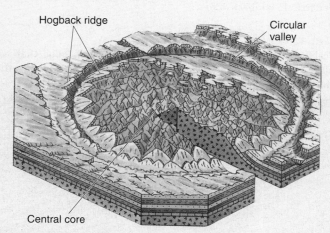

▲ **BREACHING OF THE DOME**
When the last of the strata have been removed, the ancient igneous rock is exposed in the central core of the dome. This igneous or metamorphic rock is resistant to erosion and develops into a mountainous terrain.

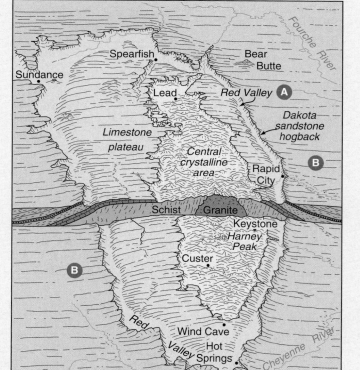

▲ **THE BLACK HILLS DOME**
The Black Hills consist of a broad, flat-topped dome deeply eroded to expose a core of igneous and metamorphic rocks.

Ⓐ Encircling the dome is the Red Valley, which is underlain by weak shale that is easily washed away.

Ⓑ On the outer side of the Red Valley is a high, sharp hogback of Dakota sandstone, rising some 150 m (about 500 ft) above the level of the Red Valley.

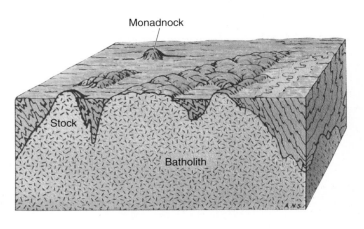

Monadnock

Stock

Batholith

A.N.S.

12.7 Batholiths

Underlying much of the Earth's continents are large expanses of igneous intrusive cooled magma, called batholiths. Surrounding the batholiths are metamorphic rocks, forged by the heat and pressure of the injection of nearby magma. After long-continued erosion and removal of thousands of meters of rocks, the resistant igneous rocks of the batholith appear at the surface, surrounded by weaker metamorphic rocks.

by long erosion, shield rocks are exposed. These rocks include highly metamorphosed rocks and crystalline igneous rocks of ancient origin, which may be related to supercontinent cycles occurring long, long, ago. Many of the igneous rocks are huge batholiths of felsic rock that melted deep within the Earth, rose as

multiple injections of magma, and cooled below the surface.

Because batholiths are typically composed of resistant rock, they are eroded into hilly or mountainous uplands. A good example is the Idaho batholith, a granite mass exposed over an area of about 40,000 sq km (about 16,000 sq mi)—a region almost as large as New Hampshire and Vermont combined. Figure 12.7 shows the features created when batholiths are exposed at the surface. Isolated mountains or hills revealed by erosion of weaker surrounding rock are called **monadnocks**. The name is taken from Mount Monadnock in southern New Hampshire (Figure 12.8).

COASTAL PLAINS

A large lithospheric plate may include a passive margin where continental crust meets oceanic crust. In this case, the coastline receives large quantities of sediments from the continental interior that are deposited in shallow waters just offshore. If the sea level falls or the passive margin rises, a gently-sloping **coastal plain** of recent sediments emerges. Beds of sand and clay alternate: clay turns into lowlands, and sandy ridges trace lines of low hills called *cuestas* (Figure 12.9). The gentle slope leads to the formation of streams in a distinctive trellis pattern. The coastal plain of the United States is a major geographical region, ranging in width from 160 to 500 km (about 100 to 300 mi) and extending for

12.8 Domes and monadnocks

Rock bodies of the ancient batholith are sometimes revealed by erosion as domes or monadnocks.

René Burri/Magnum Photos, Inc.

◄ SUGAR LOAF MOUNTAIN

This small granite dome, rising above the city of Rio de Janeiro, Brazil, is a projection of a batholith that lies below the ground and harbor.

▼ MONADNOCK

Where a larger, broader intrusive body tops a batholith, it can form an isolated mountain called a monadnock. This feature is named for Mount Monadnock in New Hampshire, shown here in fall colors.

Danita Delimont/Alamy Limited

12.9 Development of a broad coastal plain

The diagrams show a coastal zone that has recently emerged from beneath the sea, with layers of nearly flat-lying sediments.

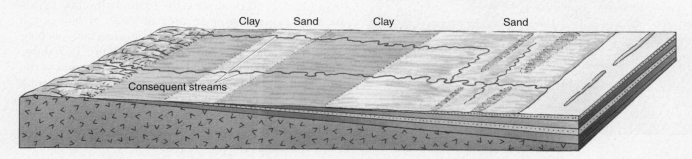

▲ EMERGENCE

As the shoreline recedes to expose the coastal plain, streams flow directly seaward on the new land surface, down the gentle slope. These are consequent streams, with courses controlled by the initial slope of the land.

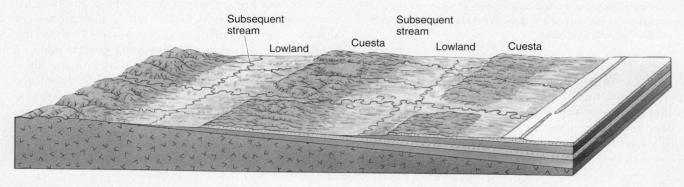

▲ EROSION

More easily eroded strata (usually clay or shale) are rapidly worn away, developing into lowlands. Between them rise broad belts of hills, called cuestas, on layers of sand, sandstone, limestone, or chalk. Subsequent streams start to flow along the lowlands, parallel with the shoreline. They take their position along any belt or zone of weak strata.

about 3000 km (about 2000 mi) along the Atlantic and Gulf coasts.

FAULTS AND FAULT LANDFORMS

As lithospheric plates move in different directions, the brittle rocks of the Earth's surface experience stress and strain. When these rocks finally break and move in different directions, they cause a fracture in the crust called a **fault**. Major fractures in the crust can be followed along the ground for many kilometers and extend down into the crust for at least several kilometers.

A single fault movement can cause slippage of as little as a centimeter or as much as 15 m (about 50 ft). Fault movements can happen many years or decades apart,

even several centuries apart. But when all these small motions are added up over long geologic time spans, they can amount to tens or hundreds of kilometers of displacement.

There are four main types of faults, characterized by how the crustal blocks move: *normal, strike-slip* or *transcurrent, reverse,* and *overthrust* faults. The type of fault can usually be identified by its relationship to the moving blocks after they move (Figure 12.10). The presence of a newly exposed cliff face,

> In a fault, rocks break apart and move in different directions. The four main types of faults are normal, strike-slip or trans-current, reverse, and overthrust.

12.10 Types of faults

A fault is fracture in the Earth's crust along which rock masses move in different directions.

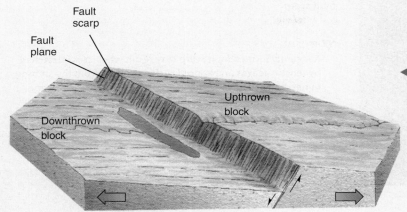

◀ NORMAL FAULT

As a result of extensional activity, the crust on one side of a normal fault drops down relative to the other side. This creates a steep, straight fault scarp.

STRIKE-SLIP OR TRANSCURRENT FAULT ▶

Movement along a strike-slip fault, also called a transcurrent fault, is mostly horizontal, so we don't see a scarp, or at most a very low one. We can usually trace only a thin fault line across the surface, although in some places the fault is marked by a narrow trench, or rift.

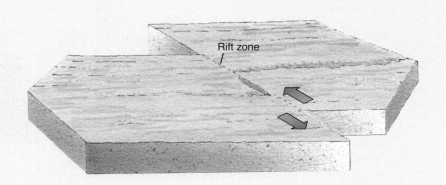

◀ REVERSE FAULT

As a result of compressional activity, the fault plane along a reverse fault is inclined such that one side rides up over the other. Reverse faults produce fault scarps similar to those of normal faults, but because the scarp tends to be overhanging there's a much greater risk of a landslide. The San Fernando, California, earthquake of 1971 was generated by slippage on a reverse fault.

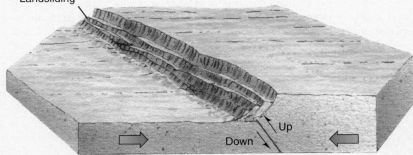

OVERTHRUST FAULT ▶

Overthrust faults involve mostly horizontal movement. One slice of rock rides over another. A thrust slice may be up to 50 km (30 mi) wide.

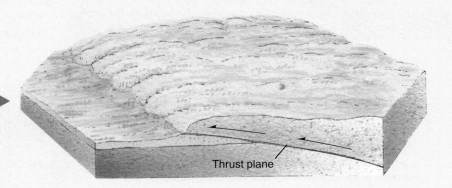

A. N. Strahler

12.11 Fault scarp

This fault scarp was formed during the Hebgen Lake, Montana, earthquake of 1959. In just a few minutes, a displacement of 6 m (20 ft) took place on a normal fault.

called a *fault scarp*, can indicate how much the fault moved (Figure 12.11).

Just as with other tectonic activity, fault movements give rise to distinctive landforms. For example, parallel normal faults can drop a block of crust down into a **graben**, or push a block up, into a **horst** (Figure 12.12).

Repeated faulting can raise a great rock cliff hundreds of meters high. Because fault planes extend hundreds of meters down into the bedrock, their landforms can persist even after millions of years of erosion. Figure 12.13 diagrams the effect of erosion on a fault scarp.

Systems of parallel faulting are typical of locations where continental spreading is occurring. The Rift Valley of East Africa is an example at a continental scale (Figure 12.14). Here, the continental lithosphere is

12.13 Fault scarp evolution

When vertical displacement at a fault occurs, the rising block can create a steep fault scarp. Once formed, the geology and weathering patterns influence the shape and erosion characteristics of the fault scarp.

▼ RECENT SCARP

A recently formed fault scarp may result when crust stretched by tension leaves an exposed surface, or cliff, from the shifting of one block relative to the other. Fault scarps create distinctive and sheer cliffs.

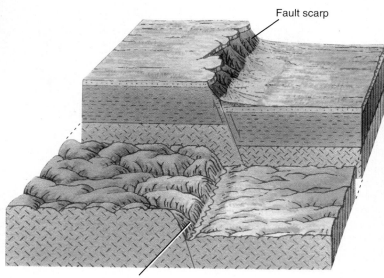

Fault scarp

Fault-line scarp

▲ FAULT-LINE SCARP

Even though the cover of sedimentary strata has been completely removed, exposing the ancient shield rock, the fault continues to produce a landform, known as a fault-line scarp.

12.12 Landforms created by parallel faults

When extensional forces cause the crust to break into sets of parallel normal faults, mountain blocks and valleys are formed.

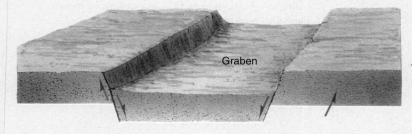

Graben

◀ GRABEN

A narrow block dropped down between two normal faults creates a graben, a trench with straight, parallel walls.

Horst

HORST ▶

A narrow block elevated between two normal faults is a horst, which produces blocklike plateaus or mountains, often with a flat top and steep, straight sides.

12.14 The Rift Valley system of East Africa

The East African Rift Valley is an example of an early stage of continental rupture.

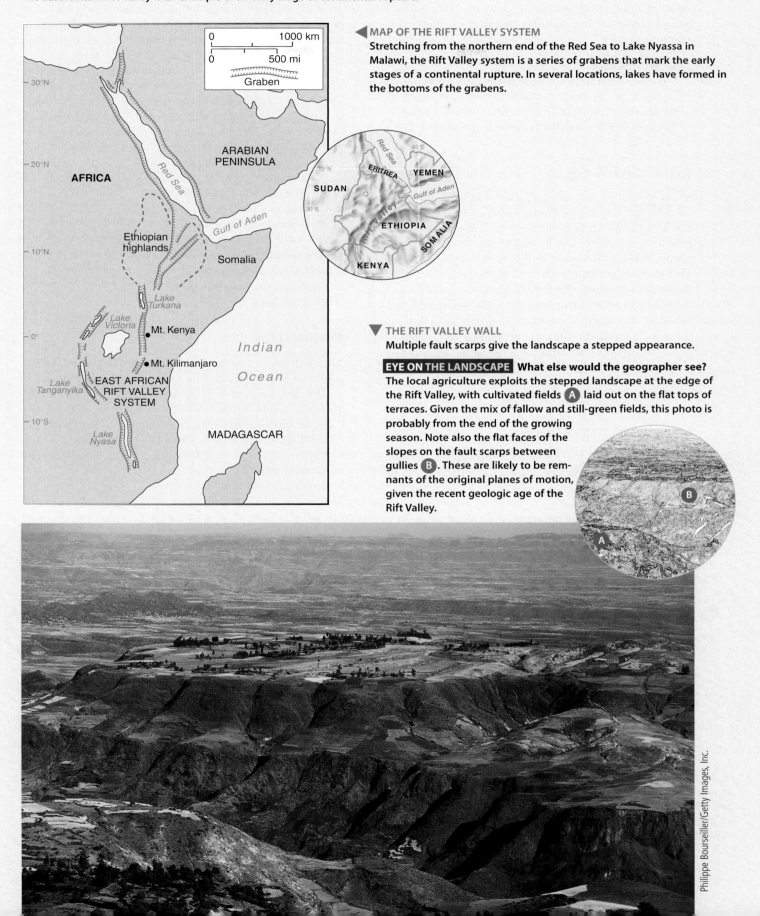

◄ **MAP OF THE RIFT VALLEY SYSTEM**
Stretching from the northern end of the Red Sea to Lake Nyassa in Malawi, the Rift Valley system is a series of grabens that mark the early stages of a continental rupture. In several locations, lakes have formed in the bottoms of the grabens.

▼ **THE RIFT VALLEY WALL**
Multiple fault scarps give the landscape a stepped appearance.

EYE ON THE LANDSCAPE What else would the geographer see?
The local agriculture exploits the stepped landscape at the edge of the Rift Valley, with cultivated fields **A** laid out on the flat tops of terraces. Given the mix of fallow and still-green fields, this photo is probably from the end of the growing season. Note also the flat faces of the slopes on the fault scarps between gullies **B**. These are likely to be remnants of the original planes of motion, given the recent geologic age of the Rift Valley.

Philippe Bourseiller/Getty Images, Inc.

beginning to rupture and split apart in the first stage of forming a new ocean basin. The Rift Valley is basically a series of linked and branching grabens, but with a more complex history that includes building volcanoes on the graben floor.

WileyPLUS Major Types of Faulting
See how crustal extension and compression create normal, reverse, and overthrust faults in this animation. Watch a transcurrent fault form at the boundary of two lithospheric plates moving in opposite directions.

Earthquakes

As rock on both sides of the fault is slowly bent over many years by tectonic forces, energy builds up in the bent rock, just as it does in a bent archer's bow. When that tension reaches a critical point, the rocks slip in different directions, relieving the strain and generating an **earthquake**. The motion generates earthquake waves, which radiate outward, traveling through the Earth's surface layer and shaking the ground. Like ripples spreading after a pebble is thrown into a quiet pond, these waves gradually lose energy as they travel outward in all directions.

> An earthquake is a seismic wave motion transmitted through the Earth. It is triggered by sudden slippage on a fault.

EPICENTER AND FOCUS

In an earthquake, the location where the fault slipped, called the *focus*, can be near the surface or deep underground. The depth of an earthquake's focus in part determines the intensity of shaking felt on the ground; shallow-focus earthquakes are likely to cause more damage than deep-focus earthquakes. The point on the Earth's surface directly above the focus is called the **epicenter** of the earthquake (Figure 12.15).

To determine where the epicenter is, and therefore where to send emergency services, scientists measure how long it takes different types of earthquake waves to reach sensors, called *seismometers*, positioned in various locations. The first waves to reach the seismometer are the faster-moving *P waves*, or *primary waves*, which shake the ground from side to side. Next come the slower-moving *S waves*, or *secondary waves*, which are characterized by up-and-down shaking. The amount of time it takes between the arrival of P and S waves at nearby seismograph stations is used to calculate the point of origin of the earthquake (Figure 12.16).

> Earthquakes generate P and S waves that radiate outward from the earthquake focus.

MAGNITUDE

The amount of energy released by an earthquake, called its *magnitude*, can be measured by the amplitude of the seismic waves produced. In 1935, the distinguished seismologist Charles F. Richter devised the scale of earthquake magnitudes that today bears his name. Figure 12.17 shows how the Richter scale relates to the energy released by some of the largest earthquakes on record.

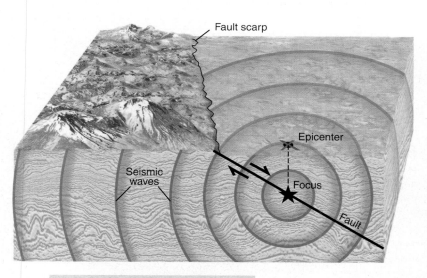

12.15 Earthquake focus and epicenter
Seismic waves emanate in all directions from the earthquake site of origin, called the focus, located along a fault line. The epicenter is the surface location directly above the subterranean focus. The earthquake focus can originate from various types of fault zones.

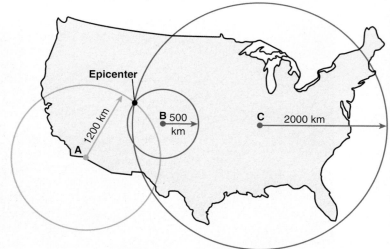

12.16 Determining earthquake epicenters
Epicenters are calculated by triangulating the readings from three different seismometer reading centers (A, B, and C), using the difference in travel times for P and S waves. The epicenter lies at the intersection of the circles.

The magnitude of an earthquake is related to the length of the fault section that broke. This means that longer faults have the potential to produce larger earthquakes. Earthquake magnitude is not, however, a direct indicator of damage severity; the extent of damage is also related to the proximity of the epicenter to populated areas, the depth of focus, the type of ground materials, and whether local buildings meet earthquake safety codes. The magnitude 9.0 Great Tohoku earthquake, which occurred 129 km (80 mi) off the coast of Japan, was 40 km (24 mi) deep and spread across a 300-km (186-mi) fault, causing the Earth to move 30 m (100 ft). Many lives were lost, but thanks to Japan's strict building codes and public education program, damages from this massive quake were curtailed.

> The Richter scale is used to measure the energy released by earthquakes.

12.17 Earthquake magnitude

Earthquake magnitude is a measure of how much energy is released by an earthquake; it is not a direct indicator of damage severity.

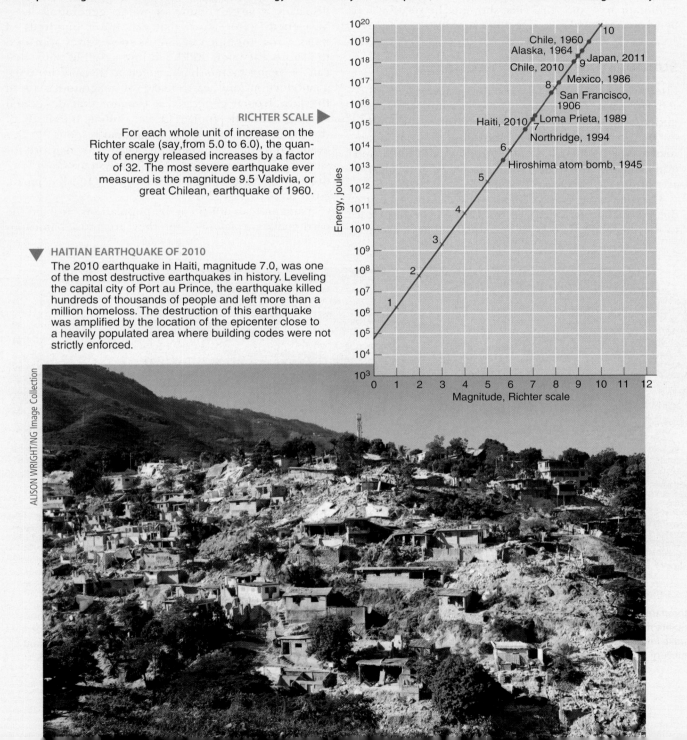

RICHTER SCALE ▶

For each whole unit of increase on the Richter scale (say, from 5.0 to 6.0), the quantity of energy released increases by a factor of 32. The most severe earthquake ever measured is the magnitude 9.5 Valdivia, or great Chilean, earthquake of 1960.

▼ HAITIAN EARTHQUAKE OF 2010

The 2010 earthquake in Haiti, magnitude 7.0, was one of the most destructive earthquakes in history. Leveling the capital city of Port au Prince, the earthquake killed hundreds of thousands of people and left more than a million homeless. The destruction of this earthquake was amplified by the location of the epicenter close to a heavily populated area where building codes were not strictly enforced.

ALISON WRIGHT/NG Image Collection

TECTONIC ENVIRONMENTS OF EARTHQUAKES

Earthquake activity is primarily (although not exclusively) associated with plate boundaries. The map in Figure 12.18 shows the centers of large earthquakes observed during a 20-year period. The pattern of earthquakes follows subduction boundaries, transform boundaries, island arcs, mountain arcs, and axial rifts, showing the relationship between earthquakes and tectonic activity.

> Earthquakes occur frequently at spreading and converging boundaries of lithospheric plates. Transcurrent faults on transform boundaries are also common earthquake sites.

SUBDUCTION-ZONE EARTHQUAKES

Intense seismic activity occurs along convergent lithospheric plate boundaries where oceanic plates are undergoing subduction. This mechanism is responsible for the greatest earthquakes, including those in Japan, Alaska, North America, Central America, and Chile, and other narrow zones close to the trenches and volcanic arcs of the Pacific Ocean Basin.

On the Pacific coast of Mexico and Central America, the subduction boundary of the Cocos plate lies close to the shoreline. The great earthquake that devastated Mexico City in 1986 was centered in the deep trench offshore. Two great shocks in close succession, the first of magnitude 8.1 and the second of 7.5, damaged cities along the coasts of the Mexican states of Michoacoán and Guerrero. Although Mexico City lies inland about 300 km (about 185 mi) from the earthquake's epicenter, it experienced intense ground shaking, killing some 10,000 people.

Tsunamis

Undersea subduction-zone earthquakes pose an additional hazard to coastal residents. When the subducted plate suddenly snaps back during an earthquake, it displaces a large volume of water, generating long-wavelength water waves that can travel hundreds of miles across the open ocean. As these waves approach the shore, they slow and build into a wall of water that can be many meters high, called a *tsunami*. In 2004, a subduction-zone earthquake of magnitude 9.0 in the Java Trench generated a tsunami that devastated areas around the Indian Ocean, killing hundreds of thousands of people. An international response to this event led to the creation of an ocean buoy network for tsunami early-warning systems. Such systems saved hundreds of thousands of lives along the Honshu, Japan, coast on March 11, 2011.

Tsunamis can affect coastal populations far from the earthquake's epicenter, but they are most hazardous

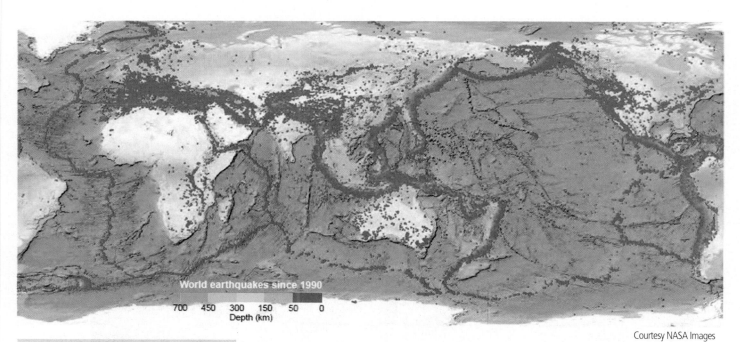

World earthquakes since 1990

700   450   300   150   50   0
Depth (km)

Courtesy NASA Images

12.18 Earthquakes and plate boundaries

If you compare the map of earthquake locations with the tectonic features, you can see that seismic activity is primarily associated with lithospheric plate motion. This world map plots hundreds of thousands of earthquake epicenters recorded over a 20-year period by international earthquake information centers. Deep-focus earthquakes are identified by yellow and green colors. They occur on subduction boundaries and are often very strong. According to geophysicists, about half of all moderate to strong quakes occur on previously unknown, or blind, faults.

when the earthquake epicenter is close to shore, leaving little time for warning between the earthquake and the arrival of the waves. Coastal residents who feel an earthquake should seek higher ground immediately, without waiting for a tsunami warning. We will discuss tsunamis in more detail in Chapter 16.

TRANSFORM BOUNDARIES

Strike-slip faults on transform boundaries that cut through the continental lithosphere cause moderate to strong earthquakes. The San Andreas Fault, which runs about 1000 km (about 600 mi) from the Gulf of California to Cape Mendocino, is an active strike-slip fault (Figure 12.19).

> Subduction-zone earthquakes are the largest and most dangerous. They can also generate deadly tsunami waves.

Movement on the San Andreas Fault generated the great San Francisco earthquake of 1906, one of the worst natural disasters in the history of the United States. Since then, this sector of the fault has been "locked"; that is, rocks on the two sides of the fault have remained in place without sudden slippage. In the meantime, the two lithospheric plates that meet along the fault have been moving steadily with respect to one another. This means that a huge amount of unrelieved strain (potential energy) has already accumulated in the crustal rock on either side of the fault.

On October 17, 1989, the San Francisco Bay area was severely shaken by an earthquake with a magnitude of 7.1 on the Richter scale. Its epicenter was located near Loma Prieta peak, about 80 km (50 mi) southeast of San Francisco, at a point only 12 km (7 mi) from the city of Santa Cruz, on Monterey Bay. Altogether, 62 lives were lost in this earthquake, with damage was estimated at $6 billion. (In comparison, the 1906 earthquake took a toll of 700 lives and caused property damage equivalent to about $30 billion in present-day dollars.) The Loma Prieta slippage, though near the San Andreas Fault, probably did not relieve more than a low level of the strain on this fault.

In Southern California, three severe earthquakes occurred in close succession in 1992, along active local faults a short distance north of the San Andreas Fault in the southern Mojave Desert. The second of these, the Landers earthquake, recorded as a powerful 7.5 on the Richter scale, occurred on a transcurrent fault trending north-northwest. It caused an 80-km (50-mi) rupture across the desert landscape.

> The San Andreas Fault and related faults in the southern and central areas of California are potential sources of great earthquakes occurring in densely populated regions.

Residents of the Los Angeles area live with the serious threat from a large number of nearby active faults, pinpointed on the map in Figure 12.20. Movements on these local faults have been felt as more than 40 damaging earthquakes since 1800, including the Long Beach earthquakes of the 1930s and the San Fernando earthquake of 1971. The San Fernando earthquake measured 6.6 on the Richter scale and severely damaged structures near the earthquake epicenter. In 1987, an earthquake of magnitude 6.1 struck the vicinity of Pasadena and Whittier, located within about 20 km (12 mi) of downtown Los Angeles. Known as the Whittier Narrows earthquake, it was generated along a local fault system that had not previously shown significant seismic activity. The Northridge earthquake of 1994, registering 6.7 on the Richter scale, produced the strongest ground motions ever recorded in an urban setting in North America, and resulted in the greatest financial losses from an earthquake in the United States since the San Francisco earthquake of 1906. Sections of three freeways were closed, including the busiest highway in the country, Interstate 5.

Spreading-Center Earthquakes

Spreading plate boundaries also generate seismic activity. Most of these boundaries are identified with the midoceanic ridge and its branches, and the earthquakes are moderate. Close examination of the midoceanic ridge displays numerous strike-slip faults that form perpendicular to the spreading plate line.

12.19 San Andreas Fault in Southern California

Throughout many kilometers of its length, the San Andreas Fault looks like a straight, narrow scar. In some places this scar is a trenchlike feature; elsewhere, it is a low scarp. It marks the active boundary between the Pacific plate and the North American plate. The Pacific plate is moving toward the northwest, carrying a great portion of the state of California and all of Lower (Baja) California with it.

Peter Essick / Aurora Photos/Alamy Limited

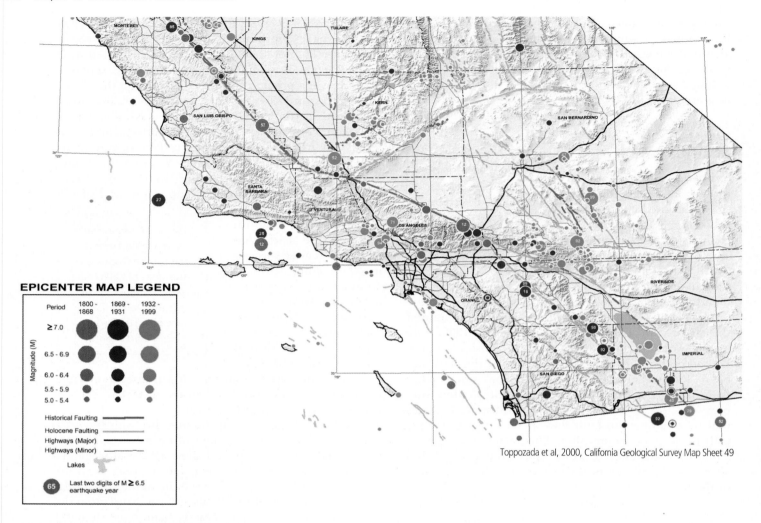

EPICENTER MAP LEGEND

Toppozada et al, 2000, California Geological Survey Map Sheet 49

12.20 Earthquake activity in Southern California

California earthquakes, identified here as circles whose size is proportional to their magnitude, are overlaid on delineated and mapped fault lines. The correlation of earthquakes and fault lines, as shown in this figure, is found around the globe.

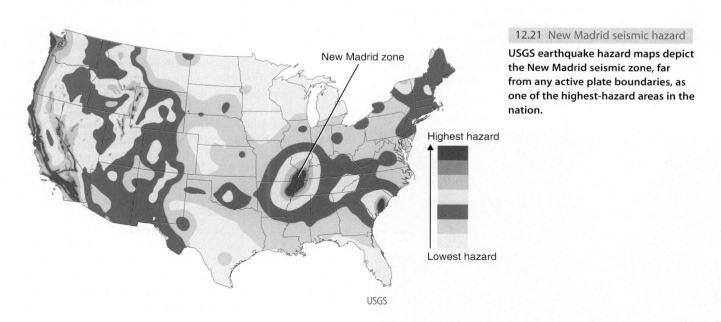

12.21 New Madrid seismic hazard

USGS earthquake hazard maps depict the New Madrid seismic zone, far from any active plate boundaries, as one of the highest-hazard areas in the nation.

New Madrid zone

Highest hazard

Lowest hazard

USGS

Isolated Earthquakes

A few earthquake epicenters are scattered over the continental plates, far from active plate boundaries. Scientists are uncertain why these occur, but suggest that spreading continental plates may be a factor. In many cases, no active fault is visible. For example, the great New Madrid earthquake of 1811 was centered in the Mississippi River floodplain in Missouri. Including three great shocks in close succession, the earthquake was rated from 8.1 to 8.3 on the Richter scale. The Earth movements caused the Mississippi River to change its course and even run backward in a few stretches for a short time. Large areas of land dropped, and new lakes were formed.

The U.S. Geological Survey (USGS) considers the present populated areas located over the New Madrid seismic zone to be in the high-risk category (Figure 12.21). Researchers conducting underwater seismic surveys have recently mapped two new unknown faults along the Mississippi River north of the city of Memphis, demonstrating an increased probability of a major earthquake, in the range of magnitude 7 on the Richter scale.

> Some earthquakes occur far from active plate boundaries, for unexplained reasons.

Earthquakes Along Blind Faults

Although scientists understand the role that faults play in earthquake risk, they don't always know where fault systems are located. Some, called *blind faults*, are not apparent on the Earth's surface. Seismographic research can help scientists identify blind faults, but they are hampered in this effort by political unrest in some areas and the geographic inaccessibility of others, which may contain fault systems that have never been mapped.

In 2003, a major earthquake on a blind strike-slip fault, previously unknown to seismologists and geoscientists, destroyed the ancient city of Bam in southern Iran (Figure 12.22). Approximately 30,000 people died and an equal number were injured and made homeless by the event, which was rated at 6.6 on the Richter scale.

This devastating earthquake reminds us that while the major plate boundaries and most seismic areas have been mapped by geologists, blind faults remain undiscovered, and as human populations grow and expand their reach, they will be increasingly exposed to unknown seismic events.

Volcanic Activity and Landforms

Like earthquakes and plate tectonics, volcanic activity creates initial landforms. Underground molten mineral matter, called *magma*, is extruded through constricted vents and fissures in the Earth's surface. When the magma reaches the surface, as *lava*, it cools and hardens, building the landform familiarly known as a **volcano**, which is typically conical or dome-shaped.

TECTONIC ENVIRONMENTS OF VOLCANOES

Volcanic activity is related to plate movement. If you compare the maps in Figure 12.23, you can see that many volcanoes are located on plate boundaries. Subduction zones around the Pacific Rim perpetuate significant volcanic activity, aptly named the Pacific Ring of Fire.

Volcanoes also occur along seafloor spreading centers. Iceland, in the North Atlantic Ocean, is an

---

12.22 The Bam, Iran, earthquake of 2003

Located on a previously unknown blind fault, Bam, Iran, experienced a 6.6 Richter magnitude earthquake in 2003. This local strike-slip fault is situated not far from the convergent boundary between the Arabian and Eurasian plates.

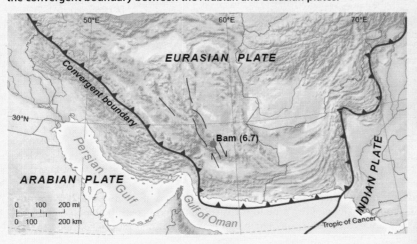

12.23 Volcanic activity of the Earth

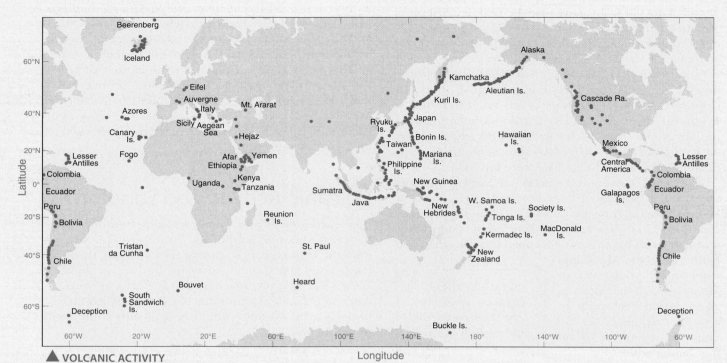

▲ VOLCANIC ACTIVITY

Dots on the top map show the locations of volcanoes known or believed to have erupted within the past 12,000 years. Each dot represents a single volcano or cluster of volcanoes.

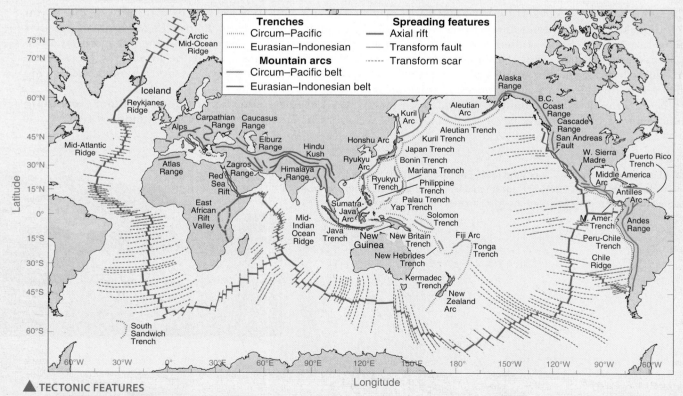

▲ TECTONIC FEATURES

On the bottom map are drawn the principal tectonic features of the world, such as midoceanic ridges, trenches, and subduction boundaries. Many volcanoes occur along these features—for example, the Ring of Fire around the Pacific Rim occurs on subduction boundaries. Spreading on the Mid-Atlantic Ridge produced Iceland's volcanoes; likewise, spreading in East Africa has also produced volcanoes.

Helgafell   Lava from Eldfell   Eldfell

Emory Kristof/NG Image Collection

12.24 Spreading-center volcano

Iceland is constructed entirely of recent eruptions of basalt emerging from the Mid-Atlantic ridge. In January 1973, the volcano beneath Heimaey Island erupted, emitting tephra at a rate of 100 m³/sec (about 3500 ft³/sec) and building a broad cone that reached 100 m (about 300 ft) above sea level. Named Edfell, the cone appears on the right, surrounded by lava flows extending into the ocean. Edfell's older sister, Helgafell, is on the left, with the town of Vestmannaeyjar behind it. Note the tongue of fresh lava from Eldfell invading the right side of the town.

outstanding example (Figure 12.24). Other islands of volcanoes located along or close to the axis of the Mid-Atlantic Ridge are the Azores, Ascension, and Tristan da Cunha.

Chains of island volcanoes can also appear away from plate boundaries, where a plume of magma, called a **hotspot**, rises through the crust (Figure 12.25). The chain of Hawaiian volcanoes, from the newest islands to the oldest seamounts, tracks the motion of the Pacific plate over the stationary hotspot.

> Volcanic activity is frequent along subduction boundaries, which accounts for the Ring of Fire around the Pacific Rim. Midocean spreading centers and hotspots also generate volcanic activity.

12.25 Hotspot volcano formation

The Hawaiian seamount chain in the northwest Pacific Ocean Basin is 2400 km (about 1500 mi) long and trends northwestward. This chain formed as the Pacific plate moved over a hotspot in the mantle. The sharp bend to the north was caused by a sudden change of direction of the Pacific plate. On the map, dots pinpoint summits, and the yellow area marks the base of the volcano at the ocean floor.

1 An upwelling plume of magma rises from the deep in the mantle to form a volcanic island.

2 Over time, plate motion carries the volcano away from its origin over the hotspot, as though it were on a conveyor belt. Once away from the hotspot—its source of magma—the volcano grows inactive and extinct.

3 As the extinct volcano is conveyed farther from the hotspot, continued erosion and subsidence reduces it below the ocean's surface to a low seamount.

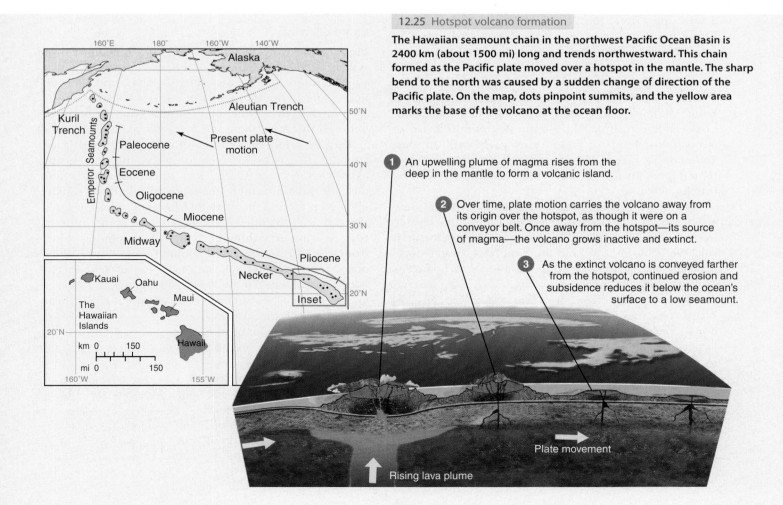

VOLCANIC ERUPTIONS

When lava is violently forced out of a volcano, in the form of a volcanic eruption, it results in one of the most severe environmental hazards on the planet. During the Mount Pelée disaster on the Caribbean island of Martinique in 1902, for example, thousands of lives were snuffed out in seconds. The destruction associated with volcanoes is wrought by a number of related effects. Clouds of incandescent gases sweep down volcano slopes like great glowing avalanches. Relentless lava flows can engulf whole cities. Showers of ash, cinders, and volcanic "bombs" (solidified masses of lava) cause terrible devastation. When hot ash melts a snowpack, it can produce a hot mudflow, called a *lahar*, that races down the sides of a volcano, burying everything in its path.

Atmospheric impacts from volcanic ash and gases have been shown to affect global weather and surface temperatures for months or years after an event. Following a major eruption, volcanic gases and ash in the stratosphere can reflect enough incoming sunlight to cause a global reduction in temperatures. For example, the year following the 1815 eruption of Mount Tambora, in Indonesia, is known as "the year without a summer," when unusually low temperatures around the world led to crop failures and, subsequently, famine.

Conversely, when volcanoes release CO_2 and other greenhouse gases into the atmosphere, the result is a warming effect. Recall from Chapter 2 that greenhouse gases in the atmosphere absorb and reradiate longwave radiation back toward the Earth in a process called the greenhouse effect. A rise in the level of greenhouse gases in the atmosphere intensifies the greenhouse effect and leads to greater warming. Some climate-change analysts have argued that volcanoes may be responsible for the documented rise in atmospheric levels of greenhouse gases over the past 150 years. In fact, volcanic sources have generated less than 1 percent of the greenhouse gases produced by human activities.

Violent earthquakes are also associated with volcanic activity. Scientists track and measure multiple seismic tremors and/or small earthquakes to better understand volcanic behavior and help them more accurately forecast major eruptions, and thus be able to warn citizens well in advance of these events. And undersea or island volcanoes put habitations along low-lying coasts in peril of tsunamis. Thanks to scientific monitoring, it is now possible to lower the death tolls and reduce the destruction caused by volcanoes. Still, not every volcano is well monitored or predictable.

> Volcanic eruptions can have extreme environmental impacts. Flows of hot gas, showers of ash, cinders, and rocks, violent earthquakes, and accompanying tsunamis can cause great loss of life.

TYPES OF VOLCANOES

The shape, size, and explosiveness of a volcano depend on the type of magma involved. Magma comes from two main types of igneous rocks: felsic and mafic (see Chapter 11).

Felsic lavas (rhyolite and andesite) are very thick, gassy, and gummy, with high viscosity. This type of lava doesn't usually flow very far from the volcano's vent and so builds up steep slopes. When the volcano erupts, ejected particles of different sizes, known collectively as *pyroclastic material*, or *tephra*, fall on the area surrounding the crater, creating a cone shape. The most common type of volcano associated with felsic magma is the **stratovolcano**, sometimes called a *composite volcano*.

In contrast to felsic lava, mafic lava (basalt) is not very viscous and holds little gas. Eruptions of basaltic lava are usually quiet, and the lava can travel long distances and spread out in thin layers. Typically, then, large basaltic volcanoes are broadly rounded domes with gentle slopes. The most common type of volcano associated with mafic lava is the **shield volcano**. The differences between felsic and mafic volcanoes are summarized graphically in Figure 12.26.

Remote sensing provides some spectacular images of volcanoes, including eruptions. Our *Focus on Remote Sensing* Figure 12.28 (overleaf) shows views of Mount Vesuvius, Mount Fuji, and Popocatepetl from several satellite sources.

Stratovolcanoes

Most of the world's active stratovolcanoes lie within the circum-Pacific mountain belt, where there is active subduction of the Pacific, Nazca, Cocos, and Juan de Fuca plates. One good example is the volcanic arc of Sumatra and Java, which lies over the subduction zone between the Australian and Eurasian plates.

Because felsic lavas from stratovolcanoes hold large amounts of gas under high pressure, they can erupt explosively (Figure 12.27). Vast quantities of ash and dust fill the atmosphere for many hundreds of square kilometers around the volcano. Clouds of white-hot gases and fine ash sometimes descend the flank of the volcanic cone, searing everything in their path. A glowing cloud from the 1902 Mount Pelée eruption spewed forth without warning, sweeping down on the city of Saint-Pierre and killing all but 2 of its 30,000 inhabitants.

These explosive eruptions can radically alter the form of a stratovolcano. During an eruption, some of the upper part of the volcano is blown outward in fragments, although most of it settles back into the cavity dug beneath the former volcano by the explosion. Sometimes the explosion of a stratovolcano is so violent,

> Stratovolcanoes are tall, steep cones built of layers of felsic lava and volcanic ash. Felsic magma can contain gases under high pressure, so felsic eruptions are often explosive.

12.26 Stratovolcanoes and shield volcanoes

Shield volcanoes and stratovolcanoes are the two most common types of volcanoes. The key difference between them is their magma source, which controls their shape, size, and explosiveness.

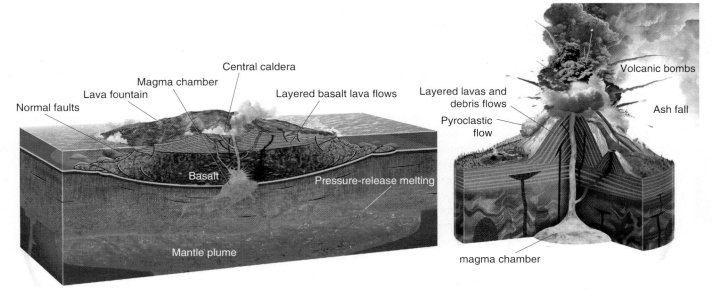

▲ SHIELD VOLCANOES

The basaltic magma that forms shield volcanoes originates in the lower mantle. It has low viscosity, so it flows smoothly and easily. It emerges under low pressure, causing mostly gentle eruptions. Repeated eruptions build broad mountains that resemble a warrior's shield, giving these volcanoes their name. They can become so massive that the crust bends under their weight. The Hawaiian Islands are a familiar example.

▲ STRATOVOLCANOES

The felsic magma that forms stratovolcanoes originates in subduction zones where descending oceanic lithosphere carries felsic sediments into the mantle, where they melt. Stratovolcanic magma is gassy and highly viscous, so it is thick and flows slowly. It often plugs the vent and then emerges under high pressure, causing violent eruptions that expel the solid plug and produce thick deposits of tephra. Lava flows then follow, thus forming the interlayered strata that give the stratovolcano its name. Examples include Mount St. Helens in Washington State and Mount Etna in Italy.

12.27 Explosive stratovolcano eruptions

Because felsic lava holds gas under high pressure, it can cause explosive eruptions.

▼ MOUNT ST. HELENS

Mount St. Helens in the Cascade Range in southwestern Washington State is a classic stratovolcano. It erupted violently on the morning of May 18, 1980, emitting a great cloud of heated gases and ash from the summit crater. Within a few minutes, the plume had risen to a height of 20 km (12 mi).

▼ SOUFRIÈRE HILLS VOLCANO

A cloud of hot, dense volcanic ash emitted by the Soufrière Hills volcano in 1997 coursed down this narrow valley on the island of Montserrat in the Lesser Antilles. It killed about 20 people in small villages. Plymouth, the island's capital, was flooded with hot ash and debris, igniting extensive fires that devastated the evacuated city. The southern two-thirds of the small island were left uninhabitable.

Chris Johns/NG Image Collection

Kevin West / Liaison/Getty Images, Inc.

Remote Sensing of Volcanoes

Volcanoes are always attractive subjects for remote sensing. As monumental landforms, they have distinctive shapes and appearances that are easy to recognize in satellite imagery. They are also very dynamic subjects with the capacity to erupt in spectacular fashion, providing smoke plumes, ash falls, and lava flows. Figure 12.32 provides four examples of volcanoes as they appear in remotely sensed images from four different sensor systems.

Mount Vesuvius This image was acquired by the ASTER instrument on NASA's Terra satellite platform. Spatial resolution is 15 by 15 m (49 by 49 ft). Vegetation appears bright red, with urban areas in blue and green tones. The magnitude of development around the volcano shows that the impact of a major eruption would be particularly catastrophic; and if sudden, thousands of deaths would be expected.

Mount Fuji This striking image of Mount Fuji was acquired by the Shuttle Radar Topography Mission (SRTM). The type of radar used for this mission sends simultaneous pulses of radio waves toward the ground from two antennas spaced 60 m (about 200 ft) apart. Very slight differences in the return signals can be related to the ground height. In this way, the radar can map elevations very precisely. The image simulates a viewpoint above and to the east of Tokyo. A color scale is assigned to elevation, ranging from white to green to brown. The vertical scale is doubled for visualization, so Mount Fuji and surrounding peaks appear twice as steep as they actually are.

Popocatepetl The close-up image of Popocatepetl (far left) was acquired by Landsat-7 at a spatial resolution of 30 by 30 m (98 by 98 ft). Snow and ice flank the summit crater. Canyons carved into the volcano lead away from the summit. The lower slopes are thickly covered with vegetation, which appears

> Remote sensing provides images of volcanoes and their settings. Spaceborne imagers can monitor eruptions in progress.

12.28 Remote sensing of volcanoes

Image courtesy NASA/GSFC/MITI/ERSDAC/JAROS and U.S./Japan ASTER Science Team

▲ **MOUNT VESUVIUS, IMAGED BY ASTER**
This famous volcano erupted in the year 79 AD, ejecting a huge cauliflower-shaped cloud of ash and debris that rapidly settled to the Earth. The nearby Roman city of Pompeii was buried in as much as 30 m (100 ft) of ash and dust, leaving an archeological treasure that was not discovered until 1748. In recent history, major eruptions were recorded in 1631, 1794, 1872, 1906, and 1944.

Image courtesy NASA/JPL/NIMA

▲ **MOUNT FUJI, IMAGED BY THE SHUTTLE RADAR TOPOGRAPHY MISSION**
Mount Fuji lies within striking distance of a large population center—Tokyo, Japan—located about 100 km (about 60 mi) to the northeast. Although it has the symmetry of a simple cone, it is actually a complex structure with two former volcanic cones buried within its outer form. Mount Fuji is considered an active volcano, with 16 eruptions since 781 AD. Its last eruption, in 1707, darkened the noontime sky and shed dust on the present-day Tokyo region.

green in this image. The surrounding plain shows intensive agricultural development.

The eruption of Popocatepetl in 2000 was captured by the Sea-viewing Wide Field-of-View Sensor (SeaWiFS) on December 19, 2000, at a spatial resolution of 1 by 1 km (0.62 by 0.62 mi). This global-scale imager pictured the width of southern Mexico from the Pacific to the Caribbean in true color. Popocatepetl (P) is shown very near the center of the image emitting a smoke plume moving south and east. A large cloud bank lies to the east and obscures much of the right-hand side of the image. High, thin cirrus clouds overlie part of the plume. Mexico City (M) is visible north and west of the volcano as a gray-brown patch flanked by north-south mountain ranges.

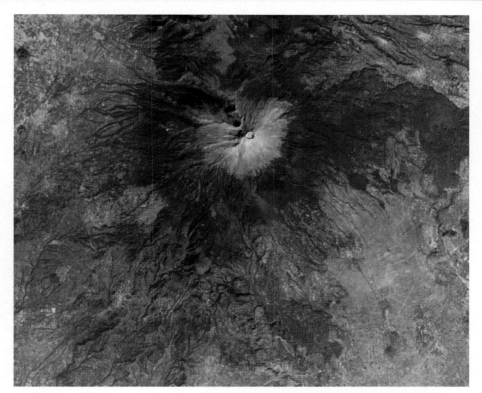

ORBIMAGES/GeoEye

▲ ERUPTION OF POPOCATEPETL, IMAGED BY SEAWIFS
On December 18, 2000, Popocatepetl came to life. Fortunately, since its last eruption in 1994, the volcano has been under constant monitoring, so about half the population in the valleys directly below the volcano was evacuated in time, and no injuries were reported.

Image courtesy Ron Beck, EROS Data Center

◀ POPOCATEPETL, IMAGED BY LANDSAT-7
Located only 65 km (40 mi) from Mexico City, Popocatepetl, "smoking mountain" in the Nahuatl language, is within the view of nearly 30 million people.

12.29 Erosion of stratovolcanoes

After a violent early history, the stratovolcanos are dissected. As the surrounding terrain succumbs to erosion, only a stub of a peak remains.

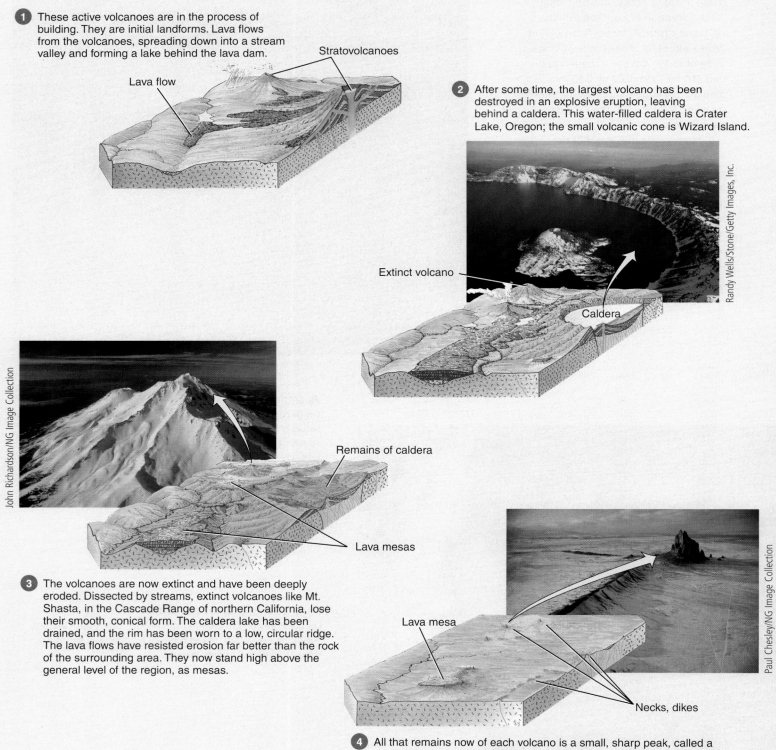

1 These active volcanoes are in the process of building. They are initial landforms. Lava flows from the volcanoes, spreading down into a stream valley and forming a lake behind the lava dam.

Stratovolcanoes

Lava flow

2 After some time, the largest volcano has been destroyed in an explosive eruption, leaving behind a caldera. This water-filled caldera is Crater Lake, Oregon; the small volcanic cone is Wizard Island.

Randy Wells/Stone/Getty Images, Inc.

Extinct volcano

Caldera

John Richardson/NG Image Collection

Remains of caldera

Lava mesas

3 The volcanoes are now extinct and have been deeply eroded. Dissected by streams, extinct volcanoes like Mt. Shasta, in the Cascade Range of northern California, lose their smooth, conical form. The caldera lake has been drained, and the rim has been worn to a low, circular ridge. The lava flows have resisted erosion far better than the rock of the surrounding area. They now stand high above the general level of the region, as mesas.

Paul Chesley/NG Image Collection

Lava mesa

Necks, dikes

4 All that remains now of each volcano is a small, sharp peak, called a volcanic neck. This is the remains of lava that solidified in the pipe of the volcano. Perhaps the finest illustration of a volcanic neck with radial dikes is Ship Rock, New Mexico.

from the pressure built up by gases captured in the felsic material, that it actually destroys the entire central portion of the volcano. The only feature remaining after the explosion is a great central depression, called a *caldera*. Over time, exogenic processes erode stratovolcanoes, creating new landscapes. Figure 12.29 summarizes the stages in the building and eroding of stratovolcanoes.

Shield volcanoes

In contrast to the explosive eruptions of felsic stratovolcanoes, eruptions of shield volcanoes are usually quiet. Basaltic lava flows smoothly over long distances and spreads out in thin layers. Most of the lava flows from fissures (long, gaping cracks) on the flanks of the volcano. Shield volcanoes are marked by erosion features that are quite different from those of stratovolcanoes (Figure 12.30).

12.30 Erosion of shield volcanoes

The gentle topography of the shield volcano gives way to deep erosion, yielding canyons flanked by steep ridges.

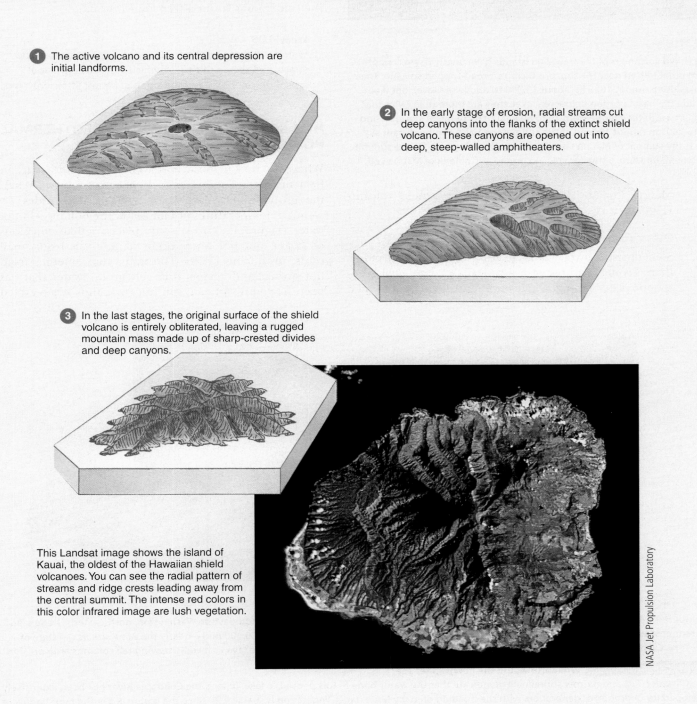

1. The active volcano and its central depression are initial landforms.

2. In the early stage of erosion, radial streams cut deep canyons into the flanks of the extinct shield volcano. These canyons are opened out into deep, steep-walled amphitheaters.

3. In the last stages, the original surface of the shield volcano is entirely obliterated, leaving a rugged mountain mass made up of sharp-crested divides and deep canyons.

This Landsat image shows the island of Kauai, the oldest of the Hawaiian shield volcanoes. You can see the radial pattern of streams and ridge crests leading away from the central summit. The intense red colors in this color infrared image are lush vegetation.

NASA Jet Propulsion Laboratory

© Frans Lanting/www.lanting.com

12.31 Hawaiian shield volcanoes

The shield volcanoes of the Hawaiian Islands have gently rising, smooth slopes that flatten near the top, producing a broad-topped volcano. Their domes rise to about 4000 m (about 13,000 ft) above sea level; but if you include the basal portion below sea level, they are more than twice that high. In width they range from 16 to 80 km (10 to 50 mi) at sea level and up to 160 km (about 100 mi) at the submerged base. On the distant skyline is the summit of Mauna Loa volcano. In the foreground is the summit cone of Mauna Kea, a smaller volcano on the north flank of Mauna Loa.

Mauna Loa in Hawaii is a distinctive shield volcano (Figure 12.31).

Other Volcanic Landforms

Sometimes, extruded lava creates a broad plateau rather than a cone. If a hotspot lies beneath a continental lithospheric plate, it can generate enormous volumes of basaltic lava that emerge from numerous vents and fissures and accumulate layer upon layer. These basalt layers, called **flood basalts**, can become thousands of meters thick and cover thousands of square kilometers (Figure 12.32).

Cinder cones are small volcanoes that form when frothy magma is ejected under high pressure from a narrow vent, producing tephra. The rain of tephra accumulates around the vent to form a roughly circular hill with a central crater. Cinder cones rarely grow more than a few hundred meters high. An exceptionally fine example of a cinder cone is Wizard Island, Oregon, which was built on the floor of Crater Lake long after the caldera was formed (see Figure 12.29).

WileyPLUS Volcanoes

Watch and compare the eruptions of shield volcanoes and stratovolcanoes in this video, which concludes with footage of the fiery, explosive eruption of Mount St. Helens.

HOT SPRINGS, GEYSERS, AND GEOTHERMAL POWER

Where hot rock material is near the Earth's surface it can heat nearby groundwater to high temperatures. When the groundwater reaches the surface, it provides *hot springs* at temperatures not far below the boiling point of water (Figure 12.33). At some places, jetlike emissions of steam and hot water occur at intervals from small vents, producing *geysers*. The water that emerges from hot springs and geysers is largely groundwater that has been heated in contact with hot rock, thus it is recycled

A. N. Strahler

12.32 Flood basalts

An important American example of flood basalts is found in the Columbia Plateau region of southeastern Washington, northeastern Oregon, and westernmost Idaho. Here, basalts of Cenozoic age cover an area of about 130,000 km² (about 50,000 mi²)—nearly the same size as the state of New York. Each set of cliffs in the photo represents a major lava flow. Vertical cracks form in the lava as it cools, leaving tall columns after erosion.

EYE ON THE LANDSCAPE What else would the geographer see?

The still water Ⓐ and narrow shoreline Ⓑ indicate that the water body in this photo is a lake. In fact, the Columbia River has been extensively developed for hydropower generation, with the building of many large dams. Vegetation is sparse Ⓒ, since the region lies in the rain shadow of the Cascades.

surface water. Little, if any, is water that was originally held in rising bodies of magma.

The heat from masses of lava close to the surface in areas of hot springs and geysers provides a source of energy for electric power generation. Here, the groundwater has been heated intensely, but because of the overlying pressure of the rock, it remains in a liquid state. To generate power, wells are drilled to tap the hot, pressurized water, which flashes into steam when it is released at the surface. The steam then drives turbines that generate electric power (Figure 12.33C).

WileyPLUS Remote Sensing and Tectonic Landforms Interactivity
Choose the Volcanoes option to inspect some of the Earth's major volcanoes located in developed areas. Identify craters and lava flows, and fly over Mount St. Helens.

A Look Ahead

In Chapters 11 and 12, we have surveyed the Earth's crust and the geologic processes that shape it. We have seen how the rocks and minerals of the Earth are formed in the cycle of rock transformation, and how the phenomenon of plate tectonics powers that cycle and results in the geographical pattern of mountain ranges, ocean basins, and continental shields visible on the Earth's surface. We have examined initial landforms of tectonic and volcanic activity, and looked at how erosion acting on rocks of different resistances has produced the landscapes that we see today.

In upcoming chapters, we will zoom in from the broad scope of landscapes to a finer scale, at which individual landform-creating processes act. In Chapter 13, we will examine the processes of weathering, which

---

12.33 Hot springs and geysers

Visitors to Yellowstone National Park enjoy many natural wonders, including hot springs and geysers that result from geothermal heating of groundwater by a body of hot rock close to the surface. In some cases, steam from hot groundwater can be harnessed to provide electric power.

GEOTHERMAL POWER PLANT ▶
Wells drilled into the hot, pressurized groundwater provide superheated water. At surface pressure, the water flashes into steam that drives turbines, producing electric power. This geothermal power plant at Svartsengi, Iceland, heats seven towns and a NATO base, and provides electric power as well. The bathers in the foreground are enjoying the warm, briny, mineral-rich waters of a runoff pond.

Images & Stories/Alamy Limited

▼ OLD FAITHFUL GEYSER
When underground crevices and fissures create just the right circulation pattern, superheated water in contact with hot subterranean rocks flashes into steam that blasts the hot water upward through pipelike crevices to the surface in the form of a geyser. This process repeats as water seeps in from surrounding groundwater and replaces the ejected geyser water.

Robert Glusic/Getty Images, Inc.

▼ MAMMOTH SPRINGS
Lacking the special underground circulation pattern, pools of hot water or hot springs form when the water reaches the surface. Small terraces ringed by mineral deposits hold steaming pools of hot water as the spring cascades down the slope.

Gordon Wiltsie/NG Image Collection

breaks rock into small particles, and mass wasting, which moves them downhill as large and small masses under the influence of gravity. Then, in Chapters 14 and 15, we will turn to running water to learn how rivers and streams shape landforms. In our two final chapters, 16 and 17, we will examine landforms created by waves, wind, and glacial ice.

WileyPLUS Web Links

Tour volcanoes, from Vesuvius to Kilauea, by following this chapter's web links. Spot new earthquakes and find out more about earthquake and tsunami hazards. Tour field sites from New York to New Mexico to view rock structures. Learn more about the Black Hills dome and Idaho batholith. It's all ready for you to explore.

IN REVIEW VOLCANIC AND TECTONIC LANDFORMS

- The motion of lithospheric plates in plate tectonics creates distinctive *rock structures* that are then eroded into characteristic landforms.
- At converging boundaries, where continental lithospheric plates collide, compression produces **folds:** *anticlines* (upbends) and *synclines* (troughs).
- Eroded fold belts create a **ridge-and-valley landscape** as weaker rocks become valleys and stronger rocks form mountain ridges. Where an anticline contains weak rocks at its core, erosion may create an *anticlinal valley*. *Synclinal mountains* occur where resistant rocks in the center of a syncline are exposed.
- More intense folding and heating shapes belts of metamorphic rock in which weak rocks (slate and marble) underlie valleys, and strong rocks (schist, gneiss, and quartzite) underlie ridges.
- A steplike bend of rock layers, or *monocline*, can erode into *hogback ridges* and *flatirons*.
- Where rock layers are warped upward into a **sedimentary dome**, erosion produces a circular arrangement of rock layers. Resistant strata form *hogbacks* flanking circular valleys underlain by weaker rocks. Igneous rocks are often revealed in the center.
- Exposed batholiths, often composed of uniform, resistant igneous rock, produce mountainous regions. **Monadnocks** of intrusive igneous rock stand up above a plain of weaker rocks.
- At passive plate margins, on gently sloping coastal plain strata, erosion removes weaker strata (clay) to form lowlands, leaving more resistant strata (sand) to become **cuestas**.
- **Faults** are fractures in crustal rock created when rocks move in different directions. The four main types of faults are *normal, strike-slip* or *transcurrent, reverse,* and *overthrust* faults.
- *Normal faults* are commonly produced by crustal rifting. They create *fault scarps,* **grabens**, and **horsts**. The Rift Valley of East Africa is an example.
- **Earthquakes** occur when rock layers, bent by tectonic activity, suddenly fracture and move, creating *seismic waves.*
- As an earthquake occurs, earthquake waves move outward through the rock from the *focus*, which lies below the **epicenter**.
- The energy released by an earthquake is measured by the Richter scale.
- Many earthquakes are associated with tectonic features such as subduction boundaries, transform boundaries, island arcs, mountain arcs, and axial rifts. But some large earthquakes occur within continental plates, far from plate boundaries.
- Subduction-zone earthquakes can be very large. In addition to shaking the ground, they can generate highly destructive tsunami waves that travel from their source in ever-widening circles.
- The San Andreas Fault is a major strike-slip fault located near two great urban areas—Los Angeles and San Francisco. The potential for a severe earthquake on this fault is high, and the probability of a major Earth movement increases every year.
- Spreading-center earthquakes are associated with mid-oceanic ridges and produce moderate earthquakes.
- Some large earthquakes occur far from active plate boundaries. The New Madrid, Missouri, earthquake of 1811 is an example. Earthquakes can also occur on *blind faults*, those not apparent at the surface.
- **Volcanoes** are landforms marking the eruption of lava at the Earth's surface. They are frequently found near subduction boundaries and on or near axial rifts. At a **hotspot**, an upwelling plume of basaltic magma creates shield volcanoes.
- Volcanic eruptions are a very severe form of environmental hazard. Lava flows, red-hot avalanches, showers of volcanic debris, and mudflows called *lahars* are destructive and deadly features of eruptions.
- Felsic lavas are thick, viscous, and gassy, and form **stratovolcanoes**. Mafic lavas are thin, less viscous, and contain little gas; they form **shield volcanoes**.
- Remote sensing of volcanoes reveals their distinctive shapes and appearances. Satellite instruments can detect and monitor volcanic eruptions.
- Stratovolcanoes, shaped by sluggish felsic lavas and showers of *tephra*, have steep slopes and tend toward explosive eruptions that can form *calderas*. Erosion of stratovolcanoes ultimately leaves a landscape of *lava mesas, volcanic necks,* and *dikes.*
- Shield volcanoes are broadly rounded domes built from flows of thin, mafic (basaltic) lavas. The Hawaiian Islands are a good example of shield volcanoes of different ages.
- Where hot rock material close to the surface heats groundwater to high temperatures, we find *hot springs* and *geysers*. Hot, pressurized groundwater can be used as a power source for generation of electricity.

KEY TERMS

REVIEW QUESTIONS

1. Define the term *rock structure* and identify three examples.
2. How does a **ridge-and-valley landscape** arise? Explain the formation of the ridges and valleys. Where in the United States can you find a ridge-and-valley landscape?
3. How are metamorphic belts formed, and in what tectonic setting do they occur? Identify weak and strong metamorphic rocks, and note whether they form lowlands, uplands, or ridges.
4. Explain how the terms *monocline, hogback ridges,* and *flatirons* are related.
5. What types of landforms are associated with **sedimentary domes?** How are they formed? Identify an example of an eroded sedimentary dome.
6. What is a **coastal plain?** Describe the strata you might find on a coastal plain. Identify two landforms found on coastal plains.
7. What is a *normal fault?* What landforms are produced by normal faults?
8. How does a *transcurrent fault* differ from a normal fault? What landforms would you expect to find along a transcurrent fault? How are transcurrent faults related to plate tectonic movements?
9. Identify and describe two types of faults produced by crustal compression.
10. What is an **earthquake**, and how does it arise? What scale is used to describe the power of an earthquake? In what plate tectonic settings do earthquakes occur?
11. How are observations of earthquake waves used to locate an earthquake?
12. Briefly summarize the geography and recent history of the San Andreas Fault system in California. What are the prospects for future earthquakes along the San Andreas Fault?
13. Identify and briefly describe three locations of earthquakes that are not related to converging boundaries.
14. How is the global pattern of volcanic activity related to plate tectonics?
15. What is a **hotspot?** What produces it? What landforms result from a hotspot in the ocean?
16. Why are volcanic eruptions environmental hazards?
17. How do *felsic* and *mafic lavas* differ? What kinds of volcanoes do they form, and how do they differ?
18. Do **shield volcanoes** erode differently from **stratovolcanoes?** Compare the types of landforms that result as they erode away.
19. How is remote sensing used to study volcanoes?
20. What are **flood basalts** and how are they related to hotspots? What is a **cinder cone?**
21. What is the origin of *hot springs* and *geysers?* What is the source of steam used in generating geothermal power?

VISUALIZING EXERCISES

1. Sketch a cross section through a fold belt showing rock layers in different colors or patterns. Label anticlines and synclines.
2. Identify and sketch a typical landform of flat-lying rock layers found in an arid region.
3. What types of landforms develop on batholiths? Sketch a batholith. What is the difference between a batholith and a monadnock?
4. Sketch a cross section through a normal fault, labeling the fault plane, upthrown side, downthrown side, and fault scarp.

ESSAY QUESTIONS

1. Imagine the following sequence of sedimentary strata: sandstone, shale, limestone, and shale. What landforms would you expect to develop in this structure if the sequence of beds is (a) flat-lying in an arid landscape; (b) slightly tilted as in a coastal plain; (c) folded into a syncline and an anticline in a fold belt; (d) fractured and displaced by a normal fault? Use sketches to illustrate your answer.
2. Write a fictional news account of a volcanic eruption. Select a type of volcano—composite or shield—and a plausible location. Describe the eruption and its effects as it was witnessed by observers. Make up any details you need, but be sure they are scientifically correct.

Chapter 13
Weathering and Mass Wasting

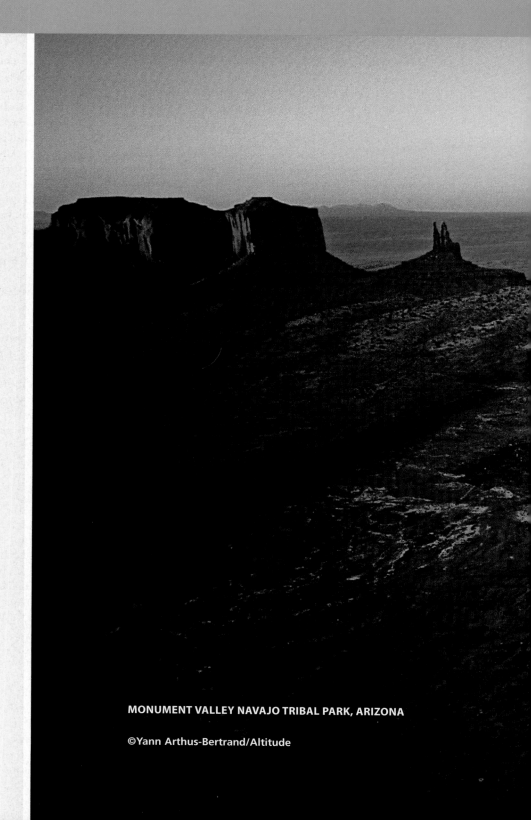

The buttes of Monument Valley, Arizona, shown here at sunset, are world famous for their unique beauty. Rising above a siltstone plain, the buttes are remnants of a resistant sandstone formation underlain by softer layers of shale and siltstone. The massive rock pillars seem to be crumbling before our very eyes. Thermal stresses and salt crystal growth fracture the sandstone, separating the rock layers along joint planes; gravity does the rest. The rock fragments spill across aprons of weaker sediments, continuing to weather as they move downslope. Here, nature's handiwork is most striking.

MONUMENT VALLEY NAVAJO TRIBAL PARK, ARIZONA

©Yann Arthus-Bertrand/Altitude

Weathering and Mass Wasting

R ock weathers into fragments that can be transported by gravity or moved by the freezing and thawing of water. What are the physical processes that turn hard rock into rubble? What are the chemical processes that make rock soft and weak? How do rock fragments move down slopes? How does gravity act on dry and wet earth materials to make them move downhill? These are some of the questions we will answer in this chapter.

The Madison Slide

For some 200 vacationers camping in a deep canyon on the Madison River just downstream from Hebgen Lake, not far west of Yellowstone National Park, the night of August 17, 1959, began quietly, with almost everyone safely bedded down in their tents or camping trailers. It was everything a great vacation should be—that is until 11:37 P.M. mountain standard time, when, in quick succession, not one but four terrifying natural disasters shocked the vacationers from their sleep: earthquake, landslide, hurricane-force wind, and raging flood. The earthquake, which measured 7.1 on the Richter scale, triggered the other three. The first shocks, lasting several minutes, rocked the campers violently in their trailers and tents. Those who managed to get outside could scarcely stand up, let alone run for safety.

Next came the landslide. A dentist and his wife watched through the window of their trailer as a mountain seemed to move across the canyon in front of them, trees flying from its surface like toothpicks in a gale. As rocks began to slam against the sides and top of their trailer, they got out and raced for safer ground. Later, they found that the slide had stopped only a few car lengths from the trailer. Propelled by the moving mountain was a vicious blast of wind. It swept upriver, tumbling trailers end over end.

The flood followed the landslide. Two women, both schoolteachers, sleeping in their car only a few feet from the riverbank were awaked by the violent shaking of the earthquake. Puzzled and frightened, they started the engine and headed for higher ground. As they did so, they were greeted by a deafening roar coming from the mountainside above and behind them. An instant later, their car was engulfed by a wall of water that surged up the riverbank; fortunately, it quickly drained back and the two women managed to drive to safety. Others were not so lucky.

After the first rush of water, set in motion when the landslide mass hit and blocked the river channel, the river began to rise rapidly. Great surges of water overtopped Hebgen Dam, located upstream, as earthquake aftershocks pitched the water of Hebgen Lake back and forth along its length. In the darkness of night, the terrified survivors of the flood had no idea what was happening, or why. The water had risen 10 m (about 30 ft) in just minutes.

The *Madison Slide*, as the huge earth movement was later named, was calculated to have a bulk of 28 million m³ (37 million yd³) of rock (Figure 13.1). Its contents consisted of a chunk of the south wall of the canyon, measuring over 600 m (about 2000 ft) in length and 300 m (about 1000 ft) in thickness. The mass descended more than half a kilometer (about a third of a mile) to the Madison River, at a speed estimated to be 160 km (100 mi) per hour. Pulverized into rock debris as it went, the slide crossed the canyon floor, its momentum carrying it over 120 m (about 400 ft) in vertical distance up the opposite canyon wall. The slide, building up into a huge dam, caused the Madison River to back up and form a new lake. In three weeks' time, the waterbody was nearly 100 m (330 ft) deep. Today, it is a permanent feature of the landscape, now known, appropriately, as Earthquake Lake.

The Madison Slide is an example of **mass wasting**, the downhill motion of rock and soil under the influence of gravity. But whereas a rock avalanche is very rapid, other types of mass wasting are slower and less dramatic. However, they all act to carve and shape the landscape into distinctive landforms.

> The Madison Slide, triggered by an earthquake, threw a huge mass of rock rubble into the canyon of the Madison River. It dammed the river, forming Earthquake Lake.

WileyPLUS The Impact of Earthquakes
Review this video from Chapter 12 and take a tour of the Madison Slide disaster site.

A. N. Strahler

◄ AERIAL VIEW
Seen from the air, the Madison Slide created a great dam of rubble across the Madison River canyon. The slide was triggered by an earthquake.

▼ THE DAMAGE
The landslide missed these cottages, but the rising lake soon floated them away.

USGS

Weathering

As a part of the lithosphere's cycle, new crust is formed and then gradually breaks down through the long-term process of *denudation*. Denudation involves, first, **weathering**, by exposure of rocks to surface elements, then mass wasting, as gravity pulls broken rock downhill, and finally *erosion* by wind and water, which transport and redeposit weathered materials. We will address weathering and mass wasting in this chapter and explore erosion and deposition by wind and water in Chapters 15 and 16.

Weathering describes the combined action of all processes that cause rock to disintegrate physically and decompose chemically due to exposure near the Earth's surface. There are two types of weathering. In **physical weathering**, rocks are fractured and broken apart. In **chemical weathering**, rock minerals are transformed from types that were stable when the rocks were formed to types that are now stable at the temperatures and pressures of the Earth's surface. Weathering produces **regolith**, a surface layer of weathered rock particles that lies above solid, unaltered rock; it also creates a number of distinctive landforms.

Weathering agents—air and water—can actively penetrate rocks through microscopic openings. Igneous and metamorphic rocks contain tiny spaces between crystals, whereas sedimentary rocks have pore spaces between grains. The type of rock openings, along with the chemical composition of the minerals exposed by the openings, determines how a rock will break down as a result of weathering.

Both air and water enter rocks through fractures called *joints*. These joints are created when rocks are exposed to heat and pressure, then cool and contract. Joints typically occur in parallel and intersecting planes, forging natural surfaces of weakness in the rock (Figure 13.2). Weathering then causes *joint-block separation*. The stratification planes, or *bedding planes*, of sedimentary rock also provide access to air and water. Joints often cut bedding planes at right angles, and relatively weak stresses will separate the joint blocks. In igneous rocks, weathering can separate individual grains to produce a fine gravel or coarse sand of single-mineral particles. This process is called *granular disintegration*.

> Weathering is the combined action of physical weathering, in which rocks are fractured and broken, and chemical weathering, in which rock minerals are transformed into softer or more soluble forms.

PHYSICAL WEATHERING

Physical weathering, also known as *mechanical weathering*, fractures rock into smaller pieces, without chemical alteration of the minerals.

Frost Action

In cold climates, an important physical weathering process is *frost cracking*, which occurs when water penetrates joints or pores in the bedrock, which weakens or breaks the rock through freezing and expansion (Figure 13.3). Once thought to be a widespread process, frost cracking in fact occurs only when there is slow and steady cooling, in the range −4°C (25°F) to −15°C (5°F). Once rocks are cracked, ground ice growing in soils and regolith can move, churn, and sort rock fragments, building up

> Under the proper conditions, when water freezes in bedrock joints and bedding planes, it can expand and split rocks apart.

13.2 Bedrock disintegration

Weathering agents break up rocks in different ways, depending on how they penetrate the rocks.

◀ **JOINT-BLOCK DISINTEGRATION**
When weathering agents penetrate along joints in the rock, joint-block disintegration breaks away large boulders.

◀ **GRANULAR DISINTEGRATION**
When weathering agents penetrate microscopic openings between rock crystals or grains, granular disintegration breaks rocks into smaller pieces.

fields of shattered rock rubble above the tree line in high-altitude and high-latitude environments.

Salt-Crystal Growth

A physical weathering process similar to frost cracking occurs in dry climates, where *salt-crystal growth* in crevices and pores builds up pressure. Crystal growth in rock pores can disintegrate rock, and it is this process that carves out many of the niches, shallow caves, rock arches, and pits seen in sandstones of arid regions. During long drought periods, groundwater rises to the rock surface by *capillary action*, a process by which the water's surface tension causes it to be drawn through fine openings and passages in the rock. The water evaporates from the sandstone pores, leaving behind tiny crystals of minerals such as halite (sodium chloride), calcite (calcium carbonate), or gypsum (calcium sulfate). Over time, the growth of these crystals breaks the sandstone apart, grain by grain.

Rock at the base of cliffs is especially susceptible to salt-crystal growth, where it can form rock niches (Figure 13.4). Salt crystallization also damages

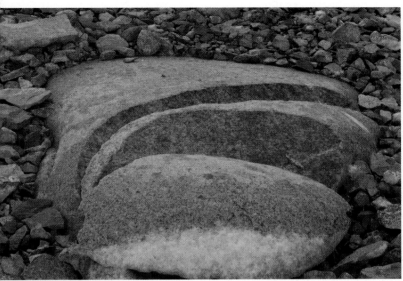

GORDON WILTSIE/NG Image Collection

13.3 Frost cracking

Unlike most other liquids, water expands when it freezes. Under proper conditions, the expansion of freezing water can fragment rocks, in a frost action process called frost cracking.

13.4 Niche formation in sandstone cliffs

In dry climates, salt-crystal growth and weathering lead to the formation of distinctive hollowed-out niches near the bases of sandstone cliffs.

WHITE HOUSE RUIN ▶

Many of the deep niches or cavelike recesses formed in this way in the southwestern United States were occupied by Native Americans. Their cliff dwellings gave them protection from the elements and safety from armed attack. This is the White House Ruin, a large niche in sandstone in the lower wall of Canyon de Chelly, Arizona.

▼ NICHE FORMATION

Groundwater flows through the porous sandstone at the top of the cliff until it reaches a layer of impermeable shale at its base, where it is forced to seep out and evaporate. Here, salt-crystal growth separates the grains of sandstone, breaking them loose and creating a niche.

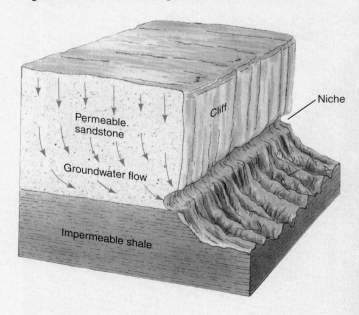

Permeable sandstone

Cliff

Niche

Groundwater flow

Impermeable shale

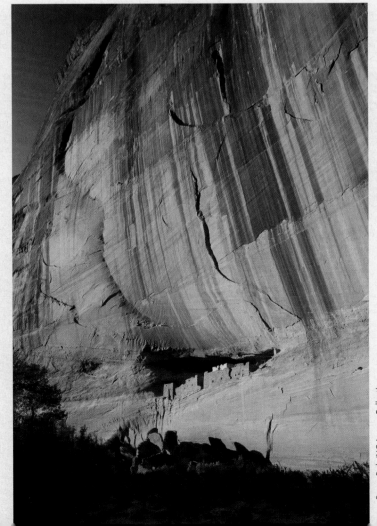

Bruce Dale/NG Image Collection

masonry buildings, concrete sidewalks, and streets. Salt-crystal growth occurs naturally in arid and semi-arid regions, but in humid climates, rainfall dissolves salts and carries them downward to groundwater.

> In arid climates, slow evaporation of groundwater from outcropping sandstone surfaces causes the growth of salt crystals that break the rock apart, grain by grain.

Thermal Action

Most rock-forming minerals expand when heated and contract when cooled, a process called *thermal action*. Because different minerals expand at different rates, internal stresses can crack rocks between mineral crystals. Intense heating of rock surfaces by the Sun during the day, followed by nightly cooling, enhances thermal action. Heat flows slowly inside rocks, intensifying the buildup of stress.

Exfoliation

Rocks are also weathered through *exfoliation*, or *unloading*, by which layers of rock are peeled away. Rock that forms deep beneath the Earth's surface is compressed by the rock above; as the upper rock is slowly worn away, the pressure is reduced, and the rock below expands slightly. This expansion makes the rock crack in layers that are more or less parallel to the surface, creating a type of jointing called *sheeting structure*. In massive rocks such as granite, thick curved layers or shells of rock peel free from the parent mass below, much like the layers of an onion. When a sheeting structure builds over the top of a single large knob or hill of massive rock, an *exfoliation dome* is the result (Figure 13.5).

> Thermal action cracks rocks when temperature changes cause minerals to expand and contract at different rates. In exfoliation, rock layers crack as the pressure of overlying rocks is reduced by erosion.

Biological Action

The active growth of plant roots can exert pressure strong enough to wedge joint blocks apart and break up rock. You have probably observed concrete sidewalk blocks that have been fractured and uplifted by the growth of tree roots. This same process occurs when roots grow between rock layers or joint blocks. Even plant seedlings on cliff surfaces can affect the physical weathering of rocks.

13.5 Exfoliation

Exfoliation occurs when underlying rock is released from the pressure of overlying rock. The rock expands, creating joint planes that follow the rock surface. In this photo of North Dome, Yosemite National Park, California, the curving joint pattern above the treetops was created by exfoliation. Weathering has loosened the rock along the joints, and several shells of rock have fallen away.

CHEMICAL WEATHERING

Chemical weathering occurs when rocks are broken down by chemical transformation. Chemical reactions can turn rock minerals into new minerals that are softer and bulkier and, therefore, easier to erode. Naturally occurring acids, produced by microorganisms digesting organic matter in soils, can dissolve some types of minerals, washing them away in runoff. Physical weathering and mechanical weathering often work in conjunction, when physical weathering exposes more rock surface area to the effects of chemical weathering.

Chemical reactions proceed more rapidly at warmer temperatures, so chemical weathering is most effective in the warm, moist climates of the equatorial, tropical, and subtropical zones. Chemical weathering in these climates, working over thousands of years, has decayed igneous and metamorphic rocks down to depths as great as 100 m (about 330 ft). Geologists conducting exploratory digs have discovered the decayed rock material to be soft, clay-rich, and easily eroded.

Chemical weathering comprises three important processes: hydrolysis, oxidation, and acid action.

Hydrolysis

In *hydrolysis*, minerals react chemically with water. For example, the mineral feldspar, a component of granite, reacts to form a soft clay mineral and silica residue that are readily washed or blown away. Through this process, granite can weather to produce interesting boulder and pinnacle forms (Figure 13.6). Thermal action or other weathering processes can then further degrade the rock.

Oxidation

In *oxidation*, oxygen and water react with metallic elements in minerals, making the minerals unstable and causing the rock to degrade in strength and, eventually, crumble into smaller particles. In time, the elements form oxides, pure compounds of elements with oxygen. Iron-bearing rocks, for example, yield rusty-red iron oxide (Fe_2O_3), which is formed by this process. Because chemical oxidation produces distinctive colors in rock, geologists can use the shades of ancient rock layers to identify periods when less oxygen existed in the atmosphere.

In chemical weathering, the minerals that make up rocks are chemically altered or dissolved. The end products are often softer and bulkier forms that are more susceptible to erosion and mass movement.

Acid Action

Acid action is the third chemical weathering process. *Carbonic acid* is a weak acid formed when carbon dioxide dissolves in water. Found in rainwater, soil water, and stream water, carbonic acid slowly dissolves some types of minerals, in a process called *carbonation*. Carbonate sedimentary rocks, such as limestone and marble, are particularly susceptible to carbonation, and produce many interesting surface forms. Carbonic acid in groundwater dissolves limestone, creating underground caverns and distinctive landscapes that develop when these caverns collapse. Soil acids, which are released as microorganisms digest organic matter, rapidly dissolve basaltic lava in wet low-latitude climates (Figure 13.7).

In urban areas, sulfur and nitrogen oxides pollute the air. When these gases dissolve in rainwater, we get acid precipitation rich in sulfuric and nitric acids, which rapidly dissolves limestone and chemically weathers other types of building stones, stone sculptures, building decorations, and tombstones (Figure 13.8).

13.6 Disintegration of granite through hydrolysis

Although there is not much rainfall in dry climates, what water there is penetrates the granite along planes between crystals of quartz and feldspar. Hydrolysis attacks the sides and edges of the feldspar crystals, breaking feldspar grains away from the main mass of rock and creating the rounded shapes shown here in Owens Valley, California.

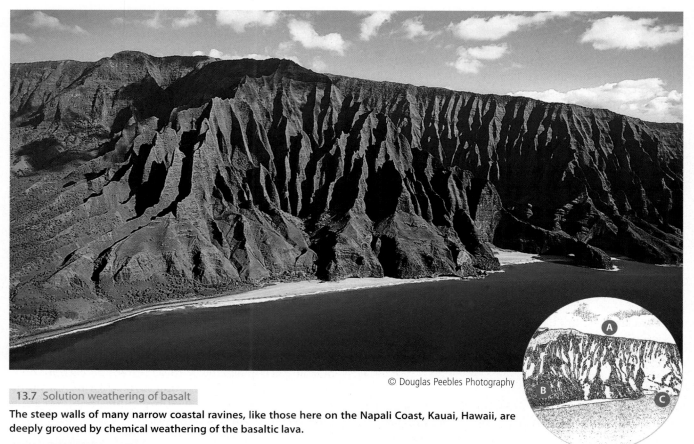

© Douglas Peebles Photography

13.7 Solution weathering of basalt

The steep walls of many narrow coastal ravines, like those here on the Napali Coast, Kauai, Hawaii, are deeply grooved by chemical weathering of the basaltic lava.

EYE ON THE LANDSCAPE What else would the geographer see?

Kauai is the oldest of the Hawaiian Islands. The rounded dome shape of the island, shown by its outline **A**, is characteristic of shield volcanoes. The red-brown colors of the soil exposed on lower slopes **B** are from iron oxides, and indicate Oxisols. Pocket beaches and an arch at **C** are products of wave action.

13.8 Chemical weathering of tombstones

These two tombstones from a burial ground in Massachusetts have weathered differently because of the susceptibility of different rock materials to acid action.

▼ SLATE

This marker is made of slate and is much more resistant to weathering. The engraved letters are clear and sharp, even though the tombstone is almost a century older than the marble example.

▼ MARBLE

This marker, carved from marble, has been strongly weathered, the lettering worn away.

Darlyne A. Murawski/NG Image Collection

Todd Gipstein/NG Image Collection

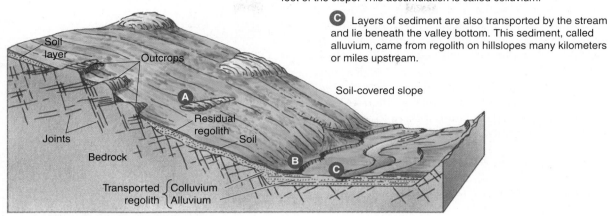

A Soil and regolith blanket the bedrock, except in a few places where the bedrock is particularly hard and projects in the form of outcrops.

B Residual regolith, formed from the rock beneath, moves very slowly down the slope toward the stream and accumulates at the foot of the slope. This accumulation is called colluvium.

C Layers of sediment are also transported by the stream and lie beneath the valley bottom. This sediment, called alluvium, came from regolith on hillslopes many kilometers or miles upstream.

13.9 Typical distribution of materials on a slope

The slope surface is covered with a soil layer, developed on underlying regolith, with some exposures of resistant bedrock. Colluvium accumulates at the foot of the slope, whereas alluvium is deposited by flowing water into the stream valley bottom.

Slopes and Slope Processes

Once rock fragments have been loosened from parent rock through physical or chemical weathering they are subjected to gravity, running water, waves, wind, and the flow of glacial ice. We concentrate here on how Earth materials are moved by gravity; we address the other effects in the following chapters.

SLOPES

The way that Earth materials move downhill is governed by the mechanics of *slopes,* patches of the land surface that are inclined from the horizontal. Slopes guide the flow of surface water downhill and fit together in stream channels. Nearly all natural surfaces slope to some degree.

Most hillslopes are covered with residual regolith, which grades downward into solid, unaltered rock, known simply as **bedrock**. *Sediment* consists of rock or mineral particles that are transported and deposited by fluids, which can be water, air, or even glacial ice. As we saw in Chapter 10, regolith and sediment are the parent materials of soil. Figure 13.9 shows a typical hillslope that makes up one wall of the valley of a small stream. Water and gravity transport regolith downhill to form *colluvium.* If water transports and deposits the debris, it becomes *alluvium.*

> Slopes are mantled with regolith, which accumulates at the foot of slopes as colluvium. Regolith that is transported by moving water is termed alluvium.

SLOPE STABILITY

Counterbalancing the downward force of gravity are the resisting force of the cohesiveness of the rock material and the internal friction holding it in place. Slopes maintain a condition of dynamic equilibrium, which means that material on a slope stays in place until one or another counterbalancing factor causes the slope system to readjust into a new state of equilibrium. Where and when materials on a slope respond to the constant forces of gravity by moving downhill is determined by the angle of the slope, the slope material, and the combination of weathering processes we have just discussed.

If a slope is too steep, the downward force of gravity will overcome the resisting force of friction and the slope materials will move until the slope is less steep and equilibrium is restored. The *angle of repose* is the maximum slope angle that allows rock, sand, clay, and debris to stay in place without movement. The angle of repose depends on the slope material (Figure 13.10) as well as its moisture content.

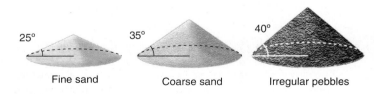

25°   Fine sand

35°   Coarse sand

40°   Irregular pebbles

13.10 Angle of repose

The angle of repose is the maximum angle at which debris, sand, or rock of a given size can rest before the force of gravity causes it to slide downslope. Coarser materials have a steeper angle of repose.

Water, vibration, and additional falling debris also shift the dynamic equilibrium, overcoming friction and allowing the slope materials to overcome the frictional resistance and move downslope. This is why landslides are often triggered by weather events such as heavy rain or by natural disasters such as earthquakes.

The stability of slopes is important for transportation routes. In designing and maintaining railway and railroad cuts, fills, and grades, transportation engineers must account for differences in earth materials, weights and pressures of loads, and vibrational forces. Geographic information systems can identify areas at high risk of slope movement, using vegetation cover, topography, drainage patterns, and rock type and other factors.

WileyPLUS Slope Stability and Mass Wasting Interactivity

Examine the factors influencing mass wasting, and view examples of earthflows, landslides, and debris flows. Check out the angle of repose for various Earth materials.

Kinds of earth materials:	Rock (dry)	Regolith, soil, alluvium, clays + water	Water + sediment
Physical properties:	Hard, brittle, solid	Plastic substance	Fluid
Kinds of motion:	Falling, rolling, sliding	Flowage within the mass	Fluid flow

13.11 Processes and types of mass wasting

Gravity is the force behind all mass wasting, but the nature of the materials and the amount of water involved determines whether movement is fast or slow, as well as the shape of the resulting landform.

Mass Wasting

Gravity pulls down continuously on all materials across the Earth's surface. Unweathered rock is usually so strong and well supported that it remains fixed in place. But when a mountain slope becomes too steep, rock masses can break free and fall or slide into valleys below. Where huge masses of bedrock are involved, towns and villages in the path of the slide can be catastrophically damaged. These events are one form of mass wasting, defined earlier as the spontaneous downhill movement of soil, regolith, and rock caused by gravity.

> Mass wasting is spontaneous movement of soil, regolith, and rock under the force of gravity. There are many types of mass wasting, depending on the speed of the motion and the amount of water involved.

The processes of mass wasting and the landforms they produce are quite varied and tend to grade one into another. Figure 13.11 shows the types of mass wasting that we will describe in this chapter, sorted by the material that is moving, the amount of water involved, and the speed of the movement.

CREEP

The slowest form of mass wasting is **soil creep**. On almost every soil-covered slope, soil and regolith are slowly and imperceptibly moving downhill. This common movement is apparent in older neighborhoods and farms, where one can see fence posts and retaining walls leaning in a downhill direction. The process is triggered when soil

and regolith are disturbed by alternate drying and wetting, growth of ice needles and lenses, heating and cooling, trampling and burrowing by animals, and shaking by earthquakes. Gravity pulls on every such rearrangement, and the particles very gradually work their way downslope. In some layered rocks, such as shales or slates, edges of the strata can appear to bend in the downhill direction (Figure 13.12).

Solifluction is a type of mass wasting that occurs when soils become thoroughly saturated with water and slowly flow downslope by gravity. It occurs where the ground underlying the soil is impermeable—a clay layer, for example—and soil water cannot drain downward. Movement is slow, perhaps a few millimeters or centimeters per day or per year. Tropical rainforest soils on slopes and tundra soils underlain by frozen ground often show this form of mass movement.

ROCKFALLS AND TALUS

The most visible form of mass wasting we are likely to encounter is a *rockfall*, in which rocks fall down steep slopes or cliff sides, often bringing soil or regolith along for the ride. The typical site of a rockfall is a cliff face where a sandstone layer is underlain by shale. Sandstone blocks, separated along joint planes, slowly

13.12 Soil creep

The slow downhill creep of soil and regolith shows up in many ways on a hillside.

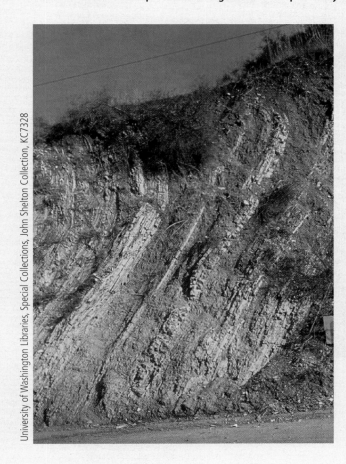

University of Washington Libraries, Special Collections, John Shelton Collection, KC7328

◀ CREEP MOTION

This photo of a roadcut near Downieville, California, shows how soil creep moves joint blocks. Although the strata appear to bend to the left, the motion is actually that of detached bedrock fragments, moving downhill along with the soil.

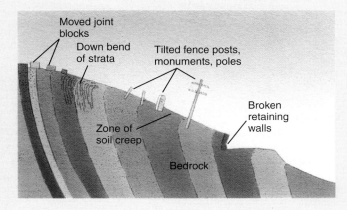

▲ SIGNS OF CREEP

Where soil creep is active, fence posts, monuments, and utility poles lean downhill. Retaining walls are broken or pushed over.

creep outward, moving on the weak and impermeable shale layer underneath, until the block falls or slides down the cliff face. Places where a road has been cut into a steep mountainside are also particularly vulnerable to rockfalls and rocky debris from the oversteepened slopes. In mountainous regions, rockfalls can be extremely dangerous to those living below steep hillsides.

Loose, fallen rocks called *talus*, or *scree*, can collect at the bottom of a slope or cliff in a cone-shaped pile (Figure 13.13). Talus slopes become picturesque boundaries along the base of many majestic mountain escarpments. On shallower slopes covered with rock rubble, gravity combined with freezing and thawing of ground ice can produce a *rock sea*, or *rock glacier*. Churned by frost action, the rubble moves slowly downhill.

SLIDES

A large mass of bedrock or regolith sliding downhill is known as a **landslide**. Landslides can be set off by earthquakes or sudden rock failures (Figure 13.14), as we saw at the beginning of this chapter. Landslides can also result when the base of a slope is made too steep by excavation or river erosion. Large, disastrous landslides are possible wherever mountain slopes are steep. In China, Switzerland, Norway, and the Canadian Rockies, for example, villages built on the floors of steep-sided valleys have been destroyed when millions of cubic meters of rock descended without warning. Landslides range from *rockslides* of jumbled bedrock fragments to *bedrock slumps* in which most of the bedrock remains more or less intact as it moves.

During a landslide, rubble travels down a mountainside at amazing speed. Geologists think this speed is reached because, somehow, the particles at the bottom of the slide translate the energy of the slide into an ultrasonic vibration mode that supports the moving mass on a fluid bed without mixing it.

In a landslide, a large mass of rock suddenly moves from a steep mountain slope to the valley below. Landslides are triggered by earthquakes or rock failures rather than heavy rains.

13.13 Rock fragments in motion

ROCK GLACIER ▶

Where blocky sandstones outcrop on slopes, frost action and gravity can set in motion moving fields of angular blocks, sometimes called rock glaciers. Although the more intense freeze/thaw cycles of the Ice Age probably shaped this example from the ridges of eastern Pennsylvania, the lack of vegetation or even lichens suggests that this rock mass is still moving.

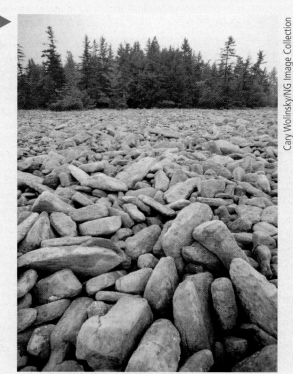

Cary Wolinsky/NG Image Collection

▼ TALUS CONES

Frost action has caused these cliffs to shed angular blocks of rock that accumulate in talus cones. In the foreground is the shore of Lake Louise in the Canadian Rockies.

EYE ON THE LANDSCAPE   What else would the geographer see?

An obvious feature is the sedimentary rock beds **A** that form the cliffs. In this case, they are ancient sedimentary rocks that were thrust over and above younger rocks of the plains during an arc-continent collision. At **B** you can see a small, shrunken glacier, coated with gray blocks of talus.

A. N. Strahler

©AP/Wide World Photos

13.14 *Santa Tecla landslide*

On January 13, 2001, an earthquake measuring 7.6 on the Richter scale triggered a landslide in Santa Tecla, El Salvador, killing hundreds. The earthquake was centered off the southern coast of El Salvador and was felt as far away as Mexico City. During the shaking, a steep slope above the neighborhood of Las Colinas collapsed, releasing a wave of earth that carved a path of destruction through the town. The landslide swept across the ordered grid of houses and streets, burying hundreds of homes and their inhabitants. The same earthquake also triggered other landslides in the region, with additional loss of life and property.

Landslides and flows can take different forms, depending on the way a slide moves (Figure 13.15). In a *rotational slide*, the material rotates along an axis that is perpendicular to the direction of the slope. This occurs when there is a downward and outward movement of the soil mass on top of a concave failure in the surface below. The rupture of a rotational slide is curved upward like a spoon, and there is no distinct gap or scarp at the surface rupture, but rather a slumping of the soil that creates a small to large cliff. When a series of curved ridges are created in the mass movement of a rotational slide, it is called a *slump*.

In a *translational slide*, the mass moves out and down the slope with little or no rotation. Translational slides can move considerable distances, often originating along geologic faults, joints, or discontinuities between rock and soil. In climates with permafrost, slides can occur along the top of the frozen ground layer.

FLOWS

Mass wasting of Earth materials with high water content results in flows. Flows occur when precipitation or snowmelt is greater than absorption into the underlying sediment or rock. In deserts, for example, thunderstorms produce rain much faster than it can be absorbed by the soil. After a wildfire, hillsides are similarly vulnerable to flows.

Earthflows

When fine-grained or clay-rich materials are saturated with water, an **earthflow** can occur. The saturated soil and regolith turns into a thick liquid that flows downhill, forming a bowl or depression at the head. The earthflow often has an hourglass shape with a spreading toe (Figure 13.16). Earthflows can range in size from a few dozen square meters to massive flows in which millions of metric tons of clay-rich or weak bedrock move like a great mass of thick mud.

Earthflows usually are not a threat to life because they move quite slowly; but during heavy rains, they can

> An earthflow is a mass of water-saturated soil that moves slowly downhill. Earthflows can block highways and railroads and severely damage or destroy buildings.

13.15 Types of slides and flows

Landslides and flows can be classified as rotational or translational.

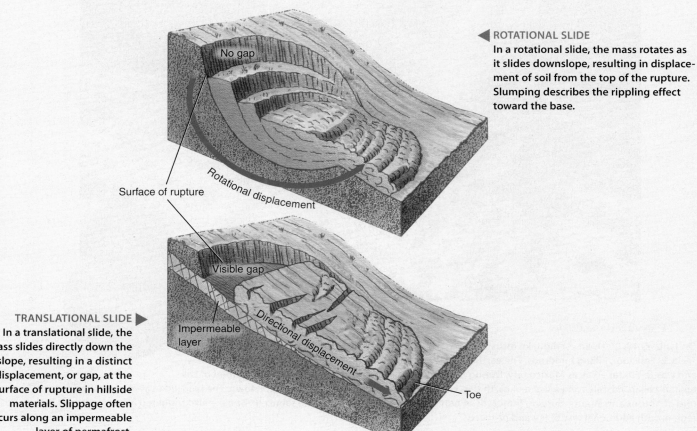

ROTATIONAL SLIDE

In a rotational slide, the mass rotates as it slides downslope, resulting in displacement of soil from the top of the rupture. Slumping describes the rippling effect toward the base.

No gap

Surface of rupture

Rotational displacement

Visible gap

TRANSLATIONAL SLIDE ▶

In a translational slide, the mass slides directly down the slope, resulting in a distinct displacement, or gap, at the surface of rupture in hillside materials. Slippage often occurs along an impermeable layer of permafrost.

Impermeable layer

Directional displacement

Toe

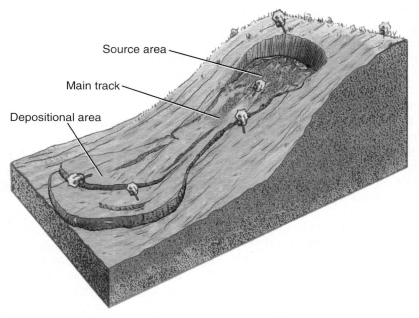

Source area

Main track

Depositional area

13.16 Earthflow

In an earthflow, saturated soil, regolith, or clay-rich bedrock moves downhill as a mass of thick, flowing mud. A bowl or depression develops at the head, and a spreading toe marks the depositional area. Between the two ends, the flow has an hourglass shape.

block highways and railroad lines, and, often, severely damage buildings, pavement, and utility lines that have been constructed on unstable slopes.

WileyPLUS Earthflows

Watch this animation to see how an earthflow creates scarps and slump terraces.

Mudflows

A **mudflow** is a rapid flow of water and soil or regolith that pours swiftly down canyons in mountainous regions. In arid regions with little or no vegetation, thunderstorm runoff picks up fine particles, becoming a thin mud that flows down to the canyon floors and then follows the stream courses. As it flows, it picks up additional sediment and becomes thicker and thicker until it is too thick to go any farther. Roads, bridges, and houses on the canyon floor are engulfed and destroyed. A mudflow can severely damage property and even cause death as it emerges from a canyon and spreads out.

Mudflows on the slopes of erupting volcanoes are called *lahars*. Lava and hot ash rapidly melt snow atop the volcano, creating flows of rock, soil, volcanic ash, and water that accelerate down the steep slopes and travel great distances (Figure 13.17). Herculaneum, a city at the base of Mt. Vesuvius, was destroyed by a lahar during the CE 79 eruption. At the same time, the neighboring city of Pompeii was buried under volcanic ash.

> Mudflows are rapid events in which water, sediment, and debris cascade down slopes and valleys to lower elevations. They are produced by very heavy rainfall or snowmelt caused by volcanic activity.

When mudflows pick up rocks and boulders, they are called *debris flows* (Figure 13.18). These fast-moving flows can carry anything from fine particles and boulders to tree trunks and limbs down steep mountain slopes.

INDUCED MASS WASTING

Human activities can induce mass wasting processes by building up unstable piles of waste soil and rock and by removing the underlying support of natural

Steve Raymer/NG Image Collection

13.17 Mudflow

More than 20,000 lives were lost when a volcanic mudflow—known as a lahar—swept through the town of Armero, Colombia. The mudflow was caused by a minor volcanic eruption of Nevado del Ruiz, which melted the snow and ice on its summit and created a cascade of mud and debris that engulfed the town. Many homes were simply swept away.

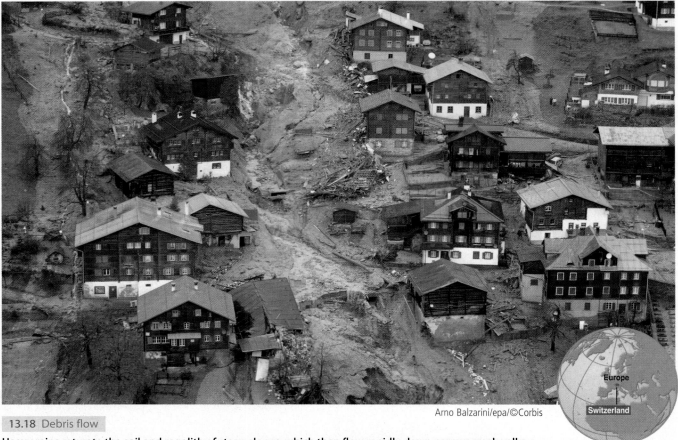

13.18 Debris flow

Arno Balzarini/epa/©Corbis

Heavy rains saturate the soil and regolith of steep slopes, which then flow rapidly down narrow creek valleys and canyons, carrying mud, rocks, and trees. This debris flow, also known as an alpine debris avalanche, struck the village of Schlans, Switzerland, in November of 2002.

masses of soil, regolith, and bedrock. Mass movements produced by human activities are called *induced mass wasting*.

For example, mass wasting is induced where roads and home sites have been bulldozed out of the deep regolith on very steep hillsides and mountainsides, steepening the slopes. The excavated regolith is piled up to construct nearby embankments, leading to further steepening. Soils receive excess water from septic tanks and lawn irrigation. This combination of raising the water content and steepening makes slopes unstable and leads to mass wasting (Figure 13.19).

Unlike natural forms of mass wasting, artificial mass wasting requires the use of machinery to raise earth materials against the force of gravity. Explosives, too, are used, and these produce disruptive forces many times more powerful than the natural forces of physical weathering. Industrial societies now transport great masses of regolith and bedrock from one place to another using such technologies. They do this to extract mineral resources or to move earth when constructing highway grades, airfields, building foundations, dams, canals, and various other large structures. These activities

destroy the preexisting ecosystems and plant and animal habitats. When the removed materials are then used to build up new land on adjacent surfaces, ecosystems and habitats are buried.

Scarification is a general term used to describe excavations and other land disturbances initiated to extract mineral resources. Strip mining is a particularly destructive scarification activity (Figure 13.20). Even though strip mining is under strict control in most locations, it is on the rise, driven by the ever-increasing human population and the growing demand for coal and other industrial minerals.

Vocal communities throughout the Appalachian region have recently raised concerns about the negative social and environmental consequences of mountaintop removal to reach coal beds. High mountain valleys near the mines are filled with rock waste and tailings, creating toxic runoff and groundwater contamination that have caused many individuals to migrate away. As more people leave the area, the local economy degrades further, causing additional social stress.

Mass wasting is often triggered by abnormally heavy rains, which are predicted to become more frequent as a

Courtesy Los Angeles County Department of Public Works

13.19 Induced earthflow

This aerial view shows houses on Point Fermun, in Palos Verdes, California, disintegrating as they slide downward toward the sea. Both large and small earthflows have been induced or aggravated by human activities in this region. The largest was this one, known as the Portuguese Bend "landslide" that affected an area of about 160 hectares (about 400 acres). It was caused when sedimentary rock layers slipped on an underlying layer of clay. It moved 20 m (about 70 ft) in three years, causing damage totaling some $10 million. Geologists believe that infiltrating water from septic tanks combined with irrigation water applied to lawns and gardens was responsible for the earthflow, weakening the clay layer enough to start and sustain the flowage.

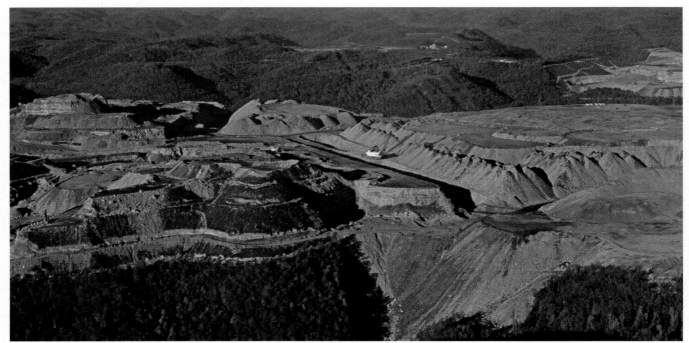

MELISSA FARLOW/NG Image Collection

13.20 Strip mining

This aerial view shows the impact of strip mining on the landscape at Samples Mine in West Virginia. Reclamation is costly and difficult. Mining companies have buried approximately 3000 km (about 2000 mi) of Appalachian streams beneath piles of toxic waste and debris, initiating decades of controversy and legal wrangling.

result of the climate change caused by rising levels of carbon emissions across the globe. The forecast increase in frequency and severity of storms should in turn increase both the frequency and severity of mass wasting events that must be endured by humans and their settlements around the world. This unexpected linkage demonstrates the complexity of global and environmental change that human society is likely to face in the future.

> Human activities can induce mass wasting by piling up unstable materials or undercutting slopes or rock masses. Cutting and filling to extract mineral resources is termed scarification.

WileyPLUS Web Quiz

Take a quick quiz on the key concepts of this chapter.

A Look Ahead

In this chapter we have examined the processes of weathering and mass wasting. In the weathering process, rock near the surface is broken up into smaller fragments and often altered in chemical composition. In mass wasting, weathered rock and soil move downhill in slow to sudden mass movements.

The landforms of mass wasting are produced by gravity acting directly on soil and regolith. Gravity also powers another landform-producing agent—running water—which we take up in the next two chapters. Chapter 14 deals with water in the hydrologic cycle, in soil, and in streams. Chapter 15 explains specifically how streams and rivers erode regolith and deposit sediment to create landforms.

WileyPLUS Web Links

Examine gravestone weathering and tour earthflows and landslides, all by visiting the web links for this chapter.

IN REVIEW WEATHERING AND MASS WASTING

- The *Madison Slide* is an impressive example of mass wasting. An earthquake triggered a huge avalanche that blocked the Madison River, creating a new lake.
- **Weathering** is the action of processes that cause rock near the surface to disintegrate and decompose into **regolith**.
- Weathering occurs when air and water penetrate spaces within rocks, between rock layers, and in fractures called *joints*.
- **Physical weathering** produces regolith from solid rock by breaking *bedrock* into pieces.
- *Frost cracking* occurs when, under proper conditions, ice crystals grow within rocks. In mountainous regions of frequent hard frost, fields of angular blocks accumulate as *rock seas* or *rock glaciers*.
- *Salt-crystal growth* in dry climates breaks individual grains of rock free, and in the process creates landforms such as niches and arches. It can also damage brick and concrete.
- *Thermal action* splits rocks when different minerals in a rock expand and contract at different rates during daily heating and cooling.
- *Exfoliation,* or *unloading* of the weight of overlying rock layers, can cause some types of rock to expand and break loose into thick shells, producing *exfoliation domes*.
- Wedging by plant roots also forces rock masses apart.
- **Chemical weathering** results from mineral alteration. Igneous and metamorphic rocks can decay to great depths, producing a regolith that is often rich in clay minerals.

- In *hydrolysis*, minerals react chemically with water. In *oxidation*, minerals combine with oxygen.
- *Acid action* occurs when weak acids attack rocks. In carbonation, acid solutions of carbon dioxide dissolve limestone and marble. Microorganisms and air pollutants also create soil acids that can slowly dissolve rocks.
- Most slopes are mantled with regolith, which lies atop unweathered **bedrock**. *Colluvium* and *alluvium* are two types of transported regolith.
- Slope stability is determined by the nature of the material and the conditions it is exposed to. The *angle of repose* is the maximum slope angle that can be held without failure.
- **Mass wasting** is the spontaneous downhill motion of soil, regolith, or rock set in motion by the force of gravity.
- **Soil creep** is a process of mass wasting by which regolith, under the influence of gravity, moves down slopes almost imperceptibly.
- In a *rockfall*, rocks fall down steep slopes or cliff sides. Cliffs shed rock fragments, which pile up as *talus* at cliff bottoms.
- A **landslide** is a rapid coursing of large masses of bedrock, sometimes triggered by an earthquake. *Rockslides* and *bedrock slumps* are forms of landslides.
- Slides and flows can be *rotational* or *translational*, depending on the movement.
- In an **earthflow**, water-saturated soil or regolith slowly flows downhill.
- A **mudflow** is much swifter than an earthflow. It follows stream courses, becoming thicker as it descends

and picks up sediment. Watery mudflows can accumulate rocks, boulders, and vegetation to become *debris flows.*

- Human activities can lead to *induced mass wasting,* mass movement of soil and regolith by oversteepening

or undercutting slopes or by building unstable piles of regolith.

- *Scarification* includes excavation and relocation of regolith to extract mineral resources. Strip mining is an example.

KEY TERMS

mass wasting, p. 446
weathering, p. 448
physical weathering, p. 448
chemical weathering, p. 448

regolith, p. 448
bedrock, p. 453
soil creep, p. 454
landslide, p. 455

earthflow, p. 457
mudflow, p. 459

REVIEW QUESTIONS

1. Describe the Madison Slide and the events that took place directly after it.
2. What is meant by the term **weathering**? What types of weathering are recognized? What is **regolith**?
3. How do weathering agents break up rock? Use the term *joint* in your answer.
4. Explain the process of *frost cracking.*
5. How does *salt-crystal growth* break up rock? Use the term *capillary action* in your answer.
6. How does *thermal action* break up rocks?
7. What is an *exfoliation dome,* and how does it arise? Refer to Figure 13.5 in your answer.
8. Identify three processes of **chemical weathering**. Describe how limestone is affected by chemical weathering.

9. What is meant by the term *angle of repose*? Provide some examples.
10. Define **mass wasting**. What factors can be used to distinguish different types of mass movements?
11. What is **soil creep**, and how does it arise?
12. Distinguish between *rockfall, talus,* and **landslide**.
13. What is an **earthflow**? What features distinguish it as a landform?
14. Contrast earthflows and **mudflows**, providing an example of each.
15. Define the term *landslide*. How does a landslide differ from an earthflow?
16. Define and describe *induced mass wasting*. Provide some examples.
17. Explain the term *scarification*. Provide an example of an activity that causes scarification.

VISUALIZING EXERCISES

1. Define the terms *regolith, bedrock, sediment,* and *alluvium*. Sketch a cross section through a part of the landscape showing these features, and label them on the sketch.
2. Using one or more sketches, distinguish between a *rotational slide, translational slide,* and an *earthflow.*

3. Copy or trace Figure 13.11. Then identify and plot on the diagram the mass movement associated with each of the following locations: Palos Verdes Hills, Madison River, Santa Tecla, Armero, Downieville, and Schlans.

ESSAY QUESTIONS

1. A landscape includes a range of lofty mountains elevated above a dry desert plain. Describe the effects of weathering and mass wasting that might be found on this landscape, and identify their location.
2. Imagine yourself as the newly appointed director of public safety and disaster planning for your state or

province. One of your first assignments is to identify locations where human populations are threatened by potential disasters, including those caused by mass wasting. Where would you look for mass wasting hazards, and why? In preparing your answer, you may want to consult maps of your state or province.

Chapter 14
Freshwater of the Continents

At the edge of the ancient basalt plateau that tops Brazil's Parana Province, the Iguazu River makes a spectacular leap, known as Iguazu Falls, into the lowlands of the Argentina's Misiones Province. Divided into 150 to 300 separate waterfalls and cataracts, depending the on the level of the water, the river thunders down 60–82 m (197–269 ft) into the plunge pool below. With an average flow about equal to that of Niagara Falls but at a height about 60 percent greater, the falls are among the most spectacular in the world. And unlike in many other regions of the world where freshwater is in short supply, it is abundant at Iguazu Falls.

IGUAZU FALLS, MISIONES PROVINCE, ARGENTINA

©Yann Arthus-Bertrand/Altitude

Freshwater of the Continents

F reshwater on the Earth is a scarce resource, regardless of where it is found—in the flowing water of streams and rivers, the still waters of lakes, or deep under the land in groundwater. How does rainfall feed streams, rivers, and lakes? How does it recharge groundwater to feed wells? What is a flood? What are the effects of urbanization on the behavior of streams and rivers? These are some of the questions we will answer in this chapter.

The Aral Sea

East of the Caspian Sea, astride the former Soviet republics of Kazakhstan and Uzbekistan, lies an immense saline lake—the Aral Sea. Fed by meltwaters of high glaciers and snowfields in the lofty Hindu Kush, Pamir, and Tien Shan ranges, the lake has endured through thousands of years, serving as an oasis for terrestrial and aquatic wildlife deep in the heart of the central Asian desert.

But since about 1960, the Aral Sea, once larger than Lake Huron, has shrunk to a shadow of its former self (Figure 14.1). The volume of its waters has decreased by more than two-thirds, and its salinity has increased from 1 to over 3 percent, making it saltier than seawater. Twenty of the lake's 24 native fish species have disappeared; and its catch of commercial fish, which once supplied 10 percent of the total for the Soviet Union, has dwindled to zero. The deltas of the Amu Darya and Syr Darya rivers, which enter the south and east sides of the lake, once were islands of great ecological diversity, teeming with fingerling fishes, birds, and their predators. Now only about half the nesting bird species remain; likewise, many species of aquatic plants, shrubs, and grasses have vanished. Commercial hunting and trapping have almost ceased.

What caused this ecological catastrophe? The answer is simple: the lake's water supply was cut off. As an inland lake with no outlet, the Aral Sea receives water from the Amu Darya and

Courtesy of Worldsat International

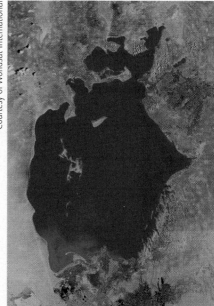

Jacques Descloitres, MODIS Land Rapid Response Team

14.1 The Aral Sea shrinks

This pair of satellite images shows the Aral Sea in 1976 (left) and 2006 (right). More than two-thirds of the sea's volume was lost in that period of 40 years. The North Aral Sea, now slowly recovering, lies in the uppermost part of the image.

Syr Darya rivers, as well as a small amount from direct precipitation. It loses water by evaporation. In the past, the lake's water gains balanced its losses, which varied from year to year, so that, until about 1960, the area, depth, and volume of the lake remained nearly constant.

That balance began to shift in the late 1950s, when the Soviet government embarked on the first phases of a vast irrigation program, using water from the Amu Darya and Syr Darya for cotton-cropping on the region's desert plain. The diversion of water soon became significant as more and more land came under irrigation, until, by the early 1980s, the inflow fell to nearly zero. The surface level of the Aral was sharply lowered, its area reduced. The sea became divided into two separate parts.

As the lake's shoreline receded, the exposed lakebed became encrusted with salts (Figure 14.2). The once-flourishing fishing port of Muynak became a ghost town, 50 km (30 mi) from the

14.2 Aral Sea gallery

As the Aral Sea slowly shrank it stranded ships and left behind blowing dust and high levels of salinization.

Panos Pictures

Shepard Sherbell/SABA/©Corbis

▲ **BLOWING DUST**
Broad expanses of lake bottom, left high and dry by the shrinking of the Aral Sea, turned into a vast dust bowl. Carried by the wind, dust has covered the area, as here along a road in the village of Kyzylkum.

▲ **GRAVEYARD OF SHIPS**
As the Aral Sea shrank, fishing vessels, useless without fish, were abandoned. Now, only their hulks remain, sad reminders of a once-vibrant industry.

▼ **SALINIZATION**
Salts accumulated in agricultural soils as the Aral Sea shrank. Salty wind-blown dust, coupled with saline groundwater drawn to the surface, concentrated salt in the top layer of soil.

Dieter Telemans/Panos Pictures

newly drawn lake shoreline. Strong winds now blow salt particles and mineral dust in great clouds southwestward over the irrigated cotton fields and westward over grazing pastures. These salts—particularly the sodium chloride and sodium sulfate components—are toxic to plants. The salt dust permanently poisons the soil and can only be flushed away with more irrigation water.

Remarkably, today, a portion of this vast ecological ruin is in rehabilitation. The flow of the Syr Darya reaching the lake is sufficient to maintain the smaller northern section of the lake, now called the North Aral Sea, in a productive state. A dike some 13 km (8 mi) separates the northern sea from its larger brother, the South Aral Sea. The dike traps the inflowing waters of the Syr Darya, and by 2010 had raised the level of the North Aral Sea from 30 m (98 ft) to 42 m (138 ft), allowing water to start to spill over the dike.

In 2006, commercial fishing returned to the North Aral Sea. Although yields are far from their former values, fish are now caught and even exported as far as the Ukraine The local population, decimated as livelihoods faded away, is coming back, with some villages doubling their numbers within a few years. And the enlarged water surface area has stimulated increased rainfall downwind of the sea, helping local agriculture.

But what of the South Aral Sea? Unfortunately, its prospects remain dim. At a salinity level of more than 85 parts per thousand (more than twice as salty as ocean water), the last of its native fish species are dying. Still, there is some hope for the South Aral Sea's fishermen. When the salinity reaches 110 parts per thousand, conditions will be favorable for brine shrimp. These tiny creatures are used as food for young fishes being raised in fish farms worldwide and thus can turn a valuable cash crop. Sadly, however, there will be no respite from the toxic dust storms that sweep salt and pesticide residues over the region, downwind from the Big Sea.

> The water of the Aral Sea has been reduced in volume and increased in salinity as its inflow was diverted to irrigation. Only a small portion, the North Aral Sea, now survives.

Freshwater and the Hydrologic Cycle

Water is essential to life. Nearly all organisms require constant access to water, or at least a water-rich environment, for survival (Figure 14.3). Humans are no exception. We need a steady supply of freshwater from

14.3 Freshwater

Humans require large quantities of freshwater. Rivers and streams are important sources of water for human uses.

precipitation over the lands. Some of this water is stored in soils, regolith, and pores in bedrock. And a small amount of water flows as freshwater in streams and rivers. In this chapter, we focus on water at the land surface and lying within the ground.

Freshwater on the continents in surface and subsurface water makes up only about 3 percent of the hydrosphere's total water supply. This freshwater is mostly locked into ice sheets and mountain glaciers. Groundwater, which can be found at almost every location on land that receives rainfall, accounts for little more than half of 1 percent of global water. Nevertheless, this small fraction is many times larger than the amount of freshwater in lakes, streams, and rivers, which only adds up to about three-hundredths of 1 percent of the total. The availability of fresh surface water varies widely across the globe, and many arid regions do not have permanent streams or rivers.

PATHS OF PRECIPITATION

Figure 14.4 shows the flow paths of water between ocean, land, and atmosphere. The study of these flows is part of the science of *hydrology*, the study of water as a complex but unified system on the Earth. In this chapter, we will trace the segment of the hydrologic cycle that arises from precipitation over land and includes both the surface and subsurface pathways of water flow.

As shown in Figure 14.5, there are three pathways for land precipitation. During light or moderate rain, most natural soil surfaces absorb rain water through **infiltration**. Infiltrating rainwater is stored temporarily in a region called the *soil-water belt*, where it is available to plants. Some water from the soil-water belt returns to the atmosphere through *evapotranspiration*, the combined effect of evaporation and transpiration through plants. The remaining water moves downward to become groundwater, which feeds streams and marshes.

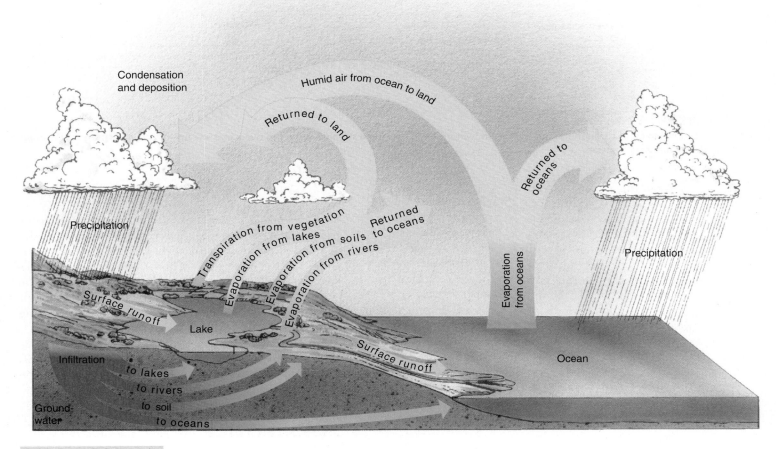

14.4 The hydrologic cycle

The hydrologic cycle traces the paths of water as it moves from oceans to the atmosphere to the land, then returns again to the ocean.

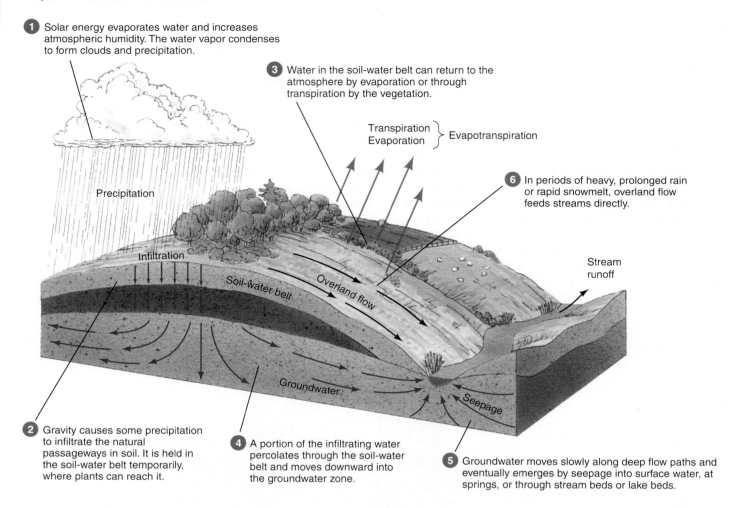

1 Solar energy evaporates water and increases atmospheric humidity. The water vapor condenses to form clouds and precipitation.

3 Water in the soil-water belt can return to the atmosphere by evaporation or through transpiration by the vegetation.

Transpiration
Evaporation } Evapotranspiration

6 In periods of heavy, prolonged rain or rapid snowmelt, overland flow feeds streams directly.

Precipitation

Infiltration

Soil-water belt

Overland flow

Stream runoff

Groundwater

Seepage

2 Gravity causes some precipitation to infiltrate the natural passageways in soil. It is held in the soil-water belt temporarily, where plants can reach it.

4 A portion of the infiltrating water percolates through the soil-water belt and moves downward into the groundwater zone.

5 Groundwater moves slowly along deep flow paths and eventually emerges by seepage into surface water, at springs, or through stream beds or lake beds.

14.5 Pathways of precipitation

Precipitation can move through the soil and groundwater belt to rivers and streams, run off to rivers and streams directly, or return to the atmosphere by evapotranspiration.

Water that drains from an area by surface, subsurface, or groundwater paths is called **runoff**. By supplying water to streams and rivers, runoff allows rivers to carve out canyons and gorges and to carry sediment to the ocean. During periods of heavy rainfall, some runoff travels over the ground surface, rather than being absorbed into the ground. This surface runoff is called **overland flow**. Overland flow moves sediment from hills to valleys, and by doing so helps shape landforms.

> Land precipitation either runs off or infiltrates into the soil. As runoff, it flows into streams. As infiltration, it returns to the air through evapotranspiration, or percolates downward to become groundwater.

Groundwater

Now let's take a closer look at how precipitation feeds groundwater. Water from precipitation sinks into the soil and flows downward through the soil-water belt under the force of gravity. This process is called *percolation*. Eventually, the percolating water fully saturates the pore spaces in bedrock, regolith, or soil, at which point it is called **groundwater** (Figure 14.6). The **water table** represents the upper limit of this *saturated zone*.

Groundwater moves slowly through pores in the rock and regolith, eventually seeping into streams, ponds, lakes, and marshes (Figure 14.7). In these places, the land surface dips below the water table.

THE WATER TABLE

The water table can be mapped in detail where there are many wells in an area, by plotting the water height in each well (Figure 14.8). Water percolating

WileyPLUS Groundwater

Watch an animation of the hydrologic cycle and trace the path of groundwater as it infiltrates the soil, percolates to the water table, and flows to streams.

14.6 Zones of subsurface water

Water in the soil-water belt is available to plants. Water in the unsaturated zone percolates downward to the saturated zone of groundwater, where all pores and spaces are filled with water.

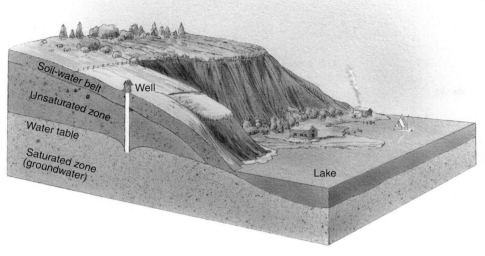

14.7 Surface waters

Streams, lakes, and ponds are example of surface waters.

Phil Schermeister/NG Image Collection

PONDS

Ponds and lakes are fed by groundwater, so their levels are determined by the water table. This pond is in Baxter State Park, Maine.

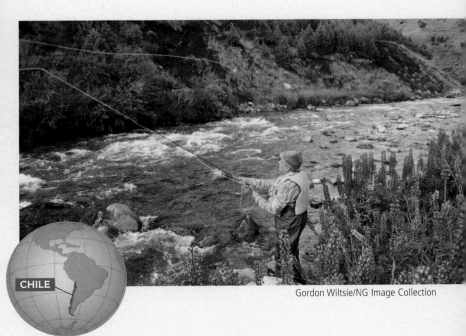

Gordon Wiltsie/NG Image Collection

STREAMS ▶

Steams and rivers carry runoff from the land surface; but in moist climates they are also fed by groundwater seeping into the bottom of the streambed. This fly fisherman is testing the waters in a stream near Coyhaique, Chile.

CHILE

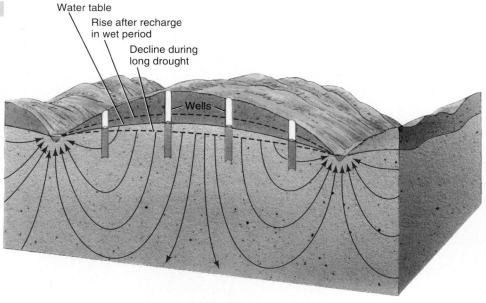

This figure shows paths of groundwater flow. It takes a long time for water to flow along the deeper paths, whereas flow near the surface moves much faster. The most rapid flow is close to the stream, where the arrows converge.

Water table
Rise after recharge in wet period
Decline during long drought
Wells

down through the unsaturated zone tends to raise the water table, while seepage into lakes, streams, and marshes draws off groundwater and tends to lower its level. When there's a large amount of precipitation, the water table rises under hilltops and divide areas. During droughts, the water table falls.

> The water table marks the top of the saturated zone of groundwater. It is highest under hilltops and divides, and it slopes to intersect the surface at lakes, marshes, and streams.

These differences in water table level are built up and maintained because groundwater moves with extreme slowness through the fine chinks and pores of bedrock and regolith. Over time, the water table level tends to remain stable, and the flow of water released to streams and lakes balances the flow of water percolating down into the water table.

AQUIFERS

The amount of groundwater that can be held in the saturated zone depends on the porosity of the sediments that make up this layer. If the layer is porous and permeable enough to hold and conduct a usable quantity of water, and the water can be easily pumped from the material, then it is called an **aquifer**. A bed of sand or sandstone is often a good aquifer because clean, well-sorted sand—such as that found in beaches, dunes, or stream deposits—can hold an amount of groundwater equal to about one-third of its bulk volume. Sandy materials have large pore spaces that allow the water to move through the sediment and be readily pumped from wells.

By contrast, layers that are relatively impermeable to groundwater are known as *aquicludes*. Clay and shale beds are examples of materials that do not conduct water in usable amounts. When an aquifer is sandwiched between two impermeable aquicludes, pressure may force the water to rise to the surface as a self-flowing *artesian well* (Figure 14.9). A fault can serve as a natural conduit for groundwater, producing artesian springs.

LIMESTONE SOLUTION BY GROUNDWATER

Carbonic acid, a weak acid produced from carbon dioxide dissolved in water, slowly dissolves limestone at the surface in moist climates in a process called *carbonation*. Similarly, limestone below the surface can be dissolved by such acid groundwater slowly flowing in the saturated zone, forming deep underground *limestone caverns*. Mammoth Cave in Kentucky and Luray Caverns in Virginia are examples of famous and spectacular caverns arising from solution of limestone. Figure 14.10 illustrates how caverns develop.

Inside caverns, water containing dissolved minerals drips through the cave ceiling and onto the floor. Where the water drips from the ceiling, calcium carbonate and other minerals are deposited to form *stalactites*, shaped like icicles hanging from the ceiling. Where the water hits the cavern floor, the deposits form an upward-pointing formation called a *stalagmite*. If the process continues for a long time, stalactites and stalagmites meet to form a column. The thickness and chemical composition of these formations gives climate scientists information about historical rainfall, enabling them to re-create past climates.

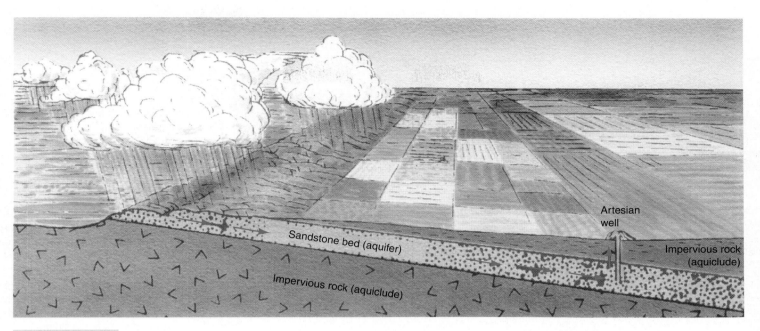

14.9 Artesian well

A porous sandstone layer (aquifer) is sandwiched between two impervious rock layers (aquicludes). Precipitation provides water that saturates the sandstone layer. The elevation of the well that taps the aquifer is below that of the range of hills feeding the aquifer, so pressure forces water to rise in the well.

14.10 Cavern development

Limestone caverns develop as limestone is dissolved by carbonic acid in groundwater at the top of the water table.

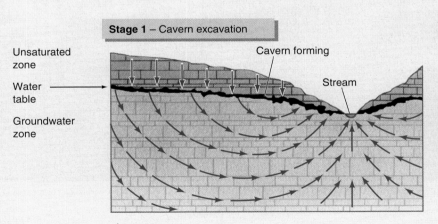

▲ STAGE 1

Carbonic acid action is concentrated in the saturated zone just below the water table. Limestone dissolves at the top of the groundwater zone, creating tortuous tubes and tunnels, great open chambers, and tall chimneys below the ground. Subterranean streams can flow in the lowermost tunnels.

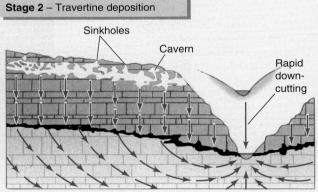

▲ STAGE 2

At this stage, the stream has deepened its valley, making the water table level drop. The previously formed cavern system now lies in the unsaturated zone. As water flows through the caverns, it deposits carbonate matter, known as travertine, on exposed rock surfaces in the caverns. Travertine encrustations take many beautiful forms–stalactites (hanging rods), stalagmites (upward-pointing rods), columns, and drip curtains.

14.11 Sinkholes

◀ **NEW MEXICO SINKHOLE**
Sinkholes in limestone are created when limestone is dissolved by carbonic acid in groundwater. These sinkholes are near Roswell, New Mexico.

EYE ON THE LANDSCAPE What else would the geographer see?
The large amount of bare rock visible in many large patches **A** suggests that the limestone is so pure that it leaves little or no residual material behind as it dissolves. Also, look at the vegetation ringing the sinkholes **B** . The plants are probably drawing on groundwater adjacent to the sinkhole.

BELIZE SINKHOLE ▶
This sinkhole, known as Nohoch Ch'en, occurs in the middle of a tropical rainforest in Belize.

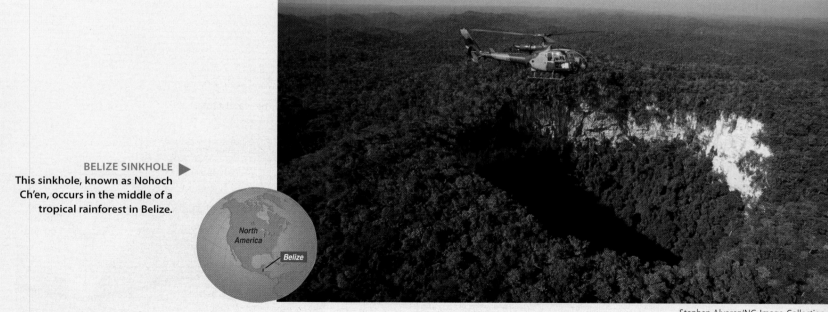

Sinkholes are surface depressions in a region of cavernous limestone (Figure 14.11). Some sinkholes are filled with soil washed from nearby hillsides, whereas others are steep-sided, deep holes. They develop where the limestone is more susceptible to solution weathering, or where an underground cavern near the surface has collapsed.

Limestone landscapes with numerous sinkholes and lacking small surface streams are called **karst**. Figure 14.12 shows how a karst landscape develops. Important regions of karst or karstlike topography are the Mammoth Cave region of Kentucky, the Yucatan Peninsula, the Dalmatian coastal area of Croatia, and parts of Cuba and Puerto Rico. The karst landscapes of southern China and western Malaysia are dominated by steep-sided, conical limestone hills or towers (Figure 14.13).

Carbonic acid action dissolves limestone, producing caverns. Cavern collapse creates sinkholes and a karst landscape.

14.12 Evolution of a karst landscape

Solution produces a landscape of unique landforms, including sinkholes (Figure 14.11) and tower karst (Figure 14.13).

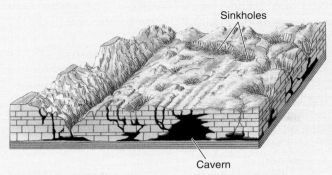

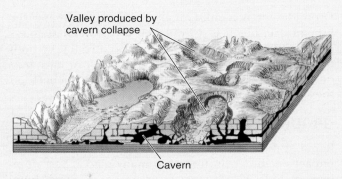

▲ **EARLY STAGE**

Over time, rainwater dissolves limestone, producing caverns and sinkholes. In warm, humid climates, a solution of pure limestone can form towers (left side of diagrams).

▲ **LATER STAGE**

Eventually, the caverns collapse, leaving open, flat-floored valleys. Surface streams flow on shale beds beneath the limestone. Some parts of the flat-floored valleys can be cultivated.

©Bruno Barbey/Magnum Photos, Inc.

14.13 Tower karst

White limestone is exposed in the almost vertical sides of these towers near Guilin (Kweilin), Guangxi Province, southern China.

EYE ON THE LANDSCAPE What else would the geographer see?

Solution weathering of certain types of bedrock in warm and wet environments can produce a landscape of steep, vertical slopes. Compare these towers Ⓐ, formed by solution of limestone, with the fins and grooves of Kauai in Figure 13.7, formed by solution of basaltic lava. Note also the flooded fields Ⓑ. They are probably rice paddies in the spring, just before planting with young rice stalks.

Groundwater Use and Management

Groundwater is a substantial source of freshwater for human use; but it is a finite resource, which requires precipitation to replenish it. It is also affected by both withdrawal and pollution, both of which can have serious and long-lasting environmental consequences on the supply of freshwater.

GROUNDWATER WITHDRAWAL

Rapid withdrawal of groundwater has had a serious impact on the environment in many places. Vast numbers of wells require powerful pumps to draw huge volumes of groundwater to the surface, greatly altering nature's balance of groundwater discharge and recharge. The yield of a single drilled well ranges from as low as a few hundred liters or gallons per day in a domestic well to many millions of liters or gallons per day for a large industrial or irrigation well (Figure 14.14).

As water is pumped from a well, the level of water in the well drops. At the same time, the surrounding water table is lowered, in the shape of a downward-pointing cone called the *cone of depression* (Figure 14.15). The difference in height between the cone tip and the original water table is known as the *drawdown*. Where many wells are in operation, their intersecting cones will lower the water table.

Groundwater is often depleted far faster than it can be replaced by water infiltrating downward to the saturated zone. As a result, we are exhausting a natural resource that is renewable only over very long periods of time. In many arid and semiarid regions, the groundwater pumped from wells today was accumulated thousands of years ago and is not being adequately replenished by present rainfall.

For example, large areas of the southwestern United States once supported large lakes when the climate was cooler and wetter, between 28,000 and 7000 years ago. When the climate changed to our present conditions, the lakes dried up, leaving behind vast groundwater reservoirs. Today, growing populations and farming communities in these regions depend upon this nonrenewable groundwater, which is part of an unsustainable system of development and agriculture.

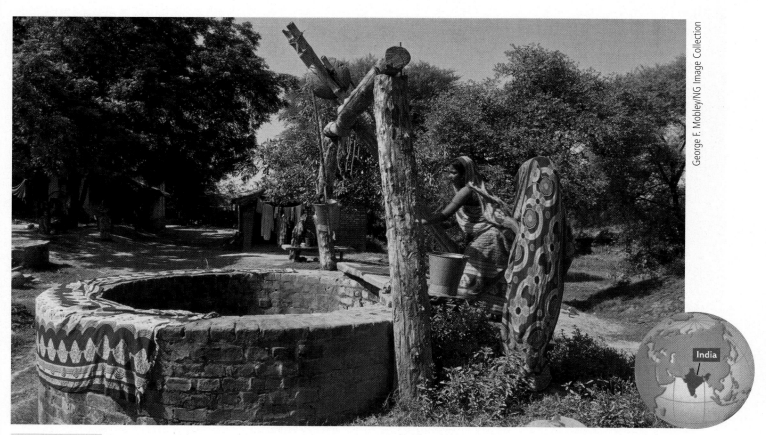

George F. Mobley/NG Image Collection

14.14 Dug well

This old-fashioned dug well supplies water for household needs in Uttar Pradesh, India. The well is lined with bricks, and groundwater seeps in around them to fill the well.

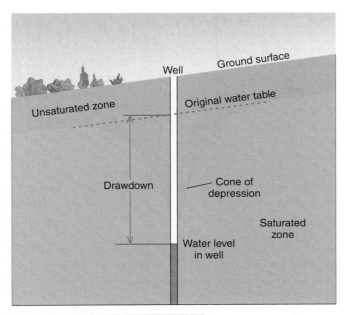

14.15 Drawdown in a pumped well

As water is drawn from the well, the water table is depressed in a cone shape centered on the well. This cone of depression may extend out as far as 15 km (9.3 mi) or more from a well where heavy pumping is continued.

SUBSIDENCE

An important environmental effect of excessive ground-water withdrawal is *subsidence*, the sinking, of the ground surface. Venice, Italy, provides a dramatic example of this side effect. Venice was built in the eleventh century on low-lying islands in a coastal lagoon, sheltered from the ocean by a barrier beach. Underlying the area are some 1000 m (about 3300 ft) of sand, gravel, clay, and silt layers, with some layers of peat. Gradual compaction of these soft layers has been going on for centuries under the heavy load of city buildings, and greatly accelerated withdrawal of groundwater in recent decades has aggravated the process.

Many ancient buildings in Venice now rest at lower levels and thus suffer more severe damage from flooding during winter storms on the adjacent Adriatic Sea (Figure 14.16). Worsening the situation is that many of the canals of Venice still receive raw sewage, meaning that the floodwater is contaminated.

Most of the subsidence was caused by the withdrawal of large amounts of groundwater from nearby industrial wells, which has now stopped. At present, the rate of subsidence is only about 1 mm (0.04

©M. Smith/Sipa Press

14.16 Flooding in Venice

Land subsidence has subjected Venice to episodes of flooding by the waters of the Adriatic Sea. Here, high water has swamped the Piazza San Marco and an outdoor café.

in.) per year. However, the threat of flooding and damage to churches and other buildings of great historical value remains. Other cities that have suffered significant land subsidence include Bangkok and Mexico City.

> Wells draw down the water table at a point, creating a cone of depression. As many wells exploit an aquifer, their cones of depression merge to cause a general lowering of the water table.

POLLUTION OF GROUNDWATER

Another major environmental problem is the contamination of groundwater by pollutants such as agricultural runoff, industrial waste, and acid rain, among other sources. For example, as rainwater filters through buried landfill waste, it picks up pollutants. If the landfill was incorrectly constructed or has failed, the percolating water carries the pollutants down to the water table (Figure 14.17). Runoff flowing over the land surface can transport excess fertilizers, agriculture pesticides, or toxic road runoff into streams and lakes, from where it gradually infiltrates groundwater. Although waves of chemical pollution in streams and rivers can move rapidly downstream, pollutants remain in relatively slow-moving groundwater for long periods of time. Chemical pollution of groundwater makes it necessary for well owners to regularly test their water for contaminants. Contamination of large urban aquifers has caused the U.S. Environmental Protection Agency to close down wells in many locations or introduce costly filtration and decontamination technologies.

When aquifers are overpumped in coastal regions, saltwater can contaminate groundwater through a process called *saltwater intrusion*. Because freshwater is less dense than saltwater, a layer of saltwater from the ocean can lie below a coastal aquifer. As the aquifer is depleted, the level of saltwater rises and eventually reaches the well from below, making the well unusable. The contamination of groundwater by saltwater intrusion also disrupts salt-intolerant ecological systems, as native vegetation is replaced by salt-tolerant species. The coastal agricultural economy may also be felt as yields are reduced and the soils become unsuitable for future farming.

> Sanitary landfills and overland flows can carry pollutants and toxic compounds to the water table, thereby contaminating groundwater.

Surface Water and Streamflow

Recall from our discussion of the hydrologic cycle earlier in this chapter that some precipitation filters through the soil to become groundwater and some travels over the surface to streams, which eventually carry the water back to the ocean. We have already seen what happens to water that infiltrates the soil. In this section we take a closer look at water that travels over the surface in *streamflow*.

14.17 Groundwater contamination

A landfill disposal site or other concentration of pollutants on the surface can contaminate groundwater below.

◀ SANITARY LANDFILL
Rainwater percolating though a landfill, like this one on the eastern shore of Maryland, can pick up contaminants and carry them to the water table, where they can reappear in streams, lakes, or marshes.

▼ MOVEMENT OF POLLUTED GROUNDWATER
Polluted water, leached from a waste disposal site, moves toward a supply well (right) and a stream (left).

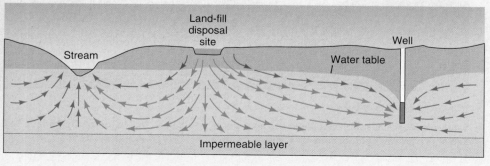

14.18 Overland flow

Overland flow, in the form of a thin sheet of water, covers this grassy field. The water converges into a small rivulet in the lower right.

OVERLAND FLOW

Water naturally travels downhill, pulled by the force of gravity, to collect in streams, which eventually carry the water back to the ocean. Streams are fed by the seepage of groundwater, melting of ice and snow, lakes, and surface runoff from precipitation.

When soils are saturated or rain falls too quickly to be absorbed into the ground, water travels directly over the surface as *overland flow*. Where the soil or rock surface is smooth, the flow may be a continuous thin film, called *sheet flow* (Figure 14.18). If the ground is rough or pitted, overland flow may be made up of a series of tiny rivulets connecting one water-filled hollow with another. On a grass-covered slope, overland flow is divided into countless tiny threads of water, passing around the stems. Even in a heavy and prolonged rain, you might not notice overland flow in progress on a sloping lawn. On heavily forested slopes, overland flow may be entirely concealed beneath a thick mat of decaying leaves.

DRAINAGE SYSTEMS

As runoff moves to lower and lower levels and eventually to the sea, it becomes organized into a branched network of stream channels. This network and the sloping ground surfaces next to the channels that contribute overland flow to the streams are together called a **drainage system**. *Drainage divides* mark the boundary between slopes that contribute water to different streams or drainage systems. The entire system is bounded by an outer drainage divide that outlines a more-or-less pear-shaped *drainage basin*, or *watershed* (Figure 14.19). Drainage systems funnel

14.19 Channel network of a stream

Smaller and larger streams merge in a network or drainage system that carries runoff downstream. Each small tributary has its own small drainage basin, bounded by drainage divides. An outer drainage divide marks the stream's watershed at any point on the stream.

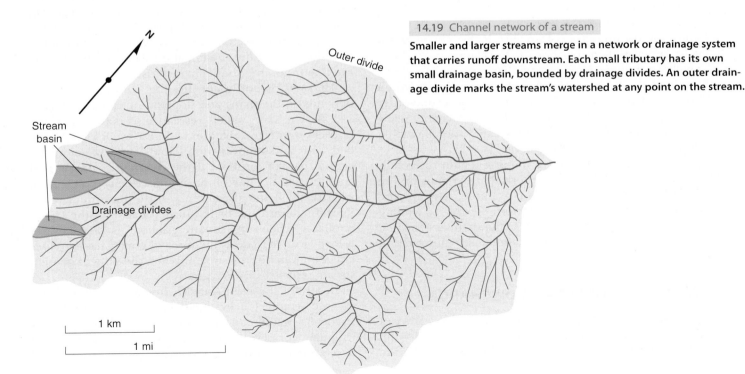

Outer divide

Stream basin

Drainage divides

1 km

1 mi

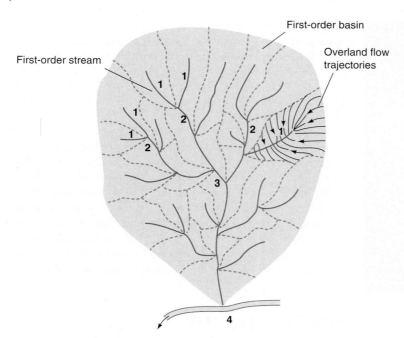

First-order basin

First-order stream

Overland flow trajectories

This schematic diagram of the drainage basin of a third-order stream shows the channel system with stream orders and the drainage divide network.

overland flow into streams, and smaller streams into larger ones.

In a channel network, the stream slopes and channel forms vary from the smaller, steeper streams at the headwaters of the network to the broad and gently sloping river channels found farther downstream. One way to study how the properties of streams change systematically is to organize them by *stream order* (Figure 14.20). The smallest tributaries are first-order streams, which receive overland flow from within their first-order basin. When two first-order streams join, the result is a second-order stream, and so forth. As you might well imagine, many stream properties are related to stream order, including slope, drainage area, and discharge.

The outline of streams draining a region often falls into a particular drainage pattern that depends on the underlying rock structure. For example, a steep stratovolcano will have a radial pattern of streams leading away from the central crater. Other patterns include dendritic, trellis, and annular, defined and illustrated in Figure 14.21.

STREAM CHANNELS AND DISCHARGE

Overland flow eventually creates streamflow. We can define a **stream** as a long, narrow body of water flowing through a channel and moving to lower levels under the

> A drainage basin, or watershed, consists of a branched network of stream channels and adjacent slopes that feed the channels. It is bounded by a drainage divide.

force of gravity. The **stream channel** is a trough shaped by the forces of flowing water. A channel may be so narrow that you can easily jump across it; or, in the case of the Mississippi River, it may be as wide as 1.5 km (about 1 mi) or more.

Scientists who study streams measure them by their cross-sectional area (A), which depends on the width (w) and depth (d) of their channels. Two other key characteristics are *stream velocity* (V), which measures how rapidly the water in the stream flows, and *discharge* (Q), which is the volume of water per unit of time passing through a cross section of the stream at that location. The slope (S) of the stream, also called the *stream gradient*, is another important characteristic.

The velocity of water flowing through a stream depends on the shape and slope of the channel (Figure 14.22). The water meets resistance as it flows because of friction with the channel walls. So, water close to the bed and banks moves more slowly than water in the central part of the flow. If the channel is straight and symmetrical, the line of maximum velocity is located in midstream. If the stream curves, the maximum velocity shifts toward the bank on the outside of the curve. Similarly, when the stream gradient is steep, water velocity will increase as the force of gravity has a greater effect on water in the stream. When the slope is gentler, stream velocity will decrease.

In all but the most sluggish streams, the movement of water is affected by *turbulence*. If we could follow a particular water molecule, we'd see it travel a highly irregular, corkscrew path as it is swept downstream. But if we were to measure the water velocity at a certain fixed

14.21 Drainage patterns

The arrangement of streams that develops on a landscape often depends on the underlying rock structure.

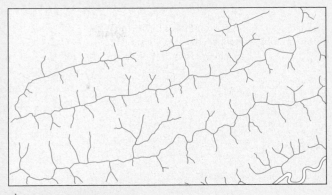

▲ TRELLIS PATTERN

As streams erode their valleys, they cut into the softer rock and leave the harder rocks behind to form the drainage divides. Where rocks are folded into anticlines and synclines, streams follow the long, parallel valleys, creating a trellis pattern. This example was taken from a folded belt in the central Appalachians.

▲ DENDRITIC PATTERN

A drainage pattern that looks much like a branching tree is known as a dendritic pattern. It usually develops on uniform rocks or on flat-lying rock layers. This example shows a dendritic stream pattern that emerged on a part of the Idaho batholith, a large area of relatively uniform granitic rocks.

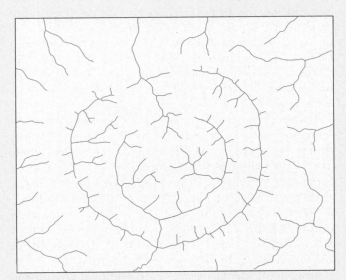

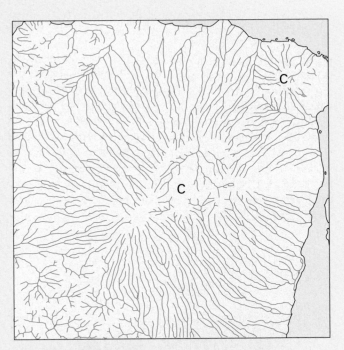

▲ ANNULAR PATTERN

If tectonic forces cause layers of sedimentary rock to bulge upward and form a dome, erosion can produce circular valleys as the dome is eroded. The streams in these valleys then follow an annular, or ring-shaped, pattern.

▲ RADIAL PATTERN

Stratovolcanoes often develop drainage patterns of streams leading away from the central crater, marked "C" on the map. This example is from a volcanic island in the East Indies. Most of the radial streams drain the large volcano in the middle. At the upper right is a smaller cone, with its own central crater.

14.22 Stream velocity

Stream velocity is determined by the shape and slope of a stream. Within the stream, it is greatest away from the bed and banks, where the flow meets frictional resistance.

PERSPECTIVE VIEW ▶
Along straight stretches, velocity is greatest in the middle of the stream.

Channel

Valley wall

V

Width, w

Area, A

d

Depth, d

Slope, S

V

Surface

Maximum shear

Bed

◀ CROSS SECTION
Velocity is greatest at the surface of the stream, where there is less frictional resistance from the streambed. Velocity is also more rapid where the stream gradient, measured by the slope, S, is steep.

point for several minutes, we would see that the average motion is in a downstream direction.

The rate of water flow in a stream is called its **discharge**, which is measured in cubic meters (cubic feet) per second. Discharge is determined by the product of the stream's cross section and the mean velocity of the water (Figure 14.23). In stretches of *rapids*, where the

> The discharge of a stream is the measure of its volume rate of flow. It is the product of the mean velocity and the cross-sectional area.

stream flows swiftly, the stream channel will be shallow and narrow. In *pools*, where the stream flows more slowly, the stream channel will be wider and deeper, to maintain the same discharge. Sequences of pools and rapids can be found along streams of all sizes. As stream discharge changes, such as during periods of flooding or drought, the stream's cross-sectional area and velocity will change accordingly.

Within a drainage system, stream discharge increases downstream, as streams and rivers combine to deliver runoff and sediment to the oceans. Figure 14.24 shows the relative discharge of major rivers of the United States. The mighty Mississippi with its tributaries dwarfs all other North American rivers. The Columbia River, draining a large segment of the Rocky Mountains in southwestern Canada and the northwestern United States, and the Great Lakes, discharging through the St. Lawrence River, also have large discharges.

In general, the larger the cross-sectional area of the stream, the lower the gradient. Great rivers, such as the Mississippi and Amazon, have gradients so low that they can be described as flat. For example, the water surface of the lower Mississippi River falls in elevation about 3 cm for each kilometer of downstream distance (1.9 in. per mi).

The discharge of a stream varies over the course of the year as a result of annual cycles of precipitation. The relationship between stream discharge and precipitation is shown by a *hydrograph*, which plots the discharge of a stream with time at a particular location

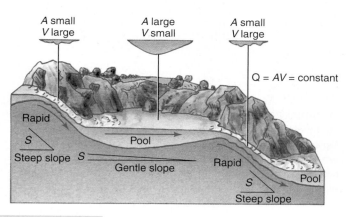

A small
V large

A large
V small

A small
V large

$Q = AV = $ constant

Rapid

S

Steep slope S

Pool

Gentle slope

Rapid

S

Pool

Steep slope

14.23 Pools and rapids

In steep areas where the velocity of the flow increases, the channel is shallow, forming rapids; whereas in flatter areas where the streamflow slows, the channel is deep, forming pools.

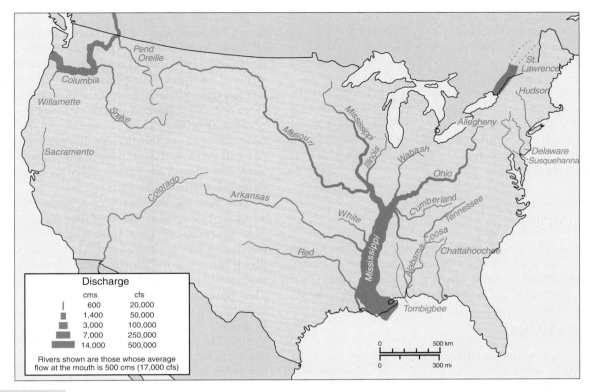

14.24 River discharge

This schematic map shows the relative magnitude of the discharge of U.S. rivers. Width of the river as drawn is proportional to mean annual discharge.

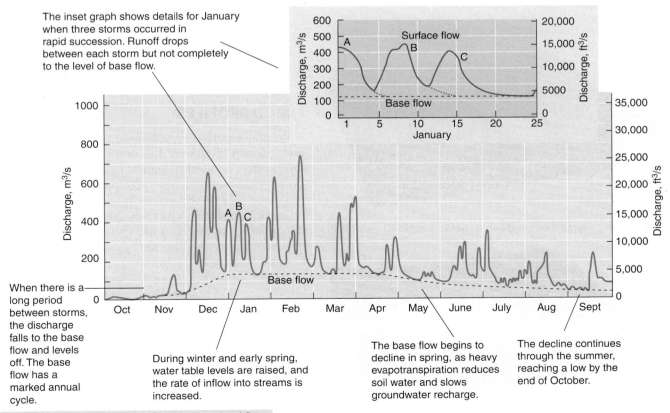

The inset graph shows details for January when three storms occurred in rapid succession. Runoff drops between each storm but not completely to the level of base flow.

When there is a long period between storms, the discharge falls to the base flow and levels off. The base flow has a marked annual cycle.

During winter and early spring, water table levels are raised, and the rate of inflow into streams is increased.

The base flow begins to decline in spring, as heavy evapotranspiration reduces soil water and slows groundwater recharge.

The decline continues through the summer, reaching a low by the end of October.

14.25 Discharge related to base flow and overland flow

This hydrograph shows the fluctuating discharge of the Chattahoochee River, Georgia, throughout a typical year. The sharp, abrupt fluctuations in discharge are produced by overland flow after rainfall periods lasting one to three days.

(Figure 14.25). Water supplied to a stream by groundwater is known as *base flow*. It increases during wet periods when the water table is high, and decreases during dry periods when the water table is low. When it rains, overland flow combines with base flow to increase the stream's discharge.

> A hydrograph plots streamflow with time. Peaks in the hydrograph occur after rainfall events. Between rains, streamflow falls to base flow, which is supplied by groundwater seepage.

Flooding

When soil is saturated by snowmelt or precipitation, runoff fills streams and rivers. When the discharge of a river cannot be accommodated within the normal channel, the water spreads over the adjacent ground, causing a **flood**.

FLOODPLAINS

Flooding is a regular occurrence in stream systems. Most low-gradient rivers of humid climates have a **floodplain**, a flat area bordering the channel on one or both sides that fills with water during a flood (Figure 14.26). As we will discuss in Chapter 15, floodplains are carved out over time as a river migrates across the valley floor. Periodic inundation

▼ LOW WATER
Migration of a river through a valley over time produces a wide floodplain.

Floodplain

▼ HIGH WATER
During flooding, a large river overflows its banks and fills its floodplain with water.

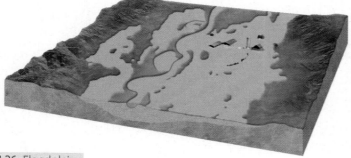

14.26 Floodplain

Most rivers are bordered by a plain of flat ground called a floodplain.

of floodplains is expected and does not prevent either crop cultivation after the flood has subsided or the growth of dense forests, which are widely distributed over low, marshy floodplains in all humid regions of the world.

Flooding on a floodplain becomes a human catastrophe only when cities and homes are situated on it (Figure 14.27). Historically, these areas were developed because they were easily accessible to river transportation and because the floodplains provided fertile agricultural ground. Today, people often build homes near a river to enjoy a picturesque view or easy access to the water for recreational activities.

The National Weather Service designates a particular water surface level as the *flood stage* for a particular river at a given place. If water rises above this critical level, the floodplain will be inundated. Over time, we see examples where even higher discharges cause rare and disastrous floods that engulf land well above the floodplain. The National Weather Service operates a River and Flood Forecasting Service through 85 offices located at strategic points along major river systems of the United States. This effort is augmented by hundreds of stations maintained by each state. When flooding potential is high, forecasters analyze precipitation patterns and the progress of high waters moving downstream. They develop specific flood forecasts after examining the flood history of the rivers and streams concerned. They then deliver these forecasts to communities that might be affected.

> A flood occurs when a river rises over its banks and covers adjacent land, which is called the floodplain. The height of the river at that time and place is called the flood stage.

FLOOD PROFILES

Following a period of increased precipitation or snowmelt, stream discharge rises and then falls over the following days or weeks (Figure 14.28). There is a delay, or *lag time*, between the precipitation event and the peak flow of a flood because it takes time for the water to move into stream channels. The length of this delay depends on several factors, including the size of the drainage basin feeding the stream. Larger drainage basins experience a longer delay.

Under some conditions, the lag time between precipitation and flooding can be quite short, leading to *flash floods* and heightening the danger to local populations. Flash floods are characteristic of streams draining small watersheds with steep slopes. These streams have short lag times, of only one or two hours, and with intense rainfall quickly rise to a high level. The flood arrives as a swiftly moving wall of turbulent water, sweeping away buildings and vehicles in its path. In arid western watersheds, great quantities of coarse rock debris are swept into the main channel and travel with the floodwater,

14.27 Mississippi flood of 2011

The spring of 2011 saw epic flooding along the Mississippi River. The discharge of the swollen river, about 65,000 m³ (2 million ft³) per second, matched the most destructive river flood in U.S. history, which occurred on the Mississippi in 1927. Many stream gauges recorded record or near-record values. At Memphis, the river peaked at a stage of 14.4 m (47.9 ft), only 0.24 m (0.8 ft) below the record, set in 1937. In Mississippi, record crests were recorded at Vicksburg and Natchez. In the city of New Orleans, the river came 1 m (3 ft) from overtopping the levees. Damage estimates for the total economic cost ranged as high as $7 to $9 billion. The flood was triggered by four major storms in April, which resulted in extensive melting of winter snow.

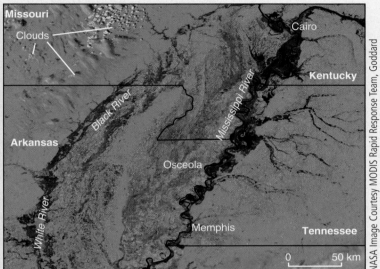

NASA Image Courtesy MODIS Rapid Response Team, Goddard Space Flight Center

NASA Image Courtesy MODIS Rapid Response Team, Goddard Space Flight Center

▲ BEFORE THE FLOOD

This satellite image, acquired by NASA's MODIS imager, shows the Mississippi Valley between Cairo, Illinois, and Memphis, Tennessee, on May 6, 2010. The brown areas are fertile agricultural lands in tributary valleys that drain into the Mississippi below Memphis.

▲ FLOOD IMAGE

On May 6, 2011, the Mississippi spilled over its banks and inundated broad areas of its floodplain. Flooding occurred as far west as the Black and White rivers in Arkansas. Two levels of clouds are visible in the upper-left part of the image.

FLOODING AT MEMPHIS ▶

Although much of Memphis escaped flooding, some areas of the city and nearby areas were immersed in flood waters. Pictured here is an inundated U.S. Navy facility in Millington, just north of Memphis. The circular feature is a recreational park with a sail-like canopy at its center.

Reuters/Landov LLC

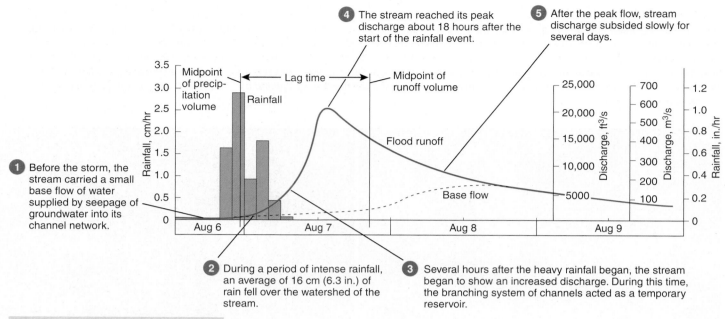

④ The stream reached its peak discharge about 18 hours after the start of the rainfall event.

⑤ After the peak flow, stream discharge subsided slowly for several days.

① Before the storm, the stream carried a small base flow of water supplied by seepage of groundwater into its channel network.

② During a period of intense rainfall, an average of 16 cm (6.3 in.) of rain fell over the watershed of the stream.

③ Several hours after the heavy rainfall began, the stream began to show an increased discharge. During this time, the branching system of channels acted as a temporary reservoir.

14.28 Stream discharge during flooding

The graph shows the discharge of Sugar Creek (blue line), a tributary of the Muskingum River in Ohio, during a four-day period marked by a heavy rainstorm.

producing debris floods. In forested landscapes, tree limbs and trunks, soil, rocks, and boulders hurtle downstream in the flood waters. Flash floods often occur too quickly to warn people, and so can cause significant loss of life.

URBANIZATION AND STREAMFLOW

The growth of cities and suburbs affects the flow of small streams in two ways. First, it is far more difficult for water to infiltrate the ground, which is more widely covered by buildings, driveways, walks, pavements, and parking lots (Figure 14.29). In a built-up residential area, 80 percent of the surface may be impervious to water. This in turn increases overland flow, making flooding more common during heavy storms for small watersheds that lie largely within these urbanized areas. It's also harder to recharge the groundwater beneath. The reduction in groundwater decreases the base flow to channels in the same area. So, in dry periods, stream discharges will tend to be lower in urban areas.

A second change caused by urbanization comes from the introduction of *storm sewers*, systems of large underground pipes designed to quickly transport storm runoff from paved areas directly to stream channels for discharge. These systems not only shorten the time it takes runoff to travel to channels, they also increase the proportion of runoff by the expansion in impervious surfaces. Together, these changes shorten the lag time of urban streams and heighten their peak discharge levels. Many suburban communities are finding that low-lying, formerly flood-free, residential areas now experience periodic flooding as a result of rapidly expanding urbanization.

14.29 Urban waterflow

Impervious surfaces like this street in Bangkok intensify runoff and hasten the flow of water into streams and rivers draining urban environments.

Jodi Cobb/NG Image Collection

WileyPLUS Surface Water

Take a video journey down the Rhine River through the heart of Europe, starting high in the Alps and ending at the Atlantic coast of The Netherlands.

Lakes

A **lake** is a body of standing water with an upper surface that is exposed to the atmosphere and does not have an appreciable gradient. Ponds, marshes, and swamps with standing water can all be included under the definition of a lake (Figure 14.30). Lakes receive water from streams, overland flow, and groundwater, and so they form part of drainage systems. Many lakes lose water at an outlet, where water drains over a dam—natural or constructed— to become an outflowing stream. Lakes also lose water by evaporation. Lakes, like streams, are landscape features but are not usually considered to be landforms.

Lakes are quite important as sources of freshwater and food, such as fish. They can also be used to generate

Volkmar K. Wentzel/NG Image Collection

14.31 Hydroelectric dam

Many lakes are artificial, created by dams built to supply water or hydroelectric power. The Picote Dam, shown here, is on the Douro River, in Portugal. It is about 100 m (328 ft) high and provides ample water pressure at its base to drive its power-generating turbines.

hydroelectric power, using dams (Figure 14.31). And, of course, lakes and ponds are sites of natural beauty. Lake basins, like stream channels, are true landforms, shaped by a number of geologic processes and ranging widely in size. For example, the tectonic process of crustal faulting creates large, deep lakes. Lava flows can form natural dams in river valleys, causing water to back up and take shape as a lake. Landslides can also create lakes spontaneously.

Where there aren't enough natural lakes, we create them by constructing dams across stream channels. Many regions that once had almost no lakes now have many. Some are small ponds built to serve ranches and farms, while others cover hundreds of square kilometers. In some areas, the number of artificial lakes is large enough to have notable effects on the region's hydrologic cycle.

On a geologic time scale, lakes are short-lived features. Lakes disappear by one of two processes, or a combination of both. First, lakes that have stream outlets will be gradually drained as the outlets are eroded to lower levels. Even when the outlet lies above strong bedrock, erosion will still occur slowly over time. Second, inorganic sediment carried by streams enters the lake and builds up, adding to organic matter produced by plants and animals within the lake. Eventually, the lake fills, becoming a boggy wetland with little or no free water surface. Many former freshwater ponds have become partially or entirely filled by organic matter from the growth and decay of water-loving plants.

14.30 Lakes and ponds

Lakes, ponds, marshes, and swamps are maintained by water from streams, overland flow, and groundwater.

Mike Theiss/NG Image Collection

▲ LAKE

Lakes, such as Lake Louise in Alberta, Canada, are usually larger and deeper than ponds. This example appeared after a rockslide dammed a stream in a narrow valley.

▼ POND

The shallow waters of this pond in Nicolet National Forest, Wisconsin, support an almost continuous cover of grasses and sedges.

James P. Blair/NG Image Collection

14.32 Water level and the water table

The water level of lakes and ponds is close to the level of the water table. In this example, retreating glaciers have left an irregular landscape of sand and gravel deposits, with depressions that are occupied by lakes.

Lakes can also disappear when the climate changes. If precipitation is reduced, or temperatures and levels of net radiation rise, evaporation can exceed input and the lake will dry up. Many former lakes of the southwestern United States that

> Lakes serve as vital reservoirs of freshwater on the land. They are formed in many different ways but are generally short-lived over geologic time.

flourished during the Ice Age have now shrunk greatly or disappeared entirely.

The water level of lakes and ponds in moist climates closely coincides with the surrounding water table (Figure 14.32). The water surface is maintained at this level as groundwater seeps into the lake and as precipitation runs off.

THE GREAT LAKES

The *Great Lakes*—Superior, Huron, Michigan, Erie, and Ontario—along with their smaller bays and connecting lakes comprise a vast network of inland waters in the heart of North America (Figure 14.33). They contain 23,000 km³ (5500 mi³) of water—about 18 percent of all the fresh, surface water on Earth. Only Lake Baikal in Siberia has a larger volume. Of the Great Lakes, Superior is by far the largest. In fact, the volume of the other Great Lakes combined would not fill its basin. The Great Lakes watershed is home to about 33 million people—22.8 million Americans and 9.2 million

14.33 The Great Lakes

The Great Lakes contain 18 percent of all the fresh surface water on Earth.

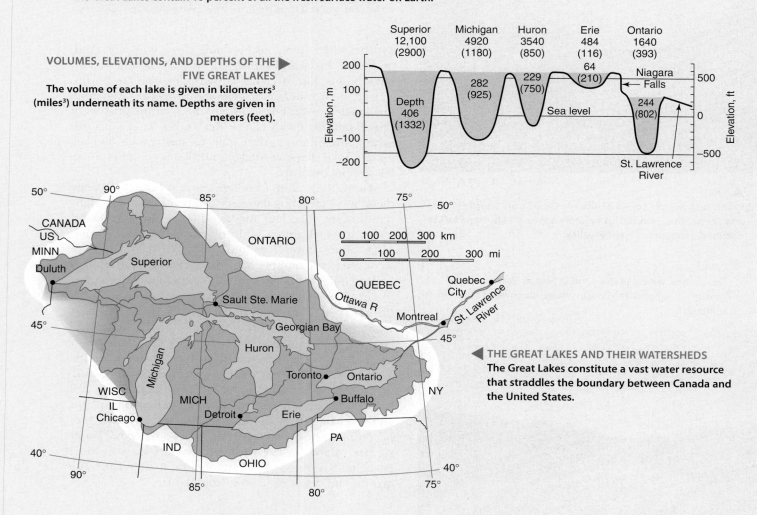

VOLUMES, ELEVATIONS, AND DEPTHS OF THE FIVE GREAT LAKES
The volume of each lake is given in kilometers³ (miles³) underneath its name. Depths are given in meters (feet).

THE GREAT LAKES AND THEIR WATERSHEDS
The Great Lakes constitute a vast water resource that straddles the boundary between Canada and the United States.

Brenda Tharp/Photo Researchers, Inc.

14.34 Salt encrustations
During a wetter climate, a lake was present in Death Valley National Park, California, but climate changed and the lake dried up, leaving a playa. Today, runoff from occasional storms and flash floods still reaches the playa, where it redissolves salts and then redeposits them as the runoff evaporates. At this location, the salt deposits take the form of white polygons.

Canadians—and the lakes are an essential resource for drinking water, fishing, agriculture, manufacturing, transportation, and power generation.

The Great Lakes are largely the legacy of Ice Age glaciation that took shape in a low interior basin of old, largely sedimentary rock. During at least four major periods in the last 2 million years, ice sheets advanced over this basin, scouring the rocks and lowering the surface by as much as 500 m (1600 ft) at some points below the surrounding terrain. As the continental ice sheets of the last glacial advance retreated, water filled these depressions, creating lakes dammed by glacial deposits and melting ice. Eventually, with the final melting of the ice, coupled with a slow, gentle uplift of the terrain, the lakes acquired their present shapes and configurations.

Situated as they are close to centers of population and agricultural development, the Great Lakes became seriously polluted in the 1960s and 1970s. Particularly hard hit was Lake Erie, which has the smallest water volume and a heavily developed coastal region. Of special concern was the threat posed by persistent organic compounds, largely of industrial origin, that are long-lasting, highly mobile in the aquatic system, and toxic in very small amounts. Many of these compounds accumulate up the food chain as predators consume contaminated prey. To address these issues, in 1987, an American and Canadian commission identified 43 "Areas of Concern," with 28 in the United States and 15 in Canada. Remedial action plans were proposed and implemented, and many of the sites have since experienced much improvement.

> The Great Lakes are a vast North American water resource, although they have suffered somewhat from water pollution.

SALINE LAKES AND SALT FLATS

In arid regions, we find lakes with no surface outlet. In these water bodies, the average rate of evaporation balances the average rate of stream inflow. When the rate of inflow increases, the lake level rises and its surface area broadens, allowing more evaporation and, thus, striking a new balance. Conversely, if the region becomes more arid, reducing input and increasing evaporation, the water will fall to a lower level.

Salt often builds up in these lakes. Streams bring dissolved solids into the lake, and since evaporation removes only pure water, the salts remain behind. The *salinity*, or "saltiness," of the water slowly increases. Eventually, the salinity level reaches a point where salts are precipitated as solids (Figure 14.34).

Sometimes the surfaces of such lakes lie below sea level. An example, shown in Figure 14.35, is the Dead Sea, along the boundary of Israel and Jordan, with a

14.35 Dead Sea
Swimmers in the Dead Sea float easily on the dense, salty water.

Priit Vesilind/NG Image Collection

Walter Meayers Edwards/NG Image Collection

14.36 Bonneville Salt Flats

Salt flats are dry lake bottoms covered with mineral salts and sediments. One of the most famous is the Bonneville Salt Flats, in Utah, which has a very uniform and smooth surface and is used as a speedway for high-speed race cars.

surface elevation of −396 m (−1299 ft). The largest of all lakes, the Caspian Sea, between Europe and Asia, has a surface elevation of −25 m (−82 ft). Both of these large lakes are saline. Another saline inland lake is the Aral Sea, described at the beginning of this chapter and shown in Figures 14.1 and 14.2.

In some cases the water is missing. In regions of high evapotranspiration and low precipitation, instead of lakes we find shallow empty basins covered with salt deposits. These are called *salt flats* or dry lakes (Figure 14.36). On rare occasions, these flats are covered by a shallow layer of water, brought by flooding streams.

DESERT IRRIGATION

Irrigating the desert is a practice as old as civilization itself. Two of the earliest civilizations—Egypt and Mesopotamia—relied heavily on large supplies of water from nondesert sources to irrigate their land. The ancient water sources for Egypt and Mesopotamia were the rivers that cross the desert but derive their flow from regions that have a water surplus. These are referred to as *exotic rivers* because their flows are derived from an outside region.

Irrigation systems in arid lands can suffer from two undesirable side effects: salinization and waterlogging of the soil (Figure 14.37). *Salinization* occurs when salts build up in

Jim Holmes/Panos Pictures

14.37 Salinization

Salinization from long-term irrigation in the Indus Valley, Sindh Province, Pakistan, has turned these once-productive fields into a barren expanse of salty earth. In addition to the Indus Valley shown here, agricultural areas of major salinization include the Euphrates valley in Syria, the Nile delta of Egypt, and the wheat belt of western Australia. In the United States, extensive regions of heavily salinized agriculture are found in the San Joaquin and Imperial valleys of California.

the soil to levels that inhibit plant growth. This happens when an irrigated area loses large amounts of soil water through evapotranspiration. Salts contained in the irrigation water remain in the soil and increase to high concentrations.

> Salinization and water-logging are undesirable side effects of long-term irrigation. Arid regions watered by exotic rivers are most affected.

Salinization can be prevented or cured by flushing the soil salts downward to lower levels by the force of more water. Clearly, however, this remedy requires greater quantities of water than for crop growth alone. In addition, new drainage systems must be installed to dispose of the excess saltwater.

Waterlogging occurs when irrigation with large volumes of water causes a rise in the water table, bringing the zone of saturation close to the surface. Most food crops cannot grow in perpetually saturated soils. When the water table rises to the point at which upward movement under capillary action can bring water to the surface, evaporation is increased and salinization is intensified.

Freshwater as a Natural Resource

Freshwater is a basic natural resource that is essential to human agricultural and industrial activities, but it is also a limited resource. Population growth and increased demand have placed a strain on freshwater supplies around the world. The supply of usable freshwater is also threatened by water pollution.

WATER ACCESS AND SUPPLY

Humans rely on clean water for food, recreation, and countless other uses (Figure 14.38). Today, however, despite the abundance of rainfall in many areas, access to freshwater is becoming a chronic problem. Our heavily industrialized societies require enormous supplies of freshwater to sustain them, and demand continues to rise. Urban dwellers in developed nations consume 150 to 400 L (50 to 100 gal) of water per person per day in their homes. We use large quantities of water in air conditioning units and power plants, much it obtained from surface water.

To increase the availability of freshwater supplies, we build dams and reservoirs that trap and store precipitation and runoff that would otherwise escape to the sea. These constructions, however, introduce a host of negative environmental ramifications, among them disruption of native fisheries, drowning of the river and creek ecosystems, displacement of local populations, and loss of nutrients that would otherwise flow into the river floodplains below the dams. Furthermore, reservoirs have finite lifetimes due to silt buildup and structural decay of their dams.

Global water is, simply, a finite resource, and in the long term, we can use only as much water as is supplied by precipitation. *Desalination*, a process that separates freshwater from seawater, offers an additional source of freshwater in locations lacking sufficient precipitation, but the high energy cost of operating desalination plants precludes their use in most of the developing world.

> Human society is heavily dependent on fresh surface water for irrigation, drinking water, and industrial usage. However, freshwater is a limited resource.

As the world's populations share dwindling supplies of this finite resource, the issue of water availability has risen to the top of the United Nations Millennium Goals and other international agendas. Climate change scientists predict a continuation of the current trends of increased flooding and droughts experienced around the globe. Clearly, more efficient use will be required to cope with the increased scarcity of freshwater and competition for it.

POLLUTION OF SURFACE WATER

Streams, lakes, bogs, and marshes provide specialized habitats for many species of plants and animals. These habitats are particularly sensitive to changes in the water balance and water chemistry. By constructing dams, irrigation systems, and canals, our industrial societies not only bring about radical changes to the flow of water, but we also pollute and contaminate our surface waters with a large variety of wastes.

There are many different sources of water pollutants. Some industrial plants dispose of toxic metals and organic compounds directly into streams and lakes. Many communities still discharge untreated or partly treated sewage wastes into surface waters. In urban and suburban areas, deicing salt and lawn conditioners (lime and fertilizers) enter and pollute streams and lakes, and contaminate groundwater. In agricultural regions, fertilizers and livestock wastes are major pollutants. Likewise, mining and processing of mineral deposits pollute water (Figure 14.39). Surface water can even be contaminated by radioactive substances released from nuclear power and processing plants.

Many chemical compounds dissolve in water by forming *ions*, charged forms of molecules or atoms. Among

14.38 Water access and use

Water is essential to support human populations. But because watersheds are not limited to single countries or political regions, water conflicts often arise, as shown on the map by red triangles.

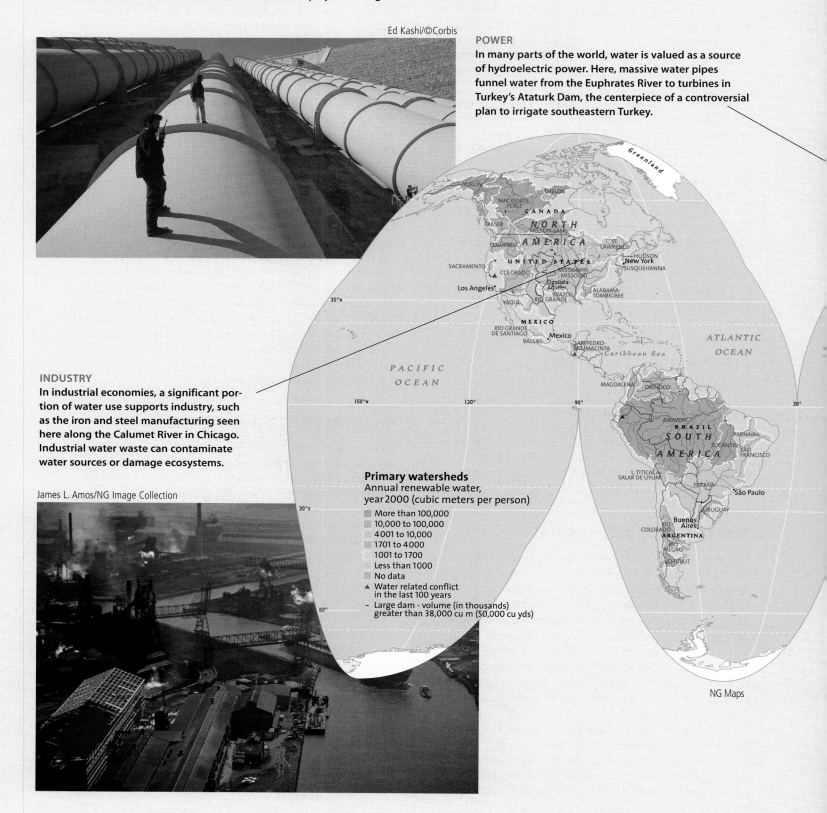

Ed Kashi/©Corbis

POWER

In many parts of the world, water is valued as a source of hydroelectric power. Here, massive water pipes funnel water from the Euphrates River to turbines in Turkey's Ataturk Dam, the centerpiece of a controversial plan to irrigate southeastern Turkey.

INDUSTRY

In industrial economies, a significant portion of water use supports industry, such as the iron and steel manufacturing seen here along the Calumet River in Chicago. Industrial water waste can contaminate water sources or damage ecosystems.

James L. Amos/NG Image Collection

Primary watersheds
Annual renewable water,
year 2000 (cubic meters per person)

- More than 100,000
- 10,000 to 100,000
- 4001 to 10,000
- 1701 to 4000
- 1001 to 1700
- Less than 1000
- No data
- ▲ Water related conflict in the last 100 years
- – Large dam - volume (in thousands) greater than 38,000 cu m (50,000 cu yds)

NG Maps

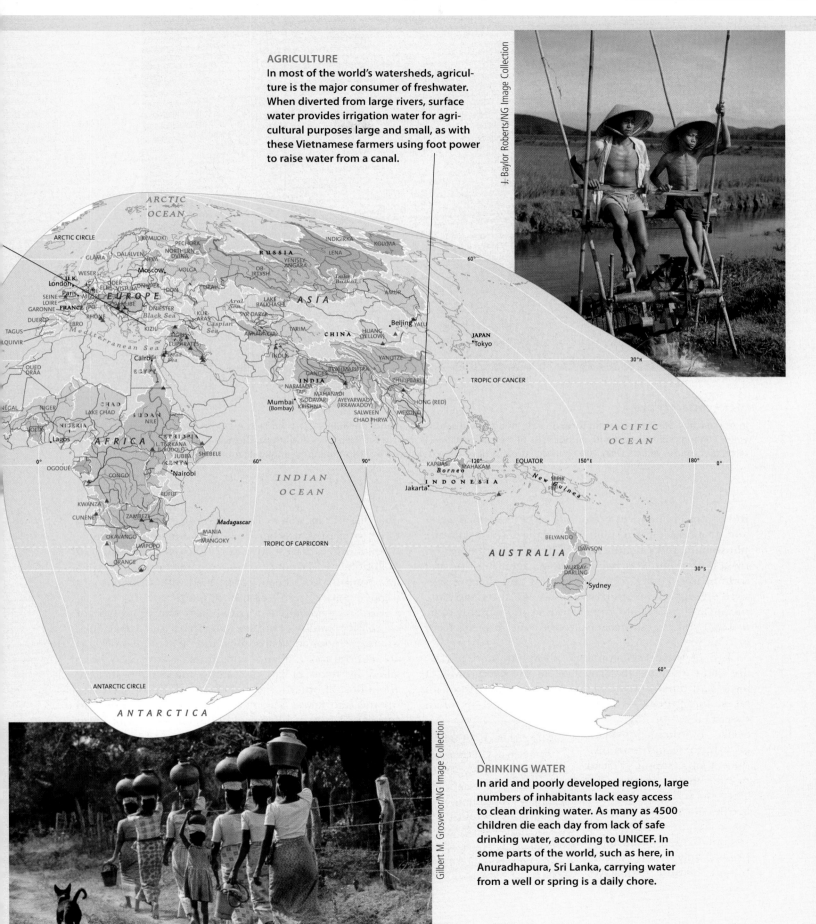

AGRICULTURE

In most of the world's watersheds, agriculture is the major consumer of freshwater. When diverted from large rivers, surface water provides irrigation water for agricultural purposes large and small, as with these Vietnamese farmers using foot power to raise water from a canal.

J. Baylor Roberts/NG Image Collection

DRINKING WATER

In arid and poorly developed regions, large numbers of inhabitants lack easy access to clean drinking water. As many as 4500 children die each day from lack of safe drinking water, according to UNICEF. In some parts of the world, such as here, in Anuradhapura, Sri Lanka, carrying water from a well or spring is a daily chore.

Gilbert M. Grosvenor/NG Image Collection

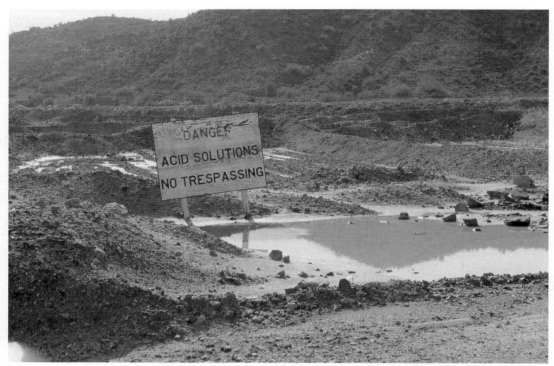

Stephen Sharnoff/NG Image Collection

14.39 Water pollution

In Appalachia, water percolating through strip-mine waste banks and abandoned mines often finds its way into streams and rivers. This water contains sulfuric acid and various salts of metals, particularly iron. In sufficient concentrations, the acid from these sources is lethal to certain species of fish. Over 2000 miles of streams have been destroyed by strip-mining fill and toxic runoff from the mines.

the common chemical pollutants of both surface water and groundwater are sulfate, chloride, sodium, nitrate, phosphate, and calcium ions. Sulfate ions enter runoff from both polluted urban air and sewage. Chloride and sodium ions come from polluted air and deicing salts used on highways. Fertilizers and sewage also contribute nitrate and phosphate ions. Nitrates can be highly toxic in large concentrations and are difficult and expensive to remove.

Two of these contaminants, phosphate and nitrate, are plant nutrients that can lead to the overgrowth of algae and other aquatic plants in streams and lakes. In lakes, this process— known as *eutrophication*—is often described as the "aging" of a lake. By stimulating plant growth, these nutrients produce a large supply of dead organic matter in the lake. The microorganisms that break down this organic matter require oxygen to do so; but oxygen is normally present only in low concentrations because it dissolves

> Water pollutants include various types of common ions and salts, as well as heavy metals, organic compounds, and acids. Excessive plant nutrients in runoff feeding lakes can lead to eutrophication.

only slightly in water. The microorganisms use up so much oxygen that other organisms, including desirable types of fish, cannot survive. After a few years of "nutrient pollution," the lake takes on the characteristics of a shallow pond that has been slowly filled with sediment and organic matter over thousands of years.

Acid mine drainage is a particularly damaging type of chemical pollution of surface water in parts of Appalachia, where abandoned coal mines and strip-mine workings are concentrated. Plants and animals have also been killed by toxic metals, including mercury, pesticides, and a host of other industrial chemicals introduced into streams and lakes. Sewage introduces live bacteria and viruses—classed as biological pollutants—that can harm humans and animals alike. Another form of pollution is *thermal pollution*, which is created when heat, generated from fuel combustion and the conversion of nuclear fuel to electricity, is discharged into the environment. Heated water entering streams, estuaries, and lakes can have drastic effects on local aquatic life, especially in a small area.

WileyPLUS Web Quiz

Take a quick quiz on the key concepts of this chapter.

A Look Ahead

This chapter has focused on processes of water flow, both within the ground and on the Earth's surface. These processes respond to the local and regional balance of the hydrologic cycle that provides precipitation to the land. The pattern of streams on the landscape acts as an efficient system to collect and carry runoff to the sea or to inland basins. This cannot happen, however, unless the gradients of slopes and streams are adjusted so that water keeps flowing downhill. This means that the landscape is shaped and organized into landforms that are an essential part of the drainage system. The shaping of landforms within the drainage system occurs as running water erodes the landscape—and that is the subject of the next chapter.

WileyPLUS Web Links
Explore caves around the world, including the famous cave paintings of Lascaux in France. Learn more about the causes and effects of floods.

IN REVIEW FRESHWATER OF THE CONTINENTS

- The Aral Sea shrank substantially because its inflow was diverted for upstream irrigation. A small northern portion, the Small Sea, is now being rehabilitated.
- The freshwater of the land accounts for only a small fraction of the Earth's water supply.
- Precipitation over land follows three paths: to the atmosphere as *evapotranspiration*, to groundwater through **infiltration**, and to streams and rivers as **runoff**.
- **Groundwater** occupies the pore spaces in rock and regolith. The **water table** marks the upper surface of the *saturated zone* of groundwater, where pores are completely full of water.
- Groundwater moves in slow paths deep underground, recharging rivers, streams, ponds, and lakes by upward seepage, thus contributing to runoff.
- The water table rises under divides, and dips down to the surface of lakes and rivers.
- Porous rock layers, such as sandstone, are good **aquifers**. Impervious layers, such as shale, are *aquicludes* that block the flow of groundwater.
- The dissolving of limestone by solution in acid groundwater can produce *limestone caverns* and create *sinkholes*. A landscape of sinkholes lacking streams is called *karst*.
- Wells draw down the water table and create a *cone of depression*. Large wells can easily lower the water table more quickly than it can be recharged.
- Land *subsidence* can occur when water is pumped out of aquifers. Venice is an example.
- Groundwater contamination can occur when precipitation percolates through contaminated soils or waste materials. Coastal wells can suffer saltwater contamination as a result of excessive withdrawals of freshwater.
- Runoff includes *overland flow*, moving as a sheet across the land surface, and *streamflow* in streams and rivers, which is confined to a *channel*.

- Rivers and streams are organized into a **drainage system** that moves runoff from slopes into channels and from smaller channels into larger ones.
- In a **stream channel**, water moves most rapidly near the center. Streamflow is turbulent and experiences friction at the bed and banks.
- Stream channels are measured by their width and depth, which create the stream's cross-sectional area. *Stream velocity* measures how rapidly the water flows. The slope of the stream is called the *stream gradient*.
- The **discharge** of a stream measures the flow rate of water moving past a given location. *Rapids* have a steeper slope, smaller cross-sectional area, and higher flow velocity than *pools*.
- Discharge increases downstream as tributary streams add more runoff. Large rivers have very low gradients.
- The *hydrograph* plots the discharge of a stream at a location through time. After a storm, the midpoint of storm discharge differs from the midpoint of precipitation by a *lag time*.
- Annual hydrographs of streams from humid regions show an annual cycle of *base flow* with superimposed discharge peaks from individual rainfall events.
- **Floods** occur when discharge increases and water spreads over the **floodplain**, inundating low fields and forests near the channel. High rates of discharge can bring damaging high waters to developed areas.
- *Flash floods* occur in small, steep watersheds and can be highly destructive. Water and debris descend rapidly along the channel, damaging structures and developments.
- Because urban surfaces are largely impervious, urban streams have shorter lag times and higher peak discharges than similar streams in more natural settings.
- **Lakes** are sources of freshwater, recreation, and hydroelectric power. Lakes shrink and disappear as their outlets are eroded and they fill with sediment.

- The *Great Lakes* are an enormous resource of freshwater for North America. But water pollution is a constant concern because of development along lake shores.
- Where lakes occur in inland basins, they are often saline. Some large saline inland lakes are below sea level. When climate changes, such lakes can dry up, creating *salt flats*.
- Irrigation is the diversion of freshwater from streams and rivers to supply the water needs of crops. In desert regions, where irrigation is most needed, *salinization* and *waterlogging* can occur, reducing productivity and eventually creating unusable land.

- Groundwater and surface water are essential natural resources, and human activities depend on abundant supplies of freshwater for many uses.
- Dams trap surface water to provide water supplies and electric power. Dams can also disrupt fisheries and river ecology.
- Water pollution arises from many sources, including industrial sites, sewage treatment plants, agricultural activities, mining, and processing of mineral deposits. Sulfate, nitrate, phosphate, chloride, sodium, and calcium ions are common contaminants.
- *Acid mine drainage* coupled with toxic metals, pesticides, and industrial chemicals are important hazards.

KEY TERMS

infiltration, p. 469
runoff, p. 470
overland flow, p. 470
groundwater, p. 470
water table, p. 470

aquifer, p. 472
karst, p. 474
drainage system, p. 479
stream, p. 480
stream channel, p. 480

discharge, p. 482
flood, p. 484
floodplain, p. 484
lake, p. 487

REVIEW QUESTIONS

1. What has happened to the Aral Sea in the last 40 years, and why? What are its future prospects?
2. What happens to precipitation falling on land? What processes are involved? Use the terms **infiltration, runoff,** and **overland flow** in your answer.
3. How are caverns formed in limestone? Describe the key features of a **karst** landscape.
4. How do wells affect the water table? What happens when pumping exceeds recharge?
5. How is **groundwater** contaminated? Describe how a well might become contaminated by a nearby landfill.
6. Why has land subsidence occurred in Venice? What are the effects?
7. What are the components of **runoff?** Which component is confined to a channel?
8. What is a **drainage system?** How are slopes and streams arranged in a drainage basin? Use the term *drainage divide* in your answer.

9. Define **discharge** (of a **stream**) and the two quantities that determine it. How does discharge vary in a downstream direction? How does *gradient* vary in a downstream direction?
10. Compare the characteristics of *rapids* and *pools*.
11. Define the term **flood**. What is a **floodplain?** What is a flash flood?
12. What are the effects of urbanization on streamflow? Describe why they occur.
13. How are **lakes** defined? What are some of their characteristics? How do lakes disappear?
14. Why are the Great Lakes important? What water-quality problems have the Great Lakes experienced?
15. Where do saline lakes occur? Why are they salty? Provide some examples of saline lakes.
16. Describe some of the problems that can arise in long-term irrigation of desert areas.
17. Discuss surface water as an important natural resource.
18. Identify common surface water pollutants and their sources.

VISUALIZING EXERCISES

1. Sketch a cross section through the land surface showing the position of the water table and indicating flow directions of subsurface water motion with arrows. Include the flow paths of groundwater. Be sure to provide a stream in your diagram. Label the saturated and unsaturated zones.

2. Why does water rise in an artesian well? Illustrate with a cross-sectional diagram the aquifer, aquicludes, and the well.

3. Sketch a hydrograph for a small stream showing both discharge and precipitation. Draw a second curve showing how the discharge would look if the stream were larger.

ESSAY QUESTIONS

1. A thundershower causes heavy rain to fall in a small region near the headwaters of a major river system. Describe the flow paths of that water as it returns to the atmosphere and ocean. What human activities influence the flows? In what ways?

2. Imagine yourself a recently elected mayor of a small city located on the banks of a large river. What issues might you be concerned with that involve the river? In developing your answer, choose and specify some characteristics for this city—such as its population, its industries, its sewage systems, and the present uses of the river for water supply or recreation.

Chapter 15
Landforms Made by Running Water

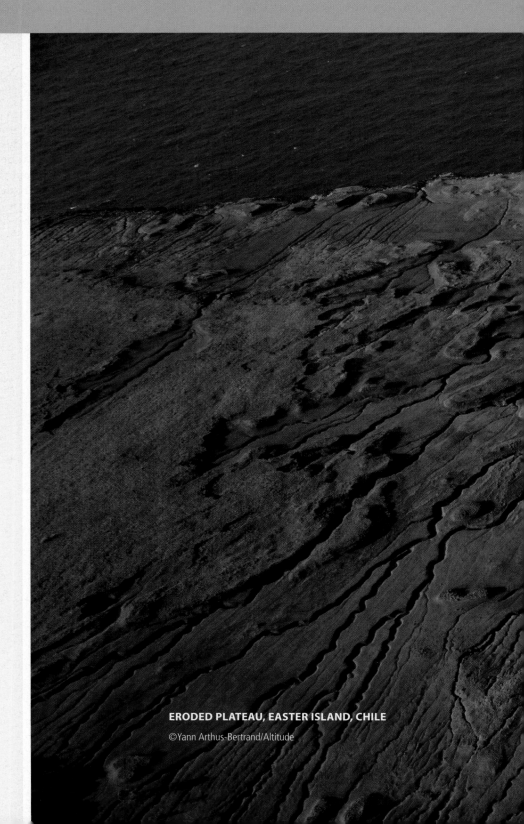

Often regarded as the most remote inhabited island in the world, Easter Island lies about 3500 km (about 2200 mi) west of the Chilean coast. First settled by Polynesians about a thousand years ago, the volcanic island has a complicated history, punctuated by famine, epidemics, civil war, slave raids, colonialism, and near deforestation. Because of forest clearing, agriculture, and sheep farming throughout most of the twentieth century, soil erosion has been severe in some island regions, where overland flow and streamflow have created a branching network of gullies in the red soil that drains an ancient lava plateau.

ERODED PLATEAU, EASTER ISLAND, CHILE
©Yann Arthus-Bertrand/Altitude

Landforms Made by Running Water

Most of the landforms we see around us are sculpted by running water as it erodes, transports, and deposits sediment. What causes slopes to erode, and what happens to eroded particles? How do streams build their beds and wear away their banks? How do stream valleys evolve over time? Under what conditions do streams form floodplains and meanders? These are some of the questions we will answer in this chapter.

Erosion, Transportation, and Deposition

Most of the world's land surface has been sculpted by running water, which acts to shape landforms through three closely related processes—*erosion*, *transportation*, and *deposition*. *Mineral* materials, from bedrock or regolith, are removed from slopes and stream channels by erosion, carving out gorges and valleys. These particles are then transported by water, either in solution as ions or as sediment of many sizes. The sediment is finally deposited downstream, where it build ups into plains, levees, fans, and deltas. Waves, glacial ice, and wind also shape unique landforms, but these processes are restricted to certain areas on the globe, as we will see in later chapters.

The landforms shaped by the progressive removal of bedrock are called **erosional landforms**. Fragments of soil, regolith, and bedrock that are removed from the parent rock mass are transported and deposited elsewhere, where they take shape as an entirely different set of surface features—the **depositional landforms** (Figure 15.1).

SLOPE EROSION

Fluvial erosion starts on the uplands as *soil erosion*. When raindrops hit bare soil, their force lifts soil particles, which fall back into new positions, causing *splash erosion* (Figure 15.2). A torrential rainstorm can disturb as much as 225 metric tons of soil per hectare (about 100 U.S. tons per acre). On a sloping ground surface, splash erosion shifts the soil slowly downhill. The soil surface also becomes much less capable of absorbing water. This important effect occurs because the natural soil openings

15.1 Erosional and depositional landforms

Steve Winter/NG Image Collection

AirPhoto-Jim Wark

become sealed by particles shifted by raindrop splash. Thus, water cannot infiltrate the soil as easily, so a much greater depth of overland flow can be triggered from a smaller amount of rain. This intensifies the rate of soil erosion.

Recall from Chapter 14 that some precipitation infiltrates the surface. When rainfall is heavy or the ground is saturated, however, water begins to flow across the surface as *overland flow*. Overland flow removes the soil in thin uniform layers through *sheet erosion*. Where land slopes are steep, runoff from torrential rains is even more destructive. *Rill erosion* scores many closely spaced channels, or *rills*, into the soil and regolith. Over time, these rills can join together into still larger channels. These deepen rapidly, turning into *gullies*, steep-walled, canyonlike trenches whose upper ends

CANYONS ▶
Plunging down the slope of South Africa's great Eastern Escarpment, the Blyde River eroded this steep, colorful canyon in flat-lying sedimentary rocks.

Digital Vision/SuperStock

◀ **PEAKS AND RAVINES**
Erosion by water, coupled with mass wasting, has carved out this ravine in a biosphere reserve on the Kamchatka Peninsula in eastern Russia.

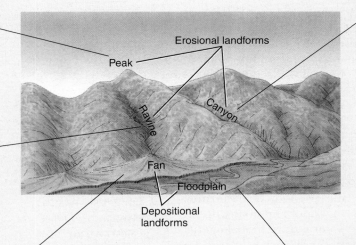

Erosional landforms

Peak

Ravine

Canyon

Fan

Floodplain

Depositional landforms

After a crustal block is uplifted by plate tectonic activity, it is shaped by running water, which erodes sediments in some places and deposits them in others. Valleys form as rock is weathered and then eroded away by fluvial agents.

▼ **FLOODPLAIN**
The Kustatan River in Alaska, carrying sediment-laden water from nearby glaciers, deposited this floodplain on its way to Cook Inlet, near Anchorage.

Alaska Stock Images/NG Image Collection

◀ **ALLUVIAL FANS**
This alluvial fan is a depositional landform, created by sediment from a steep mountain stream emerging from a narrow gap in the adjacent mountain flank. Other fans are visible in this view of Death Valley National Park, California.

15.2 Soil erosion by rain splash

A large raindrop lands on a wet soil surface, producing a miniature crater (right). Grains of clay and silt are thrown into the air, and the soil surface is disturbed.

Offical U.S. Navy Photographs

Offical U.S. Navy Photographs

grow progressively upslope (Figure 15.3). Rills and gullies are examples of *fluvial landforms*, which are created by *fluvial processes* of overland flow and streamflow.

> Fluvial landforms are shaped by the fluvial processes of overland flow and streamflow.

Soil particles picked up by overland flow are carried downslope until the surface slope meets the valley bottom. There the deposited particles accumulate in a thickening layer known as *colluvium*. Built as it is by overland flow, this deposit is distributed in sheets, making it difficult to notice unless/until it eventually buries the bases of tree trunks or fence posts.

Overland flow picks up mineral particles ranging in size from fine clay to coarse sand or even gravel. The size of particles removed depends on how fast the flow moves and how tightly plant roots and leaves hold down the soil. Under stable natural conditions in a humid climate, the erosion rate is slow enough to allow soil to develop normally. Each year, a small amount of soil is washed away, and a small amount of solid rock material is turned into regolith and soil.

In arid or semiarid climates, or where land has been cleared for cultivation, erosion can be rapid and severe. If there is no foliage to intercept rain, and no ground cover from fallen leaves and stems, raindrops fall directly on the mineral soil, eroding it much faster than it can be formed. Some natural events, such as forest fires, also speed up soil erosion.

The destruction of vegetation also lowers the ground surface's resistance to erosion under overland flow. Even deep layers of overland flow can cause only a little soil erosion on slopes that have a protective covering of grass sod. This is because the grass stems are tough and elastic, generating friction with the moving water and taking up the water's gravitational energy. Without such a cover, the water can easily dislodge soil grains and sweep them downslope. Ultimately, accelerated soil erosion leaves behind a rugged, barren topography.

> Soil erosion occurs when overland flow transports soil particles downslope. Erosion is greatest on bare slopes of fine particles, carving rills and gullies. A vegetation cover greatly reduces soil erosion.

SEDIMENT YIELD

How does erosion from vegetated land compare with erosion from open land? To compare erosion rates, we use *sediment yield*, a technical term for the rate of

15.3 Rills and gullies

Rill erosion represents the starting point for gullies.

◀ **RILLS**
Rills begin with small channels or rivulets of water following a gravitational path downslope.

Sheldan Collins/©Corbis

▼ **GULLIES**
Deforestation on these mountain slopes near Katmandu, Nepal, has led to rapid erosion and gullying.

Katmandu, Nepal

Steve McCurry/Magnum Photos, Inc.

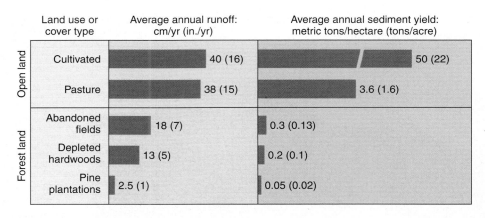

Land use or cover type	Average annual runoff: cm/yr (in./yr)	Average annual sediment yield: metric tons/hectare (tons/acre)
Open land		
Cultivated	40 (16)	50 (22)
Pasture	38 (15)	3.6 (1.6)
Forest land		
Abandoned fields	18 (7)	0.3 (0.13)
Depleted hardwoods	13 (5)	0.2 (0.1)
Pine plantations	2.5 (1)	0.05 (0.02)

15.4 Runoff and sediment yield

This bar graph shows that both runoff and sediment yield are much greater for open land than for land covered by shrubs and forest. Values are for upland surfaces in northern Mississippi, where, climate, soil, and topography are fairly uniform.

15.5 Potholes

These potholes in river bedrock were created by abrasion on the bed of a swift mountain stream. Potholes form when grinding stones are caught in shallow depressions in the streambed. Flowing water spins the stones around, carving holes in the rock.

Jeffrey Lepore/Photo Researchers, Inc.

sediment removal in metric tons per hectare per year (tons per acre per year). As shown in Figure 15.4, both surface runoff and sediment yield are much lower for vegetated surfaces. In fact, sediment yield from cultivated land under poor management can be more than 10 times greater than that of pasture and about a thousand times greater than that of a pine plantation.

STREAM EROSION

Once overland flow is channeled into a stream it can erode in various ways, depending on the nature of the channel materials, the speed and volume of the flow, and the kind of debris the streamflow is carrying. First, the flowing water drags on the bed and banks and propels particles into the bed and banks. This form of erosion, called *hydraulic action*, can, when river flow is high, rapidly wear away loose alluvial materials, such as gravel, sand, silt, and clay. In this way, erosion undermines the banks of the stream, causing large masses of loose material to slump into the river, where the particles are quickly separated and carried downstream.

Second, as the water picks up debris, it continues to break down the streambed through mechanical erosion. Where rock fragments carried by the swift current strike against bedrock channel walls and floors they knock off chips of rock. This process of mechanical wear is called *abrasion*. In bedrock that is too strong to be eroded by simple hydraulic action, abrasion is the main method of erosion. A striking example of the effects of abrasion is a *pothole* (Figure 15.5). As the rock fragments move downstream, they become rounded; cobbles and boulders roll along the streambed, crushing and grinding the smaller grains, producing a wide assortment of grain sizes.

Third, chemical weathering can remove rock from the stream channel by acid reactions and solution, a process called *corrosion*. We see this type of chemical weathering in limestone, in particular, which often develops cupped and fluted surfaces where it is being actively dissolved in the streambed. This is the same solution process that occurs when groundwater dissolves limestone by carbonation, as described in Chapter 14.

> Streambeds and banks are eroded by hydraulic action, abrasion, and corrosion. Abrasion by stones on a bedrock riverbed can dig deep depressions known as potholes.

STREAM TRANSPORTATION

The solid matter carried by a stream is the **stream load**. Stream load is carried in three ways: as *dissolved load, suspended load,* or as *bed load* (Figure 15.6). Generally, most of the stream load is carried in suspension. A large river such as the Mississippi carries as much as 90 percent of its load in suspension.

Stream capacity is the measure of the maximum solid load of debris—including bed load and suspended load—that a stream can carry. Capacity is given in units of metric tons per day passing downstream at a given location. A stream's capacity increases sharply as its velocity and discharge rise, such as during a flood, for two reasons: because there is more water volume to hold sediment, and because swifter currents are more turbulent and have more energy to hold sediment in suspension.

The capacity to move the larger particles of the bed load also increases with velocity because faster-moving water drags more powerfully against the bed. In fact, the capacity to move bed load increases according to the third to fourth power of the velocity. In other words, if a stream's velocity is doubled in times of flood, its capability to transport bed load will increase from 8 to 16 times.

> Streams carry dissolved matter, sediment in suspension, and a bed load of larger particles that bump and roll along the bottom. A stream's capacity to carry sediment increases sharply with its velocity.

DEPOSITION

A stream carries its load downslope toward a valley, a lake, or an ocean. Along that journey, whenever a stream's load exceeds its capacity, it deposits some of its load. This deposited sediment is called **alluvium**.

15.6 Stream load

Streams carry their load as dissolved, suspended, and bed load.

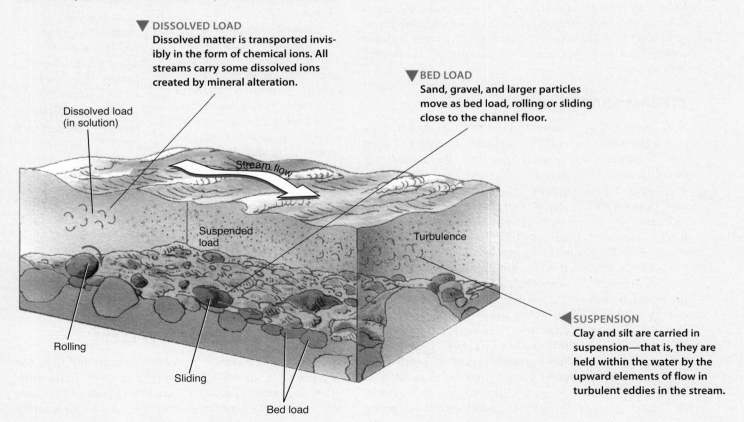

▼ DISSOLVED LOAD
Dissolved matter is transported invisibly in the form of chemical ions. All streams carry some dissolved ions created by mineral alteration.

▼ BED LOAD
Sand, gravel, and larger particles move as bed load, rolling or sliding close to the channel floor.

Dissolved load (in solution)

Stream flow

Suspended load

Turbulence

Rolling

Sliding

Bed load

◄ SUSPENSION
Clay and silt are carried in suspension—that is, they are held within the water by the upward elements of flow in turbulent eddies in the stream.

Deposition typically occurs where the velocity of streamflow decreases. For example, deposition occurs along stream banks when the streamflow slows down on the inside of a bend in the channel. During flooding, fast-moving floodwaters slow down and spread out over the valley floor, depositing alluvium in layers. Fine sediment, rich in organic matter, can improve soil fertility—although, sometimes, flooding can also leave behind sterile layers of sand or gravel.

Dams represent a special case for sediment deposition and can have a variety of negative influences on fluvial systems. Where earthen or concrete barriers block the streamflow, transported sediment quickly settles at the base of the dam. Sediment that would otherwise continue downstream and settle out during floods continuously fills in the reservoir bottom. Eventually, the sediment may displace enough water to render the original dam inoperable for water storage or electrical generation. Costly dredging is the only remedy for removing depositional sediment.

WileyPLUS Fluvial Geomorphology and Stream Processes Interactivity
Use this interactivity to master the processes by which streams erode, transport, and deposit sediment, depending on stream velocity and depth.

Stream Gradation and Evolution

Most major stream systems have experienced thousands of years of runoff, erosion, and deposition. Now that you have seen how these processes occur, let us take a closer look at the part they play in the evolution of streams over time.

STREAM GRADATION

The downhill flow of water and sediment in stream channels is governed by basic physical laws. The water falling on a mountain landscape high above the ocean, has potential energy induced by gravity; as it flows downhill in a stream, it expends the potential energy as work—in this case, done by friction on the bed and banks, which erodes sediment. Stones and boulders of bed load, pushed by the force of the water, collide, fracture, and break apart, doing more work. Energy is also expended in the internal friction of the turbulent flow of sediment-laden water.

The amount of work that a stream does within a particular length is determined largely by the potential energy available—that is, the drop in elevation within that length. But exactly how that work is carried out depends on a lot of different factors. For example, in a stretch with a deep and narrow channel of swiftly moving, clear water, a lot of energy may be expended in moving bed load. In contrast, in a stretch with a wide, shallow channel of slowly moving sediment-laden water,

the stream may expend most of its energy in internal friction and by eroding fine sediment from its bed and banks. Moreover, the amount of flowing water in the stream, and the amount and type of sediment it carries, are always changing, as precipitation causes flooding or as mass wasting and overland flow provide new inputs of sediment of different sizes. Although the laws of physics govern the motion of every particle of water or sediment, there are many possibilities for the shapes of channels and the slopes they maintain.

Through a number of feedback mechanisms, a stream can develop a more-or-less stable state, in which its long-term capacity to transport sediment is matched by the long-term rate at which it receives sediment. Channel shapes and slopes, which determine the transport rate, remain about the same over time. In this equilibrium situation, the stream is known as a **graded stream**. Streams do not reach a graded condition until they are flowing on alluvium—that is, material that the stream has already moved at least once and can move again, under the proper conditions (Figure 15.7).

A key characteristic of a graded stream is that its slope decreases fairly smoothly in a downstream direction. As the river's discharge increases downstream, it transports sediment more efficiently and so requires a lesser slope to move the sediment. In some cases, the average size of bed load particles decreases downstream, so a lesser gradient is required to move the load. The graph of the elevation of a stream against its downstream distance is known as the *stream profile* (Figure 15.8).

Graded streams often develop characteristic channel patterns. For example, wide and shallow streams often develop a channel of broad, sweeping curves, called **meanders**. The meanders move downstream, creating a broad **floodplain** that is flooded every few years. In other situations, streamflow is divided into multiple threads, called a *braided stream*. We will return to these channel forms later in the chapter.

> Over time, a stream develops a graded profile in which the gradient is just sufficient to carry the average annual load of water and sediment produced by its drainage basin.

WileyPLUS The Graded Stream
This animation demonstrates how rivers maintain a graded profile. Also watch how rivers widen their channels and deepen their beds during floods.

EVOLUTION OF STREAM VALLEYS

Through the process of gradation, streams and their valleys evolve in a predictable way (Figure 15.9). In the early stages of stream evolution, streams experience sudden changes in gradient, where faulting and tectonic uplift

15.7 Channel slope and load

Over time, a stream channel adjusts its gradient to transport its particular sediment load.

Oxford Scientific/Photolibrary/Getty Images, Inc.

Nikreates/Alamy Limited

▲ A GRADED STREAM CARRYING COARSE SEDIMENT
Riley Creek, near the entrance to Denali National Park, Alaska, has a steep slope adjusted to carrying cobbles and larger stones, which are visible on stream banks and gravel bars.

▲ A GRADED STREAM CARRYING FINE SEDIMENT
The Avon River, located in southern England, has a shallow slope adjusted to carrying sand and silt, which can be readily seen on the riverbanks at low tide near its mouth.

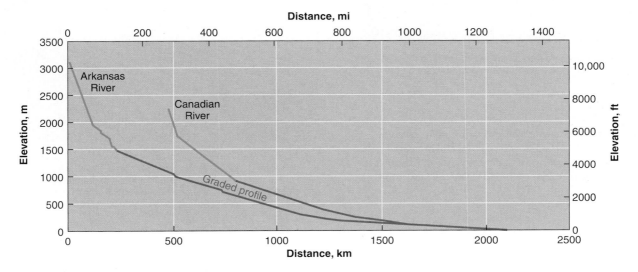

15.8 Longitudinal profiles of two river systems

This graph of profiles of elevation and distance along two rivers show the slope of each river at various points along its course. High in the Rocky Mountains, the slopes of the Arkansas River and its tributary, the Canadian River, are relatively steep and somewhat irregular. In the middle and lower portions, where the rivers are graded, the slopes are shallower and smoother. At the junction of the Arkansas with the Mississippi River, the slope is nearly flat.

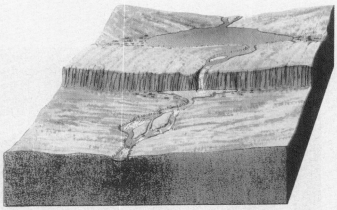

1 The flow of this ungraded stream is faster at the waterfalls and rapids, so these steeper segments of the stream are rapidly eroded. Where the stream flow slows (for example, at ponds and lakes), sediment is deposited.

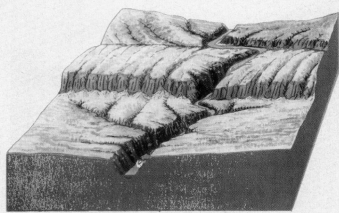

2 Waterfalls and rapids have eroded until their gradient is closer to the stream's average gradient. The lakes and ponds have drained and disappeared.

3 The river begins to wander sideways, eroding the side slopes, creating a curving path. Alluvium accumulates on the inside of each bend.

15.9 Evolution of a graded stream and its valley

Pictured here is a situation in which rapid tectonic uplift raises a highland landscape and then slows or stops. The newly created landscape is then eroded by streams and mass wasting. The process may take millions of years to shape a broad valley in a landscape of low relief.

4 As the stream flow continues to erode the banks, the channel develops sweeping meanders. Over time, the migration of these curves downstream creates a floodplain of flat land between steep bluffs. The stream has achieved a graded condition.

lead to a sudden drop-off, or where the resistance of streambed materials suddenly changes. Points at which the gradient of the stream changes abruptly, such as waterfalls and rapids, are called *nickpoints*. Streamflow is particularly fast and turbulent at nickpoints, so these segments of the stream are more rapidly eroded back to the average gradient of the stream, in a process called *downcutting*.

In some situations, nickpoints can migrate upstream as a result of erosive action. If a resistant rock layer underlies a less resistant one, a **waterfall** can form, with a cascade plunging over the resistant layer into a pool

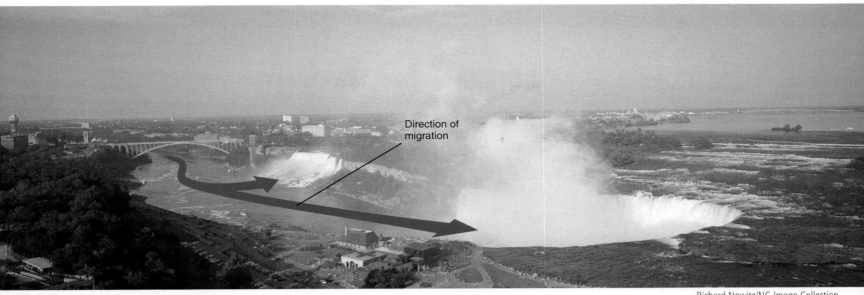

Direction of migration

Richard Nowitz/NG Image Collection

15.10 Nickpoint migration

Niagara Falls has migrated more than 11 km (7 mi) upstream from its original position of about 11,000 years ago, leaving behind a steep gorge. At the present time, the river's flow is divided into two channels. The near channel in the photo provides the horseshoe-shaped Canadian Falls, and the far channel the straighter American Falls.

Larry Dale Gordon/Getty Images, Inc.

carved from the underlying weaker rock by the tremendous energy of the falling water. The churning, oxygenated water may also enhance the chemical weathering of the weaker rock. As the resistant rock gradually weakens from weathering, and loses support from the weaker rock underneath, it breaks off in chunks and tumbles into the plunge pool, moving the waterfall farther upstream. Niagara Falls is an example of a waterfall that is migrating upstream (Figure 15.10).

Waterfalls can also form where a major river crosses a fault that has broken and fractured rock layers, making them easier to erode. Victoria Falls (Figure 15.11) and Iguazu Falls (Chapter 14 opener photo) are examples.

15.11 Victoria Falls

Located on the Zambezi River at the border of Zimbabwe and Zambia in southern Africa, Victoria Falls is one of the world's scenic wonders.

EYE ON THE LANDSCAPE **What else would the geographer see?** A fault has shattered and broken rock layers, creating a zone of weakness that has been eroded by the Zambezi River to form the gorge of Victoria Falls **A**. Note the resistant rock layer that keeps the waterfall vertical **B**. Native vegetation here, visible in the near distance **C**, is thorntree–tall grass savanna.

FOCUS ON REMOTE SENSING

Canyons from Space

Deep canyons, carved by powerful rivers crossing high terrain, are among the most dramatic features of the landscape. The Grand Canyon of the Colorado River is among the most famous in the world, spanning a length of about 450 km (about 280 mi), marked by vertical drops of up to about 1500 m (about 5000 ft). NASA's MISR imager acquired our true color image on December 31, 2000. Trace the path of the Colorado from Lake Powell, at the upper right, through narrow Marble Canyon, until the canyon broadens to the southeast of the snow-covered Kaibab Plateau. Here the Grand Canyon begins, revealing a dissected landscape between the two canyon rims as the river curves around the plateau. The canyon continues as the river flows westward to the edge of the frame.

The image acquired by the ASTER instrument on May 12, 2000, is a computer-generated perspective view of the Grand Canyon looking north up Bright Angel Canyon from the South Rim. In this image, vegetation appears green and water appears blue, rocks are not shown in their true colors.

Two canyons along the coast of southern Peru in the Arequipa region are shown in another MISR image. Here, the Pacific coastline runs nearly east-west and is obscured by low clouds and fog, visible at the bottom of the picture. The two vast canyons reaching from the sea upward into the Andean Plateau are the canyons of the Rio Ocoña (left) and Rio Camaná (right). The canyon of the middle branch of the Rio Ocoña (Rio Cotahuasi) reaches a depth of 3354 m (11,001 ft) below the plateau, more than twice as deep as the Grand Canyon. The white patch between the two canyon systems is the snow-covered Nevado Coropuna, at 6425 m (21,074 ft) elevation.

15.12 A canyon gallery

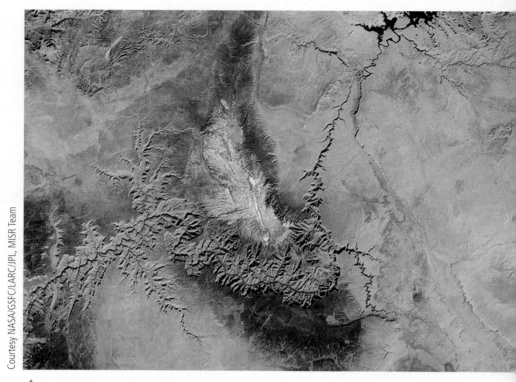

Courtesy NASA/GSFC/LARC/JPL, MISR Team

▲ GRAND CANYON, ARIZONA, IMAGED BY MISR

Courtesy NASA/GSFC/MITI/ERSDAC/JAROS and U. S. Japan ASTER Science Team

▲ GRAND CANYON PERSPECTIVE VIEW FROM ASTER

Lastly, a color-infrared ASTER image shows the Yangtze River canyon in the Three Gorges region of the provinces of Hubei and Sichuan, China. Although not a candidate for the deepest canyon, it is among the most scenic, featuring steep limestone cliffs and forest-covered slopes that separate clusters of quaint villages where tributaries meet the main channel. The inset image shows the Three Gorges Dam under construction. Completed in 2008, the dam created a reservoir 175 m (574 ft) deep and about 600 km (about 375 mi) long. The dam provides huge amounts of hydroelectric power and sharply reduces extreme flooding of the Yangtze.

Courtesy NASA/GSFC/LARC/JPL, MISR Team

▲ CANYONS OF THE ANDES, AS SEEN BY MISR

Courtesy NASA/GSFC/MITI/ERSDAC/JAROS and U. S. Japan ASTER Science Team

▲ THREE GORGES REGION OF THE YANGTZE RIVER, IMAGED BY ASTER

15.13 Entrenched meanders

Tectonic uplift of a meandering stream produces entrenched meanders.

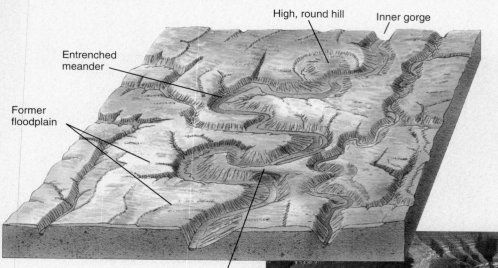

High, round hill

Inner gorge

Entrenched meander

Former floodplain

Natural bridge

◀ **HOW ENTRENCHED MEANDERS FORM**
Entrenched meanders slowly enlarge, producing cutoffs that can leave a high, round hill separated from the valley wall by the deep abandoned river channel and the shortened river course. Occasionally, if the bedrock includes a strong, massive sandstone formation, the meander cutoff can leave a natural bridge capped by the narrow meander neck.

Former floodplain level

Entrenched meanders

Joel Sartore/NG Image Collection

GOOSENECKS ▶
The Goosenecks of the San Juan River in Utah are deeply entrenched river meanders in horizontal sedimentary rock layers. The canyon, carved from sandstones and limestones, is about 370 m (1400 ft) deep.

George H.H. Huey/Alamy Limited

◀ **NATURAL BRIDGE**
Sipapu Bridge, in Natural Bridges National Monument, Utah, was created when two river meanders eroded through the meander neck separating them. Since then, weathering has enlarged the opening to form a high bridge capped by a resistant sandstone formation.

WileyPLUS Waterfalls

See some of the world's waterfalls in action, from Niagara Falls to the spectacular Iguazu Falls of Argentina. A video.

A stream may erode its streambed rapidly through downcutting in the early stages of its evolution, but eventually the balance between erosion and deposition stabilizes. There is a lower limit to how far a stream can erode its bed, called its **base level**. The hypothetical base level for all streams is sea level, because once streamflow reaches the sea, gravity can carry it no farther downhill. However, some streams have base levels much higher than sea level, such as the level of a lake or reservoir into which the stream flows and is temporarily stored.

STREAM REJUVENATION

Tectonic uplift sometimes raises the gradient of portions of a graded stream. Faulting, for example, may push up blocks of crust to form waterfalls and rapids. With new, steeper segments, the stream begins a new cycle of downcutting. Such streams are said to be *rejuvenated streams*.

When a broadly meandering stream is uplifted by rapid tectonic activity, the uplift increases the river's gradient and, in turn, its velocity, so that it cuts downward into the bedrock below. This cutting produces a steep-walled *inner gorge*. On either side of the gorge will be the former floodplain, now a flat terrace high above river level. Any river deposits left on the terrace are rapidly stripped away by runoff because floods no longer reach the terraces to restore eroded sediment. The meanders thus become impressed into the bedrock, passing on their meandering pattern to the inner gorge. We call these sinuous bends *entrenched meanders* (Figure 15.13), to distinguish them from the floodplain meanders of an alluvial river.

> Where rapid uplift causes meandering rivers to cut deeply into bedrock, entrenched meanders are formed.

As a river lowers its bed through time it can leave behind parts of its floodplain at a higher level as a series of **alluvial terraces** (Figure 15.14). Alluvial terraces attract human settlement because, unlike valley-bottom floodplains, they aren't subject to annual flooding. The terraces are easily tilled and make prime agricultural land. Causes of downcutting and terrace formation include a period of uplift or a change in climate that affects the amount of runoff and sediment the river receives.

THEORIES OF LANDSCAPE EVOLUTION

As we have seen, the landscapes sculpted by flowing water range from mountain regions of steep slopes and rugged peaks to regions of gentle hills and valleys to nearly

15.14 Alluvial terraces

Alluvial terraces are the remnants of abandoned floodplains left behind as a river carves out a new, lower floodplain.

▼ **EXCAVATION**

The meandering stream excavates alluvium from the floodplain and carries it downstream. This leaves steplike alluvial surfaces on both sides of the valley. The treads of these steps are called alluvial terraces.

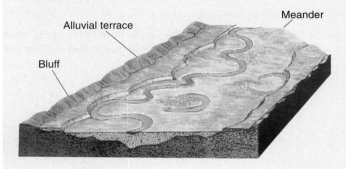

▼ **FURTHER DOWNCUTTING**

As the stream slowly lowers its floodplain, more terraces are left behind; some with bedrock outcrops sheltering the terrace slopes.

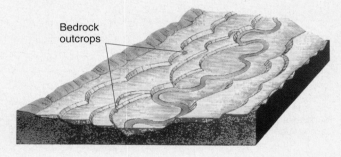

▲ **TERRACES**

These terraces line the Rakaia River gorge on the South Island of New Zealand. The flat terrace surface in the foreground is used as pasture for sheep. Two higher terrace levels can be seen on the left.

flat plains that stretch from horizon to horizon. We can think of these constantly changing landscapes as the stages in an evolutionary cycle, often called the **geomorphic cycle**. It begins with rapid uplift by plate tectonic activity and is followed by long-term erosion by streams in a graded condition until the landscape is worn down to a nearly flat surface called a *peneplain*. The surface is then uplifted again by tectonic activity (Figure 15.15). This cycle, originally called the *geographic cycle*, was first proposed by William Morris Davis, a prominent geographer and geomorphologist of the late nineteenth and early twentieth centuries.

The geomorphic cycle is useful for describing landscape evolution over very long periods of time, but

> The geomorphic cycle traces the fate of rivers and fluvial landforms from an initial uplift that creates steep slopes and canyons to a final low, gently rolling surface called a peneplain.

15.15 Idealized stages of the geomorphic cycle

The evolutionary cycle of a landmass after uplift shows how a nearly flat, eroded plain, a peneplain, is uplifted and worn away over geologic time, until it returns to a similar nearly flat surface and is ultimately uplifted again.

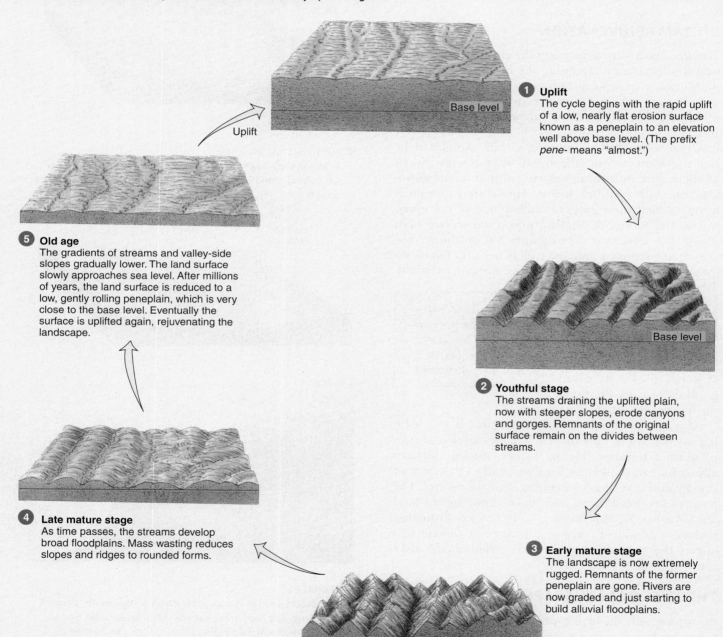

1 Uplift
The cycle begins with the rapid uplift of a low, nearly flat erosion surface known as a peneplain to an elevation well above base level. (The prefix *pene-* means "almost.")

2 Youthful stage
The streams draining the uplifted plain, now with steeper slopes, erode canyons and gorges. Remnants of the original surface remain on the divides between streams.

3 Early mature stage
The landscape is now extremely rugged. Remnants of the former peneplain are gone. Rivers are now graded and just starting to build alluvial floodplains.

4 Late mature stage
As time passes, the streams develop broad floodplains. Mass wasting reduces slopes and ridges to rounded forms.

5 Old age
The gradients of streams and valley-side slopes gradually lower. The land surface slowly approaches sea level. After millions of years, the land surface is reduced to a low, gently rolling peneplain, which is very close to the base level. Eventually the surface is uplifted again, rejuvenating the landscape.

it does little to explain the diversity of the features observed in real landscapes. Most geomorphologists think of landforms and landscapes in terms of equilibrium. This approach explains a landform as the product of forces acting upon it, including both tectonic uplift and denudation by erosion.

One strength of this viewpoint is that we can take into account the characteristics of the rock material. Thus, we find steep slopes and high relief where the underlying rock is strong and highly resistant to erosion. Even a rugged landscape that appears to be quite "young" can be in a long-lived state of equilibrium in which hillslopes and stream gradients remain steep, and thus maintain a graded condition while eroding the strong rocks of the region.

Another problem with Davis's geomorphic cycle is that it applies only where the land surface is stable over long periods of time after the initial uplift. But we know from our study of plate tectonics that crustal movements are frequent on the geologic timescale. Few regions of the land surface remain untouched by tectonic forces in the long run. Recall also that the continental lithosphere floats on a soft asthenosphere. As layer upon layer of rock is stripped from a landmass by erosion, the landmass becomes lighter and is buoyed upward.

> The equilibrium approach sees fluvial landforms as reflecting a balance between the processes of uplift and denudation acting on rocks of varying resistance to erosion.

A better model, then—referred to as the *equilibrium approach*—is one of uplift as an ongoing action to which erosional processes are constantly adjusting, rather than as a sudden event followed by denudation.

Fluvial Landforms

Now that you've seen how landscapes are shaped by water over time, let's take a closer look at several of the landforms associated with them, called **fluvial landforms**.

MEANDERING STREAMS AND FLOODPLAINS

Meandering, as described earlier, is a characteristic of graded rivers carrying substantial loads of sediment of varying sizes. In a meander bend, water moves with greater velocity on the outside of the bend, and the channel is deeper. The greater velocity erodes the floodplain sediment, creating a *cut bank* and causing the bend to grow outward or in a downstream direction (Figure 15.16). On the inside of the bend, the flow is slower, and sediment accumulates in a *point*

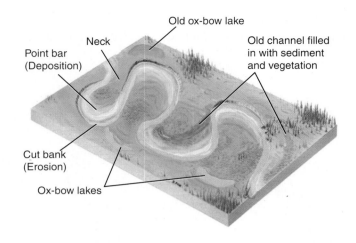

15.16 Growth and movement of meanders

A meander grows outward and migrates downstream as the riverbank is eroded on the outside of the meander, forming a cut bank, while deposition occurs at the inside bank, creating a point bar. When meanders touch, the river quickly takes the shortcut, leaving a meander scar or an ox-bow lake.

bar. Sometimes the streamflow cuts off a meander loop by eroding through the narrow portion of the meander neck. After the cutoff, silt and sand are deposited across the ends of the former channel, producing an **ox-bow lake**. Eventually, the lake fills in with sediment and becomes first a swamp and then a *meander scar*. Cutoffs are sometimes human-induced, to straighten the stream channel and make it more convenient for transport.

Over time, the erosion and deposition occurring as river meanders grow and move downstream creates a broad, flat floodplain of alluvium. The floodplain is normally flooded each year or two, when streamflow increases and the river leaves its banks. As the velocity of the water flowing away from the bank and across the floodplain decreases, the sediment begins to settle out. Sand and silt accumulate first, building up **natural levees** along the channel (Figure 15.17). Farther away, fine sediment settles out of the nearly stagnant water, accumulating between the levees and the *bluffs* that bound the floodplain—an area known as the *backswamp*.

> A graded river, with its low gradient and broad floodplain, creates characteristic landforms, including bluffs, meanders, cutoffs, ox-bow lakes, and natural levees.

Periodic flooding not only adds silt, clay, and organic matter to the floodplain soils but also infuses the floodplain with dissolved minerals. The resupply of nutrients keeps floodplain soils fertile and productive; the floodplains of some alluvial rivers—such as the Nile,

15.17 Floodplains

Meandering streams leave their banks every few years, flooding the surrounding floodplain.

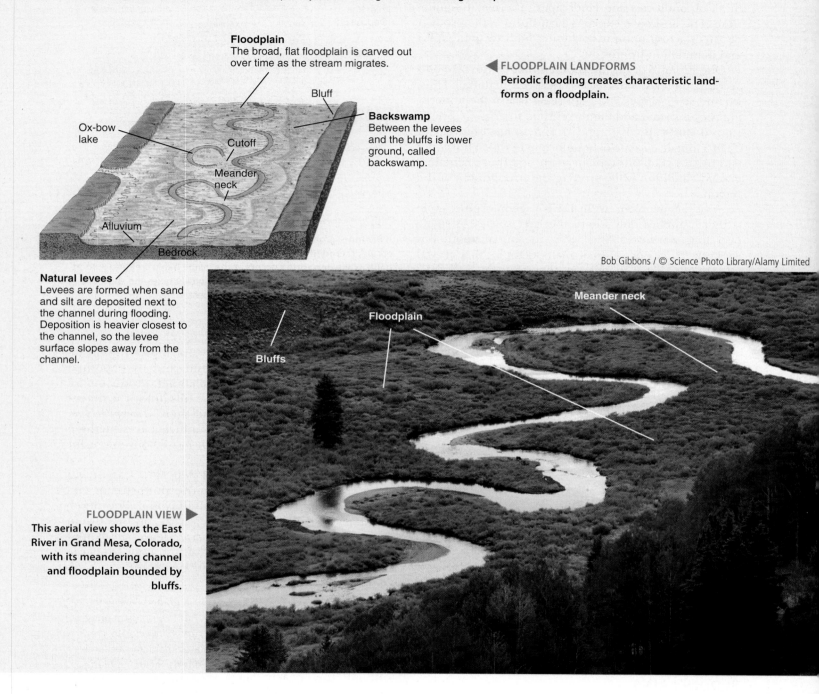

Floodplain
The broad, flat floodplain is carved out over time as the stream migrates.

Bluff

◀ **FLOODPLAIN LANDFORMS**
Periodic flooding creates characteristic land-forms on a floodplain.

Ox-bow lake

Cutoff

Backswamp
Between the levees and the bluffs is lower ground, called backswamp.

Meander neck

Alluvium

Bedrock

Natural levees
Levees are formed when sand and silt are deposited next to the channel during flooding. Deposition is heavier closest to the channel, so the levee surface slopes away from the channel.

Bob Gibbons / © Science Photo Library/Alamy Limited

Meander neck

Floodplain

Bluffs

◀ FLOODPLAIN VIEW ▶
This aerial view shows the East River in Grand Mesa, Colorado, with its meandering channel and floodplain bounded by bluffs.

Tigris, Euphrates, and Yellow—have been farmed for thousands of years.

BRAIDED STREAMS

In streams carrying a heavy sediment load, including a large proportion of bed load, the channel can take a braided form, with many individual threads of flow separated by shallow bars of sand or gravel. We refer to a stream with this type of channel as a **braided stream** (Figure 15.18). Braided channels are more likely to form in streams with great fluctuations in discharge. When the river is high, bed load is set in motion. Bars are eroded, and bed load moves downstream. As the

Braided channels are often found where fluvial sediments are accumulating and floodplains are being built up. This process, referred to as **aggradation**, can occur in a number of ways. When a stream receives a large input of sediment—too much to handle, given its current capacity—it will accumulate sediment in the floodplain. Streams that drain glaciers, which receive abundant coarse sediment coupled with high flows during the season of icemelt, are typical examples. Another cause of aggradation is change of base level. When base level rises, it reduces the slope of the river, also lowering the river's capacity to transport sediment. As a result, sediment accumulates in the floodplain.

> Braided streams develop when an aggrading stream carries a large volume of bed load and has seasonal fluctuations in discharge.

DELTAS

Where a stream enters a body of standing water, such as a lake or an ocean, it first drops its bed load, forming a bar across the mouth of the river. Coarse suspended sediment, carried near the channel bottom, then begins to settle out. The flow commonly breaks into several distributaries as it crosses the bar and the newly deposited sediment. The flow of water and fine sediment is transported into the open water, where it maintains its velocity for a short distance. Eventually, the flow mixes into the water body, and fine sediment settles out as well. If the freshwater flow mixes with saltwater, the salts cause even the finest clay particles to clot together and fall to the seafloor. The deposit built by this process is known as a **delta**.

Deltas can grow rapidly, at rates ranging from 3 m (about 10 ft) per year for the Nile River to 60 m (about 200 ft) per year for the Mississippi. Some cities and towns that were at river mouths several hundred years ago are today several kilometers inland.

Human activities have recently been linked to degradation of delta environments (Figure 15.19). In many locations, delta land is sinking due to compaction of sediment and removal of water, oil, and gas. Agricultural activities accelerate loss of land with soil runoff. Upstream river water diversion projects remove from the river sediment that would otherwise replenish the delta landscape. Millions of people living in sinking deltas will need to develop adaptations as climate change heightens the threat of flooding from storms and a rising sea level.

> Deltas are found where rivers carry sediment into lake or ocean basins. As the streamflow enters the water body, its velocity slows and it drops its load.

Frans Lanting/NG Image Collection

15.18 Braided stream

The braided channel of the Chitina River, Wrangell Mountains, Alaska, shows many channels separating and converging on a floodplain filled with coarse rock debris left by a modern valley glacier at the head of the valley.

water level falls and velocity decreases, new sand and gravel bars emerge. The low flow following the flood cannot move the bars, so the water travels in channels around them.

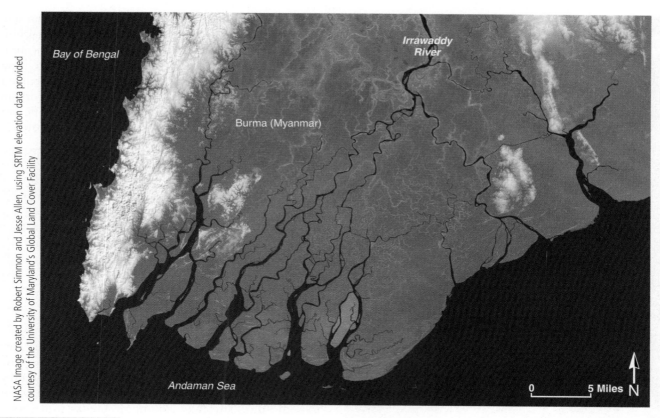

15.19 Irrawaddy Delta

Deltas form where a stream or river enters a lake or the ocean. This image of the delta of the Irrawaddy River of Myanmar was created from an elevation database acquired by NASA's Shuttle Radar Topography Mission. The lowest elevations are in tones of gray and green; much of the delta is below 5 m (16 ft). Higher elevations are shown in white (not a true color) and range up to about 2000 m (about 6600 ft). When compared to historic topographic maps, the radar data show that the Irrawaddy River Delta is sinking as a result of river diversions from upriver dams and irrigation. Erosion from farming and development is also cited as a cause of land loss in the delta, along with subsidence from oil, gas, and water drilling.

Fluvial Processes in an Arid Climate

Desert regions look strikingly different from humid regions in both vegetation and landforms. Obviously, the lower precipitation makes the difference. Vegetation in deserts is sparse or absent, and land surfaces are mantled with mineral material—sand, gravel, rock fragments, or bedrock itself.

Although deserts have low precipitation, rain falls in dry climates as well as in moist, and most landforms of desert regions are shaped by running water. A particular locality in a dry desert may experience heavy rain only once in several years, but when it does fall, stream channels carry water and perform important work as agents of erosion, transportation, and deposition.

Fluvial processes are especially effective in shaping desert landforms because of the sparse vegetation cover. The few small plants that survive offer little or no protection to soil or bedrock. Without a thick vegetative cover to shield the ground and hold back the swift downslope flow of water, large quantities of coarse rock debris are swept into the streams. A dry channel is transformed in a few minutes into a *flash flood* of muddy water heavily charged with rock fragments (Figure 15.20).

When flat-lying rocks are eroded in an arid climate—for example in the American southwest—successive rock layers are stripped or washed away, leaving behind a broad platform, or *plateau* (Figure 15.21). Such plateaus may be capped by hard rock layers, creating a sheer rock wall, or cliff, at the edge of the resistant rock layer. At the base of the cliff is an inclined slope, which flattens out into a plain beyond. As the weak clay or shale formations exposed at the cliff base are

> Although rain falls infrequently in desert environments, running water shapes desert landforms with great effectiveness because of the lack of vegetation cover.

John Cancalosi/Peter Arnold/Getty Images, Inc.

15.20 Flash flood
A flash flood has filled this river channel in the
Sonoran Desert of Arizona with raging, turbid waters.
A distant thunderstorm produced the runoff.

15.21 Landforms of flat-lying rocks in arid climates
In arid climates where there is little vegetative cover, mass wasting and overland flow carve out distinctive landforms.

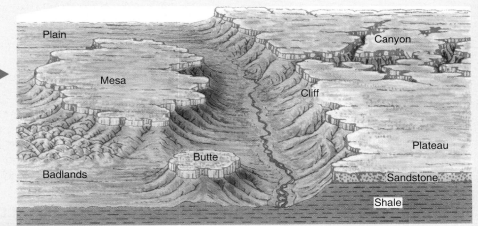

TYPICAL LANDFORMS ▶
Erosion by wind and water creates plateaus,
mesas, and buttes from resistant rock layers.

Raymond Gehman/NG Image Collection

◀ GRAND CANYON
Arizona's Grand Canyon was carved by the
Colorado River and its tributaries into a plateau
of flat-lying sedimentary rocks.

Badlands form where clay is exposed to fluvial erosion. These regions are known as badlands precisely because of the difficult terrain they presented to early travelers. The Native American Lakota tribe called them *mako sica*—literally, "bad lands"—**and French trappers called them** l*es mauvaises terres à traverser*, **or "the bad lands to cross."**

Christian Kober/Robert Harding/Getty Images, Inc.

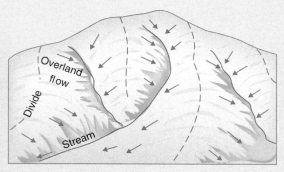

▲ **GULLY FORMATION**
Overland flow carries the fine clay particles downslope and into the channel network. Channel flow erodes the fine material rapidly, deepening the channels.

◄ **BADLANDS IN DEATH VALLEY**
These badlands at Zabriskie Point, Death Valley National Monument, California, are a classic example of these regions.

washed away, the rock in the upper cliff face repeatedly breaks away along vertical fractures. Cliff retreat produces a *mesa*, a table-topped plateau bordered on all sides by cliffs. As a mesa shrinks in area by retreat of the rimming cliffs, it maintains its flat top. Eventually, it becomes a small, steep-sided hill known as a *butte*. Further erosion may produce a single, tall column, called a *pinnacle*, before the landform is totally consumed.

Where clay is exposed at the surface, erosion is very rapid, and unstable slopes are soon dissected into *badlands* (Figure 15.22). Erosion occurs too rapidly for plants to take hold, so no soil develops. A maze of small stream channels with steep slopes results.

Many deserts have closed lowland basins of broad, arid plains. Often we find salt-rich lakes or former lake deposits in these basins. Streams heading in adjacent hills or mountains drain into these depressions, and when they are active, they contribute sediment, water, and dissolved salts to the valley bottom. But the water soon evaporates in the arid climate, leaving behind the salt and sediment. The result is

a very flat surface called a **playa** (Figure 15.23). In the southwestern United States, many playa deposits are the residue of large lakes that developed during cooler, wetter climate periods but are now completely evaporated.

ALLUVIAL FANS

Another very common arid climate landform is the **alluvial fan** (Figure 15.24). This is a low cone of alluvial sands and gravels resembling an open fan. Alluvial fans are built by streams carrying heavy loads of coarse rock waste from a mountain or an upland region. Constrained by the hard rock of the mountain, the stream flows in a narrow gorge until it emerges on the plain. The channel abruptly widens, and its capability to carry sediment

> Alluvial fans are common features of arid landscapes. They occur where streams discharge water and sediment from a narrow canyon or gorge onto an adjacent plain.

Racetrack Playa, a flat, white plain, is surrounded by rugged mountains. This desert valley lies in the northern part of the Panamint Range, west of Death Valley, California.

Nearby uplands are rugged mountain masses dissected into canyons with sleep, rocky walls.

This great uplifted fault block is the eastern face of the Inyo Mountains.

The playa occuples the floor of the central part of the basin.

A zone of coalescing alluvial fans lines the mountain front.

rapidly decreases due to loss of depth, infiltration of water into the fan sediments, and evaporation. The excess bed load settles in the channel, and the stream breaks into a few smaller channels, called *distributaries*, that spread the water and sediment over the fan surface. Alluvial fans are also found in other climates where a constrained stream emerges onto a lowland plain or the floodplain of a major river.

Alluvial fans are primary sites of groundwater reservoirs in the southwestern United States. But in many fan areas, the water table is being rapidly lowered by pumping for irrigation. In such areas, it will take an extremely long time to recharge the groundwater reserves from precipitation. Recall from Chapter 14 that one serious side effect of removing too much groundwater is subsidence.

15.24 Alluvial fans

Alluvial fans are found where streams carrying abundant rock waste emerge from canyons into a lowland.

◀ **FAN FORMATION**

The apex of an alluvial fan is firmly fixed at the canyon mouth. The streamflow breaks up into distributaries that deposit sediment, taking shape as a cone.

▼ **FAN IN DEATH VALLEY**

This large alluvial fan extends out onto the floor of Death Valley. It was built by a stream carrying rock waste from the interior of a great uplifted fault block, the Panamint Range, located to the west of the valley.

Walter Meayers Edwards/NG Image Collection

MOUNTAINOUS DESERTS

Where tectonic activity has recently produced block faulting in an area of continental desert, the assemblage of fluvial landforms is particularly diverse and interesting. The basin-and-range region of the western United States, which encompasses large parts of Nevada and Utah, southeastern California, southern Arizona, and New Mexico, is an example.

Figure 15.25 illustrates some landscape features of these mountainous deserts. The initial uplift creates two uplifted fault blocks with a downdropped block between them. Although denudation acts on the uplifted blocks as they are being raised, they are shown in the image as very little modified at the time tectonic activity has ceased. At first, the faces of the fault block are extremely steep; they are scored with deep ravines, and cones take shape at the base of talus blocks.

At a later stage of erosion, streams have carved the mountain blocks into a rugged landscape of deep canyons and high divides. Rock waste from these steep mountain slopes is carried from the mouths of canyons, where it forms large alluvial fans. The fan deposits assemble into a continuous apron that extends far out into the basins. In the central portion of the downdropped block is a playa, filling with fine sediments and precipitated salts. As erosion continues, a gently sloping rock surface, thinly veneered with alluvium, stretches from the mountain flank toward the playa. This surface is known as a *pediment*.

> Landforms of mountainous deserts include alluvial fans, dry lakes or playas, and pediments—rock platforms veneered with alluvium.

WileyPLUS Web Quiz

Take a quick quiz on the key concepts of this chapter.

15.25 Mountainous desert landforms

A desert of interleaved mountains and basins erodes to form fans, playas, and pediments.

Talus cone
Fault depression

▲ **INITIAL UPLIFT**
Shown are two blocks, with a downdropped valley between them. Talus accumulates along the mountain front.

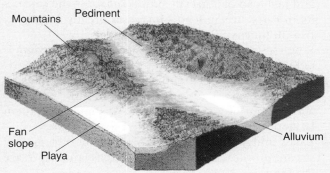

Mountains Pediment
Fan slope
Playa
Alluvium

▲ **BASIN FILLING**
Debris from the high, rugged mountain blocks fills the tectonic basins.

CONTINUED EROSION ▶
Eventually, the landscape is reduced to mountain remnants flanked by fans and pediments that extend to wide playas.

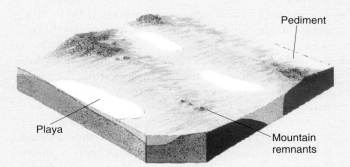

Pediment
Playa
Mountain remnants

A Look Ahead

In this chapter we have examined a number of geomorphic processes by which running water erodes, transports, and deposits sediment to create landforms. Because the landscapes of most regions on the Earth's land surface are the result of fluvial processes acting on differing rock types, running water is by far the most important agent in producing the variety of landforms we see around us. But it is not the only one; three agents of denudation remain: waves, wind, and glacial ice. These are the subjects of the last two chapters of our book.

WileyPLUS Web Links
Visit this chapter's web sites to view awesome floods and dramatic waterfalls. Tour the spectacular Grand Canyon and the scenic Hudson River Valley.

IN REVIEW LANDFORMS MADE BY RUNNING WATER

■ **Erosional landforms** are shaped by the progressive removal of bedrock. **Depositional landforms** are built from the soil, regolith, and rock fragments transported by erosion.

■ *Soil erosion* moves soil particles toward and into streams through *splash erosion* and *overland flow.*

■ *Sheet erosion* on gentler slopes and *rill erosion* on steeper slopes carry soil and regolith downhill. Eventually, *gullies* can form.

■ *Fluvial landforms* are shaped by overland flow and streamflow. Running water erodes mountains and

hills, carves valleys, and deposits sediment by *fluvial processes.*

■ Sediment yield is lowest for vegetated surfaces and highest for bare soil.

■ The work of streams includes *stream erosion, stream transportation,* and *stream deposition.*

■ Where stream channels are carved into soft materials, *hydraulic action* can generate large amounts of sediment. Where stream channels flow on bedrock, channels are deepened by the *abrasion* of bed and banks by mineral particles, large and small. *Corrosion,* or acid action, can dissolve some types of bedrock in the streambed.

- **Stream load** measures the total dissolved, suspended, and bed load of a stream. *Stream capacity*—the capability to carry stream load—increases greatly as velocity rises in times of flood.

- When a stream's load exceeds its capacity, **alluvium** is deposited.

- Over time, streams tend to become **graded streams**, in which their gradients are adjusted to move the average amount of water and sediment supplied to them by slopes. This requires a graded *stream profile*.

- Lakes and waterfalls give way to a smooth, graded stream profile. Grade is maintained as landscapes are eroded toward **base level**.

- Graded streams typically exhibit **meanders** on a broad **floodplain**.

- The gradient of a stream changes abruptly at a *nickpoint*. *Rapids* and **waterfalls** are examples of nickpoints.

- The erosion of a stream's bed is limited by its base level, the elevation at which the stream enters a lake, ocean, or inland sea.

- Remote sensing provides a way of viewing the world's deepest and most spectacular river canyons. The canyons of southern Peru are more than twice as deep as Arizona's Grand Canyon.

- When a region containing a meandering graded river is uplifted, the river cuts deeply into the bedrock below its channel, forming a *canyon* with a steep-walled *inner gorge*. *Entrenched meanders* and **alluvial terraces** can result.

- The **geomorphic cycle** organizes fluvial landscapes according to their age in a sequence of uplift that erects mountains, and subsequent erosion that produces nearly flat surfaces called *peneplains*.

- In the *equilibrium approach*, landforms are viewed as products of uplift and erosion acting as continuous processes on rocks of varying resistance.

- Meandering creates *cut banks*, *point bars*, cutoffs, **ox-bow lakes**, and *meander scars*.

- Floodplain features include *natural levees*, *backswamps*, and *bluffs*.

- Large rivers with low gradients move great quantities of sediment. The meandering of these rivers forms *bluffs*, *ox-bow lakes*, *natural levees*, and *backswamps*.

- **Aggradation** in response to an increased sediment load can create a **braided stream**.

- A **delta** forms where a stream discharges into a lake, inland sea, or ocean. The stream velocity decreases, causing sediment to settle out and bed load to accumulate.

- Desert streams are subject to *flash flooding* because vegetation is largely lacking and overland flow into stream channels is rapid.

- *Plateaus*, *mesas*, *buttes*, and *pinnacles* are all landforms of flat-lying rocks in arid climates.

- In arid and semiarid regions, *badlands* can develop on clay formations that are easily eroded by overland flow.

- Fine sediments and salts, carried by streams, accumulate in desert **playas**, from which water evaporates, leaving sediment and salt behind.

- **Alluvial fans** take shape when mountain streams carrying large amounts of rock waste drain into an adjacent lowland or valley.

- Desert mountains are eventually worn down into gently sloping rock floors called *pediments*.

KEY TERMS

erosional landforms, p. 501
depositional landforms, p. 501
stream load, p. 505
alluvium, p. 505
graded stream, p. 506
meanders, p. 506
floodplain, p. 506

waterfall, p. 508
base level, p. 513
alluvial terraces, p. 513
geomorphic cycle, p. 514
fluvial landforms, p. 515
ox-bow lake, p. 515
natural levee, p. 515

braided stream, p. 516
aggradation, p. 517
delta, p. 517
playa, p. 520
alluvial fan, p. 520

REVIEW QUESTIONS

1. Contrast **erosional landforms** and **depositional landforms**. Provide several examples of each.
2. Explain how *soil erosion* occurs. What is the role of *splash erosion*? How is sediment yield affected?
3. What is *overland flow*? How is it affected by the presence of vegetation?
4. In what ways do streams erode their beds and banks? Identify and describe three processes of stream erosion.
5. What is **stream load**? Identify its three components. In what form do large rivers carry most of their load?
6. How is velocity related to the capability of a stream to move sediment downstream? Use the term *stream capacity* in your answer.
7. Contrast the two terms *colluvium* and *alluvium*. Where on a landscape would you look to find each one?
8. Compare Arizona's Grand Canyon to the canyons of southern Peru. Where is the Three Gorges, and why is it of interest?
9. What is *stream rejuvenation*, and why does it occur? Identify some fluvial landforms associated with rejuvenation.
10. Describe the evolution of a fluvial landscape according to Davis's **geomorphic cycle**. What are the limitations of this theory? How do they lead to the *equilibrium approach*?
11. What causes **aggradation** of a stream? What are the effects of aggradation?
12. What is a **delta**? How and why does it form?
13. Why is fluvial action so effective in arid climates, considering that rainfall is scarce?
14. Identify the landforms associated with flat-lying rocks produced by fluvial processes in an arid climate.
15. Describe how an **alluvial fan** is formed. What natural resource is associated with alluvial fans? Explain.
16. Describe the evolution of the landscape in a mountainous desert. Use the terms *talus cone, fan slope*, **playa**, and *pediment* in your answer.

VISUALIZING EXERCISES

1. What is a *graded stream*? Sketch the profile of a graded stream. Identify any *nickpoints* and explain why they might occur. Identify the stream's **base level** on the graph.
2. Sketch the floodplain of a **graded river**. Identify key landforms on the sketch. How do they form?

ESSAY QUESTIONS

1. A river originates high in the Rocky Mountains, crosses the high plains, flows through the agricultural regions of the Midwest, and finally reaches the sea. Describe the fluvial processes and landforms you might expect to find on a journey along the river, from its headwaters to the ocean.
2. What effects would climate change have on a fluvial system? Choose either the effects of cooler temperatures and higher precipitation in a mountainous desert, or warmer temperatures and lower precipitation in a humid agricultural region.

Chapter 16
Landforms Made by Waves and Wind

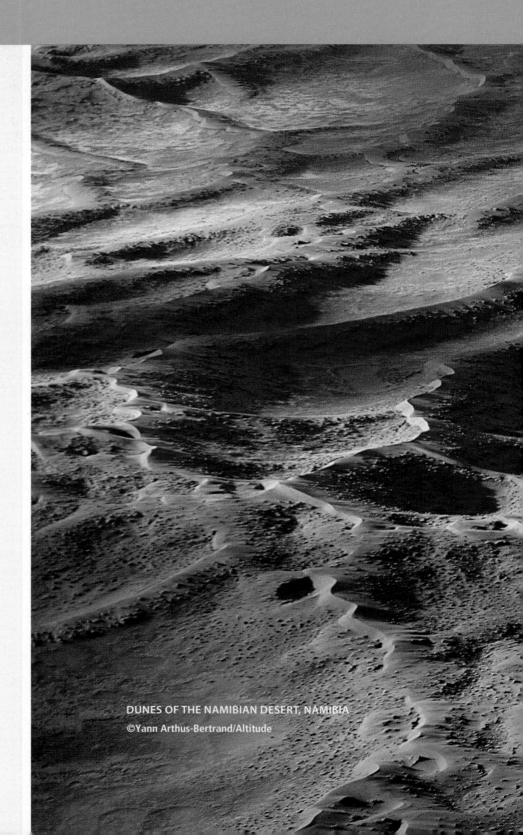

On the west coast of Africa, stretching from southern Angola to northern South Africa, is the Namibian Desert, one of the world's driest places. Here, wind has shaped a coastal band of sand seas about 80 km (about 50 mi) wide into a range of dune forms. Transverse dune crests with steep slip faces are moving from right to left, crossing some older longitudinal dune forms. In the far distance, the transverse crests grade into lines of barchan dunes. Vegetation finds a home in sheltered habitats where coastal fog provides moisture, or where deep roots can reach groundwater.

DUNES OF THE NAMIBIAN DESERT, NAMIBIA
©Yann Arthus-Bertrand/Altitude

Landforms Made by Waves and Wind

Wind creates landforms directly by moving sand and dust, and indirectly by generating waves that shape shorelines and coasts. How do waves form, and how do they move sediment along shorelines? What causes ocean tides? What is a tsunami? What distinctive landforms does wave action create? What kinds of materials does wind erode, transport, and deposit? What types of sand dunes are there, and how do they form? These are some of the questions we will answer in this chapter.

EYE ON GLOBAL CHANGE

Global Change and Coastal Environments

Global climate change over the remainder of the twenty-first century will have major impacts on coastal environments, including increases in sea-surface temperature and sea level; decreases in sea-ice cover; and changes in salinity, wave climate, and ocean circulation.

What changes have already occurred? Sea level has risen 20 cm (7.9 in.) since 1900, at a rate of 2 mm (0.08 in.) per year—that is, until the twenty-first century, when the rate climbed to 3 mm (0.12 in.) per year. Most of this rise is due to thermal expansion of seawater in response to global warming of about 0.75°C (1.35°F) since 1906, with some also due to the melting of glaciers and ice caps. Arctic sea-ice cover is decreasing at a rate of about 3 percent per decade, with summer ice cover declining by about 7 percent per decade. The incidence of extreme weather events is also escalating; there are more heat waves and more precipitation events leading to flooding; the extent of drought-affected regions is greater, as is the intensity and duration of tropical storms.

What changes are in store? Between 1990 and 2100, global average surface temperature will go up between 1.7 and 6.4°C (3.1 and 10.5°F), depending on the scenario for future economic growth and usage of fossil fuels. Sea level will rise from present levels by about 21 cm to 47 cm (8.3 to 18.5 in.), again depending on the scenario. Snow and ice cover will continue to decrease, and mountain glaciers and ice caps will carry on their retreat of the twentieth century. Tropical cyclone peak wind and precipitation intensities will increase, and El Niño extremes of flood and drought will be exaggerated.

These changes are bad news for coastal environments. Let's begin with coastal erosion (Figure 16.1). Global warming will increase the frequency of high winds and heavy precipitation events, amplifying the effects of severe coastal storms. More frequent and longer-lasting El Niños will intensify the severity of Pacific storms, leading to more rapid and damaging sea-cliff erosion along Southern California's south- and southwest-facing

Bob Jordan/©AP/Wide World Photos

16.1 Coastal erosion

The damaging effects of the waves and wind that accompanied Hurricane Dennis in 1999 are clear here on the barrier beach of North Carolina's Outer Banks. The high surf eroded the beach into a long, flat slope, undermining the supports for many beach structures. The winds then toppled them, leaving the surf to demolish the remains. The result was widespread devastation in many barrier beach communities. This homeowner was one of the lucky ones; he and his family had a house to come back to.

coastlines. During La Niña events, Atlantic hurricanes will be both more frequent and intense, and thus heighten the risk of damage to structures and coastal populations.

What will be the impact of a rising sea level on coastlines? Over the past 100 years or so, about 70 percent of sandy shorelines have retreated; 10 percent have advanced. The long-term effect of sea level rise will be to

> Rising sea level and increasing frequency and severity of storms will worsen future coastal erosion. Coastal wetlands will decrease in area and quality. Warming will stress coral reefs and speed arctic shoreline recession.

push beaches, salt marshes, and estuaries landward (Figure 16.2). Beaches will disappear and be replaced by seawalls. Salt marshes will be drained to reduce inland flooding. Estuaries will become shallower and more saline. In these ways, the most productive areas of the coast will be squeezed between a rising ocean and a water's edge that is increasingly defended.

Sea-level rise will not be uniform. Modifying factors of waves, currents, tides, and offshore topography can act to magnify the rise, depending on the location. Some models predict doubled rates of sea-level rise for portions of the eastern United States, North American Pacific coast, and the western North American arctic shoreline.

Land subsidence, too, is a contributing factor to the impact of a rising sea level. Many coastlines are fed by rivers that have now been dammed, often many times, which reduces the amount of fine sediment brought to the coast. Without new sediment, coastal wetlands slowly sink, as the older sediment that supports them compacts. This subsidence aggravates the effects of sea-level rise.

Delta coasts are especially sensitive to sediment starvation and subsidence, where rates of subsidence can reach 2 cm/yr (0.8 in./yr). The Mississippi already has lost about half of its natural sediment load; and sediment transport by such rivers as the Nile and Indus has gone down by 95 percent.

According to recent estimates, sea-level rise and subsidence could cause the loss of more than 22 percent of the world's coastal wetlands by the year 2100. Coupled with losses directly related to human activity, coastal wetlands could decrease by 30 percent or more, and have major impacts on commercially important fish and shellfish populations.

Coral reefs, like coastal wetlands, perform important ecological functions. They are highly biodiverse and serve as protective barriers to coastlines against storm waves and surges. And like coastal wetlands, they too are in jeopardy, with more than half of the total area of living coral reefs thought to be threatened by human activities coral mining. When stressed by a rise in

16.2 Beach breach

Storm waves from Hurricane Isabel breached the North Carolina barrier beach, and in the process carved out a new inlet, visible in this aerial photo from September 2003. Notice also the widespread destruction of the shoreline in the foreground, with streaks of sand carried far inland by wind and wave action. As the sea level rises, ocean waves will attack barrier beaches more frequently and more severely.

GARY O'BRIEN/KRT/NewsCom

16.3 Bleached anemones

These anemones, observed in the shallow waters of the Maldives Islands, show the effects of bleaching, a process by which coral animals expel symbiotic algae as a response to stress. The bleaching can be fatal.

Pascal Kobeh/Biosphoto

implications for ecosystems and natural resources. It will take careful management of our coastlines to minimize the risks to both human and natural systems.

The Work of Waves and Tides

This chapter brings together two different agents of erosion: wind and waves. Wind is the movement of air, a fluid of low density; waves involve the movement of water, a fluid of much higher density. This difference in fluid density means that the power of moving air to create landforms is generally much weaker than the power of moving water as waves. That said, vast desert landforms on our planet have been shaped by wind, whereas waves can act only on coastlines of oceans and large lakes. Yet the two agents are related because waves are set in motion by wind blowing across a water surface.

We will first examine how waves create landforms.

WAVES

Waves are the most influential of the agents that shape coastal landforms (Figure 16.5). When winds blow over broad expanses of water, they generate waves. Both friction between moving air and the water surface and direct wind pressure on the waves transfer energy from the atmosphere to the water.

Waves are seen and felt as a regular rising and falling of the water surface that causes a floating object to move up and down, forward and back. As shown in Figure 16.6, waves have *crests* and *troughs. Wave height* is the vertical distance between trough and crest. *Wave length* is the horizontal distance from trough to trough or from crest to crest. The wave typically travels in a forward direction

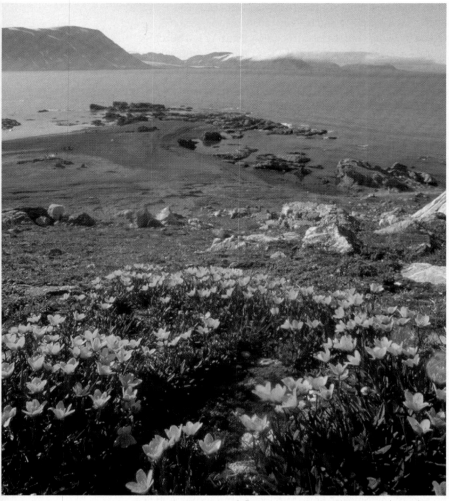

© Zoonar / Joerg Hemmer/Age Fotostock America, Inc.

16.4   Arctic shoreline

Pristine arctic shorelines, such as this beach on the coast of northern Norway, will be subjected to rapid change as the global climate warms. Thawing of ground ice will release beach sediments, and the early retreat of sheltering sea ice will expose the shoreline to fiercer attacks by summer waves.

16.5   Storm waves

High surf can rapidly reshape a beach, flattening the beach slope and moving beach sediment offshore. These large waves at Humbug Mountain State Park, Oregon, were most likely generated by an offshore storm.

Raymond Gehman/NG Image Collection

temperature, many corals respond by "bleaching" (Figure 16.3); they expel the algae that live inside their structures, stripping the coral of its color. The bleaching may be temporary if the stress subsides; but if permanent, the corals die. Major episodes of coral bleaching have been associated with higher water temperatures during strong El Niños.

Many pristine stretches of arctic shoreline are threatened by global warming (Figure 16.4). As global temperatures climb, the shoreline loses its protection from sea ice, frozen ground, and ground ice. Greater expanses of open sea make way for larger waves to attack the coast. Global warming will be especially severe at high latitudes. Rapid coastal recession has already been reported along the Beaufort Sea, in the Arctic Ocean northeast of Alaska and northwest of Canada.

It is apparent that global climate change will have major impacts on coastal environments, with very broad-ranging

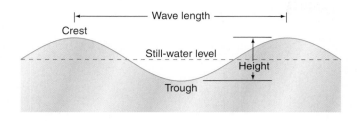

16.6 Terminology of water waves

We use the terms crest, trough, wave height, and wave length to describe water waves.

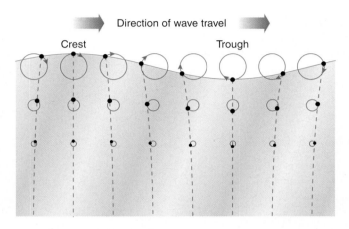

16.7 Motion in oscillatory waves

As each wave crest passes, particles move forward on the wave crest, downward as the crest passes, backward in the wave trough, and upward as the next crest approaches. At the sea surface, the orbit is of the same diameter as the wave height, but the size of the orbit decreases rapidly with depth.

as parallel fronts of crests and troughs. The *wave period* is the time in seconds between successive crests or successive troughs that pass a fixed point.

Wind-generated ocean waves are an example of *oscillatory waves*. In this type of wave, a tiny particle, such as a drop of water or a small floating object, completes one

vertical circle, or *wave orbit*, with the passage of each wave length (Figure 16.7). The circle of motion grows rapidly smaller with depth.

For lake and ocean waves pushed by the wind, particles trace an orbit in which the forward speed of each particle at the crest is slightly greater than the backward speed at the trough. As a result, there is net forward motion of water associated with most lake and ocean wave trains. The amount of motion depends on the size and steepness of the waves, and strong waves are capable of pushing large amounts of water toward the beach. This causes local sea level to rise and generates local currents of water heading back out to sea.

The height of waves is determined by wind speed, wind duration, and *fetch*—the distance that the wind blows over the water. Table 16.1 shows how these factors are interrelated. Given a sustained wind, average and peak wave heights increase with fetch and duration, until waves are fully developed.

Wave height increases rapidly with wind speed. For example, the table shows that if sustained wind speed is doubled from 25 to 50 knots (13 to 26 m/s; 29 to 58 mi/hr), average wave height increases by more than five-fold, from 2.7 m (9 ft) to 15 m (48 ft), and peak waves reach 30 m (99 ft). When you couple these values with the fact that the energy of a wave is proportional to the square of its height, you can appreciate how hurricane winds can generate waves with such enormous destructive power.

Waves retain most of their energy as they travel across the deep ocean, but when a wave reaches the shore, it begins to expend its energy. As it reaches shallow water, the drag of the bottom slows and steepens the wave. At the same time, the wave top maintains its forward velocity and eventually falls down onto the face of the wave, creating a

> Waves are driven by wind. Wave height is related to wind speed, duration, and fetch. High winds over long distances and long durations can create very large waves.

Table 16.1 Wind speed and wave characteristics

Sustained wind speed, knots (m/s, mi/hr)	Average wave height, m (ft)	Height of largest waves, m (ft)	Fetch length, km (mi)	Duration, hrs
10 (5, 12)	0.3 (0.9)	0.5 (1.8)	18 (10)	2.4
15 (8, 17)	0.8 (2.5)	1.5 (5)	63 (34)	6
20 (10, 23)	1.5 (5)	3.0 (10)	140 (75)	10
25 (13, 29)	2.7 (9)	5.5 (18)	300 (160)	16
30 (15, 35)	4.3 (14)	8.5 (28)	520 (280)	23
40 (21, 46)	8.5 (28)	17 (57)	1300 (710)	42
50 (26, 58)	15 (48)	30 (99)	2600 (1420)	69

For fully developed waves given a sustained wind speed. Wind speed is shown in knots—nautical miles per hour, where a nautical mile is one minute of longitude at the Equator. Fetch and duration are values required for full development.

16.8 Breaking waves

When waves reach shallow water, they steepen and break.

Photo Researchers/Getty Images, Inc.

◄ **BREAKERS**
A wave breaks near a rocky promontory on the coast of Oregon.

HOW A WAVE BREAKS ►
In deep water, water moves in a circular motion as the waves pass by. As a wave reaches shallower water, it begins to drag against the bottom, and the circular motions become flatter. The wave's height increases, and length decreases. Because the front of the wave is in shallower water, it is steeper and moves more slowly than the rear. Eventually, the front becomes too steep to support the advancing wave. As the rear part continues to move forward, the wave collapses, or breaks, forming turbulent surf that rushes onto shore.

Breaking wave

Wave becomes higher and steeper

Surf

Beach

Shallow water

Deep water

Turbulent water

Loops deformed by confined vertical space

Looplike motion of water

Katarina Stefanovic/Getty Images, Inc.

◄ **SWASH AND BACKWASH**
As a breaking wave hits the beach, turbulent water runs up the beach face, creating a swash. The water slows and changes direction, creating a backwash toward the next wave.

breaker (Figure 16.8). Tons of foamy turbulent water surge forward in a sheet, riding up the beach slope. This powerful *swash* moves sand and gravel on the beach landward. Once the force of the swash has been spent against the slope of the beach, the return flow, or *backwash*, pours down the beach. This undercurrent can sweep unwary bathers off their feet and carry them seaward beneath the next oncoming breaker. The backwash carries sand and gravel seaward, completing the wave cycle.

LITTORAL DRIFT

Waves breaking along the shore generate tremendous energy to move sediment along the shoreline in a process called **littoral drift**. (*Littoral* means "pertaining to a coast or shore.") Littoral drift includes two transport processes. *Beach drift* is the transport of sediment along the beach. This occurs when waves reach the shore at an angle, however slight, and the swash from the breaking waves carries sand up the beach at a similar angle. The backwash, then, is drawn downward toward the breaker zone by gravity in a straighter path. The result is that sand is carried down the beach in a series of steps, setting in motion a net movement of sand away from the angle of wave attack (Figure 16.9).

Just offshore, the net forward motion of breaking waves pushes water into the breaker zone. The water is propelled along the shoreline, away from the direction

16.9 Littoral drift

Littoral drift moves sand parallel to the beach as beach drift and longshore drift.

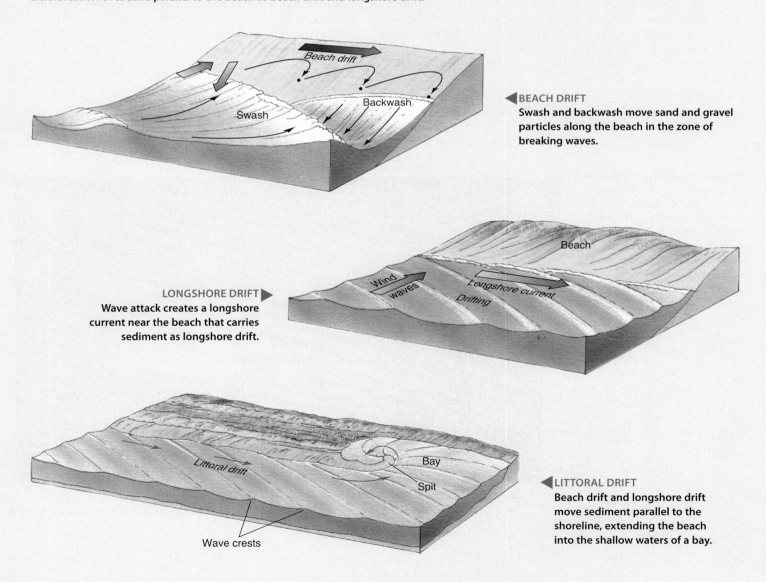

BEACH DRIFT
Swash and backwash move sand and gravel particles along the beach in the zone of breaking waves.

LONGSHORE DRIFT
Wave attack creates a longshore current near the beach that carries sediment as longshore drift.

LITTORAL DRIFT
Beach drift and longshore drift move sediment parallel to the shoreline, extending the beach into the shallow waters of a bay.

of wave attack. This produces a longshore current, which moves sand along the bottom as *longshore drift*, the second component of littoral drift. If the wave direction remains more or less the same for long periods, littoral drift can transport sediment for long distances along the coast.

> Breakers attacking the shore at an angle produce littoral drift, which includes movement of sediment along the beach, called beach drift, and just offshore, called longshore drift.

WileyPLUS The Work of Waves
Watch this video to see examples of breaking waves and to review the processes by which waves transport sediment.

WAVE REFRACTION

The process by which waves erode sediment along a shore depends on two factors: the amount of energy the waves have and the resistance of the shore materials. Where waves strike a rocky coastline, areas of softer rock are eroded more quickly, carving out bays and coves and leaving behind jutting landforms of resistant rock, called *headlands*.

When the coastline has prominent headlands that project seaward, and deep bays in between, approaching wave fronts slow when the water becomes shallow in front of the headlands. This slowing effect causes the wave front to wrap around the headland, in a process called *wave refraction* (Figure 16.10). This concentrates wave energy on the headland, enhancing erosion. The angle of the waves against the side of the headland also initiates a longshore current that moves sediment from the headland into the surrounding bays, creating crescent-shaped *pocket beaches* in the bays. Over time, the action of waves against a coast tends to have the effect of straightening the coast as it erodes sediment from headlands and deposits it into bays.

TIDES

Most marine coastlines are also influenced by the ocean tide, the rhythmic rise and fall of sea level under the influence of changing attractive forces of the Moon and Sun on the rotating Earth. In the tidal system, the Earth and Moon are coupled together by their mutual gravitational

16.10 Wave refraction

As a wave enters shallow water, its velocity decreases, causing it to become shorter and steeper.

▼ HOW WAVE REFRACTION FOCUSES
WAVE ENERGY
When a wave approaches a coastline of cliffs, its energy is concentrated at the headlands. Sediment is eroded from cliffs on the headland and carried by littoral drift along the sides of the bay. The sand is deposited at the head of the bay, forming a pocket beach.

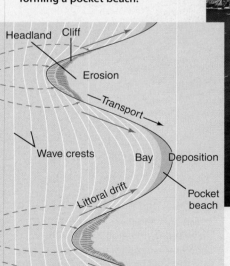

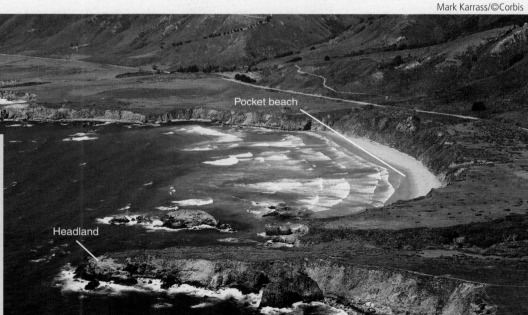

Mark Karrass/©Corbis

▲ HEADLAND AND POCKET BEACH
Steep waves attack the headland in a narrow zone, while gentler, longer waves roll into the pocket beach.

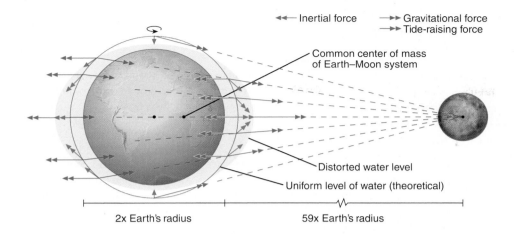

Inertial force → Gravitational force → Tide-raising force

Common center of mass of Earth–Moon system

Distorted water level

Uniform level of water (theoretical)

2x Earth's radius   59x Earth's radius

The balance between the Moon's gravitational attraction and the opposing inertial force generated by the revolution of the Earth and Moon around a common center of mass causes the Earth's oceans to bulge on the sides nearest to and farthest from the Moon. As the Earth rotates through these bulges, local areas experience two daily high tides.

attraction, which is balanced by an inertial force generated by their revolution around a common center of mass (Figure 16.11). While the inertial force is constant at all points on the globe, the gravitational attraction of the Moon is greater on the near side of the Earth; seawater responds to this attraction, causing the ocean to "bulge" toward the Moon. On the far side of the globe, the gravitational force of the Moon is weaker; but because the inertial force is the same, the ocean water is pushed away from it, and this causes a second bulge to emerge on the far side. The bulges remain essentially stationary while Earth rotates through them, creating two high tides and two low tides per day.

Ocean tides are created by the gravitational attraction of the Moon for ocean waters, as well as the rotation of the coupled Earth-Moon system around its common center of mass.

The Sun affects tides in a similar way, but its tide-producing force is only about half as strong as the Moon's. When the Sun and Moon are aligned on opposite sides of the Earth, their gravitational forces combine to produce tides with a higher range, called *spring tides*. In contrast, when the Moon and Sun are positioned at a right angle to the Earth, tides have a lower range and are called *neap tides*.

Some marine organisms time their egg laying on the beach to coincide with high spring tides, so that the eggs can incubate without subsequent exposure to wave action. Hatchlings emerge at the next spring tide, about 14 days later, and are washed out to sea. The survival of these species is dependent on close synchronization with these extreme tides.

Each day, coastal regions experience two high tides, called *flood tides*, and two low tides, called *ebb tides*; collectively, they are referred to as *semidiurnal tides*, meaning occurring approximately every half day. There are, however, two exceptions, due to unique orbital geometry: the Gulf of Mexico and South China Sea have only one high tide and one low tide each day, called *diurnal tides*. The difference between the heights of successive high and low waters is known as the *tidal range* (Figure 16.12).

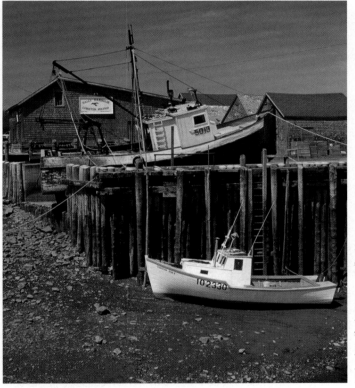

Michael Medford/NG Image Collection

The Bay of Fundy in Nova Scotia, Canada, is known for its exceptional tidal range, with a record of 17 m (55.8 ft) between high water and low water. The shape of the bay makes it resonate naturally with the rise and fall of the tide, reinforcing the tidal flood and ebb. Boats like the ones pictured here must have extra length in their mooring lines to account for the full tidal range twice each day. The boat shown in the foreground will be lifted to the height of the dock by the high tide.

If the tidal range is wide, the changing water level can play an important role in shaping coastal landforms.

Tides contribute to the erosion, transport, and deposition of sediment by ocean waters. As tides ebb and flood, ocean waters flow into and out of bays. Tidal currents pouring through narrow inlets that connect bays with the ocean are very swift and pick up sediment as they travel, eroding the inlets. This keeps the inlet open, despite littoral drift that would otherwise block the inlet with sand. Tidal currents also carry large

> Ocean tides produce tidal currents at the shoreline. These currents scour inlets and distribute fine sediment in bays and estuaries.

amounts of fine silt and clay in suspension. Some of this sediment is deposited in bays. Over time, this sediment

settles to the floors of the bays and then builds up in layers, gradually filling in the bays.

WileyPLUS Tides
Watch this animation to see how the Moon's gravity and the rotation of the Earth and Moon around their common center of mass give rise to the ocean's tides.

TSUNAMIS

Although most waves are generated by wind, waves can also be set in motion by any sudden displacement of large amounts of water, such as by an earthquake or an undersea landslide. These waves are called **tsunamis**, or seismic sea waves. (Seismic sea waves are also sometimes called tidal waves, but they are unrelated to tides.)

A tsunami originates when the sudden movement of the seafloor—resulting from an earthquake, for example—generates a train of water waves. These waves travel over the ocean at 700 to 800 km/hr (435 to 500 mi/hr), moving outward in all directions from their source. In deep ocean waters, the tsunami's motion is normally too gentle to be noticed, making it hard to detect in the open ocean. But as the wave approaches land, it is subject to the same physics as wind-driven waves; that is, as the wave drags along the bottom, and slows, it shortens and steepens to a height of 15 m (50 ft) or higher.

Ocean waters rush landward and surge far inland, destroying coastal structures and killing inhabitants as they pass, at speeds of up to 15 m/s (34 mi/hr) for several minutes. Considering that each cubic meter of water weighs about 1000 kilograms (about 2200 lbs), it is easy to understand the tremendous power of such a surge of water. Even as the waters retreat, they continue to devastate the land, pulling people and debris back out to sea with them. Tsunamis typically consist of several surging waves, one after the other—and the largest wave is not necessarily the first to arrive.

With their tremendous force, tsunamis can dramatically reshape coastal areas and inflict damage in regions on both sides of the ocean. The deadliest tsunami recorded so far struck the Indian Ocean region in December 2004, following a massive undersea earthquake—measuring 9.0 on the Richter scale—in the Java Trench, west of Sumatra. About 1000 km (600 mi) of fault ruptured, rocketing the nearby seafloor upward about 5 m (16 ft). This rapid explosion of a vast area of ocean bottom launched a giant tsunami that devastated populations in Southeast Asia, India, and Africa, reshaping coastal landscapes, wiping out cities, and killing more than 265,000 people (Figure 16.13).

> A tsunami is a gigantic ocean wave caused by an earthquake or volcanic explosion. The Indian Ocean tsunami of 2004 killed more than 200,000 people and laid waste to coastlines from Indonesia to Tanzania.

16.13 Tsunami destruction

Following the Indian Ocean tsunami of 2004, Indonesia experienced the worst devastation of any country affected.

DigitalGlobe

▲ **BEFORE THE TSUNAMI**
Banda Aceh, Indonesia, seen from space on June 23, 2004, was a city of small buildings, parks, and trees before the tsunami hit.

▼ **AFTER THE TSUNAMI**
Two days later, the devastation caused by the earthquake and tsunami was complete. All that remained of most structures was their concrete floors. Trees were uprooted or stripped of their leaves and branches.

DigitalGlobe

Where early-warning systems have been installed, tsunamis can usually be predicted far enough in advance to evacuate local populations. By monitoring seismic activity on the seafloor, it is possible to trigger tsunami warnings along coastlines that might be at risk. However, if an earthquake occurs very near the shore, or if local warning systems are not effective, the toll taken by tsunamis can be incalculable.

One such event took place on March 11, 2011, in Japan. On that date, the Pacific Tsunami Warning Center in Hawaii detected an undersea earthquake of magnitude 9.0, located 70 km (43 mi) off the east coast of Tohoku, the northeastern portion of Honshu, Japan's largest island. The Great Tohoku earthquake, the largest ever recorded in that country, occurred along a 300-km (186-mi) subduction zone at a depth of 32 km (20 mi) under the ocean floor, with its epicenter 70 km (43 mi) east of the coastline. And in spite of Japan's technologically advanced warning systems and regularly scheduled emergency training program, nothing could stave off entirely the onslaught of destruction set in motion by the earthquake and the huge tsunami it triggered— these early-warning systems did save tens of thousands of lives minutes before the 9.2 m (30 ft) wave came ashore (Figure 16.14). Unfortunately, worse was yet to come when nuclear reactors built along the coastline were flooded, and eventually experienced partial core meltdowns that released harmful radioactivity. The final death toll of the Great Tohoku earthquake and tsunami exceeded 15,000.

Coastal Landforms

As waves break upon the shore, they expend tremendous amounts of energy. This energy, along with the coastal currents it produces, shapes coastlines by eroding steep cliffs, carving out bays, and building beaches. Two terms are important to distinguish in describing coastal landforms and processes: **Shoreline** refers to the dynamic

The Asahi Shimbun/Getty Images, Inc.

16.14 Japan tsunami of 2011

Following Japan's largest underwater earthquake, tsunami warnings gave hundreds of thousands only a few minutes to flee coastal towns for higher ground along the Tohoku coastline of Japan. The disaster resulted in over $300 billion in damage and 15,000 deaths; 8000 went missing.

© 2012 Alex S. MacLean/Landslides

16.15 Retreating shoreline

This marine scarp of Ice Age sediments at Mohegan Bluffs, Block Island, Rhode Island, is undergoing rapid erosion from wave action. This historic photo shows the Southeast Lighthouse being threatened by the cliff retreat; the lighthouse has now been moved to a safer spot, about 75 m (about 250 ft) from its previous location.

EYE ON THE LANDSCAPE What else would the geographer see? The sediments exposed in the bluff were laid down by streams fed by melting stagnant continental ice sheets at the end of the Ice Age. Because the ice melted irregularly, the deposits are not very uniform, and the exposed bedding **A** gives that impression. The larger rocks in the deposits, let down from melting ice, were too large to be moved by the streams laying the sediments. A large lag deposit **B** of these stones remains at the water's edge, where it is being worked by waves.

zone of contact between water and land. **Coastline** (or **coast**) refers to the zone of shallow water and nearby land that fringes the shoreline; it is the zone in which coastal processes shape landforms.

The Earth's coastlines are widely varied. For example, along most of the East Coast of the United States, we find a coastal plain gently sloping toward the sea, with shallow lagoons and barrier islands at the coastline. On the West Coast, in contrast, we often find coastlines of rocky shores, with dramatic sea cliffs, headlands, and pocket beaches.

To explain these differences we look to plate tectonics. The eastern coastline is a passive continental margin, without tectonic activity. Abundant sediment from rivers on the continent, accumulating over millions of years, provided the material to build the coastal plain, beaches, and islands. Here, the characteristic coastal landforms are depositional—built by sediment moved by waves and currents. In contrast, the western coastline is the site of great tectonic activity, with rocks rising from the sea along subduction boundaries and transform faults. Here, the characteristic landforms are erosional, with waves and weathering slowly eroding the strong rock exposed at the shoreline.

Coastlines around the world are likewise varied, but for other reasons. During each glaciation of the recent Ice Age, sea level fell by as much as 125 m (410 ft). Rivers cut canyons to depths well below present sea level to reach the ocean. Waves broke against coastlines that are now well underwater. Now that sea level has risen, many coastal landscapes are coastlines of submergence, where fluvial landforms are now under attack by wave action. Still others are dominated by coral reefs, where ocean waters are warm enough for coral to grow.

EROSIONAL COASTAL LANDFORMS

The breaking of waves against a shoreline yields a variety of distinctive features. If the coastline is made up of weak or soft materials—various kinds of regolith, such as alluvium—the force of the forward-moving water alone easily cuts into the coastline. Here, erosion is rapid, and the shoreline may recede rapidly. Under these conditions, a steep bank, or *marine scarp*, will form and steadily erode as it is attacked by storm waves (Figure 16.15).

Where resistant rocks meet the waves, **sea cliffs** often occur. At the base of a sea cliff is a *notch*, carved largely by physical weathering. Constant splashing by waves followed by evaporation causes salt crystals to grow in tiny crevices and fissures of the rock, breaking it apart, grain by grain. Hydraulic pressure of waves, and abrasion by rock fragments thrust against the cliff, also chisel the notch. Undercut by the notch, blocks fall from the cliff face into the surf zone. As the cliff erodes, the shoreline gradually retreats shoreward. Sea cliff erosion results in a variety of erosional landforms, including *sea caves*, *sea arches*, and *sea stacks* (Figure 16.16).

16.16 Erosion of a marine cliff

Erosion of a marine cliff produces landforms such as shore platforms and sea stacks.

Chad Ehlers/Alamy Limited

Arthur N. Strahler

▲ SHORE PLATFORM

As the sea cliff at Montana d'Oro State Park, California, retreats landward, a shore platform is left behind. Any beach present in such areas is little more than a thin layer of gravel and cobblestones atop the shore platform.

◀ SEA STACKS

These sea stacks at Port Campbell National Park, Victoria, Australia, have been dubbed the "Twelve Apostles." Remnants of a collapsed arch appear between the nearest stack and the headland.

▼ EROSION OF THE CLIFF

Weathering, wave action, and gravity erode a marine cliff or headland, forming notches, sea caves, shore platforms, arches, and sea stacks.

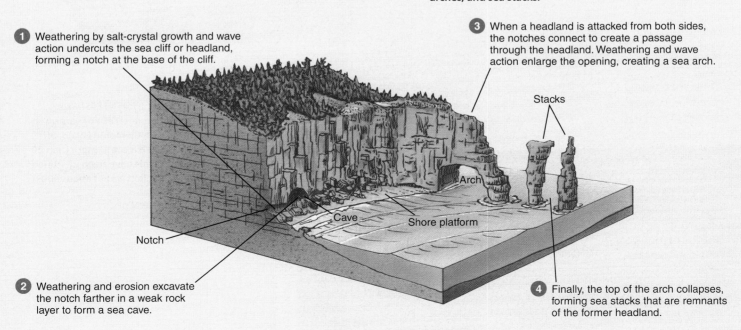

1 Weathering by salt-crystal growth and wave action undercuts the sea cliff or headland, forming a notch at the base of the cliff.

2 Weathering and erosion excavate the notch farther in a weak rock layer to form a sea cave.

3 When a headland is attacked from both sides, the notches connect to create a passage through the headland. Weathering and wave action enlarge the opening, creating a sea arch.

4 Finally, the top of the arch collapses, forming sea stacks that are remnants of the former headland.

Notch

Cave

Shore platform

Arch

Stacks

The retreat of the cliffs leaves behind a broad, gently sloping plane, called a *shore platform*, at the base of the cliffs. If tectonic activity elevates the coastline, or if sea level falls, the platform is abruptly lifted above the level of wave attack. What was a shore platform is raised up and becomes a **marine terrace**. Repeated uplifts produce a series of marine terraces in a steplike arrangement (Figure 16.17). Marine terraces are common along the continental and island coasts of the Pacific Ocean, where tectonic processes are active along the mountain and island arcs.

> Waves erode weak materials, resulting in marine scarps, and weather resistant rocks, which are reshaped as marine cliffs. Caves, arches, stacks, and abrasion platforms are landforms of marine cliffs.

DEPOSITIONAL COASTAL LANDFORMS

Most of the sediment we find along a coastline is provided by rivers that reach the ocean. Waves then transport and deposit this sediment to take shape as shoreline features such as beaches, bars, and spits. These depositional landforms are relatively transitory, however, appearing, disappearing, or migrating as a result of seasonal changes, storms, and human engineering.

Beaches

A **beach** is a wedge-shaped sedimentary deposit, usually of sand, built and worked by wave action. The face of a beach varies over time as waves either deposit or erode more sand. During short periods of storm activity, waves cut back the beach, giving it a long, flat, sloping profile. The sand moves just offshore and along the shore via longshore drift. Gentler waves return the sand to the beach, building a steeper beach face and a bench of sand at the top of it.

When sand leaves a section of beach more rapidly than it is brought in, the beach is narrowed and the shoreline moves landward. Conversely, when sand arrives at a particular section of the beach more rapidly than it is carried away, the beach is widened and

16.17 Marine terraces

Marine terraces give us a visual record of coastlines and sea levels from the Earth's past.

UPLIFT FORMS A TERRACE ▶
After a period of tectonic uplift or lowered sea level, the shore platform at the base of the cliffs is raised up to become a marine terrace.

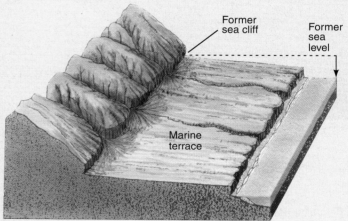

Former sea cliff

Former sea level

Marine terrace

◀ TERRACES AT SAN CLEMENTE ISLAND
This series of marine terraces appears on the western slope of San Clemente Island, off the Southern California coast. More than 20 different terrace levels have been identified; the highest has an elevation of about 400 m (about 1300 ft).

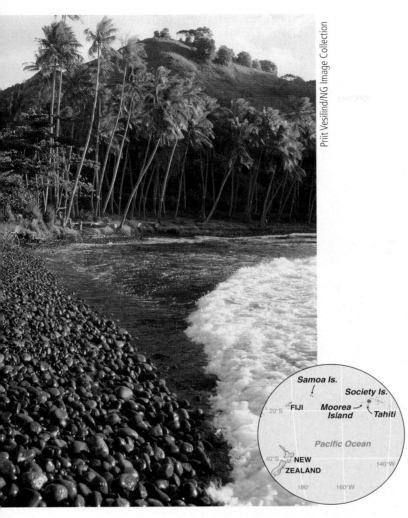

Priit Vesilind/NG Image Collection

16.18 Cobble beach

Not every beach is made of soft sand. Where sand is scarce but rocks are abundant, beaches can be composed of rounded cobbles, like these on Tahiti, Society Islands, Polynesia.

builds oceanward. During the winter, many midlatitude beaches experience more active waves and are eroded to narrow strips. This change is called *retrogradation* (cutting back). In summer, the gentler wave climate allows sediment to accumulate, and so the beach is replenished. This change is called *progradation* (building out). Retrogradation and progradation can also happen on longer time cycles, related to changes in sediment input, climate, or human activity in the coastal zone.

Although most beaches are made from particles of fine to coarse quartz sand, others are built from rounded pebbles or cobbles (Figure 16.18). Still others are composed of fragments of volcanic rock, or even shells.

Coastal Dunes

Where ample sand is available, a narrow belt of dunes, called *foredunes*, often occurs in the region landward of beaches. These dunes are usually held in place by a cover of beach grass (Figure 16.19). Although these dunes are typically built up by wind, they play an important role in maintaining a stable coastline by trapping sand blown landward from the adjacent beach.

> Coastal foredunes form a protective barrier against storm wave action, preventing waves from overwashing a beach ridge or barrier island.

As sand from the beach collects along the foredunes, the dune ridge grows upward, becoming a barrier several meters above high-tide level. This forms a protective barrier for tidal lands on the landward side of a beach ridge or barrier island. In a severe storm, the swash of storm waves chisels away the upper part of the beach. Although the foredune barrier may then be eroded by wave action and partly cut away, it will not usually give way. Between storms, the beach is rebuilt, and in due time, if a vegetative cover is maintained, wind action restores the dune ridge. If, however, the plant cover of the dune ridge is undermined by human foot and vehicular traffic, inlets may open up and allow ocean water to wash in during storms or high tides (refer back to Figure 16.2). Many coastal communities now protect their dunes by erecting raised pathways over them, to restrict foot traffic, or by planting protective vegetation.

16.19 Coastal foredunes

Beach grass growing on coastal foredunes traps drifting sand, producing a dune ridge. These foredunes are at Nauset Beach, Cape Cod National Seashore, Massachusetts.

James P. Blair/NG Image Collection

Spits And Bars

Where littoral drift moves the sand along the beach toward a bay, the sand is carried out into the open water, extending like a long finger, called a *spit*. As the spit grows, it forms a barrier, called a *baymouth bar*, across the mouth of the bay. Once the bay is isolated from the ocean, it is transformed into a lagoon. And where a spit grows to connect the mainland to a near-shore island, it forms a *tombolo* (Figure 16.20).

Barrier Islands

Much of the length of the Atlantic and Gulf coasts of North America is flanked by *barrier islands*, low ridges

16.20 Depositional landforms of littoral drift

Littoral drift deposits sand in a distinctive way, developing depositional landforms.

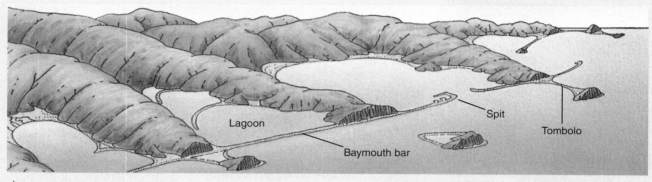

▲ **LANDFORMS**
Sand carried to the coast by rivers, or eroded from marine scarps and cliffs, is transported and deposited by littoral drift to become shoreline landforms such as spits, baymouth bars, and tombolos.

age fotostock/SuperStock

▲ **GROWTH OF A SPIT**
This large, white sand spit, at the south end of Nauset Beach, Cape Cod, Massachusetts, is growing in a direction toward the observer.

EYE ON THE LANDSCAPE **What else would the geographer see?**
The freshness of the sand here **A** indicates that Nauset Beach is growing very rapidly. The sand is supplied by the erosion of marine cliffs to the north. Note the overwash channels **B** that are cut through the vegetated sand dunes by the high waters and high surf of severe storms. To the upper left is the town of Chatham **C**, its low hills formed from rock debris shed by melting ice sheets during the Ice Age. The coastline has the "drowned" look of a coastline of submergence, with many shallow bays.

16.21 Barrier island coast

Where the gently sloping coastal plain of the Atlantic and Gulf coasts meets the ocean, the rising sea level coupled with wave and wind action has shaped a coastline of barrier islands, beaches, and lagoons.

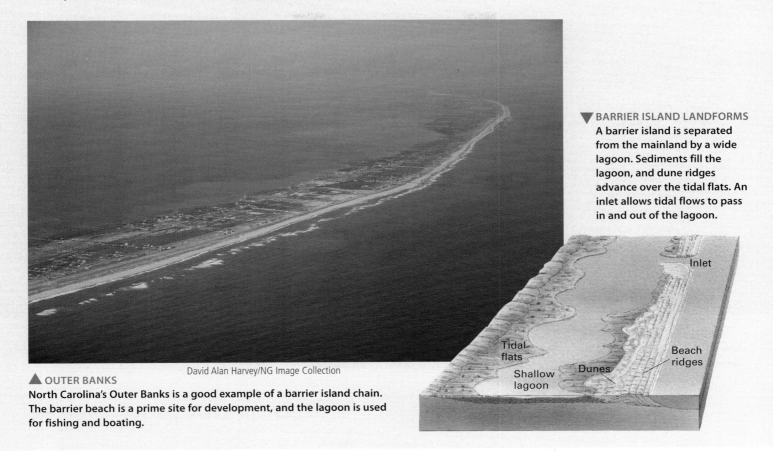

▲ OUTER BANKS

David Alan Harvey/NG Image Collection

North Carolina's Outer Banks is a good example of a barrier island chain. The barrier beach is a prime site for development, and the lagoon is used for fishing and boating.

▼ BARRIER ISLAND LANDFORMS
A barrier island is separated from the mainland by a wide lagoon. Sediments fill the lagoon, and dune ridges advance over the tidal flats. An inlet allows tidal flows to pass in and out of the lagoon.

Inlet

Tidal flats

Shallow lagoon

Dunes

Beach ridges

of sand with beaches and dunes and a landward lagoon (Figure 16.21). These islands came into being as sea level rose after the last glaciation, causing the existing beaches to grow and migrate landward with the invading shoreline. Eventually, they became too big to move, and the sea flooded the area behind them to form the lagoons. Today, barrier islands are still being sculpted by wind and waves, and they display many fine examples of depositional landforms.

Gaps in the barrier islands, called *tidal inlets,* allow tidal waters to circulate in lagoons behind the islands. Strong currents flow back and forth through the inlets as the tide rises and falls. New inlets are cut by wave action in severe storms, and they are kept open by the tidal current—though they may be closed later by longshore drift.

A *lagoon* is a shallow bay where fine sediments brought in by the tidal circulation accumulate. In time, tidal sediments can fill bays and produce *mud flats,* which are barren expanses of silt and clay. Mud flats are exposed at low tide but covered at high tide. Salt-tolerant plants start to grow on mud flats, and the plant stems trap more sediment, building up the flats into a *salt marsh*

(Figure 16.22). A thick layer of peat eventually develops on the salt marsh surface. Salt marshes are ecologically important, serving as nurseries for young ocean-going fishes.

Barrier islands and their lagoons serve to protect the mainland from hurricane waves and storm surge, although it is development on the islands that bears the brunt of the storm damage. Climate change is heightening the risk to life and property in these coastal zones as a result of greater storm frequency and intensity, as well as sea-level rise.

> Barrier islands of the coastal plain were formed as sea level rose after the last glaciation. Behind the island is a shallow lagoon that is serviced by tidal inlets crossing the barrier island.

WileyPLUS Coastal Landforms Interactivity
Review key concepts and terms of coastal landforms.
Examine photos of shorelines, from ground to satellite.
Build spits and barrier islands, varying sand supply and wind direction.

Raymond Gehman/NG Image Collection

16.22 Salt marsh

Salt marshes are abundant on the southeastern coastal plain of the United States. This marsh, photographed at high water, is near Brunswick, Georgia. At low tide, receding water will reveal the muddy bottom of the broad channel.

COASTLINES OF SUBMERGENCE

Coastlines of *submergence* are formed when rising sea level partially drowns a coast or when part of the coast sinks. At the end of the last glaciation, sea level rose by about 120 m (394 ft), submerging coastal landscapes. Climate scientists expect sea level to rise even higher over the next century as a result of global climate change. Today, many of the major river delta areas are experiencing submergence resulting from land use practices and sea-level rise.

A *ria coast* is formed when a rise of sea level or a crustal sinking (or both) brings the shoreline to rest against the sides of river valleys previously carved by streams (Figure 16.23). The shoreline rises up the sides of the stream-carved valleys, creating narrow bays. Streams that flowed through the valleys add freshwater to the bays, making them estuaries of mixed fresh- and saltwaters, thereby producing a unique habitat for many plants and animals.

A *fiord coast* is similar to a ria coast except that the bays are formed in valleys that have been scoured by glaciers (Figure 16.24). Glaciers eroded their walls and scraped away loose sediment and rock, leaving behind broad, steep-sided valleys that are now flooded with seawater. Fiords, the Scandinavian word for bays, are common features along the northern coastal countries; they can be found where glaciers occupied coastlines during the Ice Age. These deep, glacially scoured bays can be hundreds of meters deeper than the adjacent seas, and they often meander tens of kilometers inland.

> Ria and fiord coasts result from submergence of a landmass. Islands, bars, estuaries, and bays are characteristic of these "drowned" coastlines.

CORAL REEFS

So far we have been talking about coastal landforms built by sediment. **Coral reefs** are different—unique, in fact—because their origins are biological; they are made by living organisms. Growing together, corals and algae secrete rocklike deposits of carbonate minerals. As old coral colonies die, new ones grow upon them, accumulating as composite layers of limestone. When coral fragments are torn free by wave attack, the pulverized fragments accumulate on land as coral sand beaches.

For dense coral reefs to grow, water temperatures must be above 20°C (68°F); that is why coral reef coasts are usually found in warm tropical and equatorial waters between 30° N and 25° S. Furthermore, the seawater must be free of suspended sediment, and be well aerated, for vigorous coral growth to take place. For this reason, corals live near

16.23 A ria coastline

A ria coast develops where river valleys are flooded with ocean water after the submergence of a landmass or a rise in sea level.

▲ BEFORE SUBMERGENCE
Before submergence, active stream systems have carved slopes and valleys.

▲ A RIA COAST
This view of Ria Ortigueira, Coruña, Spain, shows many of the features of a ria coastline, including submerged valleys, islands, and sandbars.

▲ AFTER SUBMERGENCE
After submergence, the shoreline rests on sides of the stream valleys, creating estuaries of mixed salt- and freshwater.

Labels on A RIA COAST image: Island, Sandbar, Submerged valleys

Credit: age fotostock/SuperStock

16.24 Fiord coast

Along a fiord coast, valleys that were once scoured by glaciers are now flooded with seawater. This example is in Fiordland National Park, South Island, New Zealand.

Brian J. Skerry/NG Image Collection

the water surface and thrive in locations that are exposed to waves from the open sea. They are not found near the mouths of muddy streams because muddy water prevents coral growth. Sediment runoff from poor land-use practices and development has been identified as a primary factor in most coastal coral reef die-offs.

There are three distinctive types of coral reefs—fringing reefs, barrier reefs, and atolls. These reefs are often found around a hotspot volcano (Chapter 12) at different stages of submergence. *Fringing reefs* build up as platforms attached to shore. They are widest in front of headlands where the wave attack is strongest. *Barrier reefs* lie out from shore and are separated from the mainland by a lagoon. At intervals along barrier reefs, there are narrow gaps through which excess water from breaking waves is returned from the lagoon to the open sea.

Atolls are more-or-less circular coral reefs enclosing a lagoon; no land is inside (Figure 16.25). Most atolls are rings of coral growing on top of old, sunken hotspot volcanoes. They begin as fringing reefs surrounding a volcanic island. Then, as the volcano sinks, the reef continues to grow, until eventually only the reef remains.

Coral reefs are highly productive ecosystems that support a diversity of marine life-forms. They also perform an important role in recycling nutrients in shallow coastal environments. They serve, too, as physical barriers that dissipate the force of waves, protecting ports, lagoons, and beaches behind them. Finally, they are an important aesthetic and economic resource.

Unfortunately, because of their very specific environmental requirements, coral reefs are highly susceptible to damage from human activities, as well as from natural causes such as tropical storms. Some 50 percent of coral reefs worldwide are currently under threat of human activities. Earlier in the chapter, we noted the

Rafael Macia/Photo Researchers, Inc.

16.26 Armoring the shoreline

This seawall, under wave attack, protects these homes in Sea Bright, New Jersey, from storm surf. While protecting development, however, the seawall can also reduce biological diversity along the shoreline.

effects of global warming on corals, causing bleaching and, ultimately, the death of corals from elevated water temperatures. But geographers are also worried about the impact on corals of rising carbon dioxide levels in the atmosphere. As more CO_2 dissolves in seawater it becomes more acidic, making it harder for corals to develop their calcium carbonate skeleton structure.

> Coral reef coasts develop in warm oceans, where corals build reefs at the land-sea margin. Fringing reefs, barrier reefs, and atolls are three types of coral reef coasts.

COASTAL ENGINEERING

Around the world, human populations have always settled along coasts, drawn by ocean-related industries such as shipping, fishing, and tourism, as well as for the scenic views and coastal climates. More than 70 percent of the world's population lives on coastal plains, and 11 of the world's 15 largest cities are on coasts or estuaries. Human efforts to control dynamic coastal processes in order to protect their property and investments are often futile; worse, they can disrupt the balance of sediment supply and erosion.

Homes built on sea cliffs, beaches, or barrier islands are threatened by the erosive action of waves, especially during storms (Figure 16.26). Homeowners sometimes

16.25 Coral atoll

Kayangel Atoll, Belau, Palau Islands, is a nearly circular reef with a lagoon at the center. On the right are four islets where wind-blown sand has accumulated above the water level.

Douglas Faulkner/Photo Researchers, Inc.

construct seawalls or pile up mounds of boulders, called *rip-rap*, in an attempt to absorb some of the destructive erosive energy of the waves and protect their property. A negative side effect of these fortifications is that they can reduce the diversity and abundance of fish, crabs, and other marine species at the shoreline, disrupting marine ecosystems.

Along stretches of shoreline affected by retrogradation, the beach may be seriously depleted or even disappear entirely, destroying valuable shore property. Property owners and/or governments sometimes try to build a broad, protective beach, through a process called *artificial beach nourishment*. One method of beach nourishment is to simply pump sand from offshore onto the beach. These programs are, however, costly, and only temporarily solve the problem because waves immediately begin to carry away the newly added sand. Geographers consistently warn governments that such public works programs are futile and waste tax dollars.

Another way to protect beaches is to try to prevent the loss of sand by trapping it as it is transported down the shore by littoral drift. This is accomplished by installing walls or embankments, called *groins*, at close intervals along the beach. Groins are usually built at right angles to the shoreline and made from large rock masses, concrete, or wooden pilings. The downside of groins is that although they effectively keep sand on a specific beach, they deprive other beaches farther down the shore of sand that they would normally receive through littoral drift (Figure 16.27).

> Groins are walls or embankments built at right angles to the shoreline. They trap littoral drift and help prevent beach retrogradation.

Jetties, which are structures built to keep navigation channels open, similarly disrupt littoral drift.

In some cases, human influences on beaches result from actions taken far from the coast. Most beach sand comes from sediment that is delivered to the coast by rivers. Dams constructed far upstream can drastically reduce the sediment load of the river, starving the beach of sediment and causing retrogradation on a long stretch of shoreline.

Wind Action

Thus far, we have examined how water waves work to create landforms. Since waves are generated by wind blowing across water, we considered them to be indirectly powered by wind. But wind can also move sediment directly, by eroding, transporting, and depositing sand and finer particles.

EROSION BY WIND

Wind performs two kinds of erosional work: deflation and abrasion. **Deflation** is the removal of particles, largely silt and clay, from the ground by wind. It acts on loose soil or sediment. Dry river courses, beaches, farmlands, and areas of recently formed glacial deposits are especially susceptible to deflation. In dry climates, much of the ground surface can be deflated because vegetation is lacking to hold the soil or sediment in place.

The capability of the wind to remove particles by deflation depends on their size. Given a mixture of particles of different sizes on the ground, deflation, combined with splash erosion and overland flow, will remove the fine particles and leave the coarser particles behind. As a result, rock fragments ranging in size from pebbles to small boulders become concentrated into a surface layer known as a **desert pavement** (Figure 16.28). The large fragments become closely fitted together, sheltering the smaller particles—grains of sand, silt, and clay—that remain beneath. The pavement acts as an armor that protects the remaining fine particles from rapid removal by deflation or overland flow. This pavement is easily disturbed, however, by wheeled or tracked vehicles, which expose the finer particles underneath, leading to renewed deflation and water erosion.

In drier plains regions, deflation can scoop out a shallow basin called a **blowout**, especially where the grass cover has been broken or disturbed by grazing animals, for example (Figure 16.29). The size of the depression may range from a few meters (10 to 20 feet) to a kilometer (0.6 mile) or more in diameter, although it is usually only a few meters deep. A blowout can also form in a shallow depression in a plain where rains have filled

Doug Steley/Alamy Limited

16.27 Groins

Low walls of stone or sandbags placed at right angles to the shoreline, called groins, trap sand as it is moved along the coast by littoral drift. The groins on this beach in Maroochydore, Queensland, Australia, are trapping sand moving from left to right.

16.28 Desert pavement

After finer materials are removed by deflation, the remaining coarser materials form a desert pavement.

C. Allan Morgan/Getty Images, Inc.

▼ THE WORK OF DEFLATION
By removing fine particles, a lag deposit is left to armor the soil.

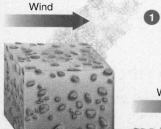

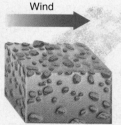

1 Newly deposited sediment contains a mixture of coarse and fine particles. Wind carries away the silt and clay and slowly reduces the volume of the deposit.

2 As the deflation continues, the surface lowers. The coarser particles—gravel, pebbles, and larger stones—are left behind and accumulate on the surface. Rare rainfall events may also remove finer particles by sheetflow.

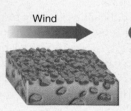

3 Eventually, the surface is covered by coarse particles, forming a desert pavement that resists further deflation or erosion.

▲ DESERT PAVEMENT
This desert pavement, photographed near Laguna San Ignacio, Baja California, Mexico, shows closely fitted rock fragments on a surface layer of sand. The larger rocks have been sandblasted by the wind into rounded and faceted shapes.

Robert and Jean Pollock/Photo Researchers, Inc.

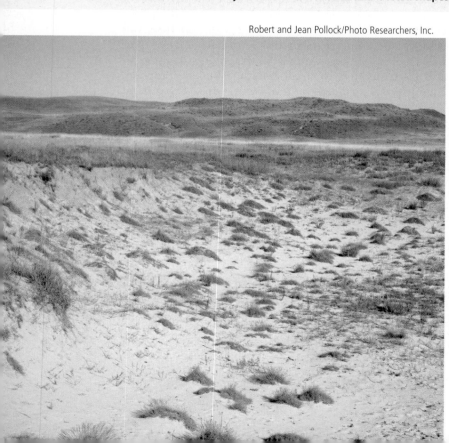

the depression and created a shallow pond or lake. As the water evaporates, the mud bottom dries out and cracks, leaving small scales or pellets of dried mud. These particles are then lifted out by the wind.

> Wind deflation can produce blowouts and help form desert pavement, a surface armor of coarse particles that minimizes further deflation.

The second process of wind erosion, **abrasion**, drives sand-sized particles against an exposed rock or soil surface, wearing down the surface by the impact of the particles. Wind abrasion is most active in a layer of about 10 to 40 cm (4 to 16 in.) above the surface. The weight of the sand grains prevents them from being lifted much higher into the air. Wooden utility poles on windswept

16.29 Blowout

A blowout forms when fine-grained soil is disturbed, breaking up the vegetation cover. Wind then picks up silt and fine particles, leaving sand and coarser particles behind. This blowout, in the Nebraska Sandhills region, has been partially revegetated.

16.30 Wind-sculpted rocks

Wind abrasion blasts sand against exposed rocks to create rock sculptures of various sizes.

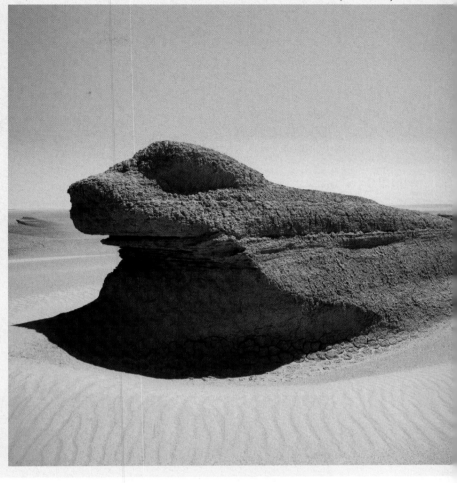

Mike P Shepherd / Alamy Limited

Mark A. Wilson/ Department of Geology/ The College of Wooster

▲ **VENTIFACT**

Wind abrasion can carve stones on a desert plain into shapes with flattened or grooved sides, creating a ventifact. In this example from the Mojave Desert, California, wind from two directions has shaped two facets, with a sharp ridge between them.

YARDANG ▶

In some types of layered rocks, wind abrasion can produce tongue- or teardrop-shaped ridges, called yardangs, oriented parallel to the prevailing wind direction. Although rare, they are quite striking. This example is near Gilf Kebir, in the western desert of Egypt.

plains often have protective metal sheathing, or heaps of large stones are arranged around their bases. Without such protection, they would quickly be cut through at the base by wind abrasion.

Wind abrasion excavates pits, grooves, and hollows in rock. Such wind-sculpted rocks are called *ventifacts* (Figure 16.30). Common examples can be found among the rocks on desert plains, which have two or three smooth or grooved sides produced by wind abrasion. Another interesting form produced by wind abrasion is the *yardang*, a low tongue- or teardrop-shaped ridge.

TRANSPORTATION BY WIND

Sediment can also be transported by wind, but how much and how far depends on the strength and turbulence of the wind as well as the size of the sediment grains. The finest particles, those of clay and silt sizes, are lifted and raised into the air—sometimes to a height of 1000 m (about 3300 ft) or more—where they are carried in suspension by turbulence.

Saltation is the process by which sand grains are moved; they are carried by the wind in low arcs, perhaps a meter long, from one point to another (Figure 16.31). Each time they hit the surface, it forces other particles into the air, which are then carried downwind. At higher wind speeds, sand grains are also dragged along the ground, in the same way that a stream carries bed load. This movement is called *surface creep*.

> Wind transports sand near the surface by saltation and surface creep. Silt and clay are carried aloft by atmospheric turbulence in dust storms.

Dust Storms

Strong, turbulent winds blowing over barren surfaces can lift large quantities of fine dust into the air and amass into a dense, high cloud called a **dust storm**.

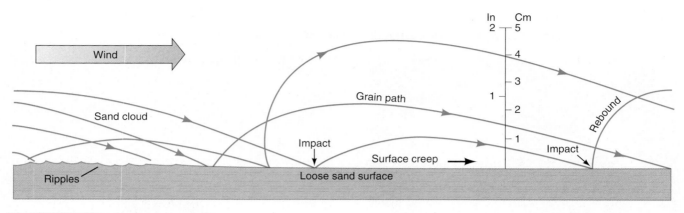

16.31　Saltation

Wind picks up individual grains of sand and carries them downwind on a low arc. Most grains travel within a few centimeters of the surface, but some reach heights of a meter or so. As each grain collides with the surface, it excavates a tiny crater, ejecting particles upward into the wind stream, which then moves them downwind. At higher wind speeds, surface sand particles are tossed and pushed along as surface creep.

Vigorous saltation of coarse sand grains, driven by the strong wind, adds to the dust. Collisions disturb the soil, breaking down soil particles and releasing fine dust that can be lofted upward by the turbulent winds.

A dust storm is seen approaching as a dark cloud, extending from the ground surface to heights of several thousand meters (Figure 16.32). Typically, the advancing cloud wall occurs along a rapidly moving cold front. Anything or anyone caught within the dust cloud is shrouded in deep gloom or even total darkness. Visibility is cut to a few meters, and a choking fine-grained dust penetrates everywhere and everything. These small particles are a health hazard for the very young and the old, as well as for persons with sensitive lungs.

©AP/Wide World Photos

16.32　Dust storm

In a front reaching up to 1000 m (3300 ft) in altitude, a dust storm approaches an American facility in Iraq. It passed over in about 45 minutes, leaving a thick layer of dust in its wake.

Severe deflation resulting from overgrazing or poor agricultural practices can cause relentless dust storms, such as those that occurred during the 1930s, in what came to be known as the American Dust Bowl. Cultivating short-grass prairie in a semiarid region makes the ground much more susceptible to deflation. Plowing disturbs the natural soil surface and grass cover, and in drought years, when vegetation dies out, the unprotected soil is easily eroded by wind action. That is why much of the Great Plains region of the United States has suffered dust storms generated by turbulent winds. Strong cold fronts frequently sweep over this area and lift dust high into the troposphere when soil moisture is low.

Human activities also are to blame for high, dense dust clouds in very dry, hot deserts. For example, in the Thar Desert, which borders the Indus River in northwest India and Pakistan, as grazing animals and humans trample the fine-textured soils, they raise dust clouds, some of which hang over the region for long periods, at heights of up to 9 km (about 30,000 ft).

Beijing, China's capital, regularly experiences severe dust storms generated from loose silt blown from the Gobi Desert and Yellow River Basin (Figure 16.33). These dust storms normally occur five to seven times each spring. The Chinese government has directed the planting of more than 40 billion trees in an attempt to help bind the soil and reduce wind strength at the ground surface. Still, however, the frequency of these seasonal dust storms is on the rise, due to a combination of poor land use practices and climate change.

Scientists calculate that Earth's atmosphere today contains the highest levels of dust particles in human history, due to deforestation, modern agricultural practices, and the growing number of droughts that result from the warming climate. The effect of dust on the global climate depends of the size and nature of the dust particles, their altitude in the atmosphere, and whether they are above a light surface, a sand desert for example, or a dark surface, such as the ocean. These complexities leave scientists still uncertain about the effects of aerosols on climate.

Eolian Landforms

Eolian landforms, named after the Greek god of wind, Aeolus, are wind-generated. Ordinarily, wind is not strong enough to dislodge mineral matter from the

Lou Linwei /Alamy Limited

16.33 Dust storm over Beijing

Springtime in Beijing means increasingly frequent dust storms, amassed from loose silt blown from desert regions of China's interior. Global warming is contributing to desertification and more common and intense seasonal winds, both of which aggravate health-threatening conditions suffered by residents of China's capital city. This severe dust storm struck the city in March 2002.

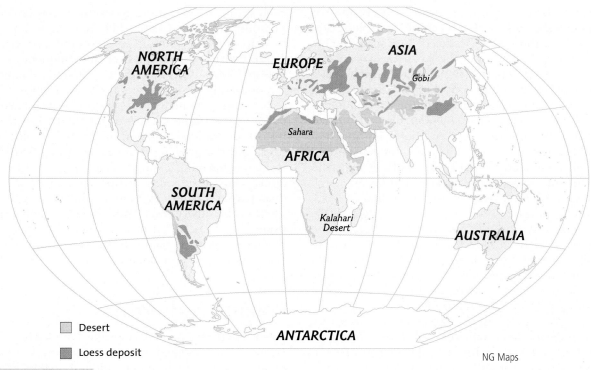

NG Maps

16.34 Eolian landforms

Desert and coastal dunes are the most common eolian landforms. During the Ice Age, strong winds carried vast clouds of silt, picked up from flooding rivers draining ice sheets, to form deep layers in continental interiors. These wind-borne deposits are known as loess.

surfaces of unweathered rock, moist, clay-rich soils, or soils bound by dense plant cover. Wind works best as a geomorphic agent when its velocity is high and moisture and vegetation cover are low—typically, in deserts and semiarid lands or steppes. In coastal environments, beaches also provide a supply of loose sand to be shaped by the wind, even where the climate is humid and the land surface inland from the coast is well protected by plant cover. Eolian landforms are also found in continental interiors where layers of silt were deposited during the retreat of glaciers at the end of the Ice Age. The map in Figure 16.34 shows the principal world regions of eolian landforms.

> Eolian landforms occur where surface mineral particles are dry and unprotected by a vegetation cover. These conditions are found in deserts and semiarid regions of the world, as well as on sandy shorelines.

SAND DUNES

A **sand dune** is any hill of loose sand shaped by wind. Dunes are one of the most common types of eolian landforms. They form where there is a source of sand—for example, a sandstone formation that weathers easily and releases individual grains, or perhaps a beach supplied with abundant sand from a nearby river mouth, or a closed interior basin in an arid region that receives abundant river sand. Active dunes constantly change shape under wind currents, but they must be nearly free of vegetation in order to move. They become inactive when stabilized by vegetation cover or when patterns of wind or sand sources change.

A dune typically begins as a sand drift in the shelter of some obstacle, such as a small hill, rock, or clump of brush that lowers wind speed, causing saltating sand to stop moving. Once a sufficient mass of sand has accumulated, it begins to move downwind (Figure 16.35). Pushed by the prevailing wind, sand blows up the windward dune slope, passes over the dune crest, and slides down the steep leeward side of the dune, called the *slip face*, moving the dune forward. The slip face maintains a more-or-less constant angle from the horizontal, known as the *angle of repose*. For loose sand, this angle is about 33° to 34°.

A vast expanse of sand dunes makes up a great desert landscape, called an *erg*, after the Arabic word for dune field. In the Sahara Desert, enormous quantities of dune sand have been weathered from

> Sand dunes move as wind-blown sand grains hop up the gentle windward side of the dune and then fall onto the steeper slip face of the leeward side.

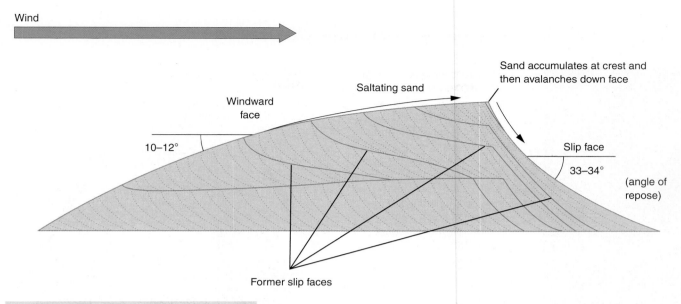

16.35 Formation and movement of sand dunes

This cross section through a sand dune shows the typical gentle windward slope (facing the wind) and the steep slip face (facing away from the wind). The solid lines inside the dune show former slip faces before the dune moved forward into its current position.

sandstone to create extensive regions of erg desert. Also common are regions covered with an aggregate of pebbles that form a vast and semi-impermeable flat surface. This desert pavement surface is called a *reg*. These two major desert landscapes can be found around the globe.

TYPES OF DUNES

Sand dunes take a variety of shapes, depending on the supply of sand, the amount and direction of wind, and the amount of vegetation cover (Figure 16.36). When sand is abundant, we find sand seas of dunes in constant motion. Where sand is in shorter supply, we find low ridges and rails of sand in parallel lines. Vegetation tends to bind the sand and minimize movement. Some plants make sand their preferred habitat, and stabilize dune forms for long time periods. Wind is needed to shape and move the dunes, and the stronger the prevailing wind, the more the sand moves.

We can recognize five basic types of dunes, shown in the figure. *Barchan dunes* are crescent-shaped, with the points of the crescent directed downwind. They require a moderate sand supply and moderate winds, and lack vegetation. *Transverse dunes* form where the sand supply is abundant and winds are strong and consistent. Their crests are perpendicular to the wind direction, and the dune field resembles a petrified sea of water waves. The moving dunes lack vegetation. *Star dunes* are large, pyramidal hills of sand built

from a moderate sand supply when winds blow regularly from several different directions. They are often fixed in position and lack vegetation.

Parabolic dunes are often found where shorelines provide a large sand supply. Coastal moisture encourages vegetation to take a foothold, anchoring the arms of the dunes in an upwind direction. *Longitudinal dunes* occur where sand is limited but winds are strong. Vegetation often stabilizes the long dune ridges, which follow the wind direction. They can cover vast areas, especially in interior Australia.

> Dune type depends on the availability of the sand supply, strength of the wind, and the amount of vegetation present. Barchan, transverse, star, parabolic, and longitudinal dunes are important types.

WileyPLUS Types of Dunes
Interact with a chart of sand dune types organized by wind, sand supply, and amount of vegetation to see photos of common types of dunes. An animation.

LOESS

In several large midlatitude areas of the world (see Figure 16.34), the surface is covered by deposits of wind-transported silt that has settled out from dust storms over many thousands of years. This material is known as **loess**, a German word pronounced somewhere between

16.36 Types of sand dunes

Dune forms are quite variable, but we can recognize at least five types that commonly occur.

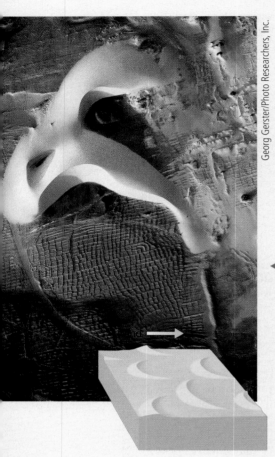

Georg Gerster/Photo Researchers, Inc.

FACTORS AFFECTING ▶ DUNE FORMS

The shape, size, and behavior of a dune depend on sand supply, wind, and amount of vegetative cover. To read the triangular diagram, project the location of a dune type onto the sides of the triangle. For example, longitudinal dunes have a low sand supply, strong winds, and a varying amount of vegetation.

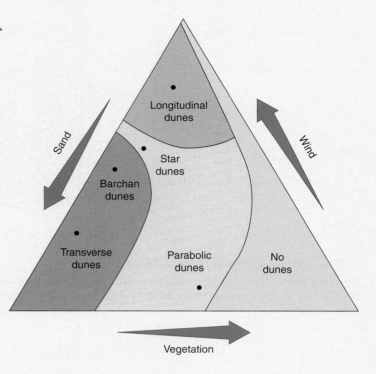

◀ BARCHAN

The barchan dune has the outline of a crescent, with the points of the crescent directed downwind. Barchan dunes form when the wind blows predominantly in one direction. They are often outliers moving away from a larger body of sand, such as a field of transverse dunes. This photo shows two or three intersecting barchan dunes moving from the upper left of the photo toward the bottom.

▼ STAR

Star dunes are elaborately shaped pyramidal mounds shaped by winds blowing from several prevailing directions. Star dunes, such as these at Naukluft Park, Namibia, are stable, almost-permanent features of the desert landscape.

▼ TRANSVERSE

When there is an abundant supply of sand, little or no vegetation, and a single dominant wind direction, transverse dunes, such as these in Baja California, Mexico, often form. Their crests are at right angles to the wind direction.

Annie Griffiths Belt/NG Image Collection

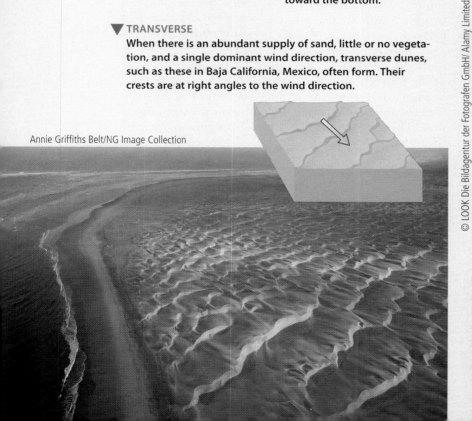

© LOOK Die Bildagentur der Fotografen GmbH/ Alamy Limited

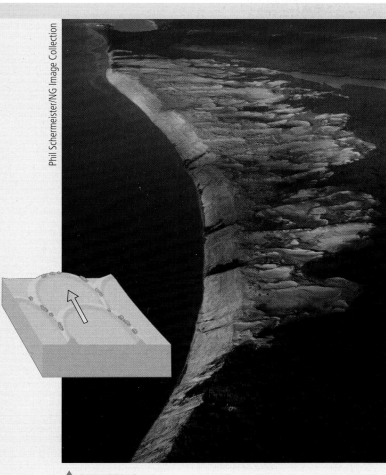

▲ PARABOLIC

Coastlines often provide abundant sand, along with a moist wind off the ocean that allows some vegetation to grow. In this environment, parabolic dunes, like these at the Grand Sable Dunes, Lake Superior, often develop. The arms of these dunes, anchored by vegetation, point upwind.

▼ LONGITUDINAL

Longitudinal dunes consist of long, narrow ridges oriented parallel with the direction of the prevailing wind. Common in deserts with meager sand supply, they are often only a few meters high but may be several kilometers long, such as these in South Australia.

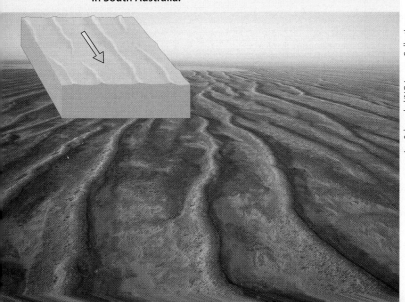

"lerse" and "luss." Loess is usually a uniform yellowish to buff color and lacks any visible layering. The dust is composed of fine silt and clay particles, which are often shaped like tiny flakes or plates. As they settle out in layers, the flat particles overlap each other, giving the deposit a weak cohesive structure. Loess often contains calcium carbonate dust, which can dissolve and recrystallize when rainwater penetrates the loess deposit, adding to its cohesiveness. As a result, loess tends to form vertical cliffs wherever it is exposed by the cutting of a stream or grading of a roadway (Figure 16.37). It is also very easily eroded by running water, and when the vegetation cover that protects it is broken, it is rapidly carved into gullies. Thick loess, a stable, easy-to-dig material, has been used for cave dwellings in some regions, including China and Central Europe.

The densest deposits of loess are in northern China, where layers over 30 m (about 100 ft) thick are common—though some have been measured to as much as 100 m (about 330 ft). These layers cover many hundreds of square kilometers and appear to have been brought as dust from the interior of Asia. Loess deposits are also common in the United States, Central Europe, Central Asia, and Argentina. In the United States, thick loess deposits can be found in the Missouri–Mississippi Valley (Figure 16.38); and eastern Washington State's famous Palouse farmland is derived from thick and fertile loess soils.

Both American and European loess deposits are directly related to the continental glaciers of the Ice Age. At the time when the ice covered much of North America and Europe, the winter climate was generally dry along the borders of the ice sheets. Strong winds blew southward and eastward over the bare ground, picking up silt from the floodplains of braided streams that discharged the meltwater from the ice. This dust settled on the ground between streams, gradually building up a smooth, level ground surface. The loess is particularly thick along the eastern sides of the valleys, the result of prevailing westerly winds carrying the loess eastward. It is well exposed along the bluffs of most streams flowing through these regions today.

Loess is an important agricultural resource thanks to its rich black soils, which are especially suited to cultivation of grains. The highly productive plains of southern Russia, the Argentine Pampas, and the rich grain region of northern China are underlain by loess. In the United States, corn is extensively cultivated on the loess plains in Kansas, Iowa, and Illinois. Wheat is grown farther west on the loess plains of Kansas and Nebraska and in the Palouse region of eastern Washington.

> Loess is a deposit of wind-blown silt that may be as thick as 30 m (about 100 ft) in some regions of North America. It forms highly productive soils.

WileyPLUS Web Quiz

Take a quick quiz on the key concepts of this chapter.

TP/Alamy Limited

16.37 Loess

This thick layer of loess, or wind-transported silt, in Roztocze, Poland, was deposited during the Ice Age. Loess has excellent cohesion and often develops vertical faces as it wastes away or is removed to build road banks.

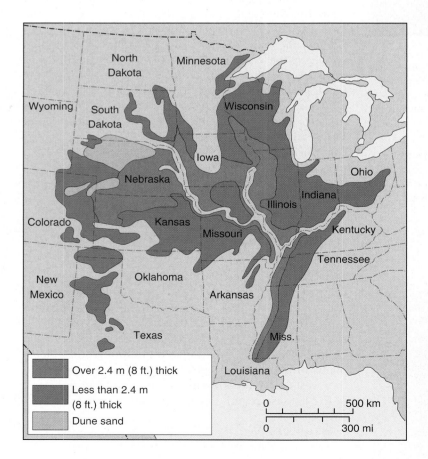

Over 2.4 m (8 ft.) thick

Less than 2.4 m (8 ft.) thick

Dune sand

0   500 km

0   300 mi

16.38 Loess distribution in the central United States

A map of loess distribution in the central United States shows that large areas of the prairie plains region of Indiana, Illinois, Iowa, Missouri, Nebraska, and Kansas are underlain by loess ranging in thickness from 1 to 30 m (about 3 to 100 ft). Extensive deposits also can be found in Tennessee and Mississippi, in areas bordering the lower Mississippi River floodplain. Still other loess deposits are in the Palouse region of northeast Washington and western Idaho.

A Look Ahead

This chapter has described the processes and landforms associated with wind action, blown either directly or as wind-driven waves. In Chapter 17, we turn to the last of the fluid agents that create landforms—glaciers. Compared to wind and water, glacial ice moves much more slowly but also much more steadily. Like a vast conveyor belt, glacial ice moves sediment forward relentlessly, depositing the sediment at the ice margin, where the ice melts. By plowing its way over the landscape, glacial

ice also shapes the local terrain, bulldozing loose rock from hillsides and plastering sediments underneath its vast bulk. This slow but steady action is very different from that of water, wind, and waves, and so produces a distinctive set of landforms, which are the subject of our last chapter.

WileyPLUS Web Links

Follow this chapter's web links to visit sites of coastal erosion from San Diego to the Carolinas. Explore dunes and deserts to see wind in action.

IN REVIEW LANDFORMS MADE BY WAVES AND WIND

- Global climate change will have major impacts on coastal environments. Sea level will rise, the size and severity of storms will increase, and flood and drought conditions will become more extreme. Coastal erosion will worsen, as well. Sea-level rise and land subsidence will cause the loss of many coastal wetlands. And as water temperatures rise, sensitive coral reefs will die off. Arctic shorelines, too, will erode as sea ice thins and ground ice and permafrost thaw.
- Wind and waves are important agents of landform sculpture. Wind moves material directly by transporting fine particles. Waves transfer wind energy to water motion that erodes, transports, and deposits coastline materials.
- Waves have *crests* and *troughs*, and are characterized by *wave height*, *wave length*, and *wave period*. Waves cause particles at or near the water surface to travel in a circular path.
- Wave height increases rapidly with increasing wind speed, duration, and *fetch*.
- Waves expend their energy as *breakers* that push a *swash*, laden with sand and gravel, onto the beach. The *backwash* returns water and sediment seaward.
- Wave action produces *beach drift* and *longshore drift*, which act together as **littoral drift** to move sediment along the shoreline. *Sand bars*, *sand spits*, and *pocket beaches* are formed.
- *Wave refraction* causes waves to wrap around a *headland*, concentrating wave energy there. Longshore drift from the headland transports sediment to adjacent *pocket beaches*.
- Sea level experiences a rhythmic rise and fall called the *ocean tide*. Tides arise from the gravitational pull of the Moon on ocean water and the push of the inertial force of the rotation of the Earth−Moon system around its center of mass.
- *Spring* and *neap* tides occur when the Sun aligns with the Earth−Moon axis or lies perpendicular to that axis.

- As water level changes with the tides, it produces *tidal currents* in bays and estuaries that move sediment. Rapid current motion keeps inlets open in spite of longshore drift.
- A **tsunami** is a great ocean wave set in motion by an earthquake or volcanic explosion. It causes a rapid and temporary rise of sea level that causes tremendous destruction, as it floods coastal regions and then retreats.
- Waves act at the **shoreline**—the boundary between water and land. The **coastline** is the zone where coastal processes operate.
- Where the shoreline consists of soft sediments, waves erode the sedimentary layers into *marine scarps*. Where the shoreline consists of hard rocks, **sea cliffs** often form.
- Intense physical weathering of a sea cliff at the water's edge carves a *notch* that undercuts the cliff. Weathering and wave action produce *caves*, *arches*, *stacks*, and *abrasion platforms*.
- **Marine terraces** occur where abrasion platforms are lifted above the shoreline by tectonic activity.
- Waves build *beaches*, which are wedge-shaped accumulations of sediment. Beach slopes are longer and shallower when wave action is more intense.
- Depending on sediment supply, beaches experience *progradation* (building out) or *retrogradation* (cutting back).
- Coastal *foredunes* are stabilized by dune grass and help protect the shoreline from storm wave action.
- Littoral drift produces *spits* of sand that project into bays. A spit can grow into a *baymouth bar* that closes a bay, providing a lagoon.
- *Barrier islands* have a flanking low ridge of sand at the edge of a tidal *lagoon*. *Tidal inlets* allow ocean water to circulate in and out of the lagoon with flood and ebb currents.
- Coastlines of *submergence* result when coastal lands sink below sea level, or sea level rises. A *ria coast* has

many drowned river valleys that develop estuaries where fresh- and saltwater mix. *Fiord coasts* form when glacial valleys and troughs are submerged.

- **Coral reefs** develop in warm oceans where corals grow in shallow water Reefs may appear as *fringing* platforms, *barrier reefs* with lagoons, or *atolls*. Reefs are highly productive ecosystems, but are also highly sensitive to human impacts of warming and acidifying ocean waters.
- *Rip-rap* and seawalls are used to protect shorelines against wave action. *Groins* act to trap sand traveling by longshore drift in order to stabilize eroding beaches. Replacing eroded sand provides *artificial beach nourishment*.
- In arid regions, **deflation** produces *desert pavement* and **blowouts** in semidesert regions. Wind **abrasion** produces *ventifacts* and *yardangs*.
- Wind transports sand by **saltation,** in which sand grains fly in low arcs, and by *surface creep* at higher wind velocities.
- Wind transports silt and clay in turbulent motion to create **dust storms**. Human activities, such as removing vegetation cover and herding animals that trample fine soils, contribute to the dust content of the air.
- *Eolian landforms* are generated by wind. They are found in deserts, semiarid lands, and steppes.

- **Sand dunes** take shape when a source, such as a sandstone outcrop or a beach, provides abundant sand that can be moved by wind action. A desert of dunes is termed an *erg*; a desert of *desert pavement* is called a *reg*.
- The type of sand dune that appears at a location depends on the sand supply, the wind strength, and the amount of vegetation cover present.
- *Barchan dunes* are shaped like crescents, with points elongated downwind.
- *Transverse dunes* form a sand sea of frozen "wave" forms arranged perpendicular to the wind direction.
- *Star dunes* are large pyramids of sand that remain fixed in place.
- *Parabolic dunes* are arc-shaped, with points elongated upwind.
- *Longitudinal dunes* parallel the wind direction and cover vast desert areas.
- **Loess** is a surface deposit of fine, wind-transported silt. It can be quite thick, and it typically forms vertical banks. Loess is very easily eroded by water and wind.
- In eastern Asia, the silt forming the loess in the region was transported by winds from extensive interior deserts located to the north and west. In Europe and North America, the silt was derived from fresh glacial deposits during the Pleistocene epoch.

KEY TERMS

littoral drift, p. 533
tsunami, p. 536
shoreline, p. 537
coastline, p. 538
sea cliff, p. 538
marine terrace, p. 540

beach, p. 540
coral reef, p. 544
deflation, p. 547
desert pavement, p. 547
blowout, p. 547
abrasion, p. 548

saltation, p. 549
dust storm, p. 549
sand dune, p. 552
loess, p. 553

REVIEW QUESTIONS

1. How has the coastal environment changed in response to climate change, and what changes are predicted for the future? What will be the impact of these changes on coastal environments?
2. How do ocean waves arise? Where do they obtain their energy, and how do they use that energy?
3. Explain how wave height is related to wind speed, duration, and *fetch* of the wind.
4. What is *littoral drift*, and how is it produced by wave action? Use the terms *beach drift* and *longshore drift* in

your answer. Identify some landforms produced by littoral drift.
5. What landforms are associated with *wave refraction*?
6. What causes *ocean tides*? Identify the terms *spring tide, neap tide, flood tide,* and *ebb tide*.
7. What causes a *tsunami*? What happens when a tsunami reaches a shoreline?
8. Contrast the terms **shoreline** and **coastline**.
9. Compare a *marine scarp* with a *marine cliff*. What landforms are associated with marine cliffs?

10. How are **marine terraces** formed?
11. What is a **beach**? Identify *progradation* and *retrogradation*. How are they related to wave activity?
12. Identify and describe three types of landforms associated with littoral drift.
13. What is a *barrier island*? How does it form?
14. Compare two types of coastlines resulting from submergence.
15. What conditions are necessary for the development of **coral reefs**? Identify three types of coral reefs, depending on their setting.
16. Identify and describe two types of erosional work done by wind.

17. What process produces **blowouts**? What is **desert pavement**, and how does it form?
18. Describe what it might be like to experience a **dust storm**. What are some sources of atmospheric dust?
19. What is a **sand dune**? What three factors affect the type of sand dune found at a location?
20. List five types of sand dunes and, on a scale of 1 to 5 (low to high), identify the need for sand, strength of wind, and amount of vegetation cover associated with each.
21. Define the term **loess**. What is the source of loess, and how are loess deposits built up? Identify several global locations of extensive loess deposits.

VISUALIZING EXERCISES

1. Sketch a profile of the surface of an ocean wave, and use it to define the following terms: *wave height, wave length,* and *wave period*. Add circles with directional arrows to show the motion of water particles at key points on the wave.

2. Take a piece of paper and let it represent a map with winds coming from the north, at the top of the page. Then sketch the shapes of the following types of dunes: barchan, transverse, parabolic, and longitudinal.

ESSAY QUESTIONS

1. Consult an atlas or Google Earth to identify a good example of each of the following coastline types: ria coast, fiord coast, barrier-island coast, and a coast of coral reefs. For each example, provide a brief description of the key features and other knowledge you used to identify each type.

2. Wind action moves sand close to the ground in a bouncing motion, whereas silt and clay are lifted and carried longer distances. Compare landforms and deposits that result from wind transportation of sand with those that result from wind transportation of silt and finer particles.

Chapter 17
Glacial and Periglacial Landforms

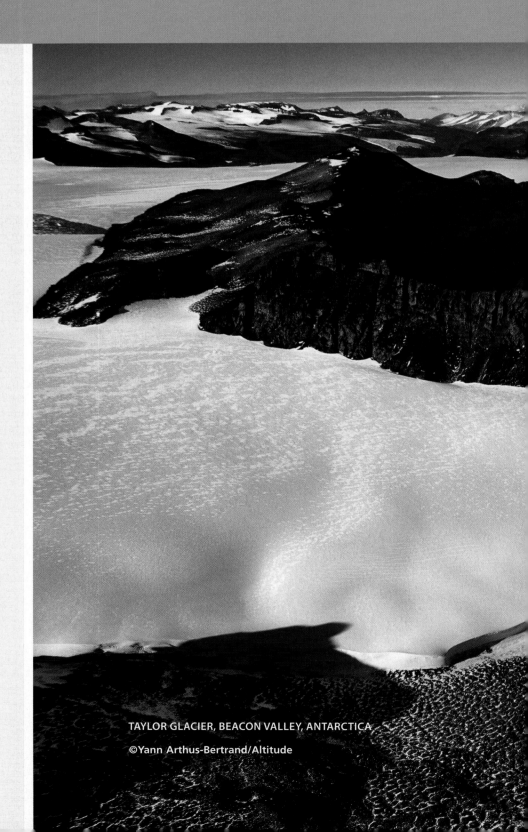

The McMurdo Dry Valleys of Antarctica, situated between the spine of the Prince Albert Mountains and the Ross Sea, are among the Earth's strangest environments. Although sheltered from the advance of the Antarctic Ice sheet by the mountains, the perpetually frozen valleys are exposed to fierce winds that drain by gravity from the icecap. The winds are warmed adiabatically as they descend, evaporating any snow cover. Glaciers heading in the mountains create alpine glacial landforms here, including horns and arêtes. Ice-wedge polygons, marked by snowy edges, cover the valley floors. These glacial landforms and others are the subjects of this chapter.

TAYLOR GLACIER, BEACON VALLEY, ANTARCTICA

©Yann Arthus-Bertrand/Altitude

Glacial and Periglacial Landforms

A bout 12,000 years ago, the Earth emerged from a glaciation that covered much of the northern hemisphere with sheets of ice. Are ice caps now receding? Is arctic sea-ice cover changing? How do glaciers form? How do they make landforms? How are alpine glaciers different from ice sheets? How does the periglacial environment provide distinctive surface features and landforms? Why has the Earth entered an ice age? What causes glaciers to come and go during an ice age? These are some of the questions we will answer in this chapter.

Ice Sheets, Sea Ice, and Global Warming

What effects will global warming have on the Earth's ice sheets and sea ice? In general, global climate models predict two types of changes in the Earth's ice sheets. First, warming on land and ocean will cause melting at the edges of the Greenland and Antarctic ice caps. More water and ice will then flow into the oceans, raising sea level. Second, warming will increase atmospheric water vapor, causing more snow to fall over the central portion of the ice caps. This will lower sea level by storing water in new ice. Both of these predictions seem to be coming true.

Let's look first at the Antarctic Ice Sheet. This huge mass holds 91 percent of the Earth's ice and is the planet's largest accumulation of freshwater. There are two different parts of the Antarctic Ice Sheet, as shown in Figure 17.1: east and west. The East Antarctic Ice Sheet is presently growing, rather than shrinking, as a result of increased snowfall. A recent study funded by NASA and the National Science Foundation (NSF) used satellite radar altimeters to measure elevation change from 1992 to 2006; researchers found that the growth corresponded well with changes in precipitation predicted by global climate models. The amount of growth, however, was quite small, and only slowed sea-level rise by about a tenth of a millimeter per year.

For the West Antarctic Ice Sheet, the story is different. This part of the ice cap contains about 10 percent of its total mass, and it is thinning significantly. In fact, the rate of loss of ice mass for the West Antarctic ice sheet increased by about 60 percent in the 10 years leading up to 2006, according to a study using satellite radar data recently published in the scientific journal *Nature Geoscience*.

Considering all parts of the Antarctic Ice Cap, ice thickness obtained from precise satellite measurements of gravity recently showed that the ice sheet's mass decreased by about 150 km³/yr (36 mi³/yr) from 2002 to 2005, accounting for about 13 percent of the rise in sea level experienced during that period.

The Greenland Ice Cap, too, has been thinning at the edges while growing at the center. Large losses from melting ice along the southeastern coast were slightly exceeded by an increase in ice thickness during the 10-year period ending in 2002. However, that situation may be changing. Since 2002, glaciers flowing into the sea from Greenland have been speeding up. More specifically,

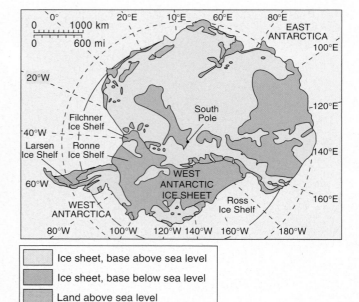

17.1 West Antarctic Ice Sheet

This map of Antarctica shows the location of the West Antarctic Ice Sheet and some of its associated ice shelves.

the rate of glacier flow escalated from 63 km³ (15.1 mi³) in 1996 to 162 km³ (38.9 mi³) in 2005, according to recently analyzed satellite radar data.

At the opposite pole, the thinning of the West Antarctic Ice Sheet, described above, has focused attention on the ice sheet's future. Much of this vast expanse of Antarctic ice is "grounded" on a bedrock base that is well below sea level; that is, a large portion of the ice sheet rests directly on deep bedrock without sea water underneath. At present, the grounded ice shelves act to hold back the flow of the main part of the ice sheet.

Geophysicists regard this as an unstable situation. Rapid melting or deterioration of the ice shelves, perhaps in response to global warming, would release the back pressure on the main part of the ice sheet, which would then move forward and thin rapidly. With the ice reduced in thickness, lower pressure at the bottom of it would allow seawater to enter, and soon most of the sheet would descend and float in ocean water. The added bulk would raise the sea level by as much as 6 m (about 20 ft). Could this actually happen?

New evidence from fossil shoreline deposits suggests that a portion of the Antarctic ice sheet collapsed into the ocean as recently as 14,200 years ago. In response, sea level rose about 20 m (65 ft) in a period of about 500 years. This rate is about 20 times faster than the slow rise of sea level measured today, and about 4 times faster than the average rate at which sea level rose following the end of the last glacial period. Whether this might occur again is still uncertain.

> Due to global warming, the Greenland and Antarctic Ice Sheets are now shrinking—they are melting at the margins—in spite of enhanced snowfall in central regions.

Meanwhile, the climatic warming of the past few decades has caused some ice shelves to thin and, thus, fracture more easily. An example is the Wilkins Ice Shelf, on the southwest side of the Antarctic Peninsula, where a major piece disintegrated in March 2008 (Figure 17.2). This incident was preceded by a collapse of the Larsen B Ice Shelf, on the opposite side of the Antarctic Peninsula, in 2002. As climate changes, it seems likely that ice shelf disintegration will become more frequent.

Global warming is also affecting Arctic sea ice. The ice pack is shrinking in size and becoming significantly thinner, according to a 2009 NASA study using satellite-borne lasers. Between 2004 and 2008, overall sea ice thickness decreased by 0.17 m/yr (7 in./yr), and the proportion of area covered by thicker, older multiyear ice dropped from 62 percent to 23 percent. Summer ice extent in 2007 set a record low. Subsequently, sea ice extent expanded again to more normal values— until 2012, which saw the smallest summer sea ice extent on record, about 49 percent of the long-term average from 1979 to 2000 and about 18 percent below the previous low value for 2007.

As the sea ice cover dissolves, more solar energy is absorbed by the ocean water that becomes exposed to the Sun, warming the Arctic Ocean further. This effect, coupled with enhanced global warming at high latitudes, has led some climate modelers to predict that the Arctic Ocean will be ice-free in the summer by 2050, although some individual predictions call for ice-free summer conditions as soon as 2020.

> According to recent satellite studies, Arctic sea ice is thinning, and summer ice cover is shrinking.

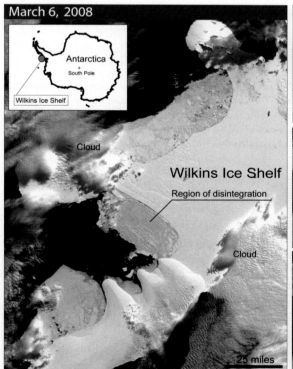

National Snow and Ice Data Center, Boulder, Co

17.2 Breakup of the Wilkins ice shelf as seen by MODIS

These four MODIS images document the disintegration of the Wilkins Ice Shelf in February and March of 2008. Over a period of about a month, 405 km² (160 mi²) of floating ice fractured and collapsed, to become thousands of individual icebergs. A remaining narrow bridge of ice connecting the shelf with nearby Charcot Island later collapsed in April 2009.

Types of Glaciers

Almost 70 percent of the Earth's freshwater is stored in the *cryosphere*, or realm of ice and snow, primarily in the form of **glaciers**. Glaciers form in regions that have low temperatures and sufficient snowfall, conditions found at both high elevations and high latitudes. In mountains, glacial ice can develop even in tropical and equatorial zones if the elevation is high enough to keep average annual temperatures below freezing. Glaciers can take a variety of forms, but they can generally be divided into two broad categories: alpine glaciers and ice sheets, which are also called *continental glaciers*.

ALPINE GLACIERS

Alpine glaciers, or mountain glaciers, are found in high mountain ranges where snow accumulates and temperatures are cold enough to maintain year-round snow cover. The specific forms of alpine glaciers depend on where they occur and the prevailing weather patterns (Figure 17.3). In high mountains, glaciers flow from high-elevation collecting grounds down to lower elevations, where temperatures are warmer. Here the ice melts and evaporates. Glacial meltwater provides freshwater sources for large portions of the globe. Himalayan glaciers, for example, feed rivers throughout China, Southeast Asia, and India. The recent decline in many alpine glaciers due to global climate change, however, presents a serious threat to the supply of freshwater for these regions.

Alpine glaciers originate from a snowfield that accumulates in a bowl-shaped depression called a *cirque*. When alpine glaciers are contained within these basins, they are called *cirque glaciers*. However, most alpine glaciers flow out of these basins to become *valley glaciers*, which occupy sloping stream valleys between steep rock walls (Figure 17.4). When a valley glacier flows out onto a surrounding plain, it appears as a *piedmont glacier*. When a valley glacier terminates in seawater, as a *tidewater glacier*, blocks of ice break off to become icebergs.

Large masses of ice often occur at the very tops of mountain ranges. *Ice caps* are continuous masses that cover mountaintops with distinctive domes. *Ice fields* consist of interconnected valley glaciers with protruding rock ridges or summits called *nunataks*. Ice caps and ice fields feed individual glaciers that travel down the mountainside to lower elevations.

> **WileyPLUS** The Cascade Range Interactivity
> Tour the mountains and glaciers of the Cascade Range and view the impact of volcanic activity on glacial activity.

ICE SHEETS

In arctic and polar regions, temperatures are low enough year-round for snow to collect over broad areas, eventually forming a vast layer of glacial ice. Snow begins to accumulate on uplands, which are eventually

17.3 Types of alpine glaciers

Alpine glaciers form in high mountain regions where snowfall is sufficient and temperatures are cold enough to maintain year-round snow cover.

Melissa Farlow/NG Image Collection

▲ CIRQUE GLACIER
A cirque glacier, such as this one in Montana's Glacier National Park, occupies a steep, bowl-shaped depression near a mountain summit. This glacier, which has shrunk in the present warm interglacial climate, would have been much larger during the last glacial period, as it flowed downhill and joined other glaciers to emerge as a valley glacier below.

▼ VALLEY GLACIER
Valley glaciers flow down steep-sided valleys once carved by streams. Several glaciers have coalesced into this valley glacier in Saint Elias National Park and Preserve, Alaska. Medial moraines of rock debris on the ice surface (see also Figure 17.11) parallel the downslope flow of the glacier.

Ron Niebrugge/Alamy Limited

Gavin Hellier/Getty Images, Inc.

▲ TIDEWATER GLACIER

A glacier that ends in a valley filled by an arm of the sea is called a tidewater glacier. By virtue of its bulk and weight, the glacier may stay grounded—resting on the valley floor—for some distance away from the shoreline before seawater begins to float the ice. Tidewater glaciers often produce icebergs that break off and float away.

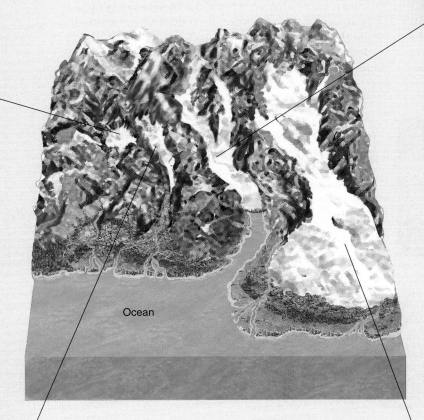

Ocean

▼ PIEDMONT GLACIER

When a mountain glacier flows out of the mountains and onto a surrounding lowland, it is called a piedmont glacier. This example is the Turnstone Glacier on Ellesmere Island, Canada, which is fed by a highland ice cap.

Bryan and Cherry Alexander/Photo Researchers, Inc.

buried under enormous volumes of ice. The layers of ice can reach a thickness of several thousand meters. The ice is thickest at the interior and thins toward the margins. The ice then spreads outward, over surrounding lowlands, and covers all landforms it encounters. We call this extensive type of ice mass an **ice sheet**, or a *continental glacier*. At some locations, ice sheets extend long tongues to reach the sea, known as *outlet glaciers*.

Ice sheets covered parts of northern North America and Eurasia many times during the most recent ice age, which began in the late-Cenozoic era. Today, the only remaining ice sheets are those of Antarctica and Greenland. The Greenland Ice Sheet is shaped like a broad, smooth dome (Figure 17.5), and the only exposed land is a narrow, mountainous coastal strip. The Antarctic Ice Sheet covers 13 million km² (about 5 million mi²) and is as much as 4000 meters (about 13,000 ft) thick—about 1000 m (about 3200 ft) thicker than the Greenland Ice Sheet. Together, these ice sheets make up over 9 percent of the world's ice. No ice sheet exists near the North Pole, which lies in the middle of the vast Arctic Ocean. Ice there occurs only as floating sea ice.

ICE SHELVES, SEA ICE, AND ICEBERGS

In Antarctica and Greenland, ice sheets meet the sea, where they become large plates of floating glacial ice, called **ice shelves**. Ice shelves are fed by the ice sheets, and they also accumulate new ice through the compaction of snow. Because much of the bulk of an ice sheet is above sea level as it reaches the shoreline, the ice rests on the sea floor at first. Here the shelf is said to be "grounded." As the sea floor descends, the shelf sinks lower, until it floats.

Free-floating ice on the sea surface takes two forms: sea ice and icebergs. **Sea ice** is molded by direct freezing of ocean water and accumulation of snow atop the ice. The surface zone of sea ice is composed of freshwater, while the deeper ice is salty.

Icebergs are masses of ice that have broken free from alpine glaciers, terminating in the ocean, or from floating

Paolo Koch/Photo Researchers, Inc.

17.4 An alpine valley glacier

This aerial view shows the Aletsch Glacier in Switzerland.

EYE ON THE LANDSCAPE **What else would the geographer see?** This alpine glacial landscape shows many of the glacial landforms and features described in this chapter. In the near distance are sharp peaks, or horns A, and knife-edge ridges, or arêtes B. In the middle foreground is a bowl-shaped cirque C that no longer bears moving glacial ice. Glacial flow features are especially evident, including medial moraines D, lateral moraines E, and crevasses F that open as the ice falls across a rock step.

Digital Vision/Getty Images, Inc.

17.5 Greenland ice sheet

The Greenland Ice Sheet, seen here in a photo from space, occupies more than 1.7 million km² (about 670,000 mi²) and covers about seven-eighths of this vast island. The view here is of the southern tip, showing the snow-covered terrain of hills and fiords around the central mass of ice.

ice shelves. This process is called *calving* of the glacier or ice shelf. Icebergs float very low in the water because they are only slightly less dense than seawater. About five-sixths of the bulk of an iceberg is submerged. A major difference between sea ice and icebergs is thickness. Sea ice is always less than 5 m (16 ft) thick, generally 2 to 3 m (7 to 10 ft), whereas icebergs may be hundreds of meters thick.

Glacial Processes

Although many parts of the world have snow and ice, only certain regions can produce glaciers. And whether glaciers survive and advance is dependent on specific climatic conditions. Scientists who study the formation and behavior of glaciers are called glaciologists.

FORMATION OF GLACIERS

Glaciers develop as snow accumulates from year to year. When the snowfall of the winter exceeds the loss of snow in the summer due to evaporation and melting, a new layer of snow is added to the surface of the glacier. Surface snow melts and refreezes on warm days and cold nights, compacting the snow and turning it into granular ice called *firn*. This intermediate ice is then compressed over a period of 25 to 100 years into hard, crystalline glacial ice by the weight of the layers above it. The *zone of accumulation* identifies the part of the glacier where snow is accumulating. Precipitation rates, temperatures, and the volume of snow layers dictate the rate of glacial ice formation.

When the ice mass becomes so thick that it begins to move under the force of gravity, it becomes an active glacier. Flowing downhill, the glacier encounters less snowfall and warmer temperatures. Evaporation and melting increase, and eventually more ice is lost than is made. This part of the glacier is the *zone of ablation*. Between the two zones is the *equilibrium line*, where the rate of snow accumulation balances the rate of evaporation and melting (Figure 17.6).

> The mass balance of a glacier depends on snow input in the zone of accumulation, and melting and evaporation in the zone of ablation.

A glacier is an open system that maintains a dynamic balance between the input of accumulated snow and the output of melting or evaporating snow. The glacier's input and output, called its *mass balance*, changes each year as a result of the annual temperature and amount of snowfall. If the mass balance remains positive for some years, the glacier grows in bulk, and its end point, called the *terminus*, moves forward. If the mass balance remains negative, the glacier shrinks. Detailed field measurements allow glaciologists to monitor how the mass balance of glaciers changes over time—for example, as a result of climate change (Figure 17.7).

WileyPLUS Glacier Mass Balance
View this animation to visualize how an alpine glacier gains mass in its upper regions and loses mass in its lower regions.

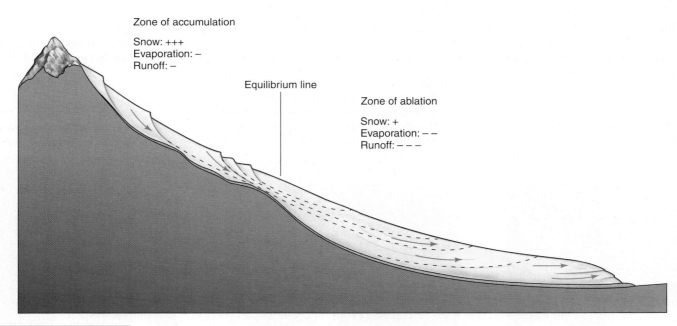

Zone of accumulation

Snow: +++
Evaporation: –
Runoff: –

Equilibrium line

Zone of ablation

Snow: +
Evaporation: – –
Runoff: – – –

17.6 Glacial mass balance

Snowfall exceeds evaporation and melting in the zone of accumulation, where new glacial ice is formed. At lower elevations, increased evaporation and melting exceed snowfall. This is the zone of ablation. The equilibrium line marks the location where accumulation and ablation are about equal. Although the amounts of snow accumulation and ablation change from year to year, the glacier remains more or less in equilibrium until climate changes. Plus and minus signs indicate the relative degree of gain (+) or loss (–) to the glacier by each process in each zone.

MOVEMENT OF GLACIERS

When we think of ice, most of us picture a brittle, crystalline solid. But large bodies of ice, with great thickness and weight, can flow in response to gravity. Flow occurs because the pressure on the ice at the bottom of an ice mass changes its physical properties, causing it to lose rigidity and become plastic. This means that a huge body of ice can move downhill or spread out over a large area. This motion typically happens at a rate of a few centimeters per day.

A glacier is always flowing outward or forward, even if the front of the glacier is retreating. Most of a glacier's movement occurs by slippage between the bottom and sides of the glacier and the rock materials that confine it. In all but the coldest glaciers, high pressure at the base, coupled with the slow flow of geothermal heat and heat caused by friction, creates a layer of liquid water that allows the glacier to slide downhill. This process is called *basal sliding*. Some plastic flow of the glacier also occurs, most rapidly along the centerline of the glacier. In contrast to the plastic ice in the deeper parts of the glacier, the surface of the glacier bears less weight and is more brittle. As the glacier moves over a ridge or cliff, this brittle surface layer may crack, as a result of the tension, and form a *crevasse* (Figure 17.8).

> When snow accumulates to a great thickness, it can turn into flowing glacial ice. Flow is always downward, forward, or outward in response to gravity.

GLACIAL EROSION AND DEPOSITION

Glacial ice normally contains rock and sediment that it has picked up as it flows. These rock fragments range from large angular boulders to pulverized rock flour. Most of this material is composed of loose rock debris and sediments found on the landscape as the ice overrides it. Alpine glaciers also carry rock debris that slides or falls from valley walls onto the surface of the ice.

But glaciers can create their own sediment, too, by eroding underlying bedrock. Glaciers don't generate enough pressure to fracture bedrock directly, but they can loosen and pluck out bedrock blocks that are already fractured and easy to split off. Glaciers also push and drag rock particles against the bedrock, creating small fractures that release rock chips and particles. As a glacier flows over an irregular bed, small volumes of water at the base of the ice commonly refreeze, picking up loose rock fragments from the bedrock and carrying them into the glacial mass. Rock fragments carried on the bottom grind along the bedrock, sometimes smoothing and polishing the rock into grooves that mark the direction of glacial flow (Figure 17.9).

The glacier finally deposits the rock debris at its terminus, where the ice melts. We use the term **glacial drift** to refer to all the varieties of rock debris that are deposited by glaciers. There are two types of drift. *Stratified drift* consists of layers of sorted and stratified clays, silts, sands, or gravels. These materials were deposited by meltwater streams or in bodies of water adjacent to the ice. **Till** is an unstratified mixture of rock fragments, ranging in size

---

17.7 Monitoring the effect of climate change on glaciers

The Grinnell Glacier, located in Glacier National Park, Montana, is one of the world's most visited and photographed glaciers. It is also shrinking rapidly in response to climate warming. This remarkable set of photos shows the history of the glacier from 1938 to 2009. Surveys in 1966 and 2005 showed a reduction in the surface area of the glacier by almost 40 percent. Glaciologists now predict that by 2030, Glacier National Park will have lost all of its glaciers by melting.

▼1938
The glacier fills its basin to a level about halfway up the rock cliff at the rear.

▼1981
The volume of the glacier has decreased very significantly, uncovering the base of the cliff. The glacier ends about halfway across the basin, providing a pond of meltwater.

▼2009
The glacier has retreated to the side of the basin and is just barely visible over the rocky slope in the foreground.

T.J. Hileman photo, Courtesy of GNP Archives

D. Fagre, USGS

Lindsey Bengtson photo, USGS

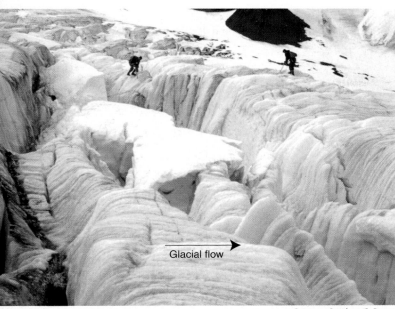

Glacial flow

Courtesy Stephen C. Porter

Deep fissures in the ice, called crevasses, open up as a result of stresses in the brittle surface layer of a glacier. Crevasses form perpendicular to the direction of glacial flow, often where the glacier descends a steep slope or ledge.

from clay to boulders, that is deposited directly from the ice, without water transport. Glaciers drop debris along their margins and where melting occurs, in the form of a *moraine*, as we will see in the following discussion.

Glacial Landforms

Glaciers create both erosional and depositional landforms in mountain ranges and in areas formerly covered by continental ice sheets. During the 2.5 million years of the present Ice Age, the Earth has experienced dozens of **glaciations**. Many parts of northern North America and Eurasia have been covered many times by massive sheets of glacial ice. As a result, glacial ice has shaped many landforms now visible in regions from the midlatitudes to subarctic zones.

> Glaciers sweep up and move loose soil, regolith, and sediments in their paths. They pluck out loose bedrock blocks and can grind and polish hard bedrock.

LANDFORMS MADE BY ALPINE GLACIERS

Alpine glaciers typically erode existing valleys down to hard bedrock, stripping them of regolith developed by weathering and mass wasting. (Figure 17.10). Valley heads are enlarged and hollowed out by glaciers, producing bowl-shaped cirques. A cirque marks the origin of the glacier and is the first landform produced. When a glacier carves a depression into the bottom of a cirque and then melts away, a small lake called a *tarn* can take shape.

Over time, glacial erosion of bedrock and mass wasting of adjacent slopes steepen the sides of the cirque. Intersecting cirques carve away the mountain, leaving peaks called *horns* and sharp ridges called *arêtes*. Where opposed cirques have intersected deeply, a notch called a *col* forms. As the glacier descends from steeper slopes into the valley, smaller tributary glaciers join together to form larger trunk glaciers occupying valleys originally carved by streams. Glacial flow strips away loose regolith and valley bottom sediment.

After the ice has finally melted, a **glacial trough** remains. Tributary glaciers also produce troughs, but they are smaller in cross section and less deeply scoured by the smaller glaciers. When the floors of these troughs lie above the level of the main trough, they are called *hanging valleys*. When streams emerge later in these valleys, they feature scenic waterfalls and rapids that cascade down steep slopes to the main trough below. Major troughs sometimes hold large, elongated trough lakes.

This grooved and polished surface marks the former path of glacial ice on the side of Tracy Arm Fiord in Alaska.

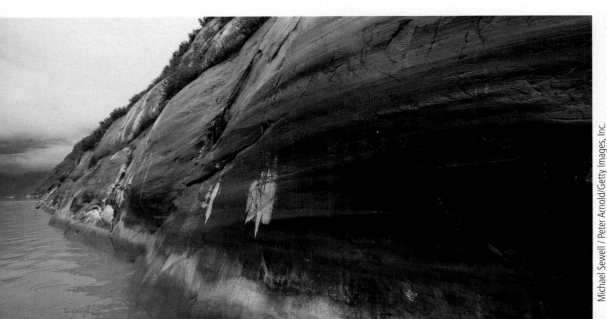

Michael Sewell / Peter Arnold/Getty Images, Inc.

17.10 Landforms produced by alpine glaciers

Alpine glaciers erode and shape mountains into distinctive landforms. Although larger alpine glaciers can widen and deepen valleys, their main work is to scrub existing slopes and valleys down to hard bedrock.

◄ **BEFORE GLACIATION**
This region has been sculptured entirely by weathering, mass wasting, and streams. The mountains look rugged, with steep slopes and ridges. Small valleys are steep and V-shaped, while larger valleys have narrow floodplains of alluvium and debris.

SNOW ACCUMULATION ►
Climate change (cooling) increases snow buildup in the higher valley heads, forming *cirques*. Their cup shapes develop as regolith is stripped from slopes and bedrock is ground by ice. Slopes above the ice undergo rapid wasting from intense frost action.

David Wrench/Leslie Garland Pic. Lib./Alamy Limited

Tarn When a glacier carves a depression into the bottom of a cirque and then melts away, a small lake called a *tarn* is formed.

Tarn Rock basin high in smaller valleys that becomes a small glacial lake.

▼ **MELTING GLACIERS**
As glaciers diminish, they reveal their handiwork in distinctive landforms.

GLACIATION ►
Thousands of years of accumulating snow and ice develops these new erosional forms.

Horn Sharp peak that develops where three or more cirques grow together.

Arête A jagged, knife-like ridge forms where two cirque walls intersect from opposite sides.

Glacial troughs When the ice disappears, a system of steep walled troughs is revealed.

Hanging valley Smaller tributary valleys that join the main glacier valley may be left "hanging" at a higher elevation as the trunk glacier deepens the main valley.

When the floor of a trough that is open to the sea lies below sea level, the seawater enters as the ice front recedes, creating a **fiord**. Fiords are opening up today along the Alaskan coast, where some glaciers are melting back rapidly and ocean waters are filling their troughs. Fiords are found largely along mountainous coasts between latitudes 50° and 70° N and S.

> Alpine glaciers strip valleys of their soil, regolith, and sediment to form glacial troughs. A glacial trough leading into the ocean is a fiord.

In addition to erosional features, alpine glaciers also leave behind distinctive depositional features. Rock waste slides or falls onto the glacier's edges, where it is worked into a ridge called a *lateral moraine* on either side of the trough (Figure 17.11). When two tributary glaciers join to form a larger trunk glacier, their lateral moraines often combine and continue together down the center of the trough, to become a *medial moraine*.

WileyPLUS Hanging Valleys
Watch alpine glaciers grow and coalesce to carve U-shaped valleys, then retreat to expose hanging valleys. An animation.

17.11 Lateral and medial moraines

When two ice streams flow together, their separate lateral moraines join together to form a medial moraine. This aerial view shows medial moraines at the junction of two ice streams in the Saint Elias Range, Wrangell-St. Elias National Park, Alaska.

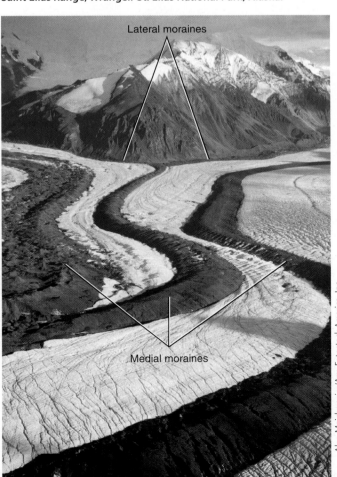

Alan Majchrowicz/Age Fotostock America, Inc.

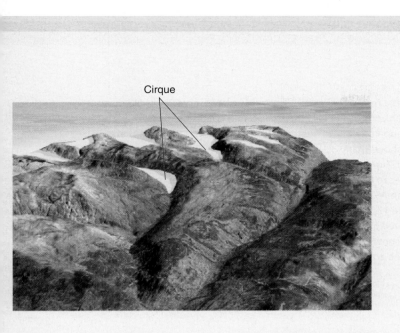

Cirque

Harvey Lloyd/Getty Images, Inc

©Marli Bryant Miller

Cirque Valley heads are enlarged and hollowed out by glaciers, producing bowl-shaped *cirques*.

Horn and Arête Intersecting cirques carve away the mountain mass, leaving peaks called *horns* and sharp ridges called *arêtes*.

Col Notch that forms where opposed cirques have intersected deeply.

Tributary glaciers flow together to form the main trunk glacier.

Lateral moraine Debris ridge formed along ice's edge next to trough wall.

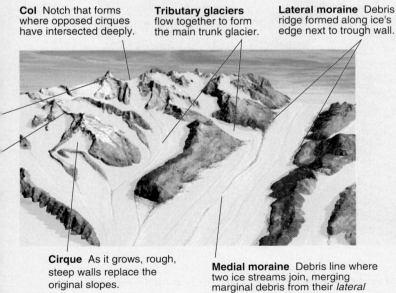

Cirque As it grows, rough, steep walls replace the original slopes.

Medial moraine Debris line where two ice streams join, merging marginal debris from their *lateral moraines*.

Remote Sensing of Glaciers

Because glaciers are often found in inaccessible terrain or in extremely cold environments, they are difficult to survey and monitor. This makes satellite remote sensing an invaluable tool for studying both continental and alpine glaciers.

Some of the world's more spectacular alpine glaciers are found in South America, along the crest of the Andes in Chile and Argentina, where mountain peaks reach as high as 3700 m (about 12,000 ft). Figure 17.12 is a true-color image of Cerro San Lorenzo (San Lorenzo Peak), acquired by astronauts aboard the International Space Station. It shows many of the glacial features described in Figure 17.10.

Northwest of Cerro San Lorenzo lies the San Quintín Glacier, shown in an ASTER image, acquired on May 2, 2000 (Figure 17.13). This color-infrared image was acquired at 15-m (49.2-ft) spatial resolution and shows the glacier in fine detail. The intense red color indicates a thick vegetation cover.

The world's largest glacier is, of course, the ice cap that covers nearly all of Antarctica. At the edges of this ice cap are *outlet glaciers*, where glacial flow into the ocean is quite rapid. Figure 17.14 shows the Lambert Glacier, one of the largest and longest of Antarctica's outlet glaciers. From the confluence of the ice streams at the left to the tip of the Amery Ice Shelf, the glacier measures about 500 km (315 mi). The image was acquired by Canada's Radarsat radar imager.

Courtesy NASA

17.12 Andean alpine glacial features

This astronaut photo shows Cerro San Lorenzo, a peak along the crest of the Andes in Chile and Argentina. The peak itself is a glacial horn (H). Leading away from the horn to the south is a long, sharp ridge, or arête (A). To the left of the peak is a cirque (C), now only partly filled with glacial ice. These features were carved during the Ice Age, when the alpine glaciers were larger and filled the now-empty glacial troughs (T) behind the ridge to the right.

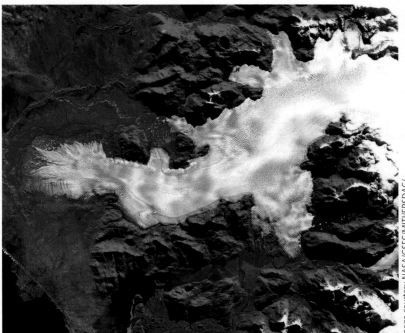

Image courtesy NASA/GSFC/MITI/ERSDAC/ JAROS and U.S./Japan ASTER Science Team

17.13 San Quintín Glacier, Chile, imaged by ASTER

This is the largest outflow glacier draining the Northern Patagonia ice field. The terminal lobe of the glacier, which is about 4.7 km (2.6 mi) wide, ends in a shallow lake of sediment-laden water that drains by fine streams into the Golfo de Penas at lower left. Note the low, semicircular ridge a short distance from the lake; this is a terminal moraine, marking a former stand of the ice tongue. A high cloud partly obscures the southern end of the snout and nearby coastline.

Image courtesy Canadian Space Agency/NASA/Ohio State University/Jet Propulsion Laboratory, Alaska SAR Facility

17.14 Lambert Glacier, Antarctica

The Lambert Glacier as imaged by Radarsat during the 2000 Antarctic Mapping Mission. This instrument allows the measurement of the velocity of glacier flow by comparing paired images acquired at different times (in this case, 24 days apart) using a technique called radar interferometry. Brown tones indicate little or no motion, and show both exposed mountains and stationary ice. Green, blue, and red tones indicate increasing velocity, with arrows showing direction. Glacial flow is most rapid at the left, where the flow is channeled through a narrow valley, and at the right, where the glacier spreads out and thins to feed the Amery Ice Shelf.

LANDFORMS MADE BY ICE SHEETS

Like alpine glaciers, ice sheets strip away surface materials and erode loose bedrock. Repeated glaciations can excavate large amounts of rock at locations where the bedrock is weak and fractured and the flow of ice is channeled by a valley along the ice flow direction. Under these

Courtesy NASA

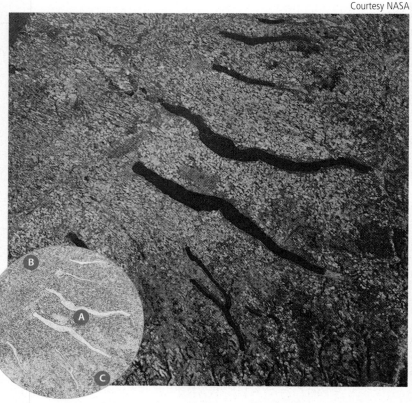

17.15 Finger Lakes

The Finger Lakes region of western New York is shown in this photo taken by astronauts aboard the Space Shuttle. The view is looking toward the northeast. The lakes occupy valleys that were eroded and deepened by glacial ice.

EYE ON THE LANDSCAPE **What else would the geographer see?**
The Finger Lakes have been called "inland fiords" for their resemblance to the fiords of glaciated regions. The lakes were eroded from preexisting stream valleys by ice action during the multiple continental glaciations of the Ice Age. The two largest lakes in the center of the image **A** are Cayuga Lake (upper) and Seneca Lake (lower). The pattern of agricultural fields in the center and upper left part of the image **B** marks the intensive agricultural development of the Lake Erie lowlands and foothills of the Appalachian Plateau, which are mantled with productive soils of glacial origin. To the south, at the bottom of the image **C**, the terrain becomes more dissected, with more pronounced valleys and ridges, as it slopes upward to the higher elevations of the plateau.

conditions, the ice sheet behaves like a valley glacier, scooping out the loose material to form a deep glacial trough (Figure 17.15).

> Moraines are piles of debris and sediment that accumulate at the front or sides of a glacier.

Ice sheets are also responsible for depositional landforms. Like huge conveyor belts, traveling ice sheets transport and deposit debris. When the glacial margin is stationary over long periods of time, as was the case during the last glaciation, deposited debris accumulates in distinctive patterns. As the glaciers retreat, they leave behind several types of depositional landforms (Figure 17.16).

Sediment deposited in a ridge along the front or side edge of a glacier is known as a **moraine** (Figure 17.17). Glacial till and water-laid sediment that accumulates at the front edge of the ice results in an irregular jumbled heap of debris, sand, and gravel called the *terminal moraine*. When the ice front retreats, it may pause for some time along a number of positions, forming *recessional moraines*.

As glacial ice moves forward, it compacts and compresses the sediment underneath it, creating *lodgment till*, a layer of particles of all sizes that is usually tough and dense (Figure 17.18) On top is a layer of *melt-out till*, which is let down as the ice retreats or stagnates and melts in place. It is a lighter and looser deposit that is easier to plow and work. Together, these two layers make up the till plain behind the terminal moraine.

The till plain can be thick, and may bury the pre-existing hills and valleys. Over parts of North America that were formerly covered by ice sheets, glacial drift thickness averages from 6 m (about 20 ft) over mountainous terrain, such as New England, to 15 m (about 50 ft) and more over the lowlands of the north-central United States. Over Iowa, drift thickness is from 45 to 60 m (about 150 to 200 ft); and over Illinois, it averages more than 30 m (about 100 ft). In some places, where deep stream valleys already existed before the glaciers advanced, such as in parts of Ohio, drift is much thicker.

Two features of the till plain are eskers and drumlins (Figure 17.19). An *esker* is a sinuous ridge of sand and gravel, developed at the bottom of an ice tunnel or channel that drains the ice sheet. A *drumlin* is a smoothly rounded, oval hill of glacial till that resembles the bowl of an inverted teaspoon.

Beyond the moraine, an outwash plain is formed from water-laid sediment left by streams carrying water away from the ice front (refer back to Figure 17.16). The plain is built up from layers of sand and gravel. Although the plain is generally smooth, it is often marked by kettles and kames. A *kettle* is a depression, originally formed as outwash sand and gravel builds up around a block of stagnant ice. When the ice block melts, the kettle hole

> Eskers, drumlins, kettles, and kames are landforms of till and outwash plains left by ice sheets.

17.16 Landforms produced by ice sheets

In glacial landscapes, a variety of features mark both the expansion and retreat of ice sheets.

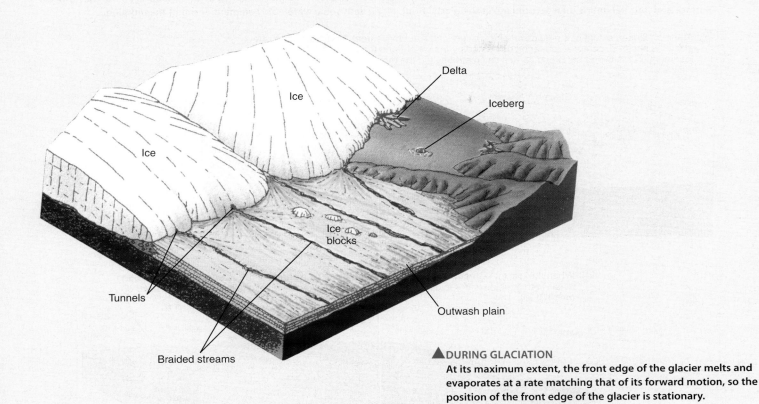

Delta

Iceberg

Ice

Ice

Ice blocks

Tunnels

Braided streams

Outwash plain

▲ **DURING GLACIATION**

At its maximum extent, the front edge of the glacier melts and evaporates at a rate matching that of its forward motion, so the position of the front edge of the glacier is stationary.

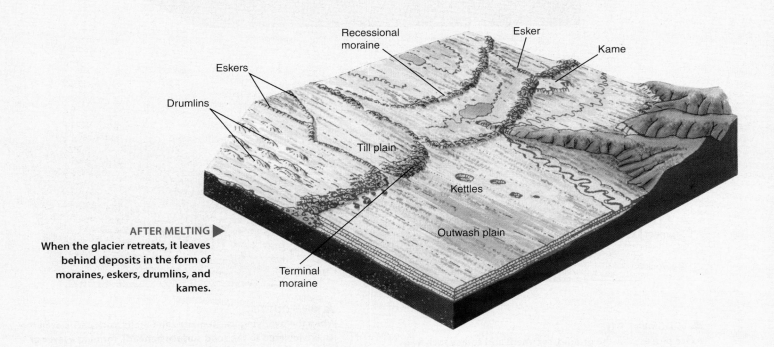

Recessional moraine

Esker

Kame

Eskers

Drumlins

Till plain

Kettles

Outwash plain

AFTER MELTING ▶

When the glacier retreats, it leaves behind deposits in the form of moraines, eskers, drumlins, and kames.

Terminal moraine

17.17 Formation of a moraine

A moraine forms at the end of an advancing glacier, which carries debris forward, like a conveyor belt, toward the glacial front. As the ice moves forward, melting and evaporation reduce the glacier's bulk and leave behind a deposit of glacial debris on the ice surface. Surface and internal debris then accumulate at the glacial front, along with some water-laid sediment, forming the moraine.

1 At the maximum extent of the glacier, debris accumulates at the ice front. Sediment under the glacier is sheared, compressed, and compacted by the weight and motion of the ice, forming lodgment till. Meltwater sorts and carries sediment beyond the moraine, forming the outwash plain.

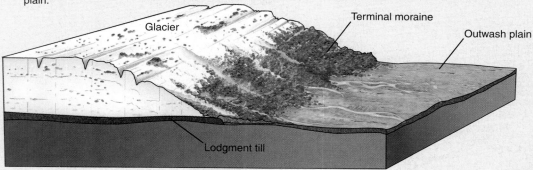

2 When the glacier retreats, the terminal moraine is left as a series of irregular piles of debris, often mixed with jumbled water-laid sediments. Debris in and on the glacier is left behind as a layer of melt-out till.

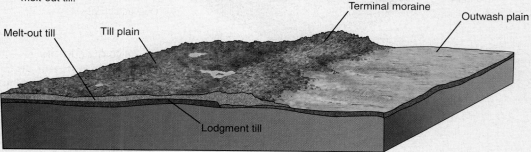

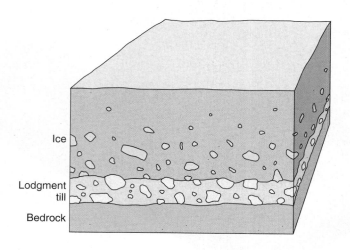

▲ LODGMENT TILL

As ice passes over the ground, sediment and coarse rock fragments of clay-rich debris that were previously dragged forward beneath the ice are pressed into a layer of lodgment till.

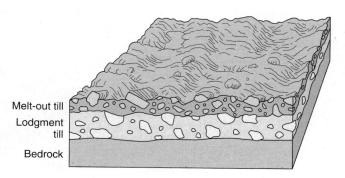

▲ MELT-OUT TILL

When the overlying ice stagnates and melts, rock particles in the ice are lowered to the solid surface beneath, forming a layer of melt-out till consisting of a mixture of sand and silt, with many angular pebbles and boulders. The layer of dense lodgment till lies below it.

17.18 Glacial till

Till is a sedimentary deposit laid down direction by glacial ice.

17.19 Landforms associated with till plains and outwash plains

Till and outwash plains often include several characteristic glacial landforms.

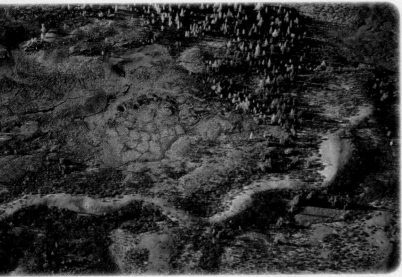

Grambo Photography/All Canada Photos/Age Fotostock America, Inc.

Arthur N. Strahler

▲ ESKER

The curving ridge of sand and gravel in this photo is an esker, marking the bed of a river of meltwater flowing underneath a continental ice sheet near its margin. Near Whitefish Lake, Northwest Territories, Canada.

▲ DRUMLIN

This small drumlin is constructed from lodgment till deposited beneath moving glacial ice. Drumlins are usually found in groups on the till plain, with the long axis of each drumlin parallel to the direction of ice movement. Drumlins often appear at rock projections. As the glacier moves forward, till piles up in front of the projection and is also deposited behind the rock in a narrowing tail. However, some drumlins lack rock projections and are composed entirely of till. This drumlin is located south of Sodus, New York. Its tapered shape indicates that it was deposited as ice moved from upper right to lower left.

▼ KAME

An isolated mound or hill of sand and gravel on the outwash plain is known as a kame. Some kames are formed as terraces between ice and a valley wall. Others are deltas built out into ice-front lakes dammed by ice. Still others accumulate when sediment is washed into openings within or between ice blocks. This is the type of kame shown in the photo, taken in Antarctica.

▼ KETTLE

The outwash plain typically contains depressions, called kettles, that mark the position of ice blocks embedded in the outwash deposits. Often, the kettle dips below the water table, creating a kettle pond. This example is in northern Minnesota, photographed with fall colors.

Carlyn Iverson/Getty Images, Inc.

USGS

in the outwash plain remains. A *kame* is a hill or mound of water-laid sediment found on the outwash plain.

Pluvial lakes are another type of landform associated with continental glaciation. During the Late-Cenozoic Ice Age, some regions experienced a cooler, moister climate with more precipitation. In the western United States, closed basins filled with water and became pluvial lakes. The largest of these, glacial Lake Bonneville, was about the size of Lake Michigan and occupied a vast area of western Utah. With the warmer and drier climate of the present interglacial period, these lakes shrank greatly in volume. Lake Bonneville became the present-day Great Salt Lake. Many other lakes dried up completely, forming desert playas.

HUMAN USE OF GLACIAL LANDFORMS

Landforms associated with glaciers are of major environmental importance. Glaciation can have both positive and negative effects on the land for agricultural development. In hilly or mountainous regions, such as New England, the glacial till is thinly distributed and extremely stony, making cultivation difficult. Till deposits built up on steep mountains or hillslopes can absorb water from melting snows and spring rains and then become earthflows. Crop cultivation is also hindered along moraine belts because of the steep slopes, the irregularity of the topography, and the number of boulders. Moraine belts are, however, well suited to pastures.

Flat till plains, outwash plains, and lake plains, on the other hand, can sometimes provide very productive agricultural land. There are fertile soils on till plains and on exposed lakebeds bordering the Great Lakes. Their fertility is enhanced by the blanket of wind-deposited silt (loess) that covers these plains.

Stratified drift provides sand and gravel deposits from outwash plains, kames, and eskers that are used for road construction and in concrete. Thick, stratified drift deposits can also serve as aquifers, which are a major source of groundwater.

Periglacial Processes and Landforms

In the arctic and alpine tundra regions, year-round cold temperatures keep the ground frozen and prevent the growth of substantial vegetation. These **periglacial** regions (*peri-* means "near") were formerly covered with ice or near a glacial front. Their present-day climate, which is exemplified by treeless, frozen ground, gives rise to a unique set of periglacial processes and landforms.

PERMAFROST

In the tundra, mean annual temperatures are below freezing, and only the top layer of soil or regolith thaws during the warmest month or two. Ground and bedrock that lie below the freezing point of freshwater (0°C;

32°F) all year-round are called **permafrost**. Permafrost includes clay, silt, sand, pebbles, and boulders, as well as solid bedrock perennially below freezing.

Distribution Of Permafrost

Permafrost is more or less extensive based on climate conditions in a particular region (Figure 17.20). In the coldest regions of the northern hemisphere, permafrost is continuous; all ground surfaces are frozen, except those beneath deep lakes. The zone of *continuous permafrost* coincides largely with the tundra climate, but it also includes a large part of the boreal forest climate of Siberia. Permafrost reaches to a depth of 300 to 450 m (about 1000 to 1500 ft) in the continuous zone near latitude 70° N. *Subsea permafrost*, formed during the last glaciation when sea level was lower, occurs at some locations on the shallow floor of the Arctic Ocean. *Discontinuous permafrost* is defined as areas where permafrost occurs only under favorable conditions, such as north-facing slopes. These areas are common in much of the boreal forest climate zone of North America and Eurasia. Zones of *alpine permafrost* occur in high-mountain areas where temperatures remain cold.

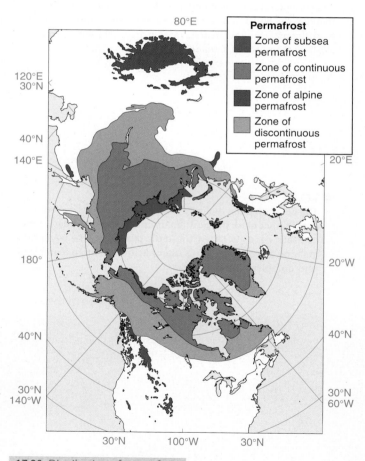

17.20 Distribution of permafrost

Most of the world's permafrost occurs in the northern hemisphere, where regions of continuous and discontinuous permafrost can be clearly distinguished. Alpine permafrost occurs in high mountain regions, and subsea permafrost occurs along continental margins at the highest latitudes.

Climate change today is raising temperatures and bringing more precipitation to much of the North American Arctic region, which in turn will alter the distribution of permafrost. Widespread warming trends are being recorded, and climate scientists predict that by the end of the century, arctic temperatures will have climbed by 4°C to 8°C (7.2°F to 14.4°F), and annual precipitation will have increased by as much as 20 percent. As the climate heats up, the boundary between continuous and discontinuous permafrost will shift poleward as increased snowfall insulates the ground under new forest. Discontinuous permafrost will be become sporadic at the southern boundary, and the isolated permafrost to the south will largely disappear.

Permafrost Processes

Permafrost terrains have a shallow surface layer, known as the **active layer**, which thaws with the changing seasons. The active layer ranges in thickness from about 15 cm (6 in.) to about 4 m (13 ft), depending on the latitude and nature of the ground. The thickness of the active layer varies with climate, getting thinner during colder periods and thicker during warmer periods. Currently, as a result of warming trends resulting in higher temperatures in the Arctic, the active layer over broad areas of permafrost is deepening.

Within even the areas of deepest continuous permafrost are isolated pockets that remain unfrozen. Called *taliks*, some of these pockets appear within permafrost as lenses of persistent unfrozen water. Other taliks occur under lakes and rivers, where water remains at or near 0°C (32°F), keeping the ground below these features unfrozen.

> Permafrost, which includes ground, water, and bedrock, is permanently below freezing. The active layer of permafrost is the shallow surface layer that freezes and thaws each year.

17.21 Ice Wedges

Ice wedges build up over many years as shrinkage cracks fill with fresh ice.

▼ ICE WEDGE FORMATION
During extreme winter cold, a crack opens when permafrost shrinks. In the spring, surface meltwater enters the crack and freezes, expanding the crack. After several hundred winters, the ice wedge will have grown quite large—and will continue to grow as this seasonal sequence repeats itself, to as wide as 3 m (about 10 ft) and as deep as 30 m (about 100 ft).

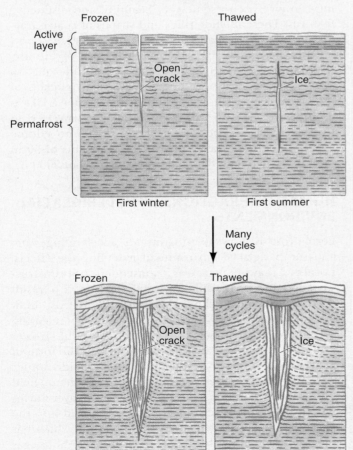

▼ ALASKAN ICE WEDGE
Exposed by river erosion, this great wedge of solid ice fills a vertical crack in organic-rich floodplain silt along the Yukon River, near Galena in western Alaska.

Courtesy Troy L. Pewe

GROUND ICE AND PERIGLACIAL LANDFORMS

Water is commonly found in pore spaces in the ground, where, in its frozen state, it is known as *ground ice*. The amount of ground ice present within and above the permafrost layer varies greatly. At great depth, pores in the rock hold little if any frozen water. Near the

17.22 Ice wedge polygons

These polygons, in the Kolyma region, Siberia, in far northeastern Russia, are formed by ice wedges. The polygons are filled with meltwater in this summer photo; green vegetation marks the high ground above the wedges.

NHPA/Photo Researchers, Inc.

17.23 Gelifluction

Bulging masses of water-saturated regolith, along with plants and soil, have slowly moved downslope, flowing atop an impermeable bed of permafrost, to form gelifluction lobes and low terraces. These examples are found at Breiddalfjellet, Breivikeidet, Troms, Norway.

Blickwinkel/Alamy Limited

surface, ground ice can take the form of a body of almost 100 percent ice.

One type of ground ice is the *ice wedge* (Figure 17.21). Ice wedges develop as repeated cycles of melting and freezing open up cracks in the ground, where ice accumulates. Ice wedges are typically interconnected into a system of polygons, called *ice-wedge polygons* (Figure 17.22). Another type of ground ice is the *ice lens*. At the end of the warm season, meltwater accumulates at the bottom of the active layer, on the permafrost table; it then freezes into a more-or-less horizontal layer of ice that can grow and persist. Ice lenses form under other conditions as well.

> Ice wedges develop in vertical cracks in permafrost opened by intense winter cold. They are often interconnected as ice-wedge polygons.

During the short summer season, when ice in the active layer thaws, the soil and regolith becomes saturated with water that cannot escape downward into solid permafrost. In a process called **gelifluction**, the saturated soil can slump or flow down shallow slopes and take shape as terraces or lobes. (Figure 17.23). With the return of cold temperatures, water in the active layer freezes. Recall from Chapter 10 that water expands when it freezes, and ice accumulation in the soil can cause the ground to shift horizontally through frost heaving, or vertically through frost thrusting. In areas that contain coarse-textured regolith, consisting of rock particles in a range of sizes, this shifting and thrusting can give rise to some very beautiful and distinctive features. These include rings and stripes of coarse fragments called **patterned ground** (Figure 17.24).

Another remarkable ice-formed feature of the arctic tundra is a conspicuous conical mound called a *pingo* (Figure 17.25). Pingos emerge when pockets of unfrozen ground under a lakebed freeze after a lake is drained. As the unfrozen groundwater is converted to ice, its volume expands and pushes upward into a dome shape. In extreme cases, pingos reach heights of 50 m (164 ft), with a base diameter of 600 m (about 2000 ft).

HUMAN INTERACTIONS WITH PERIGLACIAL ENVIRONMENTS

Permafrost is now thawing at an accelerating rate, and the periglacial environment is feeling the effects. Localized thawing occurs as a consequence of construction and industry, while regional thawing is a result of human-induced climate change. Permafrost thaw further contributes to global warming by releasing greenhouse gases into the atmosphere. The frozen ground contains vast amounts of carbon, in the form of peat, which decays to release CO_2 more rapidly under warming temperatures. Also abundant is methane, held within ice as methane hydrate. As the active layer warms and deepens, more of these gases are released into the atmosphere. Remarkably, this phenomenon, which has the potential to rapidly escalate the total greenhouse gas

17.24 Patterned ground

Frost action sorts stones into distinctive patterns.

Ralph Lee Hopkins/Alamy Limited

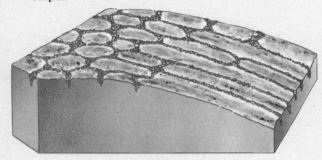

◀ **POLYGON PATTERNS**
Sorted polygons of gravel reconfigure as stone rings on this nearly flat land surface in Northeastland, Svalbard, Norway. The rings are 3 to 4 m (10 to 13 ft) across, and the gravel ridges are 20 to 30 cm (8 to 12 in,) high.

▼ **STRIPE PATTERNS**
On steep slopes, soil creep elongates these polygons in the downslope direction, turning them into parallel stone stripes.

volume in the atmosphere, is only now being included in most current climate models.

With climate warming, the upper layers of permafrost, which are often rich in ice, begin to melt and release water. The ground collapses, producing water-filled depressions. And because water conducts heat into the permafrost more effectively than vegetation-covered soil, permafrost thawing expands, and the depressions grow into shallow lakes. Now the landscape resembles limestone karst formations, so we call the new terrain *thermokarst* (Figure 17.26). Instead of being shaped by chemical solution, however, thermokarst terrain is generated by heat flow (*thermo-* means "heat").

Permafrost thaw can also be caused by burning or scraping away layers of decaying matter and plants from the surface of the tundra or arctic forest. Even the presence of a heated building or an insulated pipe directly on the frozen ground leads to thawing. As the ice under the structure

Paolo Koch/Photo Researchers, Inc.

17.25 Pingo

This low hill on the tundra landscape of Northwest Territories, Canada, is a pingo. The result of water freezing in a drained lake bed, a pingo has a core of ice that is covered with soil and regolith.

Courtesy NASA

17.26 Thermokarst

This aerial photo is used by the USGS to monitor thermokarst expansion and drainage along a section of Alaska's North Slope coastline. Thermokarst landscapes have compounded the challenges facing the oil resource extraction industry due to their unstable soils and poor drainage, the effects of which are felt on both roads and facilities.

17.27 Permafrost thaw

The permafrost has thawed under the weight of these old buildings, located in Dawson City, Yukon Territory, Canada, causing them to sag together and their foundations to settle.

imagebroker.net/SuperStock

melts, the ground subsides, causing the structure to settle (Figure 17.27). In some cases, the disturbance can set off thermokarst formation.

Responsible building practices can reduce permafrost thaw. These practices include placing buildings on piles with an insulating airspace below or, alternatively, insulating the surface with a thick pad of coarse gravel between the permafrost and the new building. In arctic settlements, steam and hot-water lines are now laid aboveground to prevent thaw of the permafrost layer.

> Clearing of natural surface layers can induce rapid thawing of ice masses in permafrost, leading to thermal erosion and the growth of shallow thermokarst lakes.

Global Climate and Glaciation

Glacial ice sheets have a major impact on our global climate. For starters, glacial ice sheets reflect much of the solar radiation they receive, and so they have a direct influence the Earth's radiation and heat balance. The temperature difference between ice sheets and warm regions near the Equator helps drive the global heat transport system: When the volume of glacial ice increases, as it does during a glaciation, sea level falls. When the ice sheets melt away, sea level rises. Today's coastal environments evolved as sea level rose in response to the melting of the last ice sheets of the Ice Age.

HISTORY OF GLACIATION

Glaciation occurs when temperatures fall in regions of ample snowfall, allowing ice to accumulate and build. *Deglaciation* happens when the ice melts at the beginning of a period of milder climate, called an **interglaciation**. An **ice age** is a period of millions of years of generally cold climate consisting of many alternating glaciations and interglaciations. Throughout its long history, the Earth has cycled through a number of ice ages.

The Late-Cenozoic Ice Age

The Cenozoic era, which began about 66 million years ago, has experienced a gradually cooling climate. By about 40 million years ago, the Antarctic Ice Sheet began to form, and by about 5 million years ago, the Arctic Ocean was ice covered. Throughout the past 2.5 million years or so, the Earth has been experiencing the **Late-Cenozoic Ice Age** (or, simply, *the Ice Age*). This period consists of alternating glacial and

> An ice age includes cycles of glaciation, deglaciation, and interglaciation. The Earth is presently in an interglaciation.

interglacial periods; five major glaciations have occurred in the past 500,000 years.

Scientists have established a record of climate cycles during this period by analyzing layers of oceanic sediment brought up in cores by deep-sea drilling. The Earth's magnetic field experienced many sudden reversals of polarity in Cenozoic time, and the absolute ages of these reversals are known with certainty. By observing these reversals in the sediments of drill cores, geoscientists were able to date the layers accurately. They also studied the composition and chemistry of the core layers, which yield a record of ancient temperature cycles in the air and ocean. These cores have revealed a long history of alternating glaciations and interglaciations going back at least 2.5 million years, with climatic oscillations beginning even earlier.

The last major glaciation, called the Wisconsin glaciation, started about 120,000 years ago. During this glaciation, ice sheets covered much of North America and Europe, as well as parts of northern Asia and southern South America (Figure 17.28); and all high mountain areas of the world developed alpine glaciers. The white surface of the expansive ice sheet progressively reflected more solar energy back to space as it covered soil and vegetation, increasing the Earth's reflectivity, or albedo. This change contributed to global cooling by reducing the amount of solar energy left to enter the lithosphere, hydrosphere, and biosphere. The maximum ice advance of the glaciation took place about 20,000 years ago, but retreat set in shortly thereafter. Once the ice began to melt, the albedo was lowered again, which contributed to more warming.

This and earlier glaciations have resculpted the face of North America's landscape. The moving ice sheet flattened and scraped it, and redefined the continent's major drainage systems. Great inland lakes were created behind tremendous ice dams. When these ice dams broke, the abrupt drainage of the lakes carved out new

17.28 Maximum glaciation

These maps show the maximum glaciation that occurred during the last advance of the ice sheets. During glaciations, sea level was much lower. The present coastline is shown for reference only.

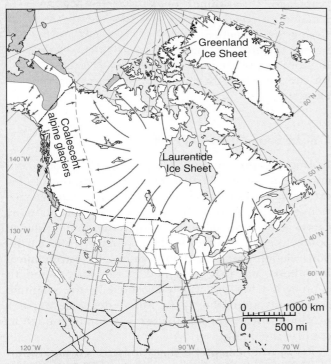

Continental glaciers of the Ice Age in North America at their maximum extent reached as far south as the present Ohio and Missouri rivers.

This area in southwestern Wisconsin escaped an ice cover and is known as the Driftless Area.

◀ **GLACIATION IN NORTH AMERICA**

The Laurentide Ice Sheet spread across eastern and central Canada and into the northern United States. Coalescent alpine glaciers extended the ice sheet to the Pacific Coast.

▼ **GLACIATION IN EUROPE**

The Scandinavian Ice Sheet dominated northern Europe during the Ice Age glaciations. The present coastline is far inland from the coastline that prevailed during glaciations.

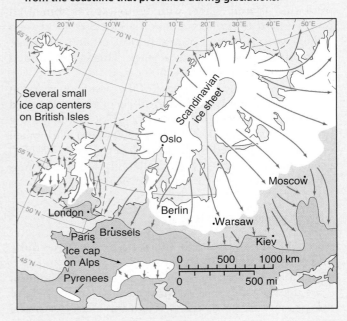

river valleys. Glaciologists note that a series of these great floods have occurred during the Ice Age.

During glaciations, much more of the Earth's water was contained in ice, so sea level was as much as 125 m (410 ft) lower than it is today, exposing large areas of the continental shelf on both sides of the Atlantic Basin. Rivers draining the continents deepened their valleys to meet the lowered sea level. The exposed continental shelf had a highly productive ecosystem and vegetated landscape, richly populated with animal life.

The weight of the continental ice sheets, covering vast areas with ice masses several kilometers thick, pushed down on the crust, depressing some regions by hundreds of meters at some locations.

> During a glaciation, sea level falls substantially. Where the ice is thick, its weight pushes down the crust.

When the ice melted, the crust began to rebound— and it is still doing so today in some locations.

THE HOLOCENE EPOCH

Since the Wisconsin glaciation ended, the Earth has been in a period of interglaciation called the **Holocene epoch**. This period began with a rapid warming of ocean-surface temperatures. Continental climate zones then quickly shifted poleward, and plants recolonized the formerly glaciated areas.

There were three major climatic periods during the Holocene epoch leading up to the last 2000 years. These periods are inferred from studies of fossil pollen and spores preserved in glacial bogs that show changes in vegetation cover over time. The earliest of the three is the Boreal stage, characterized by boreal forest vegetation in midlatitude regions. A general warming followed until the Atlantic stage, with temperatures somewhat warmer than today, was reached about 8000 years ago (–8000 years). Next came a period of below-average temperatures—the Subboreal stage. This stage spanned the age range of –5000 to –2000 years.

We can describe the climate of the past 2000 years on a finer scale, thanks to historical records and detailed evidence. A secondary warm period occurred in the period ce 1000 to 1200 (–1000 to –800 years). This warm episode was followed by the Little Ice Age, ce 1450–1850 (–550 to –150 years), when valley glaciers made new advances and extended to lower elevations.

TRIGGERING THE ICE AGE

About 50 million years ago, the Earth's climate was quite uniform and mild. Today, however, we are in an interglaciation of an ice age that began at least 2.5 million years ago, possibly earlier. Most earth scientists believe the primary cause of this ice age was plate tectonics, which changed the configuration of the continents and oceans, shifting the circulation patterns of the oceans and atmosphere.

By the middle of the Cenozoic era, North America, Eurasia, and Greenland had moved north toward the pole, creating a polar ocean largely surrounded by land. This reduced, and at times totally cut off, the flow of warm ocean currents into the polar ocean, leaving it ice-covered much of the time. The high albedo of the ice then lowered temperatures in high latitudes, setting the stage for the formation of ice sheets on the encircling continents.

In the southern hemisphere, Antarctica had taken up a position over the South Pole, where it was ideally situated to develop an ice cap. Australia, migrating toward the Equator, left a southern ocean free to circulate around the Antarctic continent, isolating it climatically from the other continents. By about 15 million years ago, both the East and West Antarctic Ice Sheets were growing. The collision of the Indian plate with the Eurasian plate pushed up the Himalaya Mountains and the Plateau of Tibet, changing atmospheric circulation patterns and adding to the world's snowy highland area. About 3 million years ago, the Isthmus of Panama emerged and the Bering Strait opened, leading to the final configuration of land, ocean, and ice for the Ice Age to begin in earnest.

> The current Ice Age was most likely triggered by the motions of continents, which provided a snow-covered Antarctic continent and an ice-covered Arctic Ocean, restricted ocean flow paths, and changed atmospheric circulation.

CYCLES OF GLACIATION

During the present Ice Age, glaciers have come and gone many times. What mechanism is responsible for the many cycles of glaciation and interglaciation experienced by the Earth during this period? Of the many causes that have been proposed, the best evidence points to cyclic changes in the geometry of the Earth–Sun relationship. This theory, called the **astronomical hypothesis**, was first presented in the early twentieth century but was not widely accepted by the scientific community until the 1970s, when deep-ocean cores provided climate records that were accurate enough to confirm the theory.

Recall from Chapter 2 that the Earth's orbit around the Sun is an ellipse, not a circle. At perihelion, the Earth is closest to the Sun and receives the strongest insolation; insolation is weakest at aphelion, when the Earth–Sun distance is greatest. Astronomers have observed that the orbit itself slowly rotates on a 108,000-year cycle, so the time of year of the strongest and weakest global

Changes in the intensity of solar radiation with latitude and the season occur in response to variations in the Earth's orbit around the Sun and variations in the tilt and position of the Earth's axis of rotation.

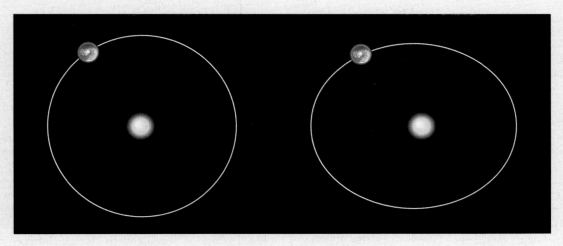

▲ ORBIT SHAPE

The shape of the Earth's orbit around the Sun varies from nearly circular to slightly elliptical, with a period of about 92,000 years. The orbit itself also slowly rotates, changing aphelion and perihelion on a cycle of 108,000 years (not shown).

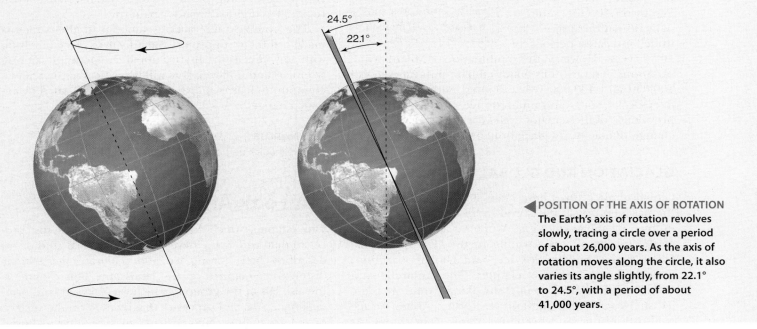

◄ POSITION OF THE AXIS OF ROTATION
The Earth's axis of rotation revolves slowly, tracing a circle over a period of about 26,000 years. As the axis of rotation moves along the circle, it also varies its angle slightly, from 22.1° to 24.5°, with a period of about 41,000 years.

insolation changes by a very small amount each year. In addition, the orbit's shape varies on a cycle of 92,000 years, becoming more elliptical and then more circular. This alters the Earth–Sun distance and, therefore, the amount of solar energy the Earth receives through the annual cycle (Figure 17.29). The Earth's axis of rotation also experiences cyclic shifts. The tilt angle of the axis varies from about 22° to 24° on a 41,000-year cycle. The axis also wobbles on a 26,000-year cycle, moving in

a slow circular motion much like a spinning top or toy gyroscope.

These cycles in solar revolution and axial rotation mean that the annual insolation experienced at each latitude changes from year to year. Glacial onsets are thought to begin with cooler summers during which snow builds up over many years in the northern hemisphere. This allows developing ice sheets to further reflect solar insolation, instigating a positive feedback

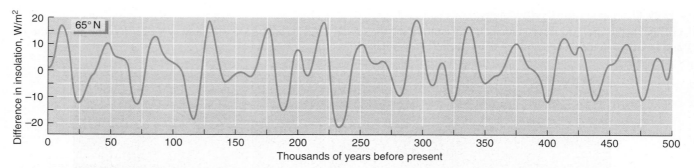

17.30 The Milankovitch curve

The vertical axis shows fluctuations in summer daily insolation at lat. 65° N for the last 500,000 years. These are calculated from mathematical models of the change in Earth–Sun distance and change in axial tilt with time. The zero value represents the present value.

loop. The cycles are plotted on a graph known as the *Milankovitch curve*, named for the engineer and mathematician Milutin Milankovitch who first calculated it in 1938. The dominant cycle shown in Figure 17.30, for summer daily insolation at 65° N latitude, indicates periods of 40,000 years. However, the combination of Milankovitch variations demonstrates major climate-shifting cycles of 100,000 and 400,000 years. Dating methods from the deep-ocean cores and antarctic ice cores confirm the prevalence of these major cycles for the onset of climate change that activates glaciation and deglaciation.

> According to the astronomical hypothesis, the timing of glaciations and interglaciations is determined by variations in insolation produced by minor cycles in the Earth's orbit and the Earth's axial rotation.

GLACIATION AND GLOBAL WARMING

Global temperatures have been slowly warming within the past century, as measured by an international community of meteorologists. We have described some of the effects of this warming in the "Eye on Global Change—Ice Sheets, Sea Ice, and Global Warming" feature that began this chapter. The Antarctic and Greenland Ice Caps are now shrinking. Some antarctic ice shelves are thinning and fracturing. Arctic sea ice is also thinning, and becoming reduced in extent. As northern ice and snow have shrunk in extent, their bright surfaces have been replaced by darker ocean water and land surfaces. The albedo of arctic regions has decreased significantly, resulting in the absorption of more solar insolation. This in turn propels a positive feedback loop that further increases warming, which then accelerates deglaciation at both poles, as well as in alpine environments.

Glacial melting is responsible for a series of events that affect the biosphere, hence humankind. In alpine regions, especially the Himalayas, the meltwater has created huge freshwater lakes that can drain suddenly,

sending cascades of water down river valleys to flood inhabited areas and affect the lives of hundreds of thousands living downslope in India and Nepal. Rising sea level, which is enhanced by the glacial melt, will affect coastal zones, where the majority of the population lives. The influx of large volumes of freshwater from melting ice in Greenland and Antarctica is projected to disrupt long-standing ocean current systems that control regional and global weather patterns.

The complex interactions among all these factors make it difficult to project the effects of glacial melting with any precision. International cooperation among scientists and policymakers will be critical as the science community learns more about the Earth's current climate change.

WileyPLUS Web Quiz

Take a quick quiz on the key concepts of this chapter.

A Look Ahead

Our chapter, which focused on glaciers and the processes that lead to the development of glacial landforms, described how flowing ice and meltwater leave their distinctive signatures on the landscape. This chapter is the also last in the group in which we reviewed landform-making agents and processes that operate on the surface of the continents, ranging from mass wasting to fluvial action, waves, wind, and glaciers. It is easier to distinguish among the wide variety and complexity of landforms by examining each agent in turn, as we have done here.

This chapter also completes our presentation of physical geography, which has covered topics from weather and climate to biogeography and landforms. By learning to observe and experience the world around you from this perspective, you can better understand and appreciate your environment and the processes that are constantly shaping it.

Human activity has changed the Earth in many ways over the last few millennia and our impact on it will continue to be felt, even more strongly, in the future. We hope that the background and context on the workings of the Earth and the environment that you have gained from the book will prove valuable to you as you face the challenges of world citizenship in the twenty-first century.

WileyPLUS Web Links
This chapter's web links lead to sites rich in photos of glaciers and glacial landforms from Alaska to Antarctica. Go exploring!

IN REVIEW GLACIAL LANDFORMS AND THE ICE AGE

- Global warming is causing an increase in the amount of snow accumulation on some ice sheets; it is also an agent of the melting and thinning of ice shelves and the edges of ice sheets.

- Although the East Antarctic Ice Sheet is holding its own currently, the West Antarctic Ice Sheet is losing ice mass, as is the Greenland Ice Cap. These combined losses are contributing to sea-level rise.

- There is a possibility that the West Antarctic Ice Sheet could collapse, raising sea level by several meters within a few hundred years.

- Arctic sea ice is thinning rapidly and shrinking in summer extent due to climate warming.

- **Glaciers** exist as **alpine glaciers** and as **ice sheets**. Alpine glaciers can take various forms—cirque, valley, piedmont, or tidewater—depending on their location on the landscape. *Ice caps* and *ice fields* can be found atop high mountain ranges.

- *Ice sheets* are huge masses of ice that cover vast areas. They are present today in Greenland and Antarctica.

- **Ice shelves** are thick plates of floating glacial ice that occur in Greenland and Antarctica where ice sheets flow into the sea. **Sea ice** is formed when ocean water freezes. **Icebergs** are masses of ice that have broken free from glaciers or ice shelves.

- The *mass balance* of a glacier is determined by the proportion of new snow accumulation in the *zone of accumulation* to melting in the *zone of ablation*.

- Glaciers move downhill by *basal sliding* on a layer of water. Some plastic flow also occurs.

- As they move, glaciers pick up and transport loose soil, regolith, and sediment. They also pluck loose bedrock and scratch and scour their beds.

- When glacial ice melts, it deposits **glacial drift**, which may be stratified by water flow (*stratified drift*) or deposited directly as **till**. *Moraines* accumulate at ice edges.

- *Alpine glaciers* develop in *cirques* in high mountain locations. Glacial erosion creates *horns, arêtes,* and *cols*. Alpine glaciers flow down valley on steep slopes, picking up rock debris and depositing it in **moraines**.

- As they remove loose soil and rock debris alpine glaciers carve **glacial troughs,** distinctive features of glaciated mountain regions. These troughs become **fiords** if they are later submerged by rising sea level.

- Alpine glaciers are often inaccessible, but they can be easily monitored by remote sensing. Images identify the locations of glaciers, and their extent; and some types of data can show the velocity and direction of flowing ice.

- Moving ice sheets can excavate and transport large amounts of loose and fractured materials. Where rocks are weak, long valleys can be excavated into troughs.

- Dense and compact *lodgment till* accumulates under glaciers as they flow, while lighter, looser *melt-out till* is let down as the ice melts in place.

- Till may be spread smooth and thick under an ice sheet, leaving a *till plain*. This plain may be studded with elongated till mounds, termed *drumlins*. Stream tunnels within the ice leave stream deposits as sinuous hills known as *eskers*.

- Outwash plains often feature *kettles* and *kames*. Kettles are holes in the drift left when a stagnant ice block melts; kames, in contrast, are isolated mounds of stratified drift formed as deltas, terraces, or crevice deposits between melting blocks.

- *Pluvial lakes* took shape during the cooler, wetter climate of the Ice Age. These lakes are now much smaller or represented by playas.

- Cultivation of stony till landscapes and moraines can be difficult, whereas flat till plains and outwash plains can be very productive. Stratified drift provides sand and gravel and serves as a source of groundwater.

- **Periglacial** regions of arctic and alpine regions are strongly influenced by past or present cold temperatures.

- Ground and bedrock at temperatures below the freezing point are called **permafrost**. *Subsea* and *continuous permafrost* occurs at the highest latitudes, flanked by bands of *discontinuous* and *sporadic permafrost*. *Alpine permafrost* can occur at high elevations.

- The **active layer** that overlies permafrost thaws each year during the warm season. *Taliks* are isolated pockets of unfrozen ground, usually occurring under lakes and rivers.

- Water in permafrost is *ground ice*. *Ice wedges* form and grow when water fills cracks in the active layer, which are caused by shrinkage due to extreme winter cold. Ice wedges occur in systems of *ice-wedge polygons*. *Ice lenses* form at the base of the active layer.

- **Gelifluction** occurs during warm periods, when the bottom of the active layer thaws, releasing water that lubricates downhill movement of the soil.

- *Pingos* are large conical mounds that build up when shallow lakes are drained and frozen over. Permafrost invades the unfrozen lake sediments, causing expansion as the water turns to ice, and creates an elevated dome.
- Human activities cause local thawing of permafrost; global warming causes more widespread regional thawing. And when frozen organic matter thaws, it yields carbon dioxide and methane, which intensifies the warming process.
- Permafrost thawing can produce a landscape of shallow lakes called *thermokarst*. Thawing also occurs when surface layers are disturbed by human construction, such as buildings, roads, and pipelines.
- An **ice age** includes alternating periods of **glaciation**, *deglaciation*, and **interglaciation**.
- During the past 2.5 million years, the Earth has experienced the **Late-Cenozoic Ice Age**.
- The most recent glaciation is the *Wisconsin glaciation*, in which ice sheets covered much of North America

and Europe, as well as parts of northern Asia and southern South America. During glaciations, sea level was as much as 125 m (410 ft) below present levels.
- The **Holocene epoch** spans the end of the Wisconsin glaciation to the present. So far, this epoch has included cool, warm, and cool stages before AD 1000. Since then, we have experienced warm, cool, and warm periods.
- The present Ice Age was triggered by the movements of continents set in motion by plate tectonics, which provided a snow-covered Antarctic continent and an ice-covered Arctic Ocean. Shifts in oceanic and atmospheric circulation patterns also cooled the planet.
- Individual cycles of glaciation are most likely caused by cyclic changes in Earth–Sun distance and axial tilt. This theory, called the **astronomical hypothesis,** was proposed by the astronomer Milutin Milankovitch.
- Global warming is shrinking ice caps and glaciers, thinning ice shelves, raising sea level, and increasing arctic albedo.

KEY TERMS

glaciers, p. 564
alpine glaciers, p. 564
ice sheet, p. 566
ice shelf, p. 566
sea ice, p. 566
icebergs, p. 566
glacial drift, p. 568
till, p. 568

glaciations, p. 569
glacial trough, p. 569
fiord, p. 571
moraine, p. 574
periglacial, p. 578
permafrost, p. 578
active layer, p. 579
gelifluction, p. 580

patterned ground, p. 580
interglaciation, p. 582
ice age, p. 582
Late-Cenozoic Ice Age, p. 582
Holocene epoch, p. 584
astronomical hypothesis, p. 584

REVIEW QUESTIONS

1. How is global warming affecting the Earth's ice sheets?
2. Why is the West Antarctic Ice Sheet considered unstable?
3. What is the effect of global warming on arctic sea ice?
4. Compare **alpine glaciers** to **ice sheets**. Where are they found?
5. Where on Earth are the great ice sheets found? What is an **ice shelf**? Where are ice shelves found?
6. Contrast **sea ice** and **icebergs**, including the processes by which they form.
7. Explain the *mass balance* of a glacier.
8. How does a glacier form? Why and how does it move?
9. Explain how glacial action erodes the landscape.
10. Define the term **glacial drift**. Identify two basic kinds of glacial drift.
11. How is a **glacial trough** related to a **fiord**?
12. How is remote sensing used to study glaciers?
13. What are **moraines**? How are they formed? What types of moraines are there?
14. Identify the landforms and deposits associated with deposition underneath a moving ice sheet.
15. Identify the landforms and deposits associated with stream action at or near the front of an ice sheet.
16. How do the landforms and deposits of a glacial landscape affect the agriculture and resources of a region?
17. What is meant by the term **periglacial**?
18. Identify the zones of **permafrost** and describe their general locations.

19. Define and describe permafrost and some of its features, including the **active layer** and *ground ice*.
20. Identify several types and features of ground ice in permafrost terrains. What is a *pingo*?
21. What is **gelifluction**? How and when does it occur?
22. What is meant by the term **patterned ground**? How is it formed?
23. How is global warming affecting permafrost? What are the consequences?
24. What is *thermokarst*? How does it form?
25. Define the terms **glaciation,** *deglaciation*, and **interglaciation**.
26. Identify the **Late-Cenozoic Ice Age**. When did it begin? Identify the most recent glaciation of the Ice Age. When did it begin and end?
27. How has climate changed during the **Holocene epoch?** What periods are recognized, and what are their characteristics?
28. Explain how plate tectonics triggered the Late-Cenozoic Ice Age.
29. What causes the glacial cycles observed within the Ice Age? What astronomical motions are involved?
30. What is the *Milankovitch curve*? What does it show about warm and cold periods during the last 500,000 years?
31. Identify some of the effects of global warming on the cryosphere, and explain their implications.

VISUALIZING EXERCISES

1. What are some typical features of an alpine glacial landscape? Sketch a mountainous landscape and identify *horns, arêtes, cols, tarns,* and **glacial troughs**.
2. Refer to Figure 17.1, which shows the Antarctic continent and its ice cap. Identify the Ross, Filchner, and Larsen ice shelves. Use the scale to measure the approximate area of each in square kilometers or miles. Consulting an atlas or almanac, identify the state or province of the United States or Canada that is nearest in land area to each.

ESSAY QUESTIONS

1. Imagine that you are planning a car trip to the Canadian Rockies. What glacial landforms would you expect to find there? Where would you look for them?
2. At some time during the latter part of the Pliocene epoch, the Earth entered an ice age. Describe the nature of this ice age and the cycles that occur within it. What explanation has been proposed for causing the Ice Age and its cycles of glaciation, deglaciation, and interglaciation? What cycles have been observed since the last ice sheets retreated?

Appendix 1
Climate Definitions and Boundaries

The following table summarizes the definitions and boundaries of climates and climate subtypes based on the soil-water balance, as described in Chapter 10 and shown on the world climate map, Figure 7.8.

Ep Water need (potential evapotranspiration)
D Soil-water shortage (deficit)
R Water surplus (runoff)
S Storage (limited to 300 mm)

GROUP I: LOW-LATITUDE CLIMATES

1. **Wet equatorial climate** ①
   $Ep \geq 100$ mm in every month, and
   $S \geq 200$ mm in 10 or more months
2. **Monsoon and trade-wind coastal climate** ②
   $Ep \geq 40$ mm in every month, or
   $Ep > 1300$ mm annual total, or both, and
   $S \geq 200$ mm in 6, 7, 8, or 9 consecutive months; or, if
   $S > 200$ mm in 10 or more months, then $Ep \leq 100$ mm in 5 or more consecutive months
3. **Wet-dry tropical climate** ③
   $D \geq 200$ mm, and
   $R \geq 100$ mm, and
   $Ep \geq 1300$ mm annual total, or $Ep \geq 40$ mm in every month, or both, and
   $S \geq 200$ mm in 5 months or fewer, or minimum monthly $S < 30$ mm
4. **Dry tropical climate** ④
   $D \geq 150$ mm, and
   $R = 0$, and
   $Ep \geq 1300$ mm annual total, or $Ep \geq 40$ mm in every month, or both

   Subtypes of dry climates (④, ⑤, ⑦, and ⑨)
   s Semiarid subtype (Steppe subtype): At least 1 month with $S > 20$ mm
   a Desert subtype: No month with $S > 20$ mm.

GROUP II: MIDLATITUDE CLIMATES

5. **Dry subtropical climate** ⑤
   $D \geq 150$ mm, and
   $R = 0$, and $Ep < 1300$ mm annual total, and
   $Ep \geq 8$ mm in every month, and
   $Ep < 40$ mm in 1 month
   (Subtypes ⑤a and ⑤s as defined under ④.)
6. **Moist subtropical climate** ⑥
   $D < 150$ mm when $R = 0$, and
   $Ep < 40$ mm in at least 1 month, and
   $Ep \geq 8$ mm in every month
7. **Mediterranean climate** ⑦
   $D \geq 150$ mm, and
   $R \geq 0$, and
   $Ep \geq 8$ mm in every month, and storage index $> 75\%$, or $P/Ea \times 100 < 40\%$
   (Subtypes ⑦a and ⑦s as defined under ④.)
8. **Marine west-coast climate** ⑧
   $D < 150$ mm, and
   $Ep < 800$ mm annual total, and
   $Ep \geq 8$ mm in every month
9. **Dry midlatitude climate** ⑨
   $D \geq 150$ mm, and
   $R = 0$, and
   $Ep \leq 7$ mm in at least 1 month, and
   $Ep > 525$ mm annual total
   (Subtypes ⑨a and ⑨s as defined under ④.)
10. **Moist continental climate** ⑩
    $D < 150$ mm when $R = 0$, and
    $Ep \leq 7$ mm in at least 1 month, and
    $Ep > 525$ mm annual total

GROUP III: HIGH-LATITUDE CLIMATES

11. **Boreal forest climate** ⑪
    525 mm $> Ep > 350$ mm annual total, and
    $Ep = 0$ in fewer than 8 consecutive months
12. **Tundra climate** ⑫
    $Ep < 350$ mm annual total, and
    $Ep = 0$ in 8 or more consecutive months
13. **Ice sheet climate** ⑬
    $Ep = 0$ in all months

Appendix 2
Conversion Factors

Metric to English

Metric Measure	Multiply by[*]	English Measure
Length		
Millimeters (mm)	0.0394	Inches (in.)
Centimeters (cm)	0.394	Inches (in.)
Meters (m)	3.28	Feet (ft)
Kilometers (km)	0.621	Miles (mi)
Area		
Square centimeters (cm^2)	0.155	Square inches (in^2)
Square meters (m^2)	10.8	Square feet (ft^2)
Square meters (m^2)	1.12	Square yards (yd^2)
Square kilometers (km^2)	0.386	Square miles (mi^2)
Hectares (ha)	2.47	Acres
Volume		
Cubic centimeters (cm^3)	0.0610	Cubic inches (in^3)
Cubic meters (m^3)	35.3	Cubic feet (ft^3)
Cubic meters (m^3)	1.31	Cubic yards (yd^3)
Milliliters (ml)	0.0338	Fluid ounces (fl oz)
Liters (l)	1.06	Quarts (qt)
Liters (l)	0.264	Gallons (gal)
Mass		
Grams (g)	0.0353	Ounces (oz)
Kilograms (kg)	2.20	Pounds (lb)
Kilograms (kg)	0.00110	Tons (2000 lb)
Tonnes (t)	1.10	Tons (2000 lb)

English to Metric

English Measure	Multiply by[*]	Metric Measure
Length		
Inches (in.)	2.54	Centimeters (cm)
Feet (ft)	0.305	Meters (m)
Yards (yd)	0.914	Meters (m)
Miles (mi)	1.61	Kilometers (km)
Area		
Square inches (in^2)	6.45	Square centimeters (cm^2)
Square feet (ft^2)	0.0929	Square meters (m^2)
Square yards (yd^2)	0.836	Square meters (m^2)
Square miles (mi^2)	2.59	Square kilometers (km^2)
Acres	0.405	Hectares (ha)
Volume		
Cubic inches (in^3)	16.4	Cubic centimeters (cm^3)
Cubic feet (ft^3)	0.0283	Cubic meters (m^3)
Cubic yards (yd^3)	0.765	Cubic meters (m^3)
Fluid ounces (fl oz)	29.6	Milliliters (ml)
Pints (pt)	0.473	Liters (l)
Quarts (qt)	0.946	Liters (l)
Gallons (gal)	3.79	Liters (l)
Mass		
Ounces (oz)	28.4	Grams (g)
Pounds (lb)	0.454	Kilograms (kg)
Tons (2000 lb)	907	Kilograms (kg)
Tons (2000 lb)	0.907	Tonnes (t)

*Conversion factors shown to three-decimal-digit precision.

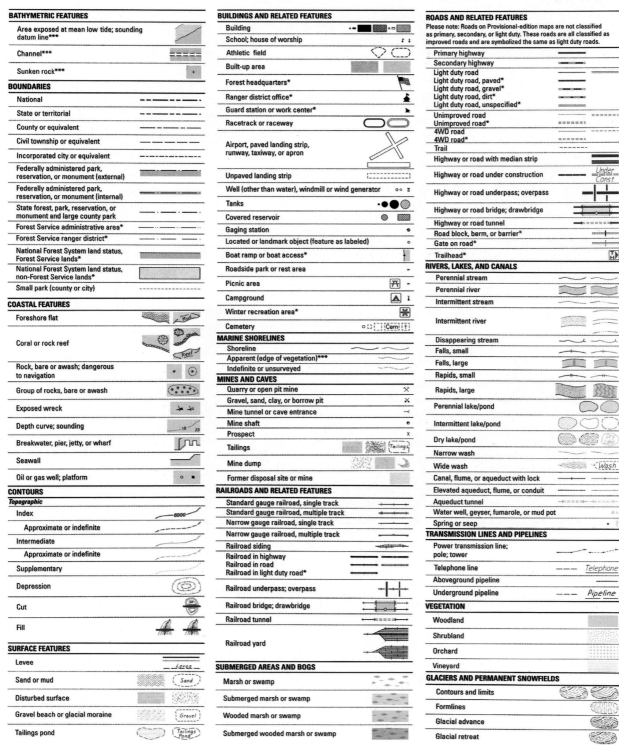

BATHYMETRIC FEATURES

Area exposed at mean low tide; sounding datum line***	
Channel***	
Sunken rock***	

BOUNDARIES

National	
State or territorial	
County or equivalent	
Civil township or equivalent	
Incorporated city or equivalent	
Federally administered park, reservation, or monument (external)	
Federally administered park, reservation, or monument (internal)	
State forest, park, reservation, or monument and large county park	
Forest Service administrative area*	
Forest Service ranger district*	
National Forest System land status, Forest Service lands*	
National Forest System land status, non-Forest Service lands*	
Small park (county or city)	

COASTAL FEATURES

Foreshore flat	
Coral or rock reef	
Rock, bare or awash; dangerous to navigation	
Group of rocks, bare or awash	
Exposed wreck	
Depth curve; sounding	
Breakwater, pier, jetty, or wharf	
Seawall	
Oil or gas well; platform	

CONTOURS

Topographic

Index	
Approximate or indefinite	
Intermediate	
Approximate or indefinite	
Supplementary	
Depression	
Cut	
Fill	

SURFACE FEATURES

Levee	
Sand or mud	
Disturbed surface	
Gravel beach or glacial moraine	
Tailings pond	

BUILDINGS AND RELATED FEATURES

Building	
School; house of worship	
Athletic field	
Built-up area	
Forest headquarters*	
Ranger district office*	
Guard station or work center*	
Racetrack or raceway	
Airport, paved landing strip, runway, taxiway, or apron	
Unpaved landing strip	
Well (other than water), windmill or wind generator	
Tanks	
Covered reservoir	
Gaging station	
Located or landmark object (feature as labeled)	
Boat ramp or boat access*	
Roadside park or rest area	
Picnic area	
Campground	
Winter recreation area*	
Cemetery	

MARINE SHORELINES

Shoreline	
Apparent (edge of vegetation)***	
Indefinite or unsurveyed	

MINES AND CAVES

Quarry or open pit mine	
Gravel, sand, clay, or borrow pit	
Mine tunnel or cave entrance	
Mine shaft	
Prospect	
Tailings	
Mine dump	
Former disposal site or mine	

RAILROADS AND RELATED FEATURES

Standard gauge railroad, single track	
Standard gauge railroad, multiple track	
Narrow gauge railroad, single track	
Narrow gauge railroad, multiple track	
Railroad siding	
Railroad in highway	
Railroad in road	
Railroad in light duty road*	
Railroad underpass; overpass	
Railroad bridge; drawbridge	
Railroad tunnel	
Railroad yard	

SUBMERGED AREAS AND BOGS

Marsh or swamp	
Submerged marsh or swamp	
Wooded marsh or swamp	
Submerged wooded marsh or swamp	

ROADS AND RELATED FEATURES

Please note: Roads on Provisional-edition maps are not classified as primary, secondary, or light duty. These roads are all classified as improved roads and are symbolized the same as light duty roads.

Primary highway	
Secondary highway	
Light duty road	
Light duty road, paved*	
Light duty road, gravel*	
Light duty road, dirt*	
Light duty road, unspecified*	
Unimproved road	
Unimproved road*	
4WD road	
4WD road*	
Trail	
Highway or road with median strip	
Highway or road under construction	Under Const
Highway or road underpass; overpass	
Highway or road bridge; drawbridge	
Highway or road tunnel	
Road block, berm, or barrier*	
Gate on road*	
Trailhead*	

RIVERS, LAKES, AND CANALS

Perennial stream	
Perennial river	
Intermittent stream	
Intermittent river	
Disappearing stream	
Falls, small	
Falls, large	
Rapids, small	
Rapids, large	
Perennial lake/pond	
Intermittent lake/pond	
Dry lake/pond	
Narrow wash	
Wide wash	Wash
Canal, flume, or aqueduct with lock	
Elevated aqueduct, flume, or conduit	
Aqueduct tunnel	
Water well, geyser, fumarole, or mud pot	
Spring or seep	

TRANSMISSION LINES AND PIPELINES

Power transmission line; pole; tower	
Telephone line	Telephone
Aboveground pipeline	
Underground pipeline	Pipeline

VEGETATION

Woodland	
Shrubland	
Orchard	
Vineyard	

GLACIERS AND PERMANENT SNOWFIELDS

Contours and limits	
Formlines	
Glacial advance	
Glacial retreat	

593

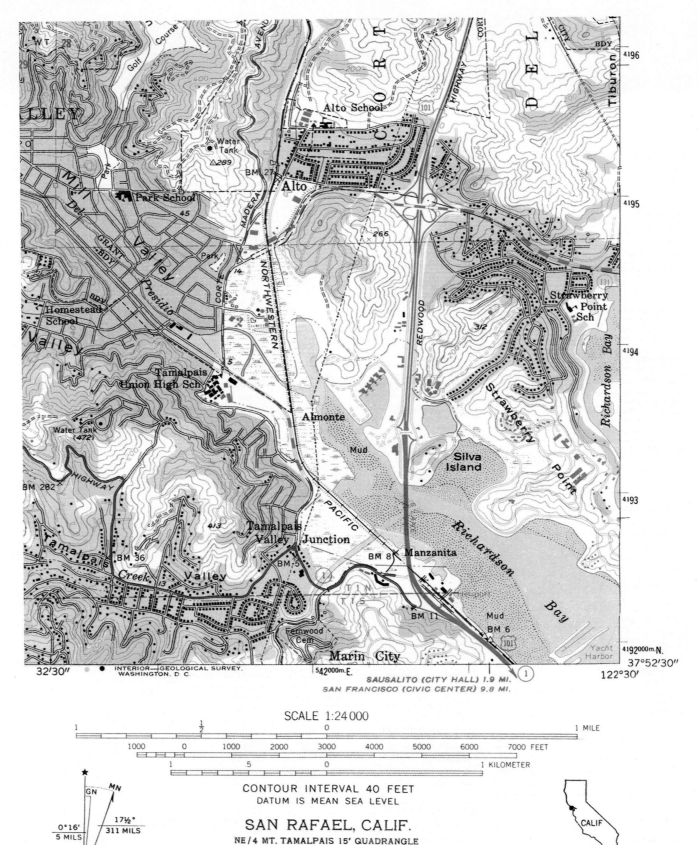

SCALE 1:24 000

CONTOUR INTERVAL 40 FEET
DATUM IS MEAN SEA LEVEL

SAN RAFAEL, CALIF.

NE/4 MT. TAMALPAIS 15' QUADRANGLE

N 3752.5—W12230/7.5

QUADRANGLE LOCATION

Glossary

This glossary contains definitions of terms shown in the text in italics or boldface, as well as other topic-relevant terms and phrases.

A

A horizon Mineral horizon of the soil, overlying the E and B horizons.

abrasion Erosion of bedrock of a stream channel by impact of particles carried in a stream and by rolling of larger rock fragments over the streambed; abrasion is also an activity of glacial ice, waves, and wind.

absorption (of radiation) Transfer of electromagnetic energy into internal energy within a gas or liquid through which the radiation is passing, or at the surface of a solid struck by the radiation.

acid action Solution of minerals by acids occurring in soil and groundwater.

acid deposition Deposition of acid raindrops and/or dry acidic dust particles on vegetation and ground surfaces.

acid mine drainage Sulfuric acid effluent from coal mines, mine tailings, or spoil ridges that are made by strip mining.

acid rain Rainwater having an abnormally low pH, between 2 and 5, as a result of air pollution.

acidic soils Soils containing an excess of hydrogen ions (low pH); typically found in cold, humid climates.

active continental margins Continental margins that coincide with tectonically active plate boundaries.

active layer Shallow surface layer subject to seasonal thawing in permafrost regions.

active pool Type of pool in the biogeochemical cycle in which the materials are in forms and places easily accessible to life processes.

active systems Remote sensing systems that emit a beam of wave energy at a source and measure the intensity of that energy reflected back to the source.

actual evapotranspiration (water use) Actual rate of evapotranspiration at a given time and place.

adiabatic lapse rate See *dry adiabatic lapse rate, moist adiabatic lapse rate.*

adiabatic principle Physical principle that a gas cools as it expands and warms as it is compressed, provided that no internal energy flows into or out of the gas during the process.

adiabatic process Change of temperature within a gas because of compression or expansion, without gain or loss of internal energy from the surroundings.

advection fog Fog produced by condensation within a moist basal air layer moving over a cold land or water surface.

aerosols Tiny particles present in the atmosphere, so small and light that the slightest movements of air keep them aloft.

aggradation Raising of stream channel elevation by continued deposition of bed load.

air Mixture of gases that surrounds the Earth.

air mass Extensive body of air within which upward gradients of temperature and moisture are fairly uniform over a large area.

air-mass thunderstorm Thunderstorm arising from daytime heating of the land surface, usually characterized by isolated cumulus and cumulonimbus clouds.

air parcel Small mass or volume of air; for example, a cubic meter or kilogram, considered in describing atmospheric processes.

air pollutant Unwanted substance injected into the atmosphere from the Earth's surface by either natural or human activities; includes aerosols, gases, and particulates.

air temperature Temperature of air, normally observed by a thermometer under standard conditions of shelter and height above the ground.

albedo Proportion or percentage of downwelling solar radiation reflected upward from a surface.

albic horizon Pale, often sandy soil horizon from which clay and free iron oxides have been removed. Found in the profile of the Spodosols.

Alfisols Soil order consisting of soils of humid and subhumid climates, with high base status and an argillic horizon.

alkaline soils Soils with an excess of hydroxide (OH^-) ions; often found in arid regions.

allele Specific version of a particular gene.

allelopathy Interaction among species in which a plant secretes substances in the soil that are toxic to other organisms.

allopatric speciation Type of speciation in which populations are geographically isolated and gene flow between the populations does not take place.

alluvial fan Gently sloping, conical accumulation of coarse alluvium deposited by a braided stream undergoing aggradation below the point of emergence of the channel from a narrow gorge or canyon.

alluvial river Stream of low gradient flowing on thick deposits of alluvium and experiencing approximately annual overbank flooding of the adjacent floodplain.

alluvial terrace Benchlike landform carved in alluvium by a stream during degradation.

alluvium Any stream-laid sediment deposit found in a stream channel and in low parts of a stream valley subject to flooding.

alpine chains High mountain ranges found along the margins of lithospheric plates, created by relatively recent volcanic or tectonic activity.

alpine glacier Long, narrow, mountain glacier on a steep downgrade.

alpine permafrost Permafrost occurring at high altitudes Equator-ward of the normal limit of permafrost.

alpine tundra   Plant formation class within the tundra biome, found at high altitudes above the limit of tree growth.

altocumulus   Cloud type in which patchy or roll clouds are found near the middle of the troposphere.

altostratus   Cloud type in which sheets or layers of clouds are found near the middle of the troposphere.

amplitude   Difference in height between a crest and the adjacent trough of a smooth, wave-like curve.

Andisols   Soil order that includes soils formed on volcanic ash; often enriched by organic matter, yielding a dark soil color.

anemometer   Weather instrument used to indicate wind speed.

angle of repose   Natural surface inclination (dip) of a slope consisting of loose, coarse, well-sorted rock or mineral fragments; for example, the slip face of a sand dune, a talus slope, or the sides of a cinder cone.

annual insolation   Average insolation rate at a location taken over the entire year; varies with latitude.

annuals   Plants that live for only a single growing season, passing the unfavorable season as a seed or spore.

Antarctic Circle   Parallel of latitude at 66½° S.

antarctic zone   Latitude zone in the latitude range 60° to 75° S (more or less), centered on the Antarctic Circle, and lying between the subantarctic zone and the polar zone.

anticlinal valley   Valley eroded in weak strata along the central line or axis of an eroded anticline.

anticline   Upfold of strata or other layered rock in an archlike structure; a class of folds.

anticyclone   Center of high atmospheric pressure.

anvil cloud   Anvil-shaped cloud extending downwind from the top of mature thunderstorms.

aphelion   Point on the Earth's elliptical orbit at which the Earth is farthest from the Sun.

aquatic ecosystem   Ecosystem of a lake, bog, pond, river, estuary, or other body of water.

Aquepts   Suborder of the soil order Inceptisols; includes Inceptisols of wet places, seasonally saturated with water.

aquiclude   Rock mass or layer that impedes or prevents the movement of groundwater.

aquifer   Rock mass or layer that readily transmits and holds groundwater.

arc   Curved line that forms a portion of a circle.

arc-continent collision   Collision of a volcanic arc with continental lithosphere along a subduction boundary.

Arctic Circle   Parallel of latitude at 66½° N.

arctic tundra   Plant formation class within the tundra biome, consisting of low, mostly herbaceous plants, but with some very small stunted trees, associated with the tundra climate ⑫.

arctic zone   Latitude zone in the latitude range 60° to 75° N (more or less), centered about on the Arctic Circle, and lying between the subarctic zone and the polar zone.

arête   Sharp, knifelike divide or crest formed between two cirques by alpine glaciation.

argillic horizon   Soil horizon, usually the B horizon, in which clay minerals have accumulated by illuviation.

arid (dry climate subtype)   Subtype of the dry climates that is extremely dry and supports little or no vegetation cover.

Aridisols   Soil order consisting of soils of dry climates, with or without argillic horizons, and with accumulations of carbonates or soluble salts.

artesian well   Drilled well in which water rises under hydraulic pressure above the level of the surrounding water table and may reach the surface.

association   Plant-animal community type identified by the typical organisms that are likely to be found together.

asthenosphere   Soft layer of the upper mantle, beneath the rigid lithosphere.

astronomical hypothesis   Theory that predicts glaciations and interglaciations, making use of cyclic variations in solar energy received by the Earth at higher latitudes.

Atlantic stage   Second climatic period of the Holocene epoch, somewhat warmer than today.

atmosphere   Envelope of gases surrounding the Earth, held by gravity.

atmospheric pressure   Pressure exerted by the atmosphere because of the force of gravity acting on the overlying column of air.

atoll   Circular or closed-loop coral reef enclosing an open lagoon with no island inside.

autogenic succession   Form of ecological succession that is self-producing—that is, resulting from the actions of plants and animals themselves.

autumnal equinox   September equinox in the northern hemisphere.

available water   Portion of soil water held by capillary tension, available for evaporation or absorption by plant rootlets.

axial rift   Narrow, trenchlike depression situated along the centerline of the mid-oceanic ridge and identified with active seafloor spreading.

axis of rotation   Centerline around which a body revolves, as the Earth's axis of rotation.

Azores high   Persistent high-pressure cell located in the eastern Atlantic Ocean south of the Azores Islands.

B

B horizon   Mineral soil horizon located beneath the A horizon, and usually characterized by a gain of mineral matter (such as clay minerals and oxides of aluminum and iron) and organic matter (humus).

backswamp   Area of low, swampy ground on the floodplain of an alluvial river between the natural levee and the bluffs.

backwash   Return flow of swash water under influence of gravity.

badlands   Rugged land surface of steep slopes, resembling miniature mountains, developed on weak clay formations or clay-rich regolith by fluvial erosion too rapid to permit plant growth and soil formation.

bar (marine or coastal)   See *sandbar*.

bar (pressure)   Unit of pressure equal to 105 Pa (pascals); approximately equal to the pressure of the Earth's atmosphere at sea level.

barchan dune   Sand dune of crescentic base outline with a sharp crest and a steep lee slip face, with crescent points (horns) pointing downwind; also known as *crescent dune*.

baroclinic instability   Property of atmospheric physics that causes small disturbances in the jet stream to grow over time.

barometer   Instrument for measuring atmospheric pressure.

barrier (to dispersal)   Zone or region that a species is unable to colonize or perhaps even occupy for a short time, thus halting diffusion.

barrier island   Long narrow island, built largely of beach sand and dune sand, parallel to the mainland and separated from it by a lagoon.

barrier reef   Coral reef separated from mainland shoreline by a lagoon.

basal sliding   Downhill movement of a glacier on a layer of liquid water found at the base.

basalt Extrusive igneous rock of mafic composition; occurs as lava.

base flow That portion of the discharge of a stream contributed by groundwater seepage.

base level Lower limiting surface or level that can ultimately be attained by a stream under conditions of stability of the Earth's crust and sea level; an imaginary surface equivalent to sea level projected inland.

base status of soils Quality of a soil as measured by the presence or absence of clay minerals capable of holding large numbers of bases. Soils of high base status are rich in base-holding clay minerals; soils of low base status are deficient in such minerals.

bases (base cations) Certain positively charged ions in the soil that are also plant nutrients; the most important are calcium, magnesium, potassium, and sodium.

batholith Large, deep-seated body of intrusive igneous rock, usually with an area of surface exposure greater than 100 km² (40 mi²).

bay Body of water sheltered from strong wave action by the configuration of the coast.

baymouth bar Sand spit crossing the mouth of a bay.

beach Thick, wedge-shaped accumulation of sand, gravel, or cobbles in the zone of breaking waves.

beach drift Transport of sand on a beach parallel with a shoreline by a succession of landward and seaward water movements at times when swash approaches obliquely.

bearing Direction angle between a line of interest and a reference line, which is usually a line pointing north.

bed load That portion of the stream load moving close to the streambed by rolling and sliding.

bedding plane Plane of stratification in a sedimentary rock, following the surface of fine layers within the stratum.

bedrock Solid rock in place with respect to the surrounding and underlying rock and relatively unchanged by weathering processes.

bedrock slump Landslide of bedrock in which most of the bedrock remains more or less intact as it moves.

Bergeron process Process of ice-crystal growth in cold clouds as water vapor evaporates from super-cooled water droplets and deposits on ice crystals.

bioclimatic frontier Geographic boundary corresponding with a critical limiting level of climate stress beyond which a species cannot survive.

biodiversity Variety of biological life on Earth or within a region.

biogeochemical cycle Total system of pathways by which a particular type of matter (a given element, compound, or ion, for example) moves through the Earth's ecosystem or biosphere; also called a *material cycle* or *nutrient cycle*.

biogeographic region Region in which the same or closely related plants and animals tend to be found together.

biogeography Study of the distributions of organisms at varying spatial and temporal scales, as well as the processes that produce these distribution patterns.

biomass Dry weight of living organic matter in an ecosystem within a designated surface area; units are kilograms of organic matter per square meter (kg/m²).

biome Largest recognizable subdivision of terrestrial ecosystems, including the total assemblage of plant and animal life interacting within the life layer.

biosphere All living organisms of the Earth and the environments with which they interact.

biota Plants and animals, referred to collectively; a list of plants and animals found at a location or in a region.

blackbody Ideal object or surface that is a perfect radiator and absorber of energy; absorbs all radiation it intercepts and emits radiation perfectly, according to physical theory.

blind fault Fracture in the crust of the Earth that is not apparent at the surface.

block mountains Class of mountains produced by block faulting, and usually bounded by normal faults.

block separation Separation of individual joint blocks during the process of physical weathering.

blowout Shallow depression produced by continued deflation.

bluffs Steeply rising ground slopes marking the outer limits of a floodplain.

bog Shallow depression filled with organic matter; for example, a glacial lake or pond basin filled with peat.

Boralfs Suborder of the soil order Alfisols; includes Alfisols of boreal forests or high mountains.

boreal forest Variety of needleleaf forest found in the boreal forest climate ⑪ regions of North America and Eurasia.

boreal forest climate ⑪ Cold climate of the subarctic zone in the northern hemisphere with long, extremely severe winters and several consecutive months of zero potential evapotranspiration (water need).

Boreal stage First climatic period of the Holocene epoch, with boreal forest in midlatitude regions.

Borolls Suborder of the soil order Mollisols; includes Mollisols of cold-winter semiarid plants (steppes) or high mountains.

braided stream Stream with shallow channel in coarse alluvium carrying multiple threads of fast flow that subdivide and rejoin repeatedly and continually shift in position.

breaker Sudden collapse of a steepened water wave as it approaches the shoreline.

broadleaf Deciduous forest type consisting of broadleaf trees that shed their leaves in the cold or dry season.

broadleaf evergreen forest Forest type consisting of broadleaf evergreen trees.

budget In flow systems, an accounting of energy and matter flows that enter, move within, and leave a system.

bush-fallow farming Agricultural system practiced in the African savanna woodland in which trees are cut and burned to provide cultivation plots.

butte Prominent, steep-sided hill or peak, often representing the final remnant of a resistant layer in a region of flat-lying strata.

C

C horizon Soil horizon lying beneath the B horizon, consisting of sediment or regolith that is the parent material of the soil.

calcification Accumulation of calcium carbonate in a soil, usually occurring in the B or C horizons.

calcite Mineral having the composition calcium carbonate.

calcium carbonate Compound consisting of calcium (Ca) and carbonate (CO_3) ions, formula $CaCO_3$, occurring naturally as the mineral calcite.

caldera Large, steep-sided circular depression resulting from the explosion and subsidence of a stratovolcano.

calving Shedding of ice into a water body by the terminus of a glacier.

Cambrian period Geologic period in which fossils first become abundant, from about 542–488 million years ago.

canyon See *gorge*.

capillarity Process by which capillary tension draws water into a small opening, such as a soil pore or a rock joint.

capillary action See *capillarity*.

capillary tension Cohesive force among surface molecules of a liquid that gives a droplet its rounded shape.

carbohydrate Class of organic compounds consisting of the elements carbon, hydrogen, and oxygen.

carbon cycle Biogeochemical cycle in which carbon moves through the biosphere; includes both gaseous and sedimentary cycles.

carbon dioxide Chemical compound CO_2, formed of oxygen and carbon. Normally a gas present in low concentration in the atmosphere, CO_2 is an important greenhouse gas.

carbon fixation See *photosynthesis*.

carbonates (carbonate minerals, carbonate rocks) Minerals that are carbonate compounds of calcium or magnesium or both; that is, calcium carbonate or magnesium carbonate.

carbonation Chemical reaction of carbonic acid in rainwater, soil water, and groundwater with minerals, which most strongly affects carbonate minerals and rocks, such as limestone and marble; an activity of chemical weathering.

carbonic acid Weak acid created when CO_2 gas dissolves in water.

cartography Science and art of making maps.

cation Positively charged mineral ions often present in soils.

Celsius scale Temperature scale on which the freezing point of water is 0° and the boiling point is 100°, measured at sea level.

Cenozoic era Last (youngest) era of geologic time.

channel See *stream channel*.

chaparral Sclerophyll scrub and dwarf forest plant formation class found throughout the coastal mountain ranges and hills of central and southern California.

chemical energy Energy stored within a substance by chemical bonds.

chemical weathering Chemical change in rock-forming minerals through exposure to atmospheric conditions in the presence of water; mainly involving oxidation, hydrolysis, carbonic acid action, or direct solution.

chemically precipitated sediment Sediment consisting of mineral matter precipitated from a water solution in which the matter has been transported in the dissolved state as ions.

chinook wind Local wind occurring at certain times to the lee of the Rocky Mountains; a very dry wind with a high capacity to evaporate snow.

chlorofluorocarbons (CFCs) Synthetic chemical compounds containing chlorine, fluorine, and carbon atoms that are widely used as coolant fluids in refrigeration systems.

chlorophyll Complex organic molecule that absorbs light energy for use by the plant cell.

choropleth map Map that identifies spatial information by categories.

cinder cone Conical hill built of coarse tephra ejected from a narrow volcanic vent; a type of volcano.

circle of illumination Great circle that divides the globe at all times into a sunlit hemisphere and a shadowed hemisphere.

circular valley Valley of roughly circular shape formed when a layer of weak rock within a geologic dome structure is eroded.

cirque Bowl-shaped depression carved in rock by glacial processes and holding the firn of the upper end of an alpine glacier.

cirque glacier Glacier confined to a cirque.

cirrocumulus Cloud type in which thin cloud patches are found in the upper troposphere.

cirrostratus Cloud type in which thin, smooth layers of clouds are found in the upper troposphere.

cirrus cloud Cloud composed of small ice crystals, typically found in the upper troposphere.

clast Rock or mineral fragment broken from a parent rock source.

clastic sediment Sediment consisting of particles broken from a parent rock source.

clay Sediment particles smaller than 0.004 mm in diameter.

clay minerals Class of minerals produced by alteration of silicate minerals and having plastic properties when moist.

claystone Sedimentary rock formed from clay; lacks a layered structure.

cliff Sheer, near-vertical rock wall formed from flat-lying resistant layered rocks, usually sandstone, limestone, or lava flows; may refer to any near-vertical rock wall.

climate Generalized statement of the prevailing weather conditions at a given place, based on statistics of a long period of record and including mean values, departures from those means, and the probabilities associated with those departures.

climatic frontier Geographical boundary that marks the limit of survival of a plant species subjected to climatic stress.

climatology Science that describes and explains the variability in space and time of the temperature and moisture states of the Earth's surface, especially its land surfaces.

climax Stable community of plants and animals reached at the end point of ecological succession.

climograph Graph on which two or more climatic variables, such as monthly mean temperature and monthly mean precipitation, are plotted for each month of the year.

cloud forest Type of low evergreen rainforest that occurs high on mountain slopes, where clouds and fog are frequent.

clouds Dense concentrations of suspended water or ice particles in the diameter range 20 to 50 μm.

coal Rock consisting of hydrocarbon compounds, formed of compacted and altered accumulations of plant remains (peat).

coarse-textured (rock) Rock with mineral crystals large enough to be visible to the naked eye or with low magnification.

coast See *coastline*.

coastal and marine geography Study of the geomorphic processes that shape shores and coastlines, and their application to coastal development and marine resource utilization.

coastal blowout dune High sand dune of the parabolic dunes class formed adjacent to a beach, usually with a deep deflation hollow (blowout) enclosed within the dune ridge.

coastal foredunes Belt of irregularly shaped sand dunes found landward of a sand beach.

coastal forest Subtype of needleleaf evergreen forest found in the humid coastal zone of the northwestern United States and western Canada.

coastal plain Coastal belt emerged from beneath the sea as a former continental shelf, underlain by strata with a gentle dip seaward.

coastline (coast) Zone in which coastal processes operate or have a strong influence.

cognitive representation In the representation perspective of geography, mental mapping of spatial relationships as they are experienced by humans.

col Natural pass or low notch in an arête between opposed cirques.

cold air outbreak Tongue of cold polar air moving from the midlatitudes into the very low latitudes.

cold front Moving weather front along which a cold air mass moves underneath a warm air mass, causing the latter to be lifted.

cold-blooded animal Animal whose body temperature passively follows the temperature of the environment.

cold-core ring Circular eddy of cold water, surrounded by warm water and lying adjacent to a warm, poleward-moving ocean current, such as the Gulf Stream.

colloids Mineral or organic particles of extremely small size, capable of remaining indefinitely in suspension in water.

colluvium Deposit of sediment or rock particles that are found on higher slopes, where sheet erosion is in progress, and accumulates from overland flow at the base of a slope.

commensalism Type of symbiosis in which one species benefits and the other is unaffected.

community Assemblage of organisms that live in a particular habitat and interact with one another.

competition Form of interaction among plant or animal species in which both draw resources from the same pool.

composite volcano Volcano composed of layers of ash and lava.

compression (tectonic) Squeezing together, as horizontal compression of crustal layers by tectonic processes.

condensation Process of change of matter in the gaseous state (water vapor) to the liquid state (liquid water) or solid state (ice).

condensation level Elevation at which an upward-moving parcel of moist air cools to the dew point and condensation begins to occur.

condensation nucleus Tiny bit of solid matter (aerosol) in the atmosphere on which water vapor condenses to form a very small water droplet.

conditionally stable air Air mass in which the environmental lapse rate is less than the dry adiabatic rate but greater than the moist adiabatic lapse rate.

conduction Transfer of internal kinetic energy of molecular motion from one substance in direct contact with another.

cone of depression Conical configuration of the lowered water table around a well from which water is being rapidly withdrawn.

conformal projection Map projection that preserves without shearing the true shape or outline of any small surface feature of the Earth.

conglomerate Sedimentary rock composed of pebbles in a matrix of finer rock particles.

conic projections Group of map projections in which the geographic grid is transformed to lie on the surface of a developed cone.

conifer Cone-bearing woody plant; nearly all are evergreen tree species with needle-shaped leaves.

consequent stream Stream that takes its course down the slope of an initial landform, such as a newly emerged coastal plain or a volcano.

consumers Animals in the food chain that live on organic matter formed by primary producers or by other consumers.

contact metamorphism Change created by heat and pressure in rock adjacent to an intrusion of magma.

continental crust Outer layer of the continents, of felsic composition in the upper part; thicker and less dense than oceanic crust.

continental drift Hypothesis, introduced by Alfred Wegener and others early in the 1900s, of the breakup of a parent continent, Pangaea, starting near the close of the Mesozoic era, and resulting in the present arrangement of continental shields and intervening ocean-basin floors.

continental glacier Large, thick plate of glacial ice moving outward in all directions from a central region of accumulation.

continental lithosphere Lithosphere bearing continental crust of felsic igneous rock.

continental margin Marginal belt of continental crust and lithosphere that is in contact with the oceanic crust and lithosphere.

continental rupture Crustal spreading apart, affecting the continental lithosphere, so as to cause a rift valley to appear and widen, eventually creating a new belt of oceanic lithosphere.

continental scale Scale of observation at which we recognize continents and other large Earth surface features, such as ocean currents.

continental shields Ancient crustal rock masses of the continents, largely igneous rock and metamorphic rock, and mostly of Precambrian age.

continental slope Steeply descending belt of seafloor between the continental shelf and the continental rise.

continental suture Long, narrow zone of crustal deformation, including underthrusting and intense folding, produced by a continental collision; for example, Himalayan Range, European Alps.

continent-continent collision Event in plate tectonics in which subduction brings two segments of the continental lithosphere into contact, leading to formation of a continental suture.

continuous permafrost Permanently frozen layer that underlies more than 90 percent of the surface area of a region.

convection (atmospheric) Air motion consisting of strong updrafts taking place within a convection cell.

convection (energy) Flow of internal energy occurring when matter moves from one place to another; for example, in the mixing of fluids of different temperatures.

convection cell Individual column of strong updrafts produced by atmospheric convection.

convection current In plate tectonics, a stream of upwelling mantle rock that rises steadily beneath a spreading plate boundary.

convection loop Circuit of moving fluid, such as air or water, created by unequal heating of the fluid.

convective precipitation Form of precipitation induced when warm, moist air is heated at the ground surface and then expands, rises, cools, and condenses to form water droplets, raindrops, and, eventually, rainfall.

convergence Horizontal motion of air creating a net inflow; causes a rising motion when it occurs at the surface, or a sinking motion when it occurs aloft.

converging boundary Boundary between two crustal plates along which subduction is occurring and lithosphere is being consumed.

Coordinated Universal Time Legal standard time, administered by the Bureau International de l'Heure, Paris.

coral reef Rocklike accumulation of carbonates secreted by corals and algae in shallow water along a marine shoreline.

core (of Earth) Spherical central mass of the Earth, composed largely of iron and consisting of an outer liquid zone and an interior solid zone.

Coriolis effect Effect of the Earth's rotation tending to turn the direction of motion of any object or fluid toward the right in the northern hemisphere and to the left in the southern hemisphere.

Coriolis force The Coriolis effect treated as a force perpendicular to the direction of motion that increases with the speed of motion but decreases with latitude.

corrosion Erosion of bedrock of a stream channel (or other rock surface) by chemical reactions between solutions in stream water and mineral surfaces.

cosmopolitan species Species found very widely.

counterradiation Longwave radiation of atmosphere directed downward to the Earth's surface.

covered shields Areas of continental shields in which the ancient rocks are covered beneath a thin layer of sedimentary strata.

crater Central summit depression associated with the principal vent of a volcano.

crescent dune See *barchan dune*.

crest In describing water waves, the highest point of the wave form.

crevasse Gaping crack in the brittle surface ice of a glacier.

crust (of Earth) Outermost solid shell or layer of the Earth, composed largely of silicate minerals.

Cryaquepts Large group within the soil suborder of Aquepts; includes Aquepts of cold-climate regions and, particularly, the tundra climate ⑫.

cryosphere Realm of the Earth's ice and snow.

cuesta Erosional landform developed on resistant strata having a low to moderate dip and taking the form of an asymmetrical low ridge or hill belt, with one side a steep slope and the other a gentle slope; usually associated with a coastal plain.

cumuliform clouds Clouds of globular shape, often with extended vertical development.

cumulonimbus cloud Large, dense cumuliform cloud yielding precipitation.

cumulus Cloud type consisting of low-lying, white cloud masses of globular shape and well separated from one another.

cumulus stage First state of development of an air-ass thunderstorm in which rising air forms cumulus clouds that carry surface energy and moisture aloft, causing instability.

cut bank Outside bank of the channel wall of a meander, where erosion is continually occurring.

cutoff Cutting-through of a narrow neck of land by a stream, bypassing the streamflow in an alluvial meander.

cycle of rock change Total cycle of changes in which rock of any one of the three major rock classes—igneous, sedimentary, metamorphic—is transformed into rock of one of the other classes.

cyclone Center of low atmospheric pressure.

cyclonic precipitation Form of precipitation that occurs as warm moist air is lifted by air motion occurring in a cyclone.

cyclonic storm Intense weather disturbance within a moving cyclone that generates strong winds, cloudiness, and precipitation.

cylindrical projection Map projection on which the geographic grid is transformed to lie on the surface of a cylinder.

D

daily insolation Average insolation rate taken over a 24-hour day; varies with latitude and time of year.

data acquisition component Element of a geographic information system (GIS) in which data are gathered together for input to the system.

data management component Element of a geographic information system (GIS) that creates, stores, retrieves, and modifies data layers and spatial objects.

daylight Period of the day during which the Sun is above the horizon at a particular location.

daylight saving time System under which time is advanced by one hour with respect to the standard time of the prevailing standard meridian.

debris flood (debris flow) Streamlike flow of muddy water heavily charged with sediment of a wide range of size grades, including boulders, generated by sporadic torrential rains on steep mountain watersheds.

decalcification Removal of calcium carbonate from a soil horizon as carbonic acid reacts with carbonate mineral matter.

December solstice Solstice that occurs on December 21 or 22, when the subsolar point is at 23½° S.

deciduous plant Tree or shrub that sheds its leaves seasonally.

declination of Sun Latitude at which the Sun is directly overhead; varies from 23½° S lat. to 23½° N lat.

decomposers Organisms that feed on dead organisms from all levels of the food chain; most are microorganisms and bacteria that feed on decaying organic matter.

deficit (soil-water shortage) In the soil-water budget, the difference between water use and water need; the quantity of irrigation water required to achieve maximum growth of agricultural crops.

deflation Lifting and transport in turbulent suspension by wind of loose particles of soil or regolith from dry ground surfaces.

deglaciation Widespread recession of ice sheets during a period of warming global climate, leading to an interglaciation.

degradation Lowering or downcutting of a stream channel by stream erosion in alluvium or bedrock.

degree of arc Measurement of the angle associated with an arc, in degrees.

delta Sediment deposit built by a stream entering a body of standing water and formed of the stream load.

denitrification Biochemical process, which is part of the nitrogen cycle, by which nitrogen in forms usable to plants is converted into molecular nitrogen in the gaseous form and returned to the atmosphere.

density of matter Quantity of mass per unit of volume, stated in kg/m^3.

denudation Total action of all processes whereby the exposed rocks of the continents are worn down and the resulting sediments are transported to the sea by the fluid agents; also includes weathering and mass wasting.

deposition (atmosphere) Change of state of a substance from a gas (water vapor) to a solid (ice). In the science of meteorology, the term *sublimation* is used to describe both this process and the change of state from solid to vapor.

deposition (of sediment) See *stream deposition*.

depositional landforms Landforms made by deposition of sediment.

derecho Violent, straight-line windstorm that precedes a squall line.

desalination Process of separating freshwater from saltwater.

desert biome Biome of the dry climates consisting of thinly dispersed plants that may be shrubs, grasses, or perennial herbs, but lacking in trees.

desert pavement Surface layer of closely fitted pebbles or coarse sand from which finer particles have been removed.

desertification See *land degradation*.

detritus Decaying organic matter on which decomposers feed.

dew point See *dew-point temperature*.

dew-point lapse rate Rate at which the dew point of an air mass decreases with elevation; typical value is 1.8°C/1000 m (1.0°F/1000 ft).

dew-point temperature Temperature of an air mass at which the air holds its full capacity of water vapor.

diagnostic horizons Rigorously defined soil horizons, that are used as diagnostic criteria in classifying soils.

differential GPS Method of application of global positioning system that reduces position error by using two receivers simultaneously.

diffuse radiation Solar radiation that has been scattered (deflected or reflected) by minute dust particles or cloud particles in the atmosphere.

diffuse reflection Solar radiation scattered back to space by the Earth's atmosphere.

diffusion In biogeography, the slow extension of the range of a species by normal processes of dispersal.

digital image Numeric representation of a picture consisting of a collection of numeric brightness values (pixels) arrayed in a fine grid pattern.

dike Thin layer of intrusive igneous rock, often near-vertical or with steep dip, occupying a widened fracture in the surrounding rock and typically cutting across older rock planes.

diploid Having two sets of chromosomes, one from each parent organism.

direct digital imaging Creation of a remotely sensed image using an instrument with large numbers of detectors arranged in a two-dimensional array.

discharge Volume of flow moving through a given cross section of a stream in a given unit of time; commonly shown in m^3/s (ft^3/s), and often symbolized by Q.

discontinuous permafrost Permanently frozen layer that underlies from 10 to 90 percent of the surface of a region.

disjunction Geographic distribution pattern of species in which one or more closely related species are found in widely separated regions.

dispersal In biogeography, the capacity of a species to move from its location of birth or origin to new sites.

dissipating stage Stage of air-mass thunderstorm development in which strong downdrafts throughout the air column inhibit the convection and latent heat release needed to sustain the thunderstorm.

distributary Branching stream channel that crosses a delta to discharge into open water.

disturbance In biogeography, an event, sometimes catastrophic, that damages or destroys an ecosystem and modifies habitats.

diurnal adjective meaning "daily."

divergence Horizontal motion of air that creates a net outflow and, when occurring aloft, causes a rising motion from below.

doldrums Belt of calm and variable winds occurring at times along the equatorial trough.

dolomite Carbonate mineral or sedimentary rock composed of calcium magnesium carbonate.

dome See *sedimentary dome.*

downcutting Process by which water flowing through a channel erodes segments of the stream back to the average gradient of the stream.

drainage basin Total land surface occupied by a drainage system, bounded by a drainage divide.

drainage divide Imaginary line following a crest of high land such that overland flow on opposite sides of the line enters different streams.

drainage pattern Plan of a network of interconnected stream channels.

drainage system Branched network of stream channels and adjacent land slopes, bounded by a drainage divide and converging into a single channel at the outlet.

drainage winds Usually cold winds that flow from higher to lower regions under the direct influence of gravity.

drawdown (of a well) Difference in height between base of cone of depression and original water table surface.

drought Occurrence of substantially lower-than-average precipitation in a season that normally has ample precipitation for the support of food-producing plants.

drumlin Hill of glacial till, oval or elliptical in basal outline and with smoothly rounded summit, formed by plastering of till beneath moving, debris-laden glacial ice.

dry adiabatic lapse rate Rate at which rising air is cooled by expansion when no condensation is occurring: 10°C per 1000 m (5.5°F per 1000 ft).

dry desert Plant formation class in the desert biome consisting of widely dispersed xerophytic plants that may be small, hard-leafed or spiny shrubs, succulent plants (cacti), or hard grasses.

dry lake Shallow basin covered with salt deposits formed when stream input to the basin is subjected to severe evaporation. May also form by evaporation of a saline lake when climate changes.

dry line Boundary separating hot, dry air from warm, moist air, and along which thunderstorms tend to form.

dry midlatitude climate ⑨ Climate of the midlatitude zone with a strong annual cycle of potential evapotranspiration (water need) and cold winters.

dry subtropical climate ⑤ Climate of the subtropical zone, transitional between the dry tropical climate ④ and the dry midlatitude climate ⑨.

dry tropical climate ④ Climate of the tropical zone with large total annual potential evapotranspiration (water need).

dune See *sand dune.*

dust bowl Western Great Plains of the United States, which suffered severe wind deflation and soil drifting during the drought years of the middle 1930s.

dust storm Heavy concentration of dust in a turbulent air mass, often associated with a cold front.

E

E horizon Soil mineral horizon lying below the A horizon and characterized by the loss of clay minerals and oxides of iron and aluminum; may show a concentration of quartz grains and is often pale in color.

earth hummock Low mound of vegetation-covered earth found in permafrost terrain, formed by cycles of ground ice growth and melting.

Earth visualization tool Web-based system for displaying images and spatial information about the Earth's surface; for example, Google Earth.

earthflow Moderately rapid downhill flowage of masses of water-saturated soil, regolith, or weak shale, typically forming a steplike terrace at the top and a bulging toe at the base.

earthquake Trembling or shaking of the ground produced by the passage of seismic waves.

earthquake focus Point within the Earth at which the energy of an earthquake is first released by rupture and from which seismic waves emanate.

Earth's crust See *crust of Earth.*

easterly wave Weak, slow-moving trough of low pressure within the belt of tropical easterlies that causes a weather disturbance with rain showers.

ebb current Oceanward flow of tidal current in a bay or tidal stream.

ebb tide Outgoing tide.

ecological biogeography Branch of biogeography focusing on how distribution patterns of organisms are related to their environment.

ecological niche Functional role played by an organism in an ecosystem; how it obtains energy and how it influences other organisms and its environment.

ecological succession Time-succession (sequence) of distinctive plant and animal communities occurring within a given area of newly formed land or land cleared of plant cover by burning, clearcutting, or other agents.

ecology Science of interactions between life-forms and their environment; science of ecosystems.

ecosystem   Group of organisms and the environment with which and in which the organisms interact.

edaphic factors   Factors relating to soil that influence a terrestrial ecosystem.

E–F scale   See *enhanced Fujita intensity scale.*

El Niño   Episodic cessation of the typical upwelling of cold deep water off the coast of Peru.   El Niño is Spanish for "the Little Boy," referring to the Christ Child, so named for its occurrence during the Christmas season once every few years.

electromagnetic radiation (electromagnetic energy)   Wavelike form of energy radiated by any substance possessing internal kinetic energy of molecular motion; travels through space at the speed of light.

electromagnetic spectrum   Total wavelength range of electromagnetic energy.

eluviation   Soil-forming process in which fine particles, particularly soil colloids (both mineral and organic), are transported from an upper soil horizon to a lower one.

emissivity   Ratio of the actual energy emitted by an object or substance to that of a blackbody at the same temperature.

endemic species   Species found only in one region or location.

endogenic processes   Internal Earth processes, such as tectonics and volcanism, that create landforms.

energy   Capacity to do work; that is, to bring about a change in the state or motion of matter.

energy balance (global)   Balance between shortwave solar radiation received by the Earth-atmosphere system and radiation lost to space by shortwave reflection and longwave radiation from the Earth-atmosphere system.

energy balance (of a surface)   Balance between the flows of energy reaching a surface and the flows of energy leaving it.

enhanced Fujita intensity scale   Measure of tornado intensity, based on damage to structures and surrounding vegetation; introduced in 2007 as an improvement to the original Fujita intensity scale.

ENSO   Abbreviation for El Niño–Southern Oscillation, referring to both of these related phenomena taken together.

Entisols   Soil order consisting of mineral soils lacking soil horizons that would persist after normal plowing.

entrenched meanders   Winding, sinuous valley produced by degradation of a stream with trenching into the bedrock by downcutting.

environmental temperature lapse rate   Rate of temperature decrease upward through the troposphere; standard value is 6.4°C/km (3°F/1000 ft).

eolian landforms   Landforms, such as dunes or loess deposits, formed by wind transport of sediment.

eon   Largest unit of geologic time, including hundreds of millions of years; subdivided into eras.

epipedon   Soil horizon that forms at the surface.

epiphytes   Plants that live above ground level out of contact with the soil, usually growing on the limbs of trees or shrubs; also called *air plants.*

epoch   Unit of geologic time up to tens of millions of years in length; a subdivision of the period time unit.

equal-angle grid   Rectangular map projection placing latitude and longitude on the axes of a grid.

equal-area projections   Class of map projections on which any given area of the Earth's surface is shown to correct relative areal extent, regardless of its position on the globe.

Equator   Parallel of latitude occupying a position midway between the Earth's poles of rotation; the largest of the parallels, designated as latitude 0°.

equatorial current   Westward-flowing ocean current in the belt of the trade winds.

equatorial easterlies   Upper-level easterly airflow over the equatorial zone.

equatorial rainforest   Plant formation class within the forest biome, consisting of tall, closely set broadleaf trees of evergreen or semideciduous habit.

equatorial trough   Atmospheric low-pressure trough centered more or less over the Equator and situated between the two belts of trade winds.

equatorial zone   Latitude zone lying between lat. 10° S and 10° N (more or less) and centered on the Equator.

equilibrium   In flow systems, a state of balance at which flow rates remain unchanged.

equilibrium approach   Model of landscape evolution in which uplift is an ongoing action to which erosional processes are constantly adjusting.

equilibrium line   Demarcation between the zone of accumulation and the zone of ablation on a glacier where the rate of snow accumulation balances the rate of evaporation and melting.

equinox   Instant in time when the subsolar point falls on the Earth's Equator and

the circle of illumination passes through both poles.

era   Major unit of geologic time, tens or hundreds of millions of years in length; subdivided into periods.

erg   Large expanse of active sand dunes in the Sahara Desert of North Africa.

erosion   Gradual wearing away of surface rocks under the influence of chemical and physical weathering and transportation by water or wind.

erosional landforms   Class of the sequential landforms shaped by the removal of regolith or bedrock by agents of erosion; for example, gorge, glacial cirque, marine cliff.

esker   Narrow, often sinuous embankment of coarse gravel and boulders deposited in the bed of a meltwater stream enclosed in a tunnel within stagnant ice of an ice sheet.

estuary   Bay that receives freshwater from a river mouth and saltwater from the ocean.

eustatic   True change in sea level, as opposed to a local change created by upward or downward tectonic motion of land.

eutrophication   Excessive growth of algae and other related organisms in a stream or lake as a result of the input of large amounts of nutrient ions, especially phosphate and nitrate.

evaporation   Process by which water in a liquid state passes into the vapor state.

evaporites   Class of chemically precipitated sediment and sedimentary rock, composed of soluble salts deposited from saltwater bodies.

evapotranspiration   Combined water loss to the atmosphere by evaporation from the soil and transpiration from plants.

evergreen plant   Tree or shrub that holds most of its green leaves or needles throughout the year.

evolution   Creation of the diversity of life-forms through the process of natural selection.

exfoliation   Process of removal of overlying rock load from bedrock by processes of denudation, accompanied by expansion, and often leading to the development of sheeting structure.

exfoliation dome   Smoothly rounded rock knob or hilltop bearing rock sheets or shells produced by the spontaneous expansion that accompanies unloading.

exogenic processes   Landform-making processes active at the Earth's surface, such as erosion by water, waves and currents, glacial ice, and wind.

exotic river Stream that flows across a region of dry climate and derives its discharge from adjacent uplands where a water surplus exists.

exponential growth Increase in number or value over time in which the increase is a constant proportion or percentage within each time unit.

exposed shields Areas of continental shields in which the ancient basement rock, usually of Precambrian age, is exposed to the surface.

extension (tectonic) Drawing apart of crustal layers by tectonic activity, resulting in faulting.

extinction Event occurring when the number of organisms of a species shrinks to zero; that is, the species no longer exists.

extratropical cyclone See *midlatitude cyclone.*

extrusion Release of molten rock magma at the surface, as in a flow of lava or shower of volcanic ash.

extrusive igneous rock Rock produced by the solidification of lava or ejected fragments of igneous rock (tephra).

eye (tropical cyclone) Cloud-free vortex of descending air that develops at the center of a tropical cyclone.

F

Fahrenheit scale Temperature scale on which the freezing point of water is 32° and the boiling point is 212°.

fair-weather system Traveling anticyclone in which the descent of air suppresses clouds and precipitation, and the weather is typically fair.

fallout Descent by gravity of pollutant particles through the atmosphere to reach the ground.

fault Sharp break in rock with a displacement (slippage) of the block on one side with respect to an adjacent block.

fault creep More or less continuous slippage on a fault plane, relieving some of the accumulated strain on the rocks crossing the fault.

fault plane Surface of slippage between two Earth blocks moving relative to each other during faulting.

fault scarp Clifflike surface feature produced by faulting and exposing the fault plane; commonly associated with a normal fault.

fault-line scarp Erosion scarp developed on an inactive fault line.

feedback In flow systems, a linkage between flow paths such that the flow in one pathway acts either to reduce or increase the flow in another pathway.

feldspar Group of silicate minerals consisting of silicate of aluminum and one or more of the metals potassium, sodium, or calcium.

felsic igneous rock Igneous rock dominantly composed of felsic minerals.

felsic minerals (felsic mineral group) Quartz and feldspars of light color and relatively low density; treated as a mineral group.

fetch Distance that wind blows over water to create a train of water waves.

fine-textured (rock) Rock with mineral crystals too small to be seen by eye or with low magnification.

fiord Narrow, deep ocean bay partially filling a glacial trough.

fiord coast Rugged coast with many deep bays formed by partial submergence of glacial troughs.

firn Granular old snow forming a surface layer in the zone of accumulation of a glacier.

flash flood Flood in which heavy rainfall causes a stream or river to rise very rapidly.

flatirons Triangular landforms created when tilted rocks are intersected by stream valleys.

flood Streamflow at a stream stage so high that it cannot be accommodated within the stream channel and must spread over the banks, thereby inundating the adjacent floodplain.

flood basalts Large-scale outpourings of basalt lava to produce thick accumulations of basalt over large areas.

flood current Landward flow of a tidal current.

flood stage Designated stream-surface level for a particular point on a stream, above which overbank flooding occurs.

flood tide Incoming tide.

floodplain Belt of low, flat ground underlain by alluvium, present on one or both sides of a stream channel, and subject to inundation by a flood about once annually.

fluvial landforms Landforms shaped by running water.

fluvial processes Geomorphic processes in which running water is the dominant fluid agent, acting as overland flow and streamflow.

focus See *earthquake focus.*

fog Cloud layer in contact with land or sea surface, or very close to that surface.

folding Process by which rock layers are bent into folds; a form of tectonic activity.

folds Wavelike corrugations of strata (or other layered rock masses) resulting from crustal compression.

food web (food chain) Organization of an ecosystem into steps or levels through which energy flows as the organisms at each level consume energy stored in the bodies of organisms at the next lower level.

forb Broad-leafed herb, as distinguished from the grasses.

foredunes Ridge of irregular sand dunes typically found adjacent to beaches on low-lying coasts and bearing a partial cover of plants.

foreland folds Folds produced by continental collision in strata of a continental margin.

forest Assemblage of trees growing close together, their crowns forming a layer of foliage that largely shades the ground.

forest biome Biome that includes all regions of forest over the lands of the Earth.

formation classes Subdivisions within a biome based on the size, shape, and structure of the plants that dominate the vegetation.

fossil fuels Naturally occurring hydrocarbon compounds that represent the altered remains of organic materials enclosed in rock; for example coal, petroleum (crude oil), and natural gas.

fractional scale See *scale fraction.*

freezing Change from liquid to solid state, accompanied by release of latent heat energy.

frictional force (atmosphere) Force opposing the motion of wind, caused by the drag of the Earth's surface; greatest close to the surface and decreasing with height.

fringing reef Coral reef directly attached to land with no intervening lagoon of open water.

front Surface of contact between two unlike air masses.

frost action Frost cracking of rocks and churning of rock rubble by the growth and melting of ice.

frost cracking Breaking or splitting of rock by freezing of water in joints or pores of bedrock.

G

gelifluction Slump or flowage of saturated soil on a layer of solid permafrost, forming low terraces or lobes.

Gelisols Soil order of cold regions, including soils underlain by permafrost with organic and mineral materials churned by frost action.

gene flow Speciation process in which evolving populations exchange alleles as individuals move among populations.

genera See *genus*.

genetic drift Speciation process by which chance mutations change the genetic composition of a breeding population until it diverges from other populations.

genus Collection of closely related species that share a similar genetic evolutionary history; plural, *genera*.

geographic cycle Cycle of stages of landscape evolution from rugged peaks to a flat lowland; described by geographer W. M. Davis.

geographic grid Complete network of parallels and meridians on the surface of the globe, used to fix the locations of surface points.

geographic information system (GIS) System for acquiring, processing, storing, querying, creating, and displaying spatial data; in the representation perspective of geography, the use of a GIS to represent and manipulate spatial data.

geographic isolation Speciation process by which a breeding population is split into parts by an emerging geographic barrier, such as an uplifting mountain range or a changing climate.

geography Study of the evolving character and organization of the Earth's surface.

geography of soils Study of the distribution of soil types and properties and the processes of soil formation.

geoid Shape of the Earth's surface coinciding with mean sea level at any point.

geologic time scale Organization of Earth history into major units of eons, eras, and periods.

geology Science of the solid Earth, including the Earth's origin and history, materials comprising the Earth, and the processes acting within the Earth and on its surface.

geomorphic factors Landform factors influencing ecosystems, such as slope steepness, slope aspect, and relief.

geomorphology Science of Earth surface processes and landforms, including their history and processes of origin.

geostationary orbit Satellite orbit in which the satellite remains nearly fixed in space above a single point on the Equator, maintaining a constant view of the same portion of the Earth.

geostrophic wind Wind at high levels above the Earth's surface blowing parallel, with a system of straight, parallel isobars.

geyser Periodic jetlike emission of hot water and steam from a narrow vent at a geothermal locality.

glacial drift General term for all varieties and forms of rock debris deposited in close association with ice sheets of the Pleistocene epoch.

glacial trough Deep, steep-sided rock trench of U-shaped cross section formed by alpine glacier erosion.

glaciation (1) General term for the total process of glacier growth and landform modification by glaciers. (2) Single episode or time period in which ice sheets formed, spread, and disappeared.

glacier Large natural accumulation of land ice affected by present or past flowage.

global positioning system (GPS) System of satellites and ground instruments that locates the global position of an observer.

global radiation balance Energy flow process by which the Earth absorbs shortwave solar radiation and emits longwave radiation; in the long run, the two flows must balance.

global scale Scale at which we are concerned with the Earth as a whole; for example, in considering Earth–Sun relationships.

glowing avalanche Cloud of white-hot gas and fine ash released by a volcanic eruption; also termed *glowing cloud*.

gneiss Variety of metamorphic rock showing banding, and commonly rich in quartz and feldspar.

GOES Acronym for Geostationary Operational Environmental Satellites, a series of Earth-imaging satellites used to monitor atmospheric and surface conditions primarily for weather forecasting.

Gondwana Supercontinent of the Permian period that encompassed much of the regions that today are South America, Africa, Antarctica, Australia, New Zealand, Madagascar, and peninsular India.

gorge (canyon) Steep-sided bedrock valley with a narrow floor limited to the width of a stream channel.

graben Trenchlike depression representing the surface of a crustal block dropped down between two opposed, inward-facing normal faults.

graded profile Smoothly descending profile displayed by a graded stream.

graded stream Stream (or stream channel) with stream gradient adjusted to achieve a balanced state in which average bed load transport is matched to average bed load input; an average condition over periods of many years' duration.

gradient Degree of slope, as the gradient of a river or a flowing glacier.

granite Intrusive igneous rock consisting largely of quartz, potash feldspar, and plagioclase feldspar, with minor amounts of biotite and hornblende; a felsic igneous rock.

granitic rock General term for rock of the upper layer of the continental crust, composed largely of felsic igneous and metamorphic rock; rock of composition similar to that of granite.

granular disintegration Grain-by-grain breakup of the outer surface of coarse-grained rock, yielding sand and gravel and leaving behind rounded boulders.

graphic scale Map scale as shown by a line divided into equal parts.

grassland biome Biome consisting largely or entirely of herbs, which may include grasses, grasslike plants, and forbs.

graupel Suspended ice pellets formed when supercooled water drops freeze around a central ice crystal.

gravel Rock particles larger than 2 mm (0.79 in.) in size.

gravitation Mutual attraction between any two masses.

gravitational water (soils) Water that drains slowly downward through the soil by gravity into the underlying lithosphere.

gravity Gravitational attraction of the Earth on any small mass near the Earth's surface.

gravity gliding Sliding of a thrust sheet away from the center of an orogen under the force of gravity.

great circle Circle formed by passing a plane through the exact center of a perfect sphere; the largest circle that can be drawn on the surface of a sphere.

greenhouse effect Accumulation of internal energy in the lower atmosphere, with accompanying increase in temperature, through the absorption of longwave radiation from the Earth's surface.

greenhouse gases Atmospheric gases such as CO_2, CH_4, N_2O, and chlorofluorocarbons (CFCs) that absorb outgoing longwave radiation, contributing to the greenhouse effect.

Greenwich meridian Meridian passing through the old Royal Observatory at Greenwich, England; taken as 0° longitude.

groin In water, wall or embankment built out into the water of a lake or ocean at right angles to the shoreline.

gross photosynthesis Total amount of carbohydrate produced by photosynthesis by a given organism or group of organisms in a given unit of time.

ground ice Frozen water within the pores of soils and regolith or as free bodies or lenses of solid ice.

ground moraine Moraine formed of till distributed beneath a large expanse of land surface covered at one time by an ice sheet.

groundwater Subsurface water occupying the saturated zone and moving under the force of gravity.

gullies Deep, V-shaped trenches carved by newly formed streams in rapid headward growth during advanced stages of accelerated soil erosion.

gust front Downdraft spreading outward in front of a severe thunderstorm, creating strong wind gusts.

gyres Large circular ocean current systems centered on the oceanic subtropical high-pressure cells.

H

habitat Subdivision of the environment according to the needs and preferences of organisms or groups of organisms.

Hadley cell Atmospheric circulation cell in low latitudes involving rising air over the equatorial trough and sinking air over the subtropical high-pressure belts.

hail Form of precipitation consisting of pellets or spheres of ice with a concentric layered structure.

hairpin dune Type of parabolic dune with long, parallel sides.

half-life Time required for an initial quantity of an unstable chemical isotope to be reduced by half in an exponential decay system.

hanging valley Stream valley truncated by glacial erosion at its intersection with a larger valley filled with ice.

hard pan Cemented layer occurring in soils, typically by accumulation of calcium carbonate.

Hawaiian high Persistent high-pressure cell located in the eastern Pacific Ocean, north of the Hawaiian Islands.

hazards assessment Field of study blending physical and human geography to focus on the perception of risk of natural hazards and, subsequently, to develop public policy to mitigate that risk.

haze Minor concentration of pollutants or natural forms of aerosols in the atmosphere causing a reduction in visibility.

headland Landform of resistant rock jutting oceanward along a shoreline.

heat Flow of internal energy in transit from one substance or location to another.

heat index Measure of apparent temperature based on actual air temperature and relative humidity, designed to account for reduced evaporative cooling of body surfaces under humid conditions.

heat island Persistent region of higher air temperatures centered over a city.

hemisphere Half of a sphere; that portion of the Earth's surface found between the Equator and a pole.

herb Tender plant, lacking woody stems, usually small or low; may be annual or perennial.

herbivory Form of interaction among species in which an animal (herbivore) grazes on herbaceous plants.

heterosphere Region of the atmosphere above about 100 km (about 60 mi) in which gas molecules tend to become increasingly sorted into layers according to molecular weight and electric charge.

hibernation Dormant state of some vertebrate animals during the winter season.

high base status See *base status of soils*.

high-latitude climates Group of climates in the subarctic zone, arctic zone, and polar zone, dominated by arctic air masses and polar air masses.

high-level temperature inversion Condition in which a high-level layer of warm air overlies a layer of cooler air, reversing the normal trend of cooling with altitude.

high-pressure cell Center of high barometric pressure; an anticyclone.

historical biogeography Branch of biogeography focusing on how spatial patterns of organisms arise over space and through time.

Histosols Soil order consisting of soils with a thick upper layer of organic matter.

hogback ridges Sharp-crested, often sawtoothed ridges formed from the upturned edge of a resistant rock layer of sandstone, limestone, or lava.

Holocene epoch Last epoch of geologic time, commencing about 10,000 years ago. The Holocene epoch followed the Pleistocene epoch and includes the present.

homosphere Lower portion of the atmosphere, below about 100 km (about 60 mi) altitude, in which atmospheric gases are uniformly mixed.

horizontal vortex Spinning circulation aligned with the ground and produced by wind shear.

horn Sharp peak in glaciated landscape formed by glacial erosion of intersecting cirques.

horse latitudes Subtropical high-pressure belt of the North Atlantic Ocean, coincident with the central region of the Azores high; a belt of weak, variable winds and frequent calms.

horst Crustal block uplifted between two normal faults.

hot springs Springs discharging heated groundwater at a temperature close to the boiling point; found in geothermal areas and thought to be related to a magma body at depth.

hotspot (biogeography) Geographic region of high biodiversity.

hotspot (plate tectonics) Center of intrusive igneous and volcanic activity thought to be located over a rising mantle plume.

human geography That part of systematic geography that deals with social, economic, and behavioral processes that differentiate places.

human-influenced vegetation Vegetation that has been affected in some way by human activity, for example, through cultivation, grazing, timber cutting, or urbanization.

humidity General term for the amount of water vapor present in the air.

humification Pedogenic process of transformation of plant tissues into humus.

humus Dark brown to black organic matter on or in the soil, consisting of fragmented plant tissues partly digested by organisms.

hurricane Tropical cyclone of the western North Atlantic and Caribbean Sea.

hydraulic action Stream erosion by impact force of flowing water on the bed and banks of a stream channel.

hydrogen bonding Weak attraction between water molecules based on the attraction between a positive hydrogen atom on one molecule and a negative oxygen atom on another.

hydrograph Graphic representation of the variation in stream discharge with elapsed time, based on data of stream gauging at a given station on a stream.

hydrologic cycle Total plan of movement, exchange, and storage of the Earth's free water in gaseous, liquid, and solid states.

hydrology Science of the Earth's water and its motions through the hydrologic cycle.

hydrolysis Chemical union of water molecules with minerals to form different, more stable mineral compounds.

hydrosphere Total water realm of the Earth's surface zone, including the oceans, surface waters of the lands, groundwater, and water held in the atmosphere.

hygrometer Instrument used to measure the water vapor content of the atmosphere; some types measure relative humidity directly.

hygroscopic water Soil water held tightly to the surfaces of soil particles by molecular attraction.

I

ice age Span of geologic time, usually on the order of 1 to 3 million years, or longer, in which glaciations alternate repeatedly with interglaciations in rhythm with cyclic global climate changes.

Ice Age (Late-Cenozoic Ice Age) Present ice age, which began in late Pliocene time, perhaps 2.5 to 3 million years ago.

ice cap Continuous mass of glacial ice covering the top of a mountain range.

ice field Mass of interconnected valley glaciers on a mountain mass with protruding rock ridges or summits.

ice lens More or less horizontal layer of segregated ice formed by capillary movement of soil water toward a freezing front.

ice sheet Large thick plate of glacial ice moving outward in all directions from a central region of accumulation.

ice sheet climate ⑬ Severely cold climate, found on the Greenland and Antarctic ice sheets, with potential evapotranspiration (water need) effectively zero throughout the year.

ice shelf Thick plate of floating glacial ice attached to an ice sheet and fed by the ice sheet and by snow accumulation.

ice storm Occurrence of heavy glaze of ice on solid surfaces.

ice wedge Vertical, wall-like body of ground ice, often tapering downward, occupying a shrinkage crack in silt of permafrost areas.

iceberg Mass of glacial ice floating in the ocean, derived from a glacier that extends into tidal water.

ice-wedge polygons Polygonal networks of ice wedges.

igneous rock Rock solidified from a high-temperature molten state; also, rock formed by cooling of magma.

illuviation Accumulation in a lower soil horizon (typically, the B horizon) of materials brought down from a higher horizon; a soil-forming process.

image processing Mathematical manipulation of digital images, for example, to enhance contrast or edges.

Inceptisols Soil order consisting of soils with weakly developed soil horizons and containing weatherable minerals.

induced deflation Loss of soil by wind erosion that is triggered by human activity such as cultivation or overgrazing.

induced mass wasting Mass wasting that is induced by human activity, such as creation of waste soil and rock piles or undercutting of slopes in construction.

infiltration Absorption and downward movement of precipitation into the soil and regolith.

infrared imagery Images formed by infrared radiation emanating from the ground surface as recorded by a remote sensor.

infrared radiation Electromagnetic energy in the wavelength range of 0.7 to about 200 µm.

initial landforms Landforms produced directly by internal Earth processes of volcanism and tectonic activity; for example, volcano, fault scarp.

inner gorge Steep-walled valley formed when a meandering stream is uplifted by rapid tectonic activity, causing the stream to cut downward into the bedrock below.

insolation Flow rate of incoming solar radiation, measured at the top of the atmosphere; varies with the Sun's position in the sky.

inspiral Horizontal inward spiral or motion, such as that found in a cyclone.

interglaciation Within an ice age, a time interval of mild global climate in which continental ice sheets were largely absent or were limited to the Greenland and Antarctic ice sheets; the interval between two glaciations.

International Date Line The 180° meridian of longitude, together with deviations east and west of that meridian, forming the time boundary between adjacent standard time zones that are 12 hours fast and 12 hours slow with respect to Greenwich standard time.

intertropical convergence zone (ITCZ) Zone of convergence of air masses of tropical easterlies (trade winds) along the axis of the equatorial trough.

intrusion Body of igneous rock injected as magma into preexisting crustal rock; for example, a dike or sill.

intrusive igneous rock Igneous rock body produced by solidification of magma beneath the surface, surrounded by preexisting rock.

invasive species Organism aggressively encroaching on a new habitat region; often, a species from a distant location imported by human activity.

inversion See *temperature inversion*.

ion Atom or group of atoms bearing an electrical charge as the result of a gain or loss of one or more electrons.

island arcs Curved lines of volcanic islands associated with active subduction zones along the boundaries of lithospheric plates.

isobars Lines on map passing through all points having the same atmospheric pressure.

isohyet Line on a map drawn through all points having the same amount of precipitation.

isopleth Line on a map or globe drawn through all points having the same value as a selected property or entity.

isostasy Principle describing the flotation of the lithosphere, which is less dense, on the plastic asthenosphere, which is more dense.

isotherm Line on a map drawn through all points having the same air temperature.

isotope Form of an element with a particular atomic mass number.

J

jet streak Localized region of very high winds embedded within a jet stream.

jet stream High-speed airflow occurring at high levels in narrow bands within the upper-air westerlies and along certain other global latitude zones.

jet stream disturbance Broad, wavelike undulation in the jet stream.

joint block separation See *block separation*.

joints Fractures within bedrock, usually occurring in parallel and intersecting sets of planes.

joule Unit of work or energy in the metric system; symbol, J.

June solstice Solstice occurring on June 21 or 22, when the subsolar point is located at 23½° N.

K

kame Hill or mound of stratified sand and gravel deposited at, or near, the terminus of a glacier.

karst Landscape or topography dominated by surface features of limestone solution and underlain by a limestone cavern system.

Kelvin scale (K) Temperature scale on which the starting point is absolute zero, equivalent to −273°C.

kettle In a glacial outwash plain, a depression formed when sand and gravel build up around a block of stagnant ice, and the ice later melts away.

kinetic energy Form of energy represented by matter (mass) in motion.

knob and kettle Terrain of numerous small hills of glacial drift and deep depressions, usually situated along the moraine belt of a former ice sheet.

knot Measure of speed used in marine and aeronautical applications equal to 1 nautical mile per hour (1 kt = 0.514 m/s = 1.15 mi/hr).

L

La Niña Period of enhanced upwelling of cold deep water off the coast of Peru. Spanish for "the Little Girl," La Niña is the complementary condition to El Niño.

lag time Interval between occurrence of precipitation and peak discharge of a stream.

lagoon Shallow body of open water lying between a barrier island or a barrier reef and the mainland.

lahar Rapid downslope or downvalley movement of a tonguelike mass of water-saturated tephra (volcanic ash) originating high up on a steep-sided volcanic cone; a variety of mudflow.

lake Body of standing water enclosed on all sides by land.

land breeze Local wind blowing from land to water during the night.

land degradation Deterioration in the quality of plant cover and soil caused by overuse by humans and their domesticated animals, especially during periods of drought.

landforms Configurations of the land surface in distinctive shapes, produced by natural processes; for example, hill, valley, plateau.

landmass Large area of continental crust lying above sea level (base level) and thus available for removal by denudation.

Landsat Series of Earth-imaging NASA satellites, 1972–present.

landslide Rapid slippage of large masses of bedrock down steep mountain slopes or from high cliffs.

lapse rate Rate at which temperature decreases with increasing altitude.

large-scale map Map with fractional scale greater than 1:100,000; usually shows a small area.

Late-Cenozoic Ice Age Period marked by a series of glaciations, deglaciations, and interglaciations during the late Cenozoic era.

latent heat Energy absorbed and held in storage in a gas or liquid during the processes of evaporation, melting, or sublimation; energy released in condensation, freezing, or deposition.

latent heat transfer Flow of latent heat that results when water absorbs energy from its surrroundings and changes from a liquid or solid to a gas and then later releases that energy to new surroundings by condensation or deposition.

lateral moraine Moraine forming an embankment between the ice of an alpine glacier and the adjacent valley wall.

laterite Rocklike layer rich in sesquioxides and iron, including the minerals bauxite and limonite, found in low latitudes in association with Ultisols and Oxisols.

latitude Arc of a meridian between the Equator and a given point on the globe.

Laurasia Supercontinent of the Permian period, including much of the continental crust that is now North America and western Eurasia.

lava Magma emerging on the Earth's solid surface, exposed to air or water.

lava mesa Mesa topped by a lava flow that is relatively resistant to erosion.

leaching Pedogenic process by which material is lost from the soil by downward washing out and removal by percolating surplus soil water.

lee-side trough Low-pressure region found on the downwind (lee) side of a mountain chain, which can subsequently generate midlatitude cyclones.

liana Climbing woody vine supported by the trunk or branches of a tree or shrub.

lichens Plant forms in which algae and fungi live together in a symbiotic relationship to create a single structure. Lichens typically form tough, leathery coatings or crusts and attach themselves to rocks and tree trunks.

lidar Acronym for LIght Detection And Ranging, a remote sensing method using light waves emitted from a laser to determine the distance between the instrument and the target.

life cycle Continuous progression of stages in a growth or development process, such as that of a living organism.

life layer Shallow surface zone containing the biosphere; a zone of interaction between atmosphere and land surface, and between atmosphere and ocean surface.

life-form Characteristic physical structure, size, and shape of a plant or of an assemblage of plants.

life zones Series of vegetation zones describing vegetation types that are encountered at increasing elevation, especially in the southwestern United States.

lightning Electric arc passing between differently charged parts of a cloud mass or between the cloud and the ground.

limestone Nonclastic sedimentary rock in which calcite is the predominant mineral, and with varying minor amounts of other minerals and clay.

limestone caverns Interconnected subterranean cavities formed in limestone by carbonic acid action occurring in slowly moving groundwater.

limnology Study of the physical, chemical, and biological processes of lakes.

limonite Mineral or group of minerals consisting largely of iron oxide and water, produced by chemical weathering of other iron-bearing minerals.

line Type of spatial object in a geographic information system (GIS) that has starting and ending nodes; may be directional.

lithosphere Strong, brittle outermost rock layer of the Earth, lying above the asthenosphere.

lithospheric plate Segment of lithosphere moving as a unit, in contact with adjacent lithospheric plates along plate boundaries.

littoral drift Transport of sediment parallel to the shoreline by the combined action of beach drift and longshore current transport.

loam Soil-texture class in which no one of the three size grades (sand, silt, clay) dominates the other two.

local scale System of observation of the Earth in which local processes and phenomena are witnessed.

local winds General term for winds generated as direct or immediate effects of the local terrain.

lodgment till Heterogeneous mixture of rock fragments deposited beneath moving glacial ice.

loess Accumulation of yellowish to buff-colored, fine-grained sediment, largely of silt size, on upland surfaces after transport in the air in turbulent suspension (i.e., carried in a dust storm).

longitude Arc of a parallel between the prime meridian and a given point on the globe.

longitudinal dunes Class of sand dunes in which the dune ridges are oriented parallel to the prevailing wind.

longshore current Current in the breaker zone, running parallel to the shoreline and set up by the oblique approach of waves.

longshore drift Littoral drift caused by action of a longshore current.

longwave radiation Electromagnetic energy emitted by the Earth, largely in the range from 3 to 50 µm.

low base status See *base status of soils.*

low-latitude climates Group of climates of the equatorial and tropical zones dominated by the subtropical high-pressure belt and the equatorial trough.

low-latitude rainforest Evergreen broadleaf forest of the wet equatorial and tropical climate zones.

low-latitude rainforest environment Low-latitude environment of warm temperatures and abundant precipitation that characterizes rainforest in the wet equatorial ① and monsoon and trade-wind coastal ② climates.

low-level temperature inversion Atmospheric condition in which temperature near the ground increases, rather than decreases, with elevation.

low-pressure trough Weak, elongated cyclone of clouds and showers resulting from surface convergence generated by divergence aloft.

M

mafic igneous rock Igneous rock dominantly composed of mafic minerals.

mafic minerals (mafic mineral group) Minerals, largely silicate minerals, rich in magnesium and iron, dark in color, and of relatively greater density.

magma Mobile, high-temperature molten state of rock, usually of silicate mineral composition and with dissolved gases.

magnitude (of earthquake) Amount of energy released by an earthquake.

manipulation and analysis component Element of a geographic information system (GIS) that responds to spatial queries and creates new data layers.

mantle Rock layer or shell of the Earth beneath the crust and surrounding the core, composed of ultramafic igneous rock of silicate mineral composition.

mantle plume Columnlike rising of heated mantle rock, thought to be the cause of a hotspot in the overlying lithospheric plate.

map Paper or digital representation of space showing point, line, or area data.

map projection Any orderly system of parallels and meridians portrayed on a flat surface to represent the Earth's curved surface.

marble Variety of metamorphic rock derived from limestone or dolomite by recrystallization under pressure.

March equinox Equinox occurring on March 20 or 21.

marine cliff Rock cliff shaped and maintained by the undermining action of breaking waves.

marine scarp Steep seaward slope appearing in poorly consolidated alluvium, glacial drift, or other forms of regolith, produced along a coastline by the undermining action of waves.

marine terrace Former shore platform elevated to become a steplike coastal landform.

marine west-coast climate ⑧ Cool moist climate of west coasts in the midlatitude zone, usually with a substantial annual water surplus and a distinct winter precipitation maximum.

mass balance (of glacier) Balance in mass between input of new snow to a glacier and output from melting.

mass wasting Spontaneous downhill movement of soil, regolith, and bedrock under the influence of gravity, rather than by the action of fluid agents.

mathematical and statistical models In the representation perspective of geography, the use of mathematical and statistical models to predict spatial phenomena.

mathematical modeling Using variables and equations to represent real processes and systems.

mature stage In geographer W. M. Davis's geographic cycle, a stage of erosion with rounded hills and gentle slopes.

mean annual temperature Mean of mean daily air temperature for a given year or succession of years.

mean daily air temperature Sum of daily maximum and minimum air temperature readings divided by 2.

mean monthly temperature Average of mean daily air temperature values for a given calendar month.

meander Sinuous bend of a graded stream flowing in the alluvial deposit of a floodplain.

meander scar Remains of a former river channel segment that has been cut off and abandoned by a meandering river.

mechanical weathering See *physical weathering.*

medial moraine Long, narrow deposit of fragments on the surface of a glacier; created by the merging of lateral moraines when two glaciers join into a single stream of ice flow.

Mediterranean climate ⑦ Climate type of the subtropical zone, characterized by the alternation of a very dry summer and a mild, rainy winter.

melting Change from solid state to liquid state, accompanied by absorption of energy from the surroundings that is stored as latent heat.

melt-out till Heterogeneous mixture of rock fragments released by the melting in place of stagnant glacial ice.

Mercator projection Conformal map projection with horizontal parallels and vertical meridians and with map scale rapidly increasing with rise in latitude.

mercury barometer Instrument using the Torricelli principle, by which atmospheric pressure counterbalances a column of mercury in a tube.

meridian of longitude North–south line on the surface of the globe, connecting the North Pole and South Pole.

mesa Table-topped plateau of comparatively small extent, bounded by cliffs and occurring in a region of flat-lying strata.

mesocyclone Vertical column of cyclonically rotating air that develops in the updraft of a severe thunderstorm cell.

mesopause Upper limit of the mesosphere.

mesoscale convective system Relatively long-lived, large, and intense convective cell or cluster of cells characterized by exceptionally strong updrafts.

mesosphere Atmospheric layer of upwardly diminishing temperature, situated above the stratopause and below the mesopause.

Mesozoic era Second of three geologic eras following Precambrian time.

metamorphic rock Rock altered in physical structure and/or chemical (mineral) composition by action of heat, pressure, shearing stress, or infusion of elements, all taking place at substantial depth beneath the surface.

meteorology Science of the atmosphere; particularly, the physics of the lower or inner atmosphere.

methane Gaseous chemical compound of carbon and hydrogen, CH_4, emitted into the atmosphere by natural and human processes; an important greenhouse gas.

microburst Brief onset of intense winds close to the ground, beneath the downdraft zone of a thunderstorm cell.

micrometer Metric unit of length equal to one-millionth of a meter (0.000001 m); abbreviated µm.

microwaves Waves of the electromagnetic radiation spectrum in the wavelength band from about 0.03 cm to about 1 cm (about 0.02 in. to about 0.34 in.).

middle-infrared radiation Electromagnetic radiation of wavelengths 3 to 6 μm.

midlatitude climates Group of climates of the midlatitude and subtropical zones, located in the polar front zone and dominated by both tropical air masses and polar air masses.

midlatitude cyclone Traveling cyclone of the midlatitudes involving interaction of cold and warm air masses along sharply defined fronts.

midlatitude deciduous forest Plant formation class within the forest biome, dominated by tall, broadleaf deciduous trees and found mostly in the moist continental climate ⑩ and marine west-coast climate ⑧.

midlatitude zones Latitude zones occupying the latitude range 35° to 55° N and S (more or less) and lying between the subtropical and subarctic (subantarctic) zones.

midoceanic ridge One of three major divisions of the ocean basins, being the central belt of submarine mountain topography with a characteristic axial rift.

Milankovitch curve Graph of summer insolation at 65° N calculated from cycles in the Earth's axial rotation and solar revolution over the last 500,000 years.

millibar Unit of atmospheric pressure equal to one-thousandth of a bar; 100 Pa.

mineral Naturally occurring inorganic substance, usually having a definite chemical composition and a characteristic atomic structure.

mineral alteration Chemical change of minerals to more stable compounds on exposure to atmospheric conditions; same as *chemical weathering*.

mineral horizon (soils) Soil horizon composed primarily of mineral particles.

mineral matter (soils) Component of soil consisting of weathered or unweathered mineral grains.

mineral oxides (soils) Secondary minerals found in soils in which original minerals have been altered by chemical combination with oxygen.

minute (of arc) One-sixtieth of a degree.

mistral wind Local drainage wind of cold air affecting the Rhône Valley of southern France.

MODIS Acronym for Moderate Resolution Imaging Spectroradiometer, a satellite-borne NASA Earth-imaging instrument.

Moho Contact surface between the Earth's crust and mantle; a contraction of Mohorovičić, the last name of the seismologist who discovered this feature.

moist adiabatic lapse rate Reduced adiabatic lapse rate when condensation is taking place in rising air; value ranges between 4 and 9°C per 1000 m (2.2 and 4.9°F per 1000 ft).

moist continental climate ⑩ Moist climate of the midlatitude zone with strongly defined winter and summer seasons, adequate precipitation throughout the year, and a substantial annual water surplus.

moist subtropical climate ⑥ Moist climate of the subtropical zone, characterized by a moderate to large annual water surplus and a strong seasonal cycle of potential evapotranspiration (water need).

mollic epipedon Relatively thick, dark-colored surface soil horizon, containing substantial amounts of organic matter (humus) and usually rich in bases.

Mollisols Soil order consisting of soils with a mollic horizon and high base status.

monadnock Prominent, isolated mountain or large hill rising conspicuously above a surrounding peneplain and composed of a rock more resistant than that underlying the peneplain; a landform of denudation in moist climates.

monocline Warping of sedimentary beds into a steplike bend without folding.

monsoon and trade-wind coastal climate ② Moist climate of low latitudes showing a strong rainfall peak in the season of high Sun and a short period of reduced rainfall.

monsoon forest Formation class within the forest biome consisting in part of deciduous trees adapted to a long, dry season in the wet–dry tropical climate ③.

monsoon system System of low-level winds blowing into a continent in summer and out of it in winter, controlled by atmospheric pressure systems developed seasonally over the continent.

montane forest Plant formation class of the forest biome found in cool upland environments of the tropical and equatorial zones.

moraine Accumulation of rock debris carried by an alpine glacier or an ice sheet and deposited by the ice to become a depositional landform.

morphology Outward form and appearance of individual organisms or species.

mottles Spots or streaks of blue or green colors in soils, produced by iron and manganese compounds that develop under cool, wet conditions.

mountain arc Curving section of an alpine chain occurring on a converging boundary between two crustal plates.

mountain roots Erosional remnants of deep portions of ancient continental sutures that were once alpine chains.

mountain winds Daytime movements of air up the gradient of valleys and mountain slopes; alternating with nocturnal valley winds.

mucks Organic soils largely composed of fine, black, sticky organic matter.

mud Sediment consisting of a mixture of clay and silt with water, often with minor amounts of sand and sometimes with organic matter.

mud flat Barren expanse of tidal sediment covered at high tide and uncovered at low tide.

mudflow Form of mass wasting consisting of the downslope flowage of a mixture of water and mineral fragments (soil, regolith, disintegrated bedrock), usually following a natural drainage line or stream channel.

mudstone Sedimentary rock formed from mud.

multipurpose map Map containing several different types of information.

multispectral image Representation consisting of two or more images, each of which is taken from a different portion of the spectrum (e.g., blue, green, red, infrared).

multispectral scanner Remote sensing instrument, flown on an aircraft or spacecraft, that simultaneously collects multiple digital images (multispectral images) of the ground. Typically, images are collected in four or more spectral bands.

mutation Change in genetic material of a reproductive cell.

mutualism Form of symbiosis in which both species benefit to such a degree that they cannot survive alone.

N

nanometer Unit of measure equal to 1/1,000,000,000 of a meter (10^{-9} m).

nappe Overturned recumbent fold of strata, usually associated with thrust sheets in a collision orogen.

natural bridge Intrinsic rock arch spanning a stream channel, formed by cutoff of an entrenched meander bend.

natural levee Belt of higher ground paralleling a meandering alluvial river on both sides of the stream channel and built up by deposition of fine sediment during periods of overbank flooding.

natural selection Selection of organisms by environment in a process similar to choice of plants or animals for breeding by agriculturalists.

natural vegetation Stable, mature plant cover characteristic of a given area of land surface, largely free from the influences and impacts of human activities.

neap tides Tides with reduced range occurring when lunar and solar tidal forces are at right angles to each other.

near-infrared radiation Electromagnetic radiation in the wavelength range 0.7 to 1.2 μm. Plant leaves reflect near-infrared light strongly.

needleleaf evergreen forest Forest composed of evergreen tree species with needle-shaped leaves, such as spruce, fir, and pine.

needleleaf forest Formation class within the forest biome, consisting largely of needleleaf trees.

needleleaf tree Tree with long, thin, or flat leaves, such as pine, fir, larch, or spruce.

negative feedback In flow systems, a linkage between flow paths such that the flow in one pathway acts to reduce the flow in another pathway.

net photosynthesis Carbohydrate production remaining in an organism after respiration has broken down sufficient carbohydrate to power the metabolism of the organism.

net primary production Rate at which carbohydrate is accumulated in the tissues of plants within a given ecosystem; units are kilograms of dry organic matter per year per square meter of surface area $(kg/yr/m^2)$.

net radiation Difference in intensity between all incoming energy (positive quantity) and all outgoing energy (negative quantity) carried by both shortwave radiation and longwave radiation.

nickpoint Portion of a stream where gradient changes abruptly, such as a waterfall or rapids.

nimbostratus Cloud type in which sheets or layers of clouds are found near the bottom of the troposphere, accompanied by precipitation.

nimbus Any cloud type in which precipitation is occurring

nitrogen cycle Biogeochemical cycle in which nitrogen moves through the biosphere by the processes of nitrogen fixation and denitrification.

nitrogen fixation Chemical process of converting gaseous molecular nitrogen of the atmosphere into compounds or ions that can be directly utilized by plants; a process carried out within the nitrogen cycle by certain microorganisms.

nitrous oxide Gas, N_2O, emitted to the atmosphere from both human and natural activity; an important greenhouse gas.

node Point marking the end of a line or the intersection of lines as spatial objects in a geographic information system (GIS).

nonrenewable resources Resources that have no practical path of renewal; for example, energy stored in fossil fuels.

noon See *solar noon.*

noon angle (of the Sun) Angle of the Sun above the horizon at its highest point during the day.

normal fault Variety of fault in which the fault plane inclines (dips) toward the downthrown block and a major component of the motion is vertical.

nortes Strong, northeasterly winds that usually accompany the intrusion of cold, dry continental polar air from the north; found in subtropical regions of North America.

North Atlantic oscillation (NAO) Atmospheric phenomenon related to variations in the surface pressure gradient between the polar sea ice cap and the midlatitudes in both the Atlantic and Pacific Ocean basins.

north polar zone Latitude zone lying between 75° and 90° N.

North Pole Point at which the northern end of the Earth's axis of rotation intersects the Earth's surface.

northeast trade winds Surface winds of low latitudes that blow steadily from the northeast.

northers See *nortes.*

notch Eroded portion of a sea cliff just above the base, carved by physical weathering.

nunatak Mountain summit or ridge protruding above valley glaciers in an ice field.

nutrient cycle See *biogeochemical cycle.*

O

O_a horizon soil horizon below the O_i horizon containing decaying organic matter that is too decomposed to recognize as specific plant parts, such as leaves or twigs.

oasis Desert area where groundwater is tapped for crop irrigation and human needs.

oblate ellipsoid Geometric solid resembling a flattened sphere, with the polar axis shorter than the equatorial diameter.

occluded front Weather front along which a moving cold front has overtaken a warm front, forcing the warm air mass aloft.

ocean current Persistent, dominantly horizontal flow of ocean water.

ocean tide Periodic rise and fall of the ocean level induced by gravitational attraction between the Earth and Moon in combination with Earth's rotation.

oceanic crust Layer of basaltic composition beneath the ocean floors, capping oceanic lithosphere.

oceanic lithosphere Lithosphere bearing oceanic crust.

oceanic trench Narrow, deep depression in the seafloor representing the line of subduction of an oceanic lithospheric plate beneath the margin of a continental lithospheric plate; often associated with an island arc.

O_i horizon Surface soil horizon containing decaying organic matter that is recognizable as leaves, twigs, or other organic structures.

old age In W. M. Davis's geographic cycle, a stage of erosion with a low, undulating landscape drained by slow and sluggish streams.

old-field succession Form of secondary succession typical of an abandoned field, such as might be found in eastern or central North America.

organic horizon Soil horizon formed largely from decaying plant and/or animal matter.

organic matter (soils) Material in soil that was originally produced by plants or animals and has been subjected to decay.

organic sediment Sediment consisting of the organic remains of plants or animals.

orogen Mass of tectonically deformed rocks and related igneous rocks produced during an orogeny.

orogenic processes Mountain-building processes generated by motions of lithospheric plates.

orogeny Major episode of tectonic activity resulting in strata being deformed by folding and faulting.

orographic Pertaining to mountains.

orographic precipitation Precipitation induced by the forced rise of moist air over a mountain barrier.

oscillatory wave Type of wave that generates an oscillating motion of mass as the wave passes through the medium; water waves are oscillatory.

outcrop Surface exposure of bedrock.

outlet glacier Tongue of glacial ice extending from an ice cap to reach the ocean.

outspiral Horizontal outward spiral or motion, such as that found in an anticyclone.

outwash Glacial deposit of stratified drift left by braided streams issuing from the front of a glacier.

outwash plain Flat, gently sloping plain built up of sand and gravel by the aggradation of meltwater streams in front of the margin of an ice sheet.

overburden Strata overlying a layer or stratum of interest, as overburden above a coal seam.

overland flow Motion of a surface layer of water over a sloping ground surface when the infiltration rate is exceeded by the precipitation rate; a form of runoff.

overrunning Rising motion of warm air as it overtakes a mass of cold air.

overthrust fault Fault characterized by the overriding of one crustal block (or thrust sheet) over another along a gently inclined fault plane; associated with crustal compression.

ox-bow lake Crescent-shaped lake representing the abandoned channel left by the cutoff of an alluvial meander.

oxidation Chemical union of free oxygen with metallic elements in minerals.

oxide Chemical compound containing oxygen; in soils, iron oxides and aluminum oxides are examples.

Oxisols Soil order consisting of very old, highly weathered soils of low latitudes with low base status.

oxygen cycle Biogeochemical cycle in which oxygen moves through the biosphere in both gaseous and sedimentary forms.

ozone Form of oxygen with a molecule consisting of three atoms of oxygen, O_3.

ozone layer Layer in the stratosphere, mostly in the altitude range 12 to 31 mi (20 to 35 km), in which a concentration of ozone is produced by the action of solar ultraviolet radiation.

P

P wave Primary earthquake wave, which shakes the ground from side to side.

Pacific Decadal Oscillation (PDO) Slowly varying change in sea-surface temperatures and sea-level pressures of the north Pacific.

pack ice Floating sea ice that completely covers the sea surface.

Paleozoic era First of three geologic eras comprising the Phanerozoic era.

pamperos Strong, southerly or southwesterly winds occurring in Argentina and Uruguay and that usually accompany the intrusion of cold maritime polar (mP) air from the south.

Pangea (Pangaea) Hypothetical parent continent, enduring until near the close of the Mesozoic era and consisting of the continental shields of Laurasia and Gondwana joined into a single unit.

parabolic dunes Isolated low sand dunes of parabolic outline, with points directed into the prevailing wind.

parallel of latitude East-west circle on the Earth's surface, lying in a plane parallel to the Equator and at right angles to the axis of rotation.

parasitism Form of negative interaction between species in which a small species (parasite) feeds on a larger one (host) without necessarily killing it.

parent material Inorganic, mineral base from which the soil is formed; usually consists of regolith.

particulates Solid and liquid particles capable of being suspended for long periods in the atmosphere.

pascal Metric unit of pressure, defined as a force of 1 newton per square meter (1 N/m^2); symbol, Pa; 100 Pa = 1 mb, 10^5 Pa = 1 bar. Named for Blaise Pascal, French mathematician and philosopher.

passive continental margin Continental margin lacking active plate boundaries where continental crust meets oceanic crust.

passive systems Electromagnetic remote sensing systems that measure radiant energy reflected or emitted by an object or surface.

patch In cartography, a map symbol denoting a particular area with a distinctive pattern or color or a line to mark its edge.

pattern In the geographical perspective, the variation in phenomena seen at a particular scale.

patterned ground General term for a ground surface that bears polygonal or ringlike features, including stone circles, nets, polygons, steps, and stripes; includes ice wedge polygons. Patterned ground is typically produced by frost action in cold climates.

peat Partially decomposed, compacted accumulation of plant remains occurring in a bog environment.

ped Individual natural soil aggregate.

pediment Gently sloping, rock-floored land surface found at the base of a mountain mass or cliff in an arid region.

pedology Science of the soil as a natural surface layer capable of supporting living plants; synonymous with *soil science*.

peneplain Land surface of low elevation and slight relief produced in the late stages of denudation of a landmass.

perched water table Surface of a lens of groundwater held above the main body of groundwater by a discontinuous impervious layer.

percolation Slow, downward flow of water by gravity through soil and subsurface layers toward the water table.

perennials Plants that live for more than one growing season.

periglacial In an environment of intense frost action, often found in cold regions or near the present or former margins of alpine glaciers or ice sheets.

perihelion Point on the Earth's elliptical orbit at which the Earth is nearest to the Sun.

period (geologic time) Unit of geologic time, about tens of millions of years in length; subdivided into epochs.

permafrost Soil, regolith, and bedrock at a temperature below 0°C (32°F), found in cold climates of arctic, subarctic, and alpine regions.

permafrost table In permafrost, the upper surface of perennially frozen ground; lower surface of the active layer.

petroleum (crude oil) Natural liquid mixture of many complex hydrocarbon compounds of organic origin, found in accumulations (oil pools) within certain sedimentary rocks.

pH Measure of the concentration of hydrogen ions in a solution; acid solutions have pH values less than 7, and basic solutions have pH values greater than 7.

photosynthesis Production of carbohydrate by the union of water with carbon dioxide while absorbing light energy.

phreatophytes Plants that draw water from the groundwater table beneath alluvium of dry stream channels and valley floors in desert regions.

phylum Highest division of higher plant and animal life.

physical geography Part of systematic geography that deals with the natural processes occurring at the Earth's surface that provide the physical setting for human activities; includes the broad fields of climatology, geomorphology, coastal and marine geography, geography of soils, and biogeography.

physical weathering Breakup of massive rock (bedrock) into small particles through the action of physical forces acting at or near the Earth's surface.

phytoplankton Microscopic plants found largely in the uppermost layer of ocean or lake water.

piedmont glacier Valley glacier that flows out onto a plain at the base of the mountain.

pingo Conspicuous conical mound or circular hill, having a core of ice, found on plains of the arctic tundra where permafrost is present.

pinnacle Single, tall rock column produced by erosion of flat-lying strata.

pioneer plants Plants that are the first to invade an environment of new land or a soil that has been cleared of vegetation cover; often, pioneer plants are annual herbs.

pioneer stage First stage of an ecological succession.

pixel One of the small discrete elements that together constitute a digital image (as on a television or computer screen).

place In geography, a location on the Earth's surface, typically a settlement or small region with unique characteristics. From the viewpoint perspective of geography, a focus on how processes are integrated at a single location or within a single region.

plane of the ecliptic Imaginary plane on which the Earth's orbit lies.

plant ecology Study of the relationships between plants and their environment.

plant nutrients Ions or chemical compounds needed for plant growth.

plate tectonics Theory of tectonic activity dealing with lithospheric plates and their activity.

plateau Upland surface, more or less flat and horizontal, upheld by resistant beds of sedimentary rock or lava flows and bounded by a steep cliff.

playa Flat land surface underlain by fine sediment or evaporite minerals deposited from shallow lake waters in a dry climate on the floor of a closed topographic depression.

Pleistocene epoch Epoch of the Cenozoic era, often identified as the Ice Age. The Pleistocene epoch preceded the Holocene epoch.

plough winds See *straight-line winds*.

plucking See *glacial plucking*.

pluton Any body of intrusive igneous rock that has solidified below the surface, enclosed in preexisting rock.

pluvial lake Lake formed during the Ice Age in a regime of precipitation greater than at present; now reduced in extent, or dry.

pocket beach Beach of crescentic outline located at a bay head.

point Spatial object in a geographic information system (GIS) with no area.

point bar Deposit of coarse bed-load alluvium accumulated on the inside of a growing alluvial meander.

polar easterlies System of easterly surface winds at high latitude centered around the pole. This system is better developed in the southern hemisphere, over Antarctica, than in the northern hemisphere.

polar front Front lying between cold polar air masses and warm tropical air masses, often situated along a jet stream within the upper-air westerlies.

polar front jet stream Jet stream found along the polar front, where cold polar air and warm tropical air are in contact.

polar front zone Broad zone in midlatitudes and higher latitudes, occupied by the shifting polar front.

polar high Persistent low-level center of high atmospheric pressure located over the polar zone of Antarctica.

polar outbreak Tongue of cold polar air, preceded by a cold front, penetrating far into the tropical zone and often reaching the equatorial zone. The polar outbreak brings rain squalls and unusual cold.

polar projection Map projection centered on Earth's North Pole or South Pole.

polar zones Latitude zones lying between 75° and 90° N and S.

pole Point at which the Earth's axis of rotation intersects the Earth's surface.

poleward energy transfer Movement of solar energy, absorbed at equatorial and tropical latitudes, toward the poles, occurring as flows of latent and sensible heat.

pollutants In air pollution studies, foreign matter injected into the lower atmosphere as particulates or chemical pollutant gases.

pollution dome Broad, low dome-shaped layer of polluted air, formed over an urban area when winds are weak or calm prevails.

pollution plume (1) Trace or path of pollutant substances, moving along the flow paths of groundwater. (2) Trail of polluted air carried downwind from a pollution source by strong winds.

polygon Type of spatial object in a geographic information system (GIS) with a closed chain of connected lines surrounding an area.

polyploidy Mechanism of speciation in which entire chromosome sets of organisms are doubled, tripled, quadrupled, and so on.

pool In flow systems, an area or location of concentration of matter.

positive feedback In flow systems, a linkage between flow paths such that the flow in one pathway acts to increase the flow in another pathway.

potential energy Energy of position; produced by gravitational attraction of the Earth's mass for a smaller mass on or near the Earth's surface.

potential evapotranspiration (water need) Ideal or hypothetical rate of evapotranspiration estimated to occur from a complete canopy of green foliage of growing plants continuously supplied with all the soil water they can use; an actual condition reached in situations where precipitation is sufficiently great or irrigation water is supplied in sufficient amounts.

pothole Cylindrical cavity in hard bedrock of a stream channel produced by abrasion of a rounded rock fragment rotating within the cavity.

power source Energy entering a flow system that can do work or increase the internal energy of a system or its parts.

prairie Plant formation class of the grassland biome, consisting of dominant tall grasses and subdominant forbs; widespread in subhumid continental climate regions of the subtropical and midlatitude zones.

Precambrian time All of geologic time older than the beginning of the Cambrian period; that is, older than about 600 million years.

precipitation Particles of liquid water or ice that fall from the atmosphere and may reach the ground.

predation Form of interaction among animal species in which one species (predator) kills and consumes the other (prey).

preprocessing component Element of a geographic information system (GIS) that prepares data for entry into the system.

pressure gradient Change of atmospheric pressure measured along a line at right angles to the isobars.

pressure gradient force Force acting horizontally, tending to move air in the direction of lower atmospheric pressure.

prevailing westerly winds (westerlies) Surface winds blowing from a generally westerly direction in the midlatitude zone, but varying greatly in direction and intensity.

primary consumers Organisms at the lowest level of the food chain that ingest primary producers or decomposers as their energy source.

primary minerals In pedology (soil science), the original, unaltered silicate minerals of igneous and metamorphic rocks.

primary producers Organisms that use light energy to convert carbon dioxide and water to carbohydrates through the process of photosynthesis.

primary succession Ecological succession that begins on newly constructed ground.

prime meridian Reference meridian of zero longitude; universally accepted as the Greenwich meridian.

process In the geographical perspective, how factors that affect a phenomenon act to produce a pattern at a particular scale.

product generation component Element of a geographic information system (GIS) that provides output products such as maps, images, or tabular reports.

progradation Shoreward building of a beach, bar, or sandspit by addition of coarse sediment carried by littoral drift or brought from deeper water offshore.

protocooperation Form of symbiosis in which the relationship benefits both species but is not essential for the survival of either.

proxy (temperature record) Indirect method of inferring air temperature, used before it was possible to make direct records of temperature; used to extend the temperature record back in time.

pyroclastic material Collective term for particles ejected from an erupting volcano.

Q

quartz Mineral of silicon dioxide composition, SiO_2.

quartzite Metamorphic rock consisting largely of the mineral quartz.

R

radar Active remote sensing system in which a pulse of electromagnetic radiation is emitted by an instrument, and the strength of the echo of the pulse is recorded.

radial drainage pattern Pattern of streams radiating outward from a central peak or highland, such as a sedimentary dome or a volcano.

radiation See *electromagnetic radiation*.

radiation balance Difference between incoming solar shortwave radiation and outgoing longwave radiation emitted by the Earth into space; taken for a particular time and place or over the Earth as a whole.

radiation fog Fog produced by radiation cooling of the basal air layer.

radioactive decay Spontaneous change in the nucleus of an atom that leads to the emission of matter and energy.

radiogenic heat Energy flow from the Earth's interior that is slowly released by the radioactive decay of unstable isotopes.

radiometric dating Method of determining the geologic age of a rock or mineral by measuring the proportions of certain of its elements in their different isotopic forms.

rain Form of precipitation consisting of falling water drops, usually 0.5 mm (0.02 in.) or larger in diameter.

rain gauge Instrument used to measure the amount of rain that has fallen.

rain shadow Belt of arid climate to lee of a mountain barrier, produced as a result of adiabatic warming of descending air.

rainforest ecosystem Ecosystem of forest marked by abundant and copious rainfall, often dense and diverse; typically refers to broadleaf rainforests of equatorial and tropical zones.

rain-green vegetation Vegetation that grows green foliage in the wet season but becomes largely dormant or deciduous in the dry season; found in the tropical zone, it includes the savanna biome and monsoon forest.

raised shoreline Former shoreline that has been lifted above the limit of wave action; also called *elevated shoreline*.

rapids Steep-gradient reaches of a stream channel in which stream velocity is high.

recessional moraine Moraine produced at the ice margin during a temporary halt in the recessional phase of deglaciation.

recombination Cause of variation in organisms arising from the free interchange of alleles of genes during the reproduction process.

recumbent In rock formations, overturned, as a folded sequence of rock layers in which the folds are doubled back on themselves.

reflection Outward scattering of radiation toward space by the atmosphere and/or Earth's surface.

reg Desert surface armored with a pebble layer, resulting from long-term deflation; found in the Sahara Desert of North Africa.

regional geography Branch of geography concerned with how the Earth's surface is differentiated into unique places.

regional metamorphism Change associated with large areas of tectonic activity from the collision of lithospheric plates.

regional scale Scale of observation at which subcontinental regions are discernible.

regolith Layer of mineral particles overlying bedrock; may be derived by weathering of underlying bedrock or be transported from other locations by fluid agents.

rejuvenated stream Uplifted stream with a steep gradient in the process of downcutting.

rejuvenation In W. M. Davis's geographic cycle, the uplift of an old-age landscape to create a new, youthful stage of erosion.

relative humidity Ratio of water vapor present in the air to the maximum quantity possible for saturated air at the same temperature.

relief (soils) Shape or configuration of the ground surface as it affects soil development.

remote sensing Measurement of some property of an object or surface by means other than direct contact; usually refers to the gathering of scientific information about the Earth's surface from great heights and over broad areas, using instruments mounted on aircraft or orbiting space vehicles.

remote sensor Instrument or device used to measure electromagnetic radiation reflected or emitted from a target body.

removal In soil science, the set of processes that result in the removal of material from a soil horizon, such as surface erosion or leaching.

representation Perspective of geography concerned with developing and manipulating tools for the display and analysis of spatial information.

representative fraction (RF) See *scale fraction*.

residual regolith Regolith formed in place by alteration of the bedrock directly beneath it.

resolution On a map, capability to distinguish small objects present on the ground.

respiration Oxidation of organic compounds by organisms that power bodily functions.

retrogradation Cutting back (retreat) of a shoreline, beach, marine cliff, or marine scarp by wave action.

reverse fault Type of fault in which one fault block rides up over another on a steep fault plane.

revolution Motion of a planet in its orbit around the Sun, or of a planetary satellite around a planet.

rhyolite Extrusive igneous rock of granite composition, occurring as lava or tephra.

ria Coastal embayment or estuary.

ria coast Deeply embayed coast formed by partial submergence of a landmass previously shaped by fluvial denudation.

Richter scale Scale of magnitude numbers describing the amount of energy released by an earthquake.

ridge (atmospheric) Elongated zone of atmospheric high pressure.

ridge-and-valley landscape Assemblage of landforms developed by denudation of a system of open folds of strata, consisting of long, narrow ridges and valleys arranged in parallel or zigzag patterns.

rift valley Trenchlike valley with steep, parallel sides; essentially, a graben between two normal faults; associated with crustal spreading.

rifting Drawing apart of crustal layers by tectonic activity, resulting in faulting.

rill erosion Form of accelerated erosion in which numerous, closely spaced miniature channels (rills) are scored into the surface of exposed soil or regolith.

rip-rap Mounds of boulders placed at the edge of the shoreline to slow wave erosion.

rock Natural aggregate of minerals in the solid state; usually hard and consisting of one, two, or more mineral varieties.

rock sea Expanse of large blocks of rock produced by joint block separation and shattering by frost action occurring at high altitudes or high latitudes.

rockfall Most visible form of mass wasting, in which rocks fall down steep slopes or cliff sides, bringing with them soil or regolith.

rockslide Landslide of jumbled bedrock fragments.

Rodinia Early supercontinent, predating Pangea, that was fully formed about 700 million years ago.

roll cloud Tube-shaped cloud whose axis is parallel to the ground, typically found above a gust front where outflowing air at the surface is in the opposite direction to the inflowing air above it.

Rossby waves Wavelike disturbances in the flow of a jet stream.

rotation Spinning of an object around an axis.

rotational slide Type of landslide in which the mass rotates as it slides downslope, resulting in displacement of soil from the top of the rupture.

runoff Flow of water from continents to oceans by way of streamflow and groundwater flow; also, a term in the water balance of the hydrologic cycle. In a more restricted sense, runoff refers to surface flow by overland flow and channel flow.

S

S wave Secondary earthquake wave, which moves the ground up and down.

Sahel region Belt of wet-dry tropical ③ and semiarid dry tropical ④ climate in Africa, located south of the Sahara Desert, where precipitation is highly variable from year to year.

salic horizon Soil horizon enriched by soluble salts.

salinity Degree of saltiness of water; refers to the abundance of such ions as sodium, calcium, potassium, chloride, fluoride, sulfate, and carbonate.

salinization Precipitation of soluble salts within the soil.

salt flat Shallow basin covered with salt deposits formed when stream input to the basin is evaporated to dryness from the basin of a lake; may also form by evaporation of a saline lake when climate changes.

salt marsh Peat-covered expanse of sediment built up to the level of high tide over a previously formed tidal mud flat.

saltation Leaping, impacting, and rebounding of sand grains transported over a sand or pebble surface by wind.

salt-crystal growth Form of weathering in which rock is disintegrated by the expansive pressure of growing salt crystals during dry weather periods, when evaporation is rapid.

saltwater intrusion Occurs in a coastal well when an upper layer of freshwater is pumped out, leaving a saltwater layer below to feed the well.

sand Sediment particles between 0.06 and 2 mm (0.0024 and 0.079 in.) in diameter.

sand dune Hill or ridge of loose, well-sorted sand shaped by wind and usually capable of downwind motion.

sand sea Field of transverse dunes.

sandbar Hill or ridge of sand found in shallow waters, often deposited by littoral drift.

sandspit Narrow, fingerlike embankment of sand constructed by littoral drift into the open water of a bay.

sandstone Sedimentary rock consisting largely of mineral particles of sand size.

sanitary landfill Facility where trash and other wastes are buried under layers of sand or soil and allowed to decompose.

Santa Ana Easterly wind, often hot and dry, that blows from the interior desert region of Southern California and passes over the coastal mountain ranges to reach the Pacific Ocean.

saturated air Air with a water vapor content equal to the saturation-specific humidity given for the temperature and pressure of the air.

saturated zone Zone beneath the land surface in which all pores of the bedrock or regolith are filled with groundwater.

saturation (atmospheric) Condition in which the specific humidity of a parcel of air is equal to the saturation-specific humidity.

saturation-specific humidity Maximum amount of water vapor an air parcel can contain based on its temperature and pressure.

savanna Vegetation cover of widely spaced trees, with grassland beneath.

savanna biome Biome that consists of a combination of trees and grassland in various proportions.

savanna woodland Plant formation class of the savanna biome consisting of a woodland of widely spaced trees and a grass layer, found throughout the wet-dry tropical climate regions in a belt adjacent to the monsoon forest and low-latitude rainforest.

scale Magnitude of a phenomenon or system, as, for example, global scale or local scale; in the viewpoint perspective of geography, a focus on examining a phenomenon at different scales.

scale fraction Ratio that relates distance on the Earth's surface to distance on a map or surface of a globe.

scanning systems Remote sensing systems that make use of a scanning beam to generate images over the frame of observation.

scarification General term for artificial excavations and other land disturbances produced for purposes of extracting and/or processing mineral resources.

scattering Turning aside of radiation by an atmospheric molecule or particle so that the direction of the scattered ray is changed.

schist Finely layered metamorphic rock in which mica flakes are typically found oriented parallel to layer surfaces.

sclerophyll forest Plant formation class of the forest biome associated with regions of Mediterranean climate ⑦ and consisting of low sclerophyll trees and often including sclerophyll woodland or scrub.

sclerophyll woodland Plant formation class of the forest biome, composed of widely spaced sclerophyll trees and shrubs.

sclerophylls Hard-leafed evergreen trees and shrubs capable of enduring long, dry summers.

scree Accumulation of loose rock fragments derived by fall of rock from a cliff; also called *talus*.

scrub Plant formation class or subclass consisting of shrubs and having a canopy coverage of about 50 percent.

sea arch Archlike landform of a rocky, cliffed coast created when weathering and waves erode through a narrow headland from both sides.

sea breeze Local wind blowing from sea to land during the day.

sea cave Cave near the base of a marine cliff, eroded by weathering and waves.

sea fog Fog layer formed at sea when warm moist air passes over a cool ocean current and is chilled to the condensation point.

sea ice Floating ice of the oceans formed by direct freezing of ocean water.

sea stack Isolated columnar mass of bedrock left standing in front of a retreating marine cliff.

second of arc One sixtieth of a minute, or 1/3600 of a degree.

secondary consumers Animals that feed on primary consumers.

secondary minerals In soil science, minerals that are stable in the surface environment, derived by mineral alteration of the primary minerals.

secondary succession Ecological succession beginning on a previously vegetated area that has been recently disturbed by such agents as fire, flood, windstorm, or humans.

sediment Finely divided mineral and organic matter derived directly or indirectly from preexisting rock and life processes.

sediment yield Quantity of sediment removed by overland flow from a land surface of a given unit area in a given unit of time.

sedimentary dome Up-arched strata forming a circular structure with domed summit and flanks with moderate to steep outward dip.

sedimentary rock Rock formed from accumulation of sediment.

seismic sea wave See *tsunami*.

seismic waves Wavelike motion of the ground, propagated through the solid Earth following a sudden slippage of rocks along an earthquake fault.

seismometer Instrument used to record ground motion caused by an earthquake.

semiarid (steppe) dry climate Subtype of the dry climates that exhibits a short wet season and supports the growth of grasses and annual plants.

semidesert Plant formation class of the desert biome, consisting of xerophytic shrub vegetation with a poorly developed herbaceous lower layer; subtypes are semidesert scrub and woodland.

semidiurnal Adjective meaning relating to or occurring every half day; also, occurring twice a day.

sensible heat Internal energy measurable by a thermometer; an indication of the intensity of kinetic energy of molecular motion within a substance.

sensible heat transfer Flow of sensible heat carried by matter as it moves from one location to another.

September equinox Equinox occurring on September 22 or 23.

sequential landforms Landforms produced by external Earth processes in the total activity of denudation; for example, gorge, alluvial fan, floodplain.

seral stage Stage in a sere.

sere In an ecological succession, the series of biotic communities that follow one another on the way to the stable stage, or climax.

sesquioxides Oxides of aluminum or iron with a ratio of two atoms of aluminum or iron to three atoms of oxygen.

severe thunderstorm Persistent, strong thunderstorm, driven by a wind shear condition that feeds cool, dry air into the storm, enhancing uplift and latent heat release.

shale Sedimentary rock of mud or clay composition, showing lamination and splitting along fine layers.

shearing (of rock) Slipping motion between very thin rock layers, similar to that of a deck of cards that has been fanned with the sweep of a hand.

sheet erosion Type of accelerated soil erosion in which thin layers of soil are removed without formation of rills or gullies.

sheet flow Overland flow taking the form of a continuous thin film of water over a smooth surface of soil, regolith, or rock.

sheeting structure Thick, subparallel layers of massive bedrock formed by spontaneous expansion accompanying unloading.

shield volcano Low, often large, dome-like accumulation of basalt lava flows emerging from long radial fissures on flanks.

shore platform Sloping, nearly flat bedrock surface extending out from the foot of a marine cliff under the shallow water of breaker zone.

shoreline Shifting line of contact between water and land.

short-grass prairie Plant formation class in the grassland biome consisting of short grasses sparsely distributed in clumps and bunches and some shrubs, widespread in areas of semiarid climate in continental interiors of North America and Eurasia; also called *steppe*.

shortwave infrared radiation Electromagnetic radiation in the wavelength range from 1.2 to 3.0 μm.

shortwave radiation Electromagnetic energy in the range from 0.2 to 3 μm, including most of the energy spectrum of solar radiation.

shrubs Woody perennial plants, usually small or low, with several low-branching stems and a foliage mass close to the ground.

Siberian high Intense and very cold high-pressure cell that forms in Siberia during the winter.

silica Silicon dioxide in any of several mineral forms.

silicate minerals (silicates) Minerals containing as silicate ions, $SiO_4^=$, within the mineral's crystalline structure.

sill Intrusive igneous rock in the form of a plate, occurring where magma was forced into a natural parting in the bedrock, such as a bedding surface in a sequence of sedimentary rocks.

silt Sediment particles between 0.004 and 0.06 mm (0.00016 and 0.0024 in.) in diameter.

sinkhole Surface depression in limestone, leading down into limestone caverns.

slash-and-burn Agricultural system practiced in the low-latitude rainforest, in which small areas are cleared and the trees burned to form plots that are cultivated for several years.

slate Compact, fine-grained variety of metamorphic rock, derived from shale, showing well-developed cleavage.

sleet Form of precipitation consisting of ice pellets, which may be frozen raindrops.

slip face Steep face of an active sand dune, receiving sand by saltation over the dune crest and repeatedly sliding because of oversteepening.

slope (1) Degree of inclination from the horizontal of an element of ground surface. 2) Any portion or element of the Earth's solid surface.

slope (fluvial)   Rate of descent of a stream's water surface to lower elevation along the length of the stream.

small circle   Circle formed by passing a plane through a sphere without going through the exact center.

small-scale map   Map with fractional scale of less than 1:100,000; usually covers a large area.

smog   Mixture of aerosols and chemical pollutants in the lower atmosphere, usually found over urban areas.

snow   Form of precipitation consisting of ice particles.

soil   Natural terrestrial surface layer containing living matter and supporting or capable of supporting plants.

soil colloids   Mineral particles of extremely small size, capable of remaining suspended indefinitely in water; typically, they take the form of thin plates or scales.

soil color   Color of soil or a soil horizon, often related to soil-forming processes.

soil creep   Extremely slow downhill movement of soil and regolith as a result of continued agitation and disturbance of the particles by such activities as frost action, temperature changes, or wetting and drying of the soil.

soil enrichment   Additions of materials to the soil body; one of the pedogenic processes.

soil erosion   Removal of material by erosion from the soil surface.

soil horizon   Distinctive layer of the soil, more or less horizontal, set apart from other soil zones or layers by differences in physical and chemical composition, organic content, structure, or a combination of those properties. The soil horizon is produced by soil-forming processes.

soil orders   Eleven soil classes forming the highest category in the classification of soils.

soil profile   Display of soil horizons on the face of a freshly cut vertical exposure through the soil.

soil science   See *pedology*.

soil structure   Presence, size, and form of aggregations (lumps or clusters) of soil particles.

soil texture   Descriptive property of the mineral portion of the soil based on varying proportions of sand, silt, and clay.

soil water   Water held in the soil and available to plants through their root systems; a form of subsurface water.

soil-water balance   Equilibrium among the component terms of the soil-water budget; namely, precipitation, evapotranspiration, change in soil water storage, and water surplus.

soil-water belt   Soil layer from which plants draw soil water.

soil-water budget   Accounting system used to evaluate the daily, monthly, or yearly amounts of precipitation, evapotranspiration, soil-water storage, water deficit, and water surplus.

soil-water recharge   Restoration of depleted soil water by infiltration of precipitation.

soil-water shortage   See *deficit*.

soil-water storage   Actual quantity of water held in the soil-water belt at any given moment; usually applied to a soil layer of given depth, such as 300 cm (about 12 in.).

solar constant   Intensity of solar radiation falling on a unit area of surface held at right angles to the Sun's rays at a point outside the Earth's atmosphere; equal to an energy flow of 1367 W/m^2.

solar day   Average time required for the Earth to complete one rotation with respect to the Sun; time elapsed between one solar noon and the next, averaged over the period of one year.

solar noon   Instant at which the subsolar point crosses the meridian of longitude of a given point on the Earth; instant at which the Sun's shadow points exactly due north or due south at a given location.

solution   Weathering process by which minerals dissolve in water; may be enhanced by the action of carbonic acid or weak organic acids.

sorting   Separation of one grade size of sediment particles from another by the action of currents of air or water.

source region   Extensive land or ocean surface over which an air mass derives its temperature and moisture characteristics.

south polar zone   Latitude zone lying between 75° and 90° S.

South Pole   Point at which the southern end of the Earth's axis of rotation intersects the Earth's surface.

southeast trade winds   Surface winds of low latitudes that blow steadily from the southeast.

Southern Oscillation   Episodic reversal of prevailing barometric pressure differences between two regions, one centered on Darwin, Australia, in the eastern Indian Ocean, and the other on Tahiti in the western Pacific Ocean; a precursor to the occurrence of an El Niño event.

southern pine forest   Subtype of needle-leaf forest dominated by pines and occurring in the moist subtropical climate ⑥; typically found on sandy soils of the Atlantic and Gulf Coast coastal plains.

space   From the viewpoint perspective of geography, a focus on how places are interdependent.

spatial data   Information associated with a specific location or area of the Earth's surface.

spatial object   Geographic area, line, or point to which information is attached.

speciation   Process by which species are differentiated and maintained.

species   Collection of individual organisms that are capable of interbreeding to produce fertile offspring.

specific heat   Physical constant of a material that describes the amount of energy in joules required to raise the temperature of 1 gram of the material by 1 Celsius degree.

specific humidity   Mass of water vapor contained in a unit mass of air.

spectral signature   In remote sensing, the reflectance of an object or a surface type in particular spectral bands.

spectrum   Entire range of wavelengths of electromagnetic radiation, or a portion of that range.

sphere   In physical geography, a great physical realm of the Earth: atmosphere, lithosphere, hydrosphere, biosphere.

spit   See *sandspit*.

splash erosion   Soil erosion caused by direct impact of falling raindrops on a wet surface of soil or regolith.

spodic horizon   Soil horizon containing precipitated amorphous materials composed of organic matter and sesquioxides of aluminum, with or without iron.

Spodosols   Soil order consisting of soils with a spodic horizon, an albic horizon, low base status, and lacking in carbonate materials.

spoil   Rock waste removed in a mining operation.

spreading plate boundary   Lithospheric plate boundary along which two plates of oceanic lithosphere are undergoing separation, while at the same time new lithosphere is being formed, by accretion.

spring tides   Tides with increased range occurring when lunar and solar tidal forces are aligned.

squall line   Line of thunderstorms and strong winds that extends for several hundred kilometers (miles).

stable air   Atmospheric condition in which the environmental temperature lapse rate is less than the dry adiabatic lapse rate, inhibiting convectional uplift and mixing.

stable air mass Air mass in which air temperature decreases with elevation so as to inhibit the rising motion of a heated air parcel.

stage Height of the surface of a river above its bed, or a fixed level near the bed.

stalactite Icicle-shaped formation created when water containing dissolved minerals drips through the ceiling of an underground cavern.

stalagmite Upward-pointing formation created when water containing dissolved minerals drips to the floor of an underground cavern.

standard meridians Standard time meridians separated by 15° of longitude and having values that are multiples of 15°—though in some cases, meridians that are multiples of 7½° are used.

standard time system Time system based on the local time of a standard meridian and applied to belts of longitude extending 7½° (more or less) on either side of that meridian.

standard time zone Zone of the Earth in which all locations keep the same time, according to a standard meridian within the zone.

star dune Large, isolated sand dune with radial ridges culminating in a peaked summit; found in the deserts of North Africa and the Arabian Peninsula.

stationary front Boundary between two differing air masses that has not moved over the last three to six hours or that has moved relatively slowly.

statistics Branch of mathematical sciences that deals with the analysis of numerical data.

steering winds Upper-level winds tending to drag cyclones in their direction.

steppe Semiarid grassland occurring largely in dry continental interiors.

steppe climate See *semiarid (steppe) dry climate subtype.*

stone polygons Linked ringlike ridges of cobbles or boulders lying at the surface of the ground in arctic and alpine tundra regions.

stone stripes Stone polygons drawn out into long stripes, usually in the direction of a gentle slope.

storage capacity Maximum capacity of soil to hold water against the pull of gravity.

storage pool Type of pool in a biogeochemical cycle in which materials are largely inaccessible to life.

storage recharge Restoration of stored soil water during periods when precipitation exceeds potential evapotranspiration (water need).

storage withdrawal Depletion of stored soil water during periods when evapotranspiration exceeds precipitation, calculated as the difference between actual evapotranspiration (water use) and precipitation.

storm sewer System of large underground pipes designed to quickly transport storm runoff from paved areas directly to stream channels for discharge.

storm surge Rapid rise of coastal water level accompanying the onshore arrival of a tropical cyclone.

storm tracks Common paths along which midlatitude cyclones tend to travel as they develop, mature, and dissolve.

straight-line winds Strong surface winds blowing from a single direction, usually arising from downburst winds of thunderstorms spreading out at the surface.

strata Layers of sediment or sedimentary rock in which individual beds are separated from one another along bedding planes.

stratified drift Glacial drift made up of sorted and layered clay, silt, sand, or gravel deposited from meltwater in stream channels or in marginal lakes close to the ice front.

stratiform clouds Clouds of layered, blanketlike form.

stratopause Upper limit of the stratosphere.

stratosphere Layer of the atmosphere lying directly above the troposphere.

stratovolcano Volcano constructed of multiple layers of lava and tephra (volcanic ash).

stratus Cloud type of the low-height family formed into a dense, dark gray layer.

stream Long, narrow body of flowing water occupying a stream channel and moving to lower levels under the force of gravity.

stream capacity Maximum stream load of solid matter that can be carried by a stream for a given discharge.

stream channel Long, narrow, troughlike depression occupied and shaped by a stream moving to progressively lower levels.

stream deposition Accumulation of transported particles on a streambed, on the adjacent floodplain, or in a body of standing water.

stream erosion Progressive removal of mineral particles from the floor or sides of a stream channel by drag force of the moving water, abrasion, or corrosion.

stream gradient Rate of descent to lower elevations along the length of a stream channel.

stream load Solid matter carried by a stream in dissolved form (as ions), in turbulent suspension and as bed load.

stream order Ranking of streams and stream segments in a drainage system, with the smallest streams first-order.

stream profile Graph of the elevation of a stream plotted against its distance downstream.

stream transportation Downvalley movement of eroded particles in a stream channel in solution, in turbulent suspension, or as bed load.

stream velocity Flow rate of water in a stream; often represented by *V.*

streamflow Water flow in a stream channel; same as *channel flow.*

strike In geology, the compass direction of the line of intersection of an inclined rock plane, and a horizontal plane of reference.

strike-slip fault Variety of fault on which the motion is dominantly horizontal along a near-vertical fault plane.

strip mining Mining method in which overburden is first removed from a seam of coal or a sedimentary ore, allowing the coal or ore to be extracted.

subantarctic low-pressure belt Persistent belt of low atmospheric pressure centered about at lat. 65° S over the Southern Ocean.

subantarctic zone Latitude zone lying between lat. 55° and 60° S (more or less) and occupying a region between the midlatitude and the antarctic zones.

subarctic zone Latitude zone between lat. 55° and 60° N (more or less), occupying a region between the midlatitude and the arctic zones.

subboreal stage Third climatic period in the Holocene epoch, marked by below-average temperatures.

subduction Descent of the downbent edge of a lithospheric plate into the asthenosphere so as to pass beneath the edge of the adjoining plate.

sublimation Process of change of ice (solid state) to water vapor (gaseous state). In meteorology, sublimation also refers to the change of state from water vapor (liquid) to ice (solid), which is referred to as *deposition* in this text.

submergence Inundation or partial drowning of a former land surface by a rise of sea level or a sinking of the crust, or both.

suborder Unit of soil classification representing a subdivision of the soil order.

subsea permafrost Permafrost lying below sea level, found in a shallow offshore zone fringing the arctic seacoast.

subsequent stream Stream that develops its course by stream erosion along a band or belt of weaker rock.

subsidence Sinking of the ground surface.

subsolar point Point on the Earth's surface at which solar rays are perpendicular to the surface.

subsurface water Water of the lands held in soil, regolith, or bedrock below the surface.

subtropical broadleaf evergreen forest Formation class of the forest biome, composed of broadleaf evergreen trees; occurs primarily in the regions of the moist subtropical climate ⑥.

subtropical evergreen forest Subdivision of the forest biome, composed of both broadleaf and needleleaf evergreen trees.

subtropical high-pressure belts Belts of persistent high atmospheric pressure trending east-west and centered on about lat. 30° N and S.

subtropical jet stream Jet stream of westerly winds forming at the tropopause, just above the Hadley cell.

subtropical needleleaf evergreen forest Formation class of the forest biome, composed of needleleaf evergreen trees occurring in the moist subtropical climate ⑥ of the southeastern United States; also referred to as *southern pine forest*.

subtropical zones Latitude zones occupying the region of lat. 25° to 35° N and S (more or less) and lying between the tropical and the midlatitude zones.

succulents Plants adapted to resist water losses by means of thickened spongy tissue in which water is stored.

summer monsoon Inflow of maritime air at low levels from the Indian Ocean toward the Asiatic low-pressure center in the season of high Sun; associated with the rainy season of the wet-dry tropical climate ③ and the Asiatic monsoon climate.

summer solstice June solstice in the northern hemisphere.

summer time Daylight saving time.

Sun-synchronous orbit Satellite orbit in which the orbital plane remains fixed in position with respect to the Sun.

supercell thunderstorm Strong, single-cell convective storm that persists for many hours and is usually accompanied by severe weather, featuring downbursts, hail, and possible tornado formation.

supercontinent Single world continent, formed when plate tectonic motions move continents together into a single, large landmass.

supercontinent cycle Plate tectonic cycle in which a supercontinent breaks apart, forming smaller continents that later reform into another supercontinent.

supercooled water Water existing in the liquid state at a temperature lower than the normal freezing point.

surface Very thin layer of a substance that receives and radiates energy and conducts energy to and away from the substance.

surface creep Dragging of sand grains by strong winds along the surface of a sand dune.

surface energy balance equation Formula expressing the balance among energy flows to and from a surface.

surface tension Tension in the surface film of water, caused by hydrogen bonding of the surface molecules to each other and to the water below; minimizes surface area, causing drops to form.

surface water Water of the lands flowing freely (as streams) or impounded (as ponds, lakes, marshes).

surges Episodes of very rapid downvalley movement within an alpine glacier.

suspended load That part of the stream load carried in turbulent suspension.

suspension See *turbulent suspension*.

suture See *continental suture*.

swash Surge of water up the beach slope (landward) following collapse of a breaker.

symbiosis Form of positive interaction between species; it is beneficial to one of the species and does not harm the other.

sympatric speciation Type of biological species formation that occurs within a larger population.

synclinal mountain Steep-sided ridge or elongate mountain developed by erosion of a syncline.

synclinal valley Valley eroded on weak strata along the central trough or axis of a syncline.

syncline Downfold of strata (or other layered rock) in a troughlike structure; a class of folds.

synthesis In geography, a perspective that focuses on collecting ideas from different fields and assembling them in new ways.

system (1) Collection of things that are somehow related or organized. (2) Scheme for naming, as in a classification system. (3) Flow system of matter and energy.

systematic geography Study of the physical, economic, and social processes that differentiate the Earth's surface into places.

systems approach Study of the interconnections among natural processes, focusing on how, where, and when natural systems are linked and interconnected.

systems theory Body of knowledge explaining how systems work.

T

taiga Plant formation class consisting of woodland with low, widely spaced trees and a ground cover of lichens and mosses, found along the northern fringes of the region of boreal forest climate; also called *cold woodland*.

tailings See *spoil*.

talik Pocket or region within permafrost that is unfrozen; ranges from small inclusions to large holes in permafrost under lakes.

tall-grass prairie Formation class of the grassland biome that consists of tall grasses with broadleafed herbs.

talus Accumulation of loose rock fragments derived by fall of rock from a cliff.

talus slope Slope formed of talus.

tar sand Sand deposit containing a combustible mixture of hydrocarbons that is highly viscous and will flow only when heated; considered a form of petroleum.

tarn Small lake occupying a rock basin in a cirque of glacial trough.

tectonic activity Process of bending (folding) and breaking (faulting) of crustal mountains, concentrated on or near active lithospheric plate boundaries.

tectonic arc Long, narrow chain of islands or mountains or a narrow submarine ridge adjacent to a subduction boundary and its trench.

tectonic crest Ridgelike summit line of a tectonic arc.

tectonics Branch of geology relating to tectonic activity and the features it produces.

temperature Measure of the amount of kinetic energy of molecular motion held by a substance.

temperature gradient Rate of temperature change along a selected line or direction.

temperature index Index indicating how air temperature is perceived, given humidity and wind conditions.

temperature inversion Upward reversal of the normal environmental temperature lapse rate, so that the air temperature increases upward.

temperature regime Distinctive type of annual temperature cycle.

tephra Collective term for all size grades of solid igneous rock particles blown out under gas pressure from a volcanic vent.

terminal moraine Moraine deposited as an embankment at the terminus of an alpine glacier or at the leading edge of an ice sheet.

terminus End point of a glacier.

terrestrial ecosystems Ecosystems of land plants and animals found on upland surfaces of the continents.

tetraploid Having four sets of chromosomes instead of the normal two sets.

thematic map Map showing a single type of information.

theme Category or class of information displayed on a map.

thermal action Cracking of rocks by expansion and contraction of minerals with change in temperature.

thermal circulation Motion of air toward a warmer region at low levels, and away from the warmer region at high levels.

thermal erosion In regions of permafrost, the physical disruption of the land surface by melting of ground ice, brought about by removal of a protective organic layer.

thermal infrared Portion of the infrared radiation wavelength band, from approximately 6 to 300 µm, in which objects at temperatures encountered on the Earth's surface (including fires) emit electromagnetic radiation.

thermal pollution Form of water pollution in which heated water is discharged into a stream or lake from the cooling system of a power plant or other industrial heat source.

thermal radiation Electromagnetic radiation in the thermal infrared portion of the electromagnetic spectrum.

thermistor Electronic device that measures (air) temperature.

thermocline Water layer of a lake or the ocean in which temperature changes rapidly in the vertical direction.

thermohaline circulation Slow flow of ocean water linking all ocean basins, driven by the sinking of cold, salty surface water at the northern edge of the Atlantic Ocean.

thermokarst In arctic environments, an uneven terrain produced by thawing of the upper layer of permafrost, with settling of sediment and related water erosion; often occurs when the natural surface cover is disturbed by fire or human activity.

thermokarst lake Shallow lake formed by the thawing and settling of permafrost, usually in response to disturbance of the natural surface cover by fire or human activity.

thermometer Instrument used to measure temperature.

thermometer shelter Louvered wooden cabinet of standard construction used to hold thermometers and other weather-monitoring equipment.

thermosphere Atmospheric layer of upwardly increasing temperature, lying above the mesopause.

thorntree semidesert Formation class within the desert biome, transitional from grassland biome and savanna biome and consisting of xerophytic trees and shrubs.

thorntree-tall-grass savanna Plant formation class, transitional between the savanna biome and the grassland biome, consisting of widely scattered trees in an open grassland.

thrust sheet Sheetlike mass of rock moving forward over a low-angle overthrust fault.

thunderstorm Intense, local convectional storm associated with a cumulonimbus cloud and yielding heavy precipitation, also with lightning and thunder, and sometimes the fall of hail.

tidal current Current set in motion by the ocean tide.

tidal inlet Narrow opening in a barrier island or baymouth bar through which tidal currents flow.

tidal range Difference between heights of successive low and high tidal waters.

tide See *ocean tide.*

tide curve Graphical presentation of the rhythmic rise and fall of ocean water caused by ocean tides.

tidewater glacier Valley glacier terminating in the ocean.

till Heterogeneous mixture of rock fragments ranging in size from clay to boulders, deposited beneath moving glacial ice or directly from the melting in place of stagnant glacial ice.

till plain Undulating, plainlike land surface underlain by glacial till.

time cycle In flow systems, a regular alternation of flow rates with time.

time zones Zones or belts of given east-west (longitudinal) extent within which standard time is applied according to a uniform system.

tombolo Sand spit connecting the mainland to a near-shore island.

topographic contour Isopleth of uniform elevation appearing on a map.

topography Shape of the Earth's surface; may be considered at any scale.

tornado Small, very intense wind vortex with extremely low air pressure in center, formed beneath a dense cumulonimbus cloud in proximity to a cold front.

trade winds (trades) Surface winds in low latitudes, representing the low-level airflow within the tropical easterlies.

trade-wind coast Coastal location where trade winds move warm, moist air from ocean to land, often generating orographic precipitation.

transcurrent fault Fault on which the relative motion is dominantly horizontal, in the direction of the strike of the fault; also called a *strike-slip fault.*

transform fault Special case of a strike-slip fault making up the boundary of two moving lithospheric plates; usually found along an offset of the midoceanic ridge where seafloor spreading is in progress.

transform plate boundary Lithospheric plate boundary along which two plates are in contact on a transform fault; the relative motion is that of a strike-slip fault.

transform scar Linear topographic feature of the ocean floor taking the form of an irregular scarp or ridge and originating at the offset axial rift of the midoceanic ridge. A transform scar represents a former transform fault but is no longer a plate boundary.

transformation (soils) Class of soil-forming processes that transform materials within the soil body; examples include mineral alteration and humification.

translational slide Landslide in which the mass moves down the slope with little or no rotation.

translocation Soil-forming process by which materials are moved within the soil body, usually from one horizon to another.

transpiration Evaporative loss of water to the atmosphere from leaf pores of plants.

transportation See *stream transportation.*

transported regolith Regolith formed of mineral matter carried by fluid agents from a distant source and deposited on bedrock or on older regolith; for example, floodplain silt, lake clay, beach sand.

transverse dunes Field of wavelike sand dunes with crests running at right angles to the direction of the prevailing wind.

traveling anticyclone Center of high pressure and outspiraling winds that travels over the Earth's surface; often associated with clear, dry weather.

traveling cyclone Center of low pressure and inspiraling winds that travels over the Earth's surface; includes wave cyclones, tropical cyclones, and tornadoes.

travertine Carbonate mineral matter, usually calcite, accumulating on limestone cavern surfaces situated in the unsaturated zone.

tree Large erect woody perennial plant typically having a single main trunk, few branches in the lower part, and a branching crown.

trellis drainage pattern Drainage pattern characterized by a dominant parallel set of major subsequent streams, joined at right angles by numerous short tributaries; typical of coastal plains and belts of eroded folds.

TRMM Abbreviation for Tropical Rainfall Measuring Mission, an American-Japanese Earth-observation satellite mission to measure and map rainfall in the tropical and equatorial zones.

Tropic of Cancer Parallel of latitude at 23½° N.

Tropic of Capricorn Parallel of latitude at 23½° S.

tropical cyclone Intense traveling cyclone of tropical and subtropical latitudes, accompanied by high winds and heavy rainfall.

tropical depression Tropical cyclone with a closed low and wind speeds lower than 17 m/s (38 mi/hr).

tropical desert Desert within the tropical zone, near the Tropic of Cancer or Capricorn.

tropical easterlies Low-latitude wind system of persistent airflow from east to west between the two subtropical high-pressure belts.

tropical easterly jet stream Upper-air jet stream of seasonal occurrence, running east to west at very high altitudes over Southeast Asia.

tropical high-pressure belt High-pressure belt occurring in tropical latitudes at a high level in the troposphere; extends downward and poleward to form the subtropical high-pressure belt, located at the surface.

tropical storm Tropical cyclone with a closed low and wind speeds between 17 to 33 m/s (38 to 74 mi/hr).

tropical zones Latitude zones centered on the Tropic of Cancer and the Tropic of Capricorn, within the latitude ranges 10° to 25° N and 10° to 25° S, respectively.

tropical-zone rainforest Plant formation class within the forest biome, similar to equatorial rainforest but occurring farther poleward in tropical regions.

tropopause Boundary between the troposphere and stratosphere.

tropophyte Plant that sheds its leaves and enters a dormant state during a dry or cold season when little soil water is available.

troposphere Lowermost layer of the atmosphere in which air temperature falls steadily with increasing altitude.

trough (atmosphere) Elongated zone of atmospheric low pressure.

trough (water waves) In describing water waves, the lowest part of the wave form.

tsunami Train of sea waves set off by an earthquake (or other seafloor disturbance) traveling over the ocean surface.

tundra See *arctic tundra*.

tundra biome Biome of the cold regions of arctic and alpine tundra, consisting of grasses, grasslike plants, flowering herbs, dwarf shrubs, mosses, and lichens.

tundra climate ⑫ Cold climate of the arctic zone, with eight or more consecutive months of zero potential evapotranspiration (water need).

turbulence In fluid flow, the motion of individual water particles in complex eddies, superimposed on the average downstream flow path.

turbulent flow Mode of fluid flow in which individual fluid particles (molecules) move in complex eddies, superimposed on the average downstream flow path.

turbulent suspension Stream transportation in which particles of sediment are held in the body of the stream by turbulent eddies. Also applies to wind transportation.

twilight Solar radiation from below the horizon that is scattered toward the ground by the atmosphere to provide illumination after the Sun has set or before it has risen.

typhoon Tropical cyclone of the western North Pacific and coastal waters of Southeast Asia.

U

Udalfs Suborder of the soil order Alfisols; includes Alfisols of moist regions, usually in the midlatitude zone, with deciduous forest as the natural vegetation.

Udolls Suborder of the soil order Mollisols; includes Mollisols of the moist soil-water regime in the midlatitude zone and with no horizon of calcium carbonate accumulation.

Ultisols Soil order consisting of soils of warm soil temperatures, with an argillic horizon and low base status.

ultramafic igneous rock Igneous rock composed almost entirely of mafic minerals, usually the olivine or pyroxene group.

ultraviolet radiation Electromagnetic energy in the wavelength range of 0.2 to 0.4 µm.

uniformitarianism Idea that the same geologic processes we can observe today have operated since the beginning of the Earth's history.

unloading See *exfoliation*.

unsaturated zone subsurface water zone in which pores are not fully saturated, except at times when infiltration is very rapid; lies above the saturated zone.

unstable air Air with substantial content of water vapor, capable of breaking into spontaneous convectional activity, leading to the development of heavy showers and thunderstorms.

unstable isotope Elemental isotope that spontaneously decays to produce one or more new isotopes.

upper-air westerlies System of westerly winds in the upper atmosphere over middle and high latitudes.

upwelling Upward motion of cold, nutrient-rich ocean waters, often associated with cool Equator-ward currents occurring along continental margins.

urban heat island Zone of warmer temperatures in the center of a city produced by enhanced absorption of solar radiation by urban surfaces and by waste heat release.

Ustalfs Suborder of the soil order Alfisols; includes Alfisols of semiarid and seasonally dry climates in which the soil is dry for a long period in most years.

Ustolls Suborder of the soil order Mollisols; includes Mollisols of the semiarid climate in the midlatitude zone, with a horizon of calcium carbonate accumulation.

V

valley glacier Alpine glacier occupying sloping stream valleys between steep rock walls.

valley winds Air movement at night down the gradient of valleys and the enclosing mountainsides; alternating with daytime mountain winds.

variation In the study of evolution, natural differences arising between parents and offspring as a result of mutation and recombination.

veins Small, irregular, branching network of intrusive rock within a preexisting rock mass.

ventifact Rock with pits, grooves, or hollows carved by wind-blown sand and dust.

verbal description From the representation perspective of geography, the use of written or oral text to describe geographic phenomena.

vernal equinox March equinox in the northern hemisphere.

Vertisols Soil order, consisting of soils of the subtropical and tropical zones with high clay content, and developing deep, wide cracks when dry, and showing evidence of movement between aggregates.

viewpoint Unique perspective of geography that considers where and how phenomena occur and how they are related to other phenomena nearby and far away.

visible light Electromagnetic energy in the wavelength range of 0.4 to 0.7 µm

visual display From the representation perspective of geography, tools such as cartography and remote sensing that display spatial information visually.

void Empty region of pore space in sediment; often occupied by water or water films.

volcanic bombs Boulder-sized, semisolid masses of lava that are ejected from an erupting volcano.

volcanic neck Isolated, narrow, steep-sided peak formed by erosion of igneous rock previously solidified in the feeder pipe of an extinct volcano.

volcanism General term for volcano building and related forms of extrusive igneous activity.

volcano Conical, circular structure built by accumulation of lava flows and tephra.

W

WAAS Abbreviation for Wide Area Augmentation System, used with GPS receivers to improve geolocation accuracy using a constant stream of differential correction information.

warm front Moving weather front along which a warm air mass is sliding up over a cold air mass, leading to production of stratiform clouds and precipitation.

warm-blooded animal Animal that possesses one or more adaptations to maintain a constant internal temperature despite fluctuations in the environmental temperature.

warm-core ring Circular eddy of warm water, surrounded by cold water and lying adjacent to a warm, poleward-moving ocean current, such as the Gulf Stream.

washout Downsweeping of atmospheric particulates by precipitation.

water gap Narrow transverse gorge cut across a narrow ridge by a stream, usually in a region of eroded folds.

water need See *potential evapotranspiration*.

water resources Field of study that couples the basic examination of the location, distribution, and movement of water with the utilization and quality of water for human use.

water surplus Water disposed of by runoff or percolation to the groundwater zone after the storage capacity of the soil is full.

water table Upper boundary surface of the saturated zone; the upper limit of the groundwater body.

water use See *actual evapotranspiration*.

water vapor Gaseous state of water.

waterfall Abrupt descent of a stream over a bedrock step in the stream channel.

waterlogging Rise of a water table in alluvium to bring the zone of saturation into the root zone of plants.

watershed See *drainage basin*.

watt Unit of power equal to the quantity of work done at the rate of 1 joule per second; symbol, W.

wave cyclone See *midlatitude cyclone*.

wave height In describing water waves, the height of the wave as measured from the top of the crest to the bottom of the trough.

wave orbit (water waves) Vertical circle or ellipse traced by a particle moved by a passing wave crest.

wave period In describing water waves, time in seconds between successive crests or successive troughs that pass a fixed point.

wave-cut notch Rock recess at the base of a marine cliff where wave impact is concentrated.

wavelength (electromagnetic radiation) Distance between successive crests or troughs of electromagnetic waves.

wavelength (water waves) Length of a wave from crest to crest or trough to trough.

weak equatorial low Weak, slow-moving low-pressure center (cyclone) accompanied by numerous convectional showers and thunderstorms. A weak equatorial low forms close to the intertropical convergence zone in the rainy season or summer monsoon.

weather Physical state of the atmosphere at a given time and place.

weather system Recurring pattern of atmospheric circulation associated with characteristic weather, such as a cyclone or anticyclone.

weathering Total of all processes acting at or near the Earth's surface to cause physical disruption and chemical decomposition of rock.

westerlies see *prevailing westerly winds, upper-air westerlies*.

west-wind drift Ocean drift current moving eastward in zone of prevailing westerlies.

wet equatorial climate ① Moist climate of the equatorial zone, marked by a large annual water surplus and with uniformly warm temperatures throughout the year.

wet-dry tropical climate ③ Climate of the tropical zone characterized by a very wet season alternating with a very dry season.

wetlands Land areas of poor surface drainage, such as marshes and swamps.

Wilson cycle Plate tectonic cycle in which continents rupture and pull apart, forming oceans and oceanic crust, then converge and collide with accompanying subduction of oceanic crust.

wilting point Quantity of stored soil water, less than which the foliage of plants not adapted to drought will wilt.

wind Air motion, dominantly horizontal relative to the Earth's surface.

wind abrasion Mechanical wearing action of wind-driven mineral particles striking exposed rock surfaces.

wind chill index Index in degrees Fahrenheit used in the United States to express how cold the air feels to the skin when the wind is blowing.

wind shear Change in wind speed or direction with height.

wind vane Weather instrument used to indicate wind direction.

window (radiation) Region of the spectrum in the thermal infrared where Earth radiation escapes to space.

Winkel tripel projection Map projection devised by Oswald Winkel (1873–1953) that minimizes the sum of distortions to area, distance, and direction; *tripel* is a German word meaning "triplet."

winter monsoon Outflow of continental air at low levels from the Siberian high, passing over Southeast Asia as a dry, cool northerly wind.

winter solstice December solstice in the northern hemisphere.

Wisconsin glaciation Last glaciation of the Pleistocene epoch.

woodland Plant formation class, transitional between forest biome and savanna biome, consisting of widely spaced trees with canopy coverage between 25 and 60 percent.

X

xenolith Fragments of surrounding rock that are incorporated into a batholith without melting.

Xeralfs Suborder of the soil order Alfisols; includes Alfisols of the Mediterranean climate ⑦.

xeric animals Animals adapted to dry conditions typical of a desert climate.

Xerolls Suborder of the soil order Mollisols; includes Mollisols of the Mediterranean climate ⑦.

xerophytes Plants adapted to a dry environment.

X-rays Electromagnetic radiation of wavelength 0.005 to 10 nm; may be hazardous to health.

Y

yardang Low, tongue- or teardrop-shaped ridge carved by wind abrasion.

youthful stage In W. M. Davis's geographic cycle, a stage of erosion with steep slopes and river gradients.

Z

zone of ablation Lower portion of an alpine glacier, where ice is lost by melting and evaporation.

zone of accumulation Upper portion of an alpine glacier, where ice is formed by accumulation of annual snowfall.

zooplankton Microscopic animals found largely in the uppermost layer of ocean or lake water.

Illustration Credits

Introduction Page 10 (bottom): From Visualzing Geology, 2e, by Barbara W. Murck, Brian J. Skinner, and Dana Mackenzie. Copyright 2010 John Wiley & Sons, Inc. Reprinted with permission of John Wiley & Sons, Inc.; Page 17: A. N. Strahler; Page 20 (left): http://www.usgs.gov/laws/info_policies. html/Public Domain/U.S. Geological Survey; Page 20 (top): Modified from U.S. Army Corps of Engineers; Page 20 (middle): Modified from U.S. Army Corps of Engineers; Page 20 (bottom): Modified from U.S. Army Corps of Engineers; Page 21: From P. Gersmehl, *Annals of the Assoc. of Amer. Geographers*, vol. 67. Copyright © Association of American Geographers. Used by permission; Page 28: Copyright © A. N. Strahler.

Chapter 1 Page 39 (top right): Courtesy Mike Sandiford, Univ. of Melbourne; Page 45: From Introducing Physical Geography, 5e by Alan Strahler. Copyright © 2011 John Wiley & Sons, Inc. Reprinted with permission of John Wiley & Sons, Inc.; Page 48: U.S. Navy Oceanographic Office.

Chapter 2 Page 62: After W. D. Sellers, *Physical Climatology*, University of Chicago Press. Used by permission; Page 63: From A. N. Strahler, "The Life Layer," *Journal of Geography*, vol. 69, Figure 2.4. Used by permission; Page 66: Copyright © A. N. Strahler. Used by permission; Page 67 (left): Copyright © A. N. Strahler. Used by permission; Page 70: Copyright © A .N. Strahler. Used by permission.

Chapter 3 Page 84: Mauna Loa Observatory; Page 92 (bottom): Courtesy NOAA; Page 93 (top): Courtesy NOAA; Page 105 (lower right): James Hansen/NASA Goddard Institute of Space Sciences; Page 106 (lower): Page 107: Climate Change 2007: Synthesis Report. Contribution of Working Groups I, II and III to the Fourth Assessment Report of the Intergovernmental Panel on Climate Change, Figure SPM.5. IPCC, Geneva, Switzerland; Page 108 (upper left): NASA - James E. Hansen; Page 108 (upper right): NOAA.

Chapter 4 Page 115: National Atmospheric Deposition Program (NRSp-3) National Trends Network, Illinois, State Water Survey; Page 122: Copyright © A. N. Strahler; Page 123: Copyright © A. N. Strahler; Page 129: Adapted from *Visualizing Weather and Climate* by Bruce T. Anderson and Alan Strahler. Copyright 2008 John Wiley & Sons Inc. Reprinted with permission of John Wiley & Sons; Page 135 (top): Adapted from *Visualizing Weather and Climate* by Bruce T. Anderson and Alan Strahler. Copyright 2008 John Wiley & Sons Inc. Reprinted with permission of John Wiley & Sons; Page 136: R. H. Skaggs, Proc. Assoc. American Geographers, vol. 6, figure 2, 1974. Used by permission; Page 139 (top): Adapted from Visualizing Weather and Climate by Bruce T. Anderson and Alan Strahler. Copyright © 2008 John Wiley & Sons Inc. Reprinted with permission of John Wiley & Sons Fig 10.5, page 275; Page 144: Data from EPA.

Chapter 5 Page 163: Data compiled by John E. Oliver; Page 165: Data compiled by John E. Oliver; Page 166: Data compiled by John E. Oliver; Page 170 (left): Copyright © A. N. Strahler; Page 170 (upper right): After National Weather Service; Page 171: Copyright © A. N. Strahler; Page 174: Copyright © A. N. Strahler; Page 176: Copyright © A. N. Strahler; Page 176 (left): National Weather Service; Page 177: Courtesy NOAA; Page 179: From Murck, Barbara W., Brian J. Skinner, and Dana Mackenzie, *Visualizing Geology*. Copyright © 2008 John Wiley & Sons, Inc. Reprinted with permission of John Wiley & Sons, Inc.

Chapter 6 Page 190: Data from U.S. Department of Commerce; Page 191: Drawn by A. N. Strahler; Page 192: Drawn by A. N. Strahler; Page 193: Drawn by A. N. Strahler; Page 202: Based on data of S. Pettersen, B. Haurwitz, and N. M. Austin, J. Namias, M. J. Rubin, and J. H. Chang; Page 203: After M. A. Garbell; Page 204: Data from Naval Research Laboratory; Page 205 (bottom): Redrawn from NOAA National Weather Service; Page 206: Courtesy NASA and National Hurricane Center. Image published on Wikipedia; Page 207: Based on data of S. Pettersen, B. Haurwitz, and N. M. Austin, J. Namias, M. J. Rubin, and J. H. Chang.

Chapter 7 Page 226: Reproduced with permission of and license from HarperCollins Publishers Ltd.; Page 232: Compiled by A. N. Strahler, from station data. Page 253: Top graph: Global Temperatures, after Hansen et. al., 2001, Proc. National Academy of Sciences. Used by permission. Top Graph, CO2 Concentration: Courtesy Scripps CO2 Program. Bottom graph: Adapted from EPA; Page 254: NASA Images; Page 255 (top left): Adapted from http://maps.grida. no/go/graphic/trends-in-natural-disasters, Centre for Research on the Epidemiology of Disasters (CRED) by Emmanuelle Bournay, UNEP/GRID-Arendal; Page 255 (middle right): NASA Images; Page 262 (top): Based on Goode Base Map.

Chapter 8 Page 272 (top right): Data courtesy of United States Environmental Protection Agency; Page 272 (middle): Compiled by the National Science Board, National Science Foundation; Page 273: Compiled by the National Science Board, National Science Foundation; Page 274 (bottom): Values are from Schlesinger, W. H. Biogeochemistry: An Analysis of Global Change, 2e, Academic Press, San Diego (1997) and based on several sources; Page 290: From J. H. Brown and M. V. Lomolino, Biogeography, 2nd edition, 1998, Sinaeur, Sunderland, Massachussetts. Used by permission; Page 292: From H. J. B. Birks, J. *Biogeography*, vol. 16, pp. 503–540; Page 296 (upper): After Goode, 1974; Page 296 (lower): After Wallace, 1876; Page 297: Data from IUCN.

Chapter 9 Page 310: Based on the maps of S. R. Eyre, 1968; Page 313: After J. S. Beard, *The Natural Vegetation of Trinidad*, Clarendon Press, Oxford. Reproduced with permission of The Bodleian Libraries, University of Oxford.; Page 315 (top): Based on maps of S. R. Eyre; Page 317: Based on maps of S. R. Eyre; Page 319: Based on maps of S. R. Eyre; Page 320: Based on maps of S. R. Eyre; Page 321: Based on maps of S. R. Eyre; Page 324: Based on maps of S. R. Eyre; Page 326: Based on maps of S. R. Eyre; Page 328: Based on maps of S. R. Eyre; Page 336: Mark Friedl, Damien Sulla-Menashe, Bin Tan, Boston University.

Chapter 10   Page 347 (top): After http://soils.usda.gov/education/resources/texture; Page 347 (bottom): from Discovering Physical Geography, 2e by Alan F. Arbogast. Copyright © 2011 John Wiley & Sons. Reprinted with permission of John Wiley & Sons; Page 348: ServiceOntario Publications; Page 350: Adapted from *Discovering Physical Geography, 2e* by Alan F. Arbogast. Copyright © 2011 John Wiley & Sons. Reprinted with permission of John Wiley & Sons; Page 351 (bottom): Copyright © A. N. Strahler. Used by permission; Page 352: Copyright © A. N. Strahler. Used by permission; Page 358: Based on data of the Natural Resources Conservation Site, USDA.

Chapter 11   Page 397: From *Visualzing Geology, 2e,* by Barbara W. Murck, Brian J. Skinner, and Dana Mackenzie. Copyright 2010 John Wiley & Sons, Inc. Reprinted with permission of John Wiley & Sons, Inc; Page 398: From *Plate Tectonics,* Arthur N. Strahler, 1998, GeoBooks Publishing. Used by permission; Page 398: From *Plate Tectonics,* Arthur N. Strahler, 1998, GeoBooks Publishing. Used by permission; Page 398: From A. Wegener, 1915, *Die Entsechung der Kontinente und Ozeane,* F. Vieweg, Braunschweig; Page 400 (top left): Based on the diagrams by A. Heim, 1922, Geologie der Schweiz, vol. II-1Tauschnitz, Leipzig; Page 400 (top right): Based on the diagrams by A. Heim, 1922, Geologie der Schweiz, vol. II-1Tauschnitz, Leipzig; Page 401 (top): Copyright © A. N. Strahler; Page 401 (middle): Copyright © A. N. Strahler; Page 404: Copyright © A. N. Strahler; Page 406: From *Plate Tectonics,* Arthur N. Strahler, 1998, Geobooks Publishing; Page 407: Copyright © A. N. Strahler; Page 408: From *Plate Tectonics,* Arthur N. Strahler, 1998, Geobooks Publishing; Page 409: From *Plate Tectonics,* Arthur N. Strahler, 1998, Geobooks Publishing.

Chapter 12   Page 417: After E. Raisz; Page 419: Drawn by A. N. Strahler; Page 420: Drawn by A. N. Strahler; Page 421: Drawn by A. N. Strahler; Page 422: Drawn by A. N. Strahler; Page 423: Copyright © A. N. Strahler; Page 424 (bottom): Copyright © A. N. Strahler; Page 424 (top right): Copyright © A. N. Strahler; Page 426 (bottom left): From *Physical Geology* by Charles Fletcher. Copyright © 2011 John Wiley & Sons. Reprinted with permission of John Wiley & Sons. FIG 12.7; Page 426 (bottom right): Adapted from *Discovering Physical Geography, 2e* by Alan Arbogast. Copyright © 2011 by Jonh Wiley & Sons, Inc. Reprinted with permission of John Wiley & Sons.; Page 428: From *Physical Geology* by Charles Fletcher. Copyright © 2011 John Wiley & Sons. Reprinted with permission of John Wiley & Sons.; Page 430 (bottom): *Physical Geology* by Charles Fletcher. Copyright © 2011 John Wiley & Sons. Reprinted with permission of John Wiley & Sons. Adapted from USGS, http://eqhazmaps.usgs.gov; Page 432: Compiled by A. N. Strahler from data from NOAA. Copyright © A. N. Strahler. Used by permission. (a) Copyright © A. N. Strahler. Used by permission. (b); Page 433: Copyright © A. N. Strahler. Used by permission. (a) Illustration by Frank Ippolito.; Page 435: Illustration by Frank Ippolito.

Chapter 13   Page 448: Drawn by A. N. Strahler; Page 453 (top): Drawn by A. N. Strahler; Page 453 (bottom): Adapted from *Discovering Physical Geography, 2e* by Alan Arbogast. Copyright © 2011 by Jonh Wiley & Sons, Inc. Reprinted with permission of John Wiley & Sons.; Page 458 (top): Adapted from USGS.

Chapter 14   Page 469: Drawn by A. N. Strahler; Page 473 (top): Copyright © A. N. Strahler; Page 473 (bottom): Copyright © A. N. Strahler; Page 474: Drawn by Erwin Raisz. Copyright © A. N. Strahler; Page 475: Copyright © A. N. Strahler; Page 477: A. H. Strahler; Page 479: Data of U. S. Geological Survey and Mark A. Melton; Page 480: Drawn by Erwin Raisz. Copyright © A. N. Strahler; Page 481: From *Introducing Physical Geography, 4e,* by Alan Strahler. Copyright © 2006 John Wiley & Sons. Reprinted with permission of John Wiley & Sons.; Page 482(top): Copyright © A. N. Strahler. Used by permission.; Page 483 (top): Data of U.S. Geological Survey and Mark A. Melton; Page 483 (bottom): Data of U. S. Geological Survey; Page 484: Adapted from *Physical Geology* by Charles Fletcher. Copyright © 2011 John Wiley & Sons. Reprinted with permission of John Wiley & Sons.; Page 486: After Hoyt and Langbein, Floods, Copyright ©1955 Princeton University Press. Used by perpmission; Page 488 (top): Redrawn from *A Geologist's View of Cape Cod,* Copyright © A. N. Strahler, 1966.Used by permission of Doubleday, a division of Bantam Doubleday Dell Publishing Group, Inc.; Page 488 (bottom): Modified from A. N. Strahler, *The Earth Sciences.* Used by permission. Harper & Row, 1971.

Chapter 15   Page 508: Drawn by Erwin Raisz. Copyright © A. N. Strahler. Used by permission.; Page 512: Drawn by Erwin Raisz. Copyright © A. N. Strahler. Used by permission.; Page 513: Drawn by A. N. Strahler; Page 514: Drawn by A. N. Strahler; Page 516: Drawn by A. N. Strahler; Page 519: Drawn by A. N. Strahler; Page 522: Copyright © A. N. Strahler. Used by permission.; Page 523: Drawn by A. N. Strahler.

Chapter 16   Page 531 (top): Copyright © A. N. Strahler; Page 531 (middle): Copyright © A. N. Strahler; Page 532: Adapted from *Visualizing Earth Science* by Zeeya Mirali and Brian J. Skiinner. Copyright © 2009 John Wiley & Sons, Inc. Reprinted with permission of John Wiley & Sons, Inc.; Page 535: From Visualizing Earth Science by Zeeya Mirali and Brian J. Skiinner. Copyright © 2009 John Wiley & Sons, Inc. Reprinted with permission of John Wiley & Sons, Inc.; Page 539: Drawn by E. Raisz. Copyright © A. N. Strahler; Page 540: Drawn by A. N. Strahler; Page 542: Drawn by A. N. Strahler; Page 543: Courtesy NASA; Page 545 (top): Drawn by A. N. Strahler; Page 553: Adapted from K. Pye and L. Tosar, *Aeolian Sand and Sand Dunes,* Figure 7.1, Chapman and Hall, 1990. Reprinted with kind permission of Springer Science and Business Media.; Page 554: Adapted from John T. Hack, The *Geographical Review,* vol. 31, fig. 19, page 260 by permission of the American Geographical Society.

Chapter 17   Page 562: Adapted with permission from C. R. Bentley, Science, Vol. 275, p. 1077; Page 567: Based on data from Dyurgerov, M. 2002, updated 2005. Glacier mass balance and regime measurements and analysis, 1945–2003, edited by M. Meier and R. Armstrong. Boulder, CO; Page 574: Courtesy NASA; Page 575: Drawn by A. N. Strahler; Page 576 (bottom): Copyright © A. N. Strahler; Page 578: Adapted from Troy L. Pewe, *Geotimes,* Vol 29, no 2, p. 11. Copyright © 1984 by the American Geological Institute.; Page 583: Based on data of R. F. Flint, Glacial and Pleistocene Geology, John Wiley & Sons, New York; Page 586: Based on calculations by A. D. Vernekar, 1968. Copyright © A. N. Strahler.

Index